VOLUME	EDITOR-IN-CHIEF	PAGES	
20	Charles F. H. Allen	113	*Out of print*
21	The Late Nathan L. Drake	120	*Out of print*
22	Lee Irvin Smith	114	*Out of print*
23	Lee Irvin Smith	124	*Out of print*
24	The Late Nathan L. Drake	119	*Out of print*
25	The Late Werner E. Bachmann	120	*Out of print*
26	The Late Homer Adkins	124	*Out of print*
27	R. L. Shriner	121	*Out of print*
28	H. R. Snyder	121	*Out of print*
29	Cliff S. Hamilton	119	*Out of print*
Collective Vol. 3	A revised edition of Annual Volumes 20–29 E. C. Horning, *Editor-in-Chief*	890	
30	Arthur C. Cope	115	
31	R. S. Schreiber	122	
32	Richard T. Arnold	119	*Out of print*
33	Charles C. Price	115	
34	William S. Johnson	121	
35	T. L. Cairns	122	
36	N. J. Leonard	120	
37	James Cason	109	
38	John C. Sheehan	120	
39	Max Tishler	114	
Collective Vol. 4	A revised edition of Annual Volumes 30–39 Norman Rabjohn, *Editor-in-Chief*	1036	
40	Melvin S. Newman	114	
41	John D. Roberts	118	
42	Virgil Boekelheide	118	
43	B. C. McKusick	124	

ORGANIC SYNTHESES

ORGANIC SYNTHESES

Collective Volume 3

A REVISED EDITION OF
ANNUAL VOLUMES 20–29

NEW YORK
JOHN WILEY & SONS, Inc.
London

FOURTH PRINTING, JANUARY, 1964

PRINTED IN THE UNITED STATES OF AMERICA

PREFACE

Collective Volume 3 follows the plan initiated by the publication of Collective Volumes 1 and 2 of *Organic Syntheses*. The material in annual volumes 20–29 has been revised and brought up to date. Errors found in the original printings have been corrected; the calculations and references have been checked; modifications and improvements which have been brought to the attention of the Editorial Board have been included; and seven new and improved checked procedures have been added. The new procedures are for the preparation of acenaphthenequinone, β-ethoxypropionaldehyde acetal, aminoacetal, 3-amino-1H-1,2,4-triazole, mesitoic acid, α-tetralone, and 2-thiophenealdehyde.

The section of Methods of Preparation under each preparation has been systematically revised to include additional methods of preparative value found in the literature covered by *Chemical Abstracts* through Volume 46 (1952). Later references are also included for a number of preparations. The *Chemical Abstracts* indexing name for each preparation is given as a subtitle when that name is different from the title.

In the interval since the publication of the annual volumes, some of the compounds whose preparation has been described have become commercially available. These compounds are noted by an asterisk after the title of the preparation.

New notes with regard to safety precautions have been included for three procedures: 2-chloromethylthiophene, 9-cyanophenanthrene, and *m*-nitrobenzaldehyde dimethyl acetal. The modifications or precautions which are recommended for each preparation were supplied by chemists who have used these preparations.

Organic Syntheses procedures frequently include Notes describing the purification of reagents or solvents, and preparative methods for reagents which are often generally useful but usually not commercially available. An index to items of this kind in Collective Volume 3 has been included along with the General Index, Type of Compound Index, Formula Index, Type of Reaction Index, and Illustration Index.

Several procedures for the preparation of hydrogenation catalysts are in this volume. Raney nickel catalysts and several types of palladium catalysts are described, and these, together with Adams' platinum oxide catalyst [Collective Volume 1, 463 (1941)] and copper chromite [Collective Volume 2, 142 (1943)], constitute the most generally used hydrogenation catalysts.

Several special items of equipment are described in detail in this volume. The ozone preparation includes specifications for the construction and use of a laboratory ozonizer. A methyl ester column is described in the procedure for methyl pyruvate. Equipment of this kind, as well as some of the other special pieces of apparatus described in annual volumes, is now commercially available.

It is now the general practice of *Organic Syntheses* to include refractive index values for all liquid products. Some of the procedures in this volume were checked and published before this practice was adopted, and consequently these data are not included at all appropriate points. Current annual volumes are complete in this respect.

The editors appreciate the help of contributors and users who have supplied corrections and improvements for this volume. Additional suggestions and corrections are welcomed and should be sent to the Secretary of the Editorial Board. Procedures for publication in annual volumes are also welcomed and should be sent, in duplicate, to the Secretary.

The preparation of this volume was aided by a number of chemists. It is a pleasure to acknowledge the help of M. G. Horning throughout this work, and of G. N. Walker, V. L. Stromberg, D. Howell, and D. Reisner for literature reviews. Dr. C. F. H. Allen also supplied much assistance for literature searches.

E. C. Horning

January, 1955

CONTENTS

ACENAPHTHENEQUINONE *

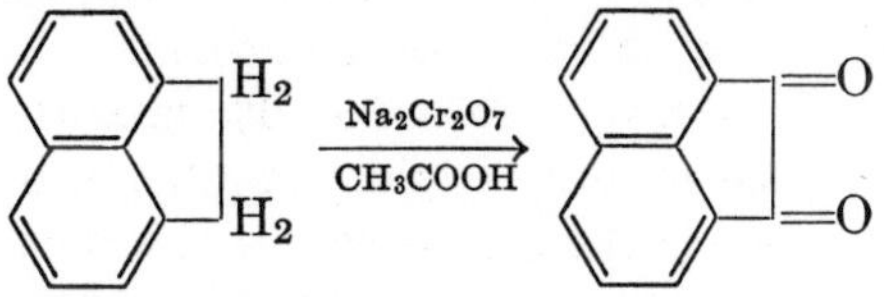

Submitted by C. F. H. ALLEN and J. A. VANALLAN.
Checked by L. MATTERNAS, H. LLOYD, and E. C. HORNING.

1. Procedure

A mixture of 100 g. (0.65 mole) of a technical grade of acenaphthene, 5 g. of ceric acetate (Note 1), and 800 ml. of glacial acetic acid is placed in a 4-l. stainless-steel beaker arranged for external cooling with cold water (Note 2). A thermometer and a powerful stirrer are inserted, and 325 g. (1.1 mole) of sodium bichromate dihydrate is added over a period of 2 hours, the temperature being kept at 40° (Note 3). Stirring is then continued at room temperature for an additional 8 hours; during this time the reaction mixture becomes quite thick, owing to the separation of the quinone and chromium salts. The suspension is diluted with 1.5 l. of cold water, and the solid is collected on a 10-in. Büchner funnel and washed free from acid.

The solid is next digested on the steam bath for 30 minutes with 500 ml. of a 10% sodium carbonate solution, and is filtered and washed. The solid is then extracted for 30 minutes at 80° with 1 l. of 4% sodium bisulfite solution; at the end of this period 15 g. each of Filtercel and Norit are added, and the suspension is filtered. The extraction is repeated, and the combined filtrates are acidified at 80° with constant stirring (Note 4), to Congo red paper, with concentrated hydrochloric acid (50–60 ml.). The temperature is maintained at 80° for 1 hour with constant stirring. The acenaphthenequinone separates as a bright yellow crystalline solid; it is collected on a Büchner funnel and washed with water until free from acid. The yield is 45–70 g. (38–60%); m.p. 256–260°.

The crude quinone (50 g.) is recrystallized from 250 ml. of *o*-dichlorobenzene without filtering (Note 5); the crystals are rinsed with methanol. The recovery is 45 g., m.p. 259–260° (Notes 6, 7).

* See Preface for explanation of asterisks.

2. Notes

1. The checkers used Eastman Kodak white label acenaphthene. Cerium salts appear to have a beneficial effect. Probably any cerous salt will be satisfactory, it being oxidized in the reaction. The submitters have used cerous chloride, ceric carbonate, and ceric acetate; the last can be obtained in the form of a 50% mixture with other rare-earth acetates from the Lindsay Light and Chemical Company, West Chicago, Illinois. The checkers used cerous chloride.

2. The checkers found no need to cool the reaction; instead the bath was used to heat the reaction to 40°. With technical grade acenaphthene, cooling is necessary.

3. If the oxidation temperature has been allowed to rise to 50°, tar formation makes it necessary to do five or six treatments with sodium bisulfite. The extractions should be continued as long as the filtrate gives a precipitate on acidification.

4. The acidification should be performed in a hood; much sulfur dioxide is evolved.

5. The quinone crystallizes so rapidly that filtration is impossible; however, there is no insoluble material if mechanical dirt has been excluded.

6. This oxidation has been run using 80 times these amounts; the yield of recrystallized material drops to 38–40%.

7. The red color of acenaphthenequinone [1] is due to biacenaphthylidenedione, m.p. 295°. It is an appreciable contaminant in a hot oxidation; the product may also contain appreciable amounts of naphthalic anhydride.

3. Methods of Preparation

Acenaphthenequinone has been prepared by oxidation of acenaphthene with chromic acid,[1–10] with calcium permanganate,[11] with air in the presence of catalysts in various solvents,[12–15] with 30% hydrogen peroxide in acetic acid,[16] by the formation of an oxime with an alkyl nitrite followed by hydrolysis,[17–19] and from oxalyl chloride and naphthalene.[20,21] This procedure is based on the disclosure in a P.B. report.[10]

[1] *Org. Syntheses*, **24**, 1 (1944).

[2] Graebe and Gfeller, *Ann.*, **276**, 4 (1893).

[3] Graebe and Gfeller, *Ber.*, **20**, 659 (1887).

[4] Graebe and Gfeller, *Ber.*, **25**, 654 (1892).

[5] Francescone and Pirazzoli, *Gazz. chim. ital.*, **33**, I, 42 (1903).

[6] Braun and Bayer, *Ber.*, **59**, 921 (1926).
[7] Dashevskii and Karishin, *Org. Chem. Ind. U.S.S.R.*, **7**, 729 (1936) [*C. A.*, **31**, 679 (1937)].
[8] Kiprianov and Dashevskii, *J. Applied Chem. U.S.S.R.*, **7**, 944 (1934) [*C. A.*, **29**, 2530 (1935)].
[9] Kalle and Co., Ger. pat. 228,698 (1910) [*Frdl.*, **10**, 198 (1910–1912)].
[10] P.B. **73485**, 1579.
[11] Morgan, *J. Soc. Chem. Ind.*, **49**, 420T (1930).
[12] Jaeger, Brit. pat. 318,617 (1928) [*C. A.*, **24**, 2145 (1930)].
[13] Duckert, *Arch. Sci. Phys. Nat.*, **15**, 244 (1933) [*C. A.*, **28**, 1255 (1934)].
[14] Paillard, *Helv. Chim. Acta*, **16**, 775 (1933).
[15] Ger. pat. 428,088 (1926) [*Frdl.*, **15**, 394 (1928)].
[16] Charrier and Moggi, *Gazz. chim. ital.*, **57**, 740 (1927).
[17] Reissert, *Ber.*, **44**, 1750 (1911).
[18] Cain, *The Manufacture of Intermediate Products for Dyes*, Macmillan Co., London, 1919, p. 242.
[19] B.I.O.S., **986**, 9; P.B. **73377**, 2201; P.B. **73719**, 2588.
[20] Lesser and Gad, *Ber.*, **60**, 243 (1927).
[21] Lesser and Gad, Ger. pat. 470,277 (1928) [*Frdl.*, **16**, 518 (1931)].

ACENAPHTHENOL-7

(1-Acenaphthenol)

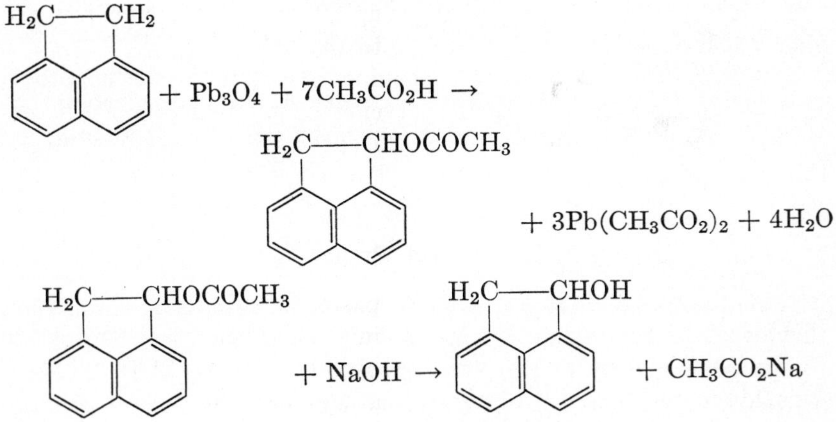

Submitted by James Cason.
Checked by R. L. Shriner and Elmer H. Dobratz.

1. Procedure

A. *Acenaphthenol acetate.* In a 2-l. round-bottomed flask are placed 154 g. (1 mole) of acenaphthene (Note 1) and 1.1 l. of glacial acetic acid (Note 2). The flask is fitted with a tantalum or Nichrome wire stirrer [1] and a thermometer extending below the surface of the liquid. The solution is stirred and heated to 60°, at which point the source of heat is removed and 820 g. of red lead (Note 3) is added in portions of about 50 g., each portion being added as soon as the color due to the previous portion has been discharged. During this operation, which requires 30–40 minutes, the temperature is maintained at 60–70° (Note 4) by external cooling. The reaction is complete when a portion of the solution gives no test for lead tetraacetate (Note 5). The dark red syrupy solution (which may contain a few suspended particles of red lead and lead dioxide) is poured into 2 l. of water contained in a 4-l. separatory funnel. The acetate is extracted with a 350-ml. portion of ether and then with a 250-ml. portion. The total extract is washed first with 100 ml. of water, then with 300 ml. of saturated sodium chloride solution and is finally dried over 50 g. of anhydrous sodium sulfate. The sodium sulfate is removed by filtration and washed colorless with three 50-ml. portions of dry ether. The combined filtrate and washings are placed in a 500-ml. Claisen flask with an inset side arm, and, after distillation of the solvent, the acetate is distilled under reduced pressure. The acetate distils almost entirely at 166–168°/5 mm. (bath temperature 180–185°, raised to 220° at the end) as a mobile yellow oil. The yield is 170–175 g. (80–82%) (Note 6).

B. *Acenaphthenol.* The acetate obtained as above is dissolved in 275 ml. of methanol in a 2-l. round-bottomed flask, and a solution of 40 g. (1.2 equiv.) of sodium hydroxide in 400 ml. of water is added (Note 7). This mixture is refluxed for 2 hours (Note 7) and then cooled below 20°. The yellow crystalline acenaphthenol is collected on a filter and washed well with about 1.5 l. of water. The crude product is air-dried (138–143 g.) and then dissolved in 2 l. of boiling benzene. The solution is treated with 6–8 g. of decolorizing carbon (Note 8) and filtered through a heated funnel. The orange-red filtrate is concentrated to about 1 l., and the acenaphthenol is allowed to crystallize. After filtering with suction and washing with cold benzene (about 500 ml.) until the wash solvent is colorless, the acenaphthenol is obtained as practically colorless needles, m.p. 144.5–145.5° (cor.) (Note 9). It weighs 117–121 g. From the filtrate may be obtained an additional quantity of material which on one recrystallization gives only 3–5 g. of pure acenaphthenol. The total yield amounts to 120–126 g. (70–74% based on the acenaphthene).

2. Notes

1. The "95% acenaphthene" sold by Reilly Tar and Chemical Corporation melts at 92.5–93.5° (cor.) and is quite satisfactory for use in this reaction. A recrystallized sample of this acenaphthene (m.p. 93–93.5°) or acenaphthene from the Gesellschaft für Teerverwertung (m.p. 93–93.5°) gives no better yield of pure acenaphthenol.

2. The glacial acetic acid should be purified by distillation from potassium permanganate. About 30–50 g. of potassium permanganate for each 1.5 l. of acetic acid should be used.

3. Mallinckrodt's analytical reagent red lead (assay 85–90%) was used. Merck's and Baker's N.F. V red lead are also quite satisfactory. Previously prepared lead tetraacetate is in no way preferable to red lead for this oxidation.

4. If the oxidation is carried out at 50° the yield is unaffected, but several hours are required to complete the addition. At 40°, the reaction is very slow and the yield is lowered.

5. A drop of the reaction mixture is placed on a moist piece of starch-iodide paper. The development of a blue color shows the presence of lead tetraacetate.

6. The acenaphthenol acetate contains small amounts of acenaphthene and acenaphthenone but is pure enough for the next step.

7. The dark violet color appearing on addition of the alkali is probably due to the presence of acenaphthenone. Crystalline acenaphthenol begins to separate almost immediately after the alkali has been added. Care must be taken in heating to refluxing because when heated too rapidly the acenaphthenol crystallizes suddenly from solution and the heat evolved may blow part of it out through the condenser.

8. If the charcoal treatment is omitted, the acenaphthenol obtained is light yellow but practically pure.

9. Marquis [2] reported the melting point as 148°; von Braun and Bayer [3] reported it as 146°.

3. Methods of Preparation

Acenaphthenol has been prepared in poor yield by the oxidation of acenaphthene with lead dioxide; [2] and it is among the products obtained by hydrogenation of acenaphthene quinone.[3] The above procedure is essentially that described more briefly in the literature.[4]

[1] Hershberg, *Ind. and Eng. Chem., Anal. Ed.*, **8**, 313 (1936); *Org. Syntheses* Coll. Vol. **2**, 117 (1943).

[2] Marquis, *Compt. rend.*, **182**, 1227 (1926).
[3] von Braun and Bayer, *Ber.*, **59**, 920 (1926).
[4] Fieser and Cason, *J. Am. Chem. Soc.*, **62**, 432 (1940).

β-(3-ACENAPHTHOYL)PROPIONIC ACID

(Acenaphthenebutyric acid, γ-oxo)

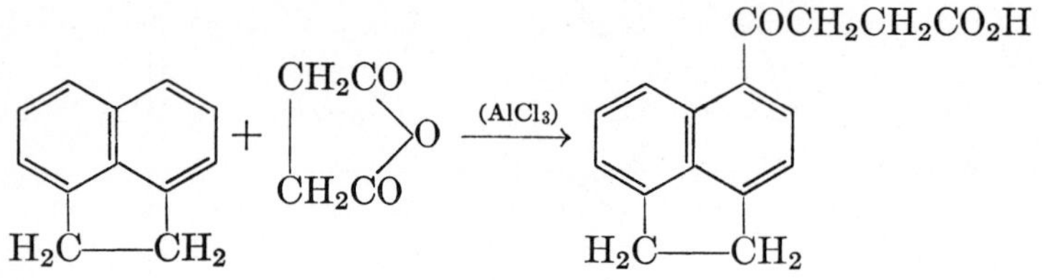

Submitted by L. F. Fieser.
Checked by W. W. Hartman and A. Weissberger.

1. Procedure

In a 3-l. round-bottomed three-necked flask (Note 1), 100 g. (0.65 mole) of pure acenaphthene (Note 2) and 72 g. (0.72 mole) of succinic anhydride are dissolved by warming in 600 ml. of nitrobenzene. The flask is clamped in a large ice bath. Through the central opening is inserted a mercury-sealed mechanical stirrer. A second opening is connected to a gas trap and also carries a thermometer; the third is for the introduction of aluminum chloride. After the mixture has been cooled to about 0°, 195 g. (1.46 moles) of aluminum chloride is added in small portions in the course of 1 hour, the temperature being kept below 5°. Stirring is continued at 0° for 4 more hours, after which time the mixture is allowed to stand for at least 12 hours so that the ice melts and the clear red solution gradually comes to room temperature.

The flask is cooled by immersion in a slush of ice and water, and the addition compound is decomposed by adding gradually 200 g. of ice, 100 ml. of water, and 100 ml. of concentrated hydrochloric acid (this is best done under a hood). The keto acid separates in the form of a stiff, grayish white paste. The solvent is removed by steam distillation, in which operation it is advisable to use a very rapid flow of steam together with an efficient condensing system, such as that illustrated (Fig. 1). The condensing flask shown need be no more than 1 l. in capacity, and the exit tube should be centered at the bottom of this flask so that it can drain the contents completely. Some condensate ordinarily remains in the flask to serve as a vapor seal, a factor which adds greatly to the efficiency of condensation. For purposes of inspection, the flask can be emptied by diverting for a moment the stream of water to one side of the flask. The stoppers which are under pressure should be secured with wire. The distilling

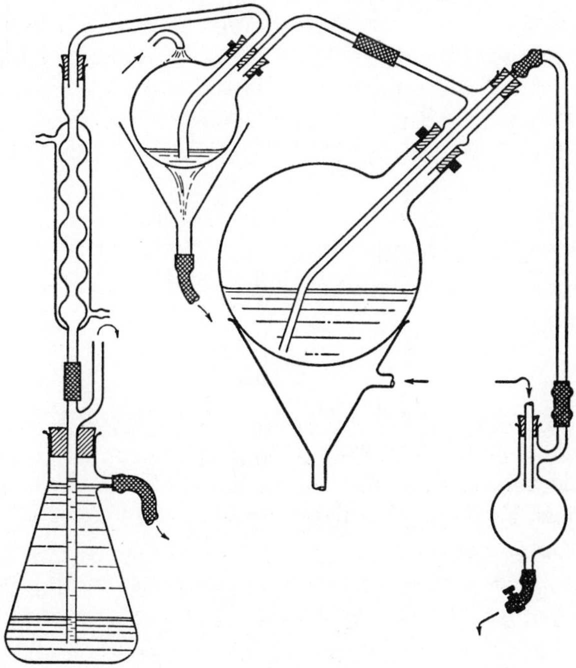

FIG. 1.

flask is heated to prevent too much condensation of steam (Note 3).

The bulk of the nitrobenzene comes over in about 1.5 hours, and the product then separates as a pasty mass which slowly disintegrates to a powder. During this process the elimination of nitrobenzene is very slow, but the steaming should be continued until only a few small lumps remain (4–5 hours), although it is not necessary to remove every trace of solvent (Note 4). The mixture is cooled with tap water, the crude acid is filtered and returned to the flask, and 115 g. of sodium carbonate decahydrate is added, together with sufficient water to make the flask a little less than half full. The mixture is heated with shaking over a free flame until most of the solid has dissolved and the frothing has diminished. A few drops of capryl alcohol may be added to dissipate the froth. The dark brown solution is steam-distilled to eliminate the last traces of nitrobenzene (about 30 minutes) and then filtered by suction from a very light residue. One hundred grams of sodium chloride is dissolved in the hot solution (volume, about 1.5 l.), which is then allowed to cool without disturbance. The sodium salt of β-(3-acenaphthoyl)propionic acid separates as colorless, fibrous needles, while the isomeric 1-acid largely remains in solution (Note 5). The product is collected on a large Büchner funnel and washed free of the dark mother liquor with half-saturated sodium chloride solution (about 150 ml.), the combined filtrates (A) being set aside. The sodium salt is crystallized once more from boiling water (1–1.5 l.) (Note 6), using Norit if required, and adding 50 g. of sodium chloride to the hot filtered solution. The mother liquor (B) is again saved. The purified salt is dissolved in 1.2 l. of hot water and the solution is acidified. The free acid separates as a white powder in a very pure condition. The yield of β-(3-acenaphthoyl)propionic acid melting at 206–208° with decomposition is 133 g. (81%) (Note 7).

2. Notes

1. By using a flask suitable for steam distillation, the loss in time and material attending a transfer is avoided.

2. Suitable material is supplied by Reilly Tar and Chemical Corporation, New York, or Gesellschaft für Teerverwertung, Duisburg-Meiderich, Germany.

3. As compared with the apparatus shown in Fig. 24 of *Org. Syntheses* Coll. Vol. **1,** 479, this arrangement requires a much smaller flask and yet offers unlimited capacity. It also enables the operator to observe more closely the nature of the distillate.

4. If the flask fills up with condensed steam, it should be cooled, the contents filtered, and the product returned to the flask. The process of disintegration can be hastened by breaking up the lumps with a flattened stirring rod.

5. The small amount of isomeric 1-acid may be obtained from the mother liquors, A and B. The first of these on acidification gives a product which is dark and tarry, but which soon solidifies on being cooled and stirred. The material is dissolved in 1 l. of water containing 25–30 g. of sodium carbonate decahydrate, and the solution is boiled for 30 minutes with Norit, filtered, cooled, and acidified. The product, which now solidifies at once and is lighter in color, is dried and combined with the material obtained by acidifying the second mother liquor, B (total amount, 23.9 g.). The crude mixture of acids

is suspended in 170 ml. of cold methanol, 8.5 ml. of concentrated sulfuric acid is added, and the mixture is heated on the steam bath for about 10 minutes, after which dissolution and esterification are complete. The dark product that crystallizes when the solution cools is largely the 1-ester, which is very much less soluble than the 3-ester. The 1-ester (13.1 g.) is washed free of acid; it crystallizes from ethanol with the use of decolorizing carbon in long needles melting at 126°; yield, 9 g. (Note 8). The ester is hydrolyzed by heating with 100 ml. of alcohol and 30 ml. of 25% sodium hydroxide solution until dissolved; the solution is then diluted with water and acidified. The β-(1-acenaphthoyl)propionic acid melts at 180° (crystallized from dilute alcohol, 181°) and weighs 8.4 g. (5%) (Note 9).

6. If, owing to hydrolysis, the sodium salt fails to dissolve completely, alkali should be added as required.

7. The 3-acid crystallizes well from glacial acetic acid, alcohol, or xylene, but large volumes of solvent are required and there is no change in the melting point.

8. The 3-ester melts at 89°.

9. Ordinarily the mother liquors from the preparation and purification of 1-ester will be discarded, but a small additional quantity of the 3-acid may be obtained by concentrating these solutions, adding alkali to hydrolyze the ester, adding water, and acidifying. The precipitated material is purified by crystallizing the sodium salt twice, and from this 8 g. (5%) of the pure 3-acid is obtained.

The ratio of 3-acid to 1-acid is dependent on the temperature, lower temperatures favoring the production of 3-acid. At −15° the yield of 3-acid is 87%, and of 1-acid, 5%. At room temperature there is some increase in the proportion of the 1-acid formed, but the product is very dark and difficult to work up, and the total yield is lower even though the aluminum chloride is added in nitrobenzene solution.

3. Methods of Preparation

This procedure is based upon a study [1] of the method outlined in the patent literature.[2] The procedure is a general one and may be used for the condensation of succinic anhydride with naphthalene and with the mono- and dimethylnaphthalenes, although in no other case are the purification and separation of isomers so easily accomplished. In this particular type of condensation, as well as in certain other types of Friedel-Crafts reactions, nitrobenzene is far superior to the solvents that are more frequently employed. This is partly because of its great solvent power and partly because it forms a molecular compound with aluminum chloride, and so decreases the activity of the catalyst in promoting side reactions.

[1] Fieser, *J. Am. Chem. Soc.*, **54**, 4350 (1932).

[2] Fr. pat. 636,065 (*Chem. Zentr.*, **1928**, I, 2751); Swiss pat. 131,959 (*Chem. Zentr.*, **1930**, I, 1539); U. S. pat. 1,759,111 (*Chem. Zentr.*, **1930**, II, 806); Ger. pat. 376,635 [*Frdl.*, **14**, 285 (1926)].

ACETOACETANILIDE *

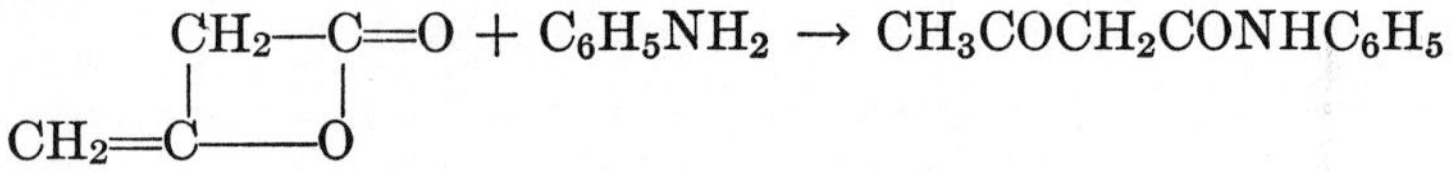

Submitted by Jonathan W. Williams and John A. Krynitsky.
Checked by Nathan L. Drake and Joseph Lann.

1. Procedure

In a 500-ml. round-bottomed three-necked flask fitted with a reflux condenser, a dropping funnel, and a mercury-sealed stirrer (Note 1) is placed a solution of 46 g. (0.5 mole) of dry aniline in 125 ml. of pure dry benzene. Stirring is started, and a solution of 42 g. (0.5 mole) of ketene dimer (p. 508) in 75 ml. of pure dry benzene is added dropwise over a period of 30 minutes. The reaction mixture is then heated under reflux on the steam bath for 1 hour. After the major portion of the benzene has been removed by distillation from the steam bath, the remainder is removed under reduced pressure. The residue is dissolved in 500 ml. of hot 50% aqueous ethanol from which the acetoacetanilide separates on cooling. The mixture is cooled to 0° before filtration. A second crop of crystals can be obtained by adding 250 ml. of water to the mother liquor and cooling again (Note 2). The total yield of product, m.p. 82–83.5°, is 65 g. (74%). Further purification by recrystallization from 300 ml. of 50% ethanol yields 55 g. of a product that melts at 84–85°.

2. Notes

1. A seal of rubber tubing lubricated by glycerol is satisfactory.

2. If the second mother liquor is evaporated to about half of its original volume, a small third crop of very impure crystals may be obtained.

3. Methods of Preparation

Acetoacetanilide has been prepared by the reaction of aniline with ethyl acetoacetate [1–5] or acetoacetyl chloride,[6] and by the reaction of ketene dimer with aniline.[7,8]

[1] Knorr, *Ann.*, **236**, 69 (1886).
[2] Roos, *Ber.*, **21**, 624 (1888).
[3] Knorr and Reuter, *Ber.*, **27**, 1169 (1894).
[4] Mizuno, *J. Pharm. Soc. Japan*, **69**, 126 (1949).
[5] U. S. pat. 2,416,738 [*C. A.*, **41**, 3485 (1947)].
[6] Hurd and Kelso, *J. Am. Chem. Soc.*, **62**, 1548 (1940).
[7] Chick and Wilsmore, *J. Chem. Soc.*, **1908**, 946.
[8] Boese, *Ind. Eng. Chem.*, **32**, 16 (1940).

ACETOBROMOGLUCOSE

(2,3,4,6-Tetraacetyl-*α*-*d*-glucopyranosyl bromide)

$$HOCH_2\underbrace{CH(CHOH)_3CHOH}_{O}\cdot H_2O + 6(CH_3CO)_2O \rightarrow$$

$$CH_3COOCH_2\underbrace{CH(CHOCOCH_3)_3CHOCOCH_3}_{O} + 7CH_3CO_2H$$

$$CH_3COOCH_2\underbrace{CH(CHOCOCH_3)_3CHOCOCH_3}_{O} + HBr \rightarrow$$

$$CH_3COOCH_2\underbrace{CH(CHOCOCH_3)_3CHBr}_{O} + CH_3CO_2H$$

Submitted by C. E. Redemann and Carl Niemann.
Checked by Lee Irvin Smith, R. T. Arnold, and Iver Lerohl.

1. Procedure

In a 1-l. round-bottomed flask are placed 66 g. (0.33 mole) of *d*-glucose monohydrate (Note 1) and 302 g. of 95% acetic anhydride (280 ml., 2.81 moles). To this mixture 3 small drops of concentrated sulfuric acid are added *from a medicine dropper* (Note 2). The glucose is kept in partial suspension by shaking the flask with a swirling motion; the reaction starts almost immediately. If the temperature of the mixture approaches the boiling point, the flask is momentarily immersed in a pan of cold water. Within 10–15 minutes nearly all the glucose will have dissolved and the temperature of the reaction mixture will have risen nearly to 100°. The flask is loosely stoppered and is heated on a steam bath for 2 hours. Then about 200 ml. of mixed acetic acid and acetic anhydride is removed by distillation under reduced pressure (Note 3).

Sixty-five grams (60 ml., 0.64 mole) of acetic anhydride is added to the warm, viscous, light-yellow syrup; the mixture is warmed slightly and is mixed by imparting a swirling motion to the flask until the solution is homogeneous. The flask is then fitted with a two-holed rubber stopper bearing an inlet tube and an exit tube, the former reaching within 5 mm. of the bottom. Dry hydrogen bromide is passed into the mixture, while it is cooled in an ice bath, until the gain in weight is 140–160 g. (Note 4). The flask is then sealed with a rubber stopper and allowed to stand at 5° overnight.

The hydrogen bromide, acetic acid, and acetic anhydride are then removed from the straw-yellow solution by distillation under reduced pressure; a water bath whose temperature does not exceed 60° should be used to heat the mixture (Note 5). During the distillation the solution becomes slightly darker. When no further distillate comes over, or when the residue crystallizes, distillation is stopped, 250–300 ml. of dry isopropyl ether is added (Note 6), and the flask is warmed carefully on a water bath to hasten solution of the product (Note 7). The hot solution is transferred to a 1-l. Erlenmeyer flask and is cooled rapidly, with cold water, to about 45°. The mixture is then allowed to cool slowly to room temperature and is finally placed in a refrigerator at 5° for 2 or more hours. The acetobromoglucose is collected on a Büchner funnel, pressed into a firm cake, and washed with about 50 ml. of dry isopropyl ether. The white crystalline material, after being dried under reduced pressure over calcium (or sodium) hydroxide, weighs 110–120 g. (80–87%) (Note 8).

2. Notes

1. Ordinary commercial glucose monohydrate (Cerelose, Clintose, etc.) was used.

2. If any uncertainty exists about the size of the drops, it is better to add 2 drops at first and then to wait at least 10 minutes before adding the third drop. If too much sulfuric acid is added the reaction may become so vigorous that it cannot be controlled.

3. The lowest pressure attainable with a good aspirator is satisfactory. The acetic acid may be removed rapidly at bath temperatures up to 100°. Two hundred milliliters of distillate is collected in about

35 minutes. Complete removal of the acetic acid is not necessary; the only purpose of removing it is to decrease the amount of hydrogen bromide required.

4. Hydrogen bromide may be generated by dropping liquid bromine into boiling tetrahydronaphthalene, or catalytically from hydrogen and bromine (*Org. Syntheses,* **15,** 35). Free bromine should be removed from the gas by passing it over red phosphorus. The hydrogen bromide may be passed in very rapidly at first, but as the solution becomes more concentrated the rate of introduction of the gas must be decreased. The absorption of 140 g. of hydrogen bromide requires about 2 hours if the solution is well stirred; otherwise a longer time is required.

5. This distillation is best conducted in the following manner: The hydrogen bromide is removed as completely as possible at a bath temperature of 40–50° under the pressure attainable with a good aspirator; this requires about 1 hour. The bath temperature is then slowly increased to 50–60° over a period of about 30 minutes, during which time considerable acetic acid and acetic anhydride are removed. The receiver is then emptied and the system is connected to a mechanical pump capable of maintaining a pressure of less than 5 mm. By keeping the temperature of the bath at 55–60°, sufficient acetic acid and anhydride are removed in 30–45 minutes. The mechanical pump must be adequately protected; the vapors should be passed through a trap cooled in carbon dioxide-ethanol, then through a 12- to 16-in. tower of flake sodium hydroxide, before they enter the pump. A somewhat higher bath temperature can be used with a good aspirator alone, but this will produce much darkening of the reaction mixture and will give a less desirable product.

6. Isopropyl ether which has been in contact with air for some time will contain peroxides. They should be removed by washing the ether, first with sodium bisulfite solution, then with sodium hydroxide solution, and finally with water. The ether is dried (24 hours over calcium chloride, then 24 hours over phosphorus pentoxide) and distilled.

7. Heating to effect solution should be as brief as possible. The water bath should be at a temperature near the boiling point, and the flask should be immersed for only short periods of time. The flask should be shaken continuously during this process.

8. The product, m.p. 87–88°, is satisfactory for most purposes. A single recrystallization from isopropyl ether gives a product having a melting point of 88–89°.

The product is best stored in a vacuum desiccator.

3. Methods of Preparation

α-Acetobromoglucose has been prepared by the action of acetyl bromide on anhydrous glucose;[1–4] by the action of hydrogen bromide in acetic acid upon β-pentaacetylglucose;[5–8] by the action of hydrogen bromide in acetic anhydride upon anhydrous glucose;[9,10] and by the action of hydrogen bromide in acetic anhydride upon starch or maltose.[11]

[1] Koenigs and Knorr, *Ber.*, **34**, 961 (1901).
[2] Colley, *Ber.*, **34**, 3206 (1901).
[3] Moll van Charante, *Rec. trav. chim.*, **21**, 43 (1902).
[4] Brauns, *J. Am. Chem. Soc.*, **47**, 1280 (1925).
[5] Fischer, *Ber.*, **49**, 584 (1916).
[6] Freudenberg, Noe, and Knopf, *Ber.*, **60**, 241 (1927).
[7] Fischer and Armstrong, *Ber.*, **34**, 2892 (1901).
[8] Fischer, *Ber.*, **44**, 1901 (1911).
[9] Dale, *J. Am. Chem. Soc.*, **38**, 2187 (1916).
[10] Levene and Raymond, *J. Biol. Chem.*, **90**, 247 (1931).
[11] Bergmann and Beck, *Ber.*, **54**, 1576 (1921).

2-ACETOTHIENONE

(Ketone, methyl 2-thienyl)

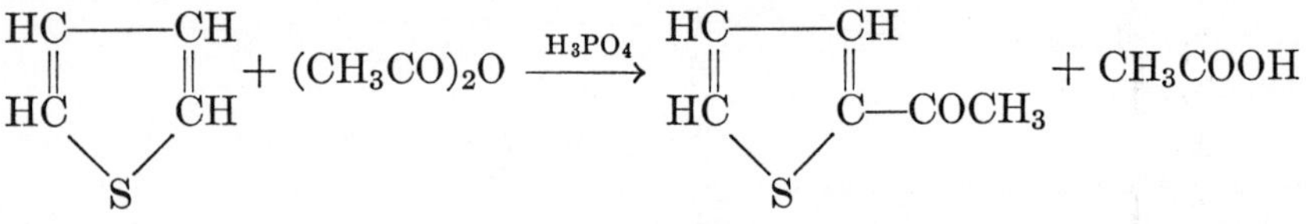

Submitted by Alvin I. Kosak and Howard D. Hartough.
Checked by George T. Gmitter, F. Lee Benton, and Charles C. Price.

1. Procedure

In a 1-l. three-necked flask fitted with a mechanical stirrer, a thermometer, and a reflux condenser are placed 168 g. (2 moles) (Note 1) of thiophene (Note 2) and 107 g. (1 mole) of 95% acetic anhydride (Note 3). The solution is heated to 70–75°, the source of heat is removed, and 10 g. (6 ml.) of 85% phosphoric acid is added with stirring. After 2–3 minutes an exothermic reaction occurs, and it is necessary to immerse the flask in a cold water bath to control the reaction.

The boiling subsides in a few minutes; heat is again applied, and the mixture is refluxed for a total of 2 hours. The cooled mixture is washed successively with one 250-ml. portion of water and two 100-ml. portions of 5% sodium carbonate and is dried over anhydrous sodium sulfate. The orange-red liquid is distilled through a short fractionating column. After the removal of 76–80 g. of unchanged thiophene (b.p. 83–84°) by distillation at atmospheric pressure the residue is distilled under reduced pressure. The yield of 2-acetothienone, b.p. 89–90°/10 mm. (m.p. 9.2–10.5°; n_D^{20} 1.5662), is 93–100 g. (74–79%).

2. Notes

1. Acetic anhydride rather than thiophene may be used in excess, but the unchanged reagent cannot be recovered by the procedure given. With a 3:1 mole ratio of thiophene to anhydride the yield is of the order of 85%.
2. Commercial 99+% thiophene was employed.
3. The use of an equivalent amount of freshly distilled 100% acetic anhydride does not improve the yield.

3. Methods of Preparation

In addition to the methods of preparation given in connection with the procedure [1] for the acetylation of thiophene with acetyl chloride in the presence of stannic chloride, 2-acetothienone has been prepared from thiophene and either acetyl chloride or acetic anhydride in the presence of iodine,[2] hydriodic acid,[2] silica-metal oxides,[3] zinc chloride,[4] inorganic oxyacids,[5,6] and boron trifluoride.[7–9] It has also been prepared from thiophene and acetic acid in the presence of hydrogen fluoride [5] or phosphorus pentoxide.[10] The acylation in the presence of phosphorus pentoxide is particularly useful with higher aliphatic acids.[10]

Procedures using acetic anhydride and stannic chloride or ferric chloride have been described.[11]

[1] *Org. Syntheses* Coll. Vol. **2,** 8 (1943).
[2] Hartough and Kosak, *J. Am. Chem. Soc.,* **68,** 2639 (1946).
[3] Hartough, Kosak, and Sardella, *J. Am. Chem. Soc.,* **69,** 1014 (1947).
[4] Hartough and Kosak, *J. Am. Chem. Soc.,* **69,** 1012 (1947).
[5] Hartough and Kosak, *J. Am. Chem. Soc.,* **69,** 3093 (1947).
[6] U. S. pat. 2,496,786 [*C. A.,* **44,** 4930 (1950)].
[7] Heid and Levine, *J. Org. Chem.,* **13,** 409 (1948).

[8] Hartough and Kosak, *J. Am. Chem. Soc.*, **70**, 867 (1948).
[9] Levine, Heid, and Farrar, *J. Am. Chem. Soc.*, **71**, 1207 (1949).
[10] Hartough and Kosak, *J. Am. Chem. Soc.*, **69**, 3098 (1947).
[11] Farrar and Levine, *J. Am. Chem. Soc.*, **72**, 4433 (1950).

ACETYLACETONE *

(Diacetylmethane; 2,4-pentanedione)

I. BORON TRIFLUORIDE METHOD

$$CH_3COCH_3 + (CH_3CO)_2O \xrightarrow{(BF_3)} CH_3COCH_2COCH_3 + CH_3COOH$$

Submitted by C. E. Denoon, Jr.
Checked by Homer Adkins and Ivan A. Wolff.

1. Procedure

One hundred and sixteen grams (2 moles) of acetone (Note 1) and 510 g. (5 moles) of reagent grade acetic anhydride are placed in a 2-l. three-necked flask and cooled in an ice-salt bath. One neck of the flask is stoppered; the second neck contains a tube for admitting boron trifluoride; and the third neck contains an outlet tube leading to an alkali trap to catch any unabsorbed boron trifluoride. Commercial grade boron trifluoride (Note 2) is passed through a Kjeldahl bulb, to prevent the reaction mixture from sucking back into the cylinder, and is then bubbled into the reaction mixture at such a rate that 500 g. is absorbed in about 5 hours (2 bubbles per second). The reaction mixture is poured into a solution of 800 g. of hydrated sodium acetate in 1.6 l. of water contained in a 5-l. flask. The mixture is then steam-distilled and the distillate collected in the following portions: 1 l., 500 ml., 500 ml., 400 ml.

A solution of reagent grade hydrated copper acetate is made by dissolving 240 g. of the salt in 3 l. of water at about 85° and filtering from any basic acetate. The copper salt of acetylacetone is then precipitated by adding 1.4 l. of the hot copper acetate solution to the first fraction of the acetylacetone, 700 ml. to the second, 500 ml. to the third, and 400 ml. to the fourth fraction. After standing for 3 hours, or better overnight, in a refrigerator the salt is filtered, washed once with water, and sucked dry. The salt is shaken in a separatory funnel with 800 ml. of 20% sulfuric acid and 800 ml. of ether, and the ether layer is removed. The aqueous layer is extracted with 400 ml. and then 200 ml. of ether. The combined extracts are dried with 250 g. of anhydrous sodium sulfate, and the ether is removed by distillation. The residue is distilled through a Widmer column (Note 3) and yields 160–170 g. of acetylacetone boiling at 134–136° (80–85% based on acetone).

2. Notes

1. Acetone is preferably dried over anhydrous potassium carbonate or anhydrous calcium sulfate, followed by phosphorus pentoxide if a very dry product is required. Calcium chloride is commonly used (100–150 g. per liter), but this is less satisfactory since it combines chemically with acetone.[1] For this preparation the checkers used acetone that had been dried over calcium chloride, followed by distillation from phosphorus pentoxide.

2. Boron trifluoride may be purchased in cylinders from Harshaw Chemical Company, Cleveland, Ohio.

3. The Widmer column used contained a spiral 15 cm. in length, 13 mm. in diameter, with 15 turns of the helix.

II. SODIUM ETHOXIDE METHOD

$$CH_3CO_2C_2H_5 + CH_3COCH_3 + NaOC_2H_5 \rightarrow CH_3C(ONa){=}CHCOCH_3 + 2C_2H_5OH$$

$$CH_3C(ONa){=}CHCOCH_3 + H_2SO_4 \rightarrow CH_3COCH_2COCH_3 + NaHSO_4$$

Submitted by Homer Adkins and James L. Rainey.
Checked by R. L. Shriner and Neil S. Moon.

1. Procedure

Sixty-nine grams (3 gram atoms) of sodium, from which all the oxide coating has been cut away, and 400 ml. of dry xylene (Note 1) are placed in a 1-l. round-bottomed flask and heated until the sodium is melted. The flask is closed with a rubber stopper (Note 2), and the sodium is finely powdered by vigorous shaking. The contents of the flask are transferred to a 3-l. three-necked flask, and the xylene decanted. The sodium is washed with two 100-ml. portions of anhydrous ether (Note 3) by decantation. One liter of anhydrous ether is added, and the flask is placed on a steam bath and fitted with a condenser,

Hershberg stirrer (*Org. Syntheses,* **17,** 31), and a 250-ml. dropping funnel. The condenser and dropping funnel are protected by drying tubes containing absorbent cotton (Note 4). One hundred and thirty-eight grams (175 ml., 3 moles) of anhydrous ethanol is placed in the dropping funnel, and the stirrer is started. The alcohol is dropped in over a period of 2–3 hours with gentle refluxing. The reaction mixture is refluxed with stirring for 6 hours (Note 5) after the addition of the alcohol. The stirrer is stopped, the condenser turned downward, and the ether distilled as completely as possible from the steam bath (Note 6).

The condenser is again arranged for refluxing, and 1.2 l. of ethyl acetate (Note 7) is added to the warm sodium ethoxide through the separatory funnel as rapidly as possible. The stirrer is started immediately, and 174 g. (220 ml., 3 moles) of acetone (Note 1, p. 17) is dropped in over a period of 15–20 minutes, refluxing being maintained by heating if necessary. Addition of the acetone must be started as soon as the ethyl acetate has been added. During the addition the solution becomes quite red, and then the mixture turns brown (Note 8). The mixture is refluxed for 1 hour; the stirrer is then stopped and the contents of the flask are allowed to stand at room temperature for 12 hours, during which time crystals of the sodium salt separate.

The liquid layer is decanted into a 5-l. flask, and the sodium salt of the diketone is dissolved and washed into the flask with 2.5 l. of ice water. After the salt is dissolved, the ester layer is separated as soon as possible (Note 9). The water layer is extracted twice with 300-ml. portions of ether, and the ether extract is discarded. To the water solution is added ice-cold dilute sulfuric acid (150 g. of concentrated sulfuric acid and 400 g. of cracked ice) until the solution is just acid to litmus. The diketone is extracted from the solution with four 300-ml. portions of ether. The combined ether extracts are dried for 24 hours over 60 g. of anhydrous sodium sulfate in the icebox. The ether solution is decanted into a 2-l. round-bottomed flask, and the sodium sulfate is extracted with 100 ml. of anhydrous ether. This extract is added to the ether solution, and the ether is distilled by means of a steam bath. The residue is transferred to a 500-ml. flask, rinsing with a little ether, and distilled through a Widmer column, the portion boiling between 130° and 139° being collected. This fraction is dried over 5 g. of anhydrous potassium carbonate for 1 hour and, after the carbonate has been removed, is redistilled through the Widmer column.

The portion boiling at 134–136° is collected; it amounts to 115–136 g. (38–45% based on acetone).

2. Notes

1. The xylene is dried by distillation from sodium.

2. Rubber stoppers should be used throughout, including the drying of reagents, as corks contain some moisture. The stoppers should be boiled in 10% sodium hydroxide solution for 2 hours, thoroughly washed with dilute acetic acid, and dried.

3. Commercial anhydrous ethyl ether and ethanol are satisfactory. If these are unavailable, the ether should be purified as for use in the Grignard reaction and the ethanol as described in *Org. Syntheses* Coll. Vol. **1,** 249 (1941).

4. Absorbent cotton is an excellent drying agent and more convenient for drying tubes than anhydrous calcium chloride.[2] It is possible to keep maleic anhydride in a flask, closed only by a plug of absorbent cotton, for 3 weeks without appreciable change in the titration value (F. P. Pingert, private communication).

5. The period of heating varies somewhat with the size of the powdered sodium. Almost all the sodium should be used up before removal of the ether. However, a few small pieces do no harm.

6. The success of the reaction depends upon the quality of the sodium ethoxide used. The product at this point should be white and very finely divided. All moisture must be excluded during its preparation in order to avoid the formation of sodium hydroxide, which markedly lowers the yield.

7. The ethyl acetate is allowed to stand over calcium chloride for 2 days, with occasional shaking. The calcium chloride is removed by filtration, and the ester is allowed to stand over phosphorus pentoxide several hours. It is then distilled directly from the phosphorus pentoxide.

8. After about half of the acetone has been added, the mixture usually sets to a solid mass. The stirrer is turned by hand and the addition of acetone continued. In a few minutes the mass can again be stirred.

9. The ethyl acetate layer is washed with water, sodium bisulfite solution, saturated calcium chloride solution, and again with water. It is further purified as in Note 6, giving 316–400 g. of recovered ester. The amount of recovered ester depends somewhat upon the length of time the two layers are allowed to remain in contact before separating.

3. Methods of Preparation

Acetylacetone has been prepared by the reaction of acetyl chloride with aluminum chloride, followed by hydrolysis; [3] by the condensation of acetone with ethyl acetate in the presence of sodium,[4] sodium amide,[5,6] sodium ethoxide,[5,7,8] and alkali or alkaline-earth hydrides; [9] by the reaction of acetone and acetic anhydride in the presence of boron trifluoride; [10] by the pyrolysis of isopropenyl acetate; [11–13] by the reaction of ethyl acetoacetate and acetic anhydride in the presence of magnesium at 140°; [11] from methyl or ethyl diacetylacetate by treatment with acids; [14] and by the dehydrogenation of 4-pentanol-2-one in the presence of Raney nickel.[15]

[1] Bagster, *J. Chem. Soc.*, **1917**, 494.
[2] Obermiller and Goertz, *Z. physik. Chem.*, **109**, 162 (1924)
[3] Combes, *Ann. chim. et phys.*, (6) **12**, 207 (1887).
[4] Claisen, *Ann.*, **277**, 168 (1893).
[5] Claisen, *Ber.*, **38**, 695 (1905).
[6] Adams and Hauser, *J. Am. Chem. Soc.*, **66**, 1220 (1944).
[7] Claisen and Ehrhardt, *Ber.*, **22**, 1010 (1889).
[8] Sprague, Beckham, and Adkins, *J. Am. Chem. Soc.*, **56**, 2665 (1934).
[9] U. S. pat. 2,158,071 [*C. A.*, **33**, 6342 (1939)].
[10] Meerwein and Vossen, *J. prakt. Chem.*, **141**, 149 (1934).
[11] U. S. pat. 2,395,800 [*C. A.*, **40**, 3130 (1946)].
[12] Brit. pat. 615,523 [*C. A.*, **43**, 7954 (1949)].
[13] Hagemeyer and Hull, *Ind. Eng. Chem.*, **41**, 2920 (1949).
[14] U. S. pat. 2,395,012 [*C. A.*, **40**, 3130 (1946)].
[15] DuBois, *Compt. rend.*, **224**, 1734 (1947).

ACETYLBENZOYL *

(1,2-Propanedione, 1-phenyl-)

$$CH_3C(=NOH)COC_6H_5 + H_2O \xrightarrow{H_2SO_4} CH_3COCOC_6H_5 + NH_2OH$$

Submitted by W. W. Hartman and L. J. Roll.
Checked by A. H. Blatt and Lewis Rothstein.

1. Procedure

In a 1-l. flask arranged for steam distillation (Note 1), 50 g. (0.31 mole) of isonitrosopropiophenone (*Org. Syntheses*, **16**, 44; Coll. Vol. **2**, 363) and 500 g. of 10% sulfuric acid are mixed, and the mixture is distilled with steam until about 2 l. of distillate is collected. During the distillation the flask is heated so that the volume of the reaction mixture is kept roughly constant. The distillation requires about 6 hours, and at the end of this time the liquid in the flask is clear (Note 2).

The lower, yellow layer of diketone in the distillate is separated; the water layer is then saturated with salt and extracted with ether, one 80-ml. and two 25-ml. portions being used for each liter of the aqueous solution. The ether extracts are combined with the diketone proper and dried over sodium sulfate. The ether is removed on the steam bath, and the residual material is distilled from a Claisen flask under reduced pressure. Acetylbenzoyl is collected at 114–116°/20 mm. (Note 3). The yield is 30–32 g. (66–70% of the theoretical amount) (Note 4).

2. Notes

1. A spray trap should be placed between the flask and the condenser; otherwise some isonitrosoketone will be carried over. The checkers used a "Kjeldahl Connecting Bulb, Cylindrical Type," illustrated as item 2020 in the Pyrex Catalog, LP–34, for 1954.

2. On being cooled, the reaction mixture deposits 3–7 g. of the dioxime of acetylbenzoyl, m.p. 234–236° dec.[1]

3. Other boiling points reported for acetylbenzoyl are 216–218°, 164–165°/116 mm., and 102–103°/12 mm.

4. According to the submitters the reaction can be carried out with about the same percentage yields using five times the amounts of material specified above.

3. Methods of Preparation

Acetylbenzoyl has been prepared by dehydrogenation of acetylphenylcarbinol over copper at 350°;[2] by oxidation of 1-methyl-2-phenylethylene glycol;[3] by action of amyl nitrite on isonitrosopropiophenone;[4] and by acid hydrolysis of isonitrosopropiophenone[5] or of isonitrosobenzyl methyl ketone.[6] The last-named reaction is reported to furnish quantitative yields of acetylbenzoyl, but the starting material is not easily accessible. Treatment of propiophenone with amyl nitrite (2 moles) without isolating the isonitroso compound has been mentioned as a method of preparing acetylbenzoyl,[4] but the yield of diketone is much better when the isonitroso compound is isolated.[7]

Acetylbenzoyl also has been prepared by the oxidation of phenyl acetone with selenium dioxide.[8]

[1] Müller and Pechmann, *Ber.*, **22**, 2128 (1889).
[2] Mailhe, *Bull. soc. chim. France*, (4) **15**, 326 (1914).
[3] Zincke and Zehn, *Ber.*, **43**, 855 (1910).
[4] Manasse, *Ber.*, **21**, 2177 (1888).
[5] Pechmann and Müller, *Ber.*, **21**, 2119 (1888).
[6] Kolb, *Ann.*, **291**, 286 (1896); Borsche, *Ber.*, **40**, 740 (1907).
[7] Coles, Manske, and Johnson, *J. Am. Chem. Soc.*, **51**, 2269 (1929).
[8] Wegmann and Dahn, *Helv. Chim. Acta*, **29**, 1247 (1946).

1-ACETYLCYCLOHEXENE

(Ketone, 1-cyclohexenyl methyl)

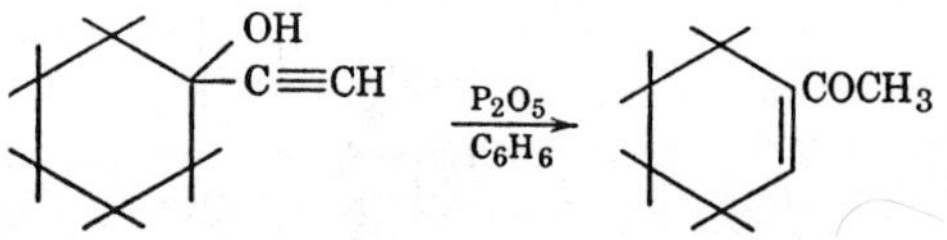

Submitted by J. H. SAUNDERS.[1]
Checked by E. L. JENNER and R. S. SCHREIBER.

1. Procedure

In a 500-ml. round-bottomed flask are placed 40 g. (0.32 mole) of 1-ethynylcyclohexanol (p. 416), 250 ml. of dry benzene, 10 g. of phosphorus pentoxide, and a boiling chip. A reflux condenser is attached to the flask, and the benzene solution is refluxed gently on a steam cone for 2.5 hours. At the end of that time the contents of the flask are cooled and the benzene is decanted from the phosphorus pentoxide, washed once with 100 ml. of 5% sodium bicarbonate solution, and dried over 15 g. of anhydrous sodium sulfate. The benzene is removed by distillation at atmospheric pressure, and the acetylcyclohexene is carefully fractionated at reduced pressure, through a 15-cm. helix-packed column. The yield of material boiling at 85–88°/22 mm., n_D^{20} 1.4892, is 22.5–28 g. (56–70%).

2. Methods of Preparation

1-Acetylcyclohexene has been prepared by treating cyclohexene with acetyl chloride and aluminum chloride,[2–5] by treating 1-ethynylcyclohexanol with oxalic acid [6] or 85% aqueous formic acid,[5,7–9] and by the dehydrohalogenation and hydrolysis of ethylidenecyclohexane nitrosochloride.[10] 1-Acetylcyclohexene and its homologs also have been prepared by the addition of a suitable diene to vinylacetylene in the presence of water and a mercury salt.[11]

[1] This investigation was carried out under the sponsorship of the Office of Rubber Reserve, Reconstruction Finance Corporation, in connection with the Government Synthetic Rubber Program.

[2] Darzens, *Compt. rend.*, **150**, 707 (1910).

[3] Christ and Fuson, *J. Am. Chem. Soc.*, **59**, 895 (1937).

[4] Nightingale, Milberger, and Tomisek, *J. Org. Chem.*, **13**, 358 (1948).

[5] Hurd and Christ, *J. Am. Chem. Soc.*, **59**, 120 (1937).

[6] Levina and Vinogradova, *J. Applied Chem. U.S.S.R.*, **9**, 1299 (1936) [*C. A.*, **31**, 2587 (1937)].

[7] Rupe, Messner, and Kambli, *Helv. Chim. Acta,* **11**, 454 (1928).

[8] Fischer and Löwenberg, *Ann.*, **475**, 203 (1929).

[9] Chanley, *J. Am. Chem. Soc.*, **70**, 246 (1948).

[10] Wallach, *Ann.*, **360**, 46 (1908).

[11] U. S. pat. 2,301,515 [*C. A.*, **37**, 2015 (1943)].

2-ACETYLFLUORENE

(Ketone, 2-fluorenyl methyl)

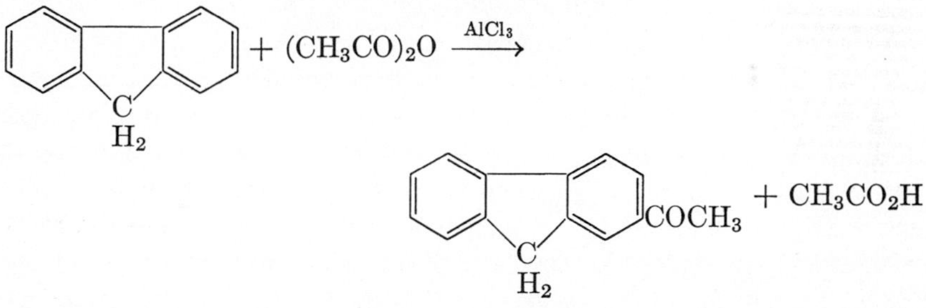

Submitted by F. E. Ray and George Rieveschl, Jr.
Checked by R. L. Shriner and Arne Langsjoen.

1. Procedure

Caution! Carbon disulfide, used as a solvent in this preparation, is highly inflammable; its vapor may ignite on contact with a hot laboratory steam line.

A 1-l. three-necked round-bottomed flask is fitted with a dropping funnel, a reflux condenser attached to a hydrogen chloride absorption trap,[1] and a very sturdy mechanical stirrer (Note 1), which may be of the mercury-sealed or rubber-sleeve type. In the flask are placed 350 ml. of dry carbon disulfide and 80 g. (0.48 mole) of fluorene (Note 2). The stirrer is started, and, after the fluorene has dissolved, 128 g. (0.96 mole) of anhydrous aluminum chloride is added in one portion. In the dropping funnel is placed 49.4 g. (0.48 mole) of redistilled acetic anhydride, and about 1 ml. of it is added dropwise to the vigorously stirred dark red reaction mixture. If the reaction does not start immediately it is initiated by warming the reaction flask in a water bath (Note 3). After the reaction has started, the balance of the acetic anhydride is added at such a rate that the carbon disulfide refluxes gently; about 45–55 minutes is required. When approximately one-half of the acetic anhydride has been added an addition complex separates as a heavy mass which makes stirring very difficult. However, stirring must be maintained to prevent excessive local reaction at the point of introduction of the acetic anhydride. The mixture is stirred and refluxed on the water bath for an hour after the addition of the acetic anhydride is complete.

The dark green mass is collected on a large Büchner funnel and transferred as quickly as possible (Note 4) to a 1-l. beaker in which it is stirred mechanically for 10 minutes with 300 ml. of carbon disulfide (Note 5). The solid is again collected and washed on the filter with two 50-ml. portions of carbon disulfide (Note 6) and with one 100-ml. portion of petroleum ether (b.p. 28–35°). The resulting granular aluminum chloride complex is decomposed by portionwise addition to a well-stirred mixture of 800 ml. of water and 30 ml. of concentrated hydrochloric acid in a 2-l. beaker under a hood. Each portion is allowed to hydrolyze before the next is added. The hydrolysis mixture should not be cooled. The crude 2-acetylfluorene is collected on a filter and washed three times with 100-ml. portions of water. After drying in an oven at 100° for 3 hours the light-orange ketone weighs 83–95 g. (83–95%) and melts over the range 113–117° (Note 7). This crude product is transferred to a 2-l. round-bottomed flask containing 800 ml. of 95% ethanol and 5 g. of decolorizing carbon. The mixture is refluxed for 1 hour and filtered hot. On cooling the filtrate deposits 71–83 g. of light-tan solid melting at 120–123°. A second recrystallization from 800 ml. of ethanol yields 55–63 g. (55–63%) of a light-cream-colored powder which melts at 124–126° and which is pure enough for most purposes (Note 8).

2. Notes

1. The stirrer, which may be of either the half-round or the propeller type, must be of heavy construction and must be driven by one of the more powerful laboratory stirring motors. Agitation must be maintained throughout the reaction period.

2. If technical fluorene (m.p. 103–107°) from Eastman Kodak Company or the Barrett Company is used, much difficulty is experienced in the purification of the product. Technical fluorene can be rendered suitable for the preparation by recrystallization from hot 95% ethanol (1 l. for 150 g.). The once-recrystallized material melts at 114–115° (lit. 116°).

3. It is necessary to make sure that the reaction has started before the addition of more acetic anhydride in order to prevent a violent reaction.

4. Exposure to the air causes the addition product to become sticky and difficult to handle.

5. Unless this operation can be conducted at a point remote from flames, hot plates, and other sources of heat, a flask should be substituted for the open beaker.

6. The carbon disulfide extracts unchanged fluorene and other impurities. Any lumps in the crude material should be crushed during the first washing. The rinsing with petroleum ether removes the last of the carbon disulfide.

7. This crude 2-acetylfluorene is completely soluble in carbon disulfide and thus is free of the insoluble 2,7-diacetylfluorene. It may be used directly for the oxidation to fluorenone-2-carboxylic acid (p. 420).

8. The pure product [2] melting at 128–129° (cor.) can be obtained in 42–45% yield by two more recrystallizations from 400-ml. portions of acetone. Pure 2-acetylfluorene has also been reported [3] as melting at 132°, but this value has not been checked.

3. Methods of Preparation

2-Acetylfluorene has been prepared by the reaction of fluorene with acetic anhydride [2,4] or with acetyl chloride [3,5] in the presence of aluminum chloride in carbon disulfide or in nitrobenzene. When nitrobenzene is employed as the solvent it must be removed by a time-consuming steam distillation, and the use of acetyl chloride as a reagent leads to the formation of considerable amounts of 2,7-diacetylfluorene.

[1] *Org. Syntheses* Coll. Vol. **2**, 4 (1943).

[2] Bachmann and Sheehan, *J. Am. Chem. Soc.*, **62**, 2687 (1940).

[3] Dziewonski and Schnayder, *Bull. intern. acad. polon. sci.*, **1930A**, 529 [*C. A.*, **25**, 5416 (1931)].

[4] Ray and Rieveschl, *J. Am. Chem. Soc.*, **65**, 836 (1943); Buu-Hoi and Cagniant, *Bull. soc. chim. France,* **1946**, 131.

[5] Ardashev, Lomovatskaya, and Kacher, *J. Applied Chem. U.S.S.R.*, **11**, 1344 (1938) [*C. A.*, **33**, 5844 (1939)].

9-ACETYLPHENANTHRENE

(Ketone, methyl 9-phenanthryl)

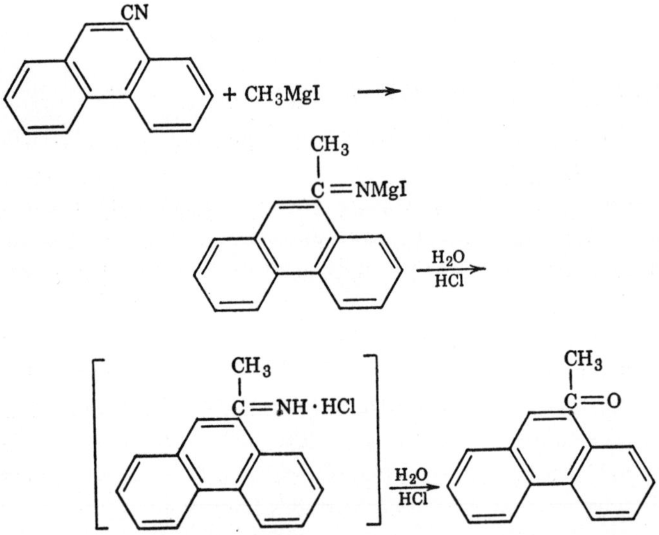

Submitted by JOSEPH E. CALLEN, CLINTON A. DORNFELD, and GEORGE H. COLEMAN.[1]
Checked by ROBERT E. CARNAHAN and HOMER ADKINS.

1. Procedure

A dry 12-l. three-necked flask is equipped with an efficient motor-driven stirrer (Note 1), a nitrogen inlet tube, a large Allihn condenser, and a 1-l. separatory funnel. Both the condenser and the funnel are provided with calcium chloride drying tubes. To the flask is added

146 g. (6 gram atoms) of magnesium turnings (Note 2), and nitrogen gas, first bubbled through concentrated sulfuric acid, is passed in to displace the air. During the reaction the nitrogen atmosphere is maintained. The magnesium is covered with 200 ml. of anhydrous ether, and a few milliliters of a solution of 852 g. (6 moles) of methyl iodide in 1 l. of anhydrous ether is added from the separatory funnel. The reaction starts spontaneously, and then the remainder of the methyl iodide solution is added slowly. When the reaction is complete (Note 3), 4 l. of dry benzene is added, a condenser is arranged for downward distillation, and about 1.2 l. of solvent is distilled (Note 4). The condenser is changed to a reflux position, 609 g. (3 moles) of 9-cyanophenanthrene (p. 212) is added quickly through a powder funnel, and the mixture is heated and stirred under reflux for 3 hours. It is then cooled in an ice bath to 0°, 3 l. of cold 6 *N* hydrochloric acid is slowly added (*Caution!*) from a separatory funnel with stirring, and the mixture is refluxed for 6 to 8 hours (Note 5).

After cooling, the layers are separated, the organic layer is washed with dilute sodium bicarbonate solution and placed in a flask equipped for distillation, and the solvent is distilled The oily residue is transferred while still warm to a 1-l. Claisen flask, and the product is distilled under reduced pressure; b.p. 190–200°/2.5 mm. (168–170°/1 mm.). The yield is 400–430 g. (61–65%). The distilled ketone is recrystallized once from ethanol (1.5–2 l.) to yield 345–390 g. (52–59%) of 9-acetylphenanthrene of m.p. 73–74°.

2. Notes

1. If a 12-l. three-necked flask is not available, a three-way adapter tube may be used in making the necessary connections. Although a mercury seal may be used, a glycerol-rubber tube seal is adequate.

2. The checkers operated on one-tenth the scale specified.

3. In several runs the Grignard reagent was filtered at this point, but the improvement in yield was not appreciable.

4. The addition of benzene and distillation of part of the solvent raises the reaction temperature.

5. The oily layer of ketimine hydrochloride usually dissolves during 6 hours' refluxing.

3. Methods of Preparation

The method described above is a modification of that of Bachmann and Boatner.[2] 9-Acetylphenanthrene has also been obtained by a

Claisen condensation of methyl phenanthrene-9-carboxylate with ethyl acetate followed by scission of the resulting phenanthroylacetic ester,[3] by the reaction of 9-phenanthrylmagnesium bromide with acetyl chloride,[4] and by dehydrogenation of 9-acetyl-1,2,3,4-tetrahydrophenanthrene by heating with sulfur.[5]

[1] Work done under contract with the Office of Scientific Research and Development.

[2] Bachmann and Boatner, *J. Am. Chem. Soc.*, **58**, 2098 (1936).

[3] Mosettig and van de Kamp, *J. Am. Chem. Soc.*, **55**, 3445 (1933).

[4] Miller and Bachman, *J. Am. Chem. Soc.*, **57**, 768 (1935).

[5] Bachmann and Struve, *J. Org. Chem.*, **4**, 476 (1939).

ACID ANHYDRIDES

$$RCOCl + C_5H_5N \rightarrow RCOCl \cdot C_5H_5N$$

$$\text{(A)} \quad RCOCl \cdot C_5H_5N + RCOOH \rightarrow (RCO)_2O + C_5H_5N \cdot HCl$$

$$\text{(B)} \quad 2RCOCl \cdot C_5H_5N + H_2O \rightarrow (RCO)_2O + 2C_5H_5N \cdot HCl$$

Submitted by C. F. H. Allen, C. J. Kibler, D. M. McLachlin, and C. V. Wilson.
Checked by Cliff S. Hamilton, Dexter Sharp, and R. Kretzinger.

1. Procedure

A. *Heptoic anhydride (enanthic anhydride).* In a 250-ml. round-bottomed three-necked flask, equipped with a stirrer, dropping funnel, and thermometer, are placed 15.8 g. (16.1 ml., 0.2 mole) of dry pyridine (Note 1) and 25 ml. of dry benzene (Note 2). Then 14.8 g. (15.5 ml., 0.1 mole) of heptoyl chloride (Note 3) is added rapidly to the stirred solution. The temperature rises only slightly, and a pyridinium complex separates. While stirring is continued, 13.0 g. (14.1 ml., 0.1 mole) of heptoic acid (Note 3) is added from the dropping funnel over a period of 5 minutes. The temperature rises rapidly to 60–65° (Note 4), and pyridine hydrochloride is formed. After stirring for 10 minutes, the solid is collected on a chilled Büchner funnel and washed twice with 25-ml. portions of dry benzene (Note 5).

The filtrate is concentrated under reduced pressure on the steam bath, and the residue is distilled using a 200-ml. modified Claisen flask.[1] The fraction boiling up to 155°/12 mm. is discarded; the anhydride is collected at 155–162°/12 mm. (170–173°/15 mm.). It amounts to 19–20 g. (78–83%).

B. *p-Chlorobenzoic anhydride (benzoic anhydride, p,p′-dichloro-)*. A mixture of 17.5 g. (0.1 mole) of *p*-chlorobenzoyl chloride (Note 6) and 50 ml. (0.6 mole) of pyridine in a loosely stoppered 200-ml. flask is warmed on the steam bath for 5 minutes and poured upon 100 g. of cracked ice and 50 ml. of concentrated hydrochloric acid (sp. gr. 1.18). The anhydride separates at once; as soon as the ice has melted sufficiently the mixture is filtered by suction. The solid is washed once with 15 ml. of methanol, then with 15 ml. of dry benzene. The yield is 14.2–14.6 g. (96–98%). Though suitable for most purposes, the crude product can be purified by recrystallization from 250 ml. of dry benzene; the recovery is 90%; it melts at 192–193°.

2. Notes

1. The pyridine was Eastman grade which was dried by long standing over potassium hydroxide for A but used without further drying in B. Adkins' studies on the mechanism of this reaction indicate that the intermediate complex may react with water or with a molar quantity of the acid to form an anhydride.[2]

2. The benzene is dried by distilling the first 10% and using the residue directly.

3. The heptoyl chloride, b.p. 173–175°, and the heptoic acid, b.p. 108–109°/9 mm., were obtained from the Eastman Kodak Company.

4. When preparing larger amounts, it would probably be better to control the temperature by external cooling as well as by the rate of addition of the acid.

5. Pyridine hydrochloride is hygroscopic; the filtration should be done rapidly, using a Büchner funnel.

6. *p*-Chlorobenzoyl chloride (m.p. 14–15°) is readily obtained by refluxing and stirring 156 g. (1.0 mole) of *p*-chlorobenzoic acid (obtained by the procedure for the *o*-isomer [3]) and 200 g. (1.7 moles) of thionyl chloride until solution is complete. The unused thionyl chloride is distilled, under slightly reduced pressure, and the product at 10 to 25 mm.; the yield of *p*-chlorobenzoyl chloride, b.p. 119–120°/22 mm., is 131–142 g. (75–81%). An additional amount can be secured by working up the fore-run and residue.

3. Methods of Preparation

These procedures are generally applicable to both aliphatic and aromatic compounds. They are reported to fail for furoic anhydride and to give poor results for *p*-nitrobenzoic anhydride (W. W. Prichard, private communication). They are superior to the common interchange method [4] in that they avoid the fractional distillation which is very troublesome in the aliphatic series. They have been used in numerous instances [2,5–12] and can be adapted to give mixed anhydrides.[13] Benzoic anhydride has been obtained, by closely related procedures, from benzoic acid and benzoyl chloride by heating under reduced pressure [14] or in the presence of zinc chloride.[15] Benzoic, acetic, and propionic anhydrides have been conveniently prepared by the action of bromine on the sodium salts of the acids in the presence of sulfur.[16]

1 *Org. Syntheses* Coll. Vol. **1,** 130 (1941).

2 Ipatieff and Friedman, *J. Am. Chem. Soc.,* **61,** 686 (1939); Adkins and Thompson, *J. Am. Chem. Soc.,* **71,** 2242 (1949).

3 *Org. Syntheses,* **10,** 20 (1930); Coll. Vol. **2,** 135 (1943).

4 *Org. Syntheses* Coll. Vol. **1,** 91 (1941).

5 Minunni, *Gazz. chim. ital.,* **22,** 213 (1892).

6 Einhorn and Holland, *Ann.,* **301,** 95 (1898).

7 Wedekind, *Ber.,* **34,** 2070 (1901).

8 Losanitsch, *Monatsh.,* **35,** 318 (1914).

9 Benary, Reiter, and Soenderop, *Ber.,* **50,** 73 (1917).

10 Rule and Paterson, *J. Chem. Soc.,* **125,** 2161 (1924).

11 Allen, Wilson, and Ball, *Can. J. Research,* **9,** 434 (1933).

12 Fuson, Corse, and Rabjohn, *J. Am. Chem. Soc.,* **63,** 2852 (1941).

13 Knoll and Company, Ger. pat. 117,267 [*Frdl.,* **6,** 146 (1900–1902)].

14 Dvornikoff, U. S. pat. 1,948,342 [*C. A.,* **28,** 2730 (1934)].

15 Doebner, *Ann.,* **210,** 278 (1881).

16 Orshansky and Bograchov, *Chemistry & Industry,* **1944,** 382.

ACRYLIC ACID *

I. PYROLYSIS METHOD

$$CH_2{=}CHCO_2C_2H_5 \xrightarrow{590°} CH_2{=}CHCO_2H + CH_2{=}CH_2$$

Submitted by W. P. Ratchford.
Checked by Arthur C. Cope, William R. Armstrong, and James J. Ryan.

1. Procedure

A 90-cm. length of 28-mm. (outside diameter) Pyrex tubing packed with pieces of Pyrex tubing (Note 1) is mounted vertically in an electric furnace (Note 2) capable of maintaining a temperature of 585–595°. A 250-ml. long-stemmed separatory funnel is connected to the upper end of the tubing with a stopper (Note 3), and the lower end is connected to a 500-ml. three-necked flask immersed in ice water. The flask, which serves as a receiver, is attached to a 50-cm. water-cooled reflux condenser, which in turn is connected by short lengths of rubber tubing to two traps in series which are immersed in a Dry Ice-trichloroethylene mixture. The exit tube of the second trap is vented to a hood. From 0.2 to 0.3 g. of hydroquinone is placed in the receiver, together with a few pieces of Dry Ice which serve to displace air from the entire apparatus. The third neck of the receiver is stoppered.

The furnace is heated to 590° (Note 4), and after the air has been displaced 200 g. (216 ml., 2 moles) of ethyl acrylate (Note 5) is placed in the separatory funnel and admitted to the reaction tube at a rate of about 90 drops a minute (Note 3), so that the addition requires about 2 hours. At the end of the addition the contents of the receiver and the small amount of liquid in the traps are combined. The total weight of crude acrylic acid containing some ethyl acrylate is 126–136 g.

The crude product is placed in a 250-ml. flask containing a capillary inlet tube through which carbon dioxide is admitted. Ten grams of hydroquinone and 15 g. of diphenyl ether are added, and the flask is attached to a suitable fractionating column (Note 6). The product is fractionated carefully (Note 7) at 135 mm. pressure. The pressure is lowered gradually when most of the ethyl acrylate has distilled, and at about 70°/90 mm. the receiver is changed. The first fraction (mostly ethyl acrylate) amounts to 9–10 g. The pressure is lowered further to 50 mm., and the acrylic acid is distilled fairly rapidly, without reflux, at 69–71°/50 mm. The acrylic acid fraction weighs 108–116 g. and is 95–97% pure according to acidimetric titration. The yield is 68–75% based upon 100% acrylic acid content (Notes 8 and 9).

If the acrylic acid is not to be used at once, it is stabilized by the addition of hydroquinone and is stored in a refrigerator.

2. Notes

1. The middle third of the Pyrex tube should be packed with 20-mm. lengths of fire-polished 7-mm. Pyrex tubing. The lower end of the tube is drawn out to a size that permits attachment to the receiver with a rubber stopper.

2. A type FD303 combustion furnace (made and sold by the Hoskins Manufacturing Company, Detroit, Michigan) or any similar furnace is satisfactory.

3. A groove filed in the stopcock of the separatory funnel aids in controlling the rate of addition. If available, a small constant-feed pump may be used to introduce the ester into the pyrolysis tube. The rate of addition of the ester is not critical, but at high rates cracking is incomplete and at low rates the yield is reduced. A stream of nitrogen (100 bubbles per minute) flowing through the tube reduces refluxing and makes the feed rate easier to observe. The nitrogen may be introduced through a tube in the stopper holding the separatory funnel or through a side arm sealed near the upper end of the pyrolysis tube.

4. The temperature is measured by a movable Chromel-Alumel thermocouple located in the furnace by the side of the tube and connected to a potentiometer or millivoltmeter. The thermocouple junction is adjusted so that during the run it is at the hottest point in the furnace. For the Hoskins Company furnace this point is about 9 in. from the top of the furnace. The temperature is controlled manually to $590 \pm 5°$ by means of an autotransformer (Variac) rated at 5 amperes, 110 volts.

5. Commercial ethyl acrylate, containing hydroquinone inhibitor, may be used directly if it is of good quality.

6. The submitter used an insulated column with a 38 by 1.1 cm. section packed with ⅛-in. copper helices made of No. 26 B & S gauge copper wire. He states that a column packed with glass helices is unsatisfactory. The checkers used a 100 by 1.7 cm. Vigreux column. Either type of column should be equipped with a total-condensation partial take-off head.

7. Ethyl acrylate and acrylic acid polymerize easily, and overheating must be avoided in the distillation. The flask is heated in an oil bath which is not permitted to rise above 115°. The diphenyl ether that is added serves to expel the acrylic acid at the end of the distillation.

8. The submitter states that the product may be purified by freezing and decanting the supernatant liquid several times. The acrylic acid may be obtained in 97% purity by this method, but it has a faint yellow color. The yield is 50–60%.

9. The submitter states that methacrylic acid may be prepared in a similar manner by pyrolyzing ethyl methacrylate. Under the same conditions of temperature and feed rate, the conversion is slightly higher and the yield is about the same.

II. ACIDOLYSIS METHOD

$$CH_2{=}CHCO_2CH_3 + HCO_2H \xrightarrow{H_2SO_4} CH_2{=}CHCO_2H + HCO_2CH_3$$

Submitted by C. E. Rehberg.
Checked by Arthur C. Cope and Elbert C. Herrick.

1. Procedure

One hundred and eighty-four grams (151 ml., 4 moles) of formic acid (Note 1), 1032 g. (1060 ml., 12 moles) of methyl acrylate (Note 2), 30 g. of hydroquinone, and 2 ml. of sulfuric acid are mixed in a 2-l. two-necked round-bottomed flask fitted with a capillary inlet tube. The flask is attached to a 100 by 1.7 cm. Vigreux column (Note 3) and is heated in an oil bath at 85–95°. The mixture is heated under total reflux until the temperature of the vapor at the still head falls to 32° (after 1–3 hours). Methyl formate then is distilled slowly at 32–35° as long as it is formed (8–10 hours). A reflux ratio of about 5 to 1 is maintained during the first part of the distillation, which is decreased to total take-off at the end. When no more methyl formate is produced, the excess methyl acrylate is distilled at 32–35°/140 mm. with the bath temperature at 60–65°. During the distillation, a slow stream of carbon dioxide is admitted through the capillary inlet. When all the methyl acrylate has been removed, the acrylic acid is distilled at 53–56°/25 mm. Upon redistillation through the same column (Note 4) acrylic acid of 97% purity (by acidimetric titration) is obtained in a yield of 220–230 g. (74–78% based upon 100% acrylic acid content), b.p. 54–56°/25 mm.

2. Notes

1. Acetic acid may be used, but it reacts much less rapidly and less completely, and fractionation of the reaction mixture is more difficult. Pure formic acid (98–100%) is preferred.
2. Commercial methyl acrylate may be used without purification if it is of good quality.
3. Either a Vigreux column or a column containing an open spiral of copper or Nichrome wire is satisfactory. The column should be jacketed and fitted with a total-condensation variable take-off head.
4. Hydroquinone or another polymerization inhibitor should be added before distillation of acrylic acid or its esters.

3. Methods of Preparation

Acrylic acid free of water has been prepared by treating lead acrylate with hydrogen sulfide;[1,2] by heating α,β-dibromopropionic acid with copper;[3] by dry distillation of a mixture of equivalent amounts of sodium acrylate and β-chloropropionic acid;[4] by pyrolysis of the polymer of β-propiolactone;[5] by heating β-chloropropionic acid with potassium fluoride;[6] by heating β-acetoxypropionic acid in the presence of hydroquinone;[7] by heating lactic acid with metal chlorosulfonates;[8] and by the two methods described here.[9] It also has been prepared by dehydration of hydracrylic acid in the presence of copper and concentrated acids,[10] by acid exchange of acrylates in the presence of polymerization inhibitors,[11] and by the reaction of β-hydroxypropionitrile with 100% sulfuric acid, followed by addition of water and distillation.[12]

[1] Caspary and Tollens, *Ann.*, **167**, 252 (1873).
[2] Wohlk, *J. prakt. Chem.*, (2) **61**, 212 (1900).
[3] Biilmann, *J. prakt. Chem.*, (2) **61**, 491 (1900).
[4] Riiber and Schetelig, *Z. physik. Chem.*, **48**, 348 (1904).
[5] Gresham, Jansen, and Shaver, *J. Am. Chem. Soc.*, **70**, 998 (1948); U. S. pat. 2,356,459.
[6] Nesmayanov, Pecherskaya, and Uretskaya, *Bull. Acad. Sci. U.S.S.R., Classe sci. chim.*, **1948**, 240 [*C. A.*, **42**, 4924 (1948)].
[7] U. S. pat. 2,459,677 [*C. A.*, **43**, 3026 (1949)].
[8] Brit. pat. 600,653 [*C. A.*, **42**, 7320 (1948)].
[9] Ratchford, Rehberg, and Fischer, *J. Am. Chem. Soc.*, **66**, 1864 (1944).
[10] U. S. pat. 2,469,701 [*C. A.*, **43**, 6221 (1949)].
[11] U. S. pat. 2,413,889 [*C. A.*, **41**, 2430 (1947)].
[12] U. S. pat. 2,425,694 [*C. A.*, **41**, 7410 (1947)].

β-ALANINE *

$$2NH_2CH_2CH_2CN + Ba(OH)_2 + 2H_2O \rightarrow (NH_2CH_2CH_2COO)_2Ba + 2NH_3$$

$$(NH_2CH_2CH_2COO)_2Ba + CO_2 + H_2O \rightarrow 2NH_2CH_2CH_2COOH + BaCO_3$$

Submitted by Jared H. Ford.
Checked by Homer Adkins and James M. Caffrey.

1. Procedure

In a 2-l. three-necked flask equipped with a mechanical stirrer, a thermometer, and a dropping funnel is placed 185 g. (0.55 mole) of technical barium hydroxide octahydrate. The flask is heated on a steam bath in a hood. When the barium hydroxide has dissolved in its water of crystallization, the stirrer is started and 70.1 g. (1.00 mole) of β-aminopropionitrile (p. 93) is added dropwise over a period of 40 minutes. The temperature is maintained at 90–95° during the addition and for 40 minutes thereafter. Forty grams of asbestos filter aid (Note 1) and 1 l. of hot water are added, and the mixture is saturated with carbon dioxide (Note 2) while the temperature is held at 85–90°. The mixture is filtered with suction, the precipitate is returned to the flask with 500 ml. of hot water, and the mixture is heated and stirred for 20 minutes. After the barium carbonate has been filtered the washing procedure is again repeated with a second 500 ml. of hot water. The combined filtrates and washings are concentrated under reduced pressure on the steam bath (Note 3) until solid material separates. To the residue are added 200 ml. of hot water and 0.5 g. of decolorizing carbon (Note 4). The resulting solution is warmed on the steam bath for a few minutes and then filtered into a weighed 500-ml. Erlenmeyer flask. The flask is heated on a steam bath, and a jet of clean compressed air is directed at the surface of the solution. When the total weight of the solution is 130 g., it is cooled to 15–20° and diluted slowly with 400 ml. of methanol. After the solution has stood for several hours in the refrigerator, the product is filtered with suction and washed with two 100-ml. portions of methanol. The yield of β-alanine melting at 197–198° (dec.) is 75–80 g. (85–90%).

2. Notes

1. Standard Super-Cel (Johns-Manville, Inc.) was used.

2. Either carbon dioxide gas or Dry Ice may be used, and the saturation may be completed in 15–20 minutes by either method. The *p*H of the saturated solution is about 8–9 when tested with a universal indicator paper, such as Alkacid or Hydrion.

3. The submitter used a special apparatus suitable for the rapid evaporation of water under reduced pressure. The checkers used standard flasks.

4. The solution is nearly colorless at this point, but the carbon aids in the removal of finely divided insoluble material.

3. Methods of Preparation

β-Alanine has been prepared by the catalytic reduction of cyanoacetic acid,[1] esters,[2] or salts;[3] by heating acrylonitrile,[4] β-aminopropionitrile,[5] bis-(β-cyanoethyl)amine,[6] β-hydroxypropionitrile,[7] β-alkoxypropionitriles,[8] bis-(β-cyanoethyl) ether,[9] or bis-(β-cyanoethyl) sulfide [9] with aqueous ammonia at 150–225°; by the hydrolysis of β-aminopropionitrile with concentrated hydrochloric acid and subsequent removal of the acid with anion-exchange resins; [10] by hydrolysis of β-phthalimidopropionitrile prepared from phthalimide and acrylonitrile; [11] from β,β'-iminodipropionic acid, β,β'-iminopropionitrile, or diethyl-β,β'-iminopropionate through preliminary conversion with phthalic anhydride at 200° to the corresponding phthalimide and subsequent hydrolysis.[12] The method as described above has been published.[13] Additional references to methods of preparation are given in connection with a procedure for making β-alanine from succinimide through the action of potassium hypobromite.[14,15]

[1] Swiss pat. 226,014 [*C. A.*, **43**, 2225 (1949)].
[2] U. S. pat. 2,365,295 [*C. A.*, **39**, 4626 (1945)].
[3] U. S. pat. 2,367,436 [*C. A.*, **39**, 3012 (1945)].
[4] U. S. pat. 2,335,997 [*C. A.*, **38**, 2972 (1944)].
[5] U. S. pat. 2,336,067 [*C. A.*, **38**, 2971 (1944)].
[6] U. S. pat. 2,334,163 [*C. A.*, **38**, 2667 (1944)].
[7] U. S. pat. 2,364,538 [*C. A.*, **39**, 3556 (1945)].
[8] U. S. pat. 2,335,605 [*C. A.*, **38**, 2970 (1944)].
[9] U. S. pat. 2,335,653 [*C. A.*, **38**, 2970 (1944)].
[10] Buc, Ford, and Wise, *J. Am. Chem. Soc.*, **67**, 92 (1945).
[11] Galat, *J. Am. Chem. Soc.*, **67**, 1414 (1945).
[12] Chodroff, Kapp, and Beckmann, *J. Am. Chem. Soc.*, **69**, 256 (1947).
[13] Ford, *J. Am. Chem. Soc.*, **67**, 876 (1945).
[14] *Org. Syntheses* Coll. Vol. **2**, 20 (1943).
[15] Parshin, *Zhur. Obshchei Khim.*, **20**, 1826 (1950) [*C. A.*, **45**, 2407 (1951)].

ALLOXAN MONOHYDRATE *

(Alloxan)

I. METHOD A

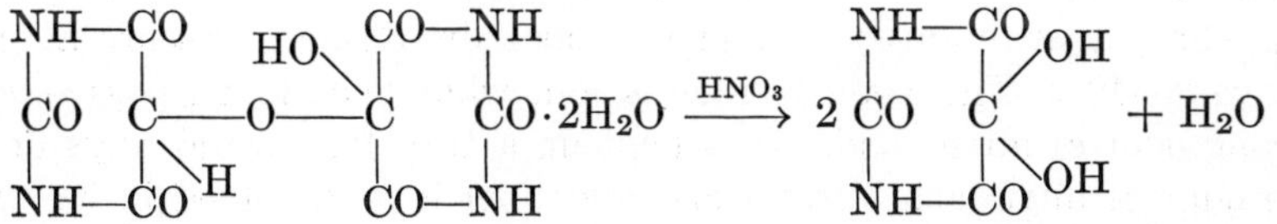

Submitted by W. W. HARTMAN and O. E. SHEPPARD.
Checked by C. S. MARVEL and B. H. WOJCIK.

1. Procedure

In a 500-ml. flask (Note 1), fitted with a mechanical stirrer, are placed 36 ml. of water and 25 g. (0.078 mole) of finely crystalline alloxantin dihydrate (p. 42). The flask and contents are heated on a steam bath to 50°, and 3.6 ml. of fuming nitric acid (sp. gr. 1.62) is added in a fine stream while vigorous stirring is maintained and the temperature is not allowed to rise above 60° (Note 2). After all the fuming nitric acid has been added, the temperature is brought to 55° and the stirrer is stopped. In a few minutes a vigorous reaction begins, and large quantities of oxides of nitrogen are evolved. The stirrer is again started, and, if the reaction becomes too violent, the mixture is cooled somewhat; otherwise, the reaction is allowed to take its course. The reaction is complete when a current of air, introduced into the flask above the mixture, does not produce much color due to formation of nitrogen tetroxide (Note 3). The mixture is then heated to 60–65° for 10–15 minutes, whereupon practically all the solid dissolves. The reaction mixture is poured into a glass (Pyrex) tray and cooled overnight at 0° or below. The large triclinic, colorless crystals of alloxan tetrahydrate are broken up, filtered with suction, washed with ice water, and pressed as dry as possible. The crystals are then added to 25–26 ml. of *hot* water (Note 4), and the mixture is shaken until solution is complete. The solution is filtered immediately, and the filtrate is cooled overnight in a tray at 0°. The crystals are broken up, filtered, washed with ice water, and pressed as dry as possible. These moist crystals of the tetrahydrate, which weigh 22–24 g., are dried to constant weight in a glass tray over concentrated sulfuric acid. The resulting product is a fine white powder (Note 5) which weighs 16 g.

The mother liquor from the recrystallization is placed in a 250-ml. flask and is concentrated to a volume of 8–10 ml. under reduced pressure at not over 30–40° (Note 6). The concentrate, when cooled overnight at 0°, deposits a solid which is filtered, recrystallized from its own weight of boiling water, and dried over sulfuric acid. This solid weighs about 2 g. The mother liquors from the two crystallizations are combined with the original mother liquor from the oxidation, and the whole is evaporated to dryness under diminished pressure at not above 30–40°. This solid residue is somewhat yellow and possesses a strong odor of nitric acid. It is kept on a tray for several days until the odor of nitric acid disappears, and then it is dissolved in its own weight (2–3 g.) of boiling water, and the solution is cooled for several days below 0° (Note 7). The solid is removed, recrystallized from water, and dried over sulfuric acid (Note 8). This crop weighs about 0.5 g. The total yield of alloxan monohydrate is 18–19 g. (72–76%) (Note 9).

2. Notes

1. The large flask is necessary because the mixture foams greatly during the reaction.

2. Very little reaction occurs during the addition of the nitric acid, and consequently there is very little rise in temperature. The acid is added during a few minutes.

3. The reaction is complete in about 30 minutes.

4. The water must not be boiled during the addition of the crystals or afterwards as this will cause decomposition of the alloxan to carbon dioxide, parabanic acid, and alloxantin.

5. The tetrahydrate effloresces readily, gradually loses part of its water, and becomes moist when allowed to stand at room temperature. For this reason it is not suitable for storage. The monohydrate results when the material is dried over sulfuric acid. Drying in an oven is likely to result in local overheating and decomposition, which starts slightly above 100°; it is also likely to result in reddening if even a trace of ammonia or amines is present in the air.

6. A higher temperature is likely to cause oxidation and to result in a violent reaction which may become explosive as the nitric acid becomes concentrated.

7. Alloxan tetrahydrate crystallizes from solution much more slowly when it is nearly free of nitric acid and when other soluble substances are present; it is, however, less soluble in nitric acid solution than in water.

8. A little alloxan still remains in solution. The mother liquor still slowly colors the skin red and on standing deposits alloxantin. The alloxantin can be removed and used in another oxidation. About 0.3–0.5 g. of it may deposit after the filtrate has stood for some time.

9. The submitters obtained the same yields (per cent) when approximately 65 times these amounts of materials were used. Thus, carrying out the oxidation of 1610 g. (5 moles) of alloxantin dihydrate in a 22-l. flask, the submitters obtained 1085 g. of the first crop, 160 g. of the second, and 28 g. of the third, a total of 1273 g.

II. METHOD B

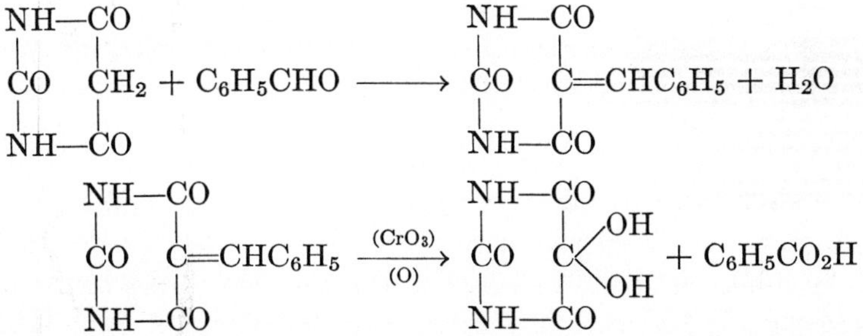

Submitted by John H. Speer and Thomas C. Dabovich.
Checked by W. E. Bachmann and R. O. Edgerton.

1. Procedure

A. *Benzalbarbituric acid.* A mixture of 128 g. (1 mole) of barbituric acid and 1250 ml. of water in a 2-l. three-necked round-bottomed flask equipped with an efficient stirrer and a reflux condenser is heated on a steam bath to effect solution (Note 1). When the acid has dissolved, 115 g. (110 ml., 1.08 moles) of benzaldehyde is added while heating and stirring are continued. The solution rapidly fills with the insoluble benzalbarbituric acid. The mixture is heated for 1 hour on the steam bath to complete the reaction, and then is filtered by suction (Note 2). The filter cake is washed with several portions of hot water and dried at 100°. The yield is 190–205 g. (88–95%) of product possessing a very pale yellow color. The substance melts at 254–256° and needs no further purification.

B. *Alloxan monohydrate.* A mixture of 730 ml. of acetic acid, 95 ml. of water, and 162 g. (1.62 moles) of chromium trioxide (Note 3) is placed in a 2-l. three-necked round-bottomed flask fitted with a stirrer and a thermometer; the stirrer is started, and the mixture is warmed to 50°. To the solution 180 g. (0.83 mole) of benzalbarbituric acid is added in small portions during the course of 30 minutes, a cold water bath being used to maintain the temperature at 50–60°. After all the acid has been added, stirring is continued and the temperature maintained at 50–60° by a warm water bath for another 30 minutes to complete the reaction. Alloxan monohydrate generally starts to crystallize from the warm solution. The mixture is cooled to 15° and filtered. The product is washed on the filter with cold glacial acetic acid until the washings are no longer green, and then is dried by washing with ether. The yield is 105–112 g. (79–84%) of yellow crystals which melt at about 254° (Note 4) with decomposition and are sufficiently pure for most purposes.

In order to obtain practically colorless alloxan monohydrate (Note 5), 25 g. of the yellow crystals is dissolved in 37 ml. of hot water, the solution is boiled with Norit, and the hot solution is filtered into a 500-ml. round-bottomed flask. About 15–20 ml. of water is removed by distillation under reduced pressure on a water bath. The colorless crystalline residue is dissolved in the minimal volume of hot water, the solution is cooled somewhat, and to it is added 250 ml. of glacial acetic acid. After the mixture has been kept cold (5–10°) for 4–6 hours, the alloxan monohydrate is filtered. The yield is 20–21 g. (80–84% recovery) (Note 6) of practically colorless crystals which melt at about 254° (Note 7) with decomposition.

2. Notes

1. The submitters report that a single attempt to use the aqueous-ethanolic solution of barbituric acid obtained in *Org. Syntheses*, **18,** 8, before recrystallization of the product gave an excellent yield of an apparently isomeric product unsuited for the preparation of alloxan.

2. This filtration may be done hot or cold at the convenience of the operator.

3. The technical grade (flakes) was found quite satisfactory.

4. This value is obtained in a Pyrex capillary tube; the solid remains yellow until about 254°, when it suddenly decomposes to a red melt with vigorous evolution of gas. When a soft-glass capillary tube is used, the solid assumes a red color at about 180–200° and melts between 240° and 250°.

5. By recrystallization of the yellow product from glacial acetic acid (12 ml. per g.), using Norit, the checkers invariably obtained yellow

crystals (75–80% recovery) instead of the practically colorless crystals reported by the submitters. By adding a volume of water equal to the weight of crystals to the hot acetic acid solution, the checkers obtained a pale yellow product.

6. This represents the first crop of crystals.

7. This value is obtained in a Pyrex capillary tube. The colorless solid begins to turn yellow at about 180° and melts at about 254° (occasionally 258–260°) to a red liquid with vigorous evolution of gas.

3. Methods of Preparation

Alloxan monohydrate has been prepared by the oxidation of uric acid with chlorine,[1,2] or potassium chlorate and hydrochloric acid;[3] by the oxidation of alloxantin,[4] xanthine,[5] uramil,[4] and thiouramil;[6] and by the hydrolysis of dibromobarbituric acid.[7] The method here described is originally due to Biilmann and Berg.[8]

[1] McElvain, *J. Am. Chem. Soc.*, **57**, 1303 (1935).

[2] Biltz and Heyn, *Ann.*, **413**, 60 (1917).

[3] Fischer and Helferich, *Anleitung zur Darstellung organischer Präparate*, 10th ed., p. 66, Braunschweig, 1922.

[4] Wöhler and Liebig, *Ann.*, **26**, 256 (1838).

[5] Fischer, *Ann.*, **215**, 310 (1882).

[6] Fischer and Ach, *Ann.*, **288**, 160 (1895).

[7] Baeyer, *Ann.*, **127**, 230 (1863); **130**, 131 (1864).

[8] Biilmann and Berg, *Ber.*, **63B**, 2201 (1930).

ALLOXANTIN DIHYDRATE

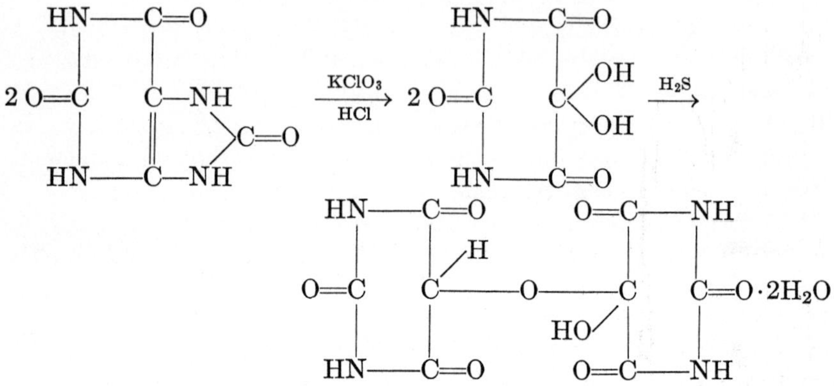

Submitted by Dorothy Nightingale.
Checked by R. L. Shriner and C. H. Tilford.

1. Procedure

In a 500-ml. three-necked flask, provided with a stirrer, are placed 15 g. (0.09 mole) of finely powdered uric acid, 30 g. (25.2 ml.) of concentrated hydrochloric acid, and 40 ml. of water. The mixture is warmed to 30°, the stirrer is started, and 4 g. (0.014 mole) of finely powdered potassium chlorate is added in small portions during a period of not less than 45 minutes (Note 1). Most of the uric acid dissolves; any undissolved material is removed by filtration through a fritted-glass filter. The clear filtrate is diluted with 30 ml. of water, and a rapid stream of hydrogen sulfide is led into it until it is saturated (about 10–15 minutes). Sulfur and alloxantin separate, and the mixture is cooled for 2–3 hours in an ice bath until the separation is complete.

The solid is collected on a Büchner funnel and washed with three 30-ml. portions of cold water. The alloxantin is dissolved by boiling the wet solid for 15 minutes with 250 ml. of water, and the *hot* solution is filtered to remove the sulfur (Note 2). Alloxantin dihydrate crystallizes from the filtrate in glistening plates which should be pressed as dry as possible on a Büchner filter, washed with about 30 ml. of ether, and dried in a vacuum desiccator (Note 3). The yield is 8–10 g. (55–69%) (Note 4). The product melts with decomposition at 234–238° (Note 5), and is pure enough for most purposes.

2. Notes

1. It is important that the temperature of the reaction mixture be kept near 30°. The potassium chlorate must be added slowly.

2. Since alloxantin is difficultly soluble, it is desirable to make a second extraction of the sulfur to be certain that all the alloxantin has been removed.

3. The compound gradually turns pink on standing in the air. It should be stored in a tightly stoppered bottle or kept in a vacuum desiccator over calcium chloride.

4. If larger amounts of material are desired it is best to oxidize several 15-g. portions of uric acid and to combine the sulfur-alloxantin mixtures for the extraction.

5. The water of crystallization may be removed by heating the dihydrate at 120–150° under reduced pressure for 2 hours. The melting points reported in the literature vary considerably. The anhydrous material turns yellow at about 225–230° and decomposes at temperatures ranging from 238–242° to 253–255°, depending on the rate of heating. The instantaneous decomposition temperatures determined on the Maquenne block were 270–275°.

3. Methods of Preparation

This procedure is essentially that described in the laboratory manual by Fischer and Helferich.[1]

Alloxantin has been obtained by the oxidation of uric acid with nitric acid, followed by reduction with hydrogen sulfide;[2] by oxidation of uric acid with potassium chlorate, followed by reduction with stannous chloride;[3] by condensation of alloxan with dialuric acid in aqueous solution;[4] and by oxidation of dialuric acid.[5] A preparation from alloxan has been published.[6]

[1] Fischer and Helferich, *Anleitung zur Darstellung organischer Präparate,* 10th ed., pp. 66, 67, Braunschweig, 1922; Doja and Mokeet, *J. Indian Chem. Soc.,* **13,** 542 (1936).

[2] Deniges, *Bull. soc. pharm. Bordeaux,* **66,** 8–12 (1928) [*C. A.,* **23,** 4160 (1929)].

[3] Davidson and Epstein, *J. Org. Chem.,* **1,** 305 (1936).

[4] Wöhler and Liebig, *Ann.,* **26,** 279 (1838); Behrend and Friederichs, *Ann.,* **344,** 1 (1906).

[5] Baeyer, *Ann.,* **127,** 11 (1863).

[6] *Org. Syntheses,* **33,** 3 (1953).

2-ALLYLCYCLOHEXANONE

(Cyclohexanone, 2-allyl-)

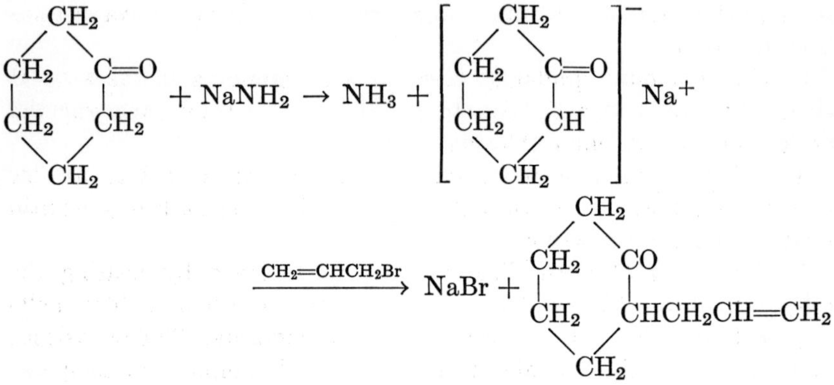

Submitted by Calvin A. Vanderwerf and Leo V. Lemmerman.
Checked by Arthur C. Cope and Theodore T. Foster.

1. Procedure

Approximately 1.5 l. of anhydrous liquid ammonia is introduced into a dry 5-l. three-necked flask fitted with a sealed mechanical stirrer and an efficient reflux condenser which is connected through a soda-lime tube to a gas-absorption trap. Freshly cut sodium (47.2 g., 2.05 gram atoms) is converted to sodium amide by addition to the liquid ammonia in the presence of a small amount of ferric nitrate, according to p. 219. A 1-l. dropping funnel and a gas inlet tube connected to a source of dry nitrogen are attached to the third neck of the flask, and, after the blue color of the solution has disappeared and a gray suspension of sodium amide remains (Note 1), 1.2 l. of dry ether is added as rapidly as the rate of vaporization of ammonia will permit. The ammonia is removed by warming the flask on a steam bath until refluxing of the ether occurs. Cyclohexanone (Note 2) (220 g., 2.24 moles) is added through the dropping funnel (Note 3), and the mixture is stirred and heated under reflux on a steam bath for 3 hours. Nitrogen is then introduced through the gas inlet tube to maintain an inert atmosphere (Note 4), and the mixture is cooled in an ice bath. A solution of 246 g. (2.03 moles) of allyl bromide (Note 5) in 1 l. of anhydrous ether is added rapidly through the dropping funnel with stirring. If the reaction does not start soon after the completion of this addition the mixture is warmed cautiously on the steam bath. When the exothermic reaction has started it is controlled by cooling in the ice bath while refluxing continues for 20–30 minutes. The mixture is finally heated under reflux on the steam bath for 3 hours.

The mixture is cooled in an ice bath, any sodium or sodium amide which may remain in the necks of the flask is scraped into the reaction mixture with a spatula, and enough water is added to dissolve the sodium bromide. The ether layer is separated and combined with five 100-ml. ether extracts of the aqueous phase, washed with 150 ml. of saturated sodium chloride solution, and dried over anhydrous sodium sulfate. The ether is removed by distillation, and the residue is fractionated carefully under reduced pressure through a 4-ft. heated column packed with glass helices and fitted with a total-condensation variable take-off head. The yield of 2-allylcyclohexanone boiling at 90–92°/17 mm. is 153–174 g. (54–62%). In addition, 28–38 g. of unchanged cyclohexanone boiling at 51–52°/17 mm., 15–35 g. of diallylcyclohexanone boiling at 123–124°/17 mm., and small intermediate fractions are obtained.

2. Notes

1. The conversion of the sodium to sodium amide requires 30–90 minutes. More liquid ammonia may be added if too much is lost by vaporization before the conversion is complete.

2. Redistilled cyclohexanone, b.p. 154–156°, was used.

3. The submitters obtained equally good results by adding 80 g. (2.05 moles) of freshly prepared finely powdered sodium amide in portions to a solution of the cyclohexanone in 1.2 l. of dry ether, heating under reflux for 3 hours, and continuing the preparation in the manner described.

4. The submitters state that the yield is increased appreciably if a nitrogen atmosphere is maintained after this point. Loss of ether may be avoided by stopping the flow of nitrogen when refluxing begins.

5. Allyl bromide was dried over calcium chloride and redistilled, b.p. 70–71.5°.

3. Methods of Preparation

2-Allylcyclohexanone has been prepared by the direct alkylation of the sodium derivative of cyclohexanone with allyl iodide, sodium amide having been used in the preparation of the sodium enolate,[1] and by ketonic hydrolysis of ethyl 1-allyl-2-ketocyclohexanecarboxylate, prepared by alkylation of ethyl 2-ketocyclohexanecarboxylate.[2,3]

[1] Cornubert, *Ann. chim.*, [9] **16,** 145 (1921).
[2] Cope, Hoyle, and Heyl, *J. Am. Chem. Soc.*, **63,** 1848 (1941).
[3] Grewe, *Ber.*, **76,** 1075 (1943).

ALLYL LACTATE

(Lactic acid, allyl ester)

$$n\text{HOCH(CH}_3\text{)COOH} \rightarrow \text{HO[CH(CH}_3\text{)COO]}_n\text{H} + (n-1)\ \text{H}_2\text{O}$$

$$\text{HO[CH(CH}_3\text{)COO]}_n\text{H} + n\text{CH}_2\text{=CHCH}_2\text{OH} \rightarrow n\text{HOCH(CH}_3\text{)COOCH}_2\text{CH=CH}_2 + \text{H}_2\text{O}$$

Submitted by CHESSIE E. REHBERG.
Checked by H. R. SNYDER and FRED E. BOETTNER.

1. Procedure

Six hundred and seventy-five grams (6 moles) of 80% lactic acid (Note 1), 300 ml. of benzene, and 5 ml. of concentrated sulfuric acid are placed in a 3-l. three-necked round-bottomed flask having a capillary ebullition tube in one neck. The flask is attached to a Vigreux column 75 cm. in length, having at its top a trap of the Dean and Stark type, preferably the Barrett modification carrying a stopcock (Note 2), and a reflux condenser. The mixture is refluxed, and the aqueous layer is withdrawn from the trap until liberation of water becomes slow or until noticeable darkening of the material begins (Note 3). Then 1394 g. (1635 ml. or 24 moles) of allyl alcohol is added (Note 4), and refluxing and removal of water are continued until production of water ceases (Note 5). The acid catalyst is then neutralized with anhydrous sodium acetate (18 g.) (Note 6), and the fractionating column is replaced by a short distillation head leading to a 3-l. round-bottomed receiving flask cooled in salt and ice. The liquid is distilled rapidly under diminished pressure, first at the water pump and finally at a pressure of about 5 mm. A viscous residue (about 130 g.) containing inorganic salts is discarded. The distillate is fractionated through the column used in the first steps. The trap at the reflux condenser is replaced with a still head arranged for controlled partial take-off. The first fraction is the benzene-allyl alcohol binary azeotrope, which distils at 76.8° and contains 17.4% of the alcohol.

Frequently a small amount of allyl ether is obtained at 94–95°. Allyl alcohol is then distilled at 97° until most of it is removed and the temperature of the residue in the pot reaches about 120°. The pressure is reduced to about 50 mm. while the remainder of the alcohol is removed. Finally the allyl lactate is distilled at 56–60°/8 mm. (Note 7). It amounts to 693–710 g. (89–91%) (Note 8).

2. Notes

1. The lactic acid used was the commercial edible grade and was almost colorless. The checkers used 635.6 g. of 85% lactic acid, U.S.P. The lower grades of commercial acid probably would give somewhat lower yields.

2. A suitable piece of apparatus is listed in Catalog LP-34, p. 43, of the Corning Glass Works.

3. Generally, about 35 ml. of water per mole of 80% lactic acid is obtained at this stage, and 4–6 hours of refluxing is required for its removal. The linear lactic acid polymer thus produced contains approximately 3 lactic acid units.

4. A large excess of alcohol is essential for a high yield of ester. Thus, when the ratio of alcohol to acid is 4:1, the yield is 90%; when it is 3:1, the yield is 85–88%; when it is 2:1, the yield is about 65%; and when it is 5:3, the yield is 58–60%.

5. The checkers found this distillation to require about 16 hours. The distillate is the ternary azeotrope. It consists of 8.6% water,[1] 9.2% allyl alcohol, and 82.2% benzene, and boils at 68.2°. The aqueous layer contains some allyl alcohol, though this loss is insignificant, since only about 15 ml. of the aqueous layer is obtained from each mole of lactic acid used.

6. It is essential to neutralize any strong acid present before distilling lactic esters; otherwise, condensation by ester interchange occurs, with liberation of alcohol and production of "polylactic acid," a linear polyester. Other neutralizing agents, such as alkali or alkaline-earth hydroxides or carbonates, doubtless could be used satisfactorily instead of sodium acetate.

7. Allyl lactate is a clear, colorless, mobile liquid boiling at 60°/8 mm., 79°/25 mm., and 175–176°/740 mm. Other properties are d_4^{20} 1.0452; n_D^{20} 1.4369.

8. This method of preparation is suitable for producing primary alkyl lactates but is unsatisfactory for β-methallyl lactate because the strong mineral acid catalyzes the rearrangement of methallyl alcohol to isobutyraldehyde. Methyl lactate [2] can be made conveniently (80–

85% yield) by heating 1 mole of lactic acid condensation polymer with 2.5–5 moles of methanol and a small quantity of sulfuric acid at 100° for 1–4 hours in a heavy-walled bottle, such as is used for catalytic hydrogenation with a platinum catalyst.

3. Methods of Preparation

Allyl lactate has been prepared by the repeated treatment of lactic acid or its polymer with allyl alcohol in the presence of mineral acid.[3]

[1] Lange, *Handbook of Chemistry,* 4th ed., p. 1217, Handbook Publishers, Sandusky, Ohio, 1941.

[2] Filachione, Fein, Fisher, and Smith, "Continuous Methods for Dehydrating Lactic Acid and Preparing Methyl Lactate and Methyl Acetoxypropionates," presented before the Division of Industrial and Engineering Chemistry at the 105th meeting of the American Chemical Society, Detroit, Michigan, April 12, 1943.

[3] Fisher, Rehberg, and Smith, *J. Am. Chem. Soc.,* **65,** 763 (1945).

ALUMINUM *tert*-BUTOXIDE *

$$3(CH_3)_3COH + Al \rightarrow Al[OC(CH_3)_3]_3 + 3(H)$$

Submitted by WINSTON WAYNE and HOMER ADKINS.
Checked by NATHAN L. DRAKE, WM. H. SOUDER, JR., and RALPH MOZINGO.

1. Procedure

In a 2-l. round-bottomed flask, bearing a reflux condenser protected by a calcium chloride tube, are placed 64 g. (2.37 gram atoms) of aluminum shavings, 200 g. (254 ml., 2.7 moles) of dry *tert*-butyl alcohol, and 5–10 g. of aluminum *tert*-butoxide (Note 1). After the mixture is heated to boiling on a steam bath, approximately 0.4 g. of mercuric chloride is added followed by vigorous shaking (Note 2). As the heating is continued the color of the reaction mixture gradually changes from clear to milky to black, and hydrogen is evolved. When the mixture has become black, the heating is interrupted.

After the reaction has been allowed to proceed for an hour without heating, an additional 244 g. (309 ml., 3.3 moles) of dry *tert*-butyl alcohol (total quantity, 6 moles) and 200 ml. of dry benzene are added. The reaction will again set in upon gentle heating and will continue vigorously without further heating. After about 2 hours the reaction subsides and the mixture is refluxed for about 10 hours.

The benzene and unchanged *tert*-butyl alcohol are removed by distillation from the steam bath, the final traces being removed under 10–30 mm. pressure. A liter of *dry* ether is added, and the solid aluminum *tert*-butoxide is dissolved by refluxing for a short period. After cooling, 35 ml. of *undried* ether is added, followed immediately by vigorous shaking (Note 3). After standing for 2 hours the solution is centrifuged for 30 minutes to remove unused aluminum, aluminum hydroxide, and mercury (Note 4).

The solvent is removed by distillation from the steam bath, the final traces under 10–30 mm. pressure. The flask is allowed to cool with a calcium chloride tube attached, and the product is crushed with a spatula and transferred to bottles sealed against moisture. The yield is 394–418 g. (80–85%) of a white or slightly gray solid.

2. Notes

1. Commercial *tert*-butyl alcohol dried over calcium oxide is suitable for this preparation. Aluminum isopropoxide or ethoxide [1,2] may be used in place of the aluminum *tert*-butoxide to remove traces of water. The grade of metal known as "fast cutting rods" has proved most satisfactory. The checkers used turnings made from aluminum cast from melted-down kitchen utensils. Aluminum *tert*-butoxide has also been prepared successfully in another laboratory from commercially pure aluminum (2S) and from rods of the alloy 17ST (communication from L. F. Fieser). The checkers were able to obtain considerably higher yields of the butoxide from pure aluminum than from a copper-bearing alloy.

2. The use of larger amounts of mercuric chloride increases the difficulty of getting the final product free from color. This difficulty may be avoided by previously amalgamating the aluminum.[3,4] The mixture is shaken to distribute the mercuric chloride and thus aid in an even amalgamation of the aluminum.

3. The small amount of water introduced with the undried ether forms aluminum hydroxide which aids in the precipitation of the black suspended material. Shaking is essential to obtain the hydroxide formation throughout the solution.

4. The centrifuging may be carried out in 250-ml. stoppered bottles at 2000 r.p.m. After centrifuging, the solution should be colorless or light tan. If it is still dark in color another 25-ml. portion of undried ether should be added and the centrifuging repeated.

3. Methods of Preparation

Aluminum *tert*-butoxide can be prepared by refluxing dry *tert*-butyl alcohol with amalgamated aluminum [1,5,6] or aluminum plus mercuric chloride.[6] The method described is that of Adkins and Cox.[6] The preparation of amalgamated aluminum has been described.[3,4] Aluminum isopropoxide can be prepared from dry isopropyl alcohol and aluminum,[1,2] the method being essentially that described for aluminum ethoxide.[7]

[1] Tischtschenko, *J. Russ. Phys. Chem. Soc.*, **31**, 694 (1899) [*Chem. Zentr.*, **71**, I, 10 (1900)].
[2] Young, Hartung, and Crossley, *J. Am. Chem. Soc.*, **58**, 100 (1936).
[3] Wislicenus and Kaufman, *Ber.*, **28**, 1325 (1895).
[4] Adkins, *J. Am. Chem. Soc.*, **44**, 2175 (1922).
[5] Oppenauer, *Rec. trav. chim.*, **56**, 137 (1937).
[6] Adkins and Cox, *J. Am. Chem. Soc.*, **60**, 1151 (1938).
[7] *Org. Syntheses* Coll. Vol. **2**, 599 (1943).

AMINOACETAL

(Acetaldehyde, amino-, diethyl acetal)

$$ClCH_2CH(OC_2H_5)_2 + 2NH_3 \rightarrow H_2NCH_2CH(OC_2H_5)_2 + NH_4Cl$$

Submitted by R. B. Woodward and W. E. Doering.
Checked by C. F. H. Allen and J. A. VanAllan.

1. Procedure

To a solution of 38.2 g. (0.25 mole) of chloroacetal (b.p. 62–64°/20 mm.) in 800 ml. of absolute methanol, cooled in a hydrogenation bomb of about 1.1-l. capacity to the temperature of a Dry Ice-acetone bath, is added approximately 300 g. (290 ml.; about 18 moles) of liquid ammonia (Note 1). The bomb is closed, connected with a pressure gauge, and heated at 140° with shaking for 10 hours.

After the bomb has cooled, the ammonia is allowed to escape (Note 2) and the solution is poured out. The bomb is rinsed with two 200-ml. portions of dry methanol (Note 3), and the combined solution and washings are filtered. The colored solution is concentrated on the steam bath to about 500 ml., 100 ml. of 5% aqueous potassium hydroxide is added, and concentration is continued until the vapors no longer burn (about 2 hours). The solution is saturated with salt (Note 4) and placed in the bulb of an automatic extractor (Note 5); 100 ml. of 50% aqueous potassium hydroxide is added, and the solution is extracted continuously with 350 ml. of ether overnight. The oil that remains after concentration of the ether extract gives on fractionation under reduced pressure (Note 6) 23–24 g. (71–74%) of aminoacetal, b.p. 99–103°/100 mm. (Notes 7, 8, and 9).

2. Notes

1. The bomb is cooled by placing it in an iron pot of about two-thirds the height of the bomb. The pot is half filled with acetone, and pieces of Dry Ice are added until the vigorous evolution of carbon dioxide accompanying the addition of each new piece is no longer observed. If liquid ammonia is added to an insufficiently cooled solution of chloroacetal and alcohol, the ammonia is volatilized so vigorously that much of the starting material is lost. It is possible to force the ammonia from a small bomb into the bomb containing the alcohol and chloroacetal, at room temperature, by the aid of compressed hydrogen.[1]

2. Some of the reaction mixture may be carried from the bomb by the escaping ammonia. If the ammonia is allowed to escape through a tube leading into a beaker or flask, this material can be collected.

3. An appreciable amount of a slimy solid collects on the walls of the bomb. It appears to consist of ammonium chloride together with a small amount of iron salts dissolved from the walls of the bomb by the action of the ammonia. The amount of product recovered by careful working of this material is less than 2 g.

4. About 70 g. is required.

5. An automatic extraction apparatus supplied by Ace Glass, Inc., Vineland, New Jersey, was used; see their catalog 40, p. 90, No. 6835 (500-ml. extraction chamber).

6. An ordinary Claisen flask having a modified side arm was used.

7. Yields of 56–70% were obtained by using larger quantities (76–100 g.) of chloroacetal but the same amounts of methanol and ammonia.

8. The residue contains diacetalylamine (b.p. 124–127°/7 mm.; 189°/100 mm.). It can be isolated by combining the residues from several runs and fractionating through an efficient column. The residues from 14 runs of double the size given above gave 83 g. of diacetalylamine.

9. The main difference between this procedure and the one published earlier [2] is that a much smaller amount of haloacetal is used in the same total volume.

3. Methods of Preparation

The most useful and general method for preparing aminoacetal consists of the action of ammonia upon the haloacetals.[3–10] It has also been prepared by the reduction of nitroacetal using sodium in alcohol,[11] and by the reduction of glycine ester hydrochloride with sodium amalgam.[12] The haloacetal-ammonia reaction has been patented.[13]

[1] *Org. Syntheses,* **23,** 69 (1943).
[2] *Org. Syntheses,* **24,** 4 (1944).
[3] Cass, *J. Am. Chem. Soc.,* **64,** 785 (1942).
[4] Wohl, *Ber.,* **21,** 617 (1888); **39,** 1953 (1906).
[5] Wolff, *Ber.,* **21,** 1482 (1888); **26,** 1832 (1893).
[6] Wolff and Marburg, *Ann.,* **363,** 179 (1908).
[7] Marckwald, *Ber.,* **25,** 2355 (1892).
[8] Hartung and Adkins, *J. Am. Chem. Soc.,* **49,** 2521 (1927).
[9] Buck and Wrenn, *J. Am. Chem. Soc.,* **51,** 3613 (1929).
[10] Woodward and Doering, *J. Am. Chem. Soc.,* **67,** 860 (1945).
[11] Losanitsch, *Ber.,* **42,** 4049 (1909).
[12] Fischer, *Ber.,* **41,** 1021 (1908).
[13] U. S. pat. 2,490,385 [*C. A.,* **44,** 6426 (1950)].

9-AMINOACRIDINE

(Acridine, 9-amino-)

$3o\text{-}C_6H_5NHC_6H_4COOH + 2POCl_3 \rightarrow$

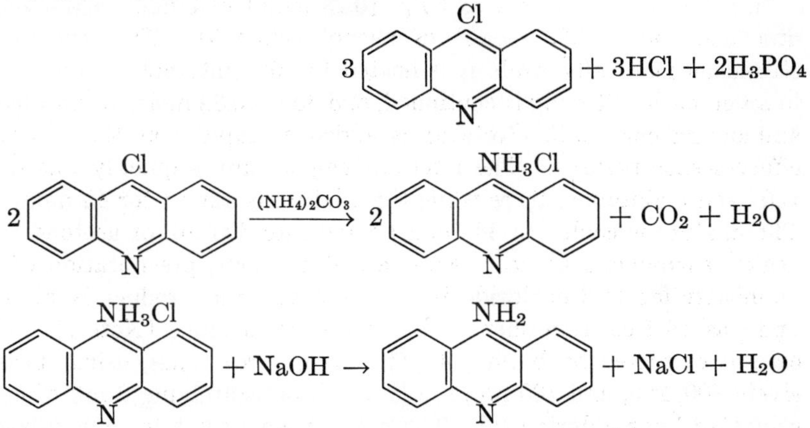

Submitted by ADRIEN ALBERT and BRUCE RITCHIE.
Checked by R. L. SHRINER and JOHN C. ROBINSON, JR.

1. Procedure

In a 500-ml. round-bottomed flask fitted with a water-cooled condenser, 50 g. (0.23 mole) of N-phenylanthranilic acid (or 46 g. of acridone) is mixed with 160 ml. (270 g., 1.76 moles) of phosphorus oxychloride (Note 1). The mixture is slowly heated (about 15 minutes) to 85–90° on a water bath. A vigorous reaction sets in, and the flask is removed at once from the hot bath. If the reaction becomes too violent, the flask is immersed in a beaker of cold water for a moment. After 5–10 minutes, when the boiling subsides somewhat, the flask is immersed in an oil bath. The temperature of the bath is then raised to 135–140°, where it is maintained for 2 hours. The excess phosphorus oxychloride is removed by distillation from an oil bath at 140–150° under a vacuum of about 50 mm. The residue, after cooling, is poured into a well-stirred mixture of 200 ml. of concentrated ammonia solution, 500 g. of ice, and 200 ml. of chloroform. The flask is rinsed by shaking with a little chloroform-ammonia mixture (about 25–30 ml.). When no more undissolved solid remains (about 30 minutes is required), the chloroform layer is separated and the aqueous layer is extracted with an additional 40 ml. of chloroform. The united chloroform extracts are dried over 10 g. of calcium chloride and filtered, and the solvent is removed by distillation. The resultant greenish gray powder is dried at 70° for 20 minutes (Notes 2 and 3). The yield of crude 9-chloroacridine is 50 g. (practically theoretical), m.p. 117–118° (Note 4).

In a 1-l. beaker are placed 50 g. (0.23 mole) of crude 9-chloroacridine and 250 g. (2.7 moles) of phenol (Note 5). The mixture is stirred mechanically while it is heated to 70° (internal temperature) in an oil bath. Stirring is continued, and 30 g. (0.38 mole) of powdered ammonium carbonate (Note 6) is added as rapidly as the vigorous effervescence permits. The internal temperature is quickly raised to 120° and maintained there while the mixture is stirred for 45 minutes. The mixture is cooled to 30° and poured into 600 ml. of acetone in a beaker surrounded by ice. After about an hour, precipitation of 9-aminoacridine hydrochloride is complete and the product is filtered and washed free from phenol with 250 ml. of acetone (Note 7). The cake is extracted by boiling it with water three times, using successively 800, 200, and 100 ml., the last portion containing 2 ml. of concentrated hydrochloric acid. The hot solutions are filtered to remove the small amount of carbonaceous matter, and the filtrates are combined. Any precipitate is redissolved by heating, and then a solution of 60 g. of sodium hydroxide in 300 ml. of water is added. The mixture is cooled and filtered; the solid is washed with 300 ml. of cold water and dried at 120°. The yield is 34–38 g. (76–85% based on phenylanthranilic acid or acridone) of bright yellow powder, m.p. 230°. This product is pure enough for most purposes, but the crude material may be purified by boiling 38 g. of it with 1.2 l. of acetone. The mixture is filtered by suction, and the filtrate is chilled in an ice-salt bath. The first crop weighs 26–27 g., and concentration of the mother liquor to 500 ml. yields an additional 6–7 g. of crystals. Both crops melt at 232–233°, and the recovery is 32–34 g. (84–89%) (Note 8).

2. Notes

1. The phosphorus oxychloride should be freshly distilled.

2. 9-Chloroacridine is readily hydrolyzed in neutral and acid solutions; hence it must not be exposed to the air after removal of the phosphorus oxychloride and before treatment with ammonia. If the drying is carried out at a higher temperature, loss results through sublimation. Care should be taken to keep 9-chloroacridine from entering the eyes, as it is distinctly irritating.

32606

3. It was found convenient to dry the chloroacridine by pouring the concentrated chloroform extract directly into the 1-l. beaker to be used for the next step, and heating the oil bath to 70° for the required 20 minutes.

4. If pure 9-chloroacridine is desired, the crude product is dissolved in a little boiling alcohol, and 0.5% ammonia is added until the solution becomes milky. About 0.5 g. of Norit is then added; the solution is quickly filtered and at once cooled in an ice bath. White crystals, melting at 119–120°, are obtained. The product keeps best in a desiccator over potassium carbonate. By warming 9-chloroacridine with various primary and secondary amines, many substituted 9-aminoacridines are readily obtained.

5. 9-Phenoxyacridine appears to be an intermediate in this reaction.

6. The ammonium carbonate should analyze for 30% ammonia. Baker's analyzed ammonium carbonate (lump) is suitable.

7. The acetone and the phenol, recovered from the filtrate by simple distillation and by vacuum distillation, respectively, may be used again.

8. 9-Aminoacridine hydrochloride is one of the most highly fluorescent of substances. The 1:1000 aqueous solution is pale yellow with only a faint green fluorescence, but the 1:100,000 solution is colorless with an intense blue fluorescence.

9-Aminoacridine is a moderately strong base; [1] the dissociation constant is 3×10^{-5} (aniline $= 5 \times 10^{-10}$).

3. Methods of Preparation

9-Chloroacridine has been prepared by heating thioacridone,[1] acridone,[2] or N-phenylanthranilic acid [3–5] with phosphorus pentachloride, phosphorus oxychloride, or a mixture of the two phosphorus halides, with and without the addition of hydrocarbon solvents. The present method is essentially that of Magidson,[3] but the troublesome filtration of the glutinous and easily hydrolyzed 9-chloroacridine has been avoided by the use of chloroform.

The method described for converting 9-chloroacridine to 9-aminoacridine was developed after experience with the methods given in the literature, namely: conversion of 9-chloroacridine, through 9,9-diphenoxyacridan, to 9-phenoxyacridine, which is then heated with ammonium chloride; [6] heating 9-chloro-, 9-ethoxy-, or 9-phenoxyacridine with ammonia and a copper compound under pressure,[7] or with phenylhydrazine followed by acid reduction.[8] 9-Aminoacridine has also been prepared by hydrolyzing 9-cyanoacridine (from acridine or 9-chloroacridine) to the corresponding acid amide, and subsequent degradation to the required amine; [9] and by the decomposition of the acid azide.[10] Some methods developed for preparing aminoethoxyacridines [11] are also of interest. 9-Aminoacridine has also been obtained by the reaction of acridine with sodium amide.[12]

[1] Edinger, *J. prakt. Chem.*, (2) **64**, 471 (1901).
[2] Graebe and Lagodinzski, *Ann.*, **276**, 48 (1893).
[3] Magidson, *Ber.*, **66**, 866 (1933).
[4] Derscherl, *Ann.*, **504**, 300 (1933).
[5] Drozdov, *J. Gen. Chem. U.S.S.R.*, **4**, 117 (1934) [*C. A.*, **28**, 5456 (1934)].
[6] Drozdov, *J. Gen. Chem. U.S.S.R.*, **5**, 1576, 1736 (1935) [*C. A.*, **30**, 2195, 3432 (1936)].
[7] Meister, Lucius, and Bruning, Ger. pat. 360,421 [*Frdl.*, **14**, 800 (1921–1925); *C. A.*, **18**, 1130 (1924)]; Ger. pat. 364,032 [*Frdl.*, **14**, 802 (1921–1925); *C. A.*, **18**, 1130 (1924)]; Ger. pat. 367,084 [*Frdl.*, **14**, 803 (1921–1925); *C. A.*, **18**, 1130 (1924)].
[8] Meister, Lucius, and Bruning, Ger. pat. 364,031 [*Frdl.*, **14**, 801 (1921–1925); *C. A.*, **18**, 1131 (1924)].
[9] Lehrnstedt, *Ber.*, **64**, 1232 (1931).
[10] Meister, Lucius, and Bruning, Ger. pat. 364,035–6 [*Frdl.*, **14**, 805–806 (1921–1925); *C. A.*, **18**, 1131 (1924)].
[11] Meister, Lucius, and Bruning, Ger. pat. 393,411 [*Frdl.*, **14**, 807 (1921–1925)]; Ger. pat. 395,683 [*Frdl.*, **14**, 812 (1921–1925)].
[12] Bauer, *Chem. Ber.*, **83**, 10 (1950).

o-AMINOBENZALDEHYDE

(Anthranilaldehyde)

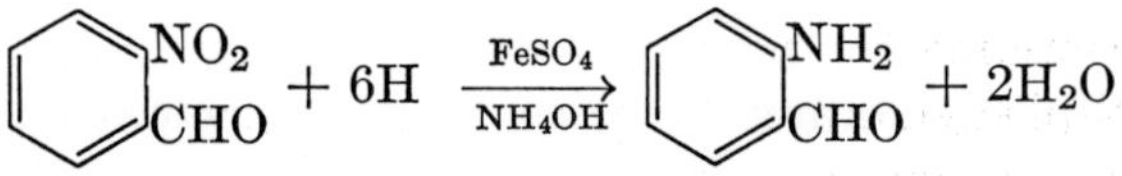

Submitted by Lee Irvin Smith and J. W. Opie.[1]
Checked by Cliff S. Hamilton, C. W. Winter, and Harry M. Walker.

1. Procedure

A 1-l. three-necked flask is employed as a reaction vessel from which the product can be steam-distilled *immediately* after completion of the reaction (Note 1). It is convenient to arrange the apparatus for the reaction and that for the steam distillation on the same steam bath, with provision for the rapid connection of the flask to the distillation assembly at the desired time. For use as a reaction vessel the flask is mounted on a steam bath and fitted with a mechanical stirrer and a reflux condenser; the third neck is closed by a cork.

In the steam-distillation assembly (Note 1) one of the small necks of the flask is fitted with a steam-inlet tube, connected through a water trap to a steam line; the other small neck is closed by a cork. The central neck is fitted to a Kjeldahl trap leading to a 50-cm. Allihn condenser set downward and connected in series to a 50-cm. Liebig condenser. The second condenser leads to a 500-ml. three-necked flask used as the receiver. The receiving flask is immersed in an ice bath and fitted with an Allihn reflux condenser.

When all the apparatus has been set up and tested the flask is connected to the reaction assembly, and 175 ml. of water, 105 g. (0.38 mole) of ferrous sulfate heptahydrate, 0.5 ml. of concentrated hydrochloric acid, and 6 g. (0.04 mole) of *o*-nitrobenzaldehyde (see p. 641) are introduced in the order given. The stirrer is then started, and the flask is heated by means of the steam bath. When the temperature of the mixture reaches 90°, 25 ml. of concentrated ammonium hydroxide is added in one portion, and at 2-minute intervals three 10-ml. portions of ammonium hydroxide are added. Stirring and heating are continued throughout. The total reaction time is 8–10 minutes.

Immediately after the addition of the last portion of ammonium hydroxide, the reflux condenser and stirrer are removed and the flask is connected to the steam-distillation assembly. The mixture is steam-distilled as rapidly as possible, and two 250-ml. fractions of distillate are collected during a period of 10–13 minutes (Note 2). The first fraction is saturated with sodium chloride, and the solution is stirred at 5° until precipitation appears to be complete. The solid is collected on a Büchner funnel and dried in the air. The product weighs 2.72–3.11 g. (57–65%) and melts at 38–39°. The second fraction of the distillate is saturated with sodium chloride and combined with the filtrate remaining from the first fraction. The combined solution is extracted with two 45-ml. portions of ether. The combined ether extract is filtered, dried over anhydrous sodium sulfate, and concentrated by distillation, finally under reduced pressure. The residue solidifies on cooling and weighs 0.6–1.0 g.; it can be purified by steam distillation from 40–50 ml. of saturated sodium chloride solution until 100 ml. of distillate is collected, saturation of the distillate with sodium chloride, cooling, and filtration. The pure product so obtained weighs 0.42–0.87 g. The total yield (Note 3) is 3.3–3.6 g. (69–75%).

2. Notes

1. Rapid removal of the product from the reaction mixture is essential to the success of this preparation. The steam-distillation assembly should be sturdily constructed, with all parts except the distilling flask

in place at the time the reaction is started. To ensure the proper fitting of the distilling flask it is convenient to construct the distillation apparatus with the flask in place and, after testing of the apparatus, to remove the flask and incorporate it into the reaction assembly.

2. The first fraction is drawn into a round-bottomed flask through a tube inserted into the third neck of the receiver (by the use of a water pump or a suction line) without interruption of the distillation.

3. The submitters have obtained yields of about 70% in runs eight times as large as that described. This amino aldehyde undergoes self-condensation on standing, especially in a desiccator, and so the product should be used immediately.

3. Methods of Preparation

o-Aminobenzaldehyde has been obtained from its oxime by the action of ferric chloride;[2] the oxime was obtained by reduction of *o*-nitrobenzaldoxime.[3] *o*-Aminobenzaldehyde has also been obtained from *o*-aminobenzyl alcohol by the action of sodium sulfide[4] or by the action of zinc dust and alkali;[5] and from *o*-nitrobenzylaniline or N-(2-nitrobenzyl)sulfanilic acid by the action of alkali sulfides and sulfur.[6] The aldehyde results in small amounts from the action of alkali alone upon *o*-nitrobenzyl alcohol[7] and from the action of ferrous sulfate and ammonia upon anthranil.[8] The only preparative methods involve the reduction of *o*-nitrobenzaldehyde with ferrous sulfate and ammonia.[8,9] Catalytic reduction of *o*-nitrobenzaldehyde has been reported to yield *o*-aminobenzaldehyde,[10,11] but the method was not successful in the hands of the submitters.

[1] Work done under contract with the Office of Scientific Research and Development.

[2] Gabriel, *Ber.*, **15**, 2004 (1882).

[3] Gabriel and Meyer, *Ber.*, **14**, 2339 (1881); **36**, 803 (1903).

[4] Ger. pat. 106,509 (*Chem. Zentr.*, **1900**, I, 1084).

[5] Freundler, *Compt. rend.*, **136**, 371 (1903); **138**, 1425 (1904); *Bull. soc. chim. France,* [3] **31**, 876 (1904).

[6] Ger. pats. 99,542, 100,968 (*Chem. Zentr.*, **1899**, I, 238, 958); Cohn and Springer, *Monatsh,* **24**, 95 (1903); Friedländer and Lenk, *Ber.*, **45**, 2084 (1912).

[7] Freundler, *Bull. soc. chim. France,* [3] **31**, 879 (1904); Carré, *Compt. rend.*, **140**, 664 (1905); *Bull. soc. chim. France,* [3] **33**, 1162, 1165 (1905); *Ann. chim.*, [8] **6**, 409, 413 (1905).

[8] Friedländer, *Ber.*, **15**, 2573 (1882).

[9] Friedländer and Göhring, *Ber.*, **17**, 456 (1884); Bamberger, *Ber.*, **60**, 319 (1927).

[10] Ruggli and Schmidt, *Helv. Chim. Acta,* **18**, 1235 (1935); Nord, *Ber.*, **52**, 1711 (1919).

[11] Borsche and Ried, *Ber.*, **76B**, 1015 (1943).

Ethyl Azodicarboxylate

WARNING

A report has been received that a sample of ethyl azodicarboxylate [*Org. Syntheses*, **28,** 59 (1948); **Coll. Vol. 3,** 375 (1955)] decomposed upon attempted distillation with sufficient violence to shatter the distillation apparatus.

It is possible that the explosion may have been due to overchlorination or to insufficient washing of the product with sodium bicarbonate solution.

It is recommended that ethyl azodicarboxylate be distilled only behind a safety shield, and protected from direct sources of light.

Please insert this sheet opposite page **59** of *Org. Syntheses*, **28** (1948) **and/or** opposite page 375 of *Org. Syntheses*, **Coll. Vol. 3** (1955).

m-AMINOBENZALDEHYDE DIMETHYLACETAL

(Benzaldehyde, *m*-amino-, dimethylacetal)

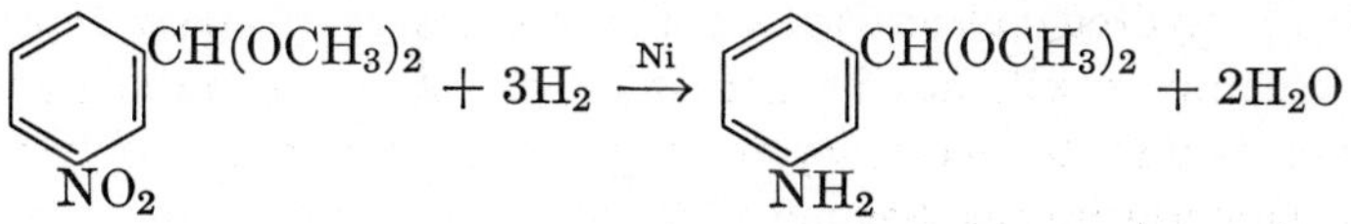

Submitted by Roland N. Icke, C. E. Redemann, Burnett B. Wisegarver, and Gordon A. Alles.
Checked by H. R. Snyder and Frank X. Werber.

1. Procedure

In a 1-l. steel bomb are placed 295 g. (1.5 moles) of *m*-nitrobenzaldehyde dimethylacetal (p. 644), 250 ml. of technical anhydrous methanol, and 1 tablespoon of Raney nickel catalyst. Hydrogen is introduced until the pressure is about 1000 lb. (Note 1). The bomb is heated to about 40°, at which point the heating is discontinued and the shaker is started. The hydrogenation soon becomes rapid as the temperature rises to about 70° (Note 2). The bomb is refilled with hydrogen as many times as necessary (Note 3). The theoretical amount of hydrogen (4.5 moles) is absorbed in about 1.5 hours.

The bomb is cooled, the remaining hydrogen is discharged, and the bomb is opened. The solution is transferred to a beaker, and the bomb is rinsed with a little methanol which is added to the solution. The catalyst is removed by filtration (*Caution! The catalyst may be pyrophoric*), and most of the filtrate is transferred to a 500-ml. Claisen flask set on a steam bath for distillation of the methanol; the remainder of the filtrate is introduced into the Claisen flask when the volume of the first portion has been reduced sufficiently by distillation. After all the methanol has been removed the aminoacetal is distilled under diminished pressure. The yield of *m*-aminobenzaldehyde dimethylacetal, a light-yellow liquid boiling at 123–124°/4 mm. or 110–112°/1.5 mm., is 168–196 g. (67–78%).

2. Notes

1. The hydrogenation is similar to one described on p. 63. As the bomb does not contain enough hydrogen to complete the reduction, more hydrogen should be admitted whenever the pressure drops below 300 lb.

2. Because of the high heat capacity of the bomb the internal temperature continues to rise (to about 70°) after the heater is turned off. As the exothermic hydrogenation begins the temperature rises to about 80°. The temperature should be kept below 85° to prevent hydrogenolysis of the acetal.

3. If the hydrogenation is started at a pressure of about 1500 lb. in a 2.5-l. bomb it will not be necessary to introduce more hydrogen. However, it may be necessary to stop the shaker occasionally to prevent a temperature rise beyond 85°.

3. Methods of Preparation

This acetal has not been described previously. The corresponding diethylacetal has been prepared by the reduction of *m*-nitrobenzaldehyde diethylacetal with sodium sulfide [1] and by the reaction of the anhydro compound of *m*-aminobenzaldehyde with ethanolic hydrogen chloride and ethyl orthoformate.[2]

[1] Haworth and Lapworth, *J. Chem. Soc.*, **121**, 76 (1922).

[2] Bottomley, Cocker, and Nanney, *J. Chem. Soc.*, **1937**, 1891.

o-AMINOBENZYL ALCOHOL

(Benzyl alcohol, *o*-amino-)

$$o\text{-}NH_2C_6H_4CO_2H + 4H \rightarrow o\text{-}NH_2C_6H_4CH_2OH + H_2O$$

Submitted by George H. Coleman and Herbert L. Johnson.
Checked by Reynold C. Fuson and E. A. Cleveland.

1. Procedure

The reduction is carried out in four cells of the type shown in Fig. 2. Each cell consists of a 1-l. beaker (B), a porous cup (P), a mechnical stirrer, and sheet lead electrodes (E_1 and E_2) each having a total surface area of 100 sq. cm. (Note 1). In the cathode space of each cell are placed 25 g. (0.18 mole) of anthranilic acid (Note 2) and 400 ml. of 15% sulfuric acid. In each porous cup is placed 200 ml. of 15% sulfuric acid. The cells are connected in series as shown in Fig. 5 with an ammeter (A) and suitable resistance (R) (Note 3) also in the circuit.

The stirrers are started, the current (110 volt d.c.) is turned on, and the resistance is so adjusted that the ammeter records 10–12 amperes.

The temperature of the solution in the cells is maintained at 20–30° by surrounding them with a bath of cool water (Note 4). The reduction is complete after 60–70 ampere-hours. This fact is indicated by the increased evolution of hydrogen and the complete solution of the anthranilic acid.

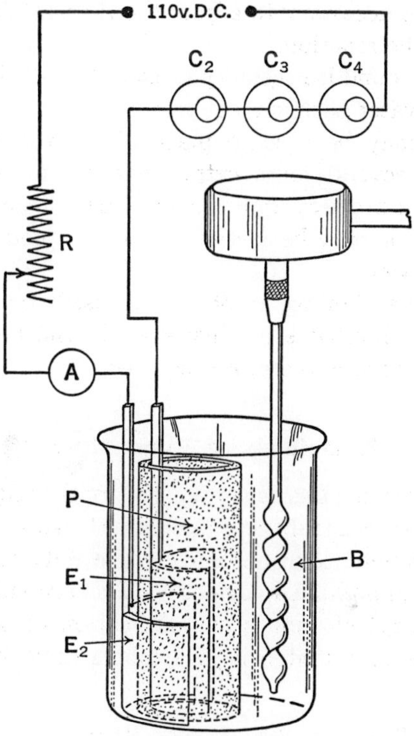

FIG. 2.

The cathode liquid is removed from the cells and neutralized with solid ammonium carbonate or concentrated aqueous ammonia. The solution is filtered to remove any resinous material, then saturated with ammonium sulfate and extracted with five 80-ml. portions of chloroform (Note 5). The chloroform solution is dried with 20 g. of anhydrous sodium or magnesium sulfate and filtered, and the chloroform is removed by evaporation on a steam bath (Note 6). The yield of *o*-aminobenzyl alcohol obtained from the four cells is 62–70 g. (69–78%).

This product is light brown and melts at 75–80°. After one recrystallization from petroleum ether the melting point is 80–81° (Note 7).

2. Notes

1. Ordinary sheet lead of 99.9% purity is satisfactory.

2. The anthranilic acid melted at 143–144°.

3. A resistance of 10–12 ohms is required for the apparatus described. It may be necessary to adjust the rheostat several times during the course of the reaction.

4. The checkers found it necessary to cool the cells by frequent addition of ice to the water bath surrounding them.

5. Ethyl ether may be used in place of chloroform but is not so satisfactory since several more extractions are necessary to remove the same amount of product from the aqueous solution.

6. The evaporation may be carried out in an apparatus that permits recovery of the solvent.

7. Petroleum ether boiling at 65–75° is used for recrystallization. The product has a limited solubility even in the hot solvent, and a relatively large volume is therefore required.

3. Methods of Preparation

o-Aminobenzyl alcohol has been prepared by the reduction of *o*-nitrobenzaldehyde [1,2] or *o*-nitrobenzyl alcohol [3,4] with zinc and hydrochloric acid in alcoholic solution, and by the reduction of anthranilic acid with lithium aluminum hydride.[5,6] The present method of preparing *o*-aminobenzyl alcohol is a modification of that described by Mettler.[7] Other substituted benzyl alcohols have been obtained by the same method.

[1] Friedländer and Henriques, *Ber.*, **15**, 2109 (1882).

[2] Paal and Senniger, *Ber.*, **27**, 1084 (1894).

[3] Gabriel and Posner, *Ber.*, **27**, 3512 (1894).

[4] Auwers, *Ber.*, **37**, 2260 (1904).

[5] Nystrom and Brown, *J. Am. Chem. Soc.*, **69**, 2548 (1947).

[6] Conover and Tarbell, *J. Am. Chem. Soc.*, **72**, 3586 (1950).

[7] Mettler, *Ber.*, **38**, 1751 (1905).

2-AMINO-*p*-CYMENE *

(Carvacrylamine)

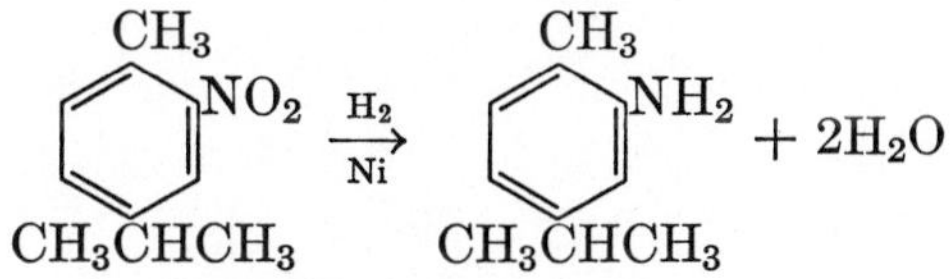

Submitted by C. F. H. ALLEN and JAMES VANALLAN.
Checked by HOMER ADKINS.

1. Procedure

A mixture of 179 g. (167 ml., 1 mole) of 2-nitro-*p*-cymene (p. 653), 300 ml. of absolute ethanol, and 3–5 g. of Raney nickel (p. 181) is placed in the steel reaction vessel of a high-pressure hydrogenation apparatus.[1] The bomb is then closed, and hydrogen is admitted until the pressure, at 25°, is about 1000 lb. While the bomb is shaken, the temperature is rapidly raised to 80–90°, and then the heater is shut off (Note 1). Owing to the strong exothermic reaction, the temperature continues to rise, reaching about 120°, while the pressure drops rapidly. The pressure in the reaction vessel is maintained at 700–1500 lb., by the introduction of hydrogen from a tank, until the rapid reaction is over (15 minutes). The reaction mixture is kept at 100–120° for 30 minutes after there is no further drop in the pressure of hydrogen. After the bomb has cooled, the hydrogen is slowly released and the catalyst is separated from the reaction mixture by centrifuging, or by filtration through a sintered-glass or Büchner funnel (Note 2). The alcohol and water are removed by distillation, and the product is fractionated in a suitable apparatus (Notes 3, 4, 5, 6, 7).

The amine distils at 239–240°/740 mm., 242° (cor.)/760 mm., 110°/10 mm., or 92–94°/2 mm. In the first distillation the portion boiling below 225°/740 mm. (or below 100°/9 mm.) is discarded. The yield is 130–135 g. (87–90%).

2. Notes

1. The hydrogenation of nitro compounds liberates so much heat, and the reaction may proceed so rapidly, that precautions must be taken against excessive reaction temperatures. Temperatures of 150–

200° may bring about the formation of secondary amines, and at higher temperatures decomposition reactions may occur with explosive violence. These dangers may be avoided by the use of small amounts of catalyst and by having an insufficient amount of hydrogen in the reaction vessel at the beginning of the reaction. After sufficient hydrogen for the hydrogenation of half of the nitro compound has been taken up there need be no fear that the reaction will proceed too rapidly. It is always desirable, however, to complete the hydrogenation as rapidly as possible while avoiding temperatures above about 120°. The use of ethanol as a solvent prevents the separation of a water layer, which would cover most of the catalyst and thus give rise to a slow and perhaps incomplete hydrogenation.

2. If the sample of nitrocymene is impure, or if the catalyst is of poor quality, the hydrogenation may not proceed to completion and a solution of the product in hydrochloric acid will be turbid. In this event, the mixture of products (after separation of the spent catalyst) may again be submitted to hydrogenation. An alternative procedure involves removal of the non-basic material by steam distillation of a mixture of the product with 350 ml. of water and 90 ml. of concentrated hydrochloric acid. The amine is then liberated from its salt by the addition of 45 g. of sodium hydroxide and is extracted with 500 ml. of ether. The ether layer may be siphoned from the water layer. The ether is removed by distillation, and the residue is fractionated as described in the procedure. For steam distillations, the submitters employ the adapter shown in Fig. 3. With this adapter the apparatus is much more compact than in the usual arrangement, for the receiver may serve as a partial condenser, as in distillation under reduced pressure.

3. The apparatus may consist of a modified Claisen flask, or of an ordinary flask connected to a Vigreux, Widmer, modified Widmer, or other column. The checker used a modified Widmer column and a distillation at 9 mm. pressure. The submitters conducted the distillation at atmospheric pressure, in the type of flask shown in Fig. 4. This type of flask, as well as the adapter described in Note 2, has been used in the laboratories of the Eastman Kodak Company for some time. In comparison with the standard Claisen flask, this flask has several advantages. It has a longer column; liquids seldom bump over, and consequently more material may be handled in a flask of given size; and material that has come into contact with the rubber stopper cannot contaminate the distillate. No dimensions are given for the flask, or for the adapter described in Note 2 above; these pieces may be constructed in various sizes to suit individual needs.

4. Nitrocymene usually contains about 8% of *p*-nitrotoluene. The complete separation of the resulting *p*-toluidine from the 2-amino-*p*-cymene requires a rather careful fractionation.

5. If air comes in contact with the hot amine vapor, the product will be colored. However, a product distilled at 9 mm. with a capillary ebullition tube was only faintly yellow.

6. The reduction may also be carried out at lower pressures in the usual laboratory apparatus [*Org. Syntheses* Coll. Vol. **1**, 53 (1932); 63 (1941)]. A mixture of 50 g. of nitrocymene, 200 ml. of 95%

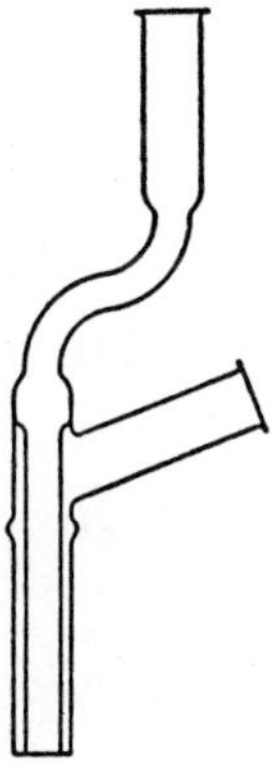

FIG. 3.

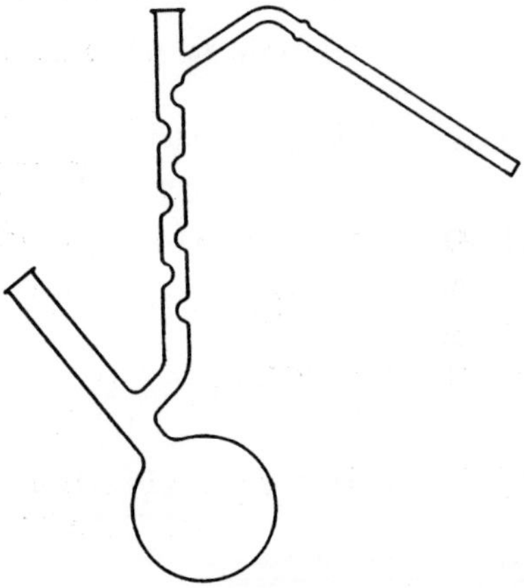

FIG. 4.

ethanol, and 5 g. of Raney nickel is reduced at 85–90°, at an initial pressure of 55–60 lb.; the time required is 6 hours. The product is processed as described above. The yield is 33–36 g. (80–85%).

7. Nitrocymene may also be reduced by the use of iron powder, essentially as described under 2,4-diaminotoluene (*Org. Syntheses*, **11**, 32; Coll. Vol. **2**, 160) (Kenneth A. Kobe, private communication).[2]

3. Methods of Preparation

Because of the accessibility of *p*-cymene, reduction of the nitro-derivative constitutes the only practical method for preparation of 2-aminocymene. Tin [3–6] and iron,[7,8] together with hydrochloric acid, have been used for this purpose. A patent [7] mentions the use of hydrogen in the presence of a nickel catalyst.

[1] American Instrument Company, Silver Springs, Maryland, or Parr Instrument Company, Moline, Illinois.

[2] Doumani and Kobe, *J. Am. Chem. Soc.*, **62**, 563 (1940).

[3] Soderbaum and Widman, *Ber.*, **21**, 2127 (1888).

[4] Wheeler and Smithey, *J. Am. Chem. Soc.*, **43**, 2613 (1921).

[5] Demonbreun and Kremers, *J. Am. Pharm. Assoc.*, **12**, 296 (1923) [*C. A.*, **17**, 3906 (1923)].

[6] Straneo, *Ann. chim. appl.*, **31**, 116 (1941).

[7] U. S. pat. 1,314,920 [*C. A.*, **13**, 2765 (1919)].

[8] Doumani and Kobe, *Ind. Eng. Chem.*, **31**, 264 (1939).

α-AMINODIETHYLACETIC ACID

(Butyric acid, α-amino-α-ethyl-)

$$C_2H_5COC_2H_5 + NaCN + NH_4Cl \rightarrow (C_2H_5)_2C(OH)CN + NaCl + NH_3$$

$$(C_2H_5)_2C(OH)CN + NH_3 \rightarrow (C_2H_5)_2C(NH_2)CN + H_2O$$

$$(C_2H_5)_2C(NH_2)CN + H_2O + HCl \xrightarrow{\text{cold}} (C_2H_5)_2C(NH_2\cdot HCl)CONH_2$$

$$(C_2H_5)_2C(NH_2\cdot HCl)CONH_2 + H_2O + HCl \xrightarrow{\text{hot}} (C_2H_5)_2C(NH_2\cdot HCl)CO_2H + NH_4Cl$$

$$(C_2H_5)_2C(NH_2\cdot HCl)CO_2H + \tfrac{1}{2}Pb(OH)_2 \rightarrow (C_2H_5)_2C(NH_2)CO_2H + \tfrac{1}{2}PbCl_2 + \tfrac{1}{2}H_2O$$

Submitted by Robert E. Steiger.
Checked by W. E. Bachmann and Yun-Tsung Chao.

1. Procedure

A solution of 50 g. (1 mole) of sodium cyanide (98% purity) in 100 ml. of water is placed in a 2-l. round-bottomed flask (Note 1) fitted with a ground-glass stopper. A solution of 58.9 g. (1.1 moles) of ammonium chloride in 140 ml. of lukewarm water is added, followed by 67 ml. (1 mole) of concentrated ammonium hydroxide (sp. gr. 0.9). This mixture is stirred mechanically and is cooled by a stream of water while a solution of 86.1 g. (1 mole) of diethyl ketone in 160 ml. of methanol (Note 2) is added. The flask is then stoppered (Note 3) and half immersed for 5 hours in a water bath, the temperature of which is kept at 55–60°. The reaction mixture is then cooled in an ice bath and poured (with precautions, i.e., under a properly ventilated hood) into 800 ml. of concentrated hydrochloric acid (sp. gr. 1.18) contained in a 5-l. round-bottomed flask which is surrounded up to the neck by ice and water. The reaction flask is rinsed with two 25-ml.

portions of water. The mixture is now saturated at 0–5° with hydrogen chloride gas. After standing overnight under a hood, the mixture is refluxed for 2.5 hours (Note 4).

The solution is evaporated to dryness under reduced pressure on a water bath. In order to remove as much hydrochloric acid as possible, the temperature of the bath is raised to 100° toward the end of the distillation. The residue of amino acid hydrochloride and inorganic salts is suspended in 500 ml. of absolute ethanol. The suspension is boiled on a steam bath for a short time, then cooled to room temperature and filtered on a Büchner funnel. The residue of inorganic salts is washed with 500 ml. of absolute ethanol. To the combined filtrates is added 400 ml. of ethyl ether (U. S. P. quality) in order to precipitate inorganic material. After several hours the mixture is filtered, and the residue is washed with a 5:2 mixture of absolute ethanol and ether. The filtrate is transferred to a 5-l. round-bottomed flask, about 200 ml. of water is added, and the liquids are removed by distillation under reduced pressure. The nearly dry residue is dissolved in 2 l. of water, and the solution is treated with an excess of freshly prepared lead hydroxide (Note 5). The suspension is diluted with water to a volume of about 3.5 l. and is then concentrated under reduced pressure, at as low a temperature as possible, to a volume of about 2 l. The suspension is then filtered with suction (Note 6), and the residue of lead salts is washed thoroughly with water. The cloudy filtrate, which still contains some free ammonia, is concentrated by distillation under reduced pressure to a volume of about 300–400 ml. The mixture is filtered, the filtrate is saturated with hydrogen sulfide gas, and the precipitate of lead sulfide is removed by filtration with suction (Note 6). The solution is now concentrated by distillation under reduced pressure on a water bath, and 1 l. of 95% ethanol is added to the nearly dry residue of the amino acid. The suspension is boiled under a reflux condenser until nearly all the amino acid is dissolved, and the mixture is then allowed to cool to room temperature. The amino acid, which separates in the form of fine needles, is collected on a Büchner funnel and washed with a little 95% ethanol. A second crop of crystals is obtained by evaporating the combined filtrates to dryness, dissolving the residue in the minimum amount of hot water, and treating the solution with 95% ethanol. The amino acid is dried in the air and then in a vacuum desiccator over phosphorus pentoxide. The total yield of product is 58.8–65 g. (39–43%, assuming that the product contains exactly one molecule of water of crystallization) (Note 7).

2. Notes

1. The checkers used a 1-l. Erlenmeyer flask fitted with a ground-glass stopper.

2. The methanol must be free from acetone since this ketone would give rise to α-aminoisobutyric acid.

3. The ground-glass stopper should be lubricated slightly with stopcock grease. It must be secured firmly by means of adhesive tape, as some pressure develops when the flask is heated.

4. The hydrolysis must be carried out under a hood. The top of the reflux condenser should be connected with the ventilation pipe by means of a piece of glass tubing. The methanol contained in the reaction mixture escapes in the form of methyl chloride, along with some hydrogen chloride.

5. The lead hydroxide is prepared by adding 1.5 l. of a 2 *N* solution of sodium hydroxide (3 moles) through a dropping funnel to a continuously stirred solution of 569 g. (1.5 moles) of lead acetate, $(CH_3CO_2)_2Pb \cdot 3H_2O$, in 1.35 l. of water. The precipitate is collected on a 13-cm. Büchner funnel and washed well with water in order to remove water-soluble impurities. The paste of lead hydroxide is transferred to the solution that is to be freed of chloride ions.

6. A Büchner funnel of adequate size is fitted with two pieces of hardened filter paper covered with a thin layer of moistened Norit.

7. The checkers obtained 65 g. of pure amino acid in the first crop; the second crop of 5 g. contained chloride ion. When heated in an open tube, the pure product sublimed at 255° without melting.

3. Methods of Preparation

α-Aminodiethylacetic acid has been prepared from α-bromodiethylacetic acid and ammonia;[1] and by hydrolysis of α-aminodiethylacetonitrile with hydrochloric acid.[1–4] The required nitrile was obtained by heating diethyl ketone cyanohydrin with one equivalent of ethanolic ammonia;[2] by heating diethyl ketone with an ethanolic solution of ammonium cyanide;[3] and by heating diethyl ketone with a solution of potassium cyanide and ammonium chloride.[1,4] α-Aminodiethylacetic acid has also been obtained from ethyl α-cyanodiethylacetate by partial hydrolysis to ethyl diethylmalonamate and subsequent degradation in alkaline hypobromite.[5]

[1] Rosenmund, *Ber.*, **42**, 4472 (1909).
[2] Tiemann and Friedländer, *Ber.*, **14**, 1973 (1881).
[3] Gulewitsch and Wasmus, *Ber.*, **39**, 1191 (1906).

[4] Freytag, *Ber.*, **48**, 649 (1915).
[5] Lin and Li, *J. Chinese Chem. Soc.*, **6**, 88 (1938) [*C. A.*, **35**, 5096 (1941)].

5-AMINO-2,3-DIHYDRO-1,4-PHTHALAZINEDIONE

(1,4-Phthalazinedione, 5-amino-2,3-dihydro-)

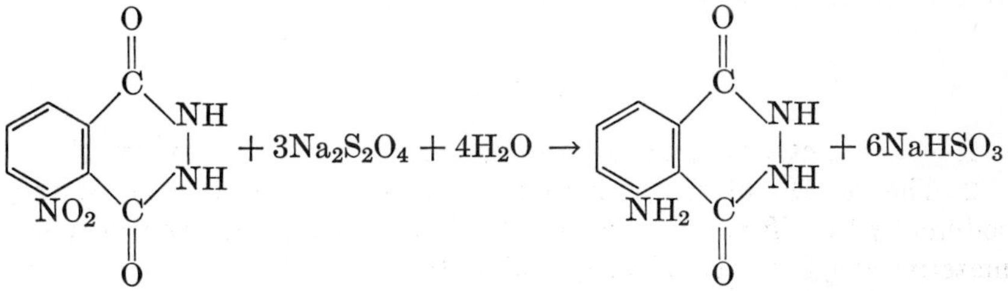

Submitted by Carl T. Redemann and C. Ernst Redemann.
Checked by Cliff S. Hamilton and C. W. Winter.

1. Procedure

In a 1-l. conical flask are placed 52 g. (about 0.15 mole) of the equimolecular mixture of 5-nitro-2,3-dihydro-1,4-phthalazinedione (p. 656) and sodium sulfate (Note 1), 200 ml. of water, and 75 ml. of 15 *N* ammonium hydroxide solution (sp. gr. 0.90). The flask is stoppered and shaken until all, or very nearly all, of the solid has dissolved, and 84 g. (0.4 mole) of sodium hydrosulfite dihydrate (Note 2) is added in three portions. The solution becomes hot, the temperature sometimes reaching the boiling point, and the dark orange-red color begins to fade. After the spontaneous reaction has subsided the solution is boiled gently for a few minutes and filtered to remove insoluble impurities. The filtrate is heated on a steam bath or over a small flame for 30 minutes. During this time the 5-amino-2,3-dihydro-1,4-phthalazinedione begins to separate as a light-yellow flocculent precipitate or as a crust adhering to the walls of the flask. The hot solution is made distinctly acid to litmus paper with glacial acetic acid and allowed to stand overnight. The yellow precipitate is separated by filtration, washed well with cold water, and dried in a hot-air oven at 110° or below. The dry material weighs 25–27 g. and melts with decomposition at 301–305° (Note 3).

This material is sufficiently pure for most purposes. The chief impurities are small amounts of inorganic salts and a trace of the unreduced nitro compound. If a purer product is desired the crude material (5 g. per 100 ml.) is dissolved in hot 3 *N* hydrochloric acid, decolorizing carbon is added, the solution is filtered promptly (Note 4), and the filtrate is made just faintly acid to Congo red paper with concentrated ammonium hydroxide. After the mixture has cooled to room temperature the pale yellow flocculent precipitate is separated by filtration, washed well with cold water, and dried in the oven at 100° or below. The recovery in the crystallization is 70–75% (Note 5), and the product melts at 329–332° (Note 4).

2. Notes

1. No advantage is gained by using the purified nitro compound.
2. The success of this reduction depends upon the quality of the sodium hydrosulfite. The reagent should be taken from a fresh bottle; material which has stood in the laboratory for a long time probably has undergone oxidation.
3. The submitters used a Kullmann copper block for the melting-point determinations. The melting point of the pure material has been reported in the literature at various values between 319° and 333°.
4. The 5-amino-2,3-dihydro-1,4-phthalazinedione should not be exposed to hot hydrochloric acid longer than necessary, since some hydrolysis appears to take place.
5. The percentage yield cannot be calculated with precision, since the exact quantity of nitro compound in the mixture taken for the reduction is unknown. The quantity of sodium hydrosulfite dihydrate employed is sufficient for the reduction of only 0.133 mole of nitro compound; the weight of the purified amino compound corresponds to about 80% of the theoretical yield calculated on the assumption that the hydrosulfite is the limiting reagent.

3. Methods of Preparation

5-Amino-2,3-dihydro-1,4-phthalazinedione, also called luminol and 3-aminophthalhydrazide, has been prepared from 5-nitro-2,3-dihydro-1,4-phthalazinedione by reduction with ammonium sulfide [1] or stannous chloride [2] and by catalytic hydrogenation over palladium on charcoal in alkaline solution [3] and by the reaction of 3-aminophthalimide [2] with hydrazine hydrate.

[1] Huntress, Stanley, and Parker, *J. Am. Chem. Soc.*, **56**, 241 (1934).
[2] Drew and Pearman, *J. Chem. Soc.*, **1937**, 30.
[3] Wegler, *J. prakt. Chem.*, **148**, 135 (1937).

4-AMINO-2,6-DIMETHYLPYRIMIDINE

(Pyrimidine, 4-amino-2,6-dimethyl-)

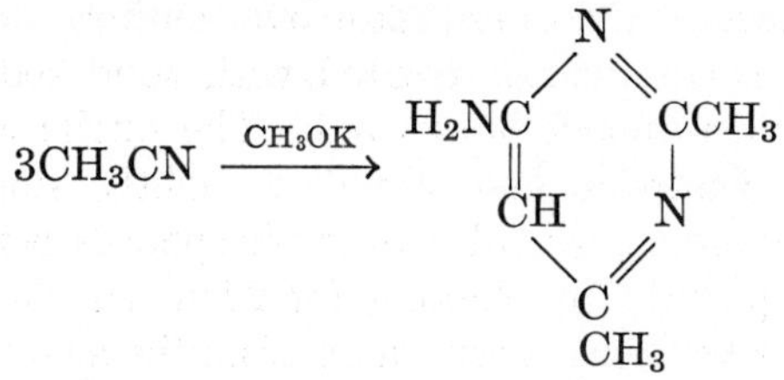

Submitted by ANTHONY R. RONZIO and WILLIAM B. COOK.
Checked by C. G. STUCKWISCH and HENRY GILMAN.

1. Procedure

Seventy grams (1 mole) of freshly prepared potassium methoxide (Note 1) and 41 g. (1 mole) of freshly purified acetonitrile (Note 2) are placed in a 500-ml. distilling flask. A cold-finger condenser which extends into the bulb of the flask is inserted through a rubber stopper in the neck of the flask, and a short piece of rubber tubing carrying a Hofmann clamp is connected to the side arm of the flask. The tubing is connected to an aspirator, and suction is applied until the acetonitrile begins to boil, whereupon the tubing is closed by means of the clamp and the flask is heated for 5 hours in an oil bath maintained at 140°.

At the end of the heating period the contents of the flask will have solidified. To the cold mixture 40 ml. of water is added to hydrolyze the potassium methoxide and precipitate the pyrimidine; the fine crystals are filtered and dried. The crude product is placed in a 500-ml. distilling flask with 250 ml. of purified kerosene (Note 3). On distilling the kerosene, the pyrimidine codistils and solidifies in the receiving flask to a snow-white mass of crystals. These are filtered, washed well with petroleum ether, and dried in an oven at 100°. The yield of pure material, melting at 182–183°, is 27.5–28.7 g. (67–70%) (Note 4).

2. Notes

1. To prepare the potassium methoxide 39 g. (1 gram atom) of metallic potassium, cut *under toluene* (*Caution!*) in 1-cm. cubes, is placed in a 1-l. three-necked flask which has been swept with nitrogen.

The flask, fitted with a reflux condenser, mechanical stirrer, and dropping funnel, is immersed in a cooling bath at −30°, and absolute methanol is added through the funnel until all the potassium has dissolved. The excess methanol is removed by heating on the steam bath, finally under reduced pressure, and the potassium methoxide is dried overnight in a vacuum desiccator over sulfuric acid.

2. Commercial acetonitrile is treated with solid sodium carbonate until no more carbon dioxide is evolved. The nitrile is then distilled over phosphorus pentoxide and stored in tightly stoppered bottles. Before use, the nitrile is redistilled over phosphorus pentoxide.

3. Kerosene is purified by shaking for 24 hours with concentrated sulfuric acid. The kerosene is separated from the acid, washed several times with dilute sodium hydroxide then with water, and finally dried over calcium chloride and distilled using an air condenser. Purified kerosene is a water-white, sweet-smelling liquid.

4. The percentage yield decreases when larger or smaller quantities of material are used.

3. Methods of Preparation

4-Amino-2,6-dimethylpyrimidine has been prepared by heating the reaction product obtained from acetic anhydride and acetamidine;[1] by treating 4-chloro-2,6-dimethylpyrimidine with ammonia;[2] and by heating acetonitrile either with sodium ethoxide in a sealed tube[3] or with sodium.[4,5]

It has been reported that this trimerization may be carried out with equally good results through the use of a small amount of sodium methoxide in place of a large amount of potassium methoxide.[6]

[1] Pinner, *Ber.*, **22**, 1600 (1889).
[2] Schmidt, *Ber.*, **35**, 1577 (1902).
[3] Schwarze, *J. prakt. Chem.*, (2) **42**, 3 (1890).
[4] Keller, *J. prakt. Chem.*, (2) **31**, 365 (1885).
[5] von Meyer, *J. prakt. Chem.*, (2) **27**, 153 (1883).
[6] Cairns, Larchar, and McKusick, *J. Am. Chem. Soc.*, **74**, 5633 (1952).

AMINOGUANIDINE BICARBONATE *

(Guanidine, amino-, bicarbonate)

$$3Zn + H_2NC({=}NH)NHNO_2 + CH_3CO_2H + 4H_2O \rightarrow H_2NC({=}NH)NHNH_2 \cdot CH_3CO_2H + 3Zn(OH)_2$$

$$H_2NC({=}NH)NHNH_2 \cdot CH_3CO_2H + NaHCO_3 \rightarrow H_2NC({=}NH)NHNH_2 \cdot H_2CO_3 + CH_3CO_2Na$$

Submitted by R. L. SHRINER and FRED W. NEUMANN.
Checked by HOMER ADKINS and M. J. CURRY.

1. Procedure

Two hundred and sixteen grams (2.07 moles) of nitroguanidine[1] and 740 g. (11.3 moles) of purified zinc dust (Note 1) are thoroughly ground together in a mortar, and then enough water (about 400 ml.) is added with stirring with the pestle to form a thick paste. The paste is transferred to a 3-l. enameled can or beaker surrounded by an ice bath. A solution of 128 g. (2.14 moles) of glacial acetic acid in 130 ml. of water is cooled to 5° in another 3-l. beaker, which is fitted with a strong mechanical stirrer and surrounded by an ice bath. The paste of nitroguanidine and zinc dust, cooled to 5°, is added slowly with mechanical stirring, the temperature of the reaction mixture being kept between 5° and 15°. A total of about 1 kg. of cracked ice is added to the mixture from time to time as the mixture becomes too warm or too thick to stir. The addition of the paste takes about 8 hours, and the final volume of the mixture is about 1.5 l. (Note 2). The mixture is then slowly warmed to 40° on a water bath with continued stirring, and this temperature is maintained for 1–5 minutes, until reduction is complete (Note 3).

The solution is immediately separated from the insoluble material by filtration on a 20-cm. Büchner funnel, and the cake is sucked as dry as possible. The residue is transferred to the 3-l. beaker, triturated well with 1 l. of water, and then separated from the liquid by filtration. In the same manner, the residue is washed twice more with two 600-ml. portions of water. The filtrates are combined and placed in a 5-l. round-bottomed flask. Two hundred grams of ammonium chloride is added, and the solution is mechanically stirred until solution is complete (Note 4). The stirring is continued, and 220 g. (2.62 moles) of sodium bicarbonate is added during a period of about 10 minutes.

The aminoguanidine bicarbonate begins to precipitate after a few minutes, and the solution is then placed in a refrigerator overnight. The precipitate is collected by filtration on a Büchner funnel. The cake is removed to a 1-l. beaker and mixed with a 400-ml. portion of a 5% solution of ammonium chloride and filtered. It is again washed with two 400-ml. portions of distilled water, the wash solution being removed each time by filtration. Finally the solid is pressed down on the Büchner funnel; the mat is broken up with a spatula and washed while on the funnel with two 400-ml. portions of 95% ethanol and then with one 400-ml. portion of ether. After air drying, the aminoguanidine bicarbonate amounts to 180–182 g. (63–64%) of a white solid, melting at 172° with decomposition (Notes 5 and 6).

2. Notes

1. The zinc is purified by stirring 1.2 kg. of commercial zinc dust with 3 l. of 2% hydrochloric acid for 1 minute. The acid is removed by filtration, and the zinc is washed in a 4-l. beaker with one 3-l. portion of 2% hydrochloric acid, three 3-l. portions of distilled water, two 2-l. portions of 95% ethanol, and finally with one 2-l. portion of absolute ether, the wash solutions being removed each time by filtration. Then the material is thoroughly dried and any lumps are broken up in a mortar.

2. The solution becomes basic to litmus after one-half to three-fourths of the paste has been added. Lower yields are obtained if a larger excess of acetic acid is employed.

3. The state of reduction can be determined by placing 3 drops of the reaction mixture in a test tube containing 5 ml. of a 10% solution of sodium hydroxide and then adding 5 ml. of a freshly prepared saturated solution of ferrous ammonium sulfate. A red coloration indicates incomplete reduction; when the reduction is complete, only a greenish precipitate is observed. The mixture should not be heated after this test shows that reduction is complete.

4. The presence of the ammonium chloride prevents the coprecipitation of zinc salts when sodium bicarbonate is added to the solution to precipitate the aminoguanidine as the bicarbonate. If the solution is not clear at this step, it should be filtered.

5. The aminoguanidine bicarbonate is pure enough for most purposes. It should not be recrystallized from hot water, since decomposition will occur.

6. W. W. Hartman and Ross Philips have submitted a procedure suitable for the preparation of aminoguanidine bicarbonate on a larger

scale. The sulfates of methylisothiourea and of hydrazine are allowed to react with the evolution of methyl mercaptan. In a 30-gal. crock are placed 10 l. of water and 5760 g. (20 moles) of methylisothiourea sulfate.[2] In a 22-l. flask, 5.2 kg. (40 moles) of hydrazine sulfate [3] is stirred with 12 l. of water, and 40% sodium hydroxide is added until all the hydrazine sulfate has dissolved and the solution is just neutral to Congo paper. The exact amount of alkali is noted and a duplicate amount added. The hydrazine solution is then added to the 30-gal. crock with stirring, as fast as possible, without allowing the foam to overflow the crock. The mixing is done out-of-doors, or in an efficient hood, since large volumes of methyl mercaptan are evolved. If the reaction is carried out on a smaller scale in 12- or 22-l. flasks, using appropriate amounts of material, the methyl mercaptan evolved may be absorbed in cold sodium hydroxide solution and isolated if desired. The solution is stirred until evolution of mercaptan stops, and then a few liters of water are distilled off under reduced pressure to free the solution entirely from mercaptan. The residual liquor is chilled in a crock, and a crop of hydrated sodium sulfate is filtered off, washed with ice water, and discarded. The filtrate is warmed to 20–25°, 25 ml. of glacial acetic acid is added, then 4 kg. of sodium bicarbonate, and the solution is stirred vigorously for 5 minutes and thereafter occasionally during an hour, or until the precipitate no longer increases. The precipitate is filtered with suction and washed with ice water and then with methanol, and is dried at a temperature not above 60–70°. The yield is 3760 g. (69% of the theoretical amount). Hydrazine sulfate may be recovered from the final filtrate, if the filtrate is strongly acidified with sulfuric acid and allowed to cool.

3. Methods of Preparation

Numerous references for the preparation of aminoguanidine bicarbonate and other salts can be found in an excellent review article by Lieber and Smith.[4] It has also been prepared by treating a cyanamide solution at 20–50° with hydrazine and carbon dioxide,[5] and by the electrolytic reduction of nitroguanidine.[6]

1 *Org. Syntheses* Coll. Vol. **1**, 399 (1941).

2 *Org. Syntheses* Coll. Vol. **2**, 411 (1943).

3 *Org. Syntheses* Coll. Vol. **1**, 309 (1941).

4 Lieber and Smith, *Chem. Revs.*, **25**, 213 (1939).

5 Ger. pat. 689,191 [*C. A.*, **35**, 3650 (1941)].

6 Shreve and Carter, *Ind. Eng. Chem.*, **36**, 423 (1944).

2-AMINO-6-METHYLBENZOTHIAZOLE *

(Benzothiazole, 2-amino-6-methyl-)

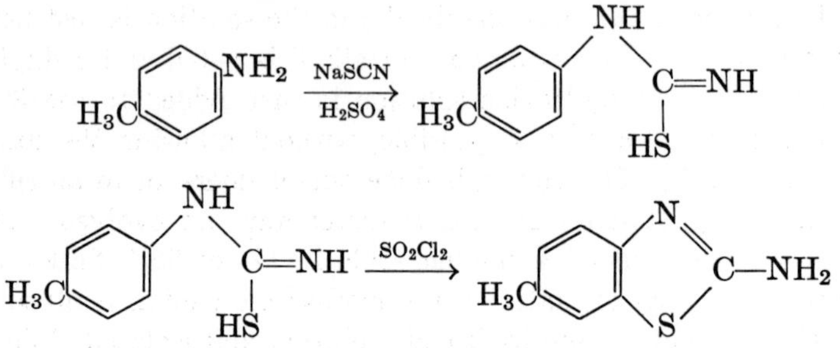

Submitted by C. F. H. ALLEN and JAMES VANALLAN.
Checked by W. E. BACHMANN and N. W. MACNAUGHTON.

1. Procedure

A solution of 107 g. (1 mole) of *p*-toluidine (Note 1) in 700 ml. of chlorobenzene is prepared in a 3-l. three-necked, round-bottom flask fitted with a stirrer, reflux condenser, thermometer, and dropping funnel. Over a period of 5 minutes, 54 g. (29.3 ml., 0.55 mole) of concentrated sulfuric acid is added dropwise. To the finely divided suspension of *p*-toluidine sulfate is added 90 g. (1.1 moles) of sodium thiocyanate, and the mixture is heated for 3 hours at 100° (inside temperature) in an oil bath (Note 2). The solution, which now contains the thiourea, is cooled to 30°, and 180 g. (108 ml., 1.34 moles) of sulfuryl chloride is added over a period of 15 minutes, with care that the temperature does not exceed 50°. The mixture is kept at 50° for 2 hours (no further evolution of hydrogen chloride), after which the chlorobenzene is removed by filtration (Note 3).

The solid residue is then dissolved in 1 l. of hot water, and the remainder of the solvent is removed by a current of steam (Note 4). The solution is filtered from a little solid and is then made alkaline to litmus by the addition of 200 ml. of concentrated ammonium hydroxide (sp. gr. 0.90). The precipitated aminomethylbenzothiazole is filtered and washed with 200 ml. of water. The solid, which melts over the range 123–128°, is dissolved in 300 ml. of hot ethanol (Note 5), 10 g. of Norit is added, and the hot suspension is filtered. The filtrate is diluted with 500 ml. of hot water, and the mixture is vigorously stirred and quickly chilled. After 30 minutes, the pale yellow granular product is filtered and washed with 150 ml. of 30% ethanol. After drying to constant weight, the product weighs 100–105 g. and melts at 135–136°, with preliminary shrinking at 130–131° (Note 6). On the addition of 200 ml. of water to the filtrate, a further 5–8 g. of product is recovered, making the total yield 105–110 g. (64–67%).

2. Notes

1. All the chemicals used were from the Eastman Kodak Company. The checkers used the practical grade of *p*-toluidine, which was distilled just before use.

2. The *p*-tolylthiourea may be isolated at this point, if desired, by filtering and washing the solid residue with ether. The dried solid (172 g.) is a mixture of the urea and sodium sulfate. It is extracted with 250 ml. of warm (50–60°) water; the residual urea, m.p. 188–189°, is completely soluble in ethanol. The yield is 139 g. (84%).

3. The recovery is about 600 ml.; it may be used again without purification.

4. Treatment of the solution with Norit at this point does not give a product of any better quality.

5. If less ethanol is used, crystallization takes place during filtration.

6. The literature gives melting points ranging from 128° to 142°, the higher value being for aminomethylbenzothiazole prepared via the hydrochloride (m.p. 250–253°). The submitters report that the product obtained above is unchanged in melting point after regeneration from the hydrochloride. By this treatment, however, the color is removed.

3. Methods of Preparation

2-Amino-6-methylbenzothiazole has been prepared by the action of cupric thiocyanate,[1–3] or of chloramine and ammonium thiocyanate,[4] on *p*-toluidine; by the action of chlorine on di-*p*-tolylthiourea; [5] or of bromine on acetyldi-*p*-tolylthiourea; [6] by treatment of *p*-tolylthiourea with halogens [7] or acid halides.[8–12] 2-Aminobenzothiazole has been prepared in excellent yield by the action of hydroxylamine upon benzothiazole,[13] and has also been obtained from 2-mercaptobenzothiazole by the action of ammonium bisulfite at 150°.[14] It has been prepared by the electrolysis of *p*-toluidine in the presence of ammonium thiocyanate.[15] Two German patents report a general method for the preparation of 2-aminobenzothiazoles [16] and 2-aminonaphthothiazoles [17] consisting of the treatment of an aromatic amine with potassium thiocyanate and acetic acid and subsequently with bromine in acetic acid.

Nearly quantitative yields are reported for a process using a modified cupric thiocyanate method.[18]

[1] Kaufmann, Oehring, and Clauberg, *Arch. Pharm.*, **266**, 197 (1928) [*C. A.*, **22**, 2166 (1928)].

[2] Kaufmann and Küchler, *Ber.*, **67**, 946 (1934).

[3] Kaufmann, Ger. pat. 579,818 [*C. A.*, **28**, 1053 (1934)].

[4] Likhosherstov and Petrov, *J. Gen. Chem. U.S.S.R.*, **3**, 759 (1933) [*C. A.*, **28**, 2690 (1934)].

[5] I. G. Farbenind. A-G., Fr. pat. 688,867 [*C. A.*, **25**, 968 (1931)].

[6] Hunter and Jones, *J. Chem. Soc.*, **1930**, 2197.

[7] Hunter, *J. Chem. Soc.*, **1926**, 1399.

[8] Herz and Schubert, U. S. pat. 2,003,444 [*C. A.*, **29**, 4774 (1935)].

[9] I. G. Farbenind. A-G., Brit. pat. 345,735 [*C. A.*, **26**, 1944 (1932)].

[10] E. I. du Pont de Nemours and Company, Fr. pat. 727,410 [*C. A.*, **26**, 5103 (1932)].

[11] Herz and Schubert, Ger. pat. 537,105 [*Frdl.*, **17**, 615 (1932); *C. A.*, **26**, 1000 (1932)].

[12] Schubert and Schütz, Ger. pat. 604,639 [*Frdl.*, **21**, 227 (1934); *C. A.*, **29**, 819 (1935)].

[13] Skraup, *Ann.*, **419**, 65 (1919).

[14] Ubaldini and Fiorenza, *Gazz. chim. Ital.*, **76**, 215 (1946).

[15] Melnikov and Cherkasova, *J. Gen. Chem. U.S.S.R.*, **14**, 113 (1944) [*C. A.*, **39**, 934 (1945)].

[16] Ger. pat. 491,223 [*Frdl.*, **16**, 2566 (1927–1929)].

[17] Ger. pat. 492,885 [*Frdl.*, **16**, 479 (1927–1929)].

[18] Fridman, *Zhur. Obshcheĭ Khim.*, **20**, 1191 (1950) [*C. A.*, **45**, 1579 (1951)].

3-AMINO-2-NAPHTHOIC ACID

(2-Naphthoic acid, 3-amino-)

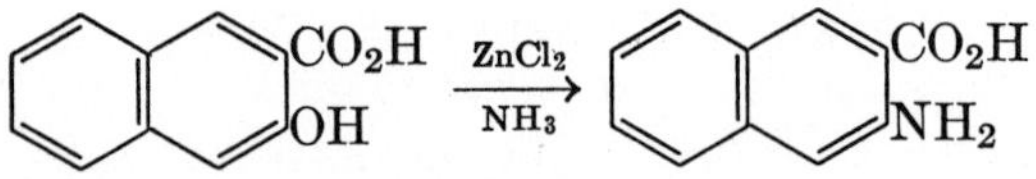

Submitted by C. F. H. Allen and Alan Bell.
Checked by Nathan L. Drake, E. W. Reeve, and J. Van Hook.

1. Procedure

An autoclave (Note 1) is charged *successively* with 600 ml. (9.5 moles) of approximately 28% aqueous ammonia, 78 g. of zinc chloride

(0.57 mole) (Notes 2 and 3), and 167 g. of 3-hydroxy-2-naphthoic acid (0.89 mole) (Note 4). The autoclave is closed, and then, with continuous stirring or shaking (Note 5), it is gradually heated, so that, at the end of 3 hours, the temperature is 195°. This temperature is maintained for 36 hours; the pressure is about 400 lb. (Note 6). The autoclave is allowed to cool to room temperature, with continuous stirring or shaking. The cover is removed, the solid is scraped off the walls, and the reaction mixture is transferred to a 5-l. flask. The autoclave is rinsed with two 700-ml. portions of water, which are added to the reaction mixture.

To the reaction mixture is added 660 g. of concentrated hydrochloric acid (sp. gr. 1.18), and the suspension is boiled for 30 minutes. It is then filtered, while hot, through a Büchner funnel. The residual cake is boiled again, for the same length of time, with a mixture of 500 ml. of water and 30 ml. of concentrated hydrochloric acid, and then filtered as before.

The combined filtrates, on cooling, deposit the hydrochloride of the aminonaphthoic acid; this is filtered with suction and is pressed as dry as possible. To obtain a second crop of the hydrochloride, the filtrate is heated to 85°, 500 g. of salt is added, and the solution is cooled.

The combined moist filter cakes are placed in a 5-l. flask with 1.3 l. of water, and, with stirring, sufficient 40% sodium hydroxide solution is added to make the solution just alkaline to Clayton yellow paper (Note 7); about 110 ml. is required. The mixture is heated to 85° and is filtered to remove a small amount of insoluble material. The filtrate is then made acid to Congo red by the addition of concentrated hydrochloric acid (about 100 ml. of acid is required). The mixture is stirred for 15 minutes, and then enough 10% sodium acetate solution is added so that Congo red paper is no longer turned blue; the amount needed depends on the acidity; usually about 15 ml. is sufficient.

The hot mixture is filtered; the aminonaphthoic acid is washed on the Büchner funnel with 400 ml. of hot water and is pressed as dry as possible. The product is dried at 50° to constant weight (24–36 hours). 3-Amino-2-naphthoic acid is a yellow, powdery substance, which melts at 214–215° (Note 8); the yield is 110–115 g. (66–70%) (Note 6). The quantities of starting materials are limited only by the size of the autoclave available; the yield (per cent) in large runs is the same as that obtained above.

2. Notes

1. An autoclave fitted for stirring or shaking is essential. An example of the former is the one furnished by the Will Corporation, catalog No. 1735. This is large enough so that three times the amounts specified may be handled. A hydrogenation autoclave of the type supplied by the American Instrument Company, catalog No. 406–21, having a capacity of about 1.4 l., may also be used. This type of autoclave is not fitted with a stirrer but is shaken by means of a "Bomb Shaker" in which the autoclave is placed. In using the hydrogenation autoclave, the long hydrogen inlet tube, which extends half way into the chamber of the autoclave, is unscrewed and removed. The inlet for hydrogen, on the outside of the autoclave, is closed by means of a solid steel rod, the end of which is finished and held in place in the same manner as the ordinary steel pressure tubing.

After making a run, the hydrogenation autoclave is poisoned for hydrogenations. It may be cleaned by filling it with 5% hydrochloric acid and rubbing the inside walls with a cloth or brush. It should then be washed with water and dried; any solid matter adhering to the walls is removed by rubbing with emery paper. Care must be taken not to scratch either the copper gasket or the groove into which the gasket fits. All parts of the hydrogenation autoclave, including the head, must be thoroughly cleaned. Traces of the reaction product will cause pitting in a few hours if they are allowed to remain in contact with the steel while the bomb is exposed to the air.

After the hydrogenation autoclave is cleaned it is best to carry out one or two ester hydrogenolyses with copper chromite in order to remove the last traces of "poison." The bomb may then be used for hydrogenations.

2. The zinc chloride should not contain appreciable amounts of the oxychloride.

3. Since zinc chloride and aqueous ammonia react with evolution of heat, the solid must be added gradually and with hand stirring. It is best to carry out this operation under a hood.

4. A technical grade of 3-hydroxy-2-naphthoic acid, m.p. 211–214°, was used.

5. Continued stirring or shaking is required throughout, i.e., from the time heating is begun until the reaction product has again attained room temperature. With a hydrogenation autoclave there will be no trouble due to leaks, but with the stirrer type of autoclave the packing around the stirrer shaft may leak. In this event the pressure will

fall and a longer period of time will be required. If a large quantity of ammonia escapes, the yield will be diminished, but the run should be finished and the unused hydroxynaphthoic acid should be recovered and used in another run.

6. In a run in which the gaskets leaked, the pressure never exceeded 200 lb. In this run, the (stirrer-type) autoclave was operated for 5 days; the yield was 67%.

7. Clayton yellow paper (thiazole paper) is paper impregnated with the dye formed by coupling diazotized primuline sulfonic acid with primuline sulfonic acid. The color change occurs at *p*H 11–12. If this paper is not available, a little less than the estimated amount of alkali should be used. The mixture is warmed to 85°, and a small test portion is removed and warmed with more alkali. If any appreciable amount of the insoluble matter dissolves, more alkali is needed.

8. This product is pure enough for most purposes. It can be recrystallized from ethanol (10 g. dissolves in 100 ml. of hot ethanol); the recovery is 78% (7.8 g.), and the purified product melts at 216–217°.

3. Methods of Preparation

3-Amino-2-naphthoic acid has been prepared by heating 3-hydroxy-2-naphthoic acid with ammonia under pressure, in the presence of catalysts. These catalysts include zinc (or calcium) chloride;[1–4] alone or with the addition of aluminum chloride.[5] Zinc oxide or carbonate and ammonium chloride,[6–8] or ferrous ammonium salts,[9,10] have also been used. When sodium hydroxynaphthoate is heated with 35% ammonia at 260–280°, the product is contaminated with considerable amounts of β-naphthol and β-naphthylamine.[11] Modifications of Henle's procedure[4] have been described.[12]

[1] Soc. anon. pour l'ind. chim. à Bâle, Brit. pat. 250,598 [*Frdl.*, **16**, 487 (1927–1929); *C. A.*, **21**, 1127 (1927)].

[2] Tobler, U. S. pat. 1,629,894 [*Frdl.*, **16**, 487 (1927–1929); *C. A.*, **21**, 2273 (1927)].

[3] I. G. Farbenind. A-G., Brit. pat. 284,998 [*Frdl.*, **16**, 484 (1927–1929); *C. A.*, **22**, 4540 (1928)].

[4] Henle and Lanz, U. S. pat. 1,871,990 [*C. A.*, **26**, 5971 (1932)].

[5] Fierz and Tobler, *Helv. Chim. Acta*, **5**, 558 (1922).

[6] Dutta, *Ber.*, **67**, 1321 (1934).

[7] I. G. Farbenind. A-G., Brit. pat. 330,941 [*Frdl.*, **16**, 2999 (1927–1929); *C. A.*, **24**, 5768 (1930)].

[8] Schweitzer, U. S. pat. 1,806,714 [*Frdl.*, **16**, 2998 (1927–1929); *C. A.*, **25**, 4012 (1931)].

[9] I. G. Farbenind. A-G., Brit. pat. 282,450 [*Frdl.*, **16**, 487 (1927–1929); *C. A.*, **22**, 3669 (1928)].

[10] Hotz and Lanz, U. S. pat. 1,690,785 [*Frdl.*, **16**, 487 (1927–1929); *C. A.*, **23**, 397 (1929)].

[11] Möhlau, *Ber.*, **28**, 3096 (1895).

[12] Yagi, *Mem. Inst. Sci. and Ind. Research Osaka Univ.*, **6**, 82 (1948).

2-AMINO-4-NITROPHENOL *

(Phenol, 2-amino-4-nitro-)

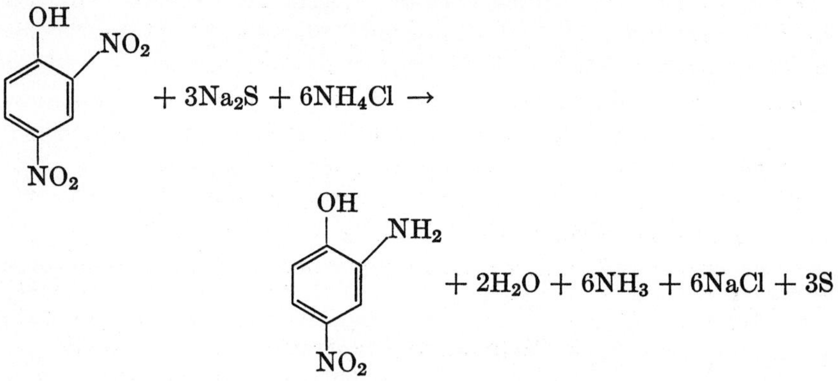

Submitted by W. W. Hartman and H. L. Silloway.
Checked by H. R. Snyder and J. Wayne Kneisley.

1. Procedure

In a 5-l. three-necked flask, suspended over a steam bath, are placed 300 g. (1.63 moles) of technical 2,4-dinitrophenol and 2.5 l. of water. The flask is fitted with an efficient stirrer, a reflux condenser, and a thermometer which dips below the surface of the mixture. After the stirrer has been started, 600 g. (11.6 moles) of ammonium chloride and 100 ml. of concentrated aqueous ammonia (about 28%) are added, and the mixture is heated to 85°. The steam is turned off, and the mixture is allowed to cool. When the temperature reaches 70° (Note 1), 700 g. (5.4 moles) of 60% fused sodium sulfide is added in portions of about 100 g. at 5-minute intervals. After two or three such additions the temperature of the reaction mixture reaches 80–85°; it is kept in this range either by adding the remaining portions of sodium sulfide at 10-minute intervals or by wrapping the flask with a wet cloth and continuing the additions at 5-minute intervals. After all the sodium sulfide has been added, the reaction mixture is heated at 85° for 15 minutes and then filtered through a heated 6-in. Büchner funnel (Note 2).

The hot filtrate is transferred to a 5-l. round-bottomed flask and cooled overnight by a stream of cold water. The mixture is filtered, and the crystals are pressed nearly dry. The solid is dissolved in 1.5 l. of boiling water, and the solution is acidified with glacial acetic acid (about 100 ml. is required, Note 3). The solution is heated with 10 g. of Norit, filtered hot, and cooled to 20°. The brown crystals are collected and dried for several hours in an oven at 65° or in a vacuum desiccator (Note 4). The yield of 2-amino-4-nitrophenol melting at 140–142° is 160–167 g. (64–67%). If a purer product is desired, the crude substance is recrystallized from 1.5 l. of hot water; 147–153 g. (58–61%) of material melting at 142–143° is obtained.

2. Notes

1. If the reaction is run at temperatures below 70° it is impossible to obtain a pure product even after several recrystallizations.

2. The Büchner funnel is preheated by inverting it over a steam bath and passing a lively current of steam through it for at least 10 minutes. During the filtration, suction is applied gently and at intervals to avoid excessive cooling by evaporation in the lower part of the funnel. If the filtration is not performed rapidly and carefully, the crude product will crystallize in the funnel. The filtration can be omitted, but the presence of insoluble materials complicates the next step by making it difficult to determine when solution of the crude product is complete.

3. The amount of acid required varies with dryness of the filter cake. The acidification can be followed by observing the color of a thin layer of the solution splashed against the side of the container. The color changes from dark red to olive brown at the end point. Both colors are so deep that they are not easily distinguished except when thin layers are viewed. After the end point is observed, an additional 10 ml. of acetic acid is added.

4. Unless properly dried the substance will melt at 80–90°, owing to the presence of water of crystalliaztion.

3. Methods of Preparation

2-Amino-4-nitrophenol has been prepared by the partial reduction of 2,4-dinitrophenol chemically [1–5] and electrolytically,[6] and by the action of sulfuric acid on 3-nitroazidobenzene.[7]

[1] Laurent and Gerhardt, *Ann.*, **75**, 68 (1850).
[2] Post and Stuckenberg, *Ann.*, **205**, 72 (1880).

[3] Auwers and Röhrig, *Ber.*, **30**, 995 (1897).
[4] Pomeranz, Ger. pat. 289,454 [*Frdl.*, **12**, 117 (1914–1916)].
[5] Gershzon, *J. Applied Chem. U.S.S.R.*, **9**, 879 (1936) [*C. A.*, **30**, 7554 (1936)].
[6] Hofer and Jacob, *Ber.*, **41**, 3196 (1908).
[7] Kehrmann and Idzkowska, *Ber.*, **32**, 1066 (1899).

dl-α-AMINOPHENYLACETIC ACID *

(Glycine, α-phenyl-)

$$C_6H_5CHO + NaCN + NH_4Cl \rightarrow C_6H_5CH(NH_2)CN + NaCl + H_2O$$

$$C_6H_5CH(NH_2)CN + 2HCl + 2H_2O \rightarrow C_6H_5CH(NH_2 \cdot HCl)CO_2H + NH_4Cl$$

$$C_6H_5CH(NH_2 \cdot HCl)CO_2H + NH_3 \rightarrow C_6H_5CH(NH_2)CO_2H + NH_4Cl$$

Submitted by Robert E. Steiger.
Checked by R. L. Shriner, S. P. Rowland, and C. H. Tilford.

1. Procedure

To a solution of 100 g. (2.0 moles) of 98% sodium cyanide in 400 ml. of water, contained in a 3-l. round-bottomed flask fitted with a Hershberg stirrer, is added 118 g. (2.2 moles) of ammonium chloride. The mixture is stirred at room temperature under a properly ventilated hood. When the ammonium chloride has dissolved, a solution of 212 g. (2.0 moles) of benzaldehyde in 400 ml. methanol is added in one portion. The reaction begins rapidly, the temperature rising to about 45°. Stirring is continued for 2 hours. The heterogeneous mixture, after dilution with 1 l. of water, is extracted with 1 l. of benzene, and the aqueous layer is discarded. The benzene layer is washed with three 50-ml. portions of water, and the aminonitrile is extracted, in the form of its hydrochloride, by shaking the benzene solution with 6 *N* hydrochloric acid, first with one 600-ml. portion and then with two 300-ml. portions.

The combined acid extracts are placed in a 3-l. round-bottomed flask and refluxed for 2 hours (Note 1). The hydrolysate is diluted with water to bring its volume to about 2 l. and is then subjected to distillation under reduced pressure (20–30 mm.) to remove all benzaldehyde and other volatile substances (Note 2). To remove some resinous matter deposited in the course of the hydrolysis, the mixture is treated with 10 g. of Norit and filtered through a Büchner funnel. The yellow filtrate is transferred to a 3-l. beaker. It is stirred by hand with a thick glass rod while ammonium hydroxide (sp. gr. 0.90) is added through a dropping funnel until the liquid is faintly alkaline to litmus (Note 3). The mixture becomes quite hot and acquires a strong odor of benzaldehyde. The amino acid separates in the form of yellow crystals. The mixture is cooled to room temperature, and the crystals are collected on a 15-cm. Büchner funnel. After washing with about 1 l. of water in small portions to remove the ammonium chloride, the solid is washed successively with 150 ml. of ethyl ether, three 50-ml. portions of hot 95% ethanol, and finally with about 500 ml. of water. The crystals, when dried by suction, weigh 220 to 240 g. (Note 4). Drying is completed in a vacuum desiccator over phosphorus pentoxide. The yield of crude amino acid is 102–116 g. (34–39%).

For purification, the product is dissolved in 800 ml. of 1 *N* sodium hydroxide, 500 ml. of ethanol is added, and the solution is filtered. The filtrate is transferred to a 2-l. beaker and is heated to the boiling point. Then 160 ml. of 5 *N* hydrochloric acid is slowly added, through a dropping funnel, while stirring by hand. The mixture is cooled to room temperature and is filtered with suction. The product is washed with 100 ml. of ethanol, then with 200 ml. of water, and is dried in a vacuum desiccator over phosphorus pentoxide. The nearly white, lustrous platelets have no definite melting point (Note 5). The yield of pure amino acid is 98–112 g. (33–37%) (Note 6).

2. Notes

1. The hydrolysis of the aminonitrile should be carried out under a hood, since some hydrocyanic acid is liberated.

2. During this process, the volume of the solution must be maintained at about 2 l. by the frequent addition of water through a dropping funnel. Otherwise the hydrochloride of *dl*-phenylglycine, which is sparingly soluble in concentrated hydrochloric acid, may separate.

3. From 375 to 425 ml. of ammonium hydroxide of sp. gr. 0.90 is required.

4. This crude product is hydrated.

5. The decomposition range was about 270–280° with sintering at 258°. A new sample, when placed in the melting-point bath at 280–300°, sintered, and then decomposed at 300–302°.

6. Increasing the quantities of cyanide and ammonium chloride to 3 moles does not markedly improve the yields.

3. Methods of Preparation

dl-Phenylglycine has been prepared by heating α-bromophenylacetic acid with three times its weight of ammonium hydroxide (sp. gr. 0.90) to 100–110°;[1] by hydrolysis of α-aminophenylacetonitrile with dilute hydrochloric acid;[2,3] and by reduction of benzoylformic acid phenylhydrazone with sodium amalgam in dilute sodium hydroxide.[4] The method described above is, with some modifications and additions, the procedure used by Marvel and Noyes.[3,5]

[1] Stöckenius, *Ber.*, **11**, 2002 (1878).

[2] Tiemann, *Ber.*, **13**, 383 (1880); Ulrich, *Ber.*, **37**, 1688 (1904); Zelinsky and Stadnikoff, *Ber.*, **39**, 1725 (1906); **41**, 2062 (1908); Ingersoll and Adams, *J. Am. Chem. Soc.*, **44**, 2933 (1922).

[3] Marvel and Noyes, *J. Am. Chem. Soc.*, **42**, 2264 (1920).

[4] Elbers, *Ann.*, **227**, 343 (1885).

[5] Dr. D. Stetten, Jr. (private communication), found that when the synthesis is carried out according to Zelinsky and Stadnikoff, *Ber.*, **41**, 2062 (1908), and the amino acid is isolated in the fashion described above, the yield is not appreciably higher than the one recorded here.

p-AMINOPHENYL DISULFIDE

(Aniline, *p,p'*-dithiodi-)

$$p\text{-}NO_2C_6H_4Cl + Na_2S \rightarrow p\text{-}NO_2C_6H_4SNa + NaCl$$

$$4p\text{-}NO_2C_6H_4SNa + 6Na_2S + 7H_2O \rightarrow 4p\text{-}NH_2C_6H_4SNa + 6NaOH + 3S + 3Na_2SO_3$$

$$2p\text{-}NH_2C_6H_4SNa + H_2O_2 \rightarrow p\text{-}NH_2C_6H_4SSC_6H_4NH_2 + 2NaOH$$

Submitted by Charles C. Price and Gardner W. Stacy.
Checked by Cliff S. Hamilton and Theo Brown.

1. Procedure

In a 5-l. round-bottomed flask equipped with a reflux condenser and a mechanical stirrer are placed 236 g. (1.5 moles) of *p*-chloronitrobenzene, 960 g. (4 moles) of sodium sulfide nonahydrate, and 2.5 l. of water. With rapid agitation, the reaction mixture is slowly heated to the reflux temperature (Note 1). Heating is continued over a period of 20 hours.

The mixture is cooled to about 15° and filtered on a large Büchner funnel to remove insoluble material, chiefly *p*-chloroaniline. The filtrate is placed in a 5-l. three-necked flask, equipped with a dropping funnel, a mechanical stirrer, and a downward condenser; it is concentrated by distillation over a flame to a volume of 1.5 l. (Note 2). The condenser is replaced by a thermometer, the stirrer is started, and the dropping funnel is charged with 230 ml. of 30% hydrogen peroxide. The temperature of the reaction mixture is maintained at 65–70° while the hydrogen peroxide is added dropwise over a period of about 2 hours (Note 3).

The reaction mixture is cooled, and the crude *p*-aminophenyl disulfide, obtained as spherical lumps amounting to 125–150 g. and melting at 73°, is collected on a Büchner funnel. It is dissolved in 1 l. of hot ethanol, and the small amount of insoluble material is removed by filtration. To the hot solution is then added 800 ml. of water containing several grams of sodium hydrosulfite to prevent discoloration of the solution by oxidation. The solution is cooled, and an additional 700 ml. of water is added with stirring to reduce the solubility of the product. The crystalline precipitate is removed by filtration and dried in a vacuum desiccator. The yield of pure product melting at 75–76° is 108–120 g. (58–64%).

2. Notes

1. The reaction mixture should not be heated too strongly as the temperature nears the boiling point, since the initial reaction is quite vigorous.

2. Some *p*-chloroaniline is removed by steam distillation during the concentration.

3. At first the solution becomes cloudy, and later large yellow globules begin to accumulate. Frequent cooling is required to maintain the temperature within the specified range.

3. Methods of Preparation

The procedure outlined is essentially that of Lantz [1] except for the use of hydrogen peroxide rather than air in the final oxidation. *p*-Aminophenyl disulfide has also been prepared by heating a mixture of aniline, aniline hydrochloride, and sulfur,[2] by the reduction of *p*-nitrophenyl disulfide with stannous chloride,[3,4] and by the action of potassium hydroxide upon *p*-acetamidobenzenesulfinic acid (autoclave) followed by hydrolysis.[5]

[1] Lantz, Fr. pat. 714,682 (*Chem. Zentr.*, **1932**, I, 1828).

[2] Hinsberg, *Ber.*, **38**, 1131 (1905).

[3] Shukla, *J. Indian Inst. Sci.*, **10**, 33 (1927).

[4] Danielsson, Christian, and Jenkins, *J. Pharm. Assoc., Sci. Ed.*, **36**, 261 (1947).

[5] Goldberg, *J. Chem. Soc.*, **1945**, 826.

dl-α-AMINO-α-PHENYLPROPIONIC ACID

(Alanine, α-phenyl-, *dl*-)

$$C_6H_5COCH_3 + NaCN + NH_4Cl \rightarrow C_6H_5C(CH_3)(NH_2)CN + H_2O + NaCl$$

$$C_6H_5C(CH_3)(NH_2)CN + H_2O + HCl \xrightarrow{\text{cold}} C_6H_5C(CH_3)(NH_3Cl)CONH_2$$

$$C_6H_5C(CH_3)(NH_3Cl)CONH_2 + H_2O + HCl \xrightarrow{\text{hot}} C_6H_5C(CH_3)(NH_3Cl)CO_2H + NH_4Cl$$

$$C_6H_5C(CH_3)(NH_3Cl)CO_2H + C_5H_5N \rightarrow C_6H_5C(CH_3)(NH_2)CO_2H + C_5H_5NHCl$$

Submitted by ROBERT E. STEIGER.
Checked by R. L. SHRINER and S. P. ROWLAND.

1. Procedure

In a 2-l. round-bottomed flask are placed, in the order mentioned, 50 g. (1 mole) of 98% sodium cyanide in 100 ml. of water, 58.9 g. (1.1 moles) of ammonium chloride in 140 ml. of lukewarm water (about 35°), and 134 ml. (2 moles) of aqueous ammonia (sp. gr. 0.90). The mixture is shaken while 120 g. (1 mole) of acetophenone in 300 ml. of 95% ethanol is added. The flask is stoppered with a rubber stopper, which is wired in place (Note 1), and is then immersed in a water bath maintained at 60°. The flask is shaken from time to time, and a homogeneous solution results within 30 minutes. The reaction mixture is heated for 5 hours at 60°, then well cooled in an ice-water mixture, and poured, with precautions (under a well-ventilated hood), into a 5-l. round-bottomed flask which is immersed up to the neck in an ice-water mixture and which contains 800 ml. of concentrated hydrochloric acid (sp. gr. 1.18–1.19). The reaction flask is rinsed with two 25-ml. portions of water, which are added to the hydrochloric acid solution. The solution of the aminonitrile is saturated at 0–5° with dry hydrogen chloride (Note 2) and is then set aside overnight in the ice-water bath, which is allowed to melt and come to the temperature of the room. The mixture of solid and liquid material is diluted with 1 l. of water and boiled vigorously for 2.5 hours under a reflux condenser in a well-ventilated hood (Note 3). The dark hydrolysate is concentrated under reduced pressure to remove the acetophenone and other volatile impurities. The solution is finally transferred to a 10-in. evaporating dish and is stirred occasionally while it is evaporated almost to dryness. Once a heavy deposit of inorganic salts has formed, very little bumping occurs. The solid residue is evaporated on a water bath twice with 100-ml. portions of water in order to remove as much hydrochloric acid as possible. The residue is crushed, and 600 ml. of absolute ethanol is added. The suspension is warmed for a short time on a steam bath; it is shaken thoroughly and then chilled in an ice-water mixture. The inorganic salts are removed by filtration through a 15-cm. Büchner funnel, and the cake is washed with small portions of absolute ethanol until 600 ml. has been used. The combined filtrates and washings are placed in a 3-l. beaker, and pyridine is added while the mixture is stirred by hand with a thick glass rod. The addition of pyridine is continued until the solution is nearly neutral to Congo red paper, and then an additional 50 ml. is added; the total amount of pyridine required is 80–100 ml. (Note 4).

The resulting rather stiff paste is chilled in an ice-water mixture for 1 hour. The solid is collected on a 15-cm. Büchner funnel and washed with 25-ml. portions of absolute ethanol until it is white and until the filtrate becomes colorless (100–200 ml. of ethanol is required). The fluffy material, after it is dried in a vacuum desiccator over flake sodium hydroxide, weighs 66 g. (40%).

The amino acid contains a small amount of ammonium chloride, which can be removed by dissolving the product in 21 times its weight of water and precipitating it by the addition of 42 times its weight of absolute ethanol (Note 5). The mixture is allowed to stand overnight in a refrigerator. The amino acid is collected on a Büchner funnel and washed with an ethanol-water mixture containing 80% by weight of ethanol. The amino acid is dried in a vacuum desiccator over flake sodium hydroxide. The recovery is about 70% (Note 6).

2. Notes

1. A flask fitted with a ground-glass stopper is convenient for carrying out this step. The ground-glass stopper should be slightly lubricated with stopcock grease. It must be firmly secured in place by means of adhesive tape, as some pressure develops when the reaction mixture is heated.

2. From 500 to 600 g. of dry hydrogen chloride gas is required. Since this gas is very rapidly absorbed, a high-speed generator can be used.

3. The hydrolysis of the aminonitrile must be carried out under a good hood. The top of the reflux condenser should be connected with the ventilating pipe by means of a piece of glass tubing.

4. The minimum amount of pyridine necessary is determined by the amount of hydrochloric acid remaining in the residue.

5. The material does not readily become wet in contact with water, and it dissolves very slowly. The finely powdered material is best suspended in a large excess of water, e.g., about 30 times its weight. The suspension is boiled under a reflux condenser and frequently shaken until the solid is dissolved. The excess water is then boiled off at atmospheric pressure until a solution of the amino acid in 21 times its weight of water is obtained. The solution is then cooled and treated with absolute ethanol.

6. The amino acid does not have a sharp melting point. It sublimes with decomposition at about 265–270°.

3. Methods of Preparation

α-Amino-α-phenylpropionitrile has been prepared by heating acetophenone cyanohydrin for 6–8 hours in a closed vessel with 1 equivalent of ethanolic ammonia at 70°; [1,2] and by heating acetophenone for 4 hours in a closed vessel with an ethanolic solution of ammonium cyanide to 80°.[3] α-Amino-α-phenylpropionitrile has been hydrolyzed by the action of, first, fuming hydrochloric acid at room temperature, and then dilute hydrochloric acid at boiling temperature, in the presence of some ethanol. This procedure gives a good yield of the amino acid hydrochloride. The amino acid has usually been liberated from its hydrochloride by means of ammonium hydroxide.[1–3] The process of isolating the amino acid by treating an ethanolic solution of its hydrochloride with pyridine is essentially the same as that developed for the preparation of glycine[4] and of α-aminoisobutyric acid.[5]

[1] Tiemann and Köhler, *Ber.,* **14,** 1981 (1881).

[2] McKenzie and Clough, *J. Chem. Soc.,* **101,** 395 (1912).

[3] Jawelow, *Ber.,* **39,** 1195, 1197 (1906).

[4] Clarke and Taylor, *Org. Syntheses,* **4,** 31 (1925); Coll. Vol. **1,** 298 (1941).

[5] Clarke and Bean, *Org. Syntheses,* **11,** 4 (1931); Coll. Vol. **2,** 29 (1943).

dl-β-AMINO-β-PHENYLPROPIONIC ACID

(Hydrocinnamic acid, β-amino-*dl*-)

$$C_6H_5CH{=}CHCO_2H + NH_2OH \rightarrow C_6H_5\underset{\textstyle NHOH}{\underset{|}{C}}HCH_2CO_2H$$

$$C_6H_5\underset{\textstyle NHOH}{\underset{|}{C}}HCH_2CO_2H + NH_2OH \rightarrow C_6H_5\underset{\textstyle NH_2}{\underset{|}{C}}HCH_2CO_2H$$

Submitted by Robert E. Steiger.
Checked by W. E. Bachmann and Yun-Tsung Chao.

1. Procedure

A hot solution of sodium ethoxide is prepared, in a 3-l. round-bottomed flask, from 46 g. (2 gram atoms) of sodium and 1.6 l. of absolute ethanol. With shaking, a solution of 139 g. (2 moles) of hydroxylamine hydrochloride in 100 ml. of hot water is added. The resulting suspension is cooled quickly by placing the flask in an ice-water mixture and is then filtered with suction through a Büchner funnel. The residue of sodium chloride is washed with small portions (total 200 ml.) of absolute ethanol. The filtrate is returned to the 3-l. flask, and to it is added 148 g. (1 mole) of cinnamic acid, whereupon a voluminous precipitate forms. The mixture is refluxed on a steam bath for 9 hours (Note 1). The amino acid begins to separate after 5–6 hours; the suspended solid causes the mixture to bump (Note 2). The suspension is allowed to remain overnight at room temperature, and the crystals are then collected on a Büchner funnel (Note 3). The product is washed with 300 ml. of absolute ethanol, then with some ice-cold water to remove all the sodium chloride, and finally with 300 ml. of absolute ethanol, always in small portions. The colorless crystals of amino acid are dried in a vacuum desiccator over flake sodium hydroxide. The yield is 56 g. (34%).

If a purer product is desired, the amino acid is dissolved in 16 times its weight of boiling water, and to the solution is added absolute ethanol (46 ml. per g. of acid). The solution is stirred mechanically while it is cooled in an ice-water mixture. After 3 hours, the snow-white crystals are collected on a Büchner funnel and are washed with 300 ml. of 95% ethanol, in small portions, and dried as before. The recovery of amino acid, melting at 221° with decomposition (Note 4), is 81%.

2. Notes

1. The solution becomes clear when the boiling point is reached. It is important that the mixture be boiled for the specified time in order to ensure complete conversion of the hydroxylamino acid into the amino acid. The solubilities of these two acids are nearly the same.

2. It may happen that the hot solution remains supersaturated, and crystallization does not take place, until the solution is cooled. The checkers obtained a poor yield when this occurred. They, therefore, seeded other runs during the boiling process in order to bring about crystallization. Posner[1] had to concentrate the solution to about half its volume in order to bring about crystallization.

3. The mother liquors from the crystallization appear to be free of amino acid if the yield mentioned is obtained. Among other products they contain at least 19 g. of acetophenoneoxime (14%), which is formed by the secondary reaction:

$$\underset{\substack{|\\ NHOH}}{C_6H_5CHCH_2CO_2H} \rightarrow \underset{\substack{\|\\ NOH}}{C_6H_5CCH_2CO_2H} \rightarrow \underset{\substack{\|\\ NOH}}{C_6H_5CCH_3} + CO_2$$

The acetophenoneoxime may be isolated by evaporating the mother liquors almost to dryness, adding water repeatedly to remove the alcohol, and then treating the oily residue with 1 *N* sodium carbonate solution.

4. The melting point varies considerably with the rate and duration of heating. Values ranging from 215° to 231° have been reported in the literature. The product obtained appears to be perfectly stable and shows no tendency to assume the pink coloration reported by Posner.[1]

3. Methods of Preparation

The procedure described is that given by Posner,[1] with some modifications and additions. The amino acid has also been prepared by boiling the oxime hydrate of β-hydroxylaminohydrocinnamohydroxamic acid with water;[2] by decarboxylation of β-amino-β-phenylethane-α,α-dicarboxylic acid;[3] and by the reaction of cinnamic acid with ammonia under pressure in the presence of anhydrous stannic chloride.[4]

[1] Posner, *Ber.*, **38**, 2320 (1905); *Ann.*, **389**, 120 (1912). See also Fischer, Scheibler, and Groh, *Ber.*, **43**, 2024 (1910).

[2] Posner, *Ber.*, **40**, 227 (1904).

[3] Rodionow and Malewinskaja, *Ber.*, **59**, 2956 (1926); Rodionow, *J. Am. Chem. Soc.*, **51**, 851 (1929); Evans and Johnson, *J. Am. Chem. Soc.*, **52**, 5001 (1930); Johnson and Livak, *J. Am. Chem. Soc.*, **58**, 301 (1936).

[4] Enkvist, Monoven, and Soderlund, *Finska Kemistsamfundets Medd.*, **53**, No. 3/4, 66 (1944); [*C. A.*, **41**, 4103 (1947)].

β-AMINOPROPIONITRILE and *bis*-(β-CYANOETHYL)-AMINE

(Propionitrile, β-amino-, and propionitrile, β,β′-iminodi-)

$$NH_3 + CH_2{=}CH{-}CN \rightarrow NH_2CH_2CH_2CN$$

$$NH_2CH_2CH_2CN + CH_2{=}CH{-}CN \rightarrow NH(CH_2CH_2CN)_2$$

Submitted by Saul R. Buc.
Checked by Homer Adkins and James M. Caffrey.

1. Procedure

Acrylonitrile is a poisonous compound. All steps in the procedure up to the distillation of the products should be carried out in a hood.

The reactions are carried out in 1-l. heavy-walled bottles provided with rubber stoppers which must be wired securely in place (Note 1). In each of four bottles are placed 400 ml. of concentrated ammonium hydroxide (28–30% ammonia) and 100 ml. (80 g., 1.5 moles) of cold acrylonitrile (Note 2). The rubber stoppers are immediately wired in place (Note 3). Each bottle is then shaken intermittently until after about 5 minutes the reaction mixture becomes homogeneous. Thereupon, the bottle, wrapped in a towel, is immediately set away under a hood (Note 4).

The reaction mixtures are allowed to stand a few hours or overnight and then are transferred to a 3-l. flask. The water and ammonia are distilled under reduced pressure as rapidly as possible until the boiling point is about 50°/20 mm. (Note 5). The higher-boiling products (395 g.) are then transferred to a 1-l. Claisen flask and fractionated under reduced pressure.

The crude primary amine (138–149 g.) is distilled over the range 75–110°/21 mm. (Note 6), and the crude secondary amine (213–226 g.) in the range 130–150°/1 mm. The primary amine, b.p. 79–81°/16 mm. or 87–89°/20 mm. (n_D^{20} 1.3496), after refractionation, is obtained in a yield of 130–140 g. (31–33%) (Note 7). The secondary amine, b.p.

134–135°/1 mm. (n_D^{20} 1.4640), is obtained in a yield of about 210 g. (57%) (Note 8).

2. Notes

1. The 1-l. centrifuge bottles (Corning No. 1280) carrying No. 6 rubber stoppers, as used for catalytic hydrogenation, are suitable for carrying out reactions under pressures up to at least 3 atm. The submitter used a heavy, selected 2-l. round-bottomed flask instead of the four bottles specified in the procedure above.

2. The acrylonitrile should be free of polymer. If there is uncertainty as to its quality, the acrylonitrile should be redistilled.

3. The temperature of the mixture does not rise during the period of solution of the acrylonitrile in the ammonium hydroxide. However, almost immediately thereafter the temperature of the solution begins to rise slowly, reaching a value of about 65° after an interval of perhaps 10 minutes. There is no significant rise in pressure within the bottle until the temperature of the reaction mixture begins to rise. The maximum pressure reached is apparently less than 2 atm.

4. The checkers placed the wrapped bottles within a 10-gal crock located under a hood. There is no danger that the bottles will be broken by the pressure developed. However, if a stopper is not firmly held, it may be pushed out, in which event a portion of the reaction mixture will foam out of the bottle.

5. The submitter used a special apparatus suitable for the rapid evaporation of water under reduced pressure. The checkers used standard flasks. Better yields result from the rapid removal of water.

6. It is not necessary to purify the crude primary amine by redistillation if it is to be used immediately for the preparation of β-alanine. However, the moist nitrile is not stable in storage, pressure being developed in a container stored at room temperature.

7. Yields of 60–80% of the primary amine [1] have been obtained by introducing the acrylonitrile below the surface of the aqueous ammonia preheated at 110° in a steel reactor suitable for pressure reactions.

8. Yields of 88.5% of the secondary amine, as well as 6% *tris*-(β-cyanoethyl)amine, have been obtained by adding the acrylonitrile dropwise to aqueous ammonia whose amount was to the amount of acrylonitrile as 0.53 mole:1 mole.[2]

3. Methods of Preparation

β-Aminopropionitrile and *bis*-(β-cyanoethyl)amine have been made by the addition of anhydrous [3] or aqueous ammonia [4] to acrylonitrile.

β-Aminopropionitrile has been made from β-chloropropionitrile and liquid ammonia (90% yield).[5] *bis*-(β-Cyanoethyl)amine may be converted to β-aminopropionitrile by distillation at atmospheric pressure.

[1] Ford, Buc, and Greiner, *J. Am. Chem. Soc.*, **69**, 844 (1947).

[2] Wiedeman and Montgomery, *J. Am. Chem. Soc.*, **67**, 1994 (1945).

[3] Whitmore, Mosher, Adams, Taylor, Chapin, Weisel, and Yanko, *J. Am. Chem. Soc.*, **66**, 725 (1944).

[4] Buc, Ford, and Wise, *J. Am. Chem. Soc.*, **67**, 92 (1945).

[5] U. S. pat. 2,443,292 [*C. A.*, **42**, 7322 (1948)].

3-AMINO-1H-1,2,4-TRIAZOLE

(1H-1,2,4-Triazole, 3-amino-)

$$NH_2C({=}NH)NHNH_2 \cdot H_2CO_3 + HCO_2H \rightarrow NH_2C({=}NH)NHNH_2 \cdot HCO_2H + CO_2 + H_2O$$

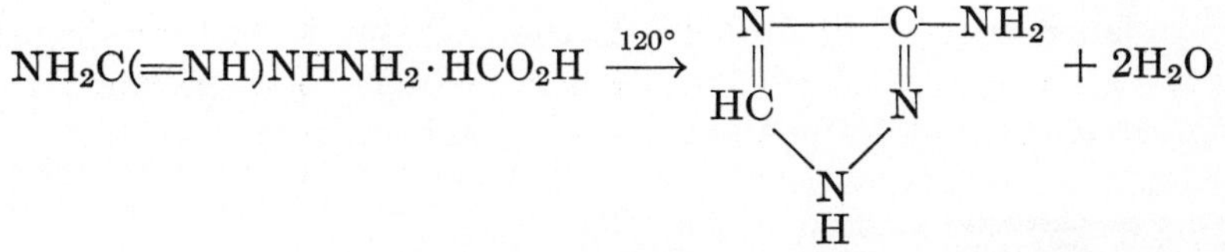

Submitted by Georg Sjostedt and Leopold Gringas.
Checked by E. C. Horning, H. Lloyd, and L. Matternas.

1. Procedure

To 136 g. (1 mole) of finely powdered aminoguanidine bicarbonate in a 500-ml. two-necked round-bottomed flask fitted with a thermometer is added 48 g. (40 ml., 1.05 moles) of 98–100% formic acid (Note 1). The foaming mixture is heated cautiously, with gentle rotation of the flask to avoid local overheating, until gas evolution ceases and the whole mass is dissolved. The solution of aminoguanidine formate is held at a temperature of 120° for 5 hours (Note 2). After the solution is cooled, 500 ml. of 95% ethanol is added, the product is dissolved by heating, and the solution is filtered while hot. The product, obtained by evaporating the ethanol solution to dryness on the steam bath, and oven-drying at 100°, is 80–81.6 g. (95–97%) of colorless crystalline 3-amino-1,2,4-triazole melting at 152–156° (Note 3). The 3-amino-1,2,4-triazole can be recrystallized from ethanol (Note 4).

2. Notes

1. The submitters report that aminoguanidine bicarbonate of practical grade is satisfactory. The checkers used Eastman Kodak white label quality.

2. The submitters and checkers used an infrared heater.

3. The main part of the ethanol can be removed by distillation. From the remaining ethanolic solution yellow crystals precipitate, which are collected on a Büchner funnel. When melted and dried they lose water and give 3-amino-1,2,4-triazole in its pure form.

4. The 3-amino-1,2,4-triazole may be crystallized from ethanol or ethanol-ether.[1] The checkers preferred ethanol (200 ml. for 40 g.). The recovery is 70–73%, m.p. 152–153°.

3. Methods of Preparation

3-Amino-1,2,4-triazole has been prepared by evaporating formylguanidine nitrate with sodium carbonate,[2] and from 5(3)-amino-1,2,4-triazolecarboxylic acid-3(5) by heating above its melting point,[3–5] or by a long digestion with acetic acid.[2] It has been prepared from the sulfate in essentially the same yield and purity.[1]

[1] *Org. Syntheses*, **26**, 11 (1946).
[2] Thiele and Manchot, *Ann.*, **303**, 45, 54 (1898).
[3] Curtius and Lang, *J. prakt. Chem.*, (2) **38**, 554 (1888).
[4] Hantzsch and Silberrad, *Ber.*, **33**, 79 (1900).
[5] Curtius, Darapsky, and Müller, *Ber.*, **40**, 818, 830 (1907).

4-AMINO-4H-1,2,4-TRIAZOLE

(4H-1,2,4-Triazole, 4-amino-)

$$HCO_2C_2H_5 + NH_2NH_2 \cdot H_2O \rightarrow HCONHNH_2 + C_2H_5OH + H_2O$$

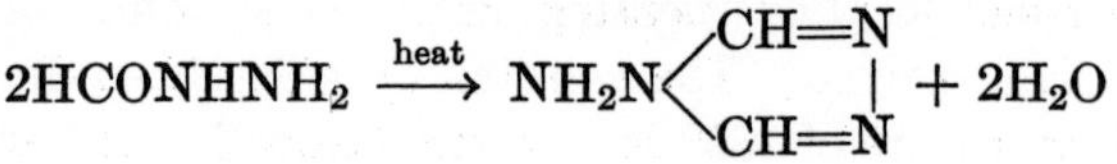

Submitted by C. F. H. Allen and Alan Bell.
Checked by W. E. Bachmann and G. Dana Johnson.

1. Procedure

In a 1-l. round-bottomed flask equipped with an efficient water condenser are placed 148 g. (2 moles) of ethyl formate (b.p. 52–53°) and

150 ml. of 95% ethanol. One hundred and twenty grams (2 moles) of 85% hydrazine hydrate is added cautiously to this solution (Note 1) with shaking over a period of 10 minutes. After the reaction has subsided, the solution is refluxed on a steam bath for 18 hours. The bulk of the water and ethanol is now removed by evaporation under reduced pressure until the volume in the flask is about 150 ml. The resulting syrup, crude formhydrazide, is heated under atmospheric pressure for 3 hours, during which time the temperature of the bath is raised from 150° to 200°. After cooling to about 100°, the oil is taken up in 50 ml. of 95% ethanol, and 5 g. of Norite is added. The filtered solution is then diluted with 75 ml. of ether and placed in an icebox to cool. The crystalline product is filtered, washed with 50 ml. of 1:2 ethanol-ether, and dried. The yield of aminotriazole, melting at 77–78°, is 55–60 g. (65–71%) (Note 2). If a purer product is desired, the crude, washed material may be recrystallized, using 2 ml. of warm 95% ethanol per gram of compound followed by the addition of 2.5 ml. of ether, and chilling. The melting point of the purified product is 81–82°.

The residual amine in the filtrate may be isolated in the form of the hydrochloride. The combined solutions are evaporated on a steam bath, 50 ml. of concentrated hydrochloric acid is added, and heating is continued for 2 hours. On cooling, the syrupy solution crystallizes. It is triturated with 50 ml. of ethanol, and the 4-amino-1,2,4-triazole hydrochloride is filtered, washed with a little ethanol, and dried. The yield of the hydrochloride is 10–18 g. (8–15%); the salt melts at 147–148° and may be recrystallized from 95% ethanol, using 10 ml. per gram; the melting point is thus raised to 151–152°.

2. Notes

1. The reaction is very vigorous, but if the hydrazine hydrate is added carefully no difficulty of control is encountered.

2. The combined yield of base and hydrochloride is always about 80–81%. When the amount of base is low, that of the hydrochloride is high.

3. Methods of Preparation

4-Amino-1,2,4-triazole has been obtained from orthoformic ester and hydrazine hydrate in a sealed tube at 120°;[1] by heating formylhydrazine at 150–210°;[2–4] by heating N,N′-diformylhydrazine at 160°;[5] by decarboxylation of 4-amino-1,2,4-triazoldicarboxylic acid;[6] by fusion of 1,2-dihydro-1,2,4,5-tetrazine;[6] and by heating 1,2-dihydro-1,2,4,5-tetrazinedicarboxylic acid above its melting point.[4,6,7]

A process in which hydrazine and carbon monoxide are heated under pressure has been patented.[8]

[1] Stolle, *J. prakt. Chem.*, (2) **68**, 467 (1903).
[2] Ruhemann and Stapleton, *J. Chem. Soc.*, **75**, 1132 (1899).
[3] Ruhemann and Merriman, *J. Chem. Soc.*, **87**, 1772 (1905).
[4] Hantzsch and Silberrad, *Ber.*, **33**, 85 (1900).
[5] Pellizzari, *Atti accad. Lincei*, (5) **8** (I), 331 (*Chem. Zentr.*, **1899**, I, 1240); *Gazz. chim. ital.*, **39** (I), 530 (1909).
[6] Curtius, Darapsky, and Müller, *Ber.*, **40**, 835, 1194 (1907).
[7] Curtius and Lang, *J. prakt. Chem.*, (2) **38**, 549 (1888).
[8] Brit. pat. 649,445 [*C. A.*, **45**, 8560 (1951)].

9-ANTHRALDEHYDE; 2-ETHOXY-1-NAPHTHALDEHYDE

$$ArH + C_6H_5N(CH_3)CHO \xrightarrow{(POCl_3)} ArCHO + C_6H_5NHCH_3$$

Submitted by (A) L. F. Fieser, J. L. Hartwell, and J. E. Jones; (B) J. H. Wood and R. W. Bost.
Checked by C. F. H. Allen and J. VanAllan.

1. Procedure

A. 9-*Anthraldehyde.* In a 2-l. round-bottomed flask fitted with a mechanical stirrer and reflux condenser are placed 35 g. (32 ml., 0.26 mole) of N-methylformanilide (p. 590), 35 g. (21 ml., 0.23 mole) of phosphorus oxychloride, 20 ml. of *o*-dichlorobenzene, and 22.5 g. (0.13 mole) of anthracene (Note 1). The flask is heated on the steam bath, with stirring, to 90–95° over a period of 20 minutes; the anthracene dissolves during this time to give a deep red solution, and hydrogen chloride is evolved (Note 2). The heating is continued for 1 hour (Note 3), after which a solution of 140 g. of crystalline sodium acetate in 250 ml. of water (Note 4) is added to the cooled mixture, and the *o*-dichlorobenzene and most of the methylaniline are rapidly distilled with steam (15–20 minutes). The residual reddish oil solidifies on cooling. The solid residue is broken up, and, after the aqueous liquor has been decanted through a Büchner funnel, it is washed by decantation with two 100-ml. portions of 6 *N* hydrochloric acid to remove amine, and then thoroughly with water (1–1.2 l.). The crude solid (22–24 g., m.p. 97–101°) is recrystallized from 50 ml. of hot glacial acetic acid; when cold, the bright yellow aldehyde is filtered by suction and washed on the filter with 30 ml. of methanol (Note 5). The yield is 20–22 g. (77–84%), and the melting point is 104.5–105° (Note 6).

B. 2-*Ethoxy*-1-*naphthaldehyde.* A mixture of 45 g. (0.33 mole) of N-methylformanilide (p. 590), 51 g. (0.33 mole) of phosphorus oxychloride, and 43 g. (0.25 mole) of β-naphthyl ethyl ether in a 500-ml. round-bottomed flask provided with an air condenser (Note 2) is heated on a steam bath for 6 hours. The hot mixture is then poured in a thin stream into 700 ml. of cold water with very vigorous stirring to avoid the formation of large lumps (Note 7); the aldehyde separates in a granular condition. It is filtered by suction and washed thoroughly, using 1 l. of water. Without drying, the crude aldehyde is dissolved in 450 ml. of ethanol and decolorized by the addition of 4 g. of Norit, boiling for 15 minutes, and filtering hot (Note 8), using a double filter paper. The filtrate is cooled, and the product is collected on a filter and washed with 40 ml. of cold ethanol; it crystallizes in pale yellow needles, m.p. 111–112°. The yield is 37–42 g. (74–84%).

2. Notes

1. The yield and purity of the anthraldehyde depend on the quality of the hydrocarbon. The figures given are attained only if the anthracene melts at 213° or higher. With anthracene, m.p. 208–210°, the yield is 19–20 g., m.p. 103–104° (Note 6).

2. The reaction may be carried on in a hood, or a gas trap may be used.

3. When run on a fourfold scale, the time of heating should be extended to 2 hours. Prolonged heating leads to the formation of tars.

4. Sodium acetate appears to decompose a product of the condensation of methylaniline with phosphorus oxyhalides; substances other than the aldehyde are largely retained in the sodium acetate solution.

5. A brighter-colored product is secured by this wash.

6. The aldehyde also exists in a low-melting form, m.p. 98.5–99.5°, and occasionally this form is obtained from the reaction. It is less stable than the high-melting form, into which it is easily converted by seeding.

7. Care must be exercised to prevent the reaction product from lumping upon being poured into water. If this happens, decomposition by the water is slow and subsequent purification is more difficult. Any lumps that are formed should be broken up, and the reaction mixture should then be permitted to stand overnight in contact with the water. The flask in which the reaction is carried out is also filled

with water to decompose the product adhering to the walls. This is purified with the rest.

8. A heated funnel is desirable. Some of the product usually crystallizes during the filtration.

3. Methods of Preparation

This aldehyde synthesis is applicable to compounds of the aromatic series having a labile hydrogen atom (phenyl ethers,[1] naphthols,[2] dialkylanilines,[3,4] naphthostyril,[2] anthrones [2]) and to certain hydrocarbons of requisite reactivity (anthracene,[5,6] 1,2-benzanthracene,[6] 3,4-benzpyrene,[3,7] pyrene,[8] styrene,[9] and α,α-diarylethylenes [9]). With polynuclear hydrocarbons the best results are secured by the use of a solvent such as *o*-dichlorobenzene. 9-Anthraldehyde has also been prepared by the action of hydrogen cyanide and aluminum chloride on anthracene in chlorobenzene.[10]

With liquid or low-melting ethers no solvent is required. 2-Ethoxy-1-naphthaldehyde has also been prepared by ethylation of the hydroxy compound,[11] and from β-naphthyl ethyl ether by the Gattermann reaction.[12]

It is possible to employ N,N-dimethylformamide in place of N-methylformanilide in some cases. A general method, applied to 9-anthraldehyde, has been described by Campaigne and Archer.[13]

1 Wood and Bost, *J. Am. Chem. Soc.*, **59**, 1722 (1937).
2 Ger. pat. 514,415 [*Frdl.*, **17**^{1}, 564 (1932)].
3 Vilsmeier and Haak, *Ber.*, **60**, 119 (1927).
4 Ger. pat. 547,108 [*Frdl.*, **18**^{2}, 2973 (1933)].
5 Ger. pat. 519,444 [*Frdl.*, **17**^{1}, 565 (1932)].
6 Fieser and Hartwell, *J. Am. Chem. Soc.*, **60**, 2556 (1938).
7 Fieser and Hershberg, *J. Am. Chem. Soc.*, **60**, 2547 (1938).
8 Vollmann, Becker, Corell, and Streeck, *Ann.*, **531**, 1 (1937).
9 Brit. pat. 504,125 [*C. A.*, **33**, 7313 (1939)].
10 Hinkel, Ayling, and Beynon, *J. Chem. Soc.*, **1936**, 344.
11 Bartsch, *Ber.*, **36**, 1975 (1903).
12 Gattermann, *Ann.*, **357**, 367 (1907).
13 Campaigne and Archer, *J. Am. Chem. Soc.*, **75**, 989 (1953).

D-ARABINOSE *

$$H(CHOCOCH_3)_5CN \xrightarrow[CH_3OH]{CH_3ONa} H(CHOH)_4\cdot CHO + NaCN + 5CH_3CO_2CH_3$$

Submitted by GÉZA BRAUN.
Checked by H. T. CLARKE and S. M. NAGY.

1. Procedure

A solution of 100 g. (0.26 mole) of pentaacetyl glucononitrile (p. 690) in 150 ml. of chloroform in a 1-l. Erlenmeyer flask is chilled to −12°. A chilled (−12°) solution of 16 g. (0.7 gram atom) of sodium in 250 ml. of anhydrous methanol is added with continual shaking and chilling to the chloroform solution of the nitrile. The mixture soon solidifies to a pale yellow gelatinous mass. After 10 minutes at −12° this is broken up with a heavy glass rod and dissolved in 600 ml. of a suspension of ice in water. The resulting solution is acidified with an ice-cold mixture of 33 g. (18 ml., 0.32 mole) of 95% sulfuric acid, 5 ml. of acetic acid, and 45 g. of ice. The aqueous layer is separated, washed once with 50 ml. of chloroform, and evaporated without delay (Note 1) under reduced pressure. The residual heavy syrup is dissolved in 300 ml. of water and again evaporated as completely as possible under reduced pressure, in order to remove residual hydrogen cyanide (Note 2). The highly viscous residue, which contains some crystals of sodium sulfate, is dissolved in 500 ml. of hot methanol. After about 10 minutes the sodium sulfate is filtered with suction and washed with two 25-ml. portions of methanol. The filtrate is concentrated under reduced pressure at 40° to a heavy syrup which is poured while warm into a 200-ml. Erlenmeyer flask. The distilling flask is rinsed twice with 20-ml. portions of hot ethanol, and this rinse is added to the filtrate. The resulting ethanol solution soon begins to deposit crystals of arabinose; it is stirred by hand during the crystallization and gradually diluted with more ethanol until 100 ml. in all has been added during the course of an hour. The mixture is allowed to stand for 4–5 hours; the crystals are then filtered, washed with two 25-ml. portions of ethanol, and dried at 40°. The yield of colorless D-arabinose, m.p. 158–158.5°, $[\alpha]_D^{20}$ −105° (final value), is 23.5–26.3 g. (61–68%) (Note 3).

2. Notes

1. After the reaction mixture has been dissolved in water and acidified, the hydrogen cyanide should be removed as soon as possible, for the arabinose tends to react with it even in dilute solution.

2. A slightly higher yield of crystalline arabinose is obtainable by removing the hydrogen cyanide with silver acetate. The procedure then consists in acidifying with acetic acid in place of sulfuric acid, adding an excess of silver acetate, shaking for an hour, filtering, saturating with hydrogen sulfide, again filtering, and adding sulfuric acid as indicated above. The rest of the procedure is the same. The yield of crystalline arabinose so obtained is 27.1 g. (70%).

3. A further quantity of arabinose may be isolated from the mother liquors by the use of diphenylhydrazine: to a solution of 22 g. of diphenylhydrazine hydrochloride in 100 ml. of absolute methanol is added a solution of 3.3 g. of sodium in 50 ml. of methanol. After 15 minutes' standing the sodium chloride is removed by filtration and washed with methanol. The filtrate, which contains approximately 18 g. of free diphenylhydrazine, is added to the alcoholic mother liquor from the arabinose, and the mixture is inoculated with diphenylhydrazone prepared from some of the crystalline arabinose. The mixture is allowed to stand overnight, and the crystalline diphenylhydrazone is filtered, washed with 95% ethanol, and dried in a vacuum desiccator. In a preparation in which the yield of crystalline arabinose had been 23.5 g., the yield of diphenylhydrazone was 16.5 g., corresponding to 7.8 g. of the sugar. Arabinose can be recovered from the diphenylhydrazone by treatment with formaldehyde in aqueous solution. In view of the high cost of diphenylhydrazine, however, it is doubtful whether its use for this purpose is profitable.

3. Methods of Preparation

D-Arabinose was first prepared by Wohl,[1] by treating pentaacetyl glucononitrile with ammoniacal silver nitrate. It has also been obtained from *d*-gluconic acid in various ways: by oxidation of the calcium salt by means of hydrogen peroxide in the presence of ferric acetate; [2–5] by boiling an aqueous solution of the mercuric salt; [6] by electrolysis; [7] by the action of sodium hypochlorite upon the amide.[8] It has also been obtained by the electrolytic reduction of D-arabonic acid lactone.[9] The present method, developed by Zemplén and Kiss,[10] furnishes better yields than that of Wohl.

A modification of the Ruff method,[2] using ion-exchange resins for the removal of salts, has been published.[11]

1 Wohl, *Ber.*, **26,** 730 (1893).
2 Ruff, *Ber.*, **32,** 553 (1899).
3 Hockett and Hudson, *J. Am. Chem. Soc.*, **56,** 1632 (1934).
4 Jones, Kent, and Stacey, *J. Chem. Soc.*, **1947,** 1341.
5 Berezovskii and Kurdyukova, *Zhur. Priklad. Khim.*, **22,** 1116 (1949) [*C. A.*, **45,** 5628 (1951)].
6 Guerbet, *Bull. soc. chim. France,* (4) **3,** 427 (1908).
7 Neuberg, *Biochem. Z.*, **7,** 527 (1908).
8 Weermann, *Rec. trav. chim.*, **37,** 16 (1918).
9 Swiss pat. 258,581 [*C. A.*, **44,** 4352 (1950)].
10 Zemplén and Kiss, *Ber.*, **60,** 165 (1927).
11 Fletcher, Diehl, and Hudson, *J. Am. Chem. Soc.*, **72,** 4546 (1950).

AZOBENZENE *

$$2C_6H_5NO_2 + 4Zn + 8NaOH \rightarrow C_6H_5N{=}NC_6H_5 + 4Na_2ZnO_2 + 4H_2O$$

Submitted by H. E. Bigelow and D. B. Robinson.
Checked by W. E. Bachmann and W. S. Struve.

1. Procedure

A 5-l. three-necked round-bottomed flask, fitted with a mercury-sealed stirrer and a reflux condenser, is placed on a steam cone. In the flask are placed 250 g. (208 ml., 2 moles) of nitrobenzene, 2.5 l. of methanol, and a solution of 325 g. (8.1 moles) of sodium hydroxide (Note 1) in 750 ml. of distilled water. To the mixture is added 265 g. (4.1 moles) of zinc dust (Note 2), the stirrer is started, and the mixture is refluxed for 10 hours (Note 3). The mixture is filtered while hot, and the precipitate of sodium zincate is washed on the filter with a little warm methanol. All the methanol is distilled from the filtrate, the residue is chilled, and the crystalline azobenzene is filtered.

In order to remove zinc salts from the crude azobenzene, the latter is added to 500 ml. of 2% hydrochloric acid, the mixture is warmed to about 70° in order to melt the azobenzene and is stirred rapidly for about 5 minutes. Stirring is continued while the mixture is chilled to solidify the azobenzene. The product is filtered, washed well with water, and recrystallized from a mixture of 720 ml. of 95% ethanol and 60 ml. of water. The yield of azobenzene melting at 66–67.5° is 156–160 g. (84–86%).

2. Notes

1. This amount assumes 100% purity. The checkers used 342 g. of 95% sodium hydroxide.

2. This amount assumes 100% purity. The checkers used 288 g. of 92% zinc dust.

3. At the end of this time, the reddish mixture should be free from the odor of nitrobenzene. If it is not, refluxing is continued for 2–3 hours longer.

3. Methods of Preparation

Azobenzene has been prepared by many different methods, of which the following are representative. It may be obtained by the reduction of nitrobenzene with iron and acetic acid;[1] with sodium amalgam;[2] with alkali sulfides;[3] with cellulose,[4] molasses,[5] or dextrose[5] in alkaline solution; and by catalytic reduction.[6] The reduction with zinc and sodium hydroxide described here is a modification of Alexejew's method.[7] Azobenzene also results from the reduction of diazotized aniline with cuprous salts.[8] Aniline has been oxidized to azobenzene by air[9] and by potassium permanganate.[10] The condensation of nitrobenzene and aniline acetate also yields azobenzene.[11]

[1] Nobel, *Ann.*, **98**, 253 (1856).
[2] Werigo, *Ann.*, **135**, 176 (1865).
[3] Lucius and Bruning, Ger. pat. 216,246 [*C. A.*, **4**, 813 (1910)].
[4] Greisheim, Ger. pat. 225,245 [*C. A.*, **5**, 592 (1911)].
[5] Opolonick, *Ind. Eng. Chem.*, **27**, 1045 (1935).
[6] Treed and Signaigo, U. S. pat. 2,344,244 [*C. A.*, **38**, 3663 (1944)]; Henke and Brown, *J. Phys. Chem.*, **26**, 324, 631 (1922).
[7] Alexejew, *Z. Chem.*, **4**, 497 (1868).
[8] Bozoslavski, *J. Gen. Chem. U.S.S.R.*, **16**, 193 (1946).
[9] Alekseevskii and Golbrakht, Russ. pat. 32,499 [*C. A.*, **28**, 3425 (1934)]; Brown and Triske, *J. Phys. & Colloid Chem.*, **51**, 1394 (1947).
[10] Glaser, *Ann.*, **142**, 364 (1867).
[11] Baeyer, *Ber.*, **7**, 1638 (1874).

BENZALACETONE DIBROMIDE

(2-Butanone, 3,4-dibromo-4-phenyl-)

$$C_6H_5CH{=}CHCOCH_3 + Br_2 \rightarrow C_6H_5CHBrCHBrCOCH_3$$

Submitted by NORMAN H. CROMWELL and RICHARD BENSON.
Checked by R. L. SHRINER and WILLIAM O. FOYE.

1. Procedure

In a 1-l. three-necked round-bottomed flask fitted with an efficient mechanical stirrer, a thermometer, and a 125-ml. dropping funnel are placed 100 g. (0.68 mole) of pure, redistilled benzalacetone[1] and 300 ml. of carbon tetrachloride. The reaction flask is immersed in an ice-water bath to maintain the reaction mixture between 10° and 20°. With stirring, a cooled solution of 109.5 g. (34.2 ml., 0.68 mole) of bromine in 60 ml. of carbon tetrachloride is run through the dropping funnel as rapidly as the color is destroyed (Note 1). During this addition the reaction flask should be shielded from direct sunlight (Note 2).

After all the bromine has been added, stirring is continued for 4–5 minutes longer and the dibromide is collected by filtration on an 11-cm. Büchner funnel, using suction. The product is washed with 100 ml. of warm 75% ethanol (Note 3). The crude product is purified by dissolving in the minimum amount of boiling methanol (800–1000 ml.) and cooling the solution in an ice bath for 4 hours. The product is collected by filtration and dried in a vacuum desiccator in the absence of light for 24 hours. The yield amounts to 110–120 g. (52–57%) of white needles which melt at 124–125° (Note 4).

2. Notes

1. Until a considerable amount of the dibromide has precipitated, the bromine solution may be run into the reaction mixture as fast as the color is discharged, within the temperature limits of 10–20°. As the mixture becomes thick with the precipitated bromide it is necessary to reduce the speed of the addition of the bromine solution considerably.

2. Strong sunlight seems to favor the substitution of the available α-hydrogen as evidenced by the strong evolution of hydrogen bromide.

3. If the crude product is dried it is found to melt at 114–117° and

to weigh 138–144 g. This is probably a mixture of the two racemic forms.

4. Evaporation and subsequent cooling of the filtrate give a second crop of white crystals, about 15 g., melting at 112–115°. This may be the lower-melting racemate.

3. Methods of Preparation

Benzalacetone dibromide has been prepared by the addition of bromine to a solution of benzalacetone in chloroform,[2] in carbon disulfide,[3] and in carbon tetrachloride;[4] and by the reaction of benzalacetone with N-bromosuccinimide.[5]

[1] *Org. Syntheses* Coll. Vol. **1,** 77 (1941).
[2] Claisen and Claparede, *Ber.,* **14,** 2463 (1881).
[3] Watson, *J. Chem. Soc.,* **85,** 464 (1904).
[4] Cromwell, *J. Am. Chem. Soc.,* **62,** 3471 (1940).
[5] Southwick, Pursglove, and Numerof, *J. Am. Chem. Soc.,* **72,** 1600 (1950).

1,2,3-BENZOTRIAZOLE *

(1H-Benzotriazole)

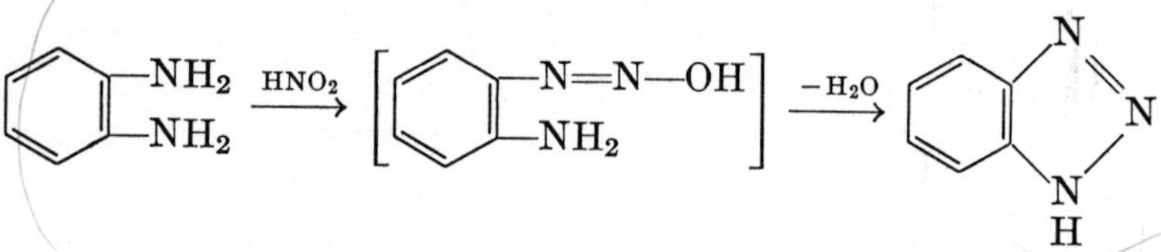

Submitted by R. E. Damschroder and W. D. Peterson.
Checked by W. E. Bachmann and W. S. Struve.

1. Procedure

In a 1-l. beaker are placed 108 g. (1 mole) of *o*-phenylenediamine, 120 g. (115 ml., 2 moles) of glacial acetic acid, and 300 ml. of water. By warming the mixture slightly a clear solution is obtained. The beaker is placed in ice water, and the contents are cooled to 5°. As soon as this temperature is reached, a cold solution of 75 g. (1.09 moles) of sodium nitrite in 120 ml. of water is added all at once, the mixture being stirred with a glass rod or by a slow mechanical stirrer. The reaction mixture turns dark green, and the temperature rises rapidly to 70–80° (Note 1). The color of the solution changes to a clear orange-red. The beaker is now removed from the cooling bath, and the contents are allowed to stand for 1 hour; as the solution cools, the benzotriazole separates as an oil. The beaker is then packed in ice, and the mixture is stirred until it sets to a solid mass. After being kept cold for 3 hours, the solid is collected on a Büchner funnel, washed with 200 ml. of ice water, and sucked as dry as possible under a rubber dam. The tan-colored product, after drying at 45–50° overnight, weighs 110–116 g.

The crude benzotriazole is placed in a 200-ml. modified Claisen flask and distilled under reduced pressure (Note 2). The yield of white solid (yellow cast) boiling at 201–204° at 15 mm. or 156–159° at 2 mm. is 92–99 g. The product in the receiver is melted over a luminous flame and poured into 250 ml. of benzene. The clear solution is stirred until crystallization sets in; after being chilled for 2 hours, the product is filtered on a Büchner funnel. The colorless benzotriazole weighs 90–97 g. (75–81%) (Note 3) and melts at 96–97°.

2. Notes

1. Too efficient cooling and stirring are to be avoided. It is essential that the temperature rise to 80°. Too rapid cooling after this temperature has been reached results in lower yields. With runs one-tenth this size it is advisable to remove the mixture from the cooling bath as soon as the sodium nitrite has been added in order to ensure the rise in temperature.

2. The crude product can be purified by repeated crystallizations from benzene or water. Greater losses accompany this tedious method of purification than a single distillation.

3. The submitters report that runs of double this size can be made with equally good results.

3. Methods of Preparation

1,2,3-Benzotriazole has been prepared directly by the action of nitrous acid on *o*-phenylenediamine [1] and by the hydrolysis of an acylated or aroylated benzotriazole which has been previously prepared by the action of nitrous acid on the corresponding mono acylated or aroylated *o*-phenylenediamine.[2–4] The above procedure is the direct method and gives better over-all yields than the methods involving several intermediate steps. Most methods described in the literature employ mineral acid. Acetic acid is much more satisfactory.

[1] Ladenburg, *Ber.*, **9**, 219 (1876).
[2] Bell and Kenyon, *J. Chem. Soc.*, **1926**, 954.
[3] Fieser and Martin, *J. Am. Chem. Soc.*, **57**, 1838 (1935).
[4] Charrier and Beretta, *Gazz. chim. ital.*, **51**, (Part 2), 267 (1921).

BENZOYLACETANILIDE

(Acetanilide, α-benzoyl-)

$$C_6H_5COCH_2CO_2C_2H_5 + C_6H_5NH_2 \rightarrow C_6H_5COCH_2CONHC_6H_5 + C_2H_5OH$$

Submitted by Charles J. Kibler and A. Weissberger.
Checked by R. L. Shriner and Fred W. Neumann.

1. Procedure

A 250-ml. three-necked round-bottomed flask is fitted with a dropping funnel, a mechanical stirrer, and a steam-jacketed column (15–20 cm. long) terminating in a still head. The still head carries a thermometer and is connected to a condenser set for downward distillation. In the flask are placed 42.2 g. (0.22 mole) of ethyl benzoylacetate (Note 1) and 50 ml. of dry xylene, and the flask is immersed in an oil bath. The bath is heated to 145–150°, stirring is begun, and 18.2 ml. (18.6 g., 0.20 mole) of aniline is added dropwise during 30 minutes. About 5 minutes after the addition has begun, the temperature in the still head rises to 75–78° as ethanol begins to distil. Approximately 12–14 ml. of distillate (Note 2) is collected in a 25-ml. graduate in about 1 hour; the end of the reaction is indicated by the fall of the temperature in the still head.

The reaction flask is removed from the oil bath, and the solution is poured into a 250-ml. beaker to crystallize; 10 ml. of benzene is used to rinse out the flask. When crystallization sets in, 50 ml. of petroleum ether (b.p. 35–55°) is added to the warm mixture with manual stirring. After the mixture has been chilled in an ice bath, the product is filtered by suction and washed with 100 ml. of a 1:1 petroleum ether-benzene mixture. It is then removed from the funnel, stirred into a slurry with 100 ml. of the mixed solvent, filtered, and again washed with 50 ml. of the solvent. These washings remove the bulk of the color, and a powdery white product remains. After standing overnight in a warm place, the product weighs 35.5–36.5 g. (74–76%) and melts at 104–105°. It may be purified further by recrystallization from benzene (3.7 ml. per g.). The yield of colorless benzoylacetanilide melting at 106–106.5° cor, is 32–34 g.

2. Notes

1. The ethyl benzoylacetate, aniline, and xylene should be redistilled before use. It is necessary that the flask and reagents be free from moisture. Impure starting materials and particularly traces of acids lower the yields. The submitters obtained 82–83% yields of benzoylacetanilide by using Eastman xylene (histological).

2. The distillate consists of a mixture of xylene and ethanol. By adding 100 ml. of water, the xylene layer can be separated and measured. Of 14 ml. of distillate, about 3.5 ml. is xylene and the balance, 10.5 ml. (89% of the theoretical amount), is ethanol.

3. Methods of Preparation

Benzoylacetanilide has been made by heating aniline and methyl benzoylacetate in an autoclave at 150° [1] and by heating the anil $C_6H_5NHCOCH_2(C{=}NC_6H_5)C_6H_5$ with dilute acid.[1]

[1] Knorr, *Ann.*, **245**, 372 (1888).

β-BENZOYLACRYLIC ACID *

(Acrylic acid, β-benzoyl-)

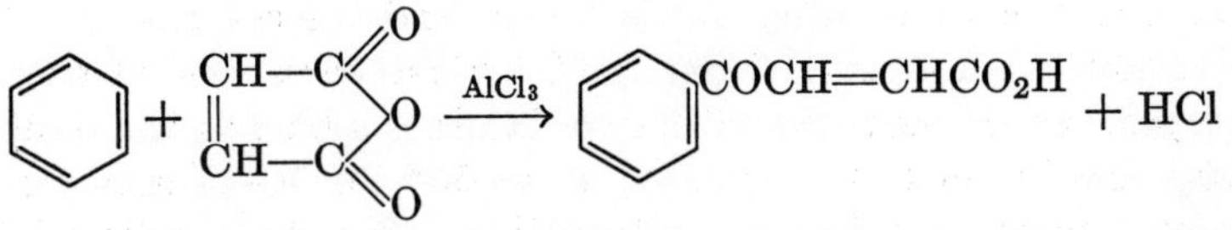

Submitted by Oliver Grummitt, E. I. Becker, and C. Miesse.
Checked by Arthur C. Cope and Claude F. Spencer.

1. Procedure

In a 1-l. three-necked round-bottomed flask fitted with a mercury-sealed stirrer and a reflux condenser are placed 34 g. (0.347 mole) of maleic anhydride (Note 1) and 175 g. (200 ml., 2.24 moles) of dry, thiophene-free benzene. Stirring is started, and, when the maleic anhydride has dissolved, 100 g. (0.75 mole) of anhydrous reagent grade aluminum chloride powder is added in 6–8 portions through the third neck of the flask at a rate so that the benzene refluxes moderately. The addition requires about 20 minutes. The mixture is then heated under reflux on a steam bath and stirred for 1 hour. The reaction flask is cooled thoroughly in an ice bath, a 250-ml. dropping funnel is attached to the third neck, and the mixture is hydrolyzed by adding 200 ml. of water with stirring and cooling (the first 50 ml. during 15–20 minutes and the balance in about 10 minutes), followed by 50 ml. of concentrated hydrochloric acid (Note 2). Stirring is continued for an additional 40 minutes, during which time it may be necessary to use a spatula to scrape adhering particles of the red-brown aluminum chloride addition compound from the walls of the flask.

The hydrolyzed mixture is transferred to a 1-l. Claisen flask, the transfer of material being completed by rinsing with about 50 ml. of warm water. The flask is heated in a water bath at 50–60°, and the benzene and some water are distilled at 20–30 mm. pressure (Note 3). While the residue is still molten, it is transferred to a 1-l. beaker, and the flask is rinsed with 50 ml. of warm water. After standing at 0–5° for 1 hour, the yellow solid is collected on a suction filter and washed with a solution of 25 ml. of concentrated hydrochloric acid in 100 ml. of water and then with 100 ml. of water. The washing is done most efficiently by suspending the solid in the wash liquid, cooling to 0–5° with stirring, and then filtering with suction. The preparation should not be interrupted before this point (Note 3), at which stage the crude acid may be air-dried overnight at room temperature if desired. The crude product is dissolved in a solution of 40 g. of anhydrous sodium carbonate in 250 ml. of water by warming to 40–50° (Note 4), 2 g. of Celite or other filter aid is added, and the solution is filtered with suction while warm. After the filter has been washed with two 30-ml. portions of warm water, 2 g. of Norit is added to the combined filtrates and the mixture is heated at 40–50° for 10–15 minutes with frequent stirring, then filtered with suction. The clear, yellow filtrate is transferred to a 1-l. beaker and cooled to 5–10°, and 70 ml. of concentrated hydrochloric acid is added dropwise with stirring. Efficient cooling and stirring are necessary to avoid the precipitation of the acid as an oil. After being cooled to 0–5°, the mixture is filtered with suction; the solid is washed with two 50-ml. portions of cold water and then is dried at 50° for 12–36 hours to give 56–63 g. of light-yellow anhydrous β-benzoylacrylic acid, m.p. 90–93° (Note 5). The crude acid may be crystallized from benzene, using 12–15 ml. of benzene per 5 g. of acid and cooling at 5–10° (Note 6) to give 44–47 g. of β-benzoylacrylic acid, m.p. 94–96° (Note 7). Concentration of the filtrate to one-fourth to one-fifth of its original volume gives an additional 3–6 g., melting in the range 92–96°. The total yield is 49–52 g. (80–85%) (Note 8).

2. Notes

1. A good grade of commercial maleic anhydride was used, m.p. 52–54°.

2. When β-benzoylacrylic acid is heated with dilute hydrochloric acid, β-benzoyllactic acid is formed, which makes the purification of the product very difficult.[1] For this reason the mixture is well cooled during hydrolysis and the hydrochloric acid is not added until the heat of the exothermic hydrolysis has been dissipated.

3. The benzene solution of β-benzoylacrylic acid is concentrated under reduced pressure to avoid overheating, and the crude product is separated from aqueous hydrochloric acid without long standing in order to minimize the possibility of forming β-benzoyllactic acid (Note 2).

4. Heating a mixture of β-benzoylacrylic acid and excess sodium carbonate causes hydrolysis to acetophenone and other products, which decreases the yield and interferes with purification of the product.[2,3]

5. Before drying, the product is the monohydrate, m.p. 64–65° when pure. It is dried until it reaches constant weight and has the melting point of the anhydrous acid.

6. The benzene solution should not be boiled longer than necessary to dissolve the acid, because prolonged heating discolors the product.

7. The evidence indicates that this is the *trans* form of β-benzoylacrylic acid.[4–6] Inhalation of β-benzoylacrylic acid dust should be avoided because of its sternutatory action.

8. It has been reported (E. L. Ringwald, private communication) that this reaction may be carried out on a large scale (170 g. of maleic anhydride, 1 kg. of dry benzene, and 1 kg. of aluminum chloride). After the first addition of 500 g. of aluminum chloride, the remaining material is added in 100-g. portions. The mixture is heated with stirring at 55–60° for 1 hour. The complex is decomposed by pouring into ice and concentrated hydrochloric acid. The crude air-dried product, separated by filtration, is combined with the residue obtained by removal of the solvent. The combined crude products are recrystallized from benzene-hexane. The yield is 252 g. (m.p. 95–96°) of first crop, and 15 g. (m.p. 90–94°) of second crop, a total yield of 87%.

3. Methods of Preparation

β-Benzoylacrylic acid has been prepared by the condensation of acetophenone and chloral to 1,1,1-trichloro-2-hydroxy-3-benzoylpropane, followed by hydrolysis to the corresponding acid and dehydration; [7] by the action of iodine, potassium iodide, and sodium carbonate on γ-phenylisocrotonic acid; [8] by bromination of β-benzoylpropionic acid and subsequent dehydrohalogenation; [1] by the action of phenylzinc chloride on maleic anhydride; [9] and by the condensation of the acid chloride of ethyl hydrogen maleate with benzene in the presence of aluminum chloride, followed by hydrolysis.[10] The present method is based on the work of von Pechmann and others.[2,11,12]

[1] Bougault, *Ann. chim. phys.*, (8) **15**, 491 (1908).
[2] von Pechmann, *Ber.*, **15**, 885 (1882).
[3] Bogert and Ritter, *J. Am. Chem. Soc.*, **47**, 526 (1925).
[4] Lutz, *J. Am. Chem. Soc.*, **52**, 3405, 3423 (1930).
[5] Rice, *J. Am. Chem. Soc.*, **52**, 2094 (1930).
[6] Lutz and Scott, *J. Org. Chem.*, **13**, 284 (1948).
[7] Koenigs and Wagstaffe, *Ber.*, **26**, 558 (1893).
[8] Bougault, *Compt. rend.*, **146**, 140 (1908).
[9] Tarbell, *J. Am. Chem. Soc.*, **60**, 215 (1938).
[10] Papa, Schwenk, Villani, and Klingsberg, *J. Am. Chem. Soc.*, **70**, 3356 (1948).
[11] Gabriel and Colman, *Ber.*, **32**, 395 (1899).
[12] Oddy, *J. Am. Chem. Soc.*, **45**, 2156 (1923).

BENZOYL CYANIDE

(Glyoxylonitrile, phenyl-)

$$C_6H_5COCl + CuCN \rightarrow C_6H_5COCN + CuCl$$

Submitted by T. S. Oakwood and C. A. Weisgerber.
Checked by R. L. Shriner and Charles R. Russell.

1. Procedure

In a 500-ml. distilling flask (Note 1) fitted with a thermometer extending to within 0.5 in. of the bottom are placed 110 g. (1.2 moles) of cuprous cyanide (Note 2) and 143 g. (118 ml., 1.02 moles) of purified benzoyl chloride (Note 3). The flask is shaken to moisten almost all the cuprous cyanide and is placed in an oil bath (Note 4) which has been previously heated to 145–150°. The temperature of the bath is raised to 220–230° and maintained between these limits for 1.5 hours. During the heating the flask is frequently removed from the bath (about every 15 minutes) and the contents are thoroughly mixed by vigorous shaking (Note 5). At the end of the 1.5 hours the flask is connected with an air-cooled condenser set for downward distillation. The temperature of the bath is slowly raised to 305–310°, and distillation is continued until no more product comes over (Note 6). About 100–112 g. of crude benzoyl cyanide boiling at 207–218°/745 mm. is obtained.

The crude benzoyl cyanide is purified by fractional distillation through a column (Note 7). The low-boiling material is taken off at a reflux ratio of 25–30 to 1 until the temperature reaches 208°; about 15 g. is collected. The benzoyl cyanide is collected at a reflux ratio of 1 to 1 at a temperature of 208–209°/745 mm. (bath temperature 260–280°). The distillate solidifies to colorless crystals which melt at 32–33°; the product weighs 80–86 g. (60–65%).

2. Notes

1. It is advisable to wrap the neck of the flask with asbestos paper or asbestos tape.

2. The cuprous cyanide is dried at 110° for 3 hours before use.

3. The commercial grade of benzoyl chloride may be purified as follows: 300 ml. (363 g.) of benzoyl chloride in 200 ml. of benzene is washed with two 100-ml. portions of cold 5% sodium bicarbonate solution. The benzene layer is separated, dried over calcium chloride, and distilled. After all the benzene has distilled, pure benzoyl chloride boiling at 196.8° (cor.)/745 mm. is collected. The recovery is 225 g.

4. Hydrogenated cottonseed oil, "Coto Flakes," obtainable from the Procter and Gamble Company, Cincinnati, Ohio, is suitable for the bath. A Wood's metal bath may also be used.

5. Upon addition of the benzoyl chloride to the cuprous cyanide thorough mixing by shaking is impossible. After the mixture is heated for about 30 minutes, the solid becomes granular and mixing is easily effected.

6. The distillate should be collected in the 200-ml. round-bottomed flask which is used in the subsequent fractionation. The distillation takes about an hour. The cake of copper salts in the flask is best removed by digestion with concentrated ammonium hydroxide solution.

7. The fractionating column was of the Whitmore-Lux type [1] and had about 14 theoretical plates. The packed section was 37 by 1.1 cm. (o.d.) and was packed with 3/32-in. single-turn glass helices. The dis-

tilling flask was a 200-ml. round-bottomed flask. A metal bath or the oil bath described in Note 4 may be used for heating.

3. Methods of Preparation

Benzoyl cyanide can be prepared by the thermal decomposition of ω-isonitrosoacetophenone,[2] from silver cyanide and benzoyl chloride,[3] from anhydrous hydrogen cyanide and benzoyl chloride in the presence of pyridine,[4] and by the thermal decomposition of phenylchloronitrocyanomethane.[5,6] It has been prepared by the vapor-phase reaction between hydrogen cyanide and benzoic anhydride in the presence of a granular catalyst (activated charcoal, alumina, or silica gel).[7]

[1] Whitmore and Lux, *J. Am. Chem. Soc.*, **54**, 3451 (1932).
[2] Claisen and Manasse, *Ber.*, **20**, 2195 (1887).
[3] Nef, *Ann.*, **287**, 307 (1895).
[4] Claisen, *Ber.*, **31**, 1024 (1898).
[5] Wislicenus and Shafer, *Ber.*, **41**, 4170 (1908).
[6] Staudinger and Kon, *Ann.*, **384**, 115 (1911).
[7] Brit. pat. 583,646 [*C. A.*, **41**, 2746 (1947)].

BENZOYLFORMIC ACID

(Glyoxylic acid, phenyl-)

$$C_6H_5COCN + 2H_2O + HCl \rightarrow C_6H_5COCO_2H + NH_4Cl$$

Submitted by T. S. Oakwood and C. A. Weisgerber.
Checked by R. L. Shriner and Charles R. Russell.

1. Procedure

In a 1-l. flask are placed 50 g. (0.38 mole) of benzoyl cyanide (p. 112) and 500 ml. of concentrated hydrochloric acid (sp. gr. 1.18). The mixture is shaken occasionally until the solid is dissolved completely and is then allowed to stand at room temperature for 5 days (Note 1). At the end of this time the clear yellow solution is poured into 2 l. of water and extracted with one 400-ml. portion and three 250-ml. portions of ether. The ether is removed by distillation from a steam bath, and the residual oil is placed in a vacuum desiccator containing phosphorus pentoxide and solid sodium hydroxide and allowed to remain there until dry (Note 2). The yield of crude solid acid melting from 57° to 64° is about 55–56 g. (96–98%). The crude acid is dissolved in

750 ml. of hot carbon tetrachloride, and 2 g. of Norit is added (Note 3). The solution is filtered and allowed to cool to room temperature and then cooled in an ice-water mixture until crystallization is complete. The solid acid is filtered with suction, and the solvent remaining on the crystals is removed by placing the product in a vacuum desiccator for about 2 days. The yield of slightly yellow benzoylformic acid melting at 64–66° is 42–44 g. (73–77%) (Note 4).

2. Notes

1. With occasional shaking about a day is necessary for complete solution of the solid benzoyl cyanide. A yellow oil separates which dissolves on shaking. At the end of this time some ammonium chloride occasionally separates.

2. If the oil does not crystallize it is cooled in an ice-water mixture until solidification is complete. The solid is returned to the desiccator for complete drying. The desiccator must be evacuated slowly or spattering will take place.

3. If decolorizing carbon is not used, the acid tends to separate as an oil which solidifies on cooling.

4. Titration with standard sodium hydroxide solution showed the acid to be about 99% pure.

3. Methods of Preparation

Benzoylformic acid can be prepared by the oxidation of acetophenone with potassium permanganate in alkaline solution,[1] by the oxidation of mandelic acid with potassium permanganate in alkaline solution,[2,3] and by the hydrolysis of benzoyl cyanide with concentrated hydrochloric acid.[4]

[1] Claus and Neukranz, *J. prakt. Chem.* (2) **44**, 80 (1891).
[2] Acree, *Am. Chem. J.*, **50**, 391 (1913).
[3] *Org. Syntheses* Coll. Vol. **1**, 241 (1941).
[4] Boeseken and Felix, *Rec. trav. chim.*, **40**, 569 (1921).

BENZOYL DISULFIDE

$$KOH + H_2S \rightarrow KSH + H_2O$$

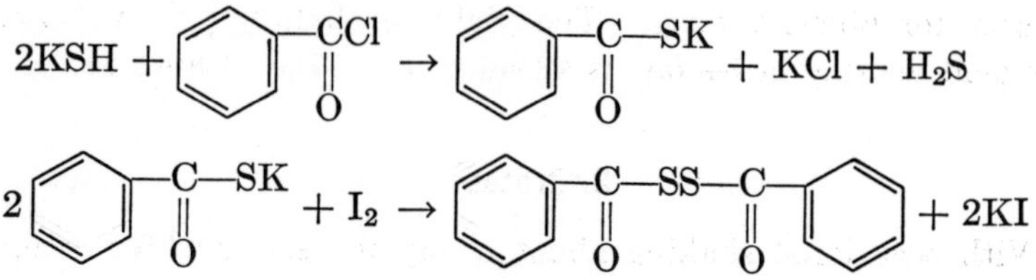

Submitted by ROBERT L. FRANK and JAMES R. BLEGEN.[1]
Checked by ARTHUR C. COPE and FRANK S. FAWCETT.

1. Procedure

A solution of 315 g. (4.76 moles) of potassium hydroxide in 3150 ml. of commercial absolute ethanol is prepared with mechanical stirring in a 5-l. three-necked round-bottomed flask. The flask is fitted with a 500-ml. dropping funnel and a gas inlet tube extending to the bottom of the flask, and hydrogen sulfide is passed in through the inlet tube with stirring until the solution is saturated and no longer gives an alkaline reaction with phenolphthalein (Notes 1, 2). The mixture is cooled to 10–15° by means of an ice bath, and 346.5 g. (2.46 moles) of redistilled benzoyl chloride is introduced dropwise with stirring while the temperature is kept below 15°. The potassium chloride which precipitates during the addition (Note 3) is separated by filtration with suction through a Büchner funnel and is washed with about 200 ml. of ethanol (Note 4). The filtrate and washings are cooled to 10–15°, and solid iodine is added slowly, with constant agitation, until a slight excess is present, as shown by a faint permanent coloration of the solution. The amount of iodine required varies from 336 g. to 407 g. (1.32–1.61 moles) (Note 5). Benzoyl disulfide precipitates during the addition. It is collected on a filter and washed with 750 ml. of 95% ethanol followed by 3 l. of water. The crude product, after drying at room temperature, or in an oven at a temperature which does not exceed 60° (Note 6), weighs 325–333 g.

The crude material is dissolved with stirring in 910 ml. of ethylene chloride heated to 60° in a water bath (Note 6). The solution is allowed to cool to room temperature, 122 ml. of a saturated aqueous solution of sodium bicarbonate is added to the resulting slurry, and the mixture is stirred for 1 hour (Note 7). The layers are then separated in a separatory funnel, and the ethylene chloride slurry is heated to 60° in a water bath. The resulting solution is filtered through a small cotton plug in a preheated funnel. Absolute ethanol (313 ml.) is added to the filtrate, and the mixture is stored in an icebox overnight while the product crystallizes. The crystals are collected on a filter and washed with 40 ml. of ether. The product is recrystallized by dissolving it in ethylene chloride (3.0 ml. per g. of product) heated to 60° in a water bath (Note 6) and cooling. The yield of white to light pink plates, m.p. 129–130°, is 230–246 g. (68–73%).

2. Notes

1. It is convenient during this step to use the dropping funnel as an exhaust tube by attaching its upper end to a gas-absorption trap. If this is not done, the preparation should be conducted in a well-ventilated hood.

2. The gas inlet tube should be of moderately large diameter or it may become plugged with crystals during the later stages of the saturation with hydrogen sulfide.

3. Stirring is discontinued after the benzoyl chloride has been added. The precipitate of potassium chloride can be separated more easily if the mixture is allowed to stand overnight before filtration.

4. After the filtration and washing, the application of suction is continued until bubbles of hydrogen sulfide no longer form in the filtrate. Much of the excess hydrogen sulfide is removed during the filtration.

5. The amount of iodine required is presumed to vary according to the presence of variable amounts of hydrogen sulfide and to the extent of oxidation by atmospheric oxygen. In one unsatisfactory preparation in which a relatively large amount of hydrogen sulfide must have remained in the solution a total of 493 g. (1.94 moles) of iodine was required and the final product contained free sulfur.

6. If the temperature exceeds 60°, discoloration occurs and the product cannot be decolorized by recrystallization or treatment with activated charcoal.

7. This operation is carried out at room temperature because heating in the presence of alkalies decomposes benzoyl disulfide.

3. Methods of Preparation

Benzoyl disulfide has been obtained by the reaction of benzoyl chloride with hydrogen sulfide,[2] hydrogen disulfide,[3] hydrogen trisulfide,[3,4] potassium sulfide,[5] sodium disulfide,[6] lead sulfide,[7] sodium hydrosulfite,[8] sodium thiosulfate,[9] sulfhydrylmagnesium bromide,[10] and thiobenzamide.[11] It is also formed by reaction of benzoic anhydride with hydrogen sulfide.[12] The better preparative methods involve the oxidation of thiobenzoic acid by means of air,[7,13,14] hydrogen peroxide,[15,16] or sulfur monochloride,[17] or of the sodium or potassium salt by means of air,[16,18] chlorine,[19] iodine,[7a,14,15,20,21] copper sulfate,[7,16] potassium ferricyanide,[7,16,22] or ferric chloride.[7,16]

[1] Work done under contract with the Office of Rubber Reserve.

[2] Szperl, *Roczniki Chem.*, **10**, 510 (1930) [*C. A.*, **25**, 503 (1931)].

[3] Bloch and Bergmann, *Ber.*, **53**, 961 (1920).

[4] Bergmann, Dreyer, and Radt, *Ber.*, **54**, 2139 (1921).

[5] Bergmann, *Ber.*, **53**, 979 (1920).

[6] Braker, U. S. pat. 2,154,488 [*C. A.*, **33**, 5415 (1939)].

[7] (*a*) Engelhardt, Latschinoff, and Malyscheff, *Z. Chem.*, **4**, 353 (1868); (*b*) Engelhardt and Latschinoff, *Z. Chem.*, **4**, 455 (1868).

[8] Binz and Marx, *Ber.*, **40**, 3855 (1907).

[9] Westlake and Dougherty, *J. Am. Chem. Soc.*, **67**, 1861 (1945).

[10] Mingoia, *Gazz. chim. ital.*, **55**, 718 (1925).

[11] Musajo and Amoruso, *Gazz. chim. ital.*, **67**, 301 (1937).

[12] Mosling, *Ann.*, **118**, 303 (1861).

[13] Cloëz, *Ann.*, **115**, 23 (1860).

[14] Weigert, *Ber.*, **36**, 1007 (1903).

[15] Moness, Lott, and Christiansen, *J. Am. Pharm. Assoc., Sci. Ed.*, **25**, 397 (1936).

[16] Shelton and Rider, *J. Am. Chem. Soc.*, **58**, 1282 (1936).

[17] Chakravarti, *J. Chem. Soc.*, **1923**, 964.

[18] Rider and Shelton, U. S. pat. 2,028,246 [*C. A.*, **30**, 1811 (1936)].

[19] Bergmann and Bloch, *Ber.*, **53**, 977 (1920).

[20] Amberg and Brunsting, *Military Surgeon*, **88**, 617 (1941) [*C. A.*, **35**, 5643 (1941)].

[21] Bloch, *Compt. rend.*, **204**, 1342 (1937).

[22] Fromm and Palma, *Ber.*, **39**, 3317 (1906); Fromm and Schmoldt, *Ber.*, **40**, 2861 (1907).

BENZYL CHLOROMETHYL KETONE

(2-Propanone, 1-chloro-3-phenyl-)

$$CH_3N(NO)COOC_2H_5 + 2KOH \rightarrow CH_2N_2 + K_2CO_3 + C_2H_5OH + H_2O$$

$$C_6H_5CH_2COCl + 2CH_2N_2 \rightarrow C_6H_5CH_2COCHN_2 + CH_3Cl + N_2$$

$$C_6H_5CH_2COCHN_2 + HCl \rightarrow C_6H_5CH_2COCH_2Cl + N_2$$

Submitted by Warren D. McPhee and Erwin Klingsberg.
Checked by Nathan L. Drake and Marvin Schwartz.

1. Procedure

Warning: Diazomethane is poisonous and explosive.

Twelve grams (0.18 mole) of ground potassium hydroxide is dissolved in 45 ml. of *n*-propyl alcohol in a three-necked 500-ml. flask set on a steam bath so that the mixture can be warmed. Two necks of the flask are fitted with dropping funnels. A Vigreux fractionating column (25 cm. in length and 12 mm. in diameter) is inserted in the third neck of the flask. The column is connected, through a condenser and adapter, to a filter flask set in ice as the receiver. The side arm of the receiver is fitted with a drying tube. About 1.5 hours is required to dissolve the hydroxide, whereupon an additional amount of alcohol is added to bring the total volume to 50 ml. The solution is then cooled, and 100 ml. of absolute ether is added. Through one dropping funnel is added dropwise 20 ml. of nitrosomethylurethan [1] (equivalent to 0.10 mole of diazomethane; Note 1). The flask is warmed on a steam bath so that diazomethane distils over with ether while the nitrosomethylurethan is being added. The addition requires about 20 minutes. During this operation more dry ether is added as needed, through the second dropping funnel, until the total volume of ether distillate is 250 ml. When the ether distilling over is colorless, the distillate is transferred to another three-necked 500-ml. flask, equipped with a mechanical stirrer, a dropping funnel, and a condenser. To the stirred solution is added, over a period of 15 minutes, a solution of 6.6 ml. (7.7 g., 0.050 mole) of phenylacetyl chloride in 10 ml. of absolute ether. A copious evolution of nitrogen accompanies this operation.

After standing for 2 hours, the solution of diazoketone is cooled in ice and treated with dry gaseous hydrogen chloride until the passage of the gas no longer causes evolution of nitrogen (Note 2). At the end of this time, the original yellow color of the solution has changed to an orange-red. About 100 ml. of water is slowly and cautiously added to the ether solution in order to cause a separation into two layers. An additional 50 ml. of water is then added to give a more complete separation. The ether layer is then washed with two 50-ml. portions of a 5% solution of sodium carbonate in water. The ether solution is dried with anhydrous calcium chloride or calcium sulfate, the solvent evaporated, and the product distilled. The yield of benzyl chloromethyl ketone, b.p. 133–135°/19 mm. or 96–98°/1 mm., is 7.0–7.1 g. (83–85%) (Notes 3 and 4).

2. Notes

1. The yield of diazomethane from nitrosomethylurethan is usually given as 0.005 mole per ml.[2] Another method for the preparation of diazomethane has been described.[3]

2. This usually requires about 25 minutes. In certain runs, the hydrogen chloride caused the deposition of a dark red flocculent solid, which was removed by filtration. The yield was then 5–10% lower. This can apparently be avoided by waiting 2 hours, as directed, and by cooling the solution before the hydrogen chloride treatment.

3. This method is general for chloromethyl ketones. Yields of 75–85% may be obtained using hydrocinnamoyl chloride or diphenylacetyl chloride.

4. *Diazomethane is poisonous and explosive.*[4] *The preparation should be carried out in a good hood.*

3. Method of Preparation

Benzyl chloromethyl ketone has been prepared by the reaction of diazomethane with phenylacetyl chloride. The method of Clibbens and Nierenstein,[5] in which one equivalent of diazomethane is added to the acyl chloride and the chloromethyl ketone obtained directly, could not be duplicated by Bradley and Schwarzenbach[6] or by the submitters.

[1] *Org. Syntheses* Coll. Vol. **2**, 464 (1943).

[2] Fieser, *Experiments in Organic Chemistry,* 2nd ed., p. 377, D. C. Heath & Company, Boston, 1941.

[3] *Org. Syntheses,* **25**, 28 (1945).

[4] Bachmann, *Organic Reactions,* **1**, 47 (1942), John Wiley & Sons, New York.

[5] Clibbens and Nierenstein, *J. Chem. Soc.,* **107**, 1491 (1915).

[6] Bradley and Schwarzenbach, *J. Chem. Soc.,* **1928**, 2904.

BIALLYL

(1,5-Hexadiene)

$$2CH_2{=}CHCH_2Cl + Mg \rightarrow CH_2{=}CH(CH_2)_2CH{=}CH_2 + MgCl_2$$

Submitted by Amos Turk and Henry Chanan.
Checked by Arthur C. Cope and Frank S. Fawcett.

1. Procedure

In a 5-l. three-necked flask fitted with a mercury-sealed stirrer (Note 1), a dropping funnel, and an efficient reflux condenser protected by a calcium chloride drying tube is placed 82 g. (3.5 gram atoms) of magnesium turnings. A solution of 459 g. (6 moles) of dry, freshly distilled allyl chloride in 2.4 l. of anhydrous ether is added to the flask through the dropping funnel in the following manner: A 100- to 200-ml. portion of the solution and a small crystal of iodine are added, and the mixture is warmed, if necessary, until the reaction starts. The remainder of the solution is added with stirring and cooling in an ice bath as rapidly as possible without loss of material through the condenser (Note 2). By sponging the upper part of the flask with ice water from the cooling bath, the addition can be completed in 1–1.5 hours. When the addition is complete, the thick slurry is allowed to stand at room temperature for 5 hours with stirring for as much of that period as is practicable (Note 1). The flask is again cooled in an ice bath, and a cold 5% solution of hydrochloric acid is added through the dropping funnel until the evolution of heat has practically ceased (Note 2) and the magnesium chloride is in solution. The mechanical stirrer is started again when the mixture becomes sufficiently fluid (Note 1).

The contents of the flask are transferred to a separatory funnel; the ether layer is separated and distilled without washing or drying through a small packed column (Note 3) until the distillation temperature begins to rise (38–40°). The residue is transferred to a separatory funnel, washed with two 500-ml. portions of water, dried over 10 g. of calcium chloride, and fractionated through the small packed column. After distillation of ether and some allyl chloride (b.p. 45°) biallyl is collected as a colorless liquid in a yield of 135–160 g. (55–65%), b.p. 59–60°/760 mm.; n_D^{20} 1.4040; n_D^{25} 1.4012.

2. Notes

1. Efficient stirring is essential during the early part of the reaction. The submitters used a double-loop-type Hershberg wire (Nichrome, Chromel, or tantalum) stirrer [1] and a motor [2] powerful enough to stir the mixture during the entire preparation. The checkers used the simpler Hershberg wire (No. 16 B and S gauge Chromel or stainless-steel) stirrer [3] and an ordinary good laboratory motor. Although the slurry became so thick that it could not be stirred with this equipment, the yield of pure biallyl obtained equaled that reported by the submitters.

2. Care must be taken to avoid loss of material (and reduction in yield) through evaporation of allyl chloride or biallyl, both of which are very volatile.

3. A simple total-condensation partial take-off column with a 2.2 by 25 cm. section packed with $\frac{3}{32}$-in. single-turn glass helices was used with a reflux ratio of 7 or 8 to 1.

3. Methods of Preparation

This procedure is a modification of one described by Cortese.[4] Allyl chloride is employed rather than allyl bromide because of its low cost. Biallyl has been prepared by the action of sodium [5] or aluminum [6] on allyl iodide; from allyl mercuric iodide by dry distillation [7] or by the action of potassium cyanide solution; [8] by the action of magnesium on allyl bromide,[4,9] allyl chloride,[4,10] allyl iodide,[11] or 1,2,3-tribromopropane; [12] and from diallyl ether by the action of sodium.[13]

[1] *Org. Syntheses* Coll. Vol. **2**, 117 (1943).

[2] Hershberg, *Ind. Eng. Chem., Anal. Ed.,* **12,** 293 (1940).

[3] Hershberg, *Ind. Eng. Chem., Anal. Ed.,* **8,** 313 (1936).

[4] Cortese, *J. Am. Chem. Soc.,* **51,** 2266 (1929).

[5] Berthelot and Luca, *Ann. chim. et phys.,* (3) **48,** 294 (1856).

[6] Domanitzkii, *J. Russ. Phys. Chem. Soc.,* **46,** 1078 (1914) [*C. A.,* **9,** 1899 (1915)].

[7] Linnemann, *Ann.,* **140,** 180 (1866).

[8] Oppenheim, *Ber.,* **4,** 670 (1871)

[9] Lespieau, *Ann. chim. et phys.,* (8), **27,** 149 (1912); Gilman and McGlumphy, *Bull. soc. chim. France,* **43,** 1322 (1928).

[10] Henne, Chanan, and Turk, *J. Am. Chem. Soc.,* **63,** 3474 (1941).

[11] Meisenheimer and Casper, *Ber.,* **54,** 1655 (1921).

[12] Krestinskii, *J. Russ. Phys. Chem. Soc.,* **58,** 1078 (1926) [*C. A.,* **22,** 1324 (1928)].

[13] U. S. pat. 2,405,347 [*C. A.,* **41,** 148 (1947)].

BROMOACETAL

(Acetaldehyde, bromo-, diethyl acetal)

$$CH_3CO_2CH{=}CH_2 + Br_2 \rightarrow CH_3CO_2CHBrCH_2Br$$

$$CH_3CO_2CHBrCH_2Br + C_2H_5OH \rightarrow BrCH_2CHO + CH_3CO_2C_2H_5 + HBr$$

$$BrCH_2CHO + 2C_2H_5OH \xrightarrow{HBr} BrCH_2CH(OC_2H_5)_2 + H_2O$$

Submitted by S. M. McElvain and D. Kundiger.
Checked by R. L. Shriner and C. H. Tilford.

1. Procedure

An apparatus is assembled as shown in Fig. 5. The 3-l. three-necked round-bottomed flask *A* is equipped with a mechanical stirrer sealed

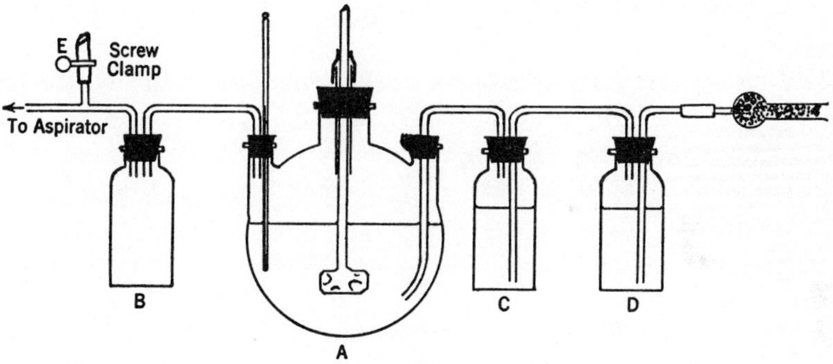

Fig. 5.

with a well-lubricated rubber sleeve. In one neck of the flask are fitted a thermometer and a glass tube leading through a safety trap *B* to a water pump. In the other neck a 7-mm. glass tube, extending to the bottom of the flask, is attached to a 500-ml. bottle *C* in which is placed 255 ml. (5 moles) of bromine (Note 1). This bottle is connected to a 500-ml. wash bottle *D* containing 250 ml. of sulfuric acid. The inlet tube of *D* is connected to a calcium chloride tube.

A solution of 430 g. (5 moles) of vinyl acetate (Note 2) in 1.5 l. (26 moles) of absolute ethanol is placed in flask *A*. The solution is cooled to about −10° by an ice-salt mixture, and stirring is started. Gentle suction is applied at the outlet tube of *B*, and the bromine is introduced into *A* by a rapid current of air. The rate of introduction of the bromine, controlled by adjustment of the clamp *E*, should be regulated so that 8–10 hours is required to volatilize all the bromine. Stirring is stopped, and the reaction mixture is allowed to stand overnight and to come to the temperature of the room. The mixture is poured into 1.7 l. of ice water (Note 3); the lower layer of bromoacetal and ethyl acetate is separated (Note 4), washed twice with 300-ml. portions of cold water and once with 300 ml. of cold 10% sodium carbonate solution, and dried over two successive 25-g. portions of anhydrous calcium chloride for 30 minutes. The crude product weighs 990–1010 g.; it is purified by distillation under diminished pressure (water pump) through a 6-in. Widmer column or a 20-cm. Vigreux column. The first fraction consists of ethyl acetate; this is followed by the pure bromoacetal which boils at 62–63°/15 mm. (84–85°/30 mm.) and which amounts to 610–625 g. (62–64%) (Note 5).

2. Notes

1. The bromine is previously washed with 100 ml. of concentrated sulfuric acid.

2. The vinyl acetate (Eastman Kodak Company) is distilled from the added preservative; the first turbid portion containing water is discarded, and the fraction boiling at 69–71°/740 mm. is used.

3. If an emulsion forms at this point, 320 g. of hydrated sodium sulfate is added.

4. Bromoacetal is a fairly strong lachrymator and is best handled under a hood.

5. A yield of 78% of bromoacetal has been obtained by a slight modification of the procedure described above. One-half of the amounts of materials specified above were used, and, after the bromination was complete, the reaction mixture was allowed to stand for 64 hours before it was processed. The yield of bromoacetal was 380 g.[1]

3. Methods of Preparation

The procedure given above is essentially a large-scale adaptation of that of Filachione.[2] Bromoacetal has been prepared by the bromination of acetal directly,[3] or in the presence of calcium carbonate;[4] by action of sodium ethoxide on α,β-dibromodiethyl ether;[5] by bromination of paraldehyde followed by action of ethanol;[6,7] by the action of ethanol on bromoacetaldehyde; [8,9] and by bromination of acetal with N-bromosuccinimide.[10]

[1] Private communication, C. F. H. Allen.
[2] Filachione, *J. Am. Chem. Soc.*, **61,** 1705 (1939).
[3] Pinner, *Ber.*, **5,** 149 (1872).
[4] Fischer and Landsteiner, *Ber.*, **25,** 2551 (1892).
[5] Wislicenus, *Ann.*, **192,** 112 (1878).
[6] Freundler and Ledru, *Bull. soc. chim. France,* (4) **1,** 75 (1907).
[7] Wizinger and Al-Attar, *Helv. Chim. Acta,* **30,** 189 (1947).
[8] Rotbart, *Ann. chim.*, (11) **1,** 451 (1934).
[9] Shchukina, *J. Gen. Chem. U.S.S.R.*, **18,** 1653 (1948).
[10] Marvell and Joncich, *J. Am. Chem. Soc.*, **73,** 973 (1951).

α-BROMOBENZALACETONE

(3-Buten-2-one, 3-bromo-4-phenyl-)

$$C_6H_5CHBrCHBrCOCH_3 + CH_3CO_2Na \rightarrow C_6H_5CH{=}CBrCOCH_3 + CH_3CO_2H + NaBr$$

Submitted by Norman H. Cromwell, Donald J. Cram, and Chas. E. Harris.
Checked by R. L. Shriner and William O. Foye.

1. Procedure

Warning: Precautions must be taken to avoid contact with α-bromobenzalacetone since it is a skin irritant (*Note* 1).

In a 500-ml. round-bottomed flask fitted with a reflux condenser are placed 100 g. (0.33 mole) of benzalacetone dibromide (p. 105), 30 g. (0.37 mole) of anhydrous sodium acetate, and 250 ml. of 95% ethanol, and the mixture is refluxed vigorously for 4 hours in the absence of direct sunlight. The precipitate of sodium bromide is removed by filtration, and the alcohol is removed from the filtrate by distillation under reduced pressure (Note 2). The residual salt-oil mixture is extracted with two 50-ml. portions of ether, and the ether solution is transferred to a 250-ml. separatory funnel (*Caution! Note* 1).

The ether solution is washed thoroughly six times with 25-ml. portions of saturated sodium chloride solution and twice with 25-ml. portions of 5% sodium bicarbonate solution (Note 3). The ether layer is allowed to dry over anhydrous sodium sulfate at room temperature for 24 hours. The ether is removed by distillation, and the residual oil is distilled from a Claisen flask under reduced pressure, using an oil bath. A yield of 47–54 g. (64–73%) of a pale yellow oil, boiling at 114–117°/1 mm. (Note 4), is obtained. On cooling, the oil crystallizes; m.p. 30–31°. The product is stored in a dark bottle in the ice chest (Notes 5 and 6).

2. Notes

1. α-Bromobenzalacetone or its solutions cause the formation of red spots on the skin. After several days these form large red blisters that are painful and take several days to heal. The affected parts should be treated with a mixture of peanut-oil and glycerol containing a little ammonia.

2. The reduced pressure produced by a water pump is satisfactory. The flask is warmed with a hot water bath (40–50°).

3. It is necessary that the product be entirely free from acetic acid before it is distilled in order to obtain the yields stated.

4. Boiling points at other pressures are: 136–138°/4 mm.; 150–151°/10 mm.

5. When stored in this manner the product is quite stable and darkens only slightly after 9 months.

6. The analogous α-bromobenzalacetophenone may be prepared by a similar procedure. In a 1-l. three-necked round-bottomed flask fitted with a mercury-sealed stirrer and a reflux condenser are placed 150 g. (0.41 mole) of benzalacetophenone dibromide,[1] 41 g. (0.50 mole) of anhydrous sodium acetate, and 250 ml. of 95% ethanol. The mixture is stirred and refluxed for 5 hours and then worked up in the same manner as described above for α-bromobenzalacetone. Distillation gives a yield of 100–110 g. (85–94%) of a pale yellow oil, boiling at 170–173°/1 mm. On cooling, the oil crystallizes and melts at 42–44°. This product should also be stored in a dark bottle in the ice chest, but it is more stable and darkens less on standing than the analogous α-bromobenzalacetone. This product is less irritating to the skin than α-bromobenzalacetone (Note 1).

3. Methods of Preparation

α-Bromobenzalacetone has been prepared from benzalacetone dibromide by heating with alcoholic potassium hydroxide[2] or with sodium acetate[3] solutions. α-Bromobenzalacetophenone is prepared by a similar procedure from benzalacetophenone dibromide.[4]

[1] *Org. Syntheses* Coll. Vol. **1**, 205 (1941).
[2] Watson, *J. Chem. Soc.*, **85**, 464 (1904).
[3] Cromwell and Cram, *J. Am. Chem. Soc.*, **65**, 305 (1943).
[4] Wislicenus, *Ann.*, **308**, 226 (1899); Cromwell, *J. Am. Chem. Soc.*, **62**, 2899 (1940).

α-BROMOHEPTALDEHYDE

(Enanthaldehyde, α-bromo-)

$$CH_3(CH_2)_4CH_2CHO + (CH_3CO)_2O \xrightarrow{CH_3COOK} CH_3(CH_2)_4CH{=}CHOCOCH_3 + CH_3COOH$$

$$CH_3(CH_2)_4CH{=}CHOCOCH_3 + Br_2 \rightarrow CH_3(CH_2)_4CHBrCHBrOCOCH_3$$

$$CH_3(CH_2)_4CHBrCHBrOCOCH_3 + 3MeOH \rightarrow CH_3(CH_2)_4CHBrCH(OMe)_2 + CH_3OCOCH_3 + HBr$$

$$CH_3(CH_2)_4CHBrCH(OMe)_2 + H_2O \xrightarrow{HCl} CH_3(CH_2)_4CHBrCHO + 2MeOH$$

Submitted by Paul Z. Bedoukian.
Checked by Joseph A. Pappalardo and Charles C. Price.

1. Procedure

Caution! Since most of the reactants are of unpleasant odor or have lachrymatory effects it is best to use a good hood.

A. *Heptaldehyde enol acetate.* A mixture of 285 g. (335 ml., 2.5 moles) of heptaldehyde (Note 1), 612 g. (566 ml., 6 moles) of acetic anhydride, and 49 g. (0.5 mole) of powdered potassium acetate is placed in a 2-l. flask fitted with a reflux condenser. The flask is heated in an oil bath kept at 155–160° for 1 hour. The mixture is then allowed to cool, placed in a 2-l. separatory funnel, washed several times with warm water (Note 2) to remove the excess acetic anhydride, and finally washed with 5% sodium carbonate solution (Note 3). The resultant oil is fractionated under reduced pressure through an efficient column (Note 4). The initial fraction consists of pure heptaldehyde followed by heptaldehyde containing heptaldehyde enol acetate. The fraction boiling at 88–90°/17 mm. is pure enol acetate, n_D^{25} 1.4295–1.4305; d_4^{25} 0.880–0.884. The yield is 175–195 g. (45–50%) (Note 5).

B. *α-Bromoheptaldehyde dimethyl acetal.* A solution of 156 g. (177 ml., 1 mole) of the enol acetate and 200 ml. of carbon tetrachloride is placed in a 1-l. flask and cooled in an ice-water bath. A mixture of 160 g. (51 ml., 1 mole) of bromine and 50 ml. of carbon tetrachloride is added slowly through a buret, the flask being constantly shaken and the rate of addition so controlled as not to allow the temperature of the brominated mixture to rise above 10° (Note 6). The addition of bromine takes from 20 minutes to 1 hour, and the end point is reached when the calculated amount is absorbed and the bromine is no longer decolorized. The brominated mixture is added to 600 ml. of anhydrous methanol (Note 7) and allowed to stand for 48 hours or longer. At the end of this period the mixture is diluted with 2 l. of water and the separated oil (lower layer) is washed with 1 l. of water and finally with 1 l. of 5% sodium carbonate (Note 8). The carbon tetrachloride and methyl acetate are removed by distillation at atmospheric pressure. The residual oil is then distilled under reduced pressure in the presence of a small amount of sodium carbonate. The fraction boiling at 117–119°/17 mm. is collected as pure α-bromoheptaldehyde dimethyl acetal, n_D^{25} 1.4510–1.4520; d_4^{25} 1.180–1.195. The yield is 191–203 g. (80–85%) (Note 9).

C. *α-Bromoheptaldehyde.* A mixture of 119.5 g. (100 ml., 0.5 mole) of α-bromoheptaldehyde dimethyl acetal and 80 ml. of concentrated hydrochloric acid is boiled gently in a 250-ml. distilling flask, and the methanol liberated is removed by distillation, which is continued slowly until the vapor temperature reaches 90°, at which point the heating is stopped and the residue and distillate are combined and diluted with 200 ml. of water. The somewhat brownish oil which separates is distilled under reduced pressure from a 250-ml. Claisen flask. The yield of pure α-bromoheptaldehyde, boiling at 87–92°/17 mm., n_D^{25} 1.4580–1.4600, d_4^{25} 1.210–1.230, is 87–92.5 g. (90–95%).

2. Notes

1. The heptaldehyde should be a freshly distilled product boiling at 151.5–153.5°.

2. Decomposition of the excess acetic anhydride takes place very slowly and with difficulty unless warm water is used. The checkers used three portions of wash water at 40–50°, totaling 1.3 l.

3. The checkers found that five or six portions of 5% sodium car-

bonate totaling 5 l. were required before rapid carbon dioxide evolution ceased.

4. The reaction mixture consists of unchanged heptaldehyde, heptaldehyde enol acetate, heptaldehyde diacetate, and a small amount of polymerized material. The proportion of free heptaldehyde and heptaldehyde diacetate depends upon the time of heating, longer periods of heating favoring the formation of the diacetate. An efficient fractionating column, preferably of the Whitmore-Fenske type, should be used in order to obtain the enol acetate free of the heptaldehyde and heptaldehyde diacetate impurities. The checkers used a Whitmore-Fenske type column of about six theoretical plates.

5. The residue consists largely of heptaldehyde diacetate, which when slowly distilled at atmospheric pressure partially decomposes to acetic anhydride and heptaldehyde. In this manner, 50–60% of the available heptaldehyde is recovered from the residue.

6. The checkers recommend a mechanical stirrer to avoid danger of contact with bromine.

7. Commercial methanol (99.5–100%) was used in all experiments.

8. The acetal must be free of acid; otherwise decomposition takes place during distillation.

9. The pure bromoacetal is a stable, colorless liquid, of mild odor. It may be kept indefinitely when stored in a dark bottle over a small amount of anhydrous sodium carbonate.

3. Methods of Preparation

The procedure described is an example of a general method of preparation of α-bromoaldehydes.[1] α-Bromoheptaldehyde has been prepared by the bromination of heptaldehyde diethyl acetal with phosphorus trichlorodibromide[2] and by direct bromination of heptaldehyde trimer with subsequent treatment with alcohol.[3]

[1] Bedoukian, *J. Am. Chem. Soc.*, **66**, 1325 (1944).

[2] Kirmann, *Compt. rend.*, **184**, 525 (1927); *Ann. chim.*, (10) **11**, 223 (1929).

[3] Dworzak and Pfifferling, *Monatsh.*, **48**, 251 (1927).

3-BROMO-4-HYDROXYTOLUENE

(*p*-Cresol, 2-bromo-)

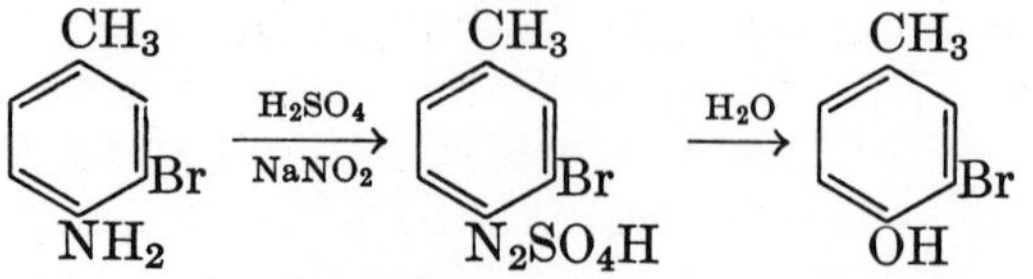

Submitted by H. E. Ungnade and E. F. Orwoll.
Checked by C. F. H. Allen and Alan Bell.

1. Procedure

In a 2-l. beaker is placed 75 g. (0.4 mole) of 3-bromo-4-aminotoluene (Note 1), and to it is added the hot diluted acid obtained by adding 72 ml. of concentrated sulfuric acid to 200 ml. of water. The clear solution is stirred and cooled to about 15°, after which 180 g. of ice is added; the amine sulfate usually separates. As soon as the temperature has dropped below +5°, a solution of 32.2 g. (0.47 mole) of sodium nitrite in 88 ml. of water is added from a dropping funnel, the stem of which extends below the surface of the liquid. The temperature of the solution is kept below +5° during the addition, which requires about 15 minutes. The solution is stirred for 5 minutes after the addition of all the sodium nitrite, and 300 g. of cold water, 3 g. of urea, and 300 g. of cracked ice are then added successively. The solution is kept in an ice bath until used.

A 1-l. Claisen flask fitted with a dropping funnel and a thermometer dipping into the liquid is attached to a condenser set for downward distillation. In the flask are placed 150 g. of anhydrous sodium sulfate, 200 g. (108 ml.) of concentrated sulfuric acid, and 100 ml. of water. The flask is heated over a wire gauze, and while the internal temperature is maintained at 130–135°, the diazonium solution, in 25-ml. portions, is added at the same rate as the distillate is collected (Note 2). When this operation has been completed, 200 ml. of water, in 25-ml. portions, is introduced and the distillation is continued until an additional 200 ml. of distillate has been collected. The complete distillation requires 3–3.5 hours.

The distillate is extracted with two 150-ml. portions of ether, and the combined extracts are washed successively with 100 ml. of water and 150 ml. of 10% sodium bicarbonate solution. The phenol is then extracted from the ether layer by use of one 200-ml. and two 50-ml. portions of 10% sodium hydroxide solution. The combined alkaline solutions are acidified, with cooling, by the addition of 100 ml. of concentrated hydrochloric acid. The phenol is extracted with one 200-ml. and two 100-ml. portions of ether, and the combined extracts are washed with 100 ml. of water and dried over 50 g. of anhydrous sodium sulfate. The mixture is filtered, and the ether is removed from the filtrate by distillation on a water bath. The residue, 65–72 g. of a brown oil (Note 3), is then distilled from a Claisen flask with modified side arm; b.p. 102–104°/20 mm. The yield is 60–69 g. (80–92%) (Note 4).

2. Notes

1. The 3-bromo-4-aminotoluene may be purchased, or prepared according to *Org. Syntheses* Coll. Vol. **1,** 111 (1941). The material used melted at 12–13°.

2. If the diazonium solution is added too rapidly, and the temperature of the liquid falls, the addition is interrupted until the temperature again exceeds 130°.

3. This crude product serves for most purposes, e.g., conversion to *o*-bromo-*p*-methylanisole.

4. The procedure has been used by the submitters for the conversion of *m*-bromoaniline into *m*-bromophenol, b.p. 234–237°/742 mm.; 105–107°/11 mm. The yield was 66%.

3. Methods of Preparation

o-Bromo-*p*-cresol has been prepared by the direct bromination of *p*-cresol in chloroform solution;[1,2] by bromination of dry sodium *p*-cresoxide;[3] and by decomposition of the diazonium sulfate of 3-bromo-4-aminotoluene;[4] the last is a general method.[5,6]

1 Vogt and Henninger, *Ber.,* **15,** 1081 (1882).
2 Zincke and Wiederhold, *Ann.,* **320,** 202 (1902).
3 Schall and Dralle, *Ber.,* **17,** 2530 (1894).
4 Cain and Norman, *J. Chem. Soc.,* **89,** 24 (1906).
5 Kalle and Company, Ger. pat. 95,339 [*Frdl.,* **4,** 124 (1894–1897)].
6 Niemann, Mead, and Benson, *J. Am. Chem. Soc.,* **63,** 609 (1941).

6-BROMO-2-NAPHTHOL

(2-Naphthol, 6-bromo-)

$$C_{10}H_7OH + 2Br_2 \rightarrow C_{10}H_5(OH)Br_2(2,1,6) + 2HBr$$

$$C_{10}H_5(OH)Br_2 + HBr + Sn \rightarrow C_{10}H_6(OH)Br(2,6) + SnBr_2$$

Submitted by C. Frederick Koelsch.
Checked by W. E. Bachmann and S. Kushner.

1. Procedure

In a 3-l. round-bottomed flask fitted with a dropping funnel and a reflux condenser (Note 1) are placed 144 g. (1 mole) of β-naphthol and 400 ml. of glacial acetic acid. Through the dropping funnel is then added a solution of 320 g. (2 moles) of bromine in 100 ml. of acetic acid. The flask is shaken gently during the addition, which requires 15–30 minutes. The β-naphthol dissolves during this period, and heat is evolved; the mixture is cooled somewhat towards the end of the addition to avoid excessive loss of hydrogen bromide. One hundred milliliters of water is then added, and the mixture is heated to boiling. It is then cooled to 100°, 25 g. of mossy tin is added (Note 2), and boiling is continued until the metal is dissolved. A second portion of 25 g. of tin is then added and dissolved by boiling, and finally a third portion of 100 g. (a total of 150 g., 1.27 gram atoms) of tin is introduced. The mixture is boiled for 3 hours, cooled to 50°, and filtered with suction. The crystalline tin salts which are thus removed are washed on the funnel with 100 ml. of cold acetic acid, the washings being added to the main portion of the filtrate.

This filtrate is stirred into 3 l. of cold water; the 6-bromo-2-naphthol which is precipitated is filtered with suction, removed from the funnel, and washed by stirring with 1 l. of cold water. After filtering again and drying at 100° there is obtained 214–223 g. (96–100%) of 6-bromo-2-naphthol. This crude product, which is pink and melts at 123–127°, contains some tin but is pure enough for most purposes.

A white product is obtained by vacuum distillation followed by crystallization of the crude product. Twenty-five grams of the crude substance on distillation (Note 3) gives 20 to 24 g. of distillate boiling at 200–205°/20 mm., and when this is crystallized from a mixture of 75 ml. of acetic acid and 150 ml. of water it gives 17.5 to 22.5 g. of 6-bromo-2-naphthol which melts at 127–129°.

2. Notes

1. An all-glass apparatus is of considerable advantage. If this is not available, the cork is covered with lead foil.

2. The reflux condenser is removed in order to add the tin, but it should be held in readiness so that it can be replaced quickly. The first two portions of tin react vigorously, and solution of the metal is

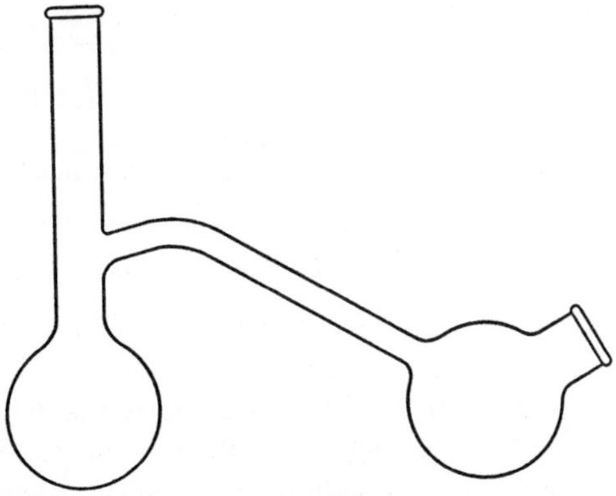

FIG. 6.

accompanied by the evolution of hydrogen and the loss of some hydrogen bromide.

3. For the distillation a sausage flask or two-bulb flask [1] (Fig. 6) is employed.

3. Methods of Preparation

6-Bromo-2-naphthol has been prepared by the reduction of 1,6-dibromo-2-naphthol with hydriodic acid,[2] with tin and hydrochloric acid,[3] or with stannous chloride and hydrochloric acid in aqueous alcohol [3] or in aqueous acetic acid.[4] It has also been obtained by the bromination of 2-naphthol with pyridinium bromide perbromide.[5]

[1] Fieser, *Experiments in Organic Chemistry*, p. 246, D. C. Heath & Company, Boston, 1935.

[2] Armstrong, *Chem. News*, **74**, 302 (1897).

[3] Franzen and Staubel, *J. prakt. Chem.*, (2) **103**, 369 (1922); Fries and Schimmelschmidt, *Ber.*, **58**, 2840 (1925).

[4] Fries and Schimmelschmidt, *Ann.*, **484**, 293 (1930).

[5] Vona and Merker, *J. Org. Chem.*, **14**, 1048 (1949).

9-BROMOPHENANTHRENE

(Phenanthrene, 9-bromo-)

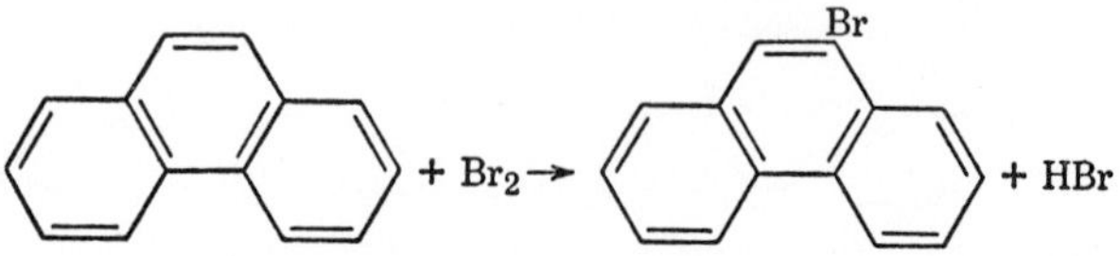

Submitted by CLINTON A. DORNFELD, JOSEPH E. CALLEN, and GEORGE H. COLEMAN.[1]
Checked by ROBERT E. CARNAHAN and HOMER ADKINS.

1. Procedure

A. *Purification of technical phenanthrene.* One and one-half kilograms of technical 90% phenanthrene is dissolved in 9 l. of ethanol in a 12-l. flask on a steam bath, and the hot solution is decanted from any insoluble material. The product crystallizes upon cooling of the solution. One kilogram of the crystallized product is dissolved in 2.2 l. of hot glacial acetic acid in a 5-l. three-necked flask provided with an efficient reflux condenser and a dropping funnel. To the boiling solution is gradually added 72 ml. of an aqueous solution containing 60 g. of chromic anhydride, and then 30 ml. of concentrated sulfuric acid is added slowly from the dropping funnel. The solution is refluxed for 15 minutes. The hot solution is then poured with vigorous stirring into 4.5 l. of water in a 12-l. round-bottomed flask. After cooling, the mixture is filtered, and the product is washed with water and air-dried. The product is distilled (Note 1) at 148–149°/1 mm. The distillate is recrystallized from ethanol to give 800–900 g. of nearly white phenanthrene, m.p. 98.7–99°.

B. 9-*Bromophenanthrene.* One kilogram (5.6 moles) of pure phenanthrene (Note 2) is dissolved in 1 l. of dry carbon tetrachloride in a 5-l. three-necked flask. A 500-ml. dropping funnel, a reflux condenser (with tube to conduct evolved hydrogen bromide to the hood), and an efficient motor-driven sealed stirrer (Note 3) are attached. The mixture is heated at gentle reflux with stirring, and 900 g. (5.64 moles) of bromine is added from the dropping funnel over a period of about 3 hours. After stirring at gentle reflux for 2 additional hours, during which most of the remaining hydrogen bromide is evolved, the reaction mixture is placed in a Claisen flask and the solvent is distilled at a pressure of 10–30 mm. The flask containing the residue is then provided with a fine capillary inlet tube, a thermometer, and a 2-l. distilling flask as receiver. The impure 9-bromophenanthrene is distilled (Note 4), and the material boiling at 177–190°/2 mm. is collected. The yield is 1300–1360 g. (90–94%), m.p. 54–56° (Notes 5 and 6).

2. Notes

1. Since the melting point of phenanthrene is relatively close to its boiling point under the pressure of the distillation, it is necessary to employ an apparatus so constructed that the solidified distillate will not clog the outlet tube. The submitters used a 2-l. modified Claisen flask attached directly, with a 14-mm. glass tube, to a 2-l. round-bottomed flask. The checkers used a similar all-glass apparatus.

2. The checkers used from 100 to 250 g. of phenanthrene with corresponding reductions of other quantities specified in the procedure.

3. A glycerol-rubber tube seal made with Neoprene rather than natural rubber is very satisfactory.

4. The flask may be supported on a wire gauze and heated directly with a Fisher or Meker burner. A modified Claisen flask equipped with a short column is desirable, but an ordinary Claisen flask can be used if the heating is carefully controlled to prevent impure bromophenanthrene from splashing over. The oil pump should be protected with the usual Dry Ice trap and with a potassium hydroxide tower to absorb hydrogen bromide. Even in runs in which the carbon tetrachloride solution was washed successively with sodium bisulfite, sodium carbonate, and water, much hydrogen bromide was evolved during distillation.

5. If the distillation is performed carefully the product is probably sufficiently pure for most purposes without recrystallization. 9-Bromophenanthrene may be recrystallized from ethanol (about 10 ml. per g.).

6. In runs of one-tenth the scale specified, the product obtained by the checkers had a melting point of 50–55° when yields above 90% were obtained. The yields were about 10% lower when 9-bromophenanthrene of melting point 54–56° was produced. The yield of recrystallized compound, melting at 65–66°, as obtained by the checkers, was 60% of the theoretical.

3. Methods of Preparation

The procedure for the purification of technical phenanthrene is that of W. E. Bachmann [2] with slight modification.

The procedure described for the preparation of 9-bromophenanthrene is an adaptation of that described by Henstock,[3] who effected the bromination at various temperatures and in different solvents but gave little experimental detail. Other methods[4] of preparation involve the formation and isolation of phenanthrene dibromide and its subsequent conversion to 9-bromophenanthrene by heating.

[1] Work done under contract with the Office of Scientific Research and Development.

[2] Bachmann, *J. Am. Chem. Soc.*, **57**, 557 (1935).

[3] Henstock, *J. Chem. Soc.*, **123**, 3097 (1923).

[4] Bachmann, *J. Am. Chem. Soc.*, **56**, 1365 (1934); Hayduck, *Ann.*, **167**, 181 (1873); Anschutz, *Ber.*, **11**, 1217 (1878); Austin, *J. Chem. Soc.*, **93**, 1763 (1908); Sandqvist, *Ann.*, **398**, 126 (1913); Miller and Bachman, *J. Am. Chem. Soc.*, **57**, 768 (1935); Mosettig and May, *J. Org. Chem.*, **11**, 15 (1946); Goldberg, Ordas, and Carsh, *J. Am. Chem. Soc.*, **69**, 260 (1947).

2-BROMOPYRIDINE *

(Pyridine, 2-bromo-)

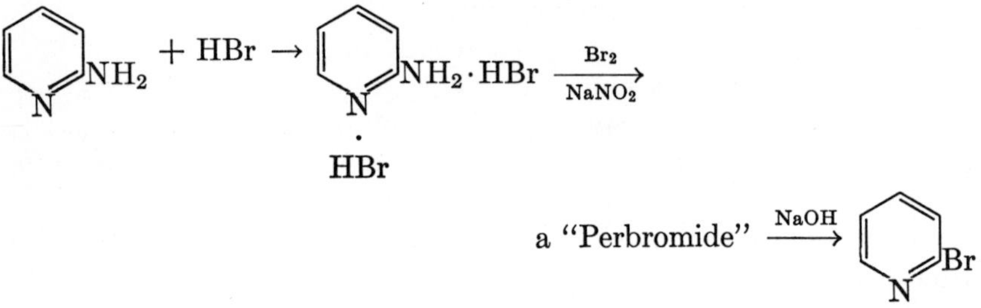

Submitted by C. F. H. Allen and John R. Thirtle.
Checked by Cliff S. Hamilton and Carol K. Ikeda.

1. Procedure

In a 5-l. three-necked flask fitted with a mechanical stirrer (Note 1), a dropping funnel, and a thermometer for reading low temperatures is placed 790 ml. (7 moles) of 48% hydrobromic acid. The flask and contents are cooled to 10–20° in an ice-salt bath, and 150 g. (1.59 moles) of 2-aminopyridine (Note 2) is added over a period of about 10 minutes. While the temperature is kept at 0° or lower, 240 ml. (4.7 moles) of bromine is added dropwise (Note 3). A solution of 275 g. (4 moles) of sodium nitrite in 400 ml. of water is added dropwise over a period of 2 hours, the temperature being carefully maintained at 0° or lower (Note 4). After an additional 30 minutes of stirring, a solution of 600 g. (15 moles) of sodium hydroxide in 600 ml. of water is added at such a rate that the temperature does not rise above 20–25° (Note 5). The nearly colorless reaction mixture is extracted with four 250-ml. portions of ether (Note 6). The extract is dried for 1 hour over 100 g. of solid potassium hydroxide and is then distilled through a Vigreux column 15 cm. in length. 2-Bromopyridine distils at 74–75°/13 mm., and the yield is 216–230 g. (86–92%) (Note 7).

2. Notes

1. A stirrer which gives efficient stirring near the walls of the flask is advisable. The fittings should not be gas tight since oxides of nitrogen and bromine are evolved during the reaction. It is advisable to work in a hood or out-of-doors.

2. The checkers used Eastman Kodak Company's practical grade of 2-aminopyridine.

3. The reaction mixture thickens, owing to formation of a yellow-orange perbromide during the addition of about one-half of the bromine; the first half of the bromine is added over a period of 30 minutes; the second half, over a period of 15 minutes.

4. The ice-salt bath is renewed before the sodium nitrite is added and once during the addition.

5. The color of the reaction mixture darkens during the addition of the alkali but becomes light yellow toward the end.

6. The separation into layers may not take place readily. The use of a "lily" may be helpful in separating the ether layer. A "lily" [1] is essentially a U-tube having unequal legs, the shorter of which has a flared top. It is easily prepared from a thistle or funnel tube by making a U bend just below the funnel. In use, a suction flask to act as a receiver is attached by means of rubber tubing. With this device it is a simple matter to draw off the upper layer when making separations, in any wide-mouth bottle or open jar. It may be necessary to filter the intermediate layer before attempting the separation.

7. This procedure can be used with 7 times the above amounts in a 22-l. flask, cooled in a half-barrel. A "Lightnin" stirrer is required (Model C-2).

3. Methods of Preparation

2-Bromopyridine has been made by direct bromination of pyridine; [2] from N-methyl-2-pyridone with phosphorus pentabromide and phosphorus oxybromide;[3] from 2-aminopyridine by diazotization with amyl nitrite in 20% hydrobromic acid;[4] from sodium 2-pyridinediazotate by solution in concentrated hydrobromic acid;[5] from 2-aminopyridine by diazotization in the presence of bromine and concentrated hydrobromic acid;[6] and from 2-aminopyridine by diazotization with nitrogen trioxide in 40% hydrobromic acid.[7] The method described here is essentially that of Craig.[6]

[1] Private communication, Emil J. Rahrs, Eastman Kodak Company.

[2] Wibaut and Den Hertog, *Rec. trav. chim.*, **51**, 385 (1932); McElvain and Goese, *J. Am. Chem. Soc.*, **65**, 2230 (1943); Wibaut, *Experientia*, **5**, 337 (1949).

[3] Fischer, *Ber.*, **32**, 1303 (1899).

[4] Tschitschibabin and Rjasanzew, *J. Russ. Phys. Chem. Soc.*, **47**, 1571 (1915) (*Chem. Zentr.*, **1916**, II, 228); *J. Chem. Soc.*, **110**, I, 224 (1916) [*C. A.*, **10**, 2898 (1916)].

[5] Tschitschibabin and Tjashelowa, *J. Russ. Phys. Chem. Soc.*, **50**, 495 (1918) (*Chem. Zentr.*, **1923**, III, 1021).

[6] Craig, *J. Am. Chem. Soc.*, **56**, 232 (1934).

[7] Newman and Fones, *J. Am. Chem. Soc.*, **69**, 1221 (1947).

4-BROMO-*o*-XYLENE

(*o*-Xylene, 4-bromo-)

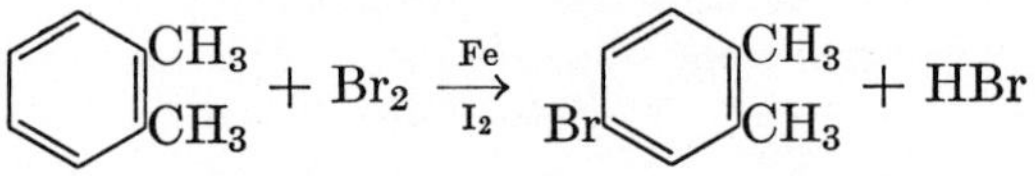

Submitted by W. A. Wisansky and S. Ansbacher.
Checked by N. L. Drake, Wilkins Reeve, and John Sterling, Jr.

1. Procedure

In a 1-l. three-necked flask having ground-glass joints are placed 500 g. (569 ml., 4.72 moles) of *o*-xylene, 12 g. of clean iron filings, and a crystal of iodine (Note 1). The flask is fitted with a dropping funnel, a stirrer (Note 2), and a condenser; a thermometer is suspended through the condenser on a platinum or Nichrome wire and arranged so that the bulb extends beneath the surface of the liquid. The top of the condenser is connected to a gas-absorption trap. The mixture is stirred and cooled in an ice-salt mixture, preferably under a hood. Six hundred and sixty grams (4.13 moles) of bromine is added dropwise over a 3-hour period; during this time the internal temperature is maintained at 0° to −5° (Note 3). After all the bromine has been added the reaction mixture is allowed to stand overnight. It is poured into water and washed successively with a 500-ml. portion of water, two 500-ml. portions of 3% sodium hydroxide solution (Note 4), and one 500-ml. portion of water. The product is then steam-distilled; about 8 l. of distillate is collected (Note 5). The organic layer is separated from the water and dried over calcium chloride. The 4-bromo-*o*-xylene is distilled through a short column under reduced pressure, and the fraction boiling at 92–94°/14–15 mm. (n_D^{22} 1.5558) is collected. The yield is 720–745 g. (94–97%, based on bromine) (Note 6).

2. Notes

1. Essentially the same yield results if only one of the catalysts (iron or iodine) is used.

2. The checkers used a glycerol-sealed stirrer.

3. In experiments in which the temperature of the reaction mixture was allowed to rise as high as +10°, the submitters noted a slight increase in the amount of dibromo-*o*-xylene formed.

4. A dilute sodium bisulfite solution may be used instead of the alkali in this washing.

5. Near the end of the steam distillation a white waxy product begins to collect in the condenser. This is apparently a dibromo-*o*-xylene.

6. This material is of sufficient purity for most uses, including the conversion to 3,4-dimethylaniline (p. 307). It is reported[1] that the boiling point of the product can be raised slightly (from 211–212° to 214–215°/760 mm.) by sulfonation, recrystallization of the barium sulfonate, and regeneration of the bromo compound by acid hydrolysis.

3. Methods of Preparation

The procedure given was developed by Ghigi[1] from an earlier preparation by Jacobsen.[2] 4-Bromo-*o*-xylene has also been prepared from 3,4-dimethylaniline by the Sandmeyer reaction.[3,4]

[1] Ghigi, *Ber.*, **71**, 684 (1938).
[2] Jacobsen, *Ber.*, **17**, 2372 (1884).
[3] Brand, Ludwig, and Berlin, *J. prakt. Chem.*, [2] **110**, 34 (1925).
[4] Kohlrausch and Pongratz, *Monatsh.*, **64**, 361 (1934).

o-n-BUTOXYNITROBENZENE

(Ether, butyl o-nitrophenyl)

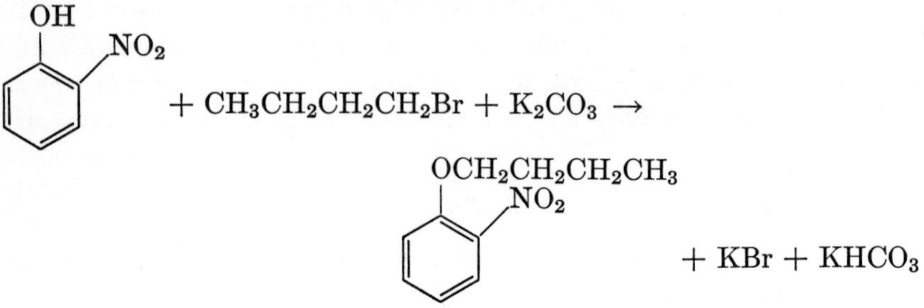

Submitted by C. F. H. Allen and J. W. Gates, Jr.
Checked by W. E. Bachmann and G. Dana Johnson.

1. Procedure

A mixture of 28 g. (0.2 mole) of *o*-nitrophenol (Note 1), 30 g. (0.22 mole) of *n*-butyl bromide, 28 g. (0.2 mole) of anhydrous potassium carbonate, and 200 ml. of dry acetone in a 1-l. round-bottomed flask is refluxed on a steam bath for 48 hours (Note 2). At the end of this time the acetone is distilled from the mixture, 200 ml. of water is added to the residue, and the product is extracted with two 100-ml. portions of benzene. The combined benzene extracts are washed with three 100-ml. portions of 10% sodium hydroxide, the benzene is removed by distillation at ordinary pressure, and the residual oil is distilled under reduced pressure. The yield of product boiling at 118–121°/1 mm. (Note 3) is 29–31 g. (75–80%) (Notes 4 and 5).

2. Notes

1. A technical grade of *o*-nitrophenol was used; the yield is no better with the pure material. In place of *n*-butyl bromide, a corresponding amount (36.8 g.) of the iodide can be used with no change in yield.

2. The checkers shook the flask occasionally during the first 1.5 hours in order to prevent caking of the contents.

3. The boiling point is 126–129°/2 mm. and 171–172°/19 mm.

4. Other nitrophenyl ethers can be prepared in a similar manner and in essentially the same yields. When the size of the run was increased to 2 moles, the yields were increased to 85–90%. The hexyl derivatives require 72 hours for reaction.

The boiling points of some ethers prepared by the present procedure are as follows: *m*-*n*-butoxynitrobenzene, 120–124°/2 mm.; *p*-*n*-butoxynitrobenzene, 150–154°/5 mm.; *p*-isopropoxynitrobenzene, 283–286°/760 mm.; *o*-*n*-hexoxynitrobenzene, 145–148°/1 mm.; *p*-*n*-hexoxynitrobenzene, 170–174°/5 mm.

5. The alkoxyanilines are obtained readily by reduction of the alkoxynitrobenzenes in alcohol in the presence of Raney nickel catalyst (see p. 63). The boiling points of some of these are as follows: *m*-*n*-butoxyaniline, 120–124°/2 mm.; *p*-*n*-butoxyaniline, 135–138°/5 mm.; *p*-isopropoxyaniline, 145–147°/20 mm.; *p*-*n*-hexoxyaniline, 155–158°/5 mm.

3. Methods of Preparation

The present procedure, which avoids the preparation of the salts of the phenols, is of general utility. It was first used by Claisen [1] for allyl ethers. *o*-*n*-Butoxynitrobenzene has been prepared in a similar manner previously.[2] The *m*-nitro- and *p*-nitrobutoxybenzenes have been obtained by alkylation of the phenol salts.[3–5] The corresponding amines have been prepared previously by reduction of the nitro compounds by means of iron and water [2] and by stannous chloride and hydrochloric acid.[4]

1 Claisen and Eisleb, *Ann.*, **401**, 39, 59 (1913).
2 Li and Adams, *J. Am. Chem. Soc.*, **57**, 1567 (1935).
3 Hodgson and Clay, *J. Chem. Soc.*, **1933**, 661.
4 Gutekunst and Gray, *J. Am. Chem. Soc.*, **44**, 1742 (1922).
5 Profft, *Deut. Chem. Ztg.*, **2**, 194 (1950).

tert-BUTYL ACETATE

(Acetic acid, *tert*-butyl ester)

I. ACETIC ANHYDRIDE METHOD

$$(CH_3)_3COH + (CH_3CO)_2O \xrightarrow{ZnCl_2} CH_3CO_2C(CH_3)_3 + CH_3CO_2H$$

Submitted by Robert H. Baker and Frederick G. Bordwell.
Checked by R. L. Shriner and Fred W. Neumann.

1. Procedure

In a 1-l. flask equipped with a reflux condenser and drying tube are placed 200 ml. (2.1 moles) of *tert*-butyl alcohol (Note 1), 200 ml. (2.1 moles) of acetic anhydride, and 0.5 g. of anhydrous zinc chloride.

After thorough shaking, the mixture is slowly heated to reflux temperature, maintained at gentle refluxing for 2 hours, and then cooled. The reflux condenser is replaced by a 20-cm. Vigreux column through which the mixture is distilled up to a temperature of 110°. The crude distillate, weighing 200–250 g., is washed with two 50-ml. portions of water, then with 50-ml. portions of 10% potassium carbonate until the ester layer is neutral to litmus; the product is finally dried over anhydrous potassium carbonate (about 20 g.).

After removal of the drying agent, the ester is fractionally distilled through an efficient fractionating column (Note 2). A fore-run of 21–37 g. is collected up to a temperature of 95°. The pure ester distils between 95° and 96° and amounts to 129–148 g. (53–60%) (Note 3).

2. Notes

1. Eastman Kodak Company's best grade of *tert*-butyl alcohol and the practical grade of acetic anhydride are satisfactory. If these are not available, the alcohol should be dried over quicklime and distilled, and the acetic anhydride should be redistilled also.
2. The fractionating column may be a 16-plate Stedman column or a 30-cm. Carborundum-packed column.
3. The corrected boiling point of the ester is given as 97.9°.[1]

II. ACETYL CHLORIDE METHOD

$$(CH_3)_3COH + CH_3COCl + C_6H_5N(CH_3)_2 \rightarrow CH_3CO_2C(CH_3)_3 + C_6H_5N(CH_3)_2 \cdot HCl$$

Submitted by C. R. Hauser, B. E. Hudson, B. Abramovitch, and J. C. Shivers.
Checked by R. L. Shriner and Fred W. Neumann.

1. Procedure

In a 2-l. flask equipped with a reflux condenser, mercury-sealed stirrer, and a dropping funnel are placed 147 ml. (114 g., 1.5 moles) of *tert*-butyl alcohol (Note 1), 212 ml. (202 g., 1.67 moles) of dimethylaniline, and 200 ml. of dry ether. The solution is heated to refluxing, and 113 ml. (124 g., 1.58 moles) of acetyl chloride is run into the stirred solution at such a rate that moderate refluxing continues after the source of heat is removed. When approximately two-thirds of the acetyl chloride has been added, dimethylaniline hydrochloride begins to crystallize and the mixture refluxes very vigorously. An ice bath is applied immediately, and, after refluxing ceases, the remainder of the acetyl chloride is added. Finally, the mixture is heated for 1 hour on a water bath. The mixture is cooled to room temperature, approximately 200 ml. of water is added, and stirring is continued until all the solid material dissolves. The ether layer is separated and extracted with 50-ml. portions of cold 10% sulfuric acid until the extract does not become cloudy when made alkaline with sodium hydroxide. After a final washing with 25 ml. of saturated sodium bicarbonate solution, the ether solution is dried by shaking it with 10 g. of anhydrous sodium sulfate. The solution is decanted and allowed to stand over 10 g. of Drierite overnight. The solution is filtered, and the ether is removed by distillation through a good fractionating column (Note 2). The residue is fractionally distilled, and 110–119 g. (63–68%) of *tert*-butyl acetate boiling at 95–98° is obtained (Note 2). Most of the ester boils at 97.0–97.5° (Note 3).

2. Notes

1. All reactants should be pure and anhydrous. The *tert*-butyl alcohol should be dried over quicklime and the dimethylaniline redistilled. Reagent grade acetyl chloride should be employed.

2. Either a 30-cm. Carborundum-packed column or a 30-cm. Widmer column may be used. The pure ester was collected over a range of 94.5–95.5°.

3. The submitters report that the following *tert*-butyl esters have been prepared in a similar manner. Eastman Kodak Company chemicals were used.

(*a*) *tert*-Butyl propionate. From 221 ml. (171 g., 2.31 moles) of *tert*-butyl alcohol, 318 ml. (303 g., 2.5 moles) of dimethylaniline, and 206 ml. (220 g., 2.38 moles) of propionyl chloride, refluxed for 3 hours, there was obtained 184 g. (61.4%) of *tert*-butyl propionate, b.p. 117–118.5°; most of the product boiled at 118.0–118.5°.

(*b*) *tert*-Butyl isobutyrate. From 94.5 ml. (73.4 g., 0.99 mole) of *tert*-butyl alcohol, 127 ml. (121 g., 1 mole) of dimethylaniline in 200 ml. of ether, and 105 ml. (108 g., 1 mole) of isobutyryl chloride in 50 ml. of ether, after standing for 15 hours at room temperature, the ether being distilled, and the residue heated for 5 hours on a water bath, there was obtained 102 g. (71%) of the ester boiling at 127–128.3°.

(*c*) *tert*-Butyl isovalerate. From 121.4 g. (1.64 moles) of *tert*-butyl alcohol, 198.5 g. (1.64 moles) of dimethylaniline, and 200.7 g. (1.66 moles) of isovaleryl chloride, after standing overnight, the ether being distilled, and the residue heated for 5 hours on a water bath, there was obtained 67.3 g. (26%) of *tert*-butyl isovalerate boiling at 154–157°.

(*d*) *tert*-Butyl cinnamate. The cinnamoyl chloride from 100 g. (0.675 mole) of cinnamic acid and 400 g. (3.37 moles) of thionyl chloride was treated with 64 ml. (49.5 g., 0.67 mole) of *tert*-butyl alcohol and 90 ml. (86.0 g., 0.71 mole) of dimethylaniline. The mixture was refluxed 12 hours and then allowed to stand 12 hours at room temperature. Distillation furnished 78.8 g. (58%) of *tert*-butyl cinnamate boiling at 144°/8 mm.

(*e*) *tert*-Butyl chloroacetate (submitted by Robert H. Baker). Thirty-five and four-tenths milliliters (0.4 mole) of *tert*-butyl alcohol was added over a period of 10 minutes to a mixture of 30.6 ml. (0.4 mole) of chloroacetyl chloride and 50 ml. (0.4 mole) of dimethylaniline, care being taken to keep the temperature below 30°. After standing 45 minutes at room temperature, the mixture was poured into water and worked up in the usual way. The ester was fractionated in a 6-in. Widmer column. The yield was 38 g. (63%) of ester which boiled at 48–49°/11 mm.; n_D^{20} 1.4259–1.4260.

III. ACETYL CHLORIDE-MAGNESIUM METHOD

$$2(CH_3)_3COH + 2CH_3COCl + Mg \rightarrow 2CH_3COOC(CH_3)_3 + MgCl_2 + H_2$$

Submitted by A. Spassow.
Checked by W. E. Bachmann and J. Korman.

1. Procedure

In a 1-l. round-bottomed flask are placed 12 g. (0.5 gram atom) of magnesium powder, 37 g. (0.5 mole) of *tert*-butyl alcohol, and 100 g. of anhydrous ether (Note 1). The flask is fitted with an addition tube, one arm of which bears a reflux condenser and the other arm a dropping funnel. While the mixture is being shaken by hand, a solution of 55 g. (0.7 mole) of acetyl chloride (Note 2) in 50 g. of anhydrous ether is added dropwise (Note 3). A lively reaction gradually ensues with evolution of hydrogen, mixed with ether vapor and a little hydrogen chloride (Note 4). After all the acetyl chloride has been added, the reaction mixture is allowed to stand in a pan of cold water for 1 hour (Note 5). After another hour at room temperature the mixture is warmed in a water bath at 40–45° for 30 minutes in order to complete the reaction.

The solid reaction product is cooled in ice water and decomposed by addition of an ice-cold solution of 20 g. of potassium carbonate in 250 ml. of water, cooling being continued throughout (Note 6). The mixture is extracted three times with 35-ml. portions of ether; the ether extract is dried over calcium chloride and then distilled (Note 7). The ester is obtained from the fraction boiling at 85–98° by fractional distillation by means of a good column (Note 8). The purified ester boils at 95–97°/740 mm. and weighs 37–45 g. (45–55%).

2. Notes

1. The *tert*-butyl alcohol must be completely anhydrous. It was dried over sodium and distilled from sodium just before use. The ether was likewise dried over sodium.

2. The acetyl chloride was distilled before use.

3. The addition of the acetyl chloride requires about 15 minutes. After the addition of about two-thirds of the acid chloride, the reaction mixture rapidly becomes semisolid.

4. The course of the esterification is best followed by the evolution of the hydrogen, the hydrogen being led by means of a tube from the top of the condenser through a wash bottle containing a small amount of water. A moderate rate of reaction is obtained throughout by judicious immersion of the reaction flask in a pan of cold water.

5. After 30 minutes the reaction becomes more lively and the mixture more fluid.

6. The potassium carbonate solution is added all at once. The carbon dioxide which is evolved does not interfere with the extraction of the ester.

7. Since the ester is quite volatile, the ether is distilled through a 60-cm. Vigreux column. The distillation is interrupted at 40°, and a 40-cm. column is employed. A fraction boiling up to 85° and one boiling from 85° to 98° are collected. The first fraction, which contains considerable amounts of the ester, is redistilled, and the portion boiling above 85° is added to the second fraction. The fraction boiling from 85° to 98° weighs 43–48 g.

8. A 40-cm. Vigreux or Widmer column is used for this distillation. The fore-run is redistilled from the same flask.

3. Methods of Preparation

tert-Butyl acetate has been prepared from the alcohol and acetyl chloride in the presence of pyridine,[1] dimethylaniline,[2] or magnesium. Acetic anhydride has been used with zinc chloride,[2] a small amount of zinc dust,[2] or anhydrous sodium acetate.[3]

[1] Bryant and Smith, *J. Am. Chem. Soc.*, **58,** 1016 (1936).
[2] Norris and Rigby, *J. Am. Chem. Soc.*, **54,** 2097 (1932).
[3] Tronow and Ssibgatullin, *Ber.*, **62,** 2850 (1929).

n-BUTYL ACRYLATE *

(Acrylic acid, *n*-butyl ester)

$$CH_2{=}CHCOOCH_3 + C_4H_9OH \rightarrow CH_2{=}CHCOOC_4H_9 + CH_3OH$$

Submitted by Chessie E. Rehberg.
Checked by H. R. Snyder and Fred E. Boettner.

1. Procedure

In a 2-l. two-necked round-bottomed flask having a capillary ebullator tube in one neck (Note 1) are placed 371 g. (5 moles) of *n*-butyl alcohol, 861 g. (10 moles) of methyl acrylate, 20 g. of hydroquinone, and 10 g. of *p*-toluenesulfonic acid (Note 2). The flask is attached to an all-glass fractionating column, preferably one without packing such as the Vigreux type (Note 3), and the solution is heated to boiling in an oil bath. The column is operated under total reflux until the temperature of the vapors at the still head falls to 62–63°, which is the boiling point of the methanol-methyl acrylate azeotrope (Note 4). This azeotrope is then distilled as rapidly as it is formed, the temperature at the still head not being allowed to exceed 65°. When the production of methanol has become very slow (6–10 hours), the excess methyl acrylate is distilled, and the butyl acrylate is then distilled, preferably at 10–20 mm. It boils at 39°/10 mm., 84–86°/101–102 mm., and at about 145° at atmospheric pressure. The yield is 500–600 g. (78–94%) (Note 5).

2. Notes

1. The capillary is used to introduce a gas to prevent bumping and superheating during the vacuum distillation of the product. As air

has some tendency to catalyze polymerization of the acrylic ester, if it is introduced through the capillary the amount must be as small as possible. The gas introduced should be an inert one, such as carbon dioxide or nitrogen. If polymeriaztion is troublesome, it may be advantageous to pass in a slow stream of carbon dioxide through the capillary during the entire reaction period.

2. Sulfuric acid is also a very satisfactory catalyst; aluminum alkoxides also are useful, especially when the alcohols would be adversely affected by strong acids. Sodium alkoxides produce undesirable side reactions and give lower yields. When alkaline catalysts are employed, an alkaline polymerization inhibitor, such as *p*-phenylenediamine or phenyl-β-naphthylamine, should be used instead of hydroquinone.

3. The fractionating column should be one that can be cleaned readily if a polymer is formed in it. A large number of plates is not required, though the column should be capable of separating the methanol-methyl acrylate azeotrope (b.p. 62–63°) from methyl acrylate (b.p. 80°), and butanol (b.p. 117°) from butyl acrylate (b.p. 145°). The necessity of effecting the latter separation can be practically eliminated by allowing the reaction to go virtually to completion, all the butanol thus being consumed. This can be done by extending the reaction period as long as reaction occurs and by adding a considerable excess of methyl acrylate. Instead of the twofold excess specified, three or four times the theoretical amount may be used with benefit. The larger amount is especially desirable when the acrylate of a relatively unreactive alcohol is being prepared.

4. The methanol-methyl acrylate azeotrope contains about 45% methyl acrylate, which can be recovered by washing out the methanol with a large volume of water or brine; the acrylate is purified by drying and distilling. An inhibitor, such as hydroquinone, should always be added to any acrylic ester before attempting to distil it, and, unless it is stored in a refrigerator, the distilled ester should not be kept more than a few hours without the addition of a small amount (0.1–1.0%) of an inhibitor.

5. Yields of the primary alkyl acrylates vary somewhat, owing to occasional losses through formation of polymer, but are usually in the range of 85–99%. Some secondary alcohols react very slowly, others readily. The method has been applied to more than fifty alcohols, some of which (with percentage yields) are listed below: ethyl, 99%; isopropyl, 37%; *n*-amyl, 87%; isoamyl, 95%; *n*-hexyl, 99%; 4-methyl-2-pentyl, 95%; 2-ethylhexyl, 95%; capryl, 80%; lauryl, 92%; myristyl, 90%; allyl, 70%; furfuryl, 86%; citronellyl, 91%; cyclohexyl, 93%;

benzyl, 81%; β-ethoxyethyl, 99%; β-(β-phenoxyethoxy)ethyl (from diethylene glycol monophenyl ether), 88%.

3. Methods of Preparation

n-Butyl acrylate has been prepared by direct esterification,[1] by debromination of *n*-butyl α,β-dibromopropionate with zinc,[2] by treatment of either butyl β-chloropropionate [1] or butyl β-bromopropionate [1] with diethylaniline, and by the pyrolysis of butyl β-acetoxypropionate.[3] Direct esterification and alcoholysis of methyl or ethyl acrylate have been recommended for the preparation of the higher alkyl acrylates.[4]

[1] Moureau, Murat, and Tampier, *Ann. chim.*, **15,** 245 (1921) [*C. A.*, **16,** 55 (1922)].

[2] Kobeko, Koton, and Florinskii, *J. Applied Chem. U.S.S.R.*, **12,** 313 (1939) [*C. A.*, **33,** 6795 (1939)].

[3] Burns, Jones, and Ritchie, *J. Chem. Soc.*, **1935,** 400.

[4] Neher, *Ind. Eng. Chem.*, **28,** 267 (1936).

tert-BUTYLAMINE *

I. HYDROGENOLYSIS OF 2,2-DIMETHYLETHYLENIMINE

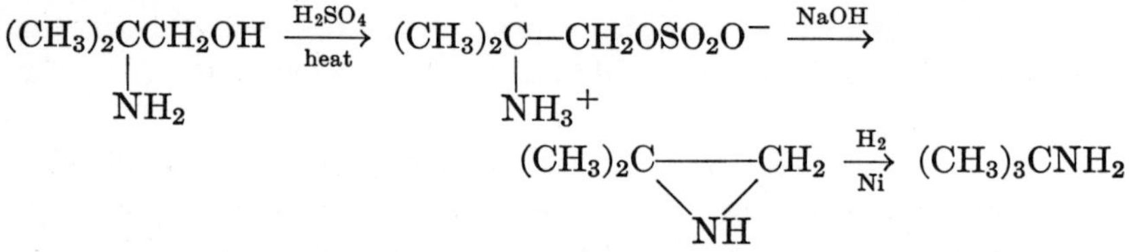

Submitted by Kenneth N. Campbell, Armiger H. Sommers, and Barbara K. Campbell.
Checked by Nathan L. Drake and Sidney Melamed.

1. Procedure

A. 2,2-*Dimethylethylenimine.* A cold mixture of 110 g. (60 ml., 1.06 moles) of concentrated sulfuric acid and 200 ml. of water is added in portions, with shaking, to a solution of 100 g. (107 ml., 1.12 moles) of 2-amino-2-methyl-1-propanol in 200 ml. of water contained in a 1-l. round-bottomed flask (Note 1). The flask is fitted with a thermometer extending into the liquid and a short still head carrying a downward condenser.

Water is distilled from the mixture at atmospheric pressure until the temperature of the solution reaches 115° (Notes 2 and 3), whereupon the liquid is transferred to a 500-ml. round-bottomed flask. This flask is connected to the distillation apparatus used previously except that the thermometer is replaced by a capillary tube. Distillation is then continued under the reduced pressure obtainable from a water aspirator. The bath temperature is raised to 175° over a period of about an hour and is held there until the mixture solidifies (usually 30–60 minutes longer), and for 1 hour thereafter. The flask is cooled and broken to remove the product.

The brown solid from the above operations is ground in a mortar and placed in a 500-ml. distilling flask equipped with a downward condenser and a receiver. A cold solution of 100 g. (2.5 moles) of technical sodium hydroxide in 150 ml. of water is added to the solid in the flask, and heat is applied by means of an oil bath whose temperature is slowly raised to 125° From 70 to 75 g. of distillate is collected; the head temperature ranges from about 70° to 101° (Note 2). The distillate is cooled in ice and saturated with technical potassium hydroxide; the organic layer that forms is separated and dried over potassium hydroxide pellets in a refrigerator for about 15 hours.

The organic layer is separated from the drying agent and distilled from a few fresh pellets of potassium hydroxide through a column of the Whitmore-Fenske type. After a 3- to 5-g. fore-run, the product distils at 71–72°; the yield is 36–41 g. (45–51%) (Notes 4 and 5).

B. *tert-Butylamine.* A hydrogenation bomb or bottle (Note 6) is charged with 100 ml. of purified dioxane,[1] 35.5 g. (0.5 mole) of freshly distilled 2,2-dimethylethylenimine, and about 9 g. (alcohol-wet weight) of Raney nickel.[2] The apparatus is filled with hydrogen, warmed to 60°, and shaken until hydrogenation is complete (Note 7). The contents of the bomb are removed and filtered to separate the catalyst, which is washed on the funnel with a little dioxane.

The filtrates from two such runs are combined and distilled through a Whitmore-Fenske column of 10–15 theoretical plates at a 5:1 reflux ratio. The yield of *tert*-butylamine is 55–60 g. (75–82%); the product boils at 44–44.5°, and its refractive index is n_D^{20} 1.3770 (Note 8).

2. Notes

1. The 2-amino-2-methyl-1-propanol used was the practical grade obtained from the Eastman Kodak Company. This aminoalcohol can also be secured from the Commercial Solvents Corporation.

2. An electric heating mantle may conveniently be substituted for the oil bath specified.

3. Approximately 285 ml. of water must be collected before the temperature of the reaction mixture reaches 115°.

4. 2,2-Dimethylethylenimine polymerizes on standing; the product should be hydrogenated within a few hours after preparation.

5. The checkers have followed the same procedure successfully using tenfold quantities; the yield of the imine was 42%.

6. The hydrogenation can be carried out equally well in metal or glass equipment. If a hydrogenator of the Parr low-pressure type is used, the bottle can be wound for electrical heating. Five turns of 24-gauge asbestos-covered Nichrome or Chromel A wire is satisfactory for the heating element. In use, current is supplied from a variable transformer, and the voltage necessary to heat the contents of the bottle to 60° is determined by experiment. The checkers used hydrogenation equipment supplied by the American Instrument Company.

7. If the hydrogenation is carried out in a Parr hydrogenation apparatus at 40–60 lb. pressure, about 2 hours is required to complete the hydrogenation as described. The low boiling point of the ethylenimine makes it impossible to remove the air from the bottle by evacuation in the usual way before hydrogenation. Instead the bottle is filled with hydrogen to 15–20 lb. pressure, the pressure is released, and the process repeated.

8. The checkers used high-pressure equipment and found that the hydrogenation of 336 g. of 2,2-dimethylethylenimine in 250 ml. of purified dioxane in the presence of 3 teaspoonfuls of Raney nickel under 3000 lb. hydrogen pressure was complete in 10–15 minutes; the temperature rise during the hydrogenation was about 50°. The yield of product was 283 g. (82%).

II. VIA *tert*-BUTYLPHTHALIMIDE

$$(CH_3)_3COH + H_2NCONH_2 \xrightarrow{H_2SO_4} (CH_3)_3CNHCONH_2 + H_2O$$

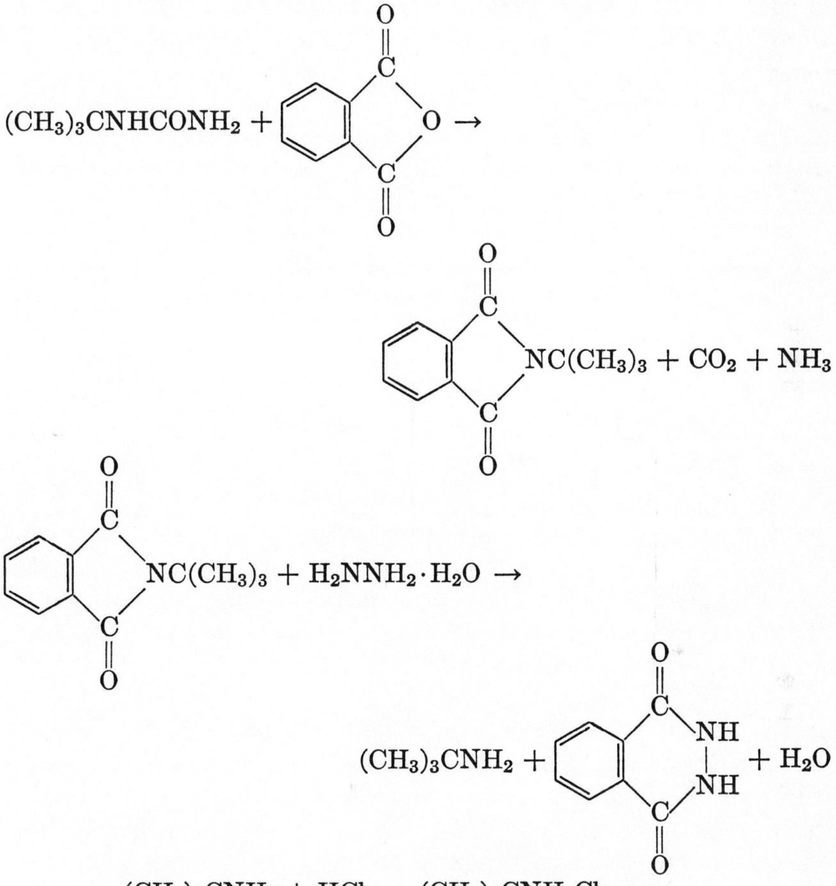

$$(CH_3)_3CNH_2 + HCl \rightarrow (CH_3)_3CNH_3Cl$$

$$(CH_3)_3CNH_3Cl + NaOH \rightarrow (CH_3)_3CNH_2 + NaCl + H_2O$$

Submitted by LEE IRVIN SMITH and OLIVER H. EMERSON.[3]
Checked by R. L. SHRINER and ARNE LANGSJOEN.

1. Procedure

A. *tert-Butylurea* (Note 1). In a 500-ml. three-necked flask equipped with a fast mechanical stirrer, a 200-ml. dropping funnel, and a thermometer is placed 193 g. (105 ml., 1.98 moles) of concentrated sulfuric acid (sp. gr. 1.84). The flask is surrounded by an ice bath, and 60 g. (1 mole) of finely powdered urea is added slowly at such a rate that the temperature remains between 20° and 25°. Then 148 g. (188 ml., 2 moles) of *tert*-butyl alcohol is added dropwise from the funnel at such a rate that the temperature is maintained between 20° and 25° (Note 2). After the addition is completed the mixture is stirred for an additional 30 minutes, allowed to stand at room temperature overnight (about 16 hours) (Note 3), and then poured with stirring on 1.5 kg. of cracked ice and water in a 4-l. beaker. Without removal of the precipitate, the mixture is made alkaline to Congo red indicator by adding slowly with stirring a solution of sodium hydroxide (160 g. in 750 ml. of water). The mixture is cooled with an ice bath to keep the temperature below 25°. The mixture is stirred in the ice bath until the temperature falls to about 15° (Note 4), at which point the precipitate is collected on a 15-cm. Büchner funnel, washed with two 100-ml. portions of cold water, and pressed and sucked as dry as possible. The cake is transferred to a 2-l. beaker, and 500 ml. of water is added. The mixture is heated to boiling and quickly filtered while hot through an 8-cm. steam-heated Büchner funnel (Note 5) with the aid of suction. The filtrate is cooled to 0–5° with occasional stirring, and the white precipitate of *tert*-butylurea is collected on a 15-cm. Büchner funnel with suction and pressed as dry as possible. After the product has been spread out on absorbent paper and air-dried overnight (Note 6) there is obtained 36–39 g. (31–33%) of *tert*-butylurea melting at 180–182° (Note 7).

B. *tert-Butylphthalimide.* Thirty-five grams (0.3 mole) of *tert*-butylurea and 100 g. (0.675 mole) of phthalic anhydride are ground together in a mortar. The mixture is placed in a 1-l. Erlenmeyer flask which is then immersed in a metal bath previously heated to 200°. The mixture melts and effervesces vigorously; after 10 minutes the temperature of the bath is raised to 240° (internal temperature 220°) and held there for 5 minutes (Note 8). The flask is removed and cooled to 60–70°, and 100 ml. of 95% ethanol is added to dissolve the contents partially. A 20% solution of sodium carbonate is added until the solution is alkaline to litmus paper, and the mixture is diluted with water to 1 l. The solid is collected on a Büchner funnel with the aid of suction and pressed as dry as possible. The filter cake is warmed on a steam bath with 500 ml. of petroleum ether (b.p. 60–70°) in a 1 l. flask, and the hot mixture is filtered. Any water layer that may separate from the filtrate is removed, and the filtrate is cooled to 25° and again filtered. The clear filtrate is concentrated by distillation to about one-third of its original volume and placed in a refrigerator overnight. The crystalline material is collected on a filter, and as much as possible of the solvent is removed by suction. After air drying, this first crop weighs 40–43 g. and melts at 58–59°. By concentration of the filtrate an additional 2–4 g. may be obtained. The total yield of *tert*-butylphthalimide is 43–46.5 g. (72–76%).

C. *tert-Butylamine hydrochloride.* In a 2-l. flask fitted with an efficient bulb-type reflux condenser are placed 203 g. (1 mole) of *tert*-butylphthalimide, 1 l. of 95% ethanol, and 59 g. (1 mole) of 85% hydrazine hydrate. The solution is refluxed for 2 hours and cooled, and concentrated hydrochloric acid (about 100 ml.) is added until the solution is strongly acid to Congo red paper, though a large excess of acid is avoided. The voluminous precipitate of phthalhydrazide is collected on a 15-cm. Büchner funnel with the aid of suction and washed with four 100-ml. portions of 95% ethanol. The combined filtrate and washings are concentrated under reduced pressure to a volume of 200 ml. About 1 l. of water is added, any insoluble material is removed by filtration, and the filtrate is concentrated to about 300–350 ml. If any additional insoluble material separates it is removed by filtration, and the solvent is completely removed by evaporation under reduced pressure. This crude amine hydrochloride may be converted to the free amine as described below, or it may be purified by solution in 500 ml. of absolute ethanol, the solution being filtered and 500 ml. of dry ether being added to the filtrate. The crude amine hydrochloride is collected on a filter; the solvent is removed by suction, and the residue is dissolved in hot absolute ethanol using 5 ml. per gram of material; absolute ether (50% of the volume of ethanol) is added, and the solution is cooled in a refrigerator for several hours. All the *tert*-butylamine hydrochloride is collected on a Büchner funnel and dried in a vacuum desiccator. It weighs 79–97 g. (72–88%) and melts with sublimation at 270–290°.

D. *tert-Butylamine.* To 150 ml. of cold 40% solution of sodium hydroxide is added 109.5 g. (1 mole) of *tert*-butylamine hydrochloride with stirring. The solution is saturated with potassium carbonate (about 100–150 g.); the layer of amine is separated and dried over 20 g. of sodium hydroxide pellets. The product is distilled using an ice-cooled receiver, and the fraction boiling at 44–46° is collected. If the purified amine hydrochloride is the starting material for this step the yield ranges from 65 g. to 69 g. (89–94%). When the crude amine salt is employed, the yield is 46–60 g. (64–83%) (Note 9).

III. HYDROLYSIS OF *tert*-BUTYLUREA

$$(CH_3)_3CNHCONH_2 + 2NaOH \rightarrow (CH_3)_3CNH_2 + NH_3 + Na_2CO_3$$

Submitted by D. E. Pearson, J. F. Baxter, and K. N. Carter.
Checked by R. L. Shriner and Calvin N. Wolf.

1. Procedure

A 1-l. round-bottomed flask is charged with 60 g. (1.5 moles) of sodium hydroxide pellets dissolved in 75 ml. of water, 70 g. (0.6 mole) of *tert*-butylurea, and 225 ml. of commercial ethylene glycol. The flask is fitted with an efficient reflux condenser, and a glass tube is led from the top of the condenser to a small flask immersed in ice water (Note 10).

The mixture is refluxed gently for 4 hours (Note 11). The flask is cooled and equipped for distillation, and the fraction boiling at 40–60° is collected in an ice-cooled receiver. The crude amine, including any collected in the trap, weighs 37–39 g. It is dried overnight with 5–7 g. of sodium hydroxide pellets (Note 12) and then distilled, a 12- to 15-cm. fractionating column (Note 13) and an ice-cooled receiver equipped with a soda-lime tube being used. The fraction boiling at 44–46° amounts to 31–34 g. (71–78%) of *tert*-butylamine; d_4^{20} 0.699; n_D^{20} 1.3800.

2. Notes

1. This step may be omitted if a practical grade of *tert*-butylurea is purchased from the Eastman Kodak Company. Since the commercial product may contain sodium sulfate and di-*tert*-butylurea it should be recrystallized from hot water, about 1 l. of solvent being used for 100 g. of the urea.

2. The temperature of the mixture should be kept in the range 20–25°. Higher temperatures lead to the formation of diisobutylene, and at lower temperatures (15°) the urea does not dissolve readily. Even at 25° the urea is usually not completely in solution. It has been found convenient to warm the *tert*-butyl alcohol to about 30–35° before placing it in the dropping funnel. This avoids solidification in the stem (the melting point of *tert*-butyl alcohol is 25.5°).

3. Longer standing does not improve the yield.

4. If the temperature is allowed to fall below 15°, large amounts of sodium sulfate decahydrate crystallize with the product. If this hap-

pens it may be necessary to recrystallize the *tert*-butylurea several times.

5. The insoluble residue weighs 10–22 g. and consists of di-*tert*-butylurea, which sublimes above 200° but melts in a sealed tube at 243°. If several runs are being made, the di-*tert*-butylurea should be saved and dried. It may be converted to *tert*-butylphthalimide in 63% yields by heating with four equivalents of phthalic anhydride under the same conditions as specified for mono-*tert*-butylurea.

6. The product should not be dried in an oven at elevated temperatures, as it sublimes.

7. This material is pure enough for the next step. Melting points of 172° (dec.) and 183° are given in the literature.[4,5] Recrystallization from dilute ethanol gives long, white needles melting at 182° (cor.), whereas use of 95% ethanol gives plates melting at the same temperature. The temperature of the melting-point bath should be raised more rapidly than usual since the product sublimes slowly above 100°.

8. Less rapid heating results in a diminished yield. The reaction is usually completed after a total time of 15 minutes, no further evolution of carbon dioxide or ammonia occurring. Prolonged heating causes the formation of a colored product and reduces the yield.

9. A similar series of reactions may be used to prepare *tert*-amylamine. The *tert*-amylurea is produced in yields of 50–58%, *tert*-amylphthalimide in yields of 63–72%, and *tert*-amylamine, b.p. 77–78°, in 87% yields (N. L. Drake and John Garman, private communication).

10. If the water in the condenser is cold enough to prevent loss of the amine the extra trap is unnecessary.

11. The *tert*-butylurea gradually dissolves, and a gelatinous mass of sodium carbonate forms.

12. If an aqueous layer forms it is separated and the amine is dried with a fresh portion of 5 g. of sodium hydroxide pellets.

13. The submitters and checkers used a Vigreux column.

3. Methods of Preparation

tert-Butylurea has been prepared by the action of *tert*-butyl bromide upon a mixture of urea and white lead;[4] by the action of *tert*-butylamine upon potassium cyanate;[5] in small amounts by heating N-*tert*-butylurethan with alcoholic ammonia.[4] The method given above is a modification of a general method for preparation of alkyl ureas, described by Harvey and Caplan in a patent.[6]

The direct hydrolysis of *tert*-butylurea is perhaps the simplest and most convenient method for preparing *tert*-butylamine.[7] In addition to the ethylene glycol method described here, it is possible to carry out the hydrolysis in a pressure vessel with aqueous sodium hydroxide (40%) at 130° (A. H. Sommers, private communication).

tert-Butylphthalimide is a new compound. The procedure given here is a modification[8] of a general procedure for the cleavage of alkylphthalimides, as developed by Ing and Manske[9] and Manske.[10]

2,2-Dimethylethylenimine has been prepared by the dehydration of 2-amino-2-methyl-1-propanol. The method described here is essentially that of Cairns.[11]

Methods for the hydrogenolysis of 2,2-dimethylethylenimine have been published.[12,13]

tert-Butylamine has also been obtained from trimethylacetamide by the Hofmann rearrangement,[14] from trimethylacetic acid by the Schmidt reaction,[15] from trimethylacetazide,[16] by the reduction of *tert*-butylhydrazide,[17] and by the reaction of *tert*-butylmagnesium chloride with O-methylhydroxylamine[18] or with monochloramine.[19]

[1] Fieser, *Experiments in Organic Chemistry,* 2nd ed., p. 369, D. C. Heath & Company, Boston, 1941. See also Hess and Frahm, *Ber.*, **71,** 2627 (1938).

[2] *Org. Syntheses* Coll. Vol. **3,** 181 (1954).

[3] Work done under a contract with the Office of Scientific Research and Development.

[4] Schneegans, *Arch. Pharm.*, **231,** 677 (1893).

[5] Brander, *Rec. trav. chim.*, **37,** 67 (1917).

[6] Harvey and Caplan, U. S. pat. 2,247,495 [*C. A.*, **35,** 6267 (1941)].

[7] Pearson, Baxter, and Carter, *J. Am. Chem. Soc.*, **70,** 2290 (1948).

[8] Smith and Emerson, *J. Am. Chem. Soc.*, **67,** 1862 (1945).

[9] Ing and Manske, *J. Chem. Soc.*, **1926,** 2348.

[10] Manske, *J. Am. Chem. Soc.*, **51,** 1209 (1929); Tingle and Brenton, *J. Am. Chem. Soc.*, **32,** 116 (1910).

[11] Cairns, *J. Am. Chem. Soc.*, **63,** 871 (1941).

[12] Campbell, Sommers, and Campbell, *J. Am. Chem. Soc.*, **68,** 140 (1946).

[13] Karabinos and Serijan, *J. Am. Chem. Soc.*, **67,** 1856 (1945).

[14] van Erp, *Rec. trav. chim.*, **14,** 16 (1895).

[15] Schuerck and Huntress, *J. Am. Chem. Soc.*, **71,** 2233 (1949).

[16] Buhler and Fierz-David, *Helv. Chim. Acta,* **26,** 2123 (1943).

[17] Klages et al., *Ann.*, **547,** 1 (1941).

[18] Brown and Jones, *J. Chem. Soc.*, **1946,** 781; Sheverdina and Kocheshkov, *J. Gen. Chem. U.S.S.R.*, **8,** 1825 (1938) [*C. A.*, **33,** 5804 (1939)].

[19] Coleman and Yager, *J. Am. Chem. Soc.*, **51,** 567 (1929).

n-BUTYLBENZENE *

(Benzene, butyl-)

$$C_6H_5Br + 2Na + n\text{-}C_4H_9Br \rightarrow n\text{-}C_4H_9C_6H_5 + 2NaBr$$

Submitted by R. R. READ, L. S. FOSTER, ALFRED RUSSELL, and V. L. SIMRIL.
Checked by C. F. H. ALLEN and JAMES VANALLAN.

1. Procedure

A dry, 3-l., three-necked, round-bottomed flask (Notes 1 and 2) is fitted with an efficient reflux condenser (Note 3) protected by a drying tube, a dropping funnel, and a thermometer which extends well into the reaction mixture (Note 4); the flask is arranged so that it can be cooled externally (Note 5). In the flask is placed 161 g. (7 gram atoms) of sodium cut into shavings 1–2 mm. in thickness (Notes 6 and 7); the sodium is just covered with dry ether (about 300 ml.) (Note 8). A mixture of 411 g. (321 ml., 3 moles) of *n*-butyl bromide and 471 g. (315 ml., 3 moles) of bromobenzene (Note 9) is added slowly from the dropping funnel over a period of about 2.5 hours, the temperature being kept as near 20° as possible; the mass acquires a bluish color.

After the flask and contents have been allowed to stand at room temperature for 2 days with occasional shaking, the liquid is decanted (Note 10). Three hundred milliliters of methanol is then added carefully, and the mixture is refluxed on a steam bath for 4 hours. Then 800 ml. of water is added to dissolve the salt, and the hydrocarbon layer is separated and added to the decanted liquid. The aqueous layer is extracted once with 250 ml. of ether (Note 11), and the combined hydrocarbon fraction, decanted solution, and ether extract are dried over 40 g. of calcium chloride. Most of the ether is removed on a steam bath, and the residual liquid is distilled through an electrically heated, jacketed column (Note 12). The fraction that boils at 180–182.5°/750 mm. is collected as *n*-butylbenzene (Notes 13 and 14); the yield is 261–281 g. (65–70%) (Note 15).

2. Notes

1. The size of the pieces of sodium, the control of temperature, and the use of an electrically heated, jacketed fractionating column are important factors in the successful preparation of the compound.

2. A copper flask and copper condenser reduce the hazard from breakage but are not essential and are less convenient with the quantities indicated.

3. Unless the reaction gets out of hand, the only function of the condenser is to prevent loss of ether. If the temperature gets above 30°, the reaction becomes violent and cannot be controlled by a single condenser.

4. A thermometer on which the scale is well above the surface of the reaction mixture is advisable; one reading from −50° to 50° is convenient.

5. The temperature range is critical. Below 15° reaction is extremely slow, but in time a vigorous reaction suddenly sets in and blows a good part of the reactants out through the condenser. Above 30° the reaction gets out of hand (Note 3).

6. This is a very tedious task. It is best accomplished by flattening the usual bars of sodium with a hammer and cutting the flattened strips with scissors. Alternatively the sodium, in 1-lb. lots, can be rolled under a heavy lawn roller (Read and Foster).

7. If the sodium is too thick, much of it fails to react, whereas sodium "sand" reacts very vigorously but gives poor yields of the desired product.

8. Larger runs require more ether. The ether may be dried over calcium chloride and used directly.

9. The *n*-butyl bromide and bromobenzene need not be redistilled.

10. Read and Foster recommend extracting the residue with benzene, using an automatic extractor.

11. Although this extract contains but 2–3 g. of hydrocarbon, its use facilitates drying later on and reduces loss through an occasional imperfect separation of layers.

12. A column such as the Whitmore-Lux or Fenske column provided with an electrically heated jacket is essential.

13. Alternatively, a crude fraction boiling at 160–185° is collected; on redistillation, the product that boils at 181–184° is collected (Read and Foster).

14. The fore-run amounts to about 45 g. The residue is largely biphenyl. Small additional amounts of product can be secured by combining these fractions from several runs and refractionating.

15. Without a heated column, the yield drops to 221 g. (54%).

3. Methods of Preparation

n-Butylbenzene has been prepared by the action of sodium (*a*) on benzyl chloride or bromide and *n*-propyl bromide without diluents,[1] or (*b*) on *n*-butyl bromide and bromobenzene without a solvent [2] or in benzene; [3] by Clemmensen [4] or Wolff-Kishner [5] reduction of *n*-butyrophenone; by the action of benzylmagnesium chloride on *n*-propyl *p*-toluenesulfonate; [6] and by the action of phenylmagnesium bromide on *n*-butyl bromide or of benzylmagnesium bromide on *n*-propyl bromide in the presence of a trace of ferric chloride.[7] It has also been prepared from benzylsodium and *n*-propyl bromide.[8]

[1] Radziszewski, *Ber.*, **9**, 261 (1876).
[2] Read and Foster, *J. Am. Chem. Soc.*, **48**, 1606 (1926).
[3] Balbiano, *Ber.*, **10**, 296 (1877); *Gazz. chim. ital.*, **7**, 343 (1877).
[4] Clemmensen, *Ber.*, **46**, 1839 (1913).
[5] Herr, Whitmore, and Schiessler, *J. Am. Chem. Soc.*, **67**, 2061 (1945).
[6] Gilman and Beaber, *J. Am. Chem. Soc.*, **47**, 523 (1925).
[7] Vavon and Mottez, *Compt. rend.*, **218**, 557 (1944).
[8] Bryce-Smith and Turner, *J. Chem. Soc.*, **1950**, 1975.

1-*n*-BUTYLPYRROLIDINE

(Pyrrolidine, 1-butyl-)

$$(n\text{-}C_4H_9)_2NH \xrightarrow{Cl_2} (n\text{-}C_4H_9)_2NCl \xrightarrow{H_2SO_4} \left.\begin{array}{l}CH_2CH_2\\ |\\ CH_2CH_2\end{array}\right\rangle N\text{—}C_4H_9\text{-}n$$

Submitted by GEORGE H. COLEMAN, GUST NICHOLS, and TED F. MARTENS.
Checked by C. F. H. ALLEN and J. VANALLAN.

1. Procedure

This preparation must be carried through the ring closure without interruption.

The apparatus is arranged as in Fig. 7. In a bottle of about 1.5-l. capacity are placed 64.5 g. (0.5 mole) of di-*n*-butylamine,[1] 350 ml. of ligroin (Note 1), and 350 ml. of 3 *N* sodium hydroxide (Note 2). The bottle is fitted with an inlet tube which passes only 1 in. through the stopper. The mixture is cooled in an ice bath (Note 3), and chlorine from a cylinder is passed in under pressure (Note 4). The bottle is kept in the ice bath and is shaken during the addition to aid in absorption of the gas. The rate of shaking and the valve on the chlorine cylinder are so regulated that the pressure as indicated on the manometer is maintained between 100 and 150 mm. (Note 5). The addition of the chlorine is continued until the white fumes of the hydrochloride, which form when chlorine comes in contact with the amine vapor, disappear, and the non-aqueous layer takes on a greenish yellow color, due to a slight excess of chlorine. This indicates that chlorination is complete (Note 6).

The ligroin solution of the chloramine is separated from the aqueous layer in a chilled, short-stemmed 1-l. separatory funnel and washed successively with 50 ml. of ice-cold 3 *N* sodium hydroxide (Note 7), 50

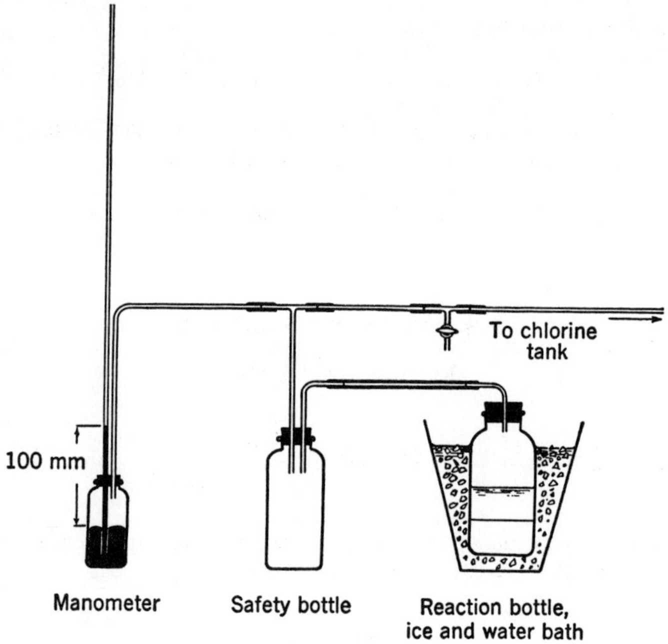

FIG. 7.

ml. of ice water, and 50 ml. of cold 2 *N* sulfuric acid. The ligroin solution becomes nearly colorless. The chloramine is extracted from the solution with sulfuric acid in the following manner. An ice-cold mixture of 200 ml. of concentrated sulfuric acid (sp. gr. 1.84) and 80 ml. of water is allowed to stand in contact with the ligroin solution for 10 minutes *without* shaking, then for 20 minutes *with* occasional shaking; the mixture is kept cool by immersion in an ice bath. After separation from the acid layer, the ligroin is extracted with two 60-ml. portions of cold concentrated sulfuric acid. The combined sulfuric acid extracts are used for ring closure (Note 8). The ligroin, which contains not more than a trace of chloramine, is discarded.

A 1-l. wide-mouthed Erlenmeyer flask, fitted with a propeller-type stirrer and a thermometer, is set in an oil bath. A mixture of 40 ml.

of concentrated sulfuric acid and 10 ml. of water is placed in the flask, the oil bath is heated to 120°, and the cold sulfuric acid solution of the chloramine is allowed to flow into the heated acid from the short-stemmed separatory funnel (Note 9). The addition, which requires 30–40 minutes (Note 10), is carried out at such a rate that the temperature of the reaction mixture in the flask is dropped from 120° rapidly to 95° by the addition of the cold chloramine solution, and thereafter at such a rate that the temperature of the reaction mixture remains at 90–100°, preferably at 95°. The reaction is exothermic, and the rise in temperature is controlled by the addition of the cold chloramine solution. To avoid an undue rise in temperature after the last addition of the chloramine, the oil bath is removed, and the flask and contents are allowed to come to room temperature.

A 5-l. flask is fitted with a separatory funnel and arranged for steam distillation (p. 64). Two liters of crushed ice is placed in the flask, and the sulfuric acid solution, cooled to 0°, is slowly added. The amines are then liberated (Note 11) by adding cold concentrated sodium hydroxide solution (650 g. in 1.5 l. of water) through the separatory funnel. The amines are steam-distilled and collected in a solution of 100 ml. of concentrated hydrochloric acid (sp. gr. 1.19) and 200 ml. of water (Note 12). Distillation is continued until all amine has passed over, as indicated by a negative test with litmus paper. About 2 l. of distillate is usually required.

The amine hydrochloride solution is evaporated nearly to dryness on a steam bath (Note 13) and transferred with about 300 ml. of water to a 1-l. three-necked flask equipped with a stirrer,[2] separatory funnel, and condenser. The stirrer is started, and the acid solution is cooled to 0° by immersing the flask in an ice bath. The amines are liberated by adding slowly 100 g. of sodium hydroxide dissolved in 250 ml. of water.

Hinsberg separation. To this solution, which should be cold (5–8°), is now added 23 g. of benzenesulfonyl chloride,[3] and the mixture is stirred vigorously for 30 minutes. The separatory funnel is replaced by a stopper bearing a thermometer, and the contents of the flask are warmed to 40° and stirred until the odor of the acid chloride is no longer noticeable. This usually requires about 30 minutes. The 1-*n*-butylpyrrolidine is separated (Note 14) from the non-volatile di-*n*-butylbenzenesulfonamide by steam distillation, the amine being collected in dilute acid (Note 12) as before. The acid solution is then evaporated to dryness, and the amine is liberated by adding 20% sodium hydroxide solution until the aqueous layer turns red litmus blue. The amine is extracted by one 200-ml. portion of ether, and the ethereal solution is dried over 15–20 pellets of potassium hydroxide. After the ether has been distilled from the decanted solution, the residue is distilled from an oil bath. The yield of 1-*n*-butylpyrrolidine boiling at 154–155°/758 mm. is 44–51 g. (70–80%); n_D^{27} 1.437 (Note 15).

2. Notes

1. Any fraction boiling within the range 60–90° may be used.

2. This is obtained by dissolving 42 g. of sodium hydroxide in 350 ml. of water.

3. All equipment, the solutions, and the chloramine solution should be kept ice-cold to prevent decomposition during the preparation. Apparatus may be stored in a refrigerator.

4. As the addition of chlorine is started, the stopper in the reaction bottle is loosened momentarily and the chlorine is allowed to replace most of the air in the system.

5. As the chlorine passes initially into the bottle, the pressure rises rapidly. Shaking greatly increases the rate of absorption, and the pressure drops. The chlorine valve is then regulated so as to maintain the proper pressure with shaking. One hundred millimeters was selected as the approximate pressure, since the rate of chlorination under this pressure is satisfactory.

6. A large excess of chlorine is undesirable and may result in greatly decreased yields. The time required varies widely, being dependent upon the chlorine pressure maintained and the vigor of the shaking. The checkers found that 20 minutes was ample under their conditions; the submitters reported that 1–1.5 hours was required.

7. If the green color, due to excess chlorine, is not removed by one washing, the operation is repeated as many times as may be necessary.

8. The sulfuric acid layers are transferred directly into the 500-ml. short-stemmed separatory funnel to be used in the next step. The solution, which usually has a light brown color, should be cold and well mixed before the ring-closure operation. The ratio of acid to water is important, and only the amounts specified should be used.

9. The funnel and contents may be supported over the reaction flask during the addition without cooling.

10. The rate of addition should be as rapid as possible provided that the proper temperature is maintained. Deviation of more than 5° from the optimum reaction temperature of 95° results in reduced yields.

11. This reaction is vigorous, and the solution of alkali should be added slowly while the flask is shaken gently with a rotary motion.

12. An adapter (rubber connector and glass tube) from the lower end of the condenser should extend below the surface of the absorbing acid in the receiver. To reduce fuming, a 3- to 4-mm. layer of ligroin is placed over the acid solution. The acid solution may be stirred occasionally.

13. Use of reduced pressure considerably diminishes the time required.

14. The amine can be separated from the sulfonamide by extracting both with ether, drying this extract over solid potassium hydroxide, and fractionating. The pyrrolidine distils smoothly, but the yield is slightly lower.

15. Several other pyrrolidines of this series can be prepared by the same general method. The temperature required for ring closure and the percentage yield of pyrrolidine vary with the different amines used. The tabulated temperatures are given as approximately correct for the ring closure of the N-chloro-derivatives of the amines listed. With lower-boiling pyrrolidines, incomplete separation from the ether may result in lower yields.

Amine	*Optimum Temperature, °C*	*Pyrrolidine Formed*
Methyl-*n*-butyl	100–110	1-Methylpyrrolidine
Ethyl-*n*-butyl	110–115	1-Ethylpyrrolidine
n-Propyl-*n*-butyl	80–85	1-*n*-Propylpyrrolidine
Methyl-*n*-amyl	90–100	1,2-Dimethylpyrrolidine
Ethyl-*n*-amyl	80–90	1-Ethyl-2-methylpyrrolidine
Methyl-*n*-octyl	60–70	1-Methyl-2-*n*-butylpyrrolidine

3. Methods of Preparation

1-*n*-Butylpyrrolidine has been prepared by heating the corresponding N-bromoamine in concentrated sulfuric acid;[4] by catalytic reduction of N-butylpyrrole;[5] from furan and *n*-butylamine in the presence of aluminum oxide at 350–450°;[6] from 1,4-dichlorobutane and *n*-butylamine with potassium carbonate.[7] The procedure described is adapted from a preparation reported earlier.[8]

[1] *Org. Syntheses* Coll. Vol. **1**, 202 (1941).
[2] *Org. Syntheses* Coll. Vol. **1**, 33, Fig. 2*B* (1941).
[3] *Org. Syntheses* Coll. Vol. **1**, 84 (1941).
[4] Britton, U. S. pat. 1,607,605 [*C. A.*, **21**, 249 (1927)].
[5] Ochiai, Tsude, and Yokoyama, *Ber.*, **68**, 2293 (1935).
[6] Yurev, Tronova, L'vova, and Bukshpan, *J. Gen. Chem. U.S.S.R.*, **11**, 1128 (1941).
[7] Elderfield and Hageman, *J. Org. Chem.*, **14**, 605 (1949).
[8] Coleman and Goheen, *J. Am. Chem. Soc.*, **60**, 730 (1938).

n-CAPROIC ANHYDRIDE

$$2CH_3(CH_2)_4CO_2H + CH_2{=}C{=}O \rightarrow [CH_3(CH_2)_4CO]_2O + CH_3CO_2H$$

Submitted by JONATHAN W. WILLIAMS and JOHN A. KRYNITSKY.
Checked by NATHAN L. DRAKE and JOSEPH LANN.

1. Procedure

One hundred and sixteen grams (126 ml., 1 mole) of *n*-caproic acid is placed in a 250-ml. gas-washing bottle. The bottle is supported in an ice bath, and 0.50 to 0.55 mole of ketene is passed into the acid at a rate of approximately 0.45 mole per hour (Notes 1 and 2).

The resulting mixture is transferred to an apparatus for fractional distillation, and carefully fractionated, an oil bath being used for heating (Note 3). A low-boiling fraction, consisting of acetone containing some ketene, acetic acid, and a small quantity of acetic anhydride, is removed at atmospheric pressure. As the distillation progresses the temperature of the oil bath is raised to 220° over a period of about 1 hour and held there until 3 hours has elapsed from the time distillation started (Note 4).

The distillation at atmospheric pressure is then discontinued, the liquid is allowed to cool somewhat, and distillation is continued at a pressure of 3–10 mm. After a fore-run of less than 20 g., *n*-caproic anhydride is collected (b.p. 109–112°/3 mm., 118–121°/6 mm.). The yield is 86–95 g. (80–87%) (Note 5).

2. Notes

1. Ketene may be generated conveniently, at the proper rate, in the apparatus described by Williams and Hurd.[1]

2. Addition of 1 mole of ketene per mole of acid does not increase the yield. Under these conditions more acetic anhydride is found in the low-boiling fraction.

3. Submitters and checkers used a column of the Whitmore-Lux type,[2] 12 mm. in diameter, 50 cm. long, packed with glass helices, and provided with the usual jackets for heating. A less efficient column will serve in the preparation of caproic anhydride but not in the preparation of propionic anhydride or butyric anhydride by the same method.

4. It is imperative to continue the distillation at atmospheric pressure until conversion of any mixed anhydride to caproic anhydride is complete. The acetic acid formed by this conversion comes off very

slowly, and approximately 3 hours is necessary to complete the distillation at atmospheric pressure.

5. According to the submitters, equally good yields can be obtained in the preparation of propionic anhydride and *n*-butyric anhydride.

3. Methods of Preparation

n-Caproic anhydride has been prepared by heating caproic acid with acetic anhydride,[3] by heating sodium caproate and acetic anhydride in a sealed tube,[4] by the action of phosphorus oxychloride on barium caproate,[5] by the action of acetyl chloride on caproic acid,[6] and by treating a mixture of sodium caproate and sulfur with chlorine.[7] The method used in the present synthesis was first described by Hurd and Dull.[8]

[1] Williams and Hurd, *J. Org. Chem.*, **5**, 122 (1940).
[2] Whitmore and Lux, *J. Am. Chem Soc.*, **54**, 3451 (1932).
[3] Autenrieth, *Ber.*, **34**, 168 (1901).
[4] Michael, *Ber.*, **34**, 918 (1901).
[5] Chiozza, *Ann.*, **86**, 359 (1853).
[6] Fournier, *Bull. soc. chim. France,* (4), **5**, 920 (1909).
[7] Brit. pat. 24,842 (1908) [*C. A.*, **4**, 2190, 2719 (1910)].
[8] Hurd and Dull, *J. Am. Chem. Soc.*, **54**, 3427 (1932).

3-CARBETHOXYCOUMARIN

(2H-1-Benzopyran-3-carboxylic acid, 2-oxo-, ethyl ester)

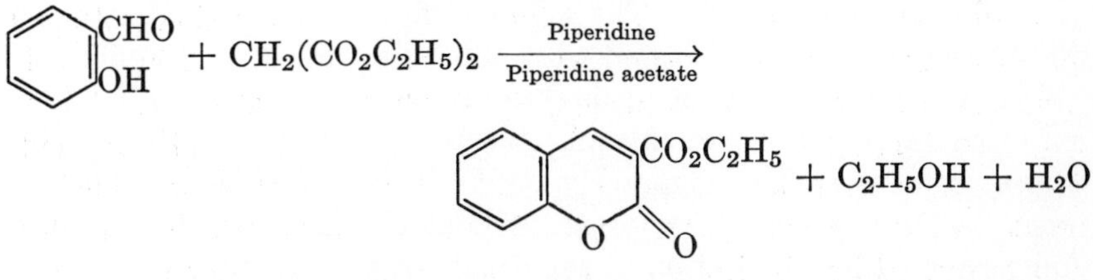

Submitted by E. C. Horning, M. G. Horning, and D. A. Dimmig.
Checked by Richard T. Arnold and Marshall Freerks.

1. Procedure

In a 500-ml. round-bottomed flask equipped with a reflux condenser are placed 61 g. (0.50 mole) of salicylaldehyde (Note 1), 88 g. (0.55 mole) of ethyl malonate, and 200 ml. of absolute ethanol. To this mixture are added 5 ml. of piperidine (Note 2) and 0.5 ml. of glacial acetic acid, and the solution is heated under reflux for 3 hours. The hot solution is transferred to a 1-l. Erlenmeyer flask, the reaction flask is rinsed with 20 ml. of ethanol, and the ethanol rinse and 330 ml. of hot water (Note 3) are added to the solution. The product crystallizes readily as the solution cools; the mixture is stirred from time to time as crystallization proceeds and is finally stored overnight in a refrigerator. The crystalline product is collected by filtration and washed with a solution made from 80 ml. of 95% ethanol and 120 ml. of water. The material is dried in the air. The yield is 85–91 g. (78–83%) of product melting at 91–93°.

The product may be recrystallized by dissolving it in 200 ml. of hot ethanol (95%), filtering, and adding 315 ml. of hot water. The recrystallized product is washed on the filter with 200 ml. of aqueous ethanol, as before, and air-dried. The yield of white 3-carbethoxycoumarin is 79–85 g. (73–78%); m.p. 92–94°.

2. Notes

1. The yield of the final product (m.p. 92–94°) may be increased to 80–84% by the use of salicylaldehyde purified in the following manner.

Preparation of pure salicylaldehyde.[1] Two hundred and fifty grams (1 mole) of copper sulfate pentahydrate is dissolved in 500 ml. of hot water in a 1-l. Erlenmeyer flask, and 244 g. (2 moles) of salicylaldehyde (Eastman practical grade) is added. A solution of 80 g. (2 moles) of sodium hydroxide in 100 ml. of water is added slowly in small portions with intermittent vigorous shaking. The mixture is permitted to cool slowly to room temperature with intermittent shaking and is finally allowed to stand overnight. The solid is collected and washed with 200 ml. of ethanol (95%), digested with 400 ml. of ether, and again collected on a filter. Without drying, the product is treated with 1 l. of water containing 108.5 g. (59 ml., 105 moles) of 95% sulfuric acid. The mixture is shaken vigorously, and 200 ml. of ether is added to break up the oily mass that forms. The aldehyde is collected in ether and recovered by distillation of the dried (over calcium sulfate) solution. It distils at 96–97°/35 mm.; the recovery is 90%.

2. Eastman practical grade is satisfactory.

3. The water should be heated to about 60°.

3. Methods of Preparation

This compound has been prepared only by the Knoevenagel[2] condensation, with a secondary amine as a catalytic agent. The corresponding methyl ester and free acid have also been prepared by using methyl malonate [3] and malonic acid,[4] respectively, in the condensation with salicylaldehyde.

[1] Claisen and Eisleb, *Ann.*, **401**, 95 (1914).
[2] Knoevenagel, *Ber.*, **31**, 2593 (1898).
[3] Werder, *Jahresber.*, **50**, 88 (1936).
[4] U. S. pat. 2,338,569 [*C. A.*, **38**, 3671 (1944)].

CARBOBENZOXY CHLORIDE AND DERIVATIVES

(Formic acid, chloro-, benzyl ester)

$$C_6H_5CH_2OH + COCl_2 \rightarrow C_6H_5CH_2OCOCl + HCl$$

$$C_6H_5CH_2OCOCl + 2NH_3 \rightarrow C_6H_5CH_2OCONH_2 + NH_4Cl$$

$$C_6H_5CH_2OCOCl + H_2NCH_2CO_2H + NaOH \rightarrow C_6H_5CH_2OCONHCH_2CO_2H + NaCl + H_2O$$

Submitted by H. E. Carter, R. L. Frank, and H. W. Johnston.
Checked by Nathan L. Drake and Charles M. Eaker.

1. Procedure

Caution! This procedure should be carried out in a hood.

A. *Benzyl chloroformate.* A 3-l. round-bottomed flask is fitted with a rubber stopper carrying an exit tube and a delivery tube extending to the bottom of the flask. The tubes are equipped with stopcocks so that the reaction flask may be disconnected. In the flask is placed 500 g. of dry toluene (Note 1), and the apparatus is weighed. The flask is then cooled in an ice bath, and phosgene (Note 2) is bubbled into the toluene until 109 g. (1.1 moles) has been absorbed (Note 3). The exit gases are passed through a flask containing toluene to remove any phosgene and then through a calcium chloride tube to a gas trap.

After the absorption of phosgene is completed the connection to the phosgene tank is replaced by a separatory funnel. The reaction flask is gently shaken while 108 g. (104 ml., 1 mole) of redistilled benzyl alcohol is added rapidly through the separatory funnel. The flask is allowed to stand in the ice bath for 30 minutes and at room temperature for 2 hours. The solution is then concentrated under reduced pressure, at a temperature not exceeding 60°, in order to remove hydrogen chloride, excess phosgene (Note 4), and the major portion of the toluene. The residue weighs 200–220 g. and contains 155–160 g.

of benzyl chloroformate (91–94% based on the benzyl alcohol) (Note 5). The amount of benzyl chloroformate present in this solution may be estimated by preparing the amide from a small aliquot portion, or it may be safely calculated by assuming a minimum yield of 90% based on the benzyl alcohol used.

B. *Benzyl carbamate.* A measured aliquot (suitably 10 ml.) of the solution of benzyl chloroformate, prepared as described above, is added slowly and with vigorous stirring to 5 volumes of cold concentrated ammonium hydroxide (sp. gr. 0.90), and the reaction mixture is allowed to stand at room temperature for 30 minutes. The precipitate is filtered with suction, washed with cold water, and dried in a vacuum desiccator. The yield of practically pure benzyl carbamate, melting at 85–86°, is 7.0–7.2 g. (91–94% based on the benzyl alcohol used in A). Pure benzyl carbamate melting at 87° is obtained by recrystallizing the slightly impure material from 2 volumes of toluene.

C. *Carbobenzoxyglycine.* A solution of 7.5 g. (0.1 mole) of glycine in 50 ml. of 2 *N* sodium hydroxide is placed in a 200-ml. three-necked flask fitted with a mechanical stirrer and two dropping funnels. The flask is cooled in an ice bath, and 17 g. (0.1 mole) of benzyl chloroformate (21–24 g. of the solution obtained in A) and 25 ml. of 4 *N* sodium hydroxide are added simultaneously to the vigorously stirred solution over a period of 20–25 minutes. The mixture is stirred for an additional 10 minutes. The toluene layer is separated, and the aqueous layer is extracted once with ether. The aqueous solution is cooled in an ice bath and acidified to Congo red with concentrated hydrochloric acid (Note 6). The precipitate is filtered, washed with small portions of cold water, and dried in the air. It is practically pure carbobenzoxyglycine; it weighs 18–19 g. (86–91%) and melts at 119–120°. The material may be recrystallized from chloroform; it then melts at 120° (Note 7).

2. Notes

1. The toluene may be dried by distillation.

2. Commercial phosgene was used; it was obtained in a tank from the Ohio Chemical Company.

3. The phosgene is absorbed rapidly for some time, then more slowly as the concentration increases. About 1 hour is required for this step. The amount of phosgene absorbed is checked by weighing the flask and delivery tubes occasionally.

4. In order to prevent the escape of phosgene, a toluene trap is inserted between the apparatus and the water pump. For this purpose

it is convenient to use a 2-l. flask immersed in an ice bath and containing about 1 l. of toluene. The flask is fitted with an inlet tube reaching almost to the bottom.

5. It is not practical to remove the toluene completely; moreover, toluene does not interfere in the preparation of the derivatives.

6. The derivative may precipitate as an oil. However, crystallization is readily induced by cooling and scratching.

7. Carbobenzoxyalanine (m.p. 114–115°) is obtained in 80–90% yield from alanine and benzyl chloroformate by the same procedure.

3. Methods of Preparation

Benzyl chloroformate has been prepared by action of phosgene on benzyl alcohol.[1–3] The methods described here for the preparation of benzyl chloroformate and the carbobenzoxy derivatives of glycine and alanine are essentially those of Bergmann and Zervas.[2] The carbobenzoxy derivatives of other amino acids are conveniently prepared in the same way. Benzyl carbamate has been prepared by the action of ammonia on benzyl chloroformate.[1,4] The present method is that of Martell and Herbst.[4]

1 Thiele and Dent, *Ann.*, **302,** 257 (1898).
2 Bergmann and Zervas, *Ber.*, **65,** 1192 (1932).
3 Farthing, *J. Chem. Soc.*, **1950,** 3213.
4 Martell and Herbst, *J. Org. Chem.*, **6,** 882 (1941).

β-CARBOMETHOXYPROPIONYL CHLORIDE

(Propionic acid, β-(chloroformyl)-, methyl ester)

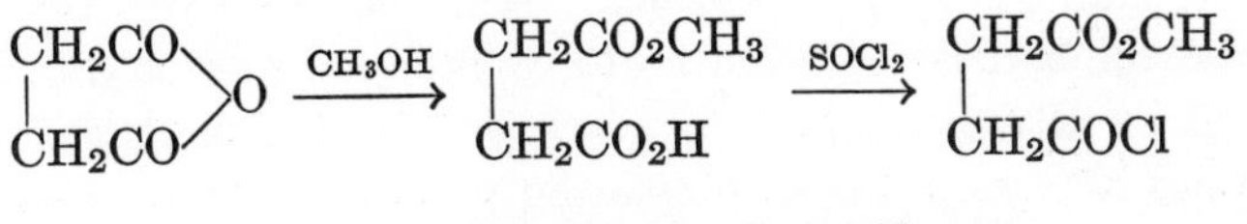

Submitted by James Cason.
Checked by C. F. H. Allen and C. V. Wilson.

1. Procedure

A. *Methyl hydrogen succinate.* A mixture of 400 g. (4 moles) of succinic anhydride (Note 1) and 194 ml. (4.8 moles) of methanol (Note 2) in a 1-l. round-bottomed flask is refluxed on a steam bath. After about 35 minutes the mixture is swirled frequently until it becomes homogeneous (this requires 15–30 minutes); the flask is then half immersed in the steam bath for an additional 30–25 minutes (Note 3).

The excess methanol is removed by distillation under reduced pressure (water pump) from a steam bath, and the residual liquid is poured into an 18- to 25-cm. evaporating dish which is cooled in a shallow pan of cold water. As the half ester crystallizes, it is stirred and scraped off the dish in order to prevent formation of a solid cake. After being dried to constant weight in a vacuum desiccator (5–8 days), the product weighs 502–507 g. (95–96%) and melts at 57–58° (Notes 4 and 5).

B. *β-Carbomethoxypropionyl chloride.* In a 1-l. flask (Note 6) bearing a reflux condenser are placed 264 g. (2 moles) of methyl hydrogen succinate and 290 ml. (4 moles) of thionyl chloride (Note 7), and the solution is warmed in a bath at 30–40° for 3 hours (Note 8). The condenser is replaced by a modified Claisen still head, the excess thionyl chloride is removed on a steam bath under reduced pressure, and the β-carbomethoxypropionyl chloride is distilled (Notes 9 and 10). The yield of colorless product is 270–278 g. (90–93%), b.p. 92–93°/18 mm. (Notes 11 and 12).

2. Notes

1. Eastman's succinic anhydride (m.p. 115–116°) was used.

2. Synthetic methanol was used. Since this anhydrous alcohol is hygroscopic, partly filled bottles that have been opened intermittently in the laboratory should be rejected.

3. Thirty minutes is allowed if solution resulted after 15 minutes of swirling; 25 minutes if 30 minutes was needed for homogeneity.

The time factor is very important. In one run in which the mixture was heated for a total of 55 minutes, a product was obtained which was shown by titration to contain about 6% of anhydride. Longer heating than that specified increases the yield of diester. Any change in quantity of materials used may necessitate a new set of conditions in order to obtain the maximum yield.

4. The checkers prefer the following procedure, which can be carried through in one day. A suspension of approximately one-half of the crude product in 750 ml. of carbon disulfide is warmed on a steam bath; two layers form, in which some solid remains in suspension. This is dissolved by the addition of 350 ml. of ether. The whole is chilled to 0°, and the solid is filtered by suction. The other half of

the crude product is now dissolved in this filtrate, the solution is again chilled to 0°, and the solid is filtered. The combined yield of acid ester, m.p. 57–58°, is 438–449 g. (83–85%). A further 32–37 g. (6–7%) of less pure material (m.p. 56–57°) can be obtained by concentrating the filtrate to half its volume and chilling to 0°.

5. The product, which is sufficiently pure for the next step, contains at least 98% methyl hydrogen succinate as shown by titration or distillation through an 18-in. Podbielniak-type column.

6. Equipment with ground-glass joints is used throughout.

7. Eastman's thionyl chloride (b.p. 75–76°) was used. The checkers obtained equivalent yields of the chloride by using only 20% excess of thionyl chloride. The mixture was heated for 1 hour at 40°, allowed to stand overnight, and heated again for 2 hours at 40°.

8. Since hydrogen chloride is evolved, it is advisable to work in a hood or employ a gas trap.

9. An electric heating mantle is convenient.

10. Other boiling points are 85.5–87°/13 mm. and 89–90°/15 mm. The use of as low a pressure as possible is advisable, since the substance tends to lose methyl chloride and form succinic anhydride.

11. The submitter obtained the same yield when phosphorus pentachloride was used instead of thionyl chloride. Consistent results were obtained with the former reagent only when the acid chloride was distilled at pressures below 3 mm. (b.p. 58–59°/2.5 mm.).

12. Methyl hydrogen glutarate, ethyl hydrogen adipate, and ethyl hydrogen sebacate may be converted to the corresponding ester acid chlorides by this procedure in about the same yields. Distillation should be carried out rapidly at a pressure of 4 mm. or lower.

3. Methods of Preparation

Methyl hydrogen succinate has been prepared by heating succinic acid with methyl succinate,[1] by treating ethyl succinate with sodium methoxide,[2] and by heating succinic anhydride with methanol.[3–6]

β-Carbomethoxypropionyl chloride has been prepared from methyl hydrogen succinate by the use of thionyl chloride[4,6] or phosphorus pentachloride.[5]

[1] Fourneau and Sabetay, *Bull. soc. chim. France,* (4) **45,** 841 (1929).

[2] Komnenos, *Monatsh.,* **32,** 77 (1911).

[3] Bone, Sudborough, and Sprankling, *J. Chem. Soc.,* **85,** 539 (1904).

[4] Nenitzescu, Ciorănescu, and Przemetsky, *Ber.,* **73B,** 313 (1940).

[5] Cason, *J. Am. Chem. Soc.,* **64,** 1106 (1942).

[6] Ruggli and Maeder, *Helv. Chim. Acta,* **25,** 936 (1942).

CARBOXYMETHOXYLAMINE HEMIHYDROCHLORIDE

(Acetic acid, aminoöxy-, hydrochloride)

$$(CH_3)_2C{=}NONa + BrCH_2CO_2Na \rightarrow (CH_3)_2C{=}NOCH_2CO_2Na + NaBr$$

$$(CH_3)_2C{=}NOCH_2CO_2Na + HCl \rightarrow (CH_3)_2C{=}NOCH_2CO_2H + NaCl$$

$$2(CH_3)_2C{=}NOCH_2CO_2H + 2H_2O + HCl \rightarrow (H_2NOCH_2CO_2H)_2HCl + 2(CH_3)_2CO$$

Submitted by H. S. Anker and H. T. Clarke.
Checked by H. R. Snyder and Peter Kovacic.

1. Procedure

A. *Acetone carboxymethoxime.* A mixture of 612 g. (4.4 moles) of bromoacetic acid (Note 1) and 500 g. of crushed ice is chilled in an ice-salt bath and made distinctly alkaline to litmus with sodium hydroxide (about 440 g. of a 40% solution). During the neutralization an additional 500 g. of ice is added. To the solution are then added 292 g. (4.0 moles) of acetoxime [1] and 440 g. of 40% sodium hydroxide (4.4 moles), the temperature being held below 20° during the addition of the alkali. The mixture is then allowed to flow dropwise, during 3–4 hours, through the inner tube of a steam-heated Liebig condenser (jacket 75 cm. long; inner tube 10-mm. diameter; angle of inclination about 20°) into a 5-l. round-bottomed flask cooled with running water (Note 2). The resulting solution is extracted three times with 500-ml. quantities of freshly distilled peroxide-free ether (Note 3), and the aqueous solution is then cooled and made strongly acid by the addition of 500 ml. of concentrated hydrochloric acid (6 moles). During the acidification the temperature should not rise above 15°. The solution is saturated with sodium chloride and immediately extracted with six successive 1.5-l. portions of peroxide-free ether. The ether is distilled from the combined ethereal extracts, and the residue, consisting of the crude acetone carboxymethoxime (333–345 g.), is used for the next step. The acetone carboxymethoxime may be purified by distillation under reduced pressure, the fraction boiling at 95–97°/1 mm. (Note 4) being collected (Note 5). The yield is 300 g. (57%) of a colorless product melting at 76°.

B. *Carboxymethoxylamine hemihydrochloride.* The crude acetone carboxymethoxime is dissolved in about twice its weight of benzene; the solution is filtered and freed of benzene by distillation under reduced pressure from a steam bath. To a solution of 200 g. (1.52 moles) of the residue in 1 l. of water, in a 5-l. flask, are added 2 mg. of hydroquinone and 1 l. of concentrated hydrochloric acid. The flask is connected with a condenser, and steam is passed through the solution until acetone no longer comes over (30–40 minutes). The solution is concentrated under reduced pressure to a volume of 180–220 ml., and 400 ml. of isopropyl alcohol is added. The solution is then stored for 12 hours in the icebox, and the crystals that separate are collected on a Büchner funnel and washed with cold isopropyl alcohol. Further crops are obtained by concentrating the mother liquors and adding isopropyl alcohol. The crude product (120–135 g.) may be recrystallized (Note 6) with very little loss by dissolving it in twice its weight of warm (50°) water (Note 7), adding 2 volumes of isopropyl alcohol, and again chilling in an icebox. Further small quantities can be recovered from the mother liquors by systematic repetition of the process described. The yield is 110–120 g. (66–72%) of white crystals which melt with decomposition at 152–153° (Notes 8 and 9).

2. Notes

1. Chloroacetic acid gives a poorer yield (46–49%) of acetone carboxymethoxime, and the crude product is more difficult to purify.[2]

2. By this procedure, the reaction takes place in a few seconds, and the formation of by-products is minimized. If the solution of the reactants is heated in bulk, the reaction temperature cannot be controlled, and a lower yield is obtained of a dark product which, however, can be purified by distillation under reduced pressure.

3. The unchanged acetoxime extracted by the ether amounts to 14–24 g. (5–8%).

4. The temperature of the vapor, during distillation under apparently comparable conditions, may differ from run to run by as much as 20°. The temperature range of 95–97° is the lowest observed for 1-mm. pressure. Boiling ranges of 110–118°/1 mm. have been reported.

5. No carbonization and only slight formation of hydrogen cyanide, which occurs extensively during the distillation of preparations from chloroacetic acid,[2] are observed.

6. The use of decolorizing carbon should be avoided, as some brands appear to contain impurities that catalyze decomposition to ammonium chloride.

7. If the resulting aqueous solution is cooled to 0° before the addition of isopropyl alcohol, about one-third of the product crystallizes in very pure form.

8. The melting point depends on the rate at which the sample is heated. When the temperature is raised in the ordinary way the material melts (with evolution of gas) at 152–153°; when the bath is heated to 150° before the sample is inserted, the melting point is 159°. On the other hand, when a sample is held between 140° and 145° it melts after about 6 minutes. Decomposition evidently plays a large part in the matter.

9. The hydrolysis of acetone carboxymethoxime may also be accomplished on a smaller scale (10 g.) by a simplified procedure as described by Lott.[3]

3. Methods of Preparation

Carboxymethoxylamine (also called hydroxylamine-O-acetic acid), which is of value for the isolation of ketones,[4] has been prepared by the hydrolysis of ethylbenzhydroximinoacetic acid [5] and of ethyl benzhydroximinoacetate.[6] The present method is a modification of that described by Borek and Clarke.[2]

[1] *Org. Syntheses* Coll. Vol. **1,** 318 (1941).
[2] Borek and Clarke, *J. Am. Chem. Soc.,* **58,** 2020 (1936).
[3] Lott, *J. Am. Chem. Soc.,* **70,** 1972 (1948).
[4] Anchel and Schoenheimer, *J. Biol. Chem.,* **114,** 539 (1936).
[5] Werner, *Ber.,* **26,** 1567 (1893); Werner and Sounenfeld, *Ber.,* **27,** 3350 (1894).
[6] Kitagawa and Takani, *J. Biochem. Japan,* **23,** 181 (1936).

o-CARBOXYPHENYLACETONITRILE

(*o*-Toluic acid, α-cyano-)

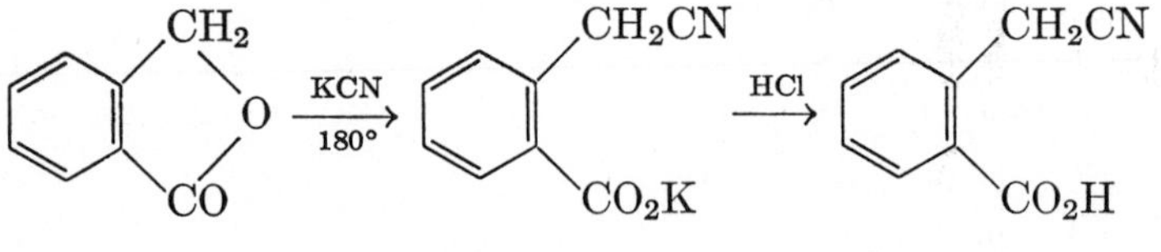

Submitted by Charles C. Price and Richard G. Rogers.
Checked by W. E. Bachmann and Richard D. Morin.

1. Procedure

A mixture of 100 g. of phthalide (Note 1) and 100 g. of powdered potassium cyanide is placed in a 2-l. round-bottomed flask fitted with a stirrer and a thermometer. The stirred mixture is heated to 180–

190° (internal temperature) for 4–5 hours in an oil bath (Note 2). One liter of water is added to the cooled mass, and the mixture is stirred until the solid salts are dissolved (Note 3). Any insoluble material that separates is removed by filtration (Note 4). Under a hood, 6 *N* hydrochloric acid (20–60 ml.) is added to the dark aqueous solution until it becomes turbid (Note 5). The solution is carefully neutralized with sodium bicarbonate (Note 6), a few grams of Norit is added, and the mixture is stirred for several minutes and filtered. The nearly colorless filtrate is acidified with 40–50 ml. of concentrated hydrochloric acid and, after cooling in an ice bath, is filtered with suction. The yield is 80–100 g. (67–83%) of white crystals which melt at 113–115° (Note 7).

2. Notes

1. The submitters used phthalide obtained from E. I. du Pont de Nemours and Company, Wilmington, Delaware. The checkers prepared it according to *Org. Syntheses* Coll. Vol. **2**, 526 (1943).

2. At the end of the reaction, the mixture should be dark brown and nearly solid. The temperature must not rise above 200°; the checkers kept it at 180°.

3. About 1 hour is required to disintegrate the mass.

4. In some runs 5–15 g. of phthalide is recovered at this point.

5. Occasionally a small amount of crystalline homophthalimide separated from the alkaline solution at this point; m.p. 235°.[1]

6. The checkers found it advisable to acidify the solution slightly at this stage in order to precipitate dark impurities.

7. This material is satisfactory for most purposes. It can be purified by recrystallization from benzene or acetic acid, though with considerable loss.

3. Methods of Preparation

o-Carboxyphenylacetonitrile has been prepared by the reaction of phthalide with potassium cyanide.[2–4] The above procedure is essentially that of Wislicenus.[2,3]

[1] Gabriel, *Ber.*, **19**, 1655 (1886).

[2] Wislicenus, *Ann.*, **233**, 102 (1886).

[3] Price, Lewis, and Meister, *J. Am. Chem. Soc.*, **61**, 2760 (1939).

[4] Johnston, Kaslow, Langsjoen, and Shriner, *J. Org. Chem.*, **13**, 477 (1948).

CATALYST, RANEY NICKEL, W-6

(With high contents of aluminum and adsorbed hydrogen)

$$NiAl_2 + 6NaOH \rightarrow Ni + 2Na_3AlO_3 + 3H_2$$

Submitted by Harry R. Billica and Homer Adkins.
Checked by Arthur C. Cope and Harold R. Nace.

Caution! The Raney nickel catalysts described below cannot be used safely under all conditions of temperature, pressure, and ratio of catalyst to hydrogen acceptor which are employed with less active nickel catalysts. They are particularly effective for low-pressure hydrogenations. No difficulty has been encountered in their use of temperatures below 100°, *or above* 100° *if the ratio of catalyst to possible hydrogen acceptor is* 5% *or less. Outside these limits their use sometimes has led to reactions proceeding with violence. In one case a hydrogenation proceeding at* 125° *and* 5000 *lb. showed a pressure rise to considerably more than* 10,000 *lb. before the reaction could be stopped or the pressure released. Several instances of sudden increases in pressure have been noted when* 10–15 *g. of catalyst was used with a similar amount of hydrogen acceptor in* 100 *ml. of ethanol in the temperature range of* 100–150° *under* 5000 *lb. of hydrogen in a bomb of* 270 *ml. void. Accordingly the catalysts should be used with caution for high-pressure hydrogenations.*

1. Procedure

W-6 *Raney nickel catalyst.* In a 2-l. Erlenmeyer flask equipped with a thermometer and a stainless-steel stirrer are placed 600 ml. of distilled water and 160 g. of c.p. sodium hydroxide pellets. The solution is stirred rapidly and allowed to cool to 50° in an ice bath equipped with an overflow siphon. Then 125 g. of Raney nickel-aluminum alloy powder is added in small portions during a period of 25–30 minutes. The temperature is maintained at 50 ± 2° by controlling the rate of addition of the alloy to the sodium hydroxide solution and the addition of ice to the cooling bath. When all the alloy has been added, the suspension is digested at 50 ± 2° for 50 minutes with gentle stirring. It is usually necessary to remove the ice bath and replace it with a hot-water bath to keep the temperature constant. After this period of digestion the catalyst is washed with three 1-l. portions of distilled water by decantation (Note 1).

A glass test tube approximately 5.1 cm. in diameter and 38 cm. in length with a side arm sealed 6 cm. from the top is used as the container for washing the catalyst. The tube is closed with a rubber stop-

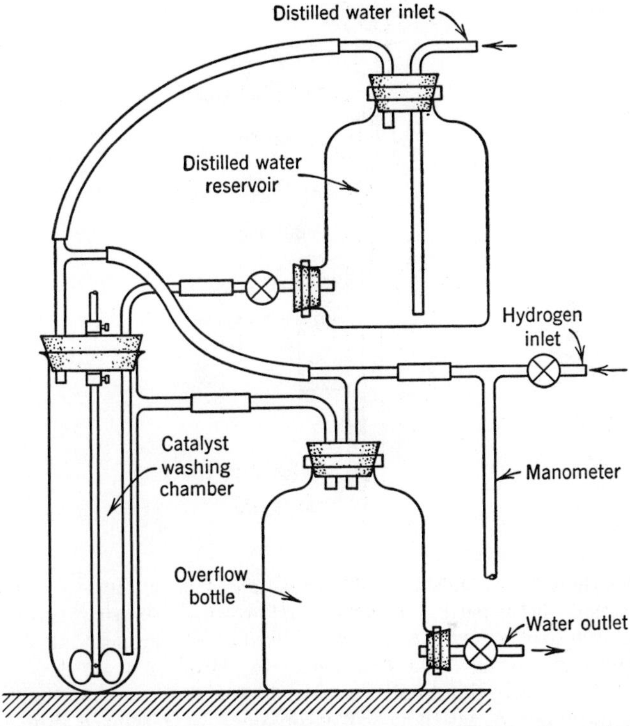

FIG. 8. Apparatus for washing catalyst. It is convenient to have the inlet tube for wash water sealed into the bottom of the chamber rather than introduced through the stopper as shown.

per clamped or wired in place tightly enough to withstand a gas pressure of 0.5 atm. The stopper contains three holes through which extend: (1) a tube 10 mm. in diameter reaching to the bottom of the test tube, for admitting distilled water; (2) a T-tube for equalizing gas pressures; (3) a gas-tight bushing through which the ¼-in. shaft of a stainless-steel stirrer projects to the bottom of the washing tube. A 5-l. aspirator bottle containing distilled water is so placed that water will flow from it through a stopcock to the bottom of the washing tube. The side arm of the test tube is connected by pressure tubing to a 5-l. aspirator bottle that serves as an overflow from which the water may be allowed to flow through a stopcock to a drain. A source of distilled water is connected to the reservoir. Other connections and the general arrangement of the apparatus are shown in Fig. 8, and details of the construction of the gas-tight brass bushing are given

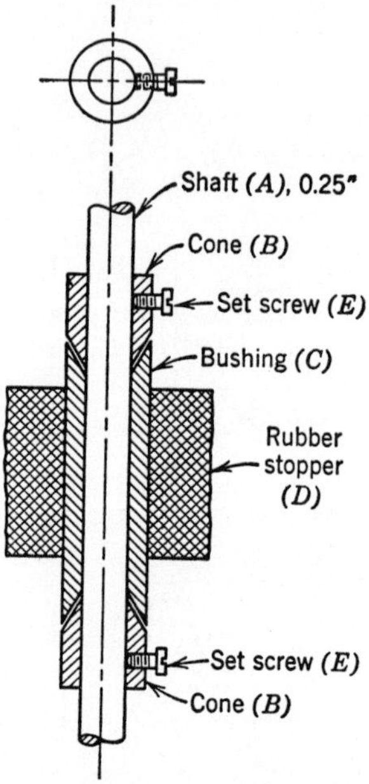

FIG. 9. Gas-tight brass bushing. The gas-tight bushing for the shaft of the stirrer consists of three parts: two cones (*B*) which fit the shaft snugly and are attached to it by set screws (*E*); and a bushing (*C*) so cut that the cones fit into it at top and bottom. The two cones are placed on shaft *A* of the stirrer, above and below the bushing and attached to the shaft so that they fit snugly against the bushing. A gas-tight seal is obtained by placing a drop or two of heavy lubricating oil between each cone and the bushing. The over-all dimensions of the bushing are approximately 13 by 65 mm., and it is held in a rubber stopper (*D*).

in Fig. 9. Rubber pressure tubing and stoppers used in making the connections should be boiled with 5% sodium hydroxide and rinsed with water to remove sulfur. All connections should be wired in place to withstand a pressure of 0.5 atm. (7.5 lb.).

The catalyst sludge is transferred to the washing tube immediately after its third washing by decantation. The last portions are rinsed from the flask into the washing tube with distilled water, and the tube, reservoir, and overflow bottle are nearly filled with distilled water. The apparatus is assembled rapidly, and hydrogen is introduced through the inlet while most of the water in the overflow bottle is displaced through the outlet. The outlet then is closed, and hydrogen is admitted until the water in the reservoir, washing tube, and overflow bottle is under a pressure about 0.5 atm. above that of the outside atmosphere. The stirrer is operated at such a speed that the catalyst is suspended to a height of 18–20 cm. Distilled water from the reservoir is allowed to flow through the suspended catalyst at a rate of about 250 ml. per minute. When the reservoir is nearly empty and the overflow bottle full, the drain cock and distilled-water inlet are opened simultaneously to an equal rate of flow such that, as the overflow bottle empties, the reservoir is filled, while the pressure in the system remains constant.

After about 15 l. of water has passed through the catalyst, the stirrer and the flow of water are stopped, the pressure is released, and the apparatus disassembled. The water is decanted from the settled sludge, which is then transferred to a 250-ml. centrifuge bottle with 95% ethanol. The catalyst is washed three times by stirring, not shaking, with 150-ml. portions of 95% ethanol, each addition being followed by centrifuging. In the same manner the catalyst is washed three times with absolute ethanol. One to two minutes' centrifugation at 1500–2000 r.p.m. is sufficient to separate the catalyst. All operations should be carried out as rapidly as possible if a catalyst of the maximum activity is desired. The catalyst should be kept in a closed bottle filled with absolute ethanol and should be stored at once in a refrigerator. The total elapsed time from the beginning of the addition of the alloy to the completion of the preparation should be not more than about 3 hours (Notes 2, 3, and 4).

2. Notes

1. The procedure in the preparation of W-7 Raney nickel, after the digestion and the three decantations, is to transfer the catalyst to a 250-ml. centrifuge bottle with 95% ethanol. It is then washed three times by stirring, not shaking, with 150-ml. portions of 95% ethanol, with centrifuging after each addition. In the same manner, the catalyst is washed three times with absolute ethanol and is at once stored in a refrigerator in a closed bottle filled with absolute ethanol. The catalyst so prepared contains alkali. This may be advantageous for various reasons in the hydrogenation of ketones, phenols, and nitriles. In certain cases the alkali may have a harmful effect upon a hydrogenation.

2. The volume of the settled catalyst (W-6) in ethanol is about 75–80 ml. containing about 62 g. of nickel and 7–8 g. of aluminum. The

catalyst is extremely pyrophoric when exposed to the air in a dry condition, and it should be kept wet with solvent at all times. It amounts to about 28 "half teaspoonfuls" if it is so measured. The catalyst loses some of its special activity upon standing but seems to be quite active for about 2 weeks when stored in a refrigerator. After this period it is similar in activity to the Raney nickel made by an earlier procedure.[1] The W-3 to W-7 catalysts all lose their special activity rather rapidly when stored under water.

3. A somewhat less active but nevertheless excellent catalyst, referred to as W-5, is made by the procedure described above for W-6 except that it is washed at atmospheric pressure without the addition of hydrogen to the system. The W-5 catalyst is similar in method of preparation and activity to the W-4 Raney nickel catalyst as prepared by Pavlic.[2]

4. Raney nickel prepared by the procedure described for W-6 will bring about the hydrogenation of alkyne and alkene linkages, of aldehydes, ketones, oximes, nitriles, nitro compounds, and benzenoid and pyridinoid nuclei under the conditions of temperature and pressure normally employed with platinum and palladium catalysts.[3] At higher pressures W-6 Raney nickel brings about reactions at a more rapid rate and at lower temperatures than Raney nickel prepared by the older procedures.

3. Methods of Preparation

The various procedures for the preparation of Raney nickel[4] from the nickel-aluminum alloy differ from one another in the method of adding the alloy, in the concentration of sodium hydroxide, in the temperature and duration of digestion, and in the method of washing the catalyst free of sodium aluminate and alkali. For convenience in reference, the Raney nickel catalysts prepared by various procedures have been designated W-1,[5] W-2,[1] W-3,[2,6] W-4,[2,6] W-5,[3] W-6,[3] and W-7.[3] They have been compared as to activity against β-naphthol.[7] A Raney nickel catalyst is available commercially from the Gilman Paint and Varnish Company, Chattanooga, Tennessee.

[1] *Org. Syntheses*, **21**, 15 (1941).
[2] Pavlic and Adkins, *J. Am. Chem. Soc.*, **68**, 1471 (1946).
[3] Adkins and Billica, *J. Am. Chem. Soc.*, **70**, 695 (1948).
[4] Murray Raney, U. S. pat. 1,628,190 [*C. A.*, **21**, 2116 (1927)].
[5] Covert and Adkins, *J. Am. Chem. Soc.*, **54**, 4116 (1932).
[6] Adkins and Pavlic, *J. Am. Chem. Soc.*, **69**, 3039 (1947).
[7] Adkins and Krsek, *J. Am. Chem. Soc.*, **70**, 412 (1948).

CATALYST, RANEY NICKEL, W-2

$$NiAl_2 + 6NaOH \rightarrow Ni + 2Na_3AlO_3 + 3H_2$$

Submitted by Ralph Mozingo.
Checked by Homer Adkins and Lawrence Richards.

1. Procedure

A solution of 380 g. of sodium hydroxide in 1.5 l. of distilled water, contained in a 4-l. beaker (Note 1) equipped with an efficient stirrer (Note 2), is cooled in an ice bath to 10°, and 300 g. of nickel-aluminum alloy (Note 3) is added to the solution in small portions, with stirring, at such a rate that the temperature does not rise above 25° (Note 4), the beaker being allowed to remain in the ice bath. When all the alloy has been added (about 2 hours is required) the stirrer is stopped, the beaker is removed from the ice bath, and the contents are allowed to come to room temperature. After the evolution of hydrogen becomes slow, the reaction mixture is allowed to stand on a steam bath until the evolution of hydrogen again becomes slow (about 8–12 hours). The heating should not be too rapid at the beginning or the solution may foam over. During this time the volume of the solution is maintained constant by adding distilled water if necessary. After heating, the nickel is allowed to settle and most of the liquid is decanted. Distilled water is then added to bring the solution to the original volume; the nickel is suspended by stirring, again allowed to settle, and the solution is decanted. The nickel is then transferred to a 2-l. beaker (Note 5) with the aid of distilled water, and the water is again decanted. A solution of 50 g. of sodium hydroxide in 500 ml. of distilled water is added; the catalyst is suspended and allowed to settle; and the alkali is decanted. The nickel is washed by suspension in distilled water and decantation until the washings are neutral to litmus and then ten times more to remove the alkali completely (twenty to forty washings are required) (Note 6). The washing process is repeated three times with 200 ml. of 95% ethanol and three times with absolute ethanol; the catalyst is then stored under absolute ethanol in bottles which are completely filled with absolute ethanol and tightly closed (Note 7). The product is highly pyrophoric and must be kept under a liquid at all times. The Raney nickel contained in the suspension weighs about 150 g. (Note 8).

To prepare the catalyst under methylcyclohexane (Note 9), the catalyst, which has been prepared as above and washed free of alkali with water, but to which no ethanol has been added, is covered with 1 l. of methylcyclohexane which is distilled from an oil bath until all the water has been codistilled with the hydrocarbon, more of the methylcyclohexane being added from time to time so that the nickel always remains covered. When the catalyst is free from water it becomes freely suspended in the liquid.

To prepare nickel under dioxane, dioxane (Note 10) is used in place of the methylcyclohexane above and the distillation is continued until the temperature of the vapor reaches 101°. (*Caution. Do not use nickel in dioxane above* 210°; *the dioxane may react almost explosively with hydrogen and Raney nickel above this temperature.*)

2. Notes

1. A Pyrex battery jar of about 10-l. capacity is also suitable and is sufficiently large for the preparation of a batch of catalyst of two to three times the size given here.

2. The stirrer should be provided with a motor which will not ignite the hydrogen. Either an induction motor or an air stirrer may be used. The stirrer blades may be made of glass, Monel, or stainless steel.

3. The alloy used is "Raney Nickel Aluminum Catalyst Powder" from the Gilman Paint and Varnish Company, Chattanooga, Tenn. It contains about 50% nickel.

4. The thermometer should not be left in the alkali or the bulb may be eaten away. The catalyst is inactivated by mercury. If the mixture foams badly at this point, 2 ml. of *n*-octyl alcohol may be added to prevent excessive foaming; however, this is not usually necessary.

5. A stoppered graduate of 2-l. capacity is somewhat more convenient than a 2-l. beaker.

6. The number of washings required may be materially reduced by allowing time for diffusion of base from the surface of the catalyst into the surrounding wash water. To this end the catalyst is stirred well with 1.5 l. of water for each washing. Diffusion is allowed to proceed for 3–10 minutes, and the mixture is then stirred again and the wash water decanted as soon as the catalyst settles to the bottom. By this method twenty washings should be sufficient to remove all traces of the alkali adsorbed on the catalyst surface.

7. The quantity of catalyst prepared should not be larger than necessary for a 6 months' supply as the catalyst may deteriorate on standing.

8. It is more convenient to measure the catalyst than to weigh it. Raney nickel in alcohol contains about 0.6 g. of the catalyst per milliliter of the settled material. The half and quarter teaspoons used in kitchens are convenient for measuring the catalyst. A level teaspoonful is about 3 g. of nickel.

9. It is necessary that the catalyst be freely suspended in the reaction mixture during a reduction. Therefore, the liquid under which the nickel is placed must be soluble in the reduction mixture at all times, i.e., in both the reactants and products.

10. The dioxane used should be dry, halogen-free, and distilled from sodium.

3. Methods of Preparation

These are summarized in the preceding procedure (p. 176). The previously published procedure [*Org. Syntheses,* **21,** 15 (1941)] which is included here results in the catalyst now known as W-2.

p-CHLOROACETYLACETANILIDE

(Acetanilide, *p*-chloroacetyl-)

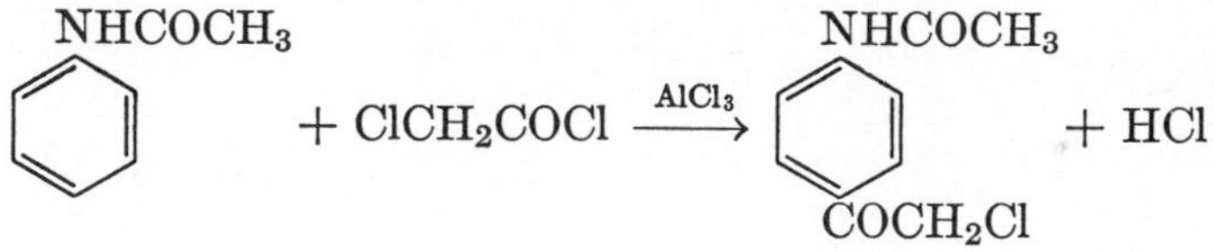

Submitted by J. L. Leiserson and A. Weissberger.
Checked by Cliff S. Hamilton and Yao-Hua Wu.

1. Procedure

Caution! Carbon disulfide, used as a solvent in this preparation, is highly inflammable; its vapor may ignite on contact with a hot laboratory steam line.

In a 5-l. three-necked flask mounted on a steam bath in the hood and equipped with a mechanical stirrer (Note 1) and a wide-bore condenser (Note 1) is placed 1.4 kg. (1.1 l.) of carbon disulfide. Through the open neck of the flask 202 g. (1.5 moles) of acetanilide and 300 g. (2.66 moles) of chloroacetyl chloride (Note 2) are introduced. The mixture is vigorously stirred while 600 g. (4.5 moles) of aluminum chloride is added in 25–50 g. portions over a period of 20–30 minutes; the neck of the flask is stoppered between additions (Note 3). After the addition of the last portion of aluminum chloride, the mixture is heated at reflux temperature for 30 minutes while stirring is continued. Heating and stirring are discontinued and the mixture is allowed to stand for 3 hours, during which time it separates into layers. The upper layer (carbon disulfide) is decanted, and the viscous red-brown lower layer is poured cautiously with stirring into about 1 kg. of finely crushed ice to which 100 ml. of concentrated hydrochloric acid has been added. After the hydrolysis of the aluminum chloride, the product crystallizes as a white solid, which is collected on a Büchner funnel and washed well with water. It is then transferred to a beaker where it is thoroughly washed by stirring with sufficient 95% ethanol to give a fluid slurry. The solid is again collected on the funnel. After drying in the air it melts at 213–214° and weighs 250–265 g. (79–83%). It can be recrystallized from 95% ethanol (about 1 l. of the solvent being required for 40 g. of the solid) as fine white crystals melting at 216°; the recovery in the recrystallization, without reworking of the mother liquor, is about 70%.

2. Notes

1. Unless a well-ventilated hood is available the stirrer should be provided with a seal and the condenser should be connected to a gas-absorption trap.

2. The chloroacetyl chloride should be weighed in the hood; it is strongly lachrymatory.

3. The addition of each portion of aluminum chloride causes vigorous boiling.

3. Methods of Preparation

The method of preparation given, devised by Kunckell,[1] is the only one reported.

[1] Kunckell, *Ber.*, **33**, 2644 (1900).

o-CHLOROBROMOBENZENE

(Benzene, 1-bromo-2-chloro-)

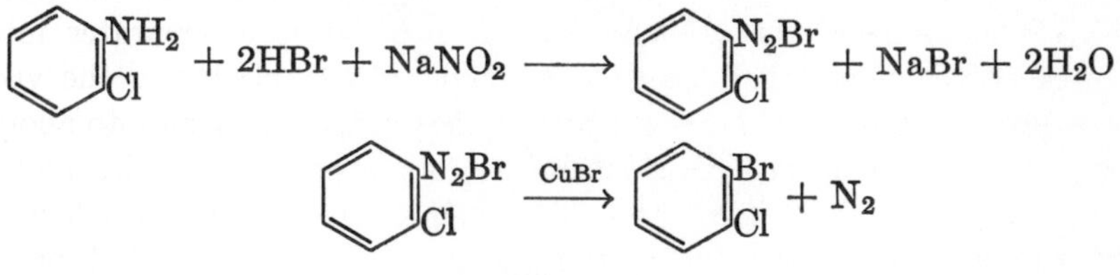

Submitted by JONATHAN L. HARTWELL.
Checked by H. R. SNYDER and ZENO WICKS, JR.

1. Procedure

A mixture of 127.5 g. (1 mole) of a good commercial grade of o-chloroaniline and 300 ml. (2.5 moles) of 48% hydrobromic acid (Note 1) in a 2-l. flask set in an ice bath is cooled to 0° by the addition of ice. A solution of 70 g. (1 mole) of sodium nitrite in 125 ml. of water is added rapidly, with stirring, the temperature being kept below 10° by the addition of small pieces of ice. When only about 5 ml. of the sodium nitrite solution remains, further additions are made cautiously until an excess of nitrous acid remains after the last addition (Note 2).

In the meantime, a mixture of 79 g. (0.55 mole) of cuprous bromide (Note 3) and 80 ml. (0.6 mole) of 48% hydrobromic acid (Note 1) is heated to boiling in a 5-l. round-bottomed three-necked flask, equipped with a condenser set for distillation and provided with a 2-l. receiving flask, a steam inlet tube closed by a screw clamp, and a separatory funnel. About one-fourth of the diazonium solution is transferred to the separatory funnel, without filtration, and immediately run into the cuprous bromide-hydrobromic acid solution, which is kept boiling over a free flame, at such a rate that boiling is continuous. When the separatory funnel is nearly empty a further portion of the cold diazonium solution is transferred to it without interrupting the addition. All the diazonium solution is added in this way over a period of about 30 minutes, during which time much of the product steam-distils. When the addition is complete, the stopcock in the separatory funnel is closed, the screw clamp in the steam line is opened, and a vigorous current of steam is passed through the mixture until no more organic material distils. About 1–1.5 l. of distillate is collected.

The heavy organic layer is separated from the distillate and washed with 10-ml. portions of concentrated sulfuric acid until the acid becomes only slightly colored during the washings; four washings usually suffice. The oil is then washed with one 100-ml. portion of water, two 50-ml. portions of 5% aqueous sodium hydroxide, and finally with one 100-ml. portion of water. The product is dried over about 3 g. of calcium chloride and distilled from a 250-ml. distilling flask. The yield of pure, colorless *o*-chlorobromobenzene, boiling at 199–201°/742 mm., is 170–183 g. (89–95%) (Notes 4 and 5).

2. Notes

1. When 40% hydrobromic acid is used in both the diazotization and Sandmeyer reaction the yield is only about 75%.

2. Free nitrous acid causes an *immediate* blue color at the point of contact with starch-iodide test paper. A delayed color or a color around the periphery of the wetted area is of no significance. At all times there must be an excess of mineral acid (blue color on Congo paper).

3. The submitter used commercial cuprous bromide. The checkers prepared cuprous bromide by dissolving 600 g. (2.4 moles) of commercial copper sulfate crystals and 350 g. (3.4 moles) of sodium bromide in 2 l. of warm water; the solution was stirred while 151 g. (1.2 moles) of solid sodium sulfite was added over a period of 10 minutes. Occasionally a little more sodium sulfite was required to discharge the blue color. The mixture was cooled, and the solid collected on an 8-in. Büchner funnel, washed once with water, pressed nearly dry, and then dried in the air overnight. The yield of cuprous bromide was 320 g. (93%).

4. Runs 3 times this size give proportional yields.

5. The checkers have prepared the following bromides by the same procedure: *m*-chlorobromobenzene (b.p. 191–194°) from *m*-chloroaniline in 91–94% yields; *m*-dibromobenzene (b.p. 215–217°) from *m*-bromoaniline in 80-87% yields; and *o*-bromoanisole (b.p. 114–116°/29 mm.) from *o*-anisidine in 88–93% yields. In the preparation of *o*-bromoanisole the washing with sulfuric acid was omitted.

3. Methods of Preparation

o-Chlorobromobenzene has been prepared by the diazotization of *o*-bromoaniline followed by replacement of the diazonium group by chlorine;[1] by the elimination of the amino group from 3-chloro-4-bromoaniline;[2] by the chlorination of bromobenzene in the presence of thallous chloride,[3] aluminum chloride,[4] or ferric chloride;[4] by the bromination of chlorobenzene without a catalyst[5] or in the presence of aluminum,[4] iron,[4] ferric bromide,[4] or aluminum-mercury couple;[6] by the diazotization of *o*-chloroaniline followed by replacement of the diazonium group with bromine;[4,7] from *o*-chlorophenylmercuric chloride by the action of bromine;[8] and by treatment of silver *o*-chlorobenzoate with bromine.[9]

1 Dobbie and Marsden, *J. Chem. Soc.*, **73**, 254 (1898).
2 Wheeler and Valentine, *Am. Chem. J.*, **22**, 266 (1899).
3 Thomas, *Compt. rend.*, **144**, 33 (1907).
4 Vander Linden, *Rec. trav. chim.*, **30**, 305 (1911).
5 Van Loon and Wibaut, *Rec. trav. chim.*, **56**, 815 (1937).
6 Sen and Bhargava, *J. Indian Chem. Soc.*, **25**, 277 (1948).
7 Narbutt, *Ber*, **52**, 1028 (1919).
8 Hanke, *J. Am. Chem. Soc.*, **45**, 1321 (1923).
9 Dauben and Tilles, *J. Am. Chem. Soc.*, **72**, 3185 (1950).

4-CHLOROBUTYL BENZOATE

(1-Butanol, 4-chloro-, benzoate)

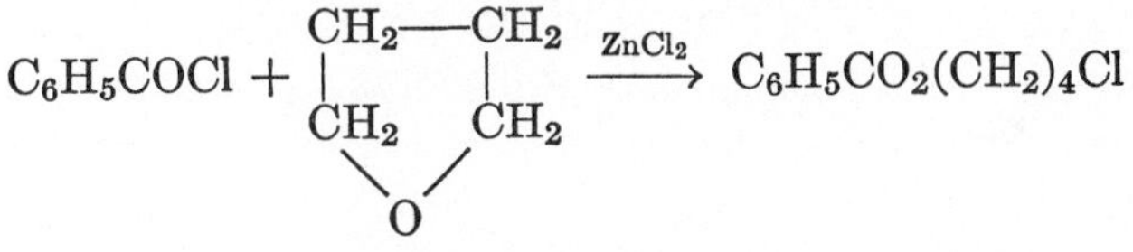

Submitted by Martin E. Synerholm.
Checked by Arthur C. Cope and James J. Ryan.

1. Procedure

In a 200-ml. round-bottomed flask fitted with an efficient reflux condenser are mixed 31.4 ml. (38 g., 0.27 mole) of freshly distilled benzoyl chloride, 28.2 ml. (25 g., 0.35 mole) of tetrahydrofuran (Note 1), and 5 g. of freshly fused zinc chloride. A vigorous reaction begins immediately, and after a few seconds, when the mixture starts to boil, external cooling is applied with an ice bath. After the initial reaction has subsided, the mixture is heated on a steam bath for 15 minutes, cooled, and dissolved in 100 ml. of benzene. The benzene solution is washed with 100 ml. of a 5% solution of sodium chloride and then with 100 ml. of a saturated solution of sodium bicarbonate. The benzene layer is dried over anhydrous sodium sulfate and fractionally distilled from a modified Claisen flask.

The product is collected at 140–143°/5 mm., 132–135°/2.5 mm.; n_D^{25} 1.5176 (Note 2). The yield is 45–48 g. (78–83%).

2. Notes

1. Good-quality commercial tetrahydrofuran may be used as received, or redistilled; b.p. 65–66°.
2. The product develops a slight yellow tint on standing.

3. Methods of Preparation

4-Chlorobutyl benzoate has been prepared by the action of benzoyl chloride on tetrahydrofuran in the presence of titanium chloride, stannic chloride,[1] or zinc chloride.[2]

[1] Gol'dfarb and Smorgonskii, *J. Gen. Chem. U.S.S.R.*, **8**, 1516 (1938) [*C. A.*, **33**, 4593 (1939)].

[2] Cloke and Pilgrim, *J. Am. Chem. Soc.*, **61**, 2667 (1939).

2-CHLOROCYCLOHEXANONE

(Cyclohexanone, 2-chloro-)

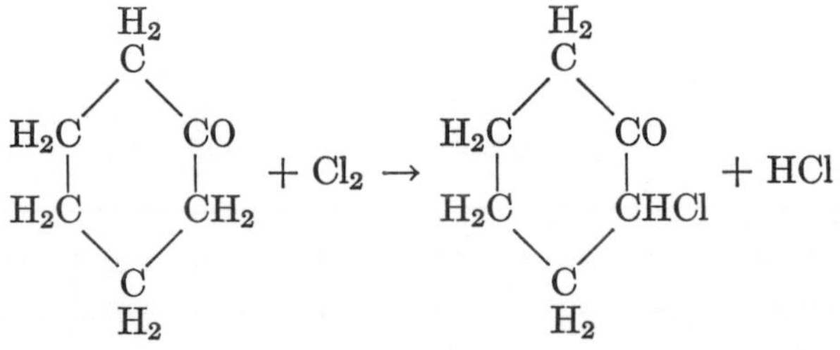

Submitted by M. S. Newman, M. D. Farbman, and H. Hipsher.
Checked by C. S. Hamilton and C. W. Whitehead.

1. Procedure

In a 3-l. three-necked round-bottomed flask, fitted with a gas inlet tube reaching almost to the bottom, a sealed mechanical stirrer (Note 1), and a gas outlet tube connected to a mercury valve (Note 2), are placed 294 g. (3 moles) of cyclohexanone (Note 3) and 900 ml. of water. After the reaction vessel has been swept out with chlorine, the gas outlet tube is connected to the mercury valve, the flask is cooled in an ice bath, the stirrer is started, and 215 g. (slightly more than 3 moles) of chlorine is bubbled in as rapidly as the gas is absorbed (about 45 minutes) (Note 4).

The heavier chlorocyclohexanone layer is separated and combined with three 150-ml. ether extracts of the aqueous phase, and washed with 150 ml. of water and then with 200 ml. of saturated sodium chloride solution. After filtration (gravity) through anhydrous sodium sulfate the ether is removed and the residue vacuum-distilled in a modified Claisen flask. The fraction (300–340 g.) boiling below 100° at 10 mm. (Note 5) is collected (Note 6). This material is then fractionated carefully under reduced pressure by means of a 42-in. modified Vigreux column (heated) with a total-condensation variable take-off head (Note 7). The yield of 2-chlorocyclohexanone boiling at 90–91°/14–15 mm. is 240–265 g. (61–66%) (Notes 8 and 9).

2. Notes

1. A large propeller-type stirrer is satisfactory. The blades should be pitched to drive the liquid upwards, and the propeller should be located just below the surface of the liquid to provide splashing. The checkers used a glycerol-lubricated rubber-tube seal.[1]

2. The mercury valve consists of a tube dipping about 0.5 in. into some mercury in a vented test tube. This allows the reaction to be carried out under a slight pressure of chlorine. The checkers found a water valve to be more satisfactory. The outlet tube was made to dip about 7 in. into water in a vented glass tube. A trap (125-ml. suction flask) was placed between the water valve and the reaction flask to prevent water from being sucked into the reaction mixture.

3. The checkers used Eastman's cyclohexanone boiling at 154–156°. The submitters report that the practical grade gives just as good yields.

4. Careful control of the temperature is unnecessary as approximately the same results are obtained when the temperature is allowed to rise to about 50° or is kept below 20°. However, at the lower temperature a greater amount of cyclohexanone is recovered.

5. Or below 110°/13 mm., or 92°/4 mm.

6. This preliminary purification is advisable before careful fractionation.

7. The head used was similar to that described by Turk and Matuszak [2] except for the stopcock, which was of the variety described by Newman.[3]

8. This material shows a long flat at 23.2° cor. in a time-temperature cooling curve. It can be stored in a refrigerator in paraffin-covered stoppered bottles for long periods of time without discoloration.

9. About 15–40 g. (5–13%) of cyclohexanone, b.p. 52°/14–15 mm., is recovered.

3. Methods of Preparation

2-Chlorocyclohexanone can be prepared by chlorinating cyclohexanone in glacial acetic acid as the solvent,[4] by passing chlorine into a mixture of cyclohexanone [5] or cyclohexanol [6] and water in the presence of powdered calcium carbonate, by the electrochemical chlorination of cyclohexanone in hydrochloric acid,[7] by the action of monochlorourea in acetic acid on the ketone,[8] by the oxidation of 2-chlorocyclohexanol,[9] and by the chlorination of the sodio derivative of cyclohexanone with ethyl hypochlorite.[10]

[1] *Org. Syntheses* Coll. Vol. **3**, 368 (1954).

[2] Turk and Matuszak, *Ind. Eng. Chem., Anal. Ed.*, **14**, 72 (1942).

[3] Newman, *Ind. Eng. Chem., Anal. Ed.*, **14**, 902 (1942).

[4] Bartlett and Rosenwald, *J. Am. Chem. Soc.*, **56**, 1992 (1934).

[5] Kotz and Grethe, *J. prakt. Chem.*, (2) **188**, 487 (1909).

[6] Ebel, *Helv. Chim. Acta,* **12,** 9 (1929); Bouveault and Chereau, *Compt. rend.*, **142,** 1086 (1906); Meyer, *Helv. Chim. Acta,* **16,** 1291 (1933); Vavon and Mitchovitch, *Bull. soc. chim. France,* (4) **45,** 961 (1929).

[7] Szper, *Bull. soc. chim. France,* (4) **51,** 653 (1932).

[8] Godchot and Mousseron, *Bull. soc. chim. France,* (4) **51,** 361 (1932).

[9] Detoeuf, *Bull. soc. chim. France,* (4) **31,** 178 (1922).

[10] Mousseron and Froger, *Bull. soc. chim. France,* **12,** 69 (1945).

ω-CHLOROISONITROSOACETOPHENONE

(Glyoxylyl chloride, phenyl-, oxime)

$$C_6H_5COCH_2Cl + n\text{-}C_4H_9ONO \xrightarrow{HCl} C_6H_5COC(NOH)Cl + n\text{-}C_4H_9OH$$

Submitted by Nathan Levin and Walter H. Hartung.
Checked by C. F. H. Allen and J. VanAllan.

1. Procedure

A 500-ml. three-necked round-bottomed flask is provided with a small dropping funnel, a sealed mechanical stirrer, a reflux condenser connected to a gas-absorption trap, and a hydrogen chloride delivery tube which extends to the bottom of the flask.

In the reaction flask are placed 15.4 g. (0.1 mole) of phenacyl chloride (Note 1) and 100 ml. of dry ether. The stirrer is started, and, after the solid has dissolved, anhydrous hydrogen chloride (Note 2) is introduced at the rate of 2–3 bubbles per second. Ten and three-tenths grams (11.8 ml., 0.1 mole) of freshly distilled *n*-butyl nitrite [1] (Note 3) is then admitted from the dropping funnel in 0.5- to 1-ml. portions. After addition of the first portion of nitrite the reaction mixture becomes orange-brown, and after several minutes, light yellow; at this point a second portion of nitrite is added and a similar color change takes place, whereupon a third portion is added; further additions are made until all the butyl nitrite has been added. The reaction mixture warms up, and the ether begins to reflux gently (Note 4). After all the nitrite has been added (about 30 minutes is required), stirring and addition of hydrogen chloride are continued an additional 15 minutes. At this point, the solution will have practically ceased boiling and will have assumed an orange color.

The reaction mixture is allowed to stand for 1–2 hours (or overnight, if more convenient); after this interval the solution will have assumed a clear, pale yellow color. The condenser is then set for downward distillation, stirring is resumed, and the solvent is removed by distillation from a steam bath (Note 5). After nearly all the ether has been removed, the distillation is continued under reduced pressure (40–50 mm.) until no further appearance of crystals is noted. The residue, which consists of yellow crystals of the crude product, is then allowed to stand until dry in a vacuum desiccator which contains concentrated sulfuric acid, soda lime, and anhydrous calcium chloride (Note 6).

The dried product is then recrystallized from 30–35 ml. of a 1:3 mixture of boiling benzene and carbon tetrachloride (Note 7). The yield of snow-white crystals is 15–15.7 g. (82–86%); the recrystallized product melts at 131–132° and is sufficiently pure for synthetic purposes. A second recrystallization gives a product which melts at 132–133° (Notes 8 and 9).

2. Notes

1. Commercial phenacyl chloride may be used; if unavailable the chloride may be prepared in 85–88% yield by a Friedel-Crafts reaction, using 234 g. (265 ml., 3 moles) of dry benzene and 79.5 g. (53 ml., 0.70 mole) of chloroacetyl chloride, in the presence of 103 g. (0.77 mole) of powdered anhydrous aluminum chloride; the product distils at 120–125°/4 mm. and melts at 56–57°. *Phenacyl chloride is a strong lachrymator and vesicant; it should be handled with care.*

2. Hydrogen chloride is now available in cylinders.

3. Any alkyl nitrite may be employed. The submitters preferred the use of isopropyl nitrite, since the low boiling point of the isopropyl alcohol formed facilitates its removal.

Isopropyl nitrite may be prepared according to the procedure given for *n*-butyl nitrite.[1]

A mixture of 147 g. (80 ml., 1.5 moles) of concentrated sulfuric acid (sp. gr. 1.84), 60 ml. of water, and 180 g. (230 ml., 3 moles) of 97% isopropyl alcohol, previously cooled to 0°, is added to a solution of 227.7 g. (3.3 moles) of 97% sodium nitrite in 1 l. of water, cooled to −5°. About 2 hours is required for the addition of the alcohol solution, during which time the temperature of the reaction mixture is maintained at −2° to 0°. The product may be isolated and purified as described under butyl nitrite. After drying over 15–20 g. of anhydrous sodium sulfate, the nitrite is distilled from a steam bath using a 20-cm. column. Practically all the isopropyl nitrite distils at 39–40°/745 mm. as a pale yellow oil; the yield of product is 191 g. (71.4%). Isopropyl nitrite, when stored in a refrigerator, has been found to be much more stable than butyl nitrite.

The checkers used commercially available *n*-butyl nitrite and experienced no difficulty in removing the *n*-butyl alcohol.

4. The rate of stirring must be kept fairly constant since an abrupt increase in speed may cause the ether to reflux at an undesirably rapid rate. The rate of addition of the nitrite is also governed by the rate of the refluxing.

5. The recovered ether may be employed without purification as the solvent in a subsequent run.

6. ω-Chloroisonitrosoacetophenone is extremely soluble in butanol, hence the alcohol should be removed as completely as possible before the crude product is recrystallized.

7. This is most conveniently done by dissolving the chloride in the benzene and then diluting.

8. This procedure works equally well in 0.5-mole runs.

9. This procedure, with minor changes, may be applied to various nuclear-substituted phenacyl chlorides. The yields vary from 74% to 92%.[2,3]

3. Methods of Preparation

The above procedure[2] is modeled after that described for the nitrosation of arylethyl ketones.[4,5] ω-Chloroisonitrosoacetophenone has been prepared by the chlorination of isonitrosoacetophenone;[6–8] by treating the ammonium salt of ω-nitroacetophenone with anhydrous hydrogen chloride,[9] or by refluxing ω-nitroacetophenone with dilute alcoholic hydrogen chloride;[10] and by the nitrosochlorination of acetophenone with nitrosyl chloride.[11] The isolation of small quantities of ω-chloroisonitrosoacetophenone from the reaction product obtained by nitrosating acetophenone in the presence of hydrogen chloride has been reported.[12,13]

[1] *Org. Syntheses* Coll. Vol. **2**, 108 (1943).
[2] Levin, Thesis, University of Maryland, 1941.
[3] Levin and Hartung, *J. Org. Chem.*, **7**, 408 (1942).
[4] Hartung and Munch, *J. Am. Chem. Soc.*, **51**, 2264 (1929).
[5] Hartung and Crosby, *Org. Syntheses* Coll. Vol. **2**, 363 (1942).
[6] Claisen and Manasse, *Ann.*, **274**, 95 (1893).
[7] Ponzio and Charrier, *Gazz. chim. ital.*, (2) **37**, 65 (1907).
[8] Ponzio, *Gazz. chim. ital.*, **61**, 946 (1931).
[9] Steinkopf and Jurgens, *J. prakt. Chem.*, (2) **84**, 712 (1911).
[10] Jakubowitsch, *J. prakt. Chem.*, **142**, 46 (1935).
[11] Rheinboldt and Schmitz-Dumont, *Ann.*, **444**, 125 (1925).
[12] Claisen, *Ber.*, **20**, 252 (1887).
[13] Claisen and Manasse, *Ber.*, **22**, 526 (1889).

2-CHLOROLEPIDINE

(Lepidine, 2-chloro-)

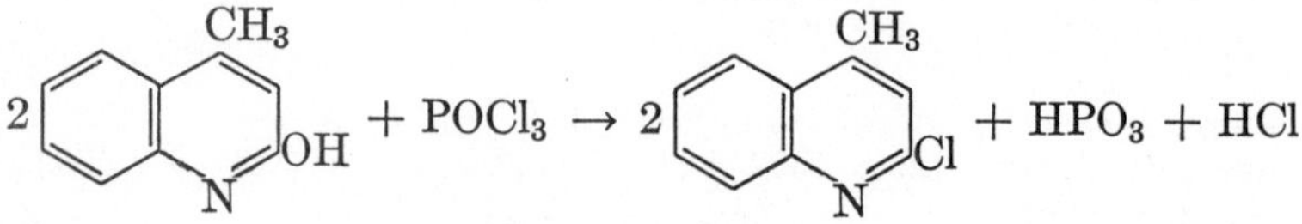

Submitted by C. E. Kaslow and W. M. Lauer.
Checked by C. F. H. Allen and H. W. J. Cressman.

1. Procedure

In a 500-ml. flask, to which is attached an air condenser whose open end is protected by absorbent cotton or calcium chloride in a drying tube, are mixed 119 g. (0.75 mole) of 4-methylcarbostyril (p. 580) and 138 g. (82.5 ml., 0.9 mole) of freshly distilled phosphorus oxychloride. The mixture is maintained at 80–85° in a water bath for about 15 minutes until most of the solid has dissolved, and then it is warmed carefully for an additional 15 minutes on a wire gauze until solution is complete. The hot reaction mixture is poured into 1 l. of water containing 1 kg. of cracked ice.

The 2-chlorolepidine is extracted, using two 750-ml. portions of ether (Note 1). The extract is shaken with two 200-ml. portions of water and then dried over 50 g. of potassium carbonate. After removal of the ether, the residual oil is distilled from a 200-ml. modified Claisen flask. The colorless distillate boils at 132–135°/3 mm. and weighs 118–122 g. (89–92%). The distillate is melted if necessary and poured into 250 ml. of petroleum ether (b.p. 40–50°); the solution is then chilled in a freezing mixture; the crystals are filtered by suction and dried in a vacuum desiccator over paraffin. The snow-white 2-chlorolepidine melts at 58–59° and weighs 114–118 g. (86–89%).

2. Note

1. An additional 5–8 g. of slightly colored material can be secured by neutralizing the aqueous solution with 200 g. of sodium carbonate, extracting with 500 ml. of ether, and distilling. The total yield of distilled product then amounts to 125–130 g. (95–97%).

3. Methods of Preparation

The preparation described is based on the method of Knorr [1] and has been used by Mikhailov,[2] Krahler and Burger,[3] and Mizuno.[4] 2-Chlorolepidine has also been prepared by the action of benzoyl chloride on 4-methylcarbostyril.[5]

[1] Knorr, *Ann.*, **236**, 98 (1886).
[2] Mikhailov, *J. Gen. Chem. U.S.S.R.*, **6**, 511 (1936) [*C. A.*, **30**, 6372 (1936)].
[3] Krahler and Burger, *J. Am. Chem. Soc.*, **63**, 2368 (1941).
[4] Mizuno, *J. Pharm. Soc. Japan*, **69**, 126 (1949) [*C. A.*, **44**, 1985 (1950)].
[5] Ellinger and Reisser, *Ber.*, **42**, 3338 (1909).

1-CHLOROMETHYLNAPHTHALENE *

(Naphthalene, 1-chloromethyl-)

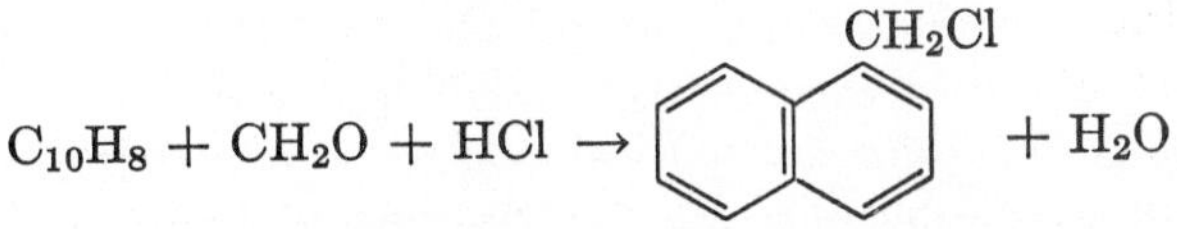

Submitted by Oliver Grummitt and Allen Buck.
Checked by C. F. H. Allen and J. VanAllan.

1. Procedure

Caution! Both chloromethylnaphthalene and the by-products are lachrymators and vesicants. Although it is not necessary to work in a hood, precautions should be taken during the handling of the substance and apparatus.

In a 3-l. three-necked flask, fitted with a reflux condenser and Hershberg stirrer, are placed 256 g. (2 moles) of naphthalene, 110 g. of paraformaldehyde (Note 1), 260 ml. of glacial acetic acid, 165 ml. of 85% phosphoric acid, and 428 g. (362 ml., 4.2 moles) of concentrated hydrochloric acid. This mixture is heated in a water bath at 80–85° and vigorously stirred for 6 hours (Note 2).

After the mixture has been cooled to 15–20° it is transferred to a 2-l. separatory funnel and the crude product is washed first with two 1-l. portions of water cooled to 5–15°, then with 500 ml. of cold 10% potassium carbonate solution, and finally with 500 ml. of cold water. The product is the lower layer in all the washings. After the addition of 200 ml. of ether, the solution is given a preliminary drying by being allowed to stand over 10 g. of anhydrous potassium carbonate, with frequent shaking, for 1 hour. The lower aqueous layer which forms is separated, and the ether solution is again dried over 20 g. of potassium carbonate for 8–10 hours (Notes 3 and 4).

The dried solution is distilled, first at atmospheric pressure to remove most of the solvent, and then under reduced pressure (Note 5). A fore-run of unused naphthalene amounting to 35–40 g. is collected at 90–110°/5 mm. (Note 6). This is followed by 195–204 g. of 1-chloromethylnaphthalene which boils at 128–133°/5 mm. or at 148–153°/14 mm. (74–77% based on naphthalene consumed) (Notes 7 and 8).

2. Notes

1. "Trioxymethylene" from the Eastman Kodak Company was used.
2. The level of the water bath should be maintained at the same height as that of the stirred reaction mixture.
3. Both the washing and drying operations must be done very carefully, because the presence of small amounts of water or acid is liable to cause the product to resinify during the final distillation.
4. The checkers added 50 ml. of dry benzene before distilling the solvent to remove traces of water by azeotropic distillation.
5. A clean, dry flask and a moderate rate of distillation help to overcome the tendency of the product to resinify.
6. Care should be taken to prevent clogging of the line by the naphthalene.
7. The oil pump should be protected from acid fumes by means of a trap containing alkali.
8. The residue left after distillation consists mainly of *bis*-(chloromethyl)naphthalene and di-1-naphthylmethane.

3. Methods of Preparation

1-Chloromethylnaphthalene has been made from naphthalene and a variety of reagents: methyl chloromethyl ether;[1] aqueous formaldehyde and hydrogen chloride with or without sulfuric acid as a condensing agent;[2–8] and paraformaldehyde with hydrogen chloride or hydrochloric acid.[9–11] Catalysts employed have been zinc chloride,[2,12–14] aluminum chloride,[13] phosphoric acid,[15–17] and *p*-toluenesulfonyl chloride.[18] Petroleum ether and glacial acetic acid have been used as solvents. The present method is a modification of that described by Cambron.[15]

The chloromethylation of aromatic hydrocarbons has been discussed by Fuson and McKeever.[19] It is reported that arsenic salts are helpful in the chloromethylation process.[20]

1 Vavon, Bolle, and Calin, *Bull. soc. chim. France,* (5) **6,** 1032 (1939).
2 Blanc, *Bull. soc. chim. France,* (4) **33,** 319 (1923).
3 Jones, U. S. pat. 2,212,099 [*C. A.,* **35,** 462 (1941)].
4 Coles and Dodds, *J. Am. Chem. Soc.,* **60,** 853 (1938).
5 Reddelien and Lange, Ger. pat. 508,890 [*C. A.,* **25,** 716 (1931)].
6 Roblin and Hechenbleikner, U. S. pat. 2,166,554 [*C. A.,* **33,** 8628 (1939)].
7 Shmuk and Guseva, *J. Applied Chem. U.S.S.R.,* **14,** 1031 (1941) [*C. A.,* **39,** 4069 (1945)].
8 Badger, Cook, and Crosbie, *J. Chem. Soc.,* **1947,** 1432.
9 Darzens and Levy, *Compt. rend.,* **202,** 74 (1936).
10 Fieser and Novello, *J. Am. Chem. Soc.,* **62,** 1856 (1940).
11 Fieser and Gates, *J. Am. Chem. Soc.,* **62,** 2338 (1940).
12 Ruggli and Burckhardt, *Helv. Chim. Acta,* **23,** 443 (1940).
13 Ger. pat. 509,149 [*C. A.,* **25,** 711 (1931)].
14 Anderson and Short, *J. Chem. Soc.,* **1933,** 485.
15 Cambron, *Can. J. Research,* **17B,** 12 (1939).
16 Manske and Ledingham, *Can. J. Research,* **17B,** 15 (1939).
17 Lock and Walter, *Ber.,* **75,** 1158 (1942).
18 Palomo, *Afinidad,* **27,** 361 (1950) [*C. A.,* **44,** 9390 (1950)].
19 Fuson and McKeever, *Organic Reactions,* **1,** 63 (1942); John Wiley & Sons.
20 U. S. pat. 2,541,408 [*C. A.,* **45,** 6662 (1951)].

2-CHLOROMETHYLTHIOPHENE

(Thiophene, 2-chloromethyl-)

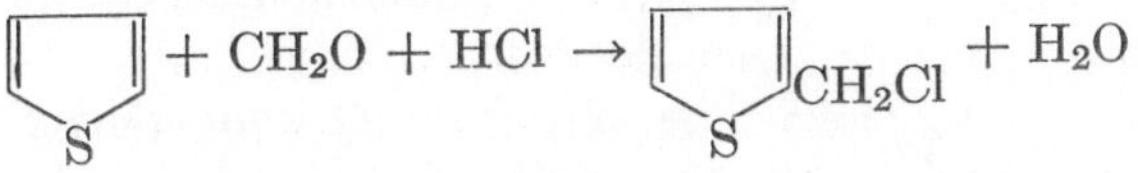

Submitted by K. B. Wiberg and H. F. McShane.
Checked by Cliff S. Hamilton and J. L. Pauley.

1. Procedure

Caution! Suitable precautions should be observed in the preparation, purification, and storage of 2-chloromethylthiophene. The product is lachrymatory, and the procedure should be carried out in an efficient hood. The distillation and storage of the material are described in Notes 6 *and* 7.

In a 2-l. beaker surrounded by an ice-salt bath (Note 1), and fitted with a mechanical stirrer and a thermometer, are placed 420 g. (392 ml., 5 moles) of thiophene (Note 2) and 200 ml. of concentrated hydrochloric acid. A rapid stream of hydrogen chloride (Note 3) is passed continuously into the mixture with vigorous stirring. When the temperature reaches 0°, 500 ml. of 37% formaldehyde solution (Note 4) is added at a rate that will permit the temperature to remain below 5°. The addition requires about 4 hours. When all the formaldehyde solution has been added, the mixture is extracted with three 500-ml. portions of ether. The ether extracts are combined, washed successively with water and saturated sodium bicarbonate solution, and then dried over anhydrous calcium chloride. The solvent is removed by distillation, and the product is distilled under reduced pressure through a 50-cm. fractionating column (Note 5). The product boiling at 73–75°/17 mm. is collected. The yield is 257–267 g. (40–41%) of a colorless oily liquid (Notes 6 and 7).

2. Notes

1. The checkers preferred to use a Dry Ice bath or to add Dry Ice to the reaction mixture as needed.

2. The thiophene used by the submitters was that supplied by the Socony-Vacuum Oil Company. It was used without any additional purification.

3. A satisfactory hydrogen chloride generator has been described earlier.[1]

4. The apparatus should be set up in a hood, for, although nearly all the hydrogen chloride is absorbed, a small amount does escape. The rate of addition of hydrogen chloride should be such that nearly complete absorption is maintained at all times.

According to the submitters, paraformaldehyde (165 g., 5.5 moles) may be substituted for the formaldehyde solution. This modification affords an easier control of the temperature. However, paraformaldehyde does not react to any appreciable extent below 0°, and so the temperature should be kept between 0° and 5°. The reaction time is 6–8 hours, and the yield is unchanged.

5. A Vigreux column was used by both the submitters and checkers.

6. Some *bis*-(2-thienyl)methane may be obtained from the residue at 125–126°/9 mm. When recrystallized from methanol it melts at 46–47°.

7. It is advisable not to store chloromethylthiophene for any length of time or in sealed containers, since it has a tendency to decompose, often with explosive violence, even when kept cold and in the dark. If the product is to be used for the preparation of 2-thiophenaldehyde

it is convenient to convert it to the more stable hexamethylenetetrammonium salt, for storage.

Instances of explosion of the product on storage have been reported; these are apparently due to generation of hydrogen chloride within a closed container. The following procedure for distillation and storage is recomended (F. C. Myer, private communication). The crude product is stabilized by the addition of 2% by weight of dicyclohexylamine and purified by rapid distillation under reduced pressure. The pot temperature should not rise above 100° during the main part of the distillation, or above 125° at the end of the distillation. The distillate is stabilized immediately with 1–2% by weight of dicyclohexylamine and stored in a glass bottle plugged loosely with a stopper wrapped with glass wool. The bottle is placed in a large beaker and stored in a refrigerator. A small amount of gelatinous salt separates on standing; this has no deleterious effect.

When stabilized properly and stored in the cold, decomposition is slow. The material should be inspected regularly to observe excessive darkening and liberation of hydrogen chloride. Restabilization and distillation may permit recovery of pure material if the decomposition is not too far advanced. Complete decomposition results in frothing, resinification, and liberation of hydrogen chloride.

3. Methods of Preparation

2-Chloromethylthiophene has been prepared by the action of hydrogen chloride on 2-thienylcarbinol [2] and by the action of hydrogen chloride and formaldehyde on thiophene.[3] The above procedure is essentially that of Blicke and Burckhalter.[3]

[1] *Org. Syntheses* Coll. Vol. **1**, 534 (1941).
[2] Biedermann, *Ber.*, **19**, 636 (1886).
[3] Blicke and Burckhalter, *J. Am. Chem. Soc.*, **64**, 477 (1942).

m-CHLOROPHENYLMETHYLCARBINOL

(Benzyl alcohol, *m*-chloro-α-methyl-)

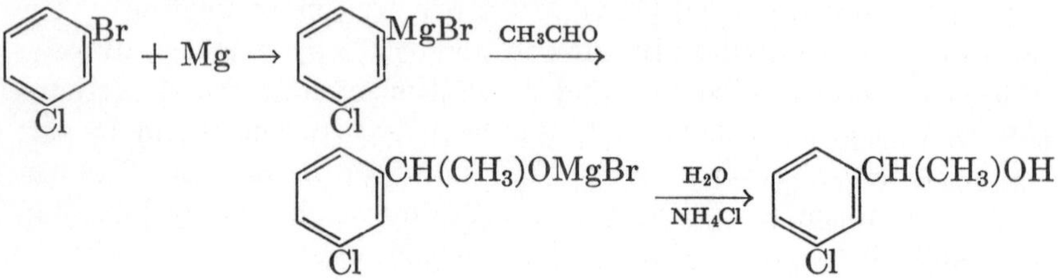

Submitted by C. G. Overberger, J. H. Saunders, R. E. Allen, and Robert Gander.
Checked by Arthur C. Cope and Theodore T. Foster.

1. Procedure

A dry 2-l. three-necked round-bottomed flask is equipped with a sealed stirrer, a 500-ml. dropping funnel (Note 1), and an efficient reflux condenser attached to a calcium chloride tube. In the flask are placed 29.1 g. (1.2 gram atoms) of magnesium turnings, a crystal of iodine, and about 50 ml. of dry ether. A solution of 229 g. (1.2 moles) of *m*-bromochlorobenzene (Note 2) in 850 ml. of dry ether is added with stirring at a rate which maintains rapid refluxing. The reaction begins after 20–50 ml. of the ether solution is added (Note 3), and the addition requires 1–3 hours. The mixture is stirred and heated on the steam bath under reflux for 1 hour after all the *m*-bromochlorobenzene has been added.

A cooled solution of 60 g. (1.365 moles) of freshly distilled acetaldehyde in 200 ml. of dry ether (Note 4) is added through the dropping funnel during 2–3.5 hours, as rapidly as the condenser capacity permits. The mixture is stirred and heated under reflux for 1 hour after the addition is completed.

The reaction mixture is cooled in ice, and the addition compound is decomposed by adding dropwise with stirring 185 ml. of a 25% solution of ammonium chloride in water (Note 5). The ether solution becomes clear, and the salts separate as a cake. The ether solution is decanted, combined with 150 ml. of ether that has been used to rinse the salt cake, and dried over anhydrous magnesium sulfate. After removal of the ether the product is distilled under reduced pressure to give 154.5–164.5 g. (82.5–88%) of *m*-chlorophenylmethylcarbinol boiling at 99–104°/4 mm.; n_D^{25} 1.5405 (Note 6).

2. Notes

1. It is convenient to use a Hershberg stirrer with a rubber slip seal protected from ether vapor with a short water-cooled condenser. The dropping funnel may be connected to the flask through an extension tube to prevent clogging, which occurs if the Grignard reagent comes in contact with the acetaldehyde in the tip of the funnel.

2. *m*-Bromochlorobenzene was prepared according to p. 185. It may be purchased from the Eastman Kodak Company.

3. If the reaction does not begin spontaneously, the mixture should be heated under reflux until the reaction starts before more than 50 ml. of the *m*-bromochlorobenzene solution is added.

4. The solution is cooled to prevent loss of acetaldehyde by vaporization. Commercial acetaldehyde may be redistilled just before it is used, or acetaldehyde may be prepared by depolymerization of paraldehyde.[1]

5. The submitters state that the complex may be decomposed by pouring the reaction mixture onto 1 kg. of crushed ice to which 50 ml. of concentrated sulfuric acid has been added. The ether layer is separated, combined with two 100-ml. ether extracts of the aqueous layer, and washed with three 250-ml. portions of water, one 250-ml. portion of 10% sodium carbonate solution, and finally with 250 ml. of water.

6. The same general method (including the procedure of Note 5) has been used by the submitters to prepare the following substituted phenylmethylcarbinols:

Carbinol	*Boiling Point*	*% Yield*
m-Trifluoromethylphenylmethylcarbinol	100–102°/17 mm.	83
m-Methylphenylmethylcarbinol	103–105°/6 mm.	71
m-tert-Butylphenylmethylcarbinol	130–134°/17 mm.	56

An alternative preparation of similar carbinols, consisting in the reaction of methylmagnesium iodide with a substituted benzaldehyde, is advantageous when the aromatic aldehyde is available. The following carbinols have been prepared by the submitters in that way:

Aldehyde	*Carbinol*	*Boiling Point*	*% Yield*
p-Chlorobenzaldehyde [Heyden Chemical Corporation; *Org. Syntheses* Coll. Vol. **2,** 133 (1943)]	*p*-Chlorophenylmethylcarbinol	98–100°/4.5 mm.	59

Aldehyde	*Carbinol*	*Boiling Point*	*% Yield*
o-Chlorobenzaldehyde (Heyden Chemical Corporation)	*o*-Chlorophenylmethylcarbinol	94°/4 mm.	69
o-Bromobenzaldehyde	*o*-Bromophenylmethylcarbinol	102–105°/2–3 mm.	73
m-Bromobenzaldehyde	*m*-Bromophenylmethylcarbinol	105–110°/2–3 mm.	74
p-Bromobenzaldehyde [*Org. Syntheses* Coll. Vol. **2**, 89, 442 (1943)]	*p*-Bromophenylmethylcarbinol	90°/1 mm.	64
2,3-Dichlorobenzaldehyde	2,3-Dichlorophenylmethylcarbinol	112–113°/2 mm. m.p. 55–57°	76
2,4-Dichlorobenzaldehyde (Heyden Chemical Corporation)	2,4-Dichlorophenylmethylcarbinol	125–126°/7 mm.	62
2,6-Dichlorobenzaldehyde (Eastman Kodak Company)	2,6-Dichlorophenylmethylcarbinol	104–107°/2.5 mm.	89
3,5-Dichlorobenzaldehyde	3,5-Dichlorophenylmethylcarbinol	126°/4 mm.	69
3,4-Dichlorobenzaldehyde (Heyden Chemical Corporation)	3,4-Dichlorophenylmethylcarbinol	125–130°/3–4 mm.	73

3. Methods of Preparation

This procedure is adapted from the preparation described by Marvel and Schertz.[2] *m*-Chlorophenylmethylcarbinol also has been prepared from *m*-chlorobenzaldehyde and methylmagnesium iodide,[3–5] and by the reduction of *m*-chloroacetophenone in the presence of copper chromite.[6]

[1] *Org. Syntheses* Coll. Vol. **2**, 407 (1943).

[2] Marvel and Schertz, *J. Am. Chem. Soc.*, **65**, 2054 (1943).

[3] Lock and Bock, *Ber.*, **70**, 916 (1937).

[4] Brooks, *J. Am. Chem. Soc.*, **66**, 1295 (1944).

[5] Ushakov and Matuzov, *J. Gen. Chem. U.S.S.R.*, **14**, 120 (1944) [*C. A.*, **39**, 916 (1945)].

[6] Emerson and Lucas, *J. Am. Chem. Soc.*, **70**, 1180 (1948).

γ-CHLOROPROPYL ACETATE

(1-Propanol, 3-chloro-, acetate)

$$CH_3COOH + HOCH_2CH_2CH_2Cl \xrightarrow{H^+} CH_3COOCH_2CH_2CH_2Cl + H_2O$$

Submitted by C. F. H. Allen and F. W. Spangler.
Checked by Millard Seeley and C. R. Noller.

1. Procedure

A 1-l. round-bottomed flask is fitted with a 30-cm. Vigreux column (Note 1) connected to a condenser. The condenser leads to an automatic separator arranged so that the lighter liquid is returned to the flask.[1] In the flask is placed a mixture of 189 g. (167 ml., 2 moles) of trimethylene chlorohydrin,[2] 180 g. (172 ml., 3 moles) of glacial acetic acid, 300 ml. of benzene, and 2 g. of *p*-toluenesulfonic acid monohydrate. The mixture is refluxed at such a rate that the temperature at the top of the column remains at 69° (boiling point of benzene-water azeotrope) during the time the greatest amount of water is distilling, and rises gradually to 80° (boiling point of benzene-acetic acid azeotrope) as the last of the water is removed. The volume of the aqueous layer collected varies with the temperature of the vapors and the length of time required for distillation, but it is approximately 50 ml. at the end of the reaction; the time required is 7–9 hours (Note 2).

The solution is cooled, then washed successively with two 500-ml. portions of 10% sodium carbonate solution, one 500-ml. portion of water, and one 100-ml. portion of saturated sodium chloride solution. The wash solutions are extracted successively with one 100-ml. portion of benzene, which is added to the main benzene solution. After removal of the benzene the ester is distilled at atmospheric pressure through a 30-cm. asbestos-wrapped Vigreux column, the fraction boiling at 166–170°, n_D^{2} 1.4295, being collected. The yield of colorless product is 257–260 g. (93–95%).

2. Notes

1. The submitters prefer to use the Clarke-Rahrs ester column,[3] for which they claim two distinct advantages: (1) dehydration requires no attention beyond occasionally noting the amount of distillate—when no more aqueous layer separates the reaction is over; (2) re-

agents do not need to be dry—any water present is removed during the heating.

2. Titration of an aliquot portion of the aqueous layer with standard alkali will give the amount of acetic acid carried over, from which the amount of water can be determined. The expected amount of water is 36 ml.

3. Methods of Preparation

γ-Chloropropyl acetate has been prepared by heating 1-bromo-3-chloropropane and potassium acetate in glacial acetic acid,[4] and by the action of acetyl chloride on trimethylene chlorohydrin.[5,6]

[1] *Org. Syntheses* Coll. Vol. **1**, 422 (1941).
[2] *Org. Syntheses* Coll. Vol. **1**, 533 (1941).
[3] *Synthetic Org. Chemicals,* **9**, No. 3 (May, 1936), Eastman Kodak Company.
[4] Henry, *Bull. acad. roy. Belg.*, **1906**, 738 [*C. A.*, **1**, 1969 (1907)].
[5] Derick and Bissel, *J. Am. Chem. Soc.*, **38**, 2483 (1916).
[6] Lespieau, *Bull. soc. chim. France,* **7**, 254 (1940).

m-CHLOROSTYRENE

(Styrene, *m*-chloro-)

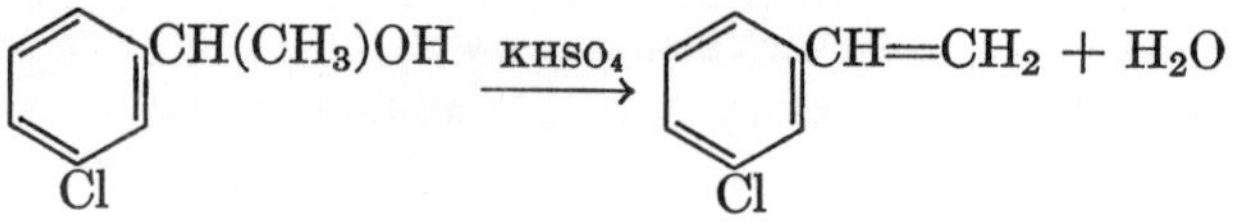

Submitted by C. G. Overberger and J. H. Saunders.
Checked by Arthur C. Cope and Theodore T. Foster.

1. Procedure

A 500-ml. three-necked round-bottomed flask is attached to a 250-ml. dropping funnel and a total-condensation variable take-off fractionating column with a 20 by 1.2 cm. section packed with glass helices (Note 1). The fractionating column is fitted to a 500-ml. receiving flask. In the reaction flask are placed 12.5 g. of powdered fused potassium acid sulfate and 0.05 g. of *p-tert*-butylcatechol. The flask is immersed in an oil bath maintained at 220–230°, and 145 g. (0.925 mole) of *m*-chlorophenylmethylcarbinol (p. 200) and 0.05 g. of *p-tert*-butylcatechol (Note 2) are placed in the dropping funnel. The system is evacuated to a pressure of 125 mm., maintained by a manostat, and the *m*-chlorophenylmethylcarbinol is added dropwise at a rate (15–20 drops per minute) which maintains a vapor temperature at 110–120° at the top of the column. The *m*-chlorostyrene and water formed are collected in a receiver. When the addition is completed (5.5–8.5 hours) the pressure is held constant until distillation stops, and it is then reduced to 20 mm. until no more liquid distils.

The distillate is rinsed into a separatory funnel with 25 ml. of ether, and the organic layer is separated and dried over 10 g. of anhydrous magnesium sulfate. The drying agent is separated and rinsed with 25 ml. of ether, and 0.1 g. of *p-tert*-butylcatechol is added to the solution. The ether is removed, and the product is distilled under reduced pressure. *m*-Chlorostyrene is obtained in a yield of 102–106 g. (80–82.5%), boiling at 55–57°/3 mm.; n_D^{20} 1.5625 (Notes 3 and 4). A small amount (3–8 g.) of *m*-chlorophenylmethylcarbinol can be recovered as a higher-boiling fraction.

2. Notes

1. Effective separation of *m*-chlorostyrene from *m*-chlorophenylmethylcarbinol is possible with the short column specified. If a fractionating column is not used, lower conversions result and the crude product contains *m*-chlorophenylmethylcarbinol, which can be separated by fractional distillation and used in a subsequent preparation.

2. The submitters state that hydroquinone is not a satisfactory polymerization inhibitor for use in this preparation but that picric acid is very effective (less than 0.01 g. is required).

3. The submitters have used a similar procedure to prepare the following substituted styrenes from arylmethylcarbinols. The reactions were conducted more rapidly (addition rate about 2 drops per second), and no fractionating column was used in the initial distillation. A lower pressure (100 mm.) was used in dehydration of the dichlorophenylmethylcarbinols. The yields cited are based on the weights of arylmethylcarbinols which reacted; varying amounts of the carbinols were recovered in the distillation of the products.

	Boiling Point	*% Yield*
o-Chlorostyrene	67–69°/3–3.5 mm.	64–66.5
p-Chlorostyrene	65°/4 mm.	57
o-Bromostyrene	64–65°/3 mm.	64.5
m-Bromostyrene	74–75°/3 mm.	56
p-Bromostyrene	87–88°/12 mm.	50
m-Fluorostyrene	30–32°/2 mm.	45
2,3-Dichlorostyrene	92–94°/4–5 mm.	44
2,4-Dichlorostyrene	81°/6 mm.	33

	Boiling Point	*% Yield*
3,4-Dichlorostyrene	95°/5 mm.	63.5
2,6-Dichlorostyrene	64–65°/3 mm.	31.5
3,5-Dichlorostyrene	59°/1 mm.	43.5

The dehydration procedure of this preparation is stated to be unsatisfactory for substituted phenylmethylcarbinols which have a strongly electronegative group, such as the nitro or trifluoromethyl group, in the *meta* position.

4. Some color due to small amounts of iodine may be present in substituted styrenes prepared by this procedure from arylmethylcarbinols synthesized from the corresponding aldehyde and methylmagnesium iodide. If present, iodine may be removed by shaking the crude product with 1 g. of powdered zinc and 25 ml. of water.

3. Methods of Preparation

m-Chlorostyrene has been prepared by dehydration of *m*-chlorophenylmethylcarbinol by modifications of this procedure,[1–3] and by the vapor-phase dehydration in the presence of alumina at reduced pressure.[4]

[1] Marvel and Schertz, *J. Am. Chem. Soc.*, **65**, 2054 (1943).

[2] Brooks, *J. Am. Chem. Soc.*, **66**, 1295 (1944).

[3] Ushakov and Matuzov, *J. Gen. Chem. U.S.S.R.*, **14**, 120 (1944) [*C. A.*, **39**, 916 (1945)].

[4] Emerson and Lucas, *J. Am. Chem. Soc.*, **70**, 1180 (1948).

CHOLESTENONE

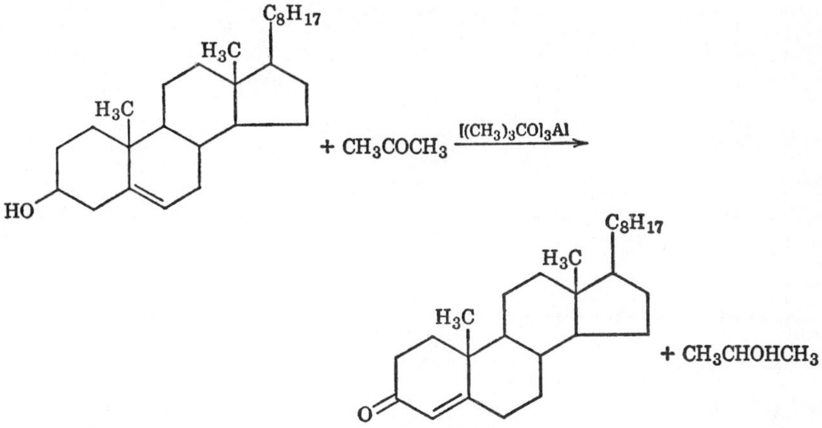

Submitted by R. V. Oppenauer.
Checked by J. Cason and L. F. Fieser.

1. Procedure

A carefully dried 5-l. round-bottomed flask equipped with a reflux condenser carrying a calcium chloride tube is charged with 100 g. of cholesterol (Note 1), 750 ml. of acetone (Note 2), and 1 l. of benzene (Note 3). A boiling tube is introduced to prevent bumping (Note 4), and the mixture is heated to boiling in an oil bath which is maintained at 75–85° during the reaction. A solution of 80 g. of aluminum *tert*-butoxide (p. 48) in 500 ml. of dry benzene is added in one portion to the boiling solution. The mixture turns cloudy and in 10–15 minutes develops a yellow color. Gentle boiling is continued at a bath temperature of 75–85° for a total of 8 hours. The mixture is then cooled, treated with 200 ml. of water and then 500 ml. of 10% sulfuric acid, shaken vigorously, and transferred to a 5-l. separatory funnel. The mixture is diluted with 1.5 l. of water and shaken for several minutes, after which the yellow aqueous layer is drawn off into a second separatory funnel and shaken out with a small amount of benzene (Note 5). The combined benzene extracts are washed thoroughly with water and dried by filtration through a layer of sodium sulfate; the solvent is evaporated, the last traces being removed by heating the residue at 60° at the water pump vacuum. The oily yellow residue solidifies when it is cooled in an ice-salt bath and scratched. For crystallization the material is dissolved in a mixture of 70 ml. of acetone and 100 ml. of methanol; the solution is allowed to cool very slowly and is seeded, for otherwise the product tends to separate as an oil. After the bulk of the material has crystallized, the mixture is allowed to stand for 1 day at 0°; the product is then collected, washed with 100 ml. of ice-cold methanol, and dried in vacuum at room temperature. The yield of almost colorless cholestenone, m.p. 77–79°, is 70–81 g. (70–81%). Recrystallization by the same method gives material melting at 78.5–80.5° with 90% recovery (Notes 6 and 7).

2. Notes

1. The material should be dried to constant weight at 80–100° in vacuum. Commercial cholesterol, m.p. 146–148.5°, gives satisfactory results; the yield is raised 5–10% by using cholesterol that has been purified by regeneration from the dibromide.

2. The acetone is distilled once from permanganate and twice from freshly fused potassium hydroxide.

3. The benzene is distilled over sodium.

4. If boiling stones are employed, fresh ones must be added at intervals; the boiling tube promotes smooth boiling at the bath temperature specified.

5. The troublesome emulsions sometimes encountered are easily broken by filtration from a trace of suspended solid.

6. For recovery of material in the mother liquors, the solutions are subjected to steam distillation to remove solvents and condensation products from the acetone. The process is continued until the distillate is odorless (1–2 hours), and the residual oil is dried by heating it under reduced pressure on the steam bath. From a solution in acetone-methanol (20 ml. and 50 ml., respectively) there is obtained after some manipulation (slow cooling, scratching) 3.3 g. of crude cholestenone, m.p. 72–80° (cloudy melt).

7. The yield is reduced by about 5% by halving the amounts of acetone and benzene specified and using only 50 g. of aluminum *tert*-butoxide.

3. Methods of Preparation

Cholestenone has been prepared by oxidation of cholesterol dibromide with chromic acid[1,2] or potassium permanganate[3] and subsequent debromination; by dehydrogenation of cholesterol over copper oxide;[4,5] and by the method described above.[6,7]

[1] Windaus, *Ber.*, **39**, 518 (1906).
[2] Bretschneider and Ajtai, *Monatsh.*, **74**, 57 (1943).
[3] Schoenheimer, *J. Biol. Chem.*, **110**, 461 (1935).
[4] Diels and Abderhalden, *Ber.*, **37**, 3099 (1904).
[5] Diels, Gädke, and Körding, *Ann.*, **459**, 21 (1927).
[6] Oppenauer, *Rec. trav. chim.*, **56**, 137 (1937).
[7] Brit. pat. 487,360 [*C. A.*, **33**, 323 (1939)].

COUMARILIC ACID *

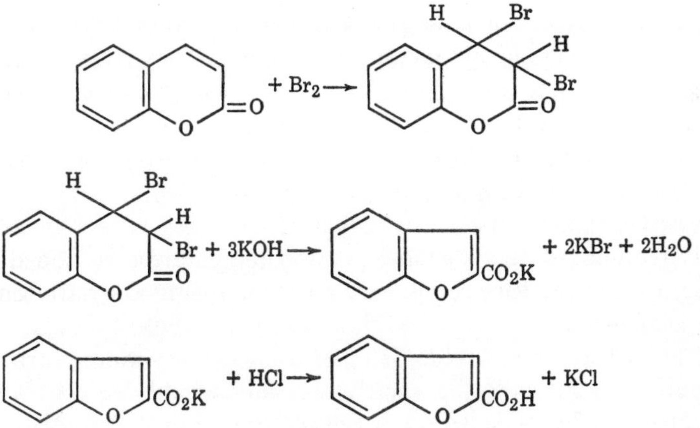

Submitted by R. C. Fuson, J. Wayne Kneisley, and E. W. Kaiser.
Checked by W. E. Bachmann and G. Dana Johnson.

1. Procedure

A. *Coumarin dibromide.* In a 1-l. three-necked flask, equipped with a mechanical stirrer, a dropping funnel, and a condenser fitted with a trap for hydrogen bromide, are placed 146 g. (1 mole) of coumarin and 200 ml. of chloroform. A solution of 160 g. (1 mole) of bromine in 85 ml. of chloroform is added dropwise to the well-stirred solution of coumarin at room temperature over a period of 3 hours (Note 1). Excess bromine is removed by adding approximately 200 ml. of a 20% solution of sodium sulfite through the dropping funnel. The colorless chloroform layer is separated, washed with two 200-ml. portions of water, and dried over magnesium sulfate (Note 2). After the mixture has stood for only a few minutes (Note 3), the magnesium sulfate is removed by filtration and the solution of the dibromide is evaporated to 100 ml. in a rapid stream of dry air (Note 4). The coumarin dibromide is collected on a filter, and the mother liquor is evaporated nearly to dryness by an air blast. This mixture is filtered, and the slightly yellow mother liquor is evaporated to dryness. The product of the third evaporation is slightly discolored, and it is washed with two 15-ml. portions of ether (Note 5). The combined precipitates weigh 215 g. (70%) and melt at 102–105° (Note 6).

B. *Coumarilic acid.* In a 5-l. three-necked flask fitted with a mechanical stirrer and reflux condenser, 450 g. (8 moles) of solid potassium hydroxide is dissolved in 700 ml. of absolute ethanol. The solution is cooled to 15° by immersing the flask in an ice bath, and 215 g. (0.7 mole) of finely divided coumarin dibromide is added in 10- to 15-g. portions to the well-stirred basic solution. The rate of addition is controlled so that the temperature never rises above 20°; the addition requires about 30 minutes. After all the dibromide has been added, the reaction mixture is refluxed, with stirring, for 30 minutes (Note 7). One and a half liters of water is added, and the resulting solution is steam-distilled until 2.5 l. of distillate has been collected (Note 8). The residue is cooled to room temperature by the addition of 1 kg. of cracked ice (Note 9) and is then acidified by the addition of 1.2 l. of 6 *N* hydrochloric acid. The crude coumarilic acid is collected on a filter and stirred with 600 ml. of cold water. The acid is separated from the water by filtration, sucked as dry as possible on the filter, and then crystallized from a mixture of 250 ml. of ethanol and 250 ml. of water (Note 10). The recrystallized coumarilic acid is colorless, weighs 93–100 g. (82–88%) (Note 11), and melts at 190–193°.

2. Notes

1. Occasional use of an ice bath is necessary to keep the temperature from rising.

2. If any coumarin dibromide separates before the magnesium sulfate is removed, enough chloroform should be added to dissolve the dibromide.

3. The mixture should be worked up immediately in order to avoid decomposition of the dibromide.

4. The checkers found that considerable decomposition occurred when this procedure was employed, perhaps because it happened to be carried out on a warm, humid day; they found it preferable to remove the chloroform by distillation under reduced pressure from a water

bath at this step and the succeeding ones. Because of the instability of the dibromide, it is advisable to convert it to coumarilic acid on the day it is made.

5. The first evaporation produces 136 g. of colorless crystals; the second, 63 g.; and the third, 16 g. of ether-washed product.

6. The first crop of crystals melts at 102–105°. The succeeeding products melt somewhat lower but are sufficiently pure for the next reaction.

7. The suspension of potassium bromide must be stirred vigorously to avoid bumping.

8. The steam distillation is carried out conveniently by fitting the reaction vessel with a steam inlet tube and a condenser set for downward distillation. The flask should be heated so that the final volume of the distilland is not more than 3 l.

9. A heavy precipitate composed of potassium coumarilate and potassium bromide may appear if the steam distillation has been very efficient. The precipitate dissolves during the addition of hydrochloric acid.

10. The crude acid is dissolved in 250 ml. of boiling absolute ethanol to which has been added 3 g. of Norit; 250 ml. of water, previously heated to 85°, is then added to the filtered solution.

11. The over-all yield, based on the amount of coumarin, is 57–62%.

3. Methods of Preparation

Coumarilic acid has been prepared by treating coumarin dibromide with alcoholic potassium hydroxide and acidifying with hydrochloric acid, a method due essentially to Perkin.[1] Coumarilic acid has also been prepared by Perkin from 3-chlorocoumarin and from 3-bromocoumarin by the action of alcoholic potassium hydroxide.[2]

Coumarilic acid has also been prepared from 2-bromobenzofuran by reaction with *n*-butyllithium and subsequent carbonation [3] and by the oxidation (NaOCl) of 2-acetylbenzofuran.[4]

Coumarin dibromide has been prepared by the addition of bromine to coumarin in chloroform solution.[1,2]

1 Perkin, *J. Chem. Soc.*, **23**, 368 (1870).
2 Perkin, *J. Chem. Soc.*, **24**, 37 (1871).
3 Gilman and Melstrom, *J. Am. Chem. Soc.*, **70**, 1655 (1948).
4 Farrar and Levine, *J. Am. Chem. Soc.*, **72**, 4433 (1950).

9-CYANOPHENANTHRENE

(9-Phenanthrenecarbonitrile)

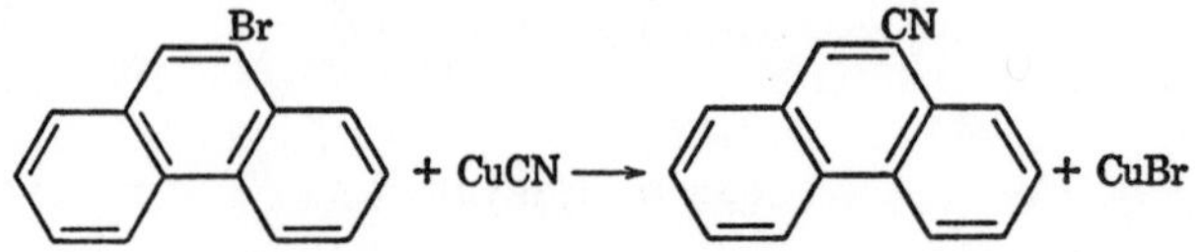

Submitted by JOSEPH E. CALLEN, CLINTON A. DORNFELD, and GEORGE H. COLEMAN.[1]
Checked by ROBERT E. CARNAHAN and HOMER ADKINS.

1. Procedure

In a 2-l. Claisen flask (Note 1) are mixed 1 kg. (3.90 moles) of 9-bromophenanthrene (p. 134) (Note 2) and 400 g. (4.46 moles) of cuprous cyanide. A small motor-driven spiral stirrer is inserted, and the flask is heated (Note 3) at 260° for 6 hours (Note 4). The flask is then provided with a fine capillary inlet tube, a thermometer, and a 2-l. distilling flask attached to the side arm of the Claisen flask as receiver. The 9-cyanophenanthrene is distilled at 190–195°/2 mm. The yield of crude product is 740 g. (93%), m.p. 94–98°. One recrystallization from about 2 l. of dry ethanol yields 690 g. (87%) of material melting at 105–107° (Notes 5, 6, and 7).

2. Notes

1. It is necessary to sacrifice the flask after the distillation, since the inorganic residue cannot be removed easily.

2. The 9-bromophenanthrene need not be recrystallized if it has been purified by distillation.

3. The heating may be readily controlled by the use of an electric heating mantle and variable transformer.

4. Increasing the heating period beyond 6 hours has no effect on the yield.

5. The melting point can be raised to 110° by further recrystallization from ethanol. The unrecrystallized product is probably pure enough for most purposes.

6. The checkers operated on one-tenth the scale and duplicated the percentage yields obtained on the larger scale. Material melting at 109–110° was obtained in a 75–77% yield.

7. It has been reported (W. R. Vaughan, private communication) that this reaction is highly exothermic, and that the flask contents may erupt when the mixture is heated to the prescribed temperature. Better control of the reaction may result if the cuprous cyanide is added in portions after the reaction has been initiated at 260–280° with about one-third or less of the total amount of cuprous cyanide.

3. Methods of Preparation

The procedure described above is an adaptation of methods of Mosettig and van de Kamp [2] and of Bachmann and Boatner.[3] 9-Cyanophenanthrene has also been prepared from 9-phenanthrenesulfonic acid [4] and from phenanthrene-9-aldoxime.[5]

[1] Work done under contract with the Office of Scientific Research and Development.

[2] Mosettig and van de Kamp, *J. Am. Chem. Soc.*, **54**, 3355 (1932).

[3] Bachmann and Boatner, *J. Am. Chem. Soc.*, **58**, 2098 (1936); Goldberg, Ordas, and Carsh, *J. Am. Chem. Soc.*, **69**, 260 (1947).

[4] Werner and Kunz, *Ann.*, **321**, 327 (1902).

[5] Shoppee, *J. Chem. Soc.*, **1933**, 40.

1,1-CYCLOBUTANEDICARBOXYLIC ACID AND CYCLOBUTANECARBOXYLIC ACID

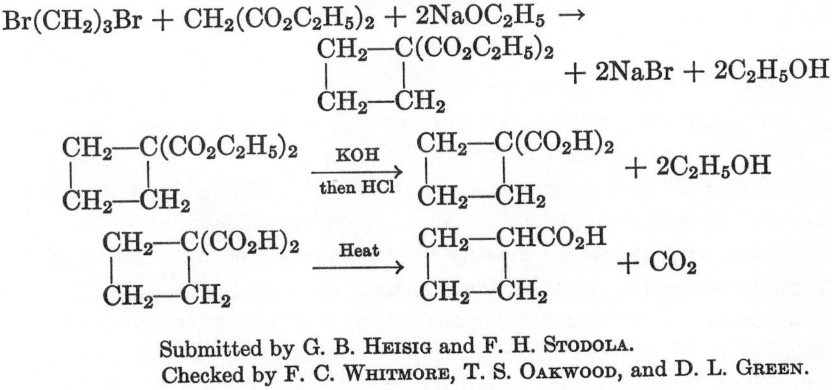

Submitted by G. B. Heisig and F. H. Stodola.
Checked by F. C. Whitmore, T. S. Oakwood, and D. L. Green.

1. Procedure

A. 1,1-*Cyclobutanedicarboxylic acid.* In a 3-l. three-necked round-bottomed flask, carrying a separatory funnel, a mechanical stirrer, a reflux condenser fitted with a calcium chloride tube, and a thermometer, are placed 160 g. (1 mole) of ethyl malonate and 212 g. (1.05 moles) of trimethylene bromide. The thermometer is adjusted so that the bulb is immersed in the liquid in the flask; the stirrer is started, and a solution of 46 g. (2 gram atoms) of sodium in 800 ml. of absolute ethanol is added through the separatory funnel while the temperature of the reaction mixture is kept at 60–65° (Note 1). During the addition of the first quarter of the ethoxide solution (20 minutes) it is necessary to cool the reaction mixture occasionally in order to maintain the proper temperature, but after this point the remainder of the ethoxide solution is added just rapidly enough to keep the reaction mixture at 60–65°; this part of the addition requires about 30 minutes.

The reaction mixture is allowed to stand until the temperature drops to 50–55°, after which it is heated on a steam bath until a sample added to water is neutral to phenolphthalein (about 2 hours). Water is added to dissolve the precipitate of sodium bromide, and the ethanol is removed by distillation. The flask is now arranged for steam distillation, and the ethyl 1,1-cyclobutanedicarboxylate and unchanged malonic ester are removed by steam distillation (Note 2); about 4 l. of distillate is collected. The ester layer in the distillate is separated, and the aqueous layer is extracted once with 1 l. of ether. The extract and the ester layer are combined, and the ether is removed on the steam bath (Note 3).

The esters are hydrolyzed by refluxing them for 2 hours with a solution of 112 g. of potassium hydroxide in 200 ml. of ethanol. Most of the ethanol is removed by distillation, and the mixture is then evaporated to dryness on a steam bath. The residue is dissolved in the minimum amount of hot water (100 to 125 ml.), and concentrated hydrochloric acid (about 90–95 ml.) is added until the solution is slightly acid (Note 4). After the solution has been boiled for a few minutes to remove carbon dioxide, it is made slightly alkaline with ammonia. To the boiling solution there is added a slight excess of barium chloride. The hot solution is filtered to remove barium malonate, the filtrate is cooled, and to it is added 100 ml. of 12 *N* hydrochloric acid. The solution is then extracted with four 250-ml. portions of ether. The extracts are combined and dried over calcium chloride, and the ether is removed by distillation on a steam bath. The residual pasty mass (about 38 g.) is pressed on a porous plate to remove adherent oil and then dissolved in 30–50 ml. of hot ethyl acetate. The solution, when cooled in an ice-salt bath, deposits the pure dicarboxylic acid. This is filtered; the filtrate, when evaporated, yields a pasty mass of acid which, in turn, is crystallized from ethyl acetate.

The yield of pure 1,1-cyclobutanedicarboxylic acid melting at 156–158° is 30–34 g. (21–23%).

B. *Cyclobutanecarboxylic acid.* The above dibasic acid is placed in a 75-ml. distilling flask carrying a thermometer and attached to a 75-ml. Claisen flask as a receiver. The receiver is cooled with running water while the flask containing the dibasic acid is heated in a metal or oil bath (bath temperature, 160–170°) until no more carbon dioxide is evolved. Then the temperature of the bath is raised to 210–220°, and the material that boils at 189–195° is collected. The crude cyclobutanecarboxylic acid (19–22 g.) is redistilled from the Claisen flask in which it was collected. The pure acid boils at 191.5–193.5°/740 mm. (Note 5) and weighs 18–21 g. (18–21% based on malonic ester). There is a small higher-boiling fraction, which comes over at 193.5–196°/740 mm.

2. Notes

1. The submitters and the checkers prepared the sodium ethoxide in the conventional manner. However, sodium ethoxide and sodium methoxide are very conveniently prepared by the "inverse" procedure, as described by Tishler in Fieser, *Experiments in Organic Chemistry,* 2nd ed., 1941, D. C. Heath and Company, Boston, p. 385 (bottom). The metal is placed in the flask, and the alcohol is added through the condenser at such a rate that rapid refluxing is maintained. It is necessary, as a precautionary measure, to clamp the flask and not to trust to the friction between a rubber stopper and the flask to hold the flask in place. When this precaution is taken, a cooling bath may be used with safety. It is necessary to cool the flask; the metal must not be allowed to melt, as this will result in the formation of one large mass with a greatly decreased metallic surface. (Private communication, C. F. H. Allen.)

2. The steam distillation separates ethyl malonate and ethyl 1,1-cyclobutanedicarboxylate from ethyl pentane-1,1,5,5-tetracarboxylate, formed in a side reaction between malonic ester (2 moles) and trimethylene bromide (1 mole). The tetraethyl ester remains in the residue from the steam distillation. In several test experiments, in which about two-thirds of the amounts specified above were used, the yield of tetraethyl ester was 30–40% of the theoretical amount, based upon the sodium ethoxide used.

3. If this extraction with ether is omitted the yield of 1,1-cyclobutanedicarboxylic acid is 3–4 g. less.

4. The submitters used 55 ml. of hydrochloric acid at this point; the checkers stated that this amount was insufficient to neutralize the mix-

ture. The purpose of acidification at this point is not to liberate the cyclobutanedicarboxylic acid, but merely to remove carbonates and excess potassium hydroxide. After the carbon dioxide has been expelled, the solution is made alkaline with ammonia; hence a great excess of hydrochloric acid should be avoided. The submitters used only enough hydrochloric acid to make the solution acid to litmus. After the solution has been made basic with ammonia, barium chloride solution is added until there is no further precipitation of barium malonate.

5. This acid is quite pure. The checkers distilled 52.5 g. of it through an analytical column; except for a fore-run of 1.5–2.0 g., all the material had a constant boiling point and index of refraction.

3. Methods of Preparation

1,1-Cyclobutanedicarboxylic acid has been prepared by hydrolysis of the ethyl ester,[1] or of the half nitrile, 1-cyano-1-carboxycyclobutane.[2] The ethyl ester has been prepared by condensation of ethyl malonate with trimethylene bromide [1,3] or chlorobromide,[4] and by the action of sodium ethoxide on diethyl γ-bromopropylmalonate.[5] The half nitrile has been prepared by condensation of trimethylene bromide with ethyl cyanoacetate followed by hydrolysis of the ester to the acid.[2]

Cyclobutanecarboxylic acid has been prepared by decarboxylation of the 1,1-dibasic acid,[1,6] and by decarboxylation of 1-cyano-1-carboxycyclobutane followed by hydrolysis of the cyano group.[2]

[1] Perkin, *J. Chem. Soc.*, **51,** 1 (1887); Rupe, *Ann.*, **327,** 183 (1903).

[2] Carpenter and Perkin, *J. Chem. Soc.*, **75,** 930 (1899).

[3] Cason and Allen, *J. Org. Chem.*, **14,** 1036 (1949).

[4] Kishner, *J. Russ. Phys. Chem. Soc.*, **37,** 507 (1905) (*Chem. Zentr.*, **1905,** II, 761).

[5] Walborsky, *J. Am. Chem. Soc.*, **71,** 2941 (1949).

[6] Perkin and Sinclair, *J. Chem. Soc.*, **61,** 40 (1892).

trans-1,2-CYCLOHEXANEDIOL

(1,2-Cyclohexanediol, *trans*-)

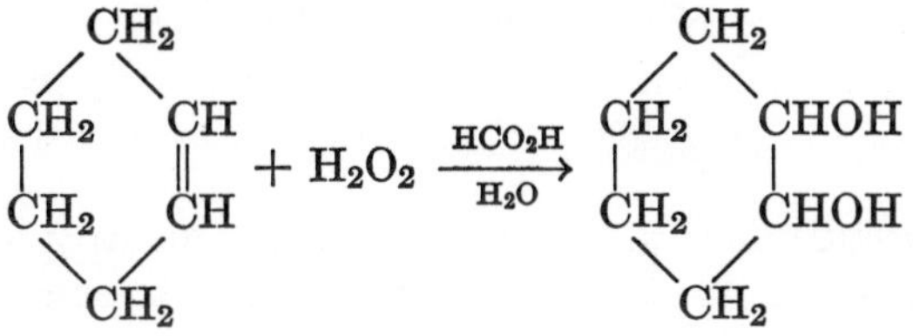

Submitted by ALAN ROEBUCK and HOMER ADKINS.
Checked by ARTHUR C. COPE and CLAUDE F. SPENCER.

1. Procedure

One hundred and forty milliliters of 30% hydrogen peroxide (1.4 moles) (Note 1) is added to 600 ml. of 88% formic acid (13.7 moles) (Note 2) in a 1-l. three-necked flask equipped with a thermometer and a motor-driven stirrer. Freshly distilled cyclohexene (82 g., 1.0 mole) (Note 3) is added slowly from a dropping funnel over a period of 20–30 minutes while the temperature of the reaction mixture is maintained between 40° and 45° by cooling with an ice bath and by controlling the rate of addition (Note 4). The reaction mixture is kept at 40° for 1 hour after all the cyclohexene has been added, and then it is allowed to stand overnight at room temperature.

The formic acid and water are removed by distillation from a steam bath under reduced pressure. An ice-cold solution of 80 g. of sodium hydroxide in 150 ml. of water is added in small portions to the residual viscous mixture of the diol and its formates, with care that the temperature of the mixture does not exceed 45°. The alkaline solution is warmed to 45°, and an equal volume (350 ml.) or more of ethyl acetate is added. After thorough extraction, the lower layer is separated and extracted at 45° six times with equal volumes of ethyl acetate. The seven ethyl acetate solutions (Note 5) are combined (total volume about 2.1 l.), and the solvent is distilled from a steam bath until the residual volume is 300–350 ml. and the solid product begins to crystallize. The mixture is cooled to 0°, and the product is separated by filtration (77–90 g., melting in the range of 90–98°). The mother liquor is concentrated on a steam bath to a volume of 65–75 ml., and more solid crystallizes. The mixture is cooled and filtered as before and yields an additional 4–15 g. of crude product melting in the range of 80–89°. *trans*-1,2-Cyclohexanediol, b.p. 120–125°/4 mm., is obtained by distillation of the combined crude products, with the use of an oil bath and a flask having a side arm and an air condenser sufficiently wide that they will not become plugged as the product solidifies (Note 6). The yield of product of m.p. 101.5–103° is 75–85 g. (65–73%) (Note 7).

2. Notes

1. The concentration of the hydrogen peroxide may be determined by titration with 0.2 *N* potassium permanganate in acid solution.

2. Less formic acid than is specified may be used, but the yields are lower and the reaction is not so easily controlled.

3. The cyclohexene used may be the commercial product or it may be prepared from cyclohexanol.[1]

4. If the temperature rises above 45°, the reaction may be stopped by discontinuing the stirring.

5. The total quantity of ethyl acetate required is less if the first three extractions are combined and the solvent required for successive extractions is distilled at a steam bath from this solution.

6. If there is an appreciable quantity of solid residue in the flask, it may be dissolved in hot water and crystallized by cooling in ice. The crystals of cyclohexanediol, after being dried, may be distilled as described above.

7. The product can be recrystallized from ethyl acetate.

3. Methods of Preparation

The method described is essentially that of Swern, Billen, Findley, and Scanlan.[2] In addition to the performic acid method,[3] *trans*-1,2-cyclohexanediol may be obtained through hydrolysis of the monoacetate [4,5] or cyclohexene oxide.[4] It has also been obtained by a peracetic acid oxidation of cyclohexene with tungsten trioxide,[6] and as a by-product in several instances.[7,8] *cis*-1,2-Cyclohexanediol has been prepared by reaction of cyclohexene with hydrogen peroxide in *tert*-butyl alcohol with osmium tetroxide as a catalyst.[6,9] Mixtures of *cis* and *trans* isomers have resulted from the hydrogenation (Raney nickel) of catechol,[10,11] and the sodium-alcohol [12] or catalytic [13] reduction of adipoin.

[1] *Org. Syntheses* Coll. Vol. **1**, 183 (1941).
[2] Swern, Billen, Findley, and Scanlan, *J. Am. Chem. Soc.*, **67**, 1786 (1945).
[3] English and Gregory, *J. Am. Chem. Soc.*, **69**, 2120 (1947).

[4] Criegee and Stanger, *Ber.*, **69**, 2753 (1936).
[5] Clarke and Owen, *J. Chem. Soc.*, **1949**, 315.
[6] Mugdan and Young, *J. Chem. Soc.*, **1949**, 2988.
[7] McKusick, *J. Am. Chem. Soc.*, **70**, 1976 (1948).
[8] Price and Mueller, *J. Am. Chem. Soc.*, **66**, 628 (1944).
[9] Milas and Sussman, *J. Am. Chem. Soc.*, **59**, 2345 (1937).
[10] Donald Robinson, Ph.D. Thesis, University of Wisconsin, 1948.
[11] English and Barber, *J. Am. Chem. Soc.*, **71**, 3310 (1949).
[12] Treibs and Bast, *Ann.*, **561**, 165 (1949).
[13] U. S. pat. 2,467,451 [*C. A.*, **43**, 6229 (1949)].

α-CYCLOHEXYLPHENYLACETONITRILE

(Cyclohexaneacetonitrile, α-phenyl-)

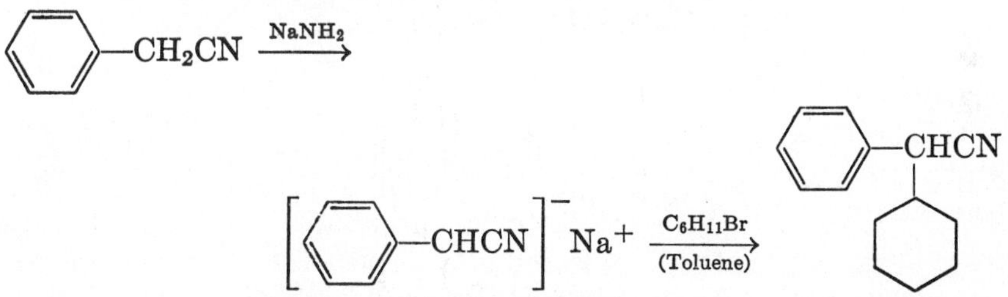

Submitted by EVELYN M. HANCOCK and ARTHUR C. COPE.
Checked by NATHAN L. DRAKE and W. MAYO SMITH.

1. Procedure

A 1-l. three-necked flask is equipped with a mercury-sealed stirrer, an inlet tube, and a reflux condenser which is connected through a soda-lime tube to a gas-absorption trap. The apparatus is dried in an oven and assembled rapidly to exclude moisture; it is advisable to set up the apparatus in a hood to vent ammonia that may escape by accident. The flask is cooled in a Dry Ice-trichloroethylene bath, and 200 ml. of anhydrous (refrigeration grade) ammonia is introduced through the inlet tube from an ammonia cylinder which is either inverted or equipped with a siphon tube. Just before the apparatus is assembled, 8.1 g. (0.35 gram atom) of sodium is cut, weighed, and kept under kerosene in a small beaker. The Dry Ice bath is removed, the inlet tube is replaced by a rubber stopper, and a crystal of hydrated ferric nitrate (about 0.2 g.) is added. A small (about 5-mm.) cube of the sodium is cut, blotted rapidly with filter paper, and added quickly to the liquid ammonia. The solution is stirred until the blue color disappears, after which the remainder of the sodium is added in narrow, thin strips about as rapidly as it can be cut and blotted, while the solution is stirred vigorously (Note 1). After the solution has turned from blue to gray, the flask is swirled by hand until the blue flecks of sodium that have spattered onto the upper part of the flask are washed into the solution.

The flask is again cooled in the Dry Ice bath, and 41 g. (0.35 mole) of benzyl cyanide (Note 2) is added during about 10 minutes through a dry dropping funnel. The Dry Ice bath is removed, and the clear solution is stirred for about 15 minutes, after which 200 ml. of dry sulfur-free toluene (Note 3) and 25 ml. of anhydrous ether are added dropwise through the funnel while the ammonia evaporates. The solution is allowed to stand or is warmed in a water bath until it comes to room temperature; the ammonia may be vaporized at any rate that does not cause gas to escape through the mercury seal on the stirrer. The remainder of the ammonia is removed by turning off the water in the reflux condenser, warming the flask in a hot water bath, and distilling most of the ether (and ammonia) through the reflux condenser.

A fresh drying tube is attached to the reflux condenser, the cooling water is turned on, and to the warm solution is added 65.2 g. (0.4 mole) of bromocyclohexane (Note 4) during approximately 20 minutes. The reaction is vigorous and may require cooling. Refluxing is continued by heating the mixture in an oil bath for 2 hours. The reaction mixture is cooled and washed with 300 ml. of water. The aqueous layer is extracted with two 50-ml. portions of benzene, and the combined benzene and toluene solutions are washed with two 50-ml. portions of water and distilled from a 500-ml. modified Claisen flask under reduced pressure. The yield of α-cyclohexylphenylacetonitrile boiling at 174–176°/13 mm. (Note 5) is 45–53 g. (65–77%); the distillate solidifies to a crystalline mass which has a melting point of 50–53.5°. The nitrile can be recrystallized from commercial pentane with approximately 15% loss; it then melts at 56–58°.

2. Notes

1. Vigorous stirring and addition of the sodium in small pieces decrease the time required for complete conversion of the sodium into sodamide.

2. Redistilled benzyl cyanide was used, b.p. 108–110°/13 mm.

3. The toluene (Merck reagent grade) was kept dry by storage over sodium wire. Dry thiophene-free benzene may be used instead but is less convenient because it freezes when added to the liquid ammonia solution.

4. Bromocyclohexane obtained from the Eastman Kodak Company was dried over calcium chloride and redistilled.

5. Another boiling point is 165–167°/9 mm.

3. Methods of Preparation

α-Cyclohexylphenylacetonitrile has been prepared by treating the sodium derivative obtained from benzyl cyanide and sodamide with bromocyclohexane in benzene,[1] in ether,[2] and in xylene.[3]

[1] Venus-Danilova and Bol'shukon, *J. Gen. Chem. U.S.S.R.*, **7**, 2823 (1937) [*C. A.*, **32**, 2925 (1938)].

[2] Vasiliu, *Bul. Soc. Chim. România,* **19A**, 75 (1937) [*C. A.*, **33**, 4207 (1939)].

[3] Cheney and Bywater, *J. Am. Chem. Soc.*, **64**, 970 (1942).

CYCLOPROPANECARBOXYLIC ACID

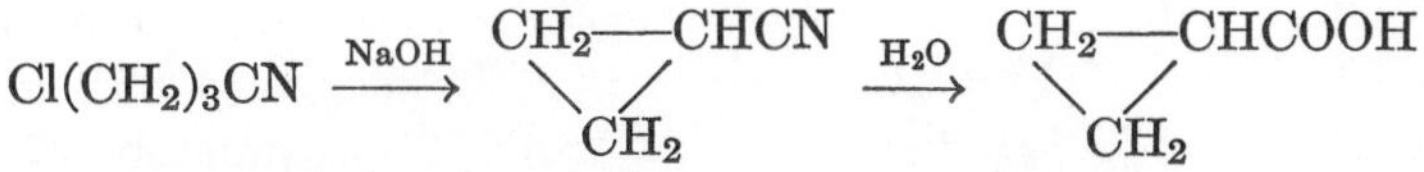

Submitted by CHESTER M. MCCLOSKEY and GEORGE H. COLEMAN.
Checked by C. F. H. ALLEN and HOMER W. J. CRESSMAN.

1. Procedure

In a 2-l. three-necked round-bottomed flask surmounted by two large condensers (Notes 1 and 2) are placed 150 g. (3.75 moles) of powdered sodium hydroxide (Note 3) and 103.5 g. (1 mole) of γ-chlorobutyronitrile.[1] The contents of the flask are well mixed by shaking, after which the mixture is heated on a steam bath; a rather vigorous reaction sets in (Note 4). The water formed in the reaction hydrolyzes some of the cyclopropyl cyanide, so that after 1 hour's heating very little liquid is apparent. The hydrolysis of the cyanide is completed by the addition of water in small portions over a period of about 2 hours and subsequent heating; 15–20 ml. of water is added at first, and portions of 60–75 ml. later at intervals of 10–15 minutes until 500 ml. in all has been added. The mixture is then heated for an additional 1.5 hours with occasional stirring; at the end of this time the oily layer will have disappeared.

The solution is then cooled in an ice bath and acidified by 200 g. of concentrated sulfuric acid (sp. gr. 1.84) previously mixed with 300 g.

of cracked ice. The solution is again cooled in an ice bath. The thick floating layer of cyclopropanecarboxylic acid and various polymers is separated and the cold aqueous solution extracted once with 1 l. of ether, using a stirrer instead of shaking (Note 5). The extract and crude acid are combined and dried over 50 g. of Drierite, and the solvent is removed in a 500-ml. modified Claisen flask on a steam bath. The residue is then distilled under reduced pressure. The yield of acid boiling at 94–95°/26 mm. or 117–118°/75 mm. is 63.5–68 g. (74–79%) (Note 6).

2. Notes

1. It is advisable to set up the apparatus in a hood, as small amounts of isocyanide are evolved.
2. The yield is slightly higher if the first vigorous reaction is allowed to take place without external cooling.
3. The commercial flakes are ground in a mortar if powder is not available. The powdered potassium hydroxide on the market can be used equally well.
4. The reaction sets in after about 15 minutes. If potassium hydroxide is used, there is only a 5-minute interval.
5. Since troublesome emulsions tend to form, it is advisable to avoid all vigorous shaking. If the mixture is stirred mechanically for 15 minutes, emulsions are avoided; if formed, they can be broken by treatment with anhydrous sodium sulfate.
6. The submitters have made runs twice this size; the yields were proportional to those given above.

3. Methods of Preparation

Cyclopropanecarboxylic acid has been prepared by the hydrolysis of cyclopropyl cyanide, although it is unnecessary to isolate the nitrile;[2–4] by heating cyclopropanedicarboxylic acid;[5,6] and by the action of alkali on ethyl γ-chlorobutyrate.[7] The last two methods do not appear to be of practical importance. The oxidation of cyclopropyl methyl ketone with sodium hypobromite is reported to be an excellent preparative method for this acid.[8]

[1] *Org. Syntheses* Coll. Vol. **1**, 156 (1941).

[2] Henry, *Bull. sci. acad. roy. Belg.*, (3) **36**, 34 (1898) (*Chem. Zentr.*, **1899**, I, 975).

[3] Kishner, *J. Russ. phys. chem. Ges.*, **37**, 304 (*Chem. Zentr.*, **1905**, I, 1703).

[4] Bruylants and Stassen, *Bull. sci. acad. roy. Belg.*, (5) **7**, 702 (1921) [*Chem. Zentr.*, **1922**, I, 1229; *C. A.*, **17**, 2872 (1923)].

[5] Perkin, *J. Chem. Soc.*, **47**, 815 (1885).
[6] Skraup and Binder, *Ber.*, **62**, 1132 (1929).
[7] Rambaud, *Bull. soc. chim. France*, (5) **5**, 1564 (1938).
[8] Jeffery and Vogel, *J. Chem. Soc.*, **1948**, 1804.

CYCLOPROPYL CYANIDE

(Cyclopropane, cyano-)

$$2Na + 2NH_3 \rightarrow 2NaNH_2 + H_2$$

$$ClCH_2CH_2CH_2CN + NaNH_2 \rightarrow \begin{array}{c} CH_2 \\ / \quad \backslash \\ CH_2 \text{—} CHCN \end{array} + NaCl + NH_3$$

Submitted by M. J. Schlatter.
Checked by R. L. Shriner and Chris Best.

1. Procedure

The apparatus shown in Fig. 10 is assembled in a good hood. One liter of liquid ammonia and 0.5 g. of hydrated ferric nitrate are placed

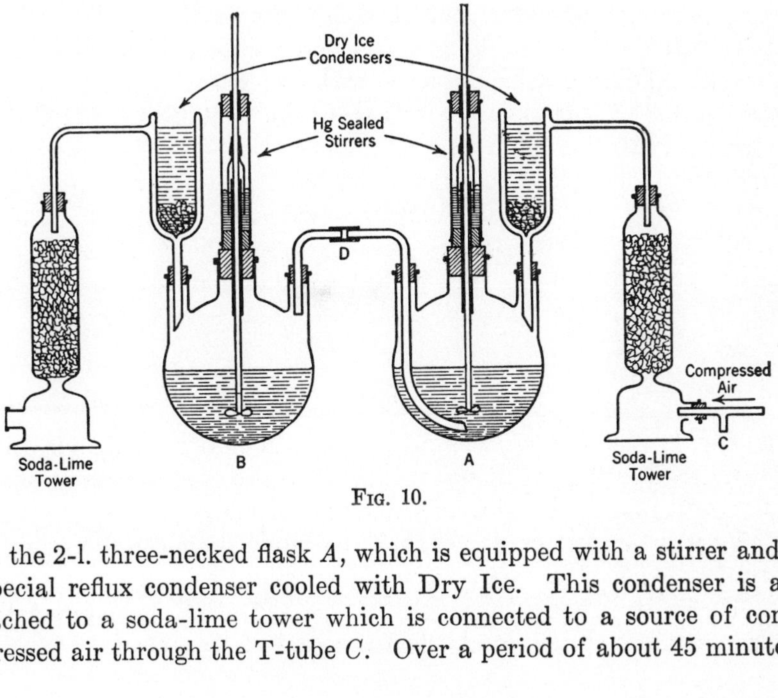

Fig. 10.

in the 2-l. three-necked flask *A*, which is equipped with a stirrer and a special reflux condenser cooled with Dry Ice. This condenser is attached to a soda-lime tower which is connected to a source of compressed air through the T-tube *C*. Over a period of about 45 minutes,

92 g. (4 gram atoms) of clean sodium shavings is added to the liquid ammonia, and the mixture is stirred until the blue color disappears (1–2 hours).

In the similarly equipped 5-l. three-necked flask *B* are placed 1.5 l. of liquid ammonia and 440 g. (4.25 moles) of γ-chlorobutyronitrile [*Org. Syntheses* Coll. Vol. **1,** 156 (1941)]. The flasks are connected by means of the 12-mm. glass tube *D* reaching to the bottom of the sodamide flask *A* and extending 1 cm. through the rubber stopper in one neck of flask *B*. Vigorous stirring is maintained in both flasks while the sodamide suspension is slowly forced over into the reaction flask in small portions (*Caution!* Note 1) by means of air pressure applied through *C*; the rate of addition is controlled by placing the finger over the by-pass in the T-tube. At first the reaction is violent, and only small amounts of the sodamide solution should be added. The addition is continued at such a rate that the total time required for addition of all the sodamide solution is 1–1.5 hours. The sodamide flask is rinsed with 300 ml. of liquid ammonia, and the washings are added to the reaction mixture, after which stirring is continued for 2 hours. During the second hour, addition of Dry Ice to the reflux condenser is discontinued, and the ammonia is permitted to evaporate slowly. At the end of the 2-hour stirring period, the inlet tube is replaced by a dropping funnel and 1 l. of dry ether is slowly added (Note 2). The reaction mixture is quickly filtered through a sintered-glass funnel, and the filter cake is washed with two 200-ml. portions of dry ether (Note 3). The ammonia and ether are removed by distillation on a water bath through a packed column (Note 4). The residue is then distilled through the column under the pressure of a water pump (Note 5). The yield of cyclopropyl cyanide boiling at 69–70°/80 mm. (75–76°/95 mm.) is 149–152 g. (52–53% based upon γ-chlorobutyronitrile) (Notes 6 and 7). The pressure is then reduced, and the unchanged γ-chlorobutyronitrile is collected at 93–96°/26 mm. It amounts to 52–62 g.

2. Notes

1. The reaction is very vigorous, and addition of the sodamide solutions in large portions must be avoided. The use of stopcocks or pinch clamps to regulate the addition is not recommended.

2. Very little cyclopropyl cyanide is obtained unless the ether is added *before* complete evaporation of the ammonia.

3. The mixture may be permitted to stand at this point and the ammonia allowed to evaporate spontaneously under a hood.

4. The Carborundum column described in *Org. Syntheses,* **20,** 96, is quite satisfactory.

5. It is desirable to interpose a Dry Ice trap between the receiver and the pump in order to prevent loss of the nitrile, and too low a pressure should not be used.

6. The submitter reports that an equivalent number of moles of γ-bromobutyronitrile or a mixture of γ-chloro- and γ-bromobutyronitrile may be substituted for γ-chlorobutyronitrile. If the bromo compound is used the reaction mixture should be refluxed 6 hours and it is necessary to filter off the sodium bromide just before the final vacuum distillation.

7. This yield is based upon the γ-chlorobutyronitrile taken. When the recovered γ-chlorobutyronitrile (52–62 g.) is taken into account, the yield of cyclopropyl cyanide is about 60%.

3. Methods of Preparation

Cyclopropyl cyanide has been prepared by the repeated distillation of γ-chlorobutyronitrile over powdered potassium hydroxide,[1–5] or over a mixture of sodium hydroxide and alumina.[6] The preparation, on a small scale, of cyclopropyl cyanide from γ-chlorobutyronitrile by action of sodium in liquid ammonia, or of sodium suspended in ether, has been described.[7] The present directions are based upon those given by Schlatter.[8]

1 Henry, *Rec. trav. chim.,* **18,** 228 (1898).

2 Dalle, *Bull. acad. roy. Belg.,* **1902,** 36–79 (*Chem. Zentr.,* **1902,** I, 913).

3 Bruylants and Stassens, *Bull. acad. roy. Belg.,* **1921,** 702–19 [*C. A.,* **17,** 2872 (1923)].

4 von Braun, Fussgänger, and Kühn, *Ann.,* **485,** 210 (1925).

5 Nicolet and Sattler, *J. Am. Chem. Soc.,* **49,** 2068 (1927).

6 Cloke, *J. Am. Chem. Soc.,* **51,** 1180 (1929).

7 Cloke, Anderson, Lachman, and Smith, *J. Am. Chem. Soc.,* **53,** 2791 (1931).

8 Schlatter, *J. Am. Chem. Soc.,* **63,** 1733 (1941).

CYSTEIC ACID MONOHYDRATE

$$(SCH_2CH(CO_2H)NH_2 \cdot HCl)_2 + 5Br_2 + 7H_2O \rightarrow$$
$$2HSO_3CH_2CH(NH_2)CO_2H \cdot H_2O + 2HCl + 10HBr$$

Submitted by H. T. Clarke.
Checked by Henry Gilman and H. J. Harwood.

1. Procedure

To a solution of 24 g. (0.1 mole) of cystine in a cold mixture of 150 ml. of water and 50 ml. of concentrated hydrochloric acid is added, dropwise, 80 g. (25 ml., 0.5 mole) of commercial bromine, with occasional stirring, during 40 minutes. The temperature of the mixture rises to about 60°. The resulting solution, which contains a little unreduced bromine, is then evaporated under reduced pressure on a steam bath. The dark-colored crystalline residue is dissolved in 100 ml. of distilled water and filtered from a small quantity of amorphous insoluble matter. The filtrate is concentrated by evaporation on a water bath to 65 ml. and allowed to crystallize by standing overnight in a refrigerator. The crystals are filtered with suction and washed well with about 100 ml. of 95% ethanol in several portions, the washings being collected separately. The crystals are dried under reduced pressure over phosphorus pentoxide. A second crop is obtained by diluting the washings with an equal volume of water, evaporating until free of ethanol (Note 1), adding the residue to the mother liquor, and evaporating the combined solution to dryness on the water bath. The residue is dissolved in 30–40 ml. of water, decolorized with 0.5–1.0 g. of charcoal, concentrated to 15 ml., and, when cold, treated with 30 ml. of 95% ethanol. The crystals so formed are collected, washed with ethanol, and dried as before (Note 2). The total yield is 30.5–33.5 g. of pure cysteic acid monohydrate (81–90%). It melts, with vigorous evolution of gas, at 278° (289° cor.) (Note 3), and shows the rotation $[\alpha]_{54}^{24.6} + 9.36°$ (6% in water).

2. Notes

1. Cysteic acid appears to esterify readily on warming with ethanol, but the resulting ester is rapidly hydrolyzed by warming with dilute mineral acid.

2. The final mother liquors, on evaporation to dryness, yield 2–3 g. of a light-brown amorphous product which is readily soluble in water but insoluble in 95% ethanol. In concentrated hydrobromic acid this by-product forms a dark solution, the color of which is discharged on dilution with water.

3. The decomposition point of 278° is obtained by placing the capillary in a bath already heated to 260–270°. If the sample is slowly heated, starting at room temperature, a decomposition point of 257–258° is observed.

3. Methods of Preparation

The most convenient oxidant for the preparation of cysteic acid from cystine is aqueous bromine.[1] Iodine[2] and hydrogen peroxide[3] also bring about the reaction, but with both substances some of the sulfur is split off as sulfuric acid.

[1] Friedmann, *Beitr. Chem. Physiol. u. Pathol.*, **3**, 1 (1903).

[2] Yamazaki, *J. Biochem. Japan,* **12**, 207 (1930); Shinohara, *J. Biol. Chem.*, **96**, 285 (1932).

[3] Schöberl, *Z. physiol. Chem.*, **216**, 193 (1933).

DECAMETHYLENE BROMIDE *

(Decane, 1,10-dibromo-)

$$HO(CH_2)_{10}OH + 2HBr \rightarrow Br(CH_2)_{10}Br + 2H_2O$$

Submitted by W. L. McEwen.
Checked by Reynold C. Fuson and E. A. Cleveland.

1. Procedure

A 2-l. three-necked flask, supported in an oil bath, is fitted with a mechanical stirrer and an inlet tube which reaches almost to the bottom of the flask. In it is placed 696 g. (4 moles) of decamethylene glycol, and, after the oil bath is heated to 95–100°, a rapid stream of dry hydrogen bromide is introduced, with stirring. When the mixture becomes saturated with the gas, as shown by vigorous fuming at the open neck of the flask (Note 1), the temperature of the oil bath is raised to 135° and a slow current of hydrogen bromide is passed in for 6 hours (Note 2).

After cooling, the crude product is transferred to a separatory funnel and the lower aqueous layer is drawn off and discarded. The upper layer is washed once with an equal volume of warm water, and then with successive portions of 10% sodium carbonate solution until all acid has been removed. It is then washed once with warm water, which is separated as completely as possible (Note 3). The product thus washed is distilled from a Claisen flask under reduced pressure. The first few drops of distillate containing some water are discarded: the main fraction distils at 139–142°/2 mm. The yield is 1080 g. (90%) (Note 4).

2. Notes

1. In larger runs it is advantageous to stop the stirrer and gas stream at this point and allow the lower aqueous layer to separate, after which it is siphoned out.

2. The total quantity of bromine used is about 950 g.

3. If difficulty is encountered with separation of layers in washing, the substance may be dissolved in an equal volume of ether.

4. This yield was almost invariably obtained by the submitter, who used quantities of glycol varying from 150 g. to 2 kg.

By the same procedure the following dibromides have been prepared.

	% Yield	*Boiling Point*
Trimethylene bromide	75	165–167°
Hexamethylene bromide	90	108–112°/8 mm.
Nonamethylene bromide	93	128–130°/2 mm.

3. Methods of Preparation

Decamethylene bromide was first prepared by heating the glycol in a sealed tube with fuming hydrobromic acid.[1] Later it was prepared in the manner here described,[2,3] and by treating the glycol with hydrobromic acid in the presence of sulfuric acid.[4]

[1] Franke and Hankam, *Monatsh.*, **31**, 181 (1910).

[2] Chuit, *Helv. Chim. Acta*, **9**, 264 (1926).

[3] Carothers, Hill, Kirby, and Jacobson, *J. Am. Chem. Soc.*, **52**, 5279 (1930).

[4] Price, Guthrie, Herbrandson, and Peel, *J. Org. Chem.*, **11**, 281 (1946).

DECAMETHYLENEDIAMINE

(1,10-Decanediamine)

$$\begin{array}{c} CN \\ | \\ (CH_2)_8 \\ | \\ CN \end{array} + 4H_2 \xrightarrow[(Ni)]{NH_3} \begin{array}{c} CH_2NH_2 \\ | \\ (CH_2)_8 \\ | \\ CH_2NH_2 \end{array}$$

Submitted by B. S. Biggs and W. S. Bishop.
Checked by C. F. H. Allen and John R. Byers, Jr.

1. Procedure

A high-pressure bomb of about 1.1-l. capacity is charged with 82 g. (0.50 mole) of sebaconitrile and about 6 g. of Raney nickel catalyst (Note 1) suspended in 25 ml. of 95% ethanol, an additional 25 ml. of ethanol being used to rinse in the catalyst. The bomb is closed (Note 2), and about 68 g. (4 moles) of liquid ammonia is introduced from a tared 5-lb. commercial cylinder (Note 3). Hydrogen is then admitted at tank pressure (1500 lb.), and the temperature is raised to 125°. The reaction starts at about 90° and proceeds rapidly at 110–125°. When hydrogen is no longer absorbed (1–2 hours) the heater is shut off and the bomb allowed to cool. The hydrogen and ammonia are allowed to escape, and the contents of the bomb are rinsed out with two 100-ml. portions of 95% ethanol. The ethanolic solution is filtered quickly through a layer of decolorizing carbon (Note 4) to remove the catalyst and transferred to a 500-ml. Claisen flask having a modified side arm and connected by ground-glass joints to a receiver (Note 5). The ethanol is removed by distillation at atmospheric pressure, the receiver is changed, and the decamethylenediamine is distilled under reduced pressure. It boils at 143–146°/14 mm. (Note 6) and solidifies, on cooling, to a white solid, freezing point 60°. The yield is 68–69 g. (79–80%) (Notes 7, 8, and 9).

2. Notes

1. Raney nickel catalyst, already prepared and suspended in water, can be obtained from the Gilman Paint and Varnish Company, Chattanooga, Tennessee.

2. The safety disk should be made of steel, nickel, or other suitable material. It must not be of copper, which is readily attacked by ammonia under pressure.

3. The ammonia may be introduced by a number of methods. A suitable one is given under α-phenylethylamine (p. 717). The amount of ammonia is not critical, but maximum yields are obtained when 6–8 moles is used per mole of dinitrile. The purpose of the ammonia is to suppress secondary amine formation.[1,2]

4. A 9-cm. Büchner funnel is used. The decolorizing carbon is deposited from a slurry in ethanol.

5. As decamethylenediamine combines with atmospheric carbon dioxide rapidly, any solutions left standing should be protected by a drying tube containing solid potassium hydroxide. When air is admitted to the apparatus at the end of the distillation it should be through such a tube.

6. Decamethylenediamine should not be allowed to solidify in a bottle or Erlenmeyer flask, since it will probably break such a container. A tared round-bottomed flask less than half filled is advisable.

7. A dermatitis is induced in susceptible individuals by decamethylenediamine.

8. The submitters reported yields of 85–90% on runs four times this size.

9. Other boiling points are 139–140°/12 mm.; 126–127°/5 mm.

3. Methods of Preparation

Decamethylenediamine has been obtained by reduction of sebaconitrile either catalytically[2] or by sodium and ethanol,[3] or by lithium aluminum hydride.[4] It has also been obtained by hydrolysis of the condensation product from decamethylene iodide and phthalimide,[5] and from decamethylene glycol in the presence of Raney nickel and ammonia at 220–260°.[6]

[1] Brit. pat. 490,922 [*C. A.*, **33**, 993 (1939)].
[2] Schwoegler and Adkins, *J. Am. Chem. Soc.*, **61**, 3499 (1939).
[3] Phookan and Krafft, *Ber.*, **25**, 2253 (1892).
[4] Nystrom and Brown, *J. Am. Chem. Soc.*, **70**, 3738 (1948).
[5] von Braun, *Ber.*, **42**, 4551 (1909).
[6] U. S. pat. 2,412,209 [*C. A.*, **41**, 1237 (1947)].

DEHYDROACETIC ACID *

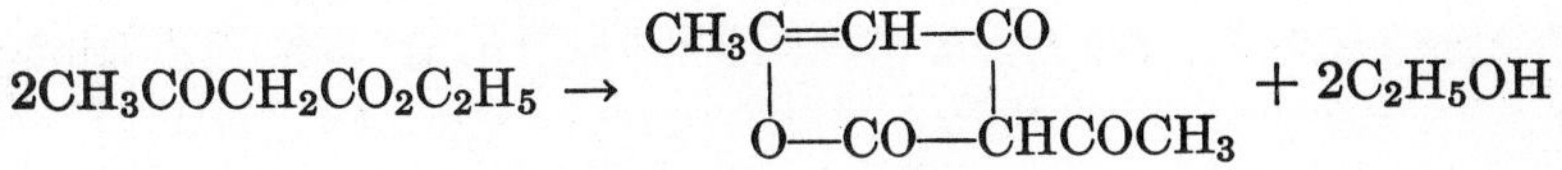

Submitted by F. Arndt.
Checked by W. W. Hartman and A. Weissberger.

1. Procedure

A 250-ml. round-bottomed flask is fitted with a thermometer reaching nearly to the bottom (Note 1) and a three- or four-bulb fractionating column without a side arm, at the upper end of which is attached a partial condenser (Fig. 11) (Note 2). The side arm (*A*) of the

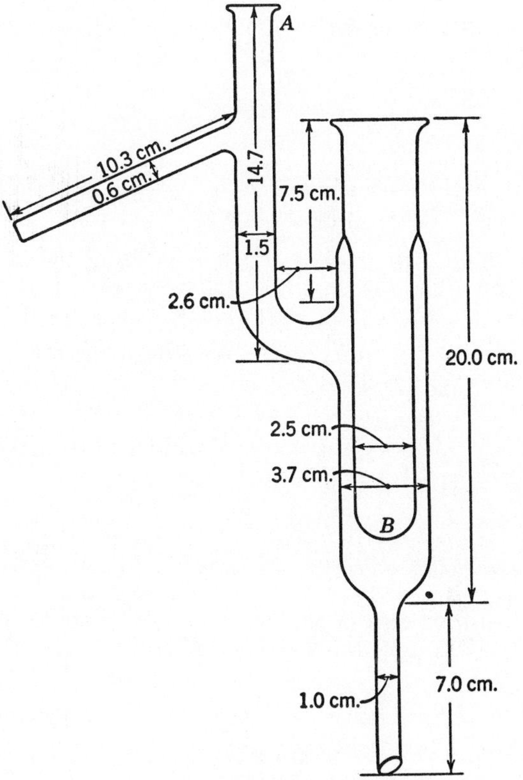

Fig. 11.

partial condenser carries a 110° thermometer and is connected to a condenser set for downward distillation. The inside container (*B*) of the partial condenser is filled halfway with toluene, a chip of porous plate is added, and the top is attached to a reflux condenser.

In the 250-ml. flask 100 g. (0.78 mole) of freshly vacuum-distilled ethyl acetoacetate and 0.05 g. of sodium bicarbonate (Note 3) are placed and heated so that the toluene is kept just boiling until the liquid in the flask has reached 200–210° (Note 4). The time for heating is usually 7–8 hours, during which period 27 g. of distillate boiling at 72° (mostly ethanol) is collected, and the color of the reaction mixture becomes dark brown. The dehydroacetic acid, while still hot (Note 5), is transferred to a 200-ml. distilling flask (Note 6) and distilled under reduced pressure. After a fore-run boiling up to 128° at 12 mm., consisting of ethyl acetoacetate, has been collected, the receiver is changed and dehydroacetic acid is collected up to 140°/12 mm. The yield of dehydroacetic acid melting at 104–110° is 34 g. (53%) (Note 7). A purer product, m.p. 108°, may be secured in 80% yield by recrystallization from ethanol, using 2 ml. per gram of material (Note 8).

2. Notes

1. The bulb of the thermometer must be immersed in the liquid.

2. This partial condenser, which is a modification of one described by Hahn,[1] is very effective in reducing the time required for fractional distillation of many mixtures. It is best constructed of Pyrex. The dimensions given are approximate and may be varied to suit individual needs. The inside container is a 30 by 140 mm. Pyrex test tube sealed at the top to standard Pyrex tubing. It is very effective in the purification of cyclohexene [*Org. Syntheses* Coll. Vol. **1**, 183 (1941)].[2] The crude hydrocarbon mixture is first put into the flask with ethanol in the partial condenser, and the whole heated as long as a distillate is obtained. The ethanol is then replaced by ethylene chloride and the cyclohexene is collected.

3. It is essential to use sodium bicarbonate to secure consistent results.

4. Above this temperature extensive decomposition sets in. The time required varies with the size of the run, being less with smaller quantities.

5. The residue solidifies on cooling. The dehydroacetic acid may be filtered and washed at this stage, but the yield is lower, owing to its solubility in the reaction mixture.

6. An ordinary 200-ml. distilling flask with a large-diameter side arm placed well down on the neck gave the best results. There was no foaming, frothing, or spattering.

7. The yield falls off with larger amounts; e.g., 500 g. of ester gave only 35% of the calculated quantity of acid.

8. The acid may be isolated through the sodium salt, but the quality is poorer and the yield less.

3. Methods of Preparation

Dehydroacetic acid has been prepared by the action of acetic anhydride on acetonedicarboxylic acid,[3] as a by-product in the pyrolysis of acetone to give ketene,[4] by treatment of ketene dimer with pyridine [5] or sodium phenoxide,[6] and by removal of ethanol from acetoacetic ester.[7]

1 Hahn, *Ber.*, **43**, 419 (1910).

2 A. W. Hutchison, private communication.

3 von Pechmann, *Ber.*, **24**, 3600 (1891); von Pechmann and Neger, *Ann.*, **273**, 194 (1893).

4 Hurd, Sweet, and Thomas, *J. Am. Chem. Soc.*, **55**, 336 (1933).

5 Chick and Wilsmore, *J. Chem. Soc.*, **93**, 946 (1908); **97**, 1987 (1910).

6 Steele, Boese, and Dull, *J. Org. Chem.*, **14**, 460 (1949).

7 Arndt and Nachtwey, *Ber.*, **57**, 1489 (1924); Arndt, Eistert, Scholz, and Aron, *Ber.*, **69**, 2373 (1936).

nor-DESOXYCHOLIC ACID

(3,12-Dihydroxy-*nor*-cholanic acid)

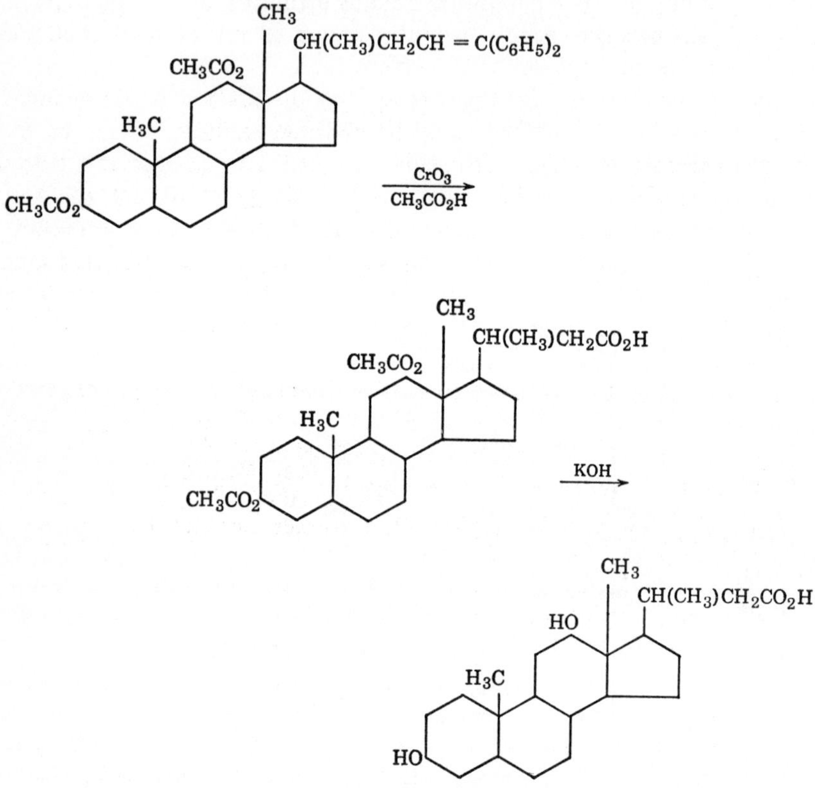

Submitted by Byron Riegel, R. B. Moffett, and A. V. McIntosh.
Checked by Richard B. Turner and Louis F. Fieser.

1. Procedure

A solution of 59.6 g. (0.1 mole) of 3,12-diacetoxy-*bisnor*-cholanyldiphenylethylene (p. 237) in 60 ml. of chloroform is prepared by warming and poured into a 1-l. round-bottomed flask containing 300 ml. of glacial acetic acid at about 40°. The flask is provided with a stirrer, a thermometer dipping below the surface of the solution, and a dropping funnel, and is surrounded by a water bath through which cold water can be circulated. A solution of 37 g. of chromium trioxide in 30 ml. of water and 200 ml. of acetic acid is added from the dropping funnel at such a rate that the temperature is kept at about 50°; the mixture is stirred and cooled. This operation should require about 10 minutes. When the temperature starts to drop, the water bath is warmed and the temperature is maintained at about 50° for an additional 20 minutes. The solution is then cooled, and the excess chromic acid is destroyed by adding carefully about 30 ml. of methanol, the temperature being kept below 50° (Note 1).

The reaction mixture is concentrated by distillation under reduced pressure. At first the solvent may be distilled rapidly, but, after the mixture becomes syrupy, the distillation should be carried out below 30° until the residue is almost solid. At a pressure of 10 mm. the concentration requires about 2 hours. The residue is diluted with 500 ml. of cold water, which should be added in several portions with thorough shaking to break up all the lumps. The product is collected on a filter and washed with dilute hydrochloric acid until the filtrate comes through colorless.

The solid crystalline cake (Note 2) is dissolved in about 400 ml. of ether and extracted with 500 ml. of 2.5% potassium hydroxide solution (Note 3). The alkaline solution is immediately acidified with 200 ml. of 10% hydrochloric acid, and the crude 3,12-diacetoxy-*nor*-cholanic acid (Note 4) is collected on a filter.

The crude diacetate is hydrolyzed by dissolving it in 350 ml. of 10% aqueous potassium hydroxide and refluxing the solution for 2 hours. The alkaline solution is diluted to about 700 ml., cooled, and filtered. The filtrate is poured into 300 ml. of 10% hydrochloric acid, and the *nor*-desoxycholic acid is separated by filtration and dried. The crude product is dissolved in about 600 ml. of acetone and filtered, while hot, to remove small amounts of salt. On cooling, 25–30 g. (57–68%) of white crystals which melt at 209–211° is obtained in two crops (Note 5). This material contains 1 molecule of acetone of crystallization. The fully purified substance softens at about 160° (loss of acetone) and melts at 213.5–214.5°.

2. Notes

1. Solid chromic acid and methanol will ignite spontaneously; care should be taken that the methanol does not come in contact with any of the chromic acid solution that may have dried around the edges of the dropping funnel.

2. If the cake is not nearly colorless it should be stirred with 250 ml. of 10% hydrochloric acid and the mixture extracted with 400 ml. of

ether. The ether solution is then extracted with the 2.5% potassium hydroxide as described.

3. Vigorous mixing should be avoided in order to prevent the formation of an emulsion.

4. The 3,12-diacetoxy-*nor*-cholanic acid may be purified by crystallization from acetone. It melts at 207–208°.

5. This procedure, coupled with the procedure described on p. 237, illustrates the Barbier-Wieland method for systematically degrading carboxylic acids. *bisnor*-Desoxycholic acid may be prepared from *nor*-desoxycholic acid by repetition of this procedure. If the chromic acid oxidation product is not sufficiently solid to filter after dilution with water, the mixture must be extracted with ether and washed with dilute hydrochloric acid before the alkaline extraction. *bisnor*-Desoxycholic acid may be crystallized from ethyl alcohol. It melts at 239–241.°

3. Methods of Preparation

nor-Desoxycholic acid has been prepared from desoxycholic acid by modification of the Barbier-Wieland degradation.[1]

[1] Hoehn and Mason, *J. Am. Chem. Soc.*, **60,** 1493 (1938); Sawlewicz, *Roczniki Chem.*, **18,** 250, 755 (1938); Kazuno and Simizu, *J. Biochem. Japan,* **29,** 421 (1939); Reichstein and Arx, *Helv. Chim. Acta,* **23,** 747 (1940).

3,12-DIACETOXY-*bisnor*-CHOLANYLDIPHENYLETHYLENE

(Bisnorcholanyldiphenylethylene, 3,12-diacetoxy-)

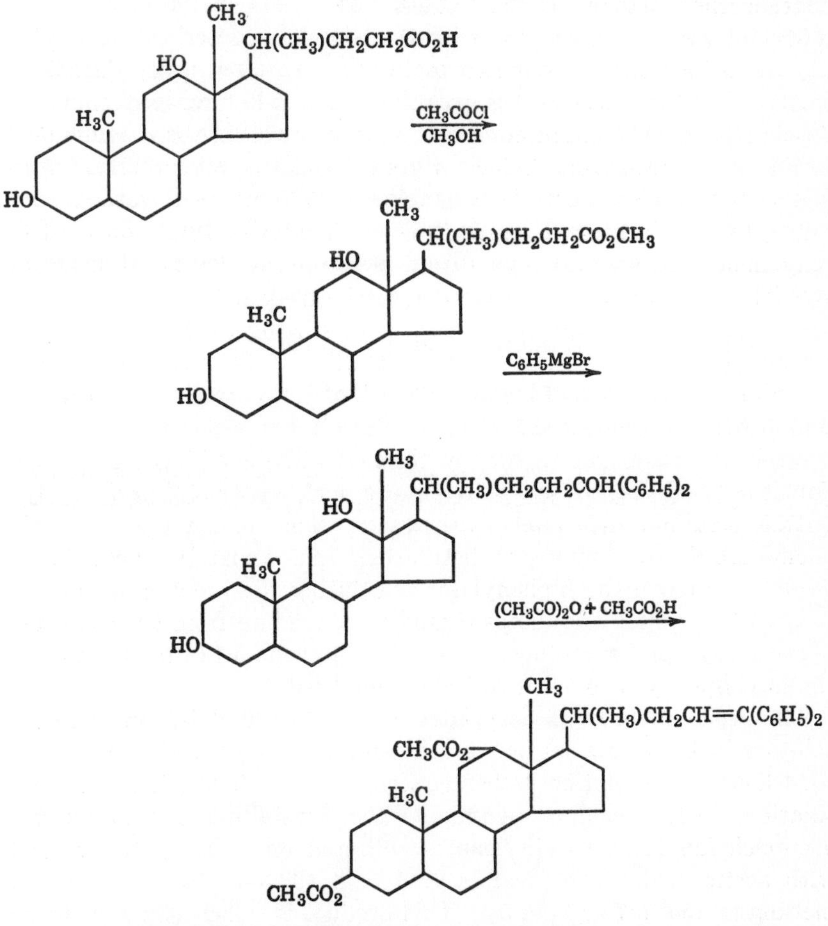

Submitted by Byron Riegel, R. B. Moffett, and A. V. McIntosh.
Checked by Richard B. Turner and Louis F. Fieser.

1. Procedure

A. *Methyl desoxycholate.* To a cooled solution of 100 g. (0.255 mole) of desoxycholic acid (Note 1) in 1 l. of methanol is added carefully 50 ml. of acetyl chloride. The solution is allowed to stand overnight at room temperature (Note 2) and is then diluted with cold water until just turbid. Crystallization is induced by scratching and seeding, if necessary. When much of the ester has crystallized, the mixture is further diluted to about 2.5 l. and allowed to stand for 30 minutes until crystallization is complete. The ester is collected on a filter, washed with water, and dried. The yield is 100–103 g. (97–100%) of material which melts at 95–100° (Note 3).

B. 3,12-*Diacetoxy-bisnor-cholanyldiphenylethylene.* A solution of phenylmagnesium bromide is prepared in a 5-l. three-necked flask, fitted with a dropping funnel, an efficient reflux condenser, and a mechanical stirrer, from 97.2 g. (4 gram atoms) of magnesium, 675 g. (450 ml., 4.3 moles) of bromobenzene, and 1250 ml. of dry ether.

To the solution of the Grignard reagent is added a solution of 102 g. (0.25 mole) of methyl desoxycholate in 700 ml. of dry benzene (Note 4). The desoxycholate is washed into the flask with 500 ml. of dry benzene, and the mixture is refluxed with stirring for 3 hours. After cooling, the complex is decomposed by pouring its benzene solution into a mixture of about 4 l. of ice and 700 ml. of concentrated hydrochloric acid. After thorough shaking, the layers are separated, and the aqueous layer is extracted twice with ether. The combined ether solution is washed with dilute hydrochloric acid, water, 5% sodium hydroxide solution, and finally with water. The solvent and some biphenyl are removed by steam distillation; approximately 5 hours is required to remove the biphenyl formed during the preparation of the Grignard reagent. The lumps should be broken up from time to time, if necessary. After cooling, the residue of crude 3,12-dihydroxy-*nor*-cholanyldiphenylcarbinol is collected and dried.

The crude carbinol is acetylated and dehydrated by refluxing its solution in 1 l. of glacial acetic acid and 500 ml. of acetic anhydride for 1 hour. The solution is then concentrated to about 500 ml. by distillation. After cooling overnight, the crystalline 3,12-diacetoxy-*bisnor*-cholanyldiphenylethylene is collected on a filter and washed with acetic acid. The yield is 95–105 g. (63.5–70.0%) of material melting at 154–157° (Note 5). This product is sufficiently pure to be used for the preparation of *nor*-desoxycholic acid (p. 234); one crystallization from acetone gives white crystals which melt at 156–157.5°; fully purified material melts at 159.5–160.5°.

2. Notes

1. A good grade of desoxycholic acid should be used. The product from Wilson Laboratories has been found to be satisfactory.

2. At this point the solution should be filtered if any insoluble material is present.

3. This methyl desoxycholate is pure enough for most purposes, but if desired it may be recrystallized from methanol or from a mixture of ether and petroleum ether.

4. The solution of methyl desoxycholate in dry benzene is conveniently prepared by dissolving the ester in 900 ml. of ordinary benzene and distilling the excess solvent.

5. An additional 5–7 g. of product may be obtained by concentrating the filtrate and recrystallizing the product from acetone. If the material is to be used for the preparation of *nor*-desoxycholic acid (p. 234), it is more convenient to oxidize the filtrate directly with chromic acid and isolate the product as the acid.

3. Methods of Preparation

The literature on methods of preparation is the same as that given for *nor*-desoxycholic acid (p. 234).

4,4'-DIAMINODIPHENYLSULFONE

(Aniline, 4,4'-sulfonylbis-)

$$p\text{-}CH_3CONHC_6H_4SO_2Na + p\text{-}NO_2C_6H_4Cl \rightarrow p\text{-}CH_3CONHC_6H_4SO_2C_6H_4NO_2\text{-}p'$$

$$p\text{-}CH_3CONHC_6H_4SO_2C_6H_4NO_2\text{-}p' \xrightarrow[HCl_2]{SnCl_2} p\text{-}NH_2C_6H_4SO_2C_6H_4NH_2\text{-}p'$$

Submitted by C. W. Ferry, J. S. Buck, and R. Baltzly.
Checked by C. F. H. Allen and James VanAllan.

1. Procedure

In a 1-l. beaker on a steam bath, 155 g. (0.78 mole) of *p*-acetaminobenzenesulfinic acid is suspended in 310 ml. of water. The suspension is stirred by hand and warmed gently while it is neutralized by the addition of a solution of 31.5 g. of sodium hydroxide in 125 ml. of water (Note 1). The solution is then transferred to a 2-l. flask and the bulk of the water is removed under the vacuum of a good water pump; 330–350 ml. of water is removed during 1.5 hours. During this time, the solid salt begins to separate. When bumping becomes severe, the flask is cooled by immersion in an ice bath for 30 minutes, and the salt is filtered with suction. The filtrate is concentrated as before, the product being added to the first crop. The second filtrate, on concentration, yields a less pure salt which may be worked up separately. The yields are 107–124 g. (63–73%) of the first crop, 25–34 g. (15–20%) of the second crop, and 12–17 g. (7–10%) of material of inferior grade.

In a 500-ml. three-necked round-bottomed flask fitted with a reflux condenser, a mechanical stirrer, and a thermometer are placed 48.5 g. (0.22 mole) of the sodium sulfinate and 60 ml. of a mixture of 75 ml. of ethylene glycol and 120 ml. of carbitol or methyl carbitol. The mixture is stirred and heated in an oil bath until solution is complete, after which 31.5 g. (0.2 mole) of 4-chloronitrobenzene (m.p. 76–78°) is added. The mixture is heated for 3.5 hours at 141–143° (thermometer in the mixture), with continued stirring, and then is allowed to cool overnight. After the addition of 20 ml. of water, the pasty lumps are broken up, and the solid is filtered with suction and washed with 50–75 ml. of hot water. The solid is then transferred to a 1-l. flask and refluxed with 250 ml. of 95% ethanol for 15 minutes. After cooling, the *p*-nitro-*p'*-acetylaminodiphenylsulfone is filtered with suction and washed on the funnel, first with 25 ml. of ethanol and then with 25 ml. of ether. After drying in the air, the tan-colored solid weighs 32–33 g. (50–52%) and melts at 226–228° (Note 2).

To a solution of 300 g. of stannous chloride dihydrate in 300 ml. of concentrated hydrochloric acid (sp. gr. 1.19) in a 1-l. beaker there is added 96 g. of the sulfone, and the mixture is stirred occasionally. Evolution of heat is sufficiently great so that, after 10–15 minutes, external cooling is necessary to prevent violent boiling with possible loss of material (Note 3). After nearly all the solid has dissolved, the mixture is heated on the steam bath for 2 hours. The solution is then cooled and added to 1.35 l. of 40% sodium hydroxide solution contained in a 3-l. beaker; it is necessary to stir mechanically and to add about 1.5 kg. of ice during the operation. The final temperature should be about 10°. After standing for 30 minutes the crude amine is filtered with suction (Note 4) and washed with water (200–250 ml.) until free from alkali.

The amine is recrystallized by dissolving it in 250 ml. of 95% ethanol, boiling a few minutes with 5 g. of Norit, and filtering. The clear filtrate is either concentrated to a small volume or is diluted by the addition of 200–250 ml. of water (Note 5) and allowed to stand overnight in the icebox. The crystalline amine (m.p. 176°) is then filtered and air-dried. The yield is 55–57 g. (74–77%) (Note 6).

2. Notes

1. Frequently the solution develops a blue color. If this persists, a little additional sodium hydroxide solution is required.

2. The yield (per cent) is the same if twice the amounts given are used. It is convenient to combine several runs for the reduction.

3. Sometimes the reduction does not start readily. In this event the beaker may be cautiously heated, but provision for rapid cooling should be made.

4. Filtration is facilitated by use of a Pyrex glass or Vinyon filter fabric; a 30-cm. funnel is advisable.

5. A part of the amine usually crystallizes during the filtration. It is dissolved by warming, after which water is added until the solution is just cloudy.

6. If the melting point is low, a second recrystallization is needed.

3. Methods of Preparation

4,4′-Diaminodiphenylsulfone and/or 4,4′-diacetylaminodiphenylsulfone have been prepared by various procedures, starting with 4,4′-dinitrodiphenylsulfide;[1] or 4,4′-dichlorodiphenylsulfone.[2,3] It has also been made from a sulfinate and a halonitrobenzene;[4] from 4-acetylaminobenzenesulfonyl chloride and acetanilide;[5] from acetanilide and thionyl chloride;[6] from 4-nitro-4′-aminodiphenylsulfide;[7] and by acetylation of thioaniline (4,4′-diaminodiphenylsulfide) followed by hydrogen peroxide oxidation.[8] The 4,4′-diacetylaminodiphenylsulfone obtained by any of these procedures is readily deacetylated.[7,8] A method somewhat similar to the one described above has been patented.[9]

[1] Fromm and Wittmann, *Ber.*, **41**, 2270 (1908).

[2] I. G. Farbenind. A-G., Brit. pat. 506,227 [*C. A.*, **33**, 9328 (1939)].

[3] I. G. Farbenind. A-G., Fr. pat. 829,926 [*C. A.*, **33**, 1760 (1939)].

[4] Schering A-G., Brit. pat. 510,127 [*C. A.*, **34**, 4079 (1940)].

[5] Kereszty and Wolf, Hung. pat. 120,021 [*C. A.*, **33**, 4600 (1939)].

[6] Sugasawa and Sakurai, *J. Pharm. Soc. Japan*, **60**, 22 (1940 [*C. A.*, **34**, 3704 (1940)].

[7] Raiziss, Clemence, Severac, and Moetsch, *J. Am. Chem. Soc.*, **61**, 2763 (1939).

[8] Van Arendonk and Kleiderer, *J. Am. Chem. Soc.*, **62**, 3521 (1940).

[9] Roblyn and Williams, U. S. pat. 2,227,400 [*C. A.*, **35**, 2531 (1941)].

1,2-DIAMINO-4-NITROBENZENE

(o-Phenylenediamine, 4-nitro-)

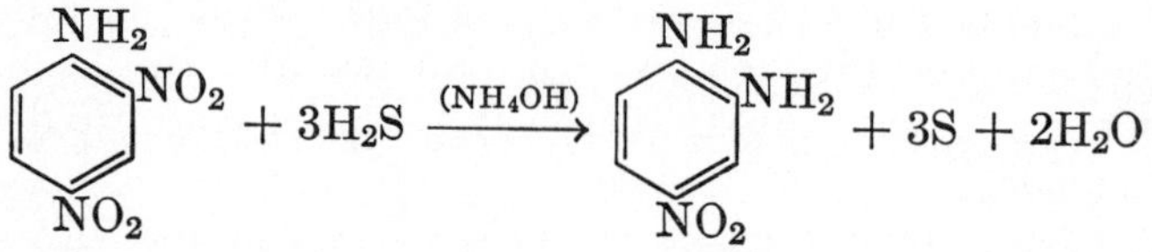

Submitted by K. P. Griffin and W. D. Peterson.
Checked by Nathan L. Drake, Ralph Mozingo, and Jonathan Williams.

1. Procedure

A 5-l. three-necked flask is fitted with a mechanical stirrer, a reflux condenser, a thermometer, and an inlet tube extending to the bottom of the flask (Note 1). In the flask is placed a mixture of 238 g. (1.3 moles) of 2,4-dinitroaniline, 2.4 l. of 95% ethanol, and 1.2 l. of concentrated ammonium hydroxide (sp. gr. 0.90).

The mixture is heated to 45°, and with good stirring hydrogen sulfide is passed into the reaction mixture while its temperature is maintained between 45° and 55° (Notes 2 and 3). The yellow suspended particles of 2,4-dinitroaniline dissolve slowly to form an intensely red-colored solution. The reaction is complete when all the *yellow* particles have disappeared; reduction should be complete in 30–60 minutes (Note 4).

The reaction mixture is allowed to stand in an icebox for 16–18 hours to complete the separation of the product, which forms small, well-defined, deeply red-colored crystals. The 1,2-diamino-4-nitrobenzene is filtered by suction, washed with 150–250 ml. of cold water, and sucked dry on the funnel (Note 5).

The crude product is purified by dissolving it in a boiling mixture of 900 ml. of water and 110 ml. of concentrated hydrochloric acid (sp. gr. 1.19), and filtering the hot solution through a Norit bed using suction. The filter bed is washed with a boiling mixture of 90 ml. of water and 10 ml. of concentrated hydrochloric acid, and the washings are added to the main body of the solution. The filtered solution, while still hot, is treated with 100 ml. of concentrated ammonia (sp. gr. 0.90). The precipitated 1,2-diamino-4-nitrobenzene is filtered hot on a Büchner funnel, washed on the funnel with 150 ml. of water, and dried in an oven at 40–50°.

The purified material melts at 197–198° and weighs 105–115 g. (52–58%).

2. Notes

1. The inlet tube should have a diameter of about 15 mm.
2. Enough heat is generated so that it is necessary to play a stream of cold water on the flask from time to time to maintain this temperature.
3. The rate of the gas stream should be as rapid as is consistent with complete absorption.
4. The product may begin to separate during the last few minutes.
5. Concentration of the mother liquor yields only a few grams of material (8–10 g.), a mixture of 1,4-diamino-2-nitrobenzene and 1,2-diamino-4-nitrobenzene.

3. Methods of Preparation

1,2-Diamino-4-nitrobenzene can be prepared by the partial reduction of 2,4-dinitroaniline in alcohol solution using sodium hydrosulfide [1] or ammonium sulfide.[2,3] The method described here is a modification of that given by Kehrmann.[3] 2-N,N-Dialkyl derivatives of 1,2-diamino-4-nitrobenzene have been prepared by acylation, nitration, and hydrolysis of substituted phenylenediamines.[4]

1 Brand, *J. prakt. Chem.*, (2) **74**, 471 (1907).
2 Heim, *Ber.*, **21**, 2305 (1888).
3 Kehrmann, *Ber.*, **28**, 1707 (1895).
4 Ger. pat. 653,259 [*C. A.*, **32**, 1946 (1938)].

DIAZOMETHANE

(Methane, diazo-)

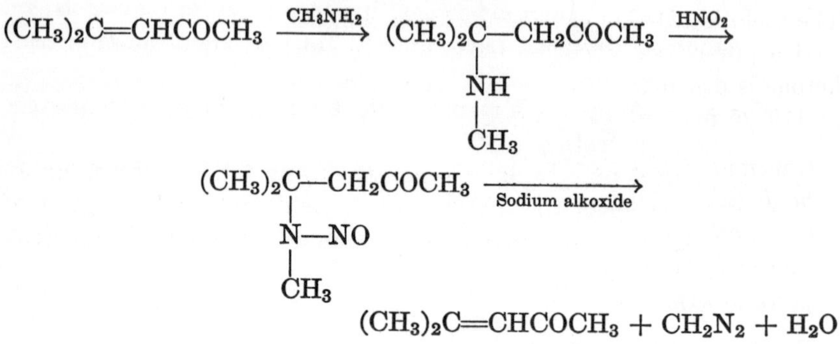

Submitted by C. Ernst Redemann, F. O. Rice, R. Roberts, and H. P. Ward.
Checked by Nathan L. Drake, Charles M. Eaker, Robert K. Preston, and W. Mayo Smith.

1. Procedure

A. *N-Nitroso-β-methylaminoisobutyl methyl ketone.* In a 2-l. three-necked flask, fitted with a mechanical stirrer, a thermometer, and a dropping funnel, is placed 250 ml. (2.1 moles) of 30% aqueous methylamine (Note 1). The flask is surrounded by an ice bath, and the stirrer is started. When the temperature of the solution has dropped to 5°, 196 g. (2 moles) of mesityl oxide is added through the dropping funnel at such a rate that the temperature remains below 20° (Notes 2 and 3). After the mesityl oxide has been added, the mixture is allowed to stand without cooling for 30 minutes.

The solution is then cooled to 10° by means of the ice bath, and 125 ml. of glacial acetic acid is added through the dropping funnel at such a rate that the temperature remains below 15°; then an additional 75 ml. of the acid is added rapidly.

The ice bath is removed, and 300 ml. of 8 *N* sodium nitrite (Note 4) is added in 20–30 minutes to the stirred solution, which is kept at 25–35° by intermittent cooling with the ice bath. Stirring is then discontinued, and the mixture is allowed to stand for 6 hours or longer (Note 5).

The oily layer is separated from the aqueous layer, the aqueous layer is extracted with two 200-ml. portions of ether, and the combined extracts and oil are dried over calcium chloride. The drying agent is removed by filtration, and the ether is distilled on a water bath; finally, all low-boiling material is removed from the mixture on a boiling water bath under the lowest pressure obtainable with a water pump (Note 6). The nitrosoaminoketone which remains in the flask is sufficiently pure for the preparation of diazomethane (Note 7). The yield is 221–257 g. (70–80% based on mesityl oxide).

If a purer product is desired for other purposes, the nitrosoaminoketone is distilled at low pressure (see *Caution*). The substance boils at 119°/5 mm., at 111°/3 mm., or at 101°/1.5 mm.

Caution! Only a small quantity of the nitrosoaminoketone should be distilled at a time. It is reported that the substance occasionally undergoes violent decomposition. If the distillation is carried out, the operator should be protected by a suitable screen or plate of safety glass (Note 8).

B. *Diazomethane. Sodium isopropoxide method* (Note 9).

Caution! Diazomethane is very toxic; its preparation should be carried out only in a well-ventilated hood. Individuals differ in their susceptibility; some develop symptoms similar to asthma from very small concentrations, whereas others notice no ill effects from much larger quantities. The use of a safety screen is also recommended.

Thirty milliliters of a solution of sodium isopropoxide, prepared from 1 g. of sodium and 100 ml. of isopropyl alcohol, is placed in a 250-ml. Claisen flask arranged for heating in a water bath. The flask is provided with a dropping funnel, a condenser, and a receiver cooled in a Dry Ice bath; this receiver is connected to a second one containing 20 ml. of anhydrous ether; the inlet tube of the second receiver should dip below the surface of the ether.

The water bath is heated to 70–75°, and one-half of a solution prepared by dissolving 15.8 g. (0.1 mole) of N-nitroso-β-methylaminoisobutyl methyl ketone in a mixture of 80 ml. of anhydrous ether and 12 ml. of isopropyl alcohol is added through the dropping funnel at a rate slightly greater than that of distillation. When the separatory funnel is empty, an additional 15 ml. of the solution of sodium isopropoxide is added; the remainder of the solution of nitroso compound is then added as before. Anhydrous ether is then added gradually through the dropping funnel until the condensing ether becomes colorless. The diazomethane comes over with the ether as a golden yellow distillate. The process can be continued until all the nitrosoaminoketone has been decomposed. The yield of diazomethane, which varies with the purity of the nitrosoaminoketone, is 1.9–2.5 g (45–60%).

C. *Diazomethane. Sodium cyclohexoxide method* (Note 10). A solution of sodium cyclohexoxide is prepared from 4 g. of sodium and 100 ml. of cyclohexanol (Note 11) in a 2-l. flask; the cyclohexanol is heated to boiling under reflux to speed up the formation of the alkoxide. As soon as the sodium has disappeared, heating is discontinued and the condenser is removed. When solid begins to separate from the solution, the mixture is stirred with a stout rod to prevent the formation of a hard cake. The flask is surrounded by an ice bath, and, when the temperature of the mixture has dropped to 10°, 300 ml. of dry ether and a solution of 49 g. (0.33 mole) of N-nitroso-β-methylaminoisobutyl methyl ketone dissolved in 600 ml. of dry ether are added. The flask is connected to a 25-cm. fractionating column and an efficient water-cooled condenser (Note 12). The delivery end of the condenser should be connected to an adapter which dips below the surface of 50 ml. of ether contained in a 1-l. Erlenmeyer flask immersed in an ice bath. The mixture is heated by a water bath whose temperature is maintained at 50–55°. Distillation is continued until the condensing ether is colorless; usually 700–750 ml. of distillate containing 10–11 g. (77–84% yield) (Note 13) of diazomethane is collected (Notes 14 and 15). The diazomethane-ether solution may be purified further with only slight loss by drying over potassium hydroxide pellets (Note 16) and redistillation through a fractionating column.

2. Notes

1. Methylamine is available from several manufacturers. The Commercial Solvents Corporation and Rohm and Haas Company offer an aqueous solution containing 30% to 33% of amine. Solutions of other strengths are satisfactory if appropriate quantities are employed. The amine solution may be prepared from the hydrochloride by adding slowly 210 ml. of 10 *N* sodium hydroxide solution to an ice-cold solution of 142 g. (2.1 moles) of methylamine hydrochloride in 250 ml. of water. The resulting solution is used without further treatment.

2. The mesityl oxide should be freshly distilled.

3. The length of time needed for this addition varies with the efficiency of the cooling bath; a period of 30 minutes to 1 hour is usually required.

4. The nitrite solution is very nearly saturated. The sodium nitrite need not be of C.P. grade; if it is less pure, appropriate allowance must be made when the solution is prepared.

5. It is preferable to allow the mixture to stand overnight; care should be taken, however, that its temperature does not exceed 35° at any time.

6. The pressure should not be above 30 mm.; a pressure of 20 mm. is desirable.

7. Because of the low cost of the starting materials and the stability in storage of the nitrosoaminoketone, this substance is an excellent intermediate for preparing diazomethane. One of the submitters (Redemann) reports that a sample of the crude material was kept in the laboratory in a brown bottle for more than 2 years without evident decomposition; the trace of acetic acid remaining in the ketone tends to stabilize it.

8. One of the submitters (Redemann) reports that he has never experienced difficulty during distillation of this ketone.

9. Submitted by F. O. Rice, R. Roberts, and H. P. Ward.

10. Submitted by C. Ernst Redemann.

11. The cyclohexanol may be replaced by benzyl alcohol. Benzyl alcohol is reported to give a slightly increased rate of formation of diazomethane and a slightly more concentrated ethereal solution; however, the over-all yield is somewhat lower. If only a small amount of diazomethane is wanted, concentrated aqueous sodium hydroxide may replace the solution of the alkoxide of cyclohexanol. The yield under these conditions is 40–50%.

12. A column with smooth packing or smooth inner surface should be used; rough surfaces catalyze the decomposition of diazomethane. A column helps to remove mesityl oxide.

13. The diazomethane content of the ethereal solution may be determined by the method described previously.[1]

14. A solution of diazomethane in dry ether may be stored in a smooth flask or bottle in a refrigerator for several weeks. Since slow decomposition of the diazomethane occurs, the concentration of the solution should be checked before use.

15. Other alkyl diazo compounds can be prepared readily by the same general method with some alterations because of the lower volatility of the higher homologs.[2]

16. Pellets are recommended because their smooth surface causes little decomposition of diazomethane.

3. Methods of Preparation

Diazomethane has been prepared from nitrosomethylurea,[1,3] from nitrosomethylurethane,[4,5] from a mixture of chloroform and hydrazine hydrate by reaction with potassium hydroxide,[6] and from N-methyl-N-nitroso-N′-nitroguanidine with potassium hydroxide.[7] The methods described above are adapted from the work of Kenner.[8,9] Two reviews of the reactions and use of diazomethane have appeared.[10,11]

1 *Org. Syntheses* Coll. Vol. **2**, 165 (1943).
2 Adamson and Kenner, *J. Chem. Soc.*, **1937**, 1551.
3 Owen, *Current Sci. India,* **12**, 228 (1943).
4 von Pechmann, *Ber.*, **27**, 1888 (1894); **28**, 855 (1895).
5 Meerwein and Burneleit, *Ber.*, **61**, 1845 (1928).
6 Staudinger and Kupfer, *Ber.*, **45**, 505 (1912).
7 McKay, *J. Am. Chem. Soc.*, **70**, 1974 (1948).
8 Jones and Kenner, *J. Chem. Soc.*, **1933**, 363.
9 Adamson and Kenner, *J. Chem. Soc.*, **1935**, 286; **1937**, 1551.
10 Smith, *Chem. Revs.*, **23**, 193 (1938).
11 Eistert, *Angew. Chem.*, **54**, 99, 124 (1941).

trans-DIBENZOYLETHYLENE *

(2-Butene-1,4-dione, 1,4-diphenyl-)

$$\begin{array}{c}\text{HCCOCl}\\ \| \\ \text{ClCOCH}\end{array} + 2C_6H_6 \xrightarrow{(AlCl_3)} \begin{array}{c}\text{HCCOC}_6\text{H}_5\\ \| \\ \text{C}_6\text{H}_5\text{COCH}\end{array} + 2HCl$$

Submitted by R. E. Lutz.
Checked by C. F. H. Allen and F. P. Pingert.

1. Procedure

Eighteen hundred milliliters of benzene (Note 1) and 350 g. of finely ground anhydrous aluminum chloride (2.6 moles) (Note 2) are placed in a 3-l. three-necked flask, fitted with a mechanical stirrer, a dropping funnel (Note 3) containing 153 g. (1 mole) of fumaryl chloride (p. 422), and a reflux condenser (Note 4). A trap for absorbing hydrogen chloride is attached to the condenser.

The mixture is well stirred and heated externally by hot water (50–60°) (Note 5), the water removed, and the fumaryl chloride admitted at a *brisk* rate, moderated only enough to avoid a too rapid evolution of hydrogen chloride; this requires 15–25 minutes (Notes 6 and 7). The mixture turns dark red and soon reaches the boiling point; hydrogen chloride is rapidly given off. The mixture is then refluxed gently for 10 minutes with stirring.

The pasty red mixture is then poured portionwise upon 4 kg. of cracked ice to which has been added 75 ml. of concentrated hydrochloric acid, the reaction mixture being thoroughly stirred before each pouring so that the aluminum chloride complex does not settle out and become concentrated at the bottom of the flask. The residue in the flask is decomposed by adding some of the ice and water. After standing 20–30 minutes, very hot water (Note 8) is added to melt any ice or frozen benzene and raise the temperature generally. The bulk of the aqueous layer is discarded by drawing it off with a glass tube connected to a suction flask. The benzene layer is next washed at least four times with hot water (Notes 8 and 9). Finally the warm benzene layer is transferred to a large separatory funnel, any crusts of dibenzoylethylene adhering to the various pieces of apparatus are dissolved in hot benzene, and the solutions combined; the small water layer is separated and discarded (Note 10). The hot benzene layer is then filtered, either by gravity using a large glass funnel or by suction on a Büchner funnel, into a 3-l. round-bottomed flask. After a few porcelain chips have been added, the bulk of the solvent is distilled, using a steam or boiling water bath. Most of the residual solvent is removed under diminished pressure, using a water pump, and heating until the syrupy liquid begins to crystallize suddenly (Note 11). At this point, the heating and suction are discontinued and 125 ml. of 95% ethanol is added rapidly, stirring with a wooden paddle to break up any lumps. The flask is cooled a few minutes under the tap, and the bright yellow product is collected on a 127-mm. Büchner funnel. The solid is triturated on the funnel with cold ethanol for 10 minutes to remove adhering mother liquor, and sucked as dry as possible. The yield is 186–197 g. (78–83%) (Note 11), and the melting point is 109–110° (Notes 12 and 13).

2. Notes

1. The checkers used thiophene-free benzene, m.p. 5°; it gave a product of better quality than the commercial hydrocarbon. A large excess is used to facilitate stirring.

2. Resublimed aluminum chloride is suitable; it requires no further grinding. The excess over the required 2 moles assures a complete reaction and a product of good color.

3. A dropping funnel is preferred to the ordinary separatory funnel, since the rate of addition of the fumaryl chloride is important.

4. A wide-bore condenser permits a more rapid reaction, favoring an increased yield and better product.

5. If the mixture is not heated before the fumaryl chloride is added the reaction is slow, and when the temperature finally rises the ac-

cumulated chloride and intermediates react so vigorously that frothing and boiling over occur.

6. If for any reason stirring is interrupted, the *addition* of fumaryl chloride *must be stopped immediately,* and the stirrer started again *very cautiously.*

7. The product and yield are better with the shorter time of addition.

8. Very hot water is desirable; unless the benzene layer is really warm (50–60°), the separation into layers is poor and the product does not readily dissolve. For the same reasons, hot water is used in the subsequent washings.

9. The first and second wash waters should be acidulated with 25 ml. of concentrated hydrochloric acid; this facilitates the formation of layers.

10. If the water used has been hot enough to keep the benzene really warm (50–60°), there are no crusty deposits.

11. Ordinarily there is but 8 g. in the second crop; however, in the event that not enough solvent has been removed, there may be a larger amount, with a correspondingly smaller first crop. The total yield is 194–205 g. (82–86%). If the first crop weighs over 190 g., it is not economical to work up the mother liquor.

12. In some runs, the product sinters slightly at 106°, but the melting point is unaltered.

13. Lower-melting material, whatever its source, can be recrystallized from 95% ethanol, using 3 ml. per gram.

3. Methods of Preparation

trans-Dibenzoylethylene has been prepared by the present method,[1] by heating dibenzoylmalic acid,[2] by condensing benzoylformaldehyde and acetophenone,[3] and by the Friedel-Crafts reaction on benzoylacrylyl chloride.[4] Analogs may be prepared using other aromatics such as mesitylene,[1,4,5,9] with carbon disulfide as the solvent, and also using other acid chlorides such as mesaconyl chloride,[6] dibromofumaryl chloride,[7] and dimethylfumaryl chloride.[8]

A mixture of stereoisomers has been prepared by the action of alcoholic potassium hydroxide on phenacyl chloride.[10]

[1] Conant and Lutz, *J. Am. Chem. Soc.,* **45,** 1305 (1923); Oddy, *J. Am. Chem. Soc.,* **45,** 2159 (1923).

[2] Paal and Schulze, *Ber.,* **33,** 3798 (1900).

[3] Smedley, *J. Chem. Soc.,* **95,** 219 (1909).

[4] Lutz, *J. Am. Chem. Soc.,* **52,** 3432 (1930).

[5] Conant and Lutz, *J. Am. Chem. Soc.,* **47,** 891 (1925).

[6] Lutz and Taylor, *J. Am. Chem. Soc.*, **55**, 1177 (1933).
[7] Lutz, *J. Am. Chem. Soc.*, **52**, 3421 (1930).
[8] Lutz and Taylor, *J. Am. Chem. Soc.*, **55**, 1599 (1933).
[9] Weygand and Lanzendorf, *J. prakt. Chem.*, **151**, 209 (1938).
[10] Bogoslovskii, *J. Gen. Chem. U.S.S.R.*, **14**, 993 (1944) [*C. A.*, **39**, 4600 (1945)].

DIBENZOYLMETHANE *

(1,3-Propanedione, 1,3-diphenyl-)

$$C_6H_5CO_2C_2H_5 + CH_3COC_6H_5 + NaOC_2H_5 \rightarrow C_6H_5\overset{\overset{\displaystyle ONa}{|}}{C}{=}CHCOC_6H_5 + 2C_2H_5OH$$

$$C_6H_5\overset{\overset{\displaystyle ONa}{|}}{C}{=}CHCOC_6H_5 + H_2SO_4 \rightarrow C_6H_5COCH_2COC_6H_5 + NaHSO_4$$

Submitted by ARTHUR MAGNANI and S. M. McELVAIN.
Checked by R. L. SHRINER and F. J. WOLF.

1. Procedure

In a dry 2-l. three-necked flask are placed 600 g. (4 moles) of freshly distilled ethyl benzoate and 60 g. (0.5 mole) of freshly distilled acetophenone (Note 1). The flask is fitted with a mercury-sealed stirrer which must be rugged enough to stir the reaction mixture even after it becomes very viscous (Note 2). A condenser for downward distillation is attached to one of the necks, and a 500-ml. filter flask is used as the receiver. This receiver is connected to a water pump through a suction flask carrying a two-holed rubber stopper, one hole of which is left open. The reaction flask is heated in an oil bath kept at 150–160°, and, after the mixture is hot, 44 g. (0.65 mole) of sodium ethoxide (Note 3) is added through the third arm of the flask in 1- to 2-g. portions. The ethoxide addition can be conveniently accomplished with a spoon shaped to enter the arm of the flask or by placing a very short-stemmed large funnel in the third neck and pushing the ethoxide through the funnel with a small glass rod or wire spatula. The reaction mixture becomes orange immediately; ethanol distils after the first few additions, and thereafter the evolution of ethanol is quite vigorous. The additions are made as rapidly as evolution of the ethanol permits. During the ethoxide addition a gentle stream of air is kept passing through the flask by means of a water pump attached to the receiver in order to prevent the ethanol vapor from rising in the third arm of the flask and interfering with the addition (Note 4).

After all the ethoxide has been added (20–30 minutes) the gelatinous reaction mixture is stirred until no more distillate comes over (15–30 minutes). The weight of this distillate amounts to 38–45 g. (Note 5).

The oil bath is removed, and, with the stirring maintained, the reaction mixture is cooled to room temperature by running cold water over the flask. Then 150 ml. of water is added to dissolve the reaction mass, and both layers of the mixture are transferred to a separatory funnel. An ice-cold solution of 25 ml. of concentrated sulfuric acid in 200 ml. of water is added, and the mixture is shaken vigorously. The ester layer is separated and washed with 200 ml. of water; it is then shaken with successive 200-ml. portions of 5% sodium bicarbonate solution until the evolution of carbon dioxide ceases, after which it is washed with 200 ml. of water. The bicarbonate solution is separated and extracted with 100 ml. of ether (Note 6). The ether extract, after washing with 50 ml. of water, is combined with the ester layer and the resulting ethereal solution dried with 40 g. of calcium chloride. The ether is removed by distillation from a water bath, and the excess ethyl benzoate is removed by distillation under reduced pressure. The recovered ester, b.p. 80–83°/8 mm., amounts to 475–490 g. After the ester is removed the temperature of the oil bath is slowly raised to 180–185° while the system is maintained under 8 mm. pressure. A small amount of liquid distilling higher than the ester is thus removed. When no more distillate comes over at this temperature, the remaining oil, while still warm, is poured into a 500-ml. Erlenmeyer flask and allowed to crystallize. This crude dibenzoylmethane weighs 92–108 g. and is usually brown. It is recrystallized by dissolving in 150 ml. of hot methanol, adding 1 g. of Norit, filtering, and cooling the filtrate to 0° (Note 7). The yield of yellow crystals of dibenzoylmethane, m.p. 77–78° (Note 8), thus obtained is 70–80 g. (62–71% based on the acetophenone).

2. Notes

1. The apparatus and all reagents must be carefully dried.

2. A thick glass rod stirrer bent to fit around the inside of the flask works well. The mixture is too viscous to be stirred with a Hershberg stirrer.

3. The sodium ethoxide should be freshly prepared according to the directions under acetylacetone (p. 16). It must be kept in a tightly stoppered bottle and handled quickly.

4. The water flow through the pump is so regulated that no appreciable vacuum is allowed to build up in the reaction flask when the side arm through which the ethoxide is added is closed.

5. The amount of this distillate depends somewhat upon the rate of flow of air through the apparatus. If an appreciable vacuum builds up, this distillate may contain some acetophenone.

6. By acidification of the aqueous bicarbonate layer, 20–25 g. of benzoic acid may be obtained.

7. If the product is still dark colored, it may be purified by recrystallization from 400 ml. of 90% methanol and the color removed with Norit. In either method of crystallization 4–6 g. of inferior product may be obtained from the mother liquor.

8. Occasionally crystals melting at 71–72° may be obtained which upon standing change to the higher-melting form.

3. Methods of Preparation

In addition to the methods listed in *Org. Syntheses* Coll. Vol. **1,** 205 (1941), dibenzoylmethane has been prepared by the action of sodium amide [1] and lithium amide [2] on acetophenone and ethyl benzoate, by the action of sodium ethoxide on certain alkyl benzoates,[3] by the reaction of ethyl benzoate and sodium acetylbenzoylmethane,[4] and by the hydrolysis of N-(1,3-dimethylbutyl)acetophenone imide.[5]

[1] Levine, Conroy, Adams, and Hauser, *J. Am. Chem. Soc.,* **67,** 1510 (1945).

[2] Zellars and Levine, *J. Org. Chem.,* **13,** 160 (1948).

[3] Magnani and McElvain, *J. Am. Chem. Soc.,* **60,** 813 (1938).

[4] McElvain and Weber, *J. Am. Chem. Soc.,* **63,** 2192 (1941).

[5] Haury, Cerrito, and Ballard, U. S. pat. 2,418,173 [*C. A.,* **41,** 4510 (1947)].

β-DI-*n*-BUTYLAMINOETHYLAMINE

(Ethylenediamine, N,N-dibutyl-)

$$(C_4H_9)_2NCH_2CH_2Br \cdot HBr + 2NH_3 \rightarrow (C_4H_9)_2NCH_2CH_2NH_2 \cdot HBr + NH_4Br$$

$$(C_4H_9)_2NCH_2CH_2NH_2 \cdot HBr + NaOH \rightarrow (C_4H_9)_2NCH_2CH_2NH_2 + NaBr + H_2O$$

Submitted by LAWRENCE H. AMUNDSEN, KARL W. KRANTZ, and JAMES J. SANDERSON.
Checked by R. L. SHRINER and JOHN L. RENDALL.

1. Procedure

A rubber tube from an inverted cylinder of liquid ammonia is attached to a glass tube which extends beneath the surface of 325 ml. of 90% ethanol in a 500-ml. Erlenmeyer flask. Ammonia is passed into the ethanol until the weight increases by 41 g. (2.4 moles). The flask is cooled in ice water occasionally in order to hasten the absorption of the ammonia.

The ethanolic ammonia is added to 49 g. (0.16 mole) of di-*n*-butylaminoethyl bromide hydrobromide [*Org. Syntheses* Coll. Vol. **2,** 92 (1943)] (Note 1) in a 1-l. round-bottomed flask. The flask is stoppered tightly and allowed to stand for 6 days at room temperature. The solution is then transferred to an evaporating dish and stirred while it is evaporated to a crystalline paste under a hood. The crystalline paste is transferred to a 400-ml. beaker, and to it is added a solution of 16 g. of sodium hydroxide in 18 ml. of water. The solution separates into an upper layer of brown oil and a lower layer which contains a considerable amount of undissolved sodium bromide. About 10 ml. of water is added to dissolve most of the precipitate. The whole is then extracted with three 50-ml. portions of benzene. The combined extracts are placed in a 250-ml. flask over 15 g. of potassium carbonate and set aside to dry overnight.

The dried benzene extract is placed in a 250-ml. flask fitted with a 30-cm. fractionating column, and the benzene is removed by distillation at atmospheric pressure. Fractionation of the residue is carried out under reduced pressure using a 50-ml. Claisen flask with a built-in 12- to 15-cm. Vigreux column. The fraction boiling at 100–103°/13 mm. is collected; it weighs 11 to 14.7 g. (41–55%) (Note 2).

2. Notes

1. It is not necessary to isolate the di-*n*-butylaminoethyl bromide hydrobromide. The crude di-*n*-butylaminoethyl bromide hydrobromide prepared from 405 g. (2.34 moles) of di-*n*-butylaminoethanol is taken up in 600 ml. of water, and the tarry impurities are removed by extraction with three 200-ml. portions of benzene. About 400 ml. of water is evaporated from this solution, and to it is then added 2.8 l. of ethanolic ammonia prepared from absolute ethanol; the concentration of the ethanol in the reaction mixture should be 90% with respect to water. The yield is 114 g. (28% based on 405 g. of the aminoethanol). The over-all yield obtained by the procedure described above is 27% based on the aminoethanol.

2. The submitter reports that diethylaminoethylamine and di-*n*-propylaminoethylamine have also been prepared by this method.

3. Method of Preparation

Di-*n*-butylaminoethylamine has been prepared by the action of ethanolic ammonia on di-*n*-butylaminoethyl bromide hydrobromide,[1] and by the reduction of di-*n*-butylaminoacetonitrile with sodium in butanol.[2]

[1] Amundsen and Krantz, *J. Am. Chem. Soc.*, **63**, 305 (1941).

[2] Bloom, Breslow, and Hauser, *J. Am. Chem. Soc.*, **67**, 539 (1945).

γ-DI-*n*-BUTYLAMINOPROPYLAMINE

(1,3-Propanediamine, N,N-di-*n*-butyl-)

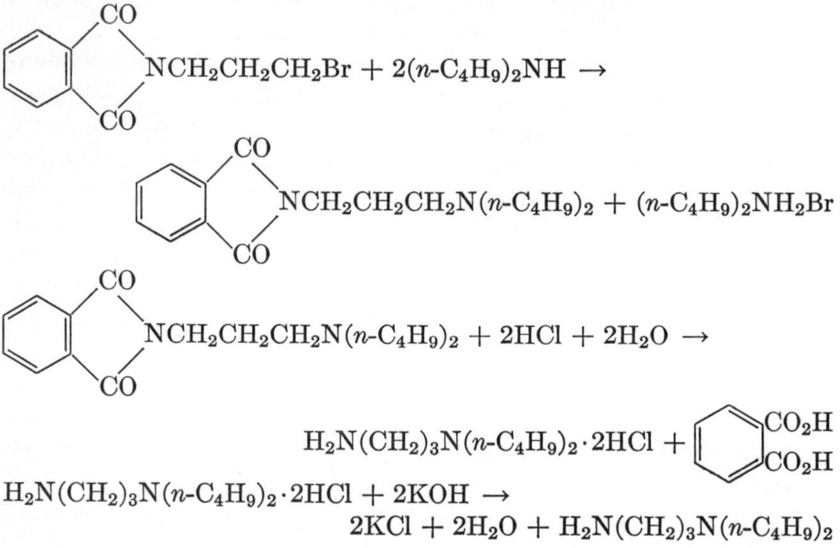

Submitted by Lawrence H. Amundsen and James J. Sanderson.
Checked by R. L. Shriner and John L. Rendall.

1. Procedure

In a 500-ml. distilling flask are placed 107 g. (0.40 mole) of γ-bromopropylphthalimide (Note 1) and 240 ml. of xylene. Solution is effected by heating, and 24 ml. of xylene is distilled to remove traces of moisture. After cooling, the solution is transferred to a 1-l. round-bottomed flask with a ground-glass joint and treated with 107 g. (140 ml., 0.83 mole) of di-*n*-butylamine. The flask is fitted with a reflux condenser and heated with occasional shaking for 10 hours in an oil bath maintained at 140–150°.

The solution is allowed to cool, and the di-*n*-butylamine hydrobromide, which separates in the course of the reaction, is removed by means of a suction filter (Note 2). The filtrate is transferred to a 500-ml. distilling flask, and the xylene is removed by distillation. The crude γ-di-*n*-butylaminopropylphthalimide, a brown oil, is transferred to a 500-ml. round-bottomed flask with a ground-glass joint and treated with 20 ml. of water and 120 ml. of 12 *N* hydrochloric acid. The flask is fitted with a reflux condenser, and the solution is heated for 6 hours in an oil bath maintained at 140–150°. After the solution has cooled to room temperature, the precipitated phthalic acid is removed by filtration and washed with four 25-ml. portions of cold water. The combined filtrates are transferred to a 600-ml. beaker and heated on a steam bath to evaporate the water and hydrochloric acid.

The residue, a thick brown oil, which is crude γ-di-*n*-butylaminopropylamine dihydrochloride, is treated with a solution of 80 g. of potassium hydroxide in 80 ml. of water. A brown oil separates above the aqueous layer, which contains a considerable amount of solid potassium chloride. The whole is extracted with one 200-ml. portion and two 50-ml. portions of benzene (Note 3). The combined extracts are then placed in a 500-ml. Erlenmeyer flask over 50 g. of potassium hydroxide and allowed to dry overnight.

The dried benzene extract is placed in a 500-ml. round-bottomed flask with a ground-glass joint. The flask is fitted with a Vigreux column, and the benzene is distilled from the solution at atmospheric pressure, an oil bath maintained at 100–110° being the source of heat. The crude γ-di-*n*-butylaminopropylamine is fractionated under reduced pressure from a 250-ml. Claisen flask; heat is supplied by an oil bath maintained at 170–180°. The yield of product boiling at 108–110.5°/5–6 mm. or 98–100°/2 mm. amounts to 57–60 g. (77–80%) (Note 4).

2. Notes

1. The γ-bromopropylphthalimide was prepared from potassium phthalimide and trimethylene bromide in 78% yield, using the conditions and molar quantities specified for the preparation of β-bromoethylphthalimide.[1] γ-Bromopropylphthalimide can also be prepared from phthalimide, potassium carbonate, and trimethylene bromide.[2]

2. About 55–58 g. of di-*n*-butylamine hydrobromide is recovered.

3. The increase in volume of the first portion is about 75 ml. There is no noticeable increase in the volumes of the two 50-ml. portions.

4. This same procedure has been followed in the preparation of γ-diethylaminopropylamine and γ-di-*n*-propylaminopropylamine. When these substances are prepared by the method described, yields of 60% and 67% respectively are obtained.

3. Methods of Preparation

The above procedure is based on the directions of Sanderson.[3] γ-Di-*n*-butylaminopropylamine has also been prepared from β-di-*n*-butylaminopropionitrile by the action of sodium and alcohol [4] and by catalytic reduction.[5] (β-Di-*n*-butylaminopropionitrile has been prepared from di-*n*-butylamine and acrylonitrile.)

[1] *Org. Syntheses* Coll. Vol. **1**, 119 (1941).
[2] *Org. Syntheses* Coll. Vol. **2**, 84, note 5 (1943).
[3] Sanderson, M.S. Thesis, University of Connecticut, 1942.
[4] Singh and Singh, *J. Indian Chem. Soc.*, **23**, 224 (1946).
[5] Burckhalter, Jones, Holcomb, and Sweet, *J. Am. Chem. Soc.*, **65**, 2012 (1943).

DI-β-CARBETHOXYETHYLMETHYLAMINE

(Propionic acid, β,β′-(methylimino)di-, diethyl ester)

$$CH_3NH_2 + 2CH_2{=}CHCO_2C_2H_5 \rightarrow CH_3N(CH_2CH_2CO_2C_2H_5)_2$$

Submitted by Ralph Mozingo and J. H. McCracken.
Checked by C. F. H. Allen and Joseph Dec.

1. Procedure

In a 2-l. flask, fitted with a rubber stopper carrying a soda-lime drying tube and a glass tube reaching almost to the bottom, is placed 325 g. (420 ml.) of absolute ethanol. The whole is then tared and placed in an ice bath, and the inlet tube is connected to a source of dry methylamine.

The methylamine generator consists of a 1-l. flask fitted with a dropping funnel and outlet tube, which in turn is connected to a 25-cm. drying tower containing soda-lime, followed by a 1-l. safety trap (Note 1). In the flask is placed 200 g. of technical sodium hydroxide flakes, and 263 g. of a 33–35% solution of methylamine in water (Note 2) is dropped in slowly at such a rate that an even current of gas is evolved. When the addition has been completed and the gas bubbles very slowly into the ethanol, the flask and contents are removed and weighed; the increase should be 84–86 g. (2.71–2.77 moles) (Note 3).

The flask is reimmersed in the ice bath, and 900–925 g. (5.4–5.5 moles), depending upon the weight of methylamine, of a 60% solution of ethyl acrylate in ethanol (Note 4) is added in portions of about 100 ml. at such a rate that the temperature of the mixture does not rise above 40°; this requires about 10 minutes. After each addition the flask is stoppered to exclude moisture. When all the ester has been added the flask is closed by a rubber stopper and allowed to stand 6 days (Note 5).

Most of the ethanol is removed by distillation (1.5 hours) on a steam bath, and the residue is transferred to a modified Claisen flask and distilled. The fraction boiling below 50°/20 mm. (Note 6) is removed with the aid of a water pump, and the residue is distilled using an oil pump (at 3 mm. pressure) (Note 7). The portion boiling at 105–108°/3 mm. (Note 8) is collected as di-β-carbethoxyethylmethylamine (Note 9); it amounts to 519–550 g. (83–86%) (Note 10).

2. Notes

1. The safety trap is important, since the ethanol tends to suck back easily.

2. Solutions of methylamine in water are obtainable from Eastman Kodak Company (33%) and Rohm and Haas Company (35%).

3. It is much more convenient to use the amount of methylamine secured than to try to collect a predetermined weight.

4. The ethyl acrylate (60% solution in ethanol) is available from the Rohm and Haas Company. The solution, containing a polymerization inhibitor, is used without treatment. The inhibitor must not be removed.

5. If the time is shortened, the yield is correspondingly decreased.

6. Most of this distillate comes over with the oil-bath temperature at 75°; the temperature is finally raised to 150°. The fractionating arm of the Claisen flask used was wrapped with asbestos cord.

7. Any ethyl N-methyl-β-aminopropionate formed comes over in the first or middle fraction (6–8 ml.). The middle fraction is obtained as the fraction boiling up to 105°/3 mm., on changing from the water pump to an oil pump. The oil bath is cooled to 100° before applying the higher vacuum.

8. The observed boiling points vary with the particular apparatus used.

9. Since the ester decomposes very slightly during distillation, the oil pump should be protected from ethyl acrylate and methylamine by a Dry Ice trap. The distillation should not be interrupted.

10. Di-β-carbomethoxyethylmethylamine may be prepared in a similar yield by substituting an equivalent quantity of methyl acrylate in methanol in the above procedure.

3. Methods of Preparation

This ester has been prepared by the action of ethyl β-bromopropionate on methylamine hydrochloride in the presence of silver oxide,[1] by the addition of methylamine to ethyl acrylate,[2] and by heating ethyl β-chloropropionate, methylamine, and benzene in an autoclave.[3]

[1] McElvain, *J. Am. Chem. Soc.*, **46**, 1724 (1924).
[2] Morsch, *Monatsh.*, **63**, 229 (1933).
[3] Ger. pat. 491,877 [*Frdl.*, **16**, 2908 (1931)] [*C. A.*, **24**, 2468 (1930)].

α,α-DICHLOROACETAMIDE

(Acetamide, α,α-dichloro-)

$$CCl_3CH(OH)_2 + 2NH_4OH \xrightarrow{(KCN)} CHCl_2CONH_2 + NH_4Cl + 3H_2O$$

Submitted by John R. Clark, W. J. Shibe, and Ralph Connor.
Checked by R. L. Shriner and Neil S. Moon.

1. Procedure

A solution of 134 g. (0.81 mole) of chloral hydrate (Note 1) in 400 ml. of ether (Note 2) is placed in a 2-l. round-bottomed three-necked flask equipped with a dropping funnel, a reflux condenser, and an efficient mercury-sealed stirrer (Note 3). A solution of 12 g. of potassium cyanide (Note 4) in 220 ml. of concentrated ammonium hydroxide (sp. gr. 0.9) is added through the dropping funnel over the course of 15 minutes at a rate sufficient to cause the ether to reflux vigorously. Stirring is continued for an additional 20 minutes (Note 5). The layers are separated, and the ether layer is washed once with 75 ml. of water and once with 75 ml. of 10% aqueous sulfuric acid solution (Note 6). (These washings are retained and used again later.) The aqueous layer from the reaction mixture is extracted with three 75-ml. portions of ether, and each ether extract is washed successively with the same water and sulfuric acid solutions used previously. The combined ether extracts are dried with 40 g. of sodium sulfate, the ether is removed by distillation, and the residue is recrystallized from 200 ml. of benzene. The solid is removed by filtration with suction and washed with two 25-ml. portions of cold benzene. The yield is 66–76 g. melting at 97.5–99.5° (cor.). Concentration of the filtrate gives 1–5 g. of material with a slightly lower (96–97° cor.) melting point, making the total yield 67–81 g. (65–78%).

2. Notes

1. The chloral hydrate was of U.S.P. XI quality.
2. Ether decreases the amount of charring, presumably by controlling the temperature of the reaction mixture.
3. It is very difficult to prevent the escape of ammonia and ether. The reaction should be carried out in a hood.
4. Baker's potassium cyanide, 94–96%, was used.
5. A decided increase in reaction time will cause charring and give a product that is difficult to purify. The reaction should not be interrupted until the ethereal extracts have been washed as described.
6. The ethereal extracts of the reaction mixture contain impurities that cause charring when the solvent is removed. The water and acid treatments remove these impurities. Equally good yields may be obtained by omitting these washings, but then it is necessary to decolorize with Norit in the recrystallization from benzene, and a second recrystallization may be necessary to obtain a white product.

3. Methods of Preparation

Dichloroacetamide has been prepared from ethyl dichloroacetate with alcoholic ammonia [1] or aqueous ammonium hydroxide; [2] from ethyl dichloromalonate and ethanolic ammonia; [3] by the action of ammonia on pentachloroacetone,[4] chloral cyanohydrin,[5] and hexachloro-1,3,5-cyclohexanetrione; [6] from chloral ammonia and potassium cyanide; [7] by the action of hydrogen chloride on dichloroacetonitrile; [8] from the reaction of asparagine with the sodium salt of N-chloro-*p*-toluenesulfonamide; [9] and by the action of an alkali cyanide and ammonia on chloral hydrate.[10]

1 Geuther, *Jahresbericht der Chemie,* **1864,** 317.

2 d'Ouville and Connor, *J. Am. Chem. Soc.,* **60,** 33 (1938).

3 Conrad and Brückner, *Ber.,* **24,** 2993 (1891); Dootson, *J. Chem. Soc.,* **75,** 169 (1899).

4 Cloez, *Compt. rend.,* **53,** 1122 (1864).

5 Pinner and Fuchs, *Ber.,* **10,** 1058 (1877).

6 Zincke and Kegel, *Ber.,* **23,** 230 (1890).

7 Schiff and Speciale, *Gazz. chim. ital.,* **9,** 335 (1879).

8 Steinkopf and Malinowski, *Ber.,* **44,** 2898 (1911).

9 Dakin, *Biochem. J.,* **11,** 79 (1917).

10 Chattaway and Irving, *J. Chem. Soc.,* **1929,** 1038.

2,6-DICHLOROANILINE AND 2,6-DIBROMOANILINE

(Aniline, 2,6-dichloro- and 2,6-dibromo-)

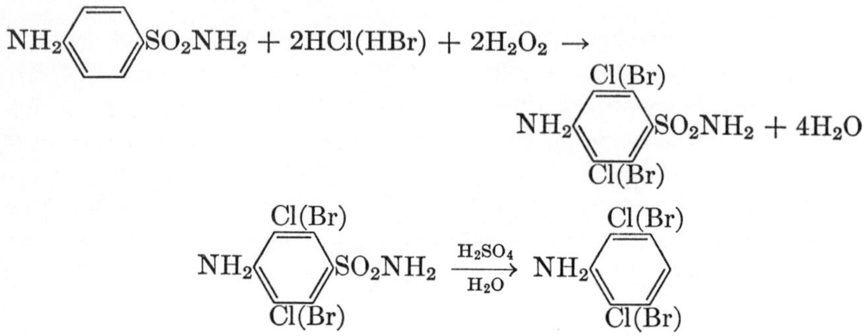

Submitted by MARGARET K. SEIKEL.
Checked by H. R. SNYDER and A. B. SPRADLING.

1. Procedure

A. 3,5-*Dichlorosulfanilamide.* In a 2-l. round-bottomed flask, fitted with a two-holed stopper carrying a mechanical stirrer and a thermometer, are placed 50 g. (0.29 mole) of sulfanilamide and 500 ml. of water. About 50 ml. of a 500-ml. portion (approximately 6 moles) of pure concentrated hydrochloric acid is added, and the mixture is stirred until a clear solution results (Note 1). The remainder of the 500 ml. of hydrochloric acid is then added. If the internal temperature does not rise to 45°, the stirred solution should be warmed gently with a free flame until this temperature is reached. At this point 65 g. (59 ml., 0.58 mole) of 30% hydrogen peroxide (sp. gr. 1.108) is added and rapid stirring is initiated (Note 2). The heat of reaction causes a progressively faster rise in temperature. After 5 minutes the solution fills with a white precipitate which increases rapidly in amount and becomes delicately colored. When the temperature has reached 60°, about 10 minutes after adding the peroxide (Note 3), any further rise is preferably prevented by judicious cooling (Note 4). The reaction is allowed to proceed for 15 minutes more at 60°, and then an ice bath is raised about the flask while stirring is continued. When the temperature has fallen to 25–30°, the mixture is filtered at once. The yield of 3,5-dichlorosulfanilamide is 45–50 g. (65–71%), and the crude dusky pink product melts over a range of 1–2° in the region 200–205° (Note 5).

B. 2,6-*Dichloroaniline.* The crude 3,5-dichlorosulfanilamide is added to 200–250 ml. (5 ml./g.) of 70% sulfuric acid (50 ml. of concentrated acid diluted with 31 ml. of water) in a 500-ml. flask, and the mixture is boiled gently for 2 hours; heat is supplied by an oil bath kept at 165–195° (Note 6). The dark mixture is then poured into 1 l. of water in a 2-l. round-bottomed flask, whereupon partial separation of a suspension of black oil occurs. The flask is attached to a condenser by a gooseneck of glass tubing; the tubing entering the flask is bent at a slight angle and cut so that its opening is at right angles to the level of the solution (Note 7). The product is steam-distilled by boiling the mixture with a free flame; the collecting vessel should be immersed in an ice bath. The solid product is separated from the distillate and dried in the air (Note 8); the white 2,6-dichloroaniline weighs 23.5–26 g. (75–80% based on 3,5-dichlorosulfanilamide or 50–55% based on sulfanilamide) and melts at 39–40°. If the product is colored it can be purified by a second steam distillation; recovery is over 90%.

C. 3,5-*Dibromosulfanilamide.* The bromination of sulfanilamide is carried out in much the same way as the chlorination. The stirrer must be more efficient (Note 9); a glass stirrer with two sets of blades is satisfactory if run at high speed. Fifty grams (0.29 mole) of sulfanilamide is dissolved in a mixture of 850 ml. of water (Note 9) and 100 ml. (0.68 mole) of 40% hydrobromic acid (Note 10). The solution is heated as above, but to 70–75°, and 65 g. (59 ml., 0.58 mole) of 30% hydrogen peroxide is added (Notes 2 and 11). A precipitate settles in 2–3 minutes, and the solution becomes yellow. The heat of reaction causes the internal temperature to rise without further application of heat to a maximum of 85–90° after 10 minutes; by the end of the reaction time the temperature will have fallen to about 70° (Note 12). After a total reaction time of 30 minutes (Note 13), during which the mixture has become almost solid and is very difficult to stir, the material is filtered hot (Note 14). The yield of 3,5-dibromosulfanilamide is 85–90 g. (90–94%), and the crude tan product (Note 15) melts over a range of 1–2° in the region 230–237° (Note 16).

D. 2,6-*Dibromoaniline.* In a 2-l. flask, equipped with a two-holed stopper carrying an exit tube to a condenser and an entrance tube for steam, 50 g. of crude 3,5-dibromosulfanilamide (Notes 17 and 18) and 250 ml. (5 ml./g.) of 70% sulfuric acid are heated in an oil bath; when the temperature of the bath reaches 175–180°, steam is passed rapidly through the mixture (Note 19). The hydrolysis is continued in this way for 2 hours; small amounts of the dibromoaniline distil (Note 20). The bath is then allowed to cool to 105–110°. At this temperature the main mass of the product is steam-distilled. The slightly colored 2,6-dibromoaniline melts at 84–86° and weighs 25–30 g. (66–79% based on 3,5-dibromosulfanilamide) (Note 21). It may be purified by recrystallization from 70% alcohol (7 ml./g.); after recrystallization the product is obtained as long colorless needles which melt at 87–88°. The recovery is 85–90%.

2. Notes

1. Solution of sulfanilamide is more readily obtained in this way than by treating the amide at once with 6 *N* acid for, in the latter case, the salt precipitates at first and redissolves only slowly. If acid stronger than 6 *N* is used, a larger volume is required to dissolve the sulfanilamide hydrochloride, and the product is partly held in solution as dichlorosulfanilamide hydrochloride unless the solution is diluted. If 4 *N* acid is used, the reaction is much slower without any worthwhile decrease in the color of the product.

2. Equivalent quantities of sulfanilamide and hydrogen peroxide are used in order to minimize the cost since no better yields are obtained with either in excess. If sulfanilamide is in excess, the product is tinged more orange-tan than pink and never becomes red from subsequent oxidation. If hydrogen peroxide is in excess, the reaction is faster but rapid oxidation giving dark-colored materials occurs toward the end.

3. When smaller amounts of reactants are employed, it is necessary to apply heat to reach and maintain a temperature of 60°.

4. Without cooling, the temperature rises above 70° and the precipitate becomes deep rose from oxidation by-products. The color, however, does not seem to affect the yield or purity of the dichloroaniline to any great extent.

5. The color of the product varies; higher temperatures, larger excesses of peroxide, longer reaction times, higher acidity, and higher concentrations of the reactants lead to the formation of more deeply colored materials. The color may be removed by several recrystallizations from glacial acetic acid with decolorizing carbon. As a preparative method for dichlorosulfanilamide, when a pure product is desired, the following procedure is recommended. For each 10 g. (0.058 mole) of sulfanilamide, 200 ml. of water, 200 ml. of concentrated hydrochloric acid, and 24 ml. (0.23 mole or 2 equivalents) of

30% hydrogen peroxide are used and the reaction is run at 25° for not more than 2 hours. A practically white product is obtained in 45–60% yield. It may be recrystallized from a very large volume of water, from 95% ethanol, or from glacial acetic acid; the recrystallized product melts at 205–205.5°.

6. Desulfonamidation is brought about faster by 75% sulfuric acid, but the yield is lower. The reaction is inconveniently slow with 65% sulfuric acid.

7. This bend prevents the mechanical carry-over of dark materials from the spray rising from the boiling dark solution. A distilling flask is not employed because dark oil creeps out the side arm.

8. It is difficult to measure the yield accurately; the material must be air-dried on account of its low melting point; the substance volatilizes to a significant extent on standing.

9. The larger amount of product obtained requires a more dilute initial solution coupled with violent stirring to maintain sufficient agitation.

10. Speedier reactions can be run at lower temperatures, producing lighter-colored crude materials if the amount of hydrobromic acid is doubled (30 minutes at 45°) or quadrupled (20 minutes at 25°); the use of the minimum quantity recommended above (approximately 1.15 times the required amount) produces as high yields of as high-melting materials with an attendant reduction in cost.

The 100 ml. of 40% hydrobromic acid can be replaced by 75 ml. of 48% hydrobromic acid.

11. The use of excess peroxide in the bromination also causes the formation of tribromoaniline.[1–4] If only 0.29 mole of hydrogen peroxide is used, the monobromo derivative is obtained.

12. When small runs are made, it is necessary to use a bath to attain and maintain an internal temperature of 80–85° during the reaction time of 30 minutes.

13. The reaction is considered complete when a sudden increase in or appearance of the brownish yellow color of free bromine is noted. This may or may not be evident under the above conditions, for the solution is deeply colored because of the high temperature and a very slight excess of hydrobromic acid is used; it is clearly evident whenever the excess is greater or the temperature lower.

14. Hot filtration removes traces of the monobromo compound, which is soluble in water and acids.

15. When the reaction is run at lower temperature with excess hydrobromic acid, the crude product is lighter colored. However, it melts

no higher and is purified with no greater ease and with only slightly better recovery.

16. The purification of the crude material is effected by recrystallization from glacial acetic acid or, preferably, 95% ethanol (25 ml./g.), using decolorizing carbon; colorless needles, which melt at 239–240°, are obtained after two recrystallizations.

17. The crude material to be used for desulfonamidation should be tested for absence of *sym*-tribromoaniline by ascertaining its solubility in 1 *N* alkali. A clear, though colored, solution should result. If the solution is cloudy, purification of the impure material by dissolving it in alkali, filtering the solution, and reprecipitating the amide is essential.

18. Fifty-gram lots seem to be the maximum size if consistently high-melting pure material is to be obtained under the conditions employed. If a very pure product is desired, less starting material should be used.

19. The use of a current of steam in the desulfonamidation of the dibromo compound as recommended by earlier investigators [1,3,5] was found to be absolutely necessary in this instance. Without the steam, considerable amounts of *sym*-tribromoaniline are formed; [6] superheated steam is not necessary. The purity of the crude product varies directly with the ratio of the amount of steam to the size of the reaction mixture; the current of steam should be very rapid or the amount of material in a single operation should be small.

20. One and a half liters of distillate was collected in the time given.

21. If the crude product melts lower, it contains too much tribromoaniline to permit satisfactory purification. Recrystallization effects no separation; repeated fractional steam distillation is very slowly effective.

3. Methods of Preparation

2,6-Dichloroaniline was prepared by Beilstein [7] and by Körner [8] by reduction of 2,6-dichloronitrobenzene. A better method, more recently reported,[9] involves the chlorination of sulfanilic acid in 1% solution with free chlorine, subsequent evaporation of the solution, and desulfonation as above. The present method (reported earlier in slightly different form and for small amounts [10]), although requiring more expensive raw materials, is more convenient and gives higher yields.

2,6-Dibromoaniline has been prepared many times [1,3,5,6,11] by bromination of sulfanilic acid and desulfonation of the product or its salts. Fuchs [12] brominated sulfanilamide with free bromine and desulfonamidated it in the usual manner. Reduction of the corresponding nitro compound[13] and other methods[14] also have been employed.

1 Heinichen, *Ann.*, **253**, 274 (1889).
2 Heinichen, *Ann.*, **253**, 269 (1889).
3 Orton and Pearson, *J. Chem. Soc.*, **93**, 735 (1908).
4 Sudborough and Lakhumalani, *J. Chem. Soc.*, **111**, 45 (1917).
5 Montagne and Moll Van Charante, *Rec. trav. chim.*, **31**, 334 (1912).
6 Limpricht, *Ber.*, **10**, 1541 (1877).
7 Beilstein and Kurbatow, *Ann.*, **196**, 219, 228 (1879).
8 Körner and Contardi, *Atti accad. Lincei*, (5) **18**, I, 100 [*C. A.*, **4**, 1619 (1910)].
9 Dyson, George, and Hunter, *J. Chem. Soc.*, **1926**, 3043.
10 Seikel, *J. Am. Chem. Soc.*, **62**, 1214 (1940).
11 Willgerodt, *J. prakt. Chem.*, (2) **71**, 562 (1905).
12 Fuchs, *Monatsh.*, **36**, 124 (1915).
13 Claus and Weil, *Ann.*, **269**, 220 (1892).
14 Buning, *Rec. trav. chim.*, **40**, 345, 347 (1921).

2,6-DICHLOROPHENOL

(Phenol, 2,6-dichloro-)

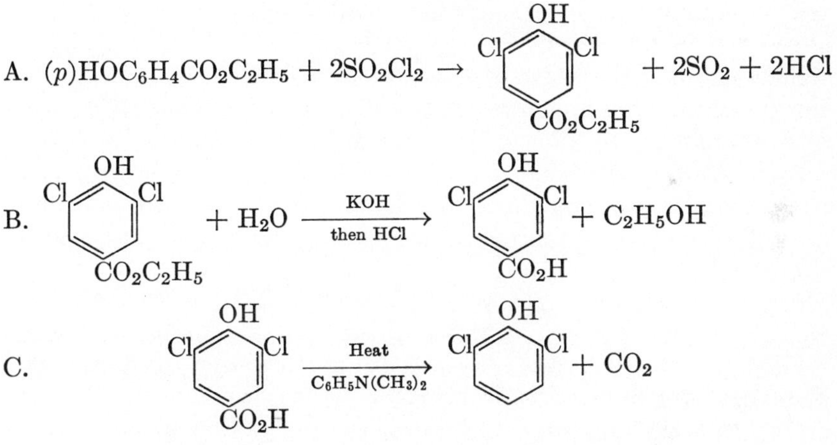

Submitted by D. S. Tarbell, J. W. Wilson, and Paul E. Fanta.
Checked by R. L. Shriner and Arne Langsjoen.

1. Procedure

A. *Ethyl* 3,5-*dichloro*-4-*hydroxybenzoate.* A tared 2-l. round-bottomed flask is equipped with an efficient reflux condenser connected through a calcium chloride drying tube to a gas-absorption trap, and is set up on a steam bath in a hood. In the flask are placed 250 g. (1.5 moles) of ethyl 4-hydroxybenzoate (Note 1) and 444 g. (266 ml., 3.3 moles) of sulfuryl chloride. The mixture is warmed on the steam bath (gently at first) until gas is no longer evolved (about 1 hour). Then 50 ml. of sulfuryl chloride is added to the reaction flask, and the warming is continued until gas is no longer evolved. This entire chlorination procedure requires about 1.5 hours. The excess sulfuryl chloride is then removed by attaching a water pump through an empty safety flask to the reaction flask and warming on the steam bath until no evidence of vapor can be detected above the white solid in the reaction flask. The residue in the flask should weigh about 350 g. (Note 2). Recrystallization from a mixture of 600 ml. of ethanol and 140 ml. of water gives 315–334 g. (83–88%) of ethyl 3,5-dichloro-4-hydroxybenzoate hydrate (Note 3) melting at 110–114° with decomposition.

B. 3,5-*Dichloro-4-hydroxybenzoic acid.* In a 2-l. round-bottomed flask equipped with a reflux condenser and set upon a steam bath are placed 315 g. (1.25 moles) of ethyl 3,5-dichloro-4-hydroxybenzoate hydrate and 600 ml. of Claisen's alkali (Note 4). Saponification is brought to completion by heating on the steam bath for 1 hour. A yellow homogeneous solution results which is diluted with 400 ml. of water and acidified to Congo red paper by pouring it into a rapidly stirred solution of 320 ml. of concentrated hydrochloric acid and 380 ml. of water in a 4-l. beaker. The thick white slurry is then cooled to 0–10°; the acid is collected on a 20-cm. Büchner funnel and washed with two 250-ml. portions of cold water (Note 5). After being freed from as much water as possible by suction, the filter cake is broken up and dissolved in a boiling mixture of 1 l. of ethanol and 350 ml. of water. Cooling to 0° gives the first crop of crystals, weighing about 200 g. Concentration of the mother liquors to 750 ml. and cooling to 0° yields an additional 40–50 g. of product of equal purity. After drying in an oven at 85–95° for 8 hours, the white crystalline acid weighs 240–250 g. (93–97%) and melts at 266–268°.

C. 2,6-*Dichlorophenol.* A mixture of 250 g. (1.2 moles) of dry 3,5-dichloro-4-hydroxybenzoic acid and 575 g. (600 ml., 4.8 moles) of redistilled dimethylaniline is placed in a 2-l. round-bottomed flask provided with a thermometer and a short air-cooled condenser and is heated slowly in an oil bath. Evolution of gas commences at 130° and is vigorous at 150°. The solution is heated at 190–200° for 2 hours or until the evolution of gas has ceased. After cooling, the solution is poured by portions into 600 ml. of concentrated hydrochloric acid in a

3-l. separatory funnel, with cooling from time to time by holding the funnel under a stream of cold water. When the solution is thoroughly cooled and acid to Congo red paper, the phenol is extracted with three 250-ml. portions and three 100-ml. portions of ether (Note 6). The combined ether extracts are washed with 15 ml. of 6 *N* hydrochloric acid, dried overnight over 20 g. of anhydrous sodium sulfate, and filtered into a 2-l. Claisen flask, and the ether is removed by distillation. After cooling, the residue begins to solidify, and 500 ml. of petroleum ether (40–60°) is added. After the phenol has been brought into solution by refluxing gently on the steam bath it is poured into a 1-l. beaker and cooled to 0°. A crop of white crystals (130–140 g.) is collected, having a melting point of 64.5–65.5°. By concentration of the mother liquor to 200 ml., a second crop (25–40 g.) melting at 64–65° can be collected. The total yield amounts to 157–180 g. (80–91%) (Note 7).

2. Notes

1. The ethyl 4-hydroxybenzoate was obtained from the Eastman Kodak Company and melted at 115.5–116°.

2. If the product weighs much less than 350 g. at this point, an additional 50 ml. of sulfuryl chloride should be added and the heating continued until gas is no longer evolved.

3. The ester crystallizes from dilute ethanol as the monohydrate; after long drying in a vacuum desiccator over phosphorus pentoxide, it melts at 111–112°.

4. Claisen's alkali is prepared by dissolving 350 g. of potassium hydroxide in 250 ml. of water, cooling, and diluting to 1 l. with methanol.

5. If the acid is dried completely at this point it contains about 5% potassium chloride. The submitters state that recrystallization is unnecessary if the acid is to be used in Part C.

6. The aqueous layer should be saved. Low yields are due to incomplete extraction of the 2,6-dichlorophenol. If the final yield is low, the aqueous layer should be extracted with three 100-ml. portions of ether and the product recovered by following the procedure outlined.

7. If the product is colored or low melting it may be purified further by recrystallization from 600 ml. of petroleum ether (40–60°) containing about 4 g. of Norit. By cooling and concentrating, 170–175 g. may be recovered (m.p. 65–66°) from 180 g. of crude product.

3. Methods of Preparation

2,6-Dichlorophenol has been prepared by the chlorination of phenol with chlorine gas in the presence of nitrobenzene and fuming sulfuric acid,[1] by the decomposition of the diazotate of 2,6-dichloro-4-aminophenol,[2] and by the decarboxylation of 3,5-dichloro-4-hydroxybenzoic acid in quinoline [3] or dimethylaniline.[4]

[1] Huston and Neeley, *J. Am. Chem. Soc.*, **57**, 2177 (1935).
[2] Seifart, *Ann.*, *Spl.*, **7**, 203 (1870).
[3] Blicke, Smith, and Powers, *J. Am. Chem. Soc.*, **54**, 1467 (1932).
[4] Tarbell and Wilson, *J. Am. Chem. Soc.*, **64**, 1066 (1942).

DI-(*p*-CHLOROPHENYL)ACETIC ACID

(Acetic acid, di-(*p*-chlorophenyl))

$$(p\text{-}ClC_6H_4)_2CHCCl_3 + KOH \rightarrow (p\text{-}ClC_6H_4)_2C{=}CCl_2 + KCl + H_2O$$

$$(p\text{-}ClC_6H_4)_2C{=}CCl_2 + 3KOH \rightarrow (p\text{-}ClC_6H_4)_2CHCOOK + 2KCl + H_2O$$

$$(p\text{-}ClC_6H_4)_2CHCOOK + H_2SO_4 \rightarrow (p\text{-}ClC_6H_4)_2CHCOOH + KHSO_4$$

Submitted by Oliver Grummitt, Allen Buck, and Richard Egan.
Checked by Lee Irvin Smith, R. T. Arnold, and Paul N. Craig.

1. Procedure

A mixture of 400 ml. of diethylene glycol (Note 1) and 49.5 g. (0.14 mole) of 1,1-di-(*p*-chlorophenyl)-2,2,2-trichloroethane (Note 2) is placed in a 1-l. three-necked flask fitted with a stirrer, a reflux condenser, and a thermometer. To this is added a solution of 63 g. (1.12 mole) of potassium hydroxide in 35 ml. of water. The mixture is stirred and refluxed for 6 hours at such a rate that the temperature is maintained at 134–137° (Notes 3 and 4). The mixture is allowed to cool and, with vigorous stirring, is poured into 1 l. of cold water. The insoluble material is filtered and washed twice with 50-ml. portions of warm water (Note 5). The filtrate is then boiled gently for 5 minutes with 2 g. of Norit; the carbon is removed, the filtrate is acidified to litmus with 20% sulfuric acid (approximately 120 ml.), and then an additional 30 ml. of acid is added. The mixture is cooled to 0–5°; the precipitate is collected by filtration under suction, washed free of sulfate ions with water, and dried at 100–110°. The product is di-

(p-chlorophenyl)acetic acid melting at 163–165°; it weighs 27–28.5 g. (69–73%). For purification, this material is dissolved in 100 ml. of boiling 95% ethanol, the solution is filtered, and approximately 45 ml. of water is added to the boiling filtrate until it just becomes turbid. The solution is then cooled to 0–5°, and the solid is removed by suction filtration. The material weighs 25–26 g. and melts at 164–166°.

2. Notes

1. Diethylene glycol with a boiling range of 230–270°, from Carbide and Carbon Chemicals Corporation, is satisfactory.

2. A purified grade of 1,1-di-(p-chlorophenyl)-2,2,2-trichloroethane (DDT) melting at 105–106° should be used. It can be obtained by crystallizing the technical material from ethanol. Thus, 100 g. of technical DDT melting at 81–96°, when crystallized from 550 ml. of 95% ethanol, gave about 70 g. of material melting at 105–106°.

3. The intermediate product, 1,1-di-(p-chlorophenyl)-2,2-dichloroethylene,[1,2] can be isolated readily. The mixture is gradually (7–8 minutes) heated to reflux temperature and is then refluxed 2 minutes. The mixture is cooled and poured into 1 l. of cold water, and the product is removed by suction filtration, washed well with water, and dried. The yield of crude product melting at 83–86° is 43.5 g. (97%). This may be purified by crystallization from 200 ml. of 95% ethanol; the product weighs 39.5 g. and melts at 86.5–88°.

4. Longer reaction times or higher temperatures favor the formation of di-(p-chlorophenyl)methane; see Note 5.

5. The insoluble material consists of silica, 1,1-di-(p-chlorophenyl) ethylene and a small amount of di-(p-chlorophenyl)methane[3] (melting at 54–55° when purified), formed by decarboxylation of the acid.

3. Methods of Preparation

Di-(p-chlorophenyl)acetic acid has been made by the action of alcoholic potassium hydroxide on 1,1-di-(p-chlorophenyl)-2,2-dichloroethane;[2,4] by the action of barium hydroxide on DDT in ethylene glycol;[4] and by the condensation of chlorobenzene with glyoxylic acid.[4]

[1] Zeidler, *Ber.*, **7**, 1181 (1874).

[2] Grummitt, Buck, and Stearns, *J. Am. Chem. Soc.*, **67**, 156 (1945).

[3] Montagne, *Rec. trav. chim.*, **25**, 379 (1906).

[4] White and Sweeney, *Public Health Repts. U. S.*, **60**, 66 (1945).

4,7-DICHLOROQUINOLINE

(Quinoline, 4,7-dichloro-)

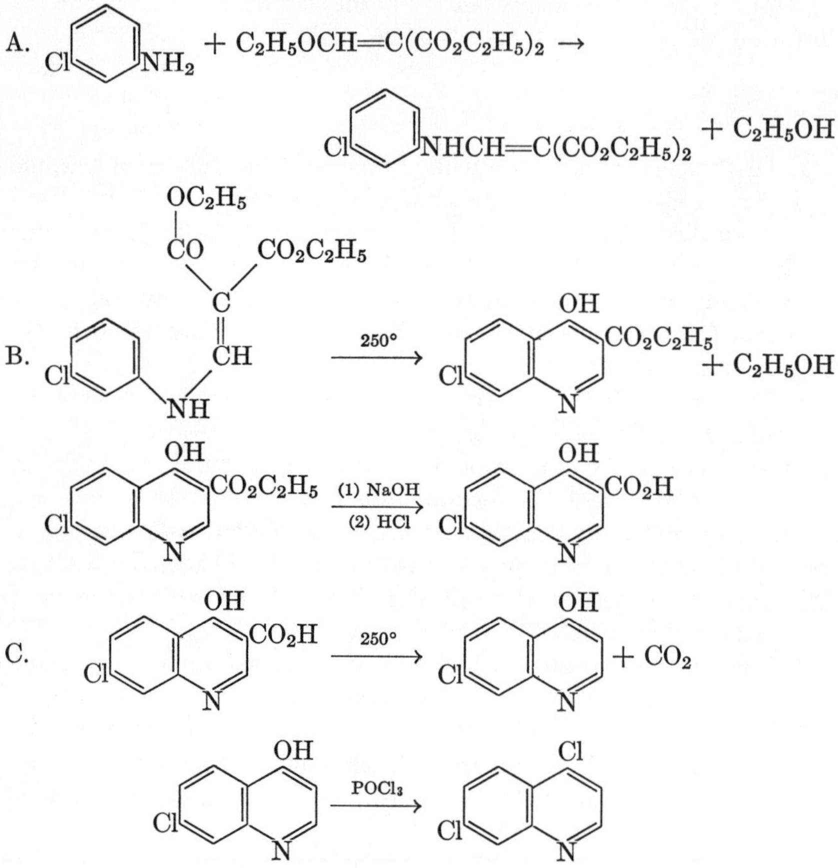

Submitted by Charles C. Price and Royston M. Roberts.[1]
Checked by J. R. Roland and R. S. Schreiber.

1. Procedure

A. *Ethyl α-carbethoxy-β-m-chloroanilinoacrylate.* A few boiling chips are added to a mixture of 127.5 g. (1.0 mole) of *m*-chloroaniline (Note 1) and 233 g. (1.1 moles) of ethyl ethoxymethylenemalonate (p. 395) (Note 2) in an open 500-ml. round-bottomed flask. The mixture is heated on a steam bath for 1 hour, the evolved ethanol being allowed to escape. The warm product is used directly in the next step (Note 3).

B. 7-*Chloro*-4-*hydroxy*-3-*quinolinecarboxylic acid.* In a 5-l. round-bottomed flask equipped with an air condenser 1 l. of Dowtherm A (Note 4) is heated to vigorous boiling, and the product of the above step is poured in through the condenser. Heating is continued for 1 hour, during which time a large proportion of the cyclization product crystallizes. The mixture is cooled, filtered, and washed with two 400-ml. portions of Skellysolve B (b.p. 61–70°) to remove the major portion of colored impurities. The air-dried filter cake (Note 5) is mixed with 1 l. of 10% aqueous sodium hydroxide, and the mixture is refluxed vigorously until all the solid ester dissolves (about 1 hour). The saponification mixture is cooled, and the aqueous solution is separated from any oil that may be present. The solution is acidified to Congo red paper with concentrated hydrochloric acid (ca. 270 ml. of the 38% acid, sp. gr. 1.19) or 10% sulfuric acid. The 7-chloro-4-hydroxy-3-quinolinecarboxylic acid, weight 190–220 g. (85–98%), is collected by filtration and washed thoroughly with water. The dry acid melts at about 266° with effervescence (Note 6).

C. 7-*Chloro*-4-*quinolinol and* 4,7-*dichloroquinoline.* The above air-dried acid (Note 7) is suspended in 1 l. of Dowtherm A in a 2-l. flask equipped with a stirrer and a reflux condenser. The mixture is boiled for 1 hour under a stream of nitrogen to assist in the removal of the water (Note 8). The clear, light-brown solution is cooled to room temperature, and 90 ml. (150 g., 0.98 mole) of phosphorus oxychloride is added. The temperature is raised to 135–140°, and the mixture is stirred for 1 hour. The reaction mixture is cooled and poured into a separatory funnel. The portion of the mixture adhering to the flask is rinsed into the funnel with ether, and the solution is washed with three 500-ml. portions of 10% hydrochloric acid. The combined acid extracts are cooled in ice and neutralized with 10% sodium hydroxide to precipitate the 4,7-dichloroquinoline. The solid is collected, washed thoroughly with water, and dried; it weighs 130–145 g. (66–73%) and melts at 80–82°. The pure product is obtained by one recrystallization from Skellysolve B (b.p. 61–70°); weight 110–120 g. (55–60%), m.p. 84–85°.

2. Notes

1. The checkers found it desirable to distil the *m*-chloroaniline through a 20-plate column, the fraction boiling at 64–66°/18 mm. being collected for use in this preparation.

2. A good criterion of the purity of the ethyl ethoxymethylenemalonate is the refractive index; material of $n_D^{2} > 1.4600$ is satisfactory. The checkers redistilled the ethyl ethoxymethylenemalonate through a 1 by 12 in. bead-packed column just before use and employed the fraction boiling at 112–115°/0.1 mm., n_D^{25} 1.4604.

3. The anilinoacrylate can be recrystallized from low-boiling petroleum ether as slender white needles, m.p. 55–56°.

4. Dowtherm A, a mixture of biphenyl and diphenyl ether, may be replaced by diphenyl ether. The high-boiling solvent is most conveniently heated to its boiling point (ca. 250°) by an electric heating mantle.

5. The yield of ester isolated at this point is 215–240 g. (85–95%), m.p. 295–297°.

6. The acid can be recrystallized from ethanol as fine white needles melting with decomposition at 273–274°.

7. The acid need not be dry if care is taken to remove water during the decarboxylation in boiling Dowtherm.

8. If the 7-chloro-4-quinolinol is desired, it is more convenient to effect the decarboxylation without a solvent.[2]

3. Methods of Preparation

4,7-Dichloroquinoline has been prepared through a somewhat similar scheme from *m*-chloroaniline and oxaloacetic ester[3] or formylacetic ester.[4] The synthesis outlined above can be modified in various ways.[2,5–7]

The procedure has been utilized on a large scale in the preparation of several thousand pounds of 4,7-dichloroquinoline, essentially as described above. It has also been applied successfully to many other aromatic amines, including aniline,[8] *o*-,[9] *m*-,[9] and *p*-anisidine,[3] 3,4-dimethylaniline,[3] *o*-nitroaniline,[8] *p*-chloroaniline,[8] *m*- and *p*-phenoxyaniline,[8] *p*-dimethylaminoaniline,[8] 3,4-dimethoxyaniline,[8] 3-aminopyridine,[3] *o*-phenylenediamine,[10] and 8-aminoquinoline.[10]

[1] Work done under contract with the Office of Scientific Research and Development.

[2] Price and Roberts, *J. Am. Chem. Soc.*, **68**, 1206 (1946).

[3] Surrey and Hammer, *J. Am. Chem. Soc.*, **68**, 115 (1946).

[4] Price, Leonard, and Reitsema, *J. Am. Chem. Soc.*, **68**, 1256 (1946).

[5] Snyder and Jones, *J. Am. Chem. Soc.*, **68**, 1253 (1946).

[6] Price and Roberts, *J. Am. Chem. Soc.*, **68**, 1255 (1946).

[7] Price, Leonard, and Herbrandson, *J. Am. Chem. Soc.*, **68**, 1251 (1946); Price, Roberts, and Herbrandson, Brit. pat. 627,297 [*C. A.*, **44**, 2572 (1950)].

[8] Riegel, Lappin, Adelson, Jackson, Albisetti, Dodson, and Baker, *J. Am. Chem. Soc.*, **68**, 1264 (1946).
[9] Lauer, Arnold, Tiffany, and Tinker, *J. Am. Chem. Soc.*, **68**, 1268 (1946).
[10] Snyder and Freier, *J. Am. Chem. Soc.*, **68**, 1320 (1946).

DIETHYLAMINOACETONITRILE *

(Acetonitrile, diethylamino-)

$$C_2H_5)_2NH + NaCN + CH_2O + NaHSO_3 \rightarrow (C_2H_5)_2NCH_2CN + Na_2SO_3 + H_2O$$

Submitted by C. F. H. Allen and J. A. VanAllan.
Checked by Cliff S. Hamilton, A. F. Harris, and C. W. Winter.

1. Procedure

This preparation should be carried out under a good hood since poisonous hydrogen cyanide may be evolved.

To a solution of 312 g. (3 moles) of sodium bisulfite in 750 ml. of water in a 3-l. beaker is added 225 ml. of a 37–40% formaldehyde solution, and the mixture is warmed to 60°. After cooling to 35°, 219 g. (309 ml., 3 moles) of diethylamine is added with hand stirring, and the mixture is allowed to stand for 2 hours. The beaker containing the reaction mixture is placed under a good hood, and to it is added a solution of 147 g. (3 moles) of sodium cyanide dissolved in 400 ml. of water with efficient stirring so that the two layers are thoroughly mixed. After 1.5 hours the upper nitrile layer is separated and dried over 25 g. of Drierite; it weighs 299–309 g. (90–92%). The crude product is purified by distillation; the portion boiling at 61–63°/14 mm., n_D^{25} 1.4230, amounts to 298–302 g. (88–90%) (Note 1).

2. Note

1. Higher homologs have been prepared from other aldehydes.

3. Method of Preparation

This procedure is essentially that recorded in the literature.[1]

[1] Knoevenagel and Mercklin, *Ber.*, **37**, 4089 (1904).

2,3-DIHYDROPYRAN *

(Pyran, dihydro-)

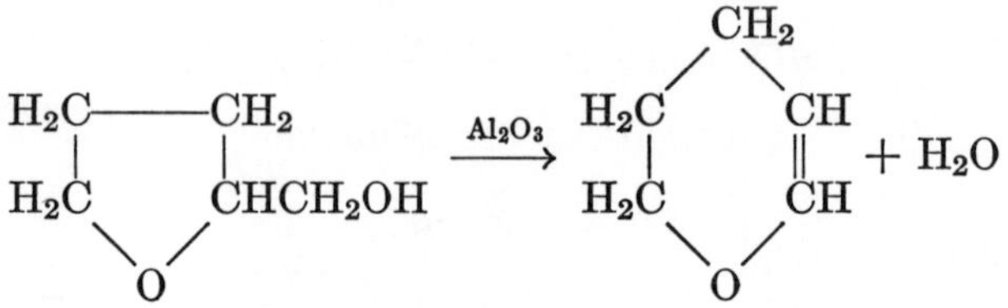

Submitted by R. L. SAWYER and D. W. ANDRUS.
Checked by NATHAN L. DRAKE and CHARLES M. EAKER.

1. Procedure

The reaction is carried out in a heated tube similar to that described by Herbst and Manske [*Org. Syntheses* Coll. Vol. **2,** 389 (1943)] except that the receiving chamber B (*loc. cit.,* Fig. 2) does not have a side arm, and that a 10-cm. water-cooled condenser is attached to the exit end of the tube leading directly to the receiver. The tube is packed with activated alumina (Note 1) held in place at the ends by plugs of glass wool.

The furnace is heated to 300–340° (Note 2), and 204 g. (195 ml., 2 moles) of tetrahydrofurfuryl alcohol (Note 3) is introduced from the dropping funnel at the rate of 50 ml. per hour. The product, collected in an Erlenmeyer flask which contains 30 g. of anhydrous potassium carbonate, consists of a light-brown oil and a lower aqueous layer. When the reaction tube has drained, the lower aqueous layer is separated and discarded. The upper layer is fractionated through a short column, and a fraction boiling at 70–86° is collected. This consists of a mixture of water and dihydropyran, most of which distils at 83–86°. The residue (25–35 g.) is mainly unchanged tetrahydrofurfuryl alcohol (Note 4).

The water-dihydropyran fraction separates into two layers. The lower aqueous layer is separated and discarded. The upper layer, consisting of fairly pure dihydropyran, is dried over 5–6 g. of anhydrous potassium carbonate, decanted, refluxed for 1 hour with 2–3 g. of metallic sodium, and then distilled from sodium (Note 5). The yield is 110–118 g. of dihydropyran, boiling at 84–86° (66–70%).

2. Notes

1. Several varieties of technical activated alumina were used. After three or four runs the catalyst becomes covered with a brown tar and the yield of dihydropyran decreases. The catalyst may be regenerated by igniting it at red heat until the tar is burned off. The checkers used 8–14 mesh activated alumina from the Aluminum Ore Company of America. The catalyst was reactivated at 450° by drawing a slow stream of air through it until the tar was burned off.

2. The temperature of the furnace is measured by a thermometer placed alongside the glass tube inside the furnace. The temperature should be 330–340° except in the regions about 10 cm. from each end of the furnace; here the temperatures will be 300–340°, depending on the construction of the furnace.

3. According to the submitters, Eastman's practical grade of tetrahydrofurfuryl alcohol must be purified by distillation; the fraction boiling at 79–80°/20 mm. was used. The checkers used, without purification, tetrahydrofurfuryl alcohol obtained from the Quaker Oats Company. The yields were equally good.

4. The recovered tetrahydrofurfuryl alcohol turns yellow on standing and is unsuitable for further runs. If it is used, the yield of dihydropyran drops to 36–38%, and the catalyst must be regenerated after each run.

5. Dihydropyran is very difficult to dry. Even after this treatment the product often contains traces of water.

3. Methods of Preparation

The procedure given above is essentially that of Paul.[1,2] Aluminum silicate, titanium oxide, and basic aluminum phosphate have also been used as catalysts in this rearrangement.[3] The action of sodium amide upon tetrahydrofurfuryl bromide gives chiefly 1,4-epoxy-4-pentene and a small amount of dihydropyran. 1,4-Epoxy-4-pentene undergoes rearrangement at 380° in the presence of alumina to yield dihydropyran.[4]

[1] Paul, *Bull. soc. chim. France,* (4) **53,** 1489 (1933).
[2] Schniepp and Geller, *J. Am. Chem. Soc.,* **68,** 1646 (1946).
[3] Wilson, *J. Am. Chem. Soc.,* **69,** 3004 (1947).
[4] Paul, *Bull. soc. chim. France,* (5) **2,** 745 (1935).

DIHYDRORESORCINOL

(1,3-Cyclohexanedione)

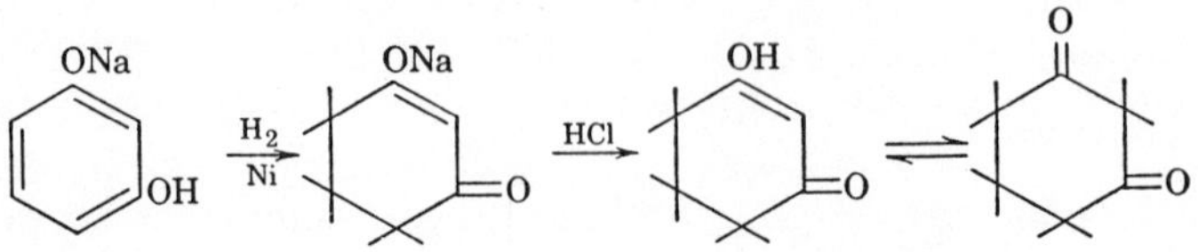

Submitted by R. B. Thompson.
Checked by Nathan L. Drake, G. Forrest Woods, and I. W. Tucker.

1. Procedure

A solution of 24 g. (0.6 mole) of sodium hydroxide, 100 ml. of water, and 55 g. (0.5 mole) of resorcinol is placed in an apparatus for high-pressure hydrogenation together with 10 g. of reduced Universal Oil Products hydrogenation catalyst (Note 1) or Raney nickel (see p. 181). The pressure in the bomb is raised to 1000–1500 lb. with hydrogen, and the temperature is adjusted to 50° (Note 2). The bomb is shaken and the reaction allowed to proceed for 10 to 12 hours, during which time 0.5 mole of hydrogen is absorbed (Note 3).

The apparatus is allowed to cool to room temperature, the pressure is released, and the catalyst is removed by filtration. The filtrate is made acid to Congo red with concentrated hydrochloric acid, and the solution is cooled to 0° in an ice-salt bath and held at that temperature for 30 minutes before filtration. The dihydroresorcinol which crystallizes is separated by filtration and dried; 50–60 g. of crude dry product containing sodium chloride is obtained.

The crude dihydroresorcinol is dissolved in 125–150 ml. of hot benzene, filtered to remove the sodium chloride, and allowed to crystallize. The solid is separated by filtration and dried overnight in a vacuum desiccator. The product melts at 103–104° and weighs 48–53 g. (85–95%) (Note 4).

2. Notes

1. The Universal Oil Products hydrogenation catalyst consists of a mixture of nickel, nickel oxides, and kieselguhr compressed into pills containing 50–55% nickel. Before use in a liquid-phase hydrogenation of this sort, the catalyst must be reduced in a stream of hydrogen at 430°. The reduced catalyst is cooled in the stream of hydrogen and

may be kept under alcohol. It also may be saturated with carbon dioxide and kept in a sealed bottle. The pilled material should be pulverized before use in this preparation. The pelleted Universal Oil Products hydrogenation catalyst may be purchased from Universal Oil Products Company, 310 S. Michigan Ave., Chicago 4, Illinois.

2. It is particularly important that the temperature should not rise above 50°; at higher temperatures complex condensation products result.

3. It is important that hydrogenation be complete; incompletely hydrogenated material yields an oily product which is exceedingly difficult to crystallize. The usual time of hydrogen absorption is 4 to 5 hours with Universal Oil Products catalyst and about 4 hours with Raney nickel. The extra time mentioned in the procedure avoids any chance of incomplete reduction.

4. Dihydroresorcinol is unstable; it can be stored only a short time. If it is not used immediately, it should be stored under an inert gas in a brown bottle in a refrigerator.

3. Methods of Preparation

Dihydroresorcinol has been prepared by the reduction of resorcinol with sodium amalgam,[1] by reduction of hydroxyhydroquinone or its carboxylic acid with sodium amalgam,[2] by hydrolysis of its dioxime,[3] or by cyclization of ethyl γ-acetylbutyrate.[4] The present method of preparation is essentially that of Klingenfuss.[5]

[1] Merling, *Ann.*, **278**, 28 (1894).

[2] Thiele and Jaeger, *Ber.*, **34**, 2841 (1901).

[3] Kötz and Grethe, *J. prakt. Chem.*, (2) **80**, 502 (1909).

[4] Vorländer, *Ann.*, **294**, 270 (1897); Schilling, *Ann.*, **308**, 190 (1899).

[5] Klingenfuss (to Hoffmann-La Roche, Inc.), U. S. pat. 1,965,499 [*C. A.*, **28**, 5476 (1934)].

2,5-DIHYDROXYACETOPHENONE

(Acetophenone, 2,5-dihydroxy-; quinacetophenone)

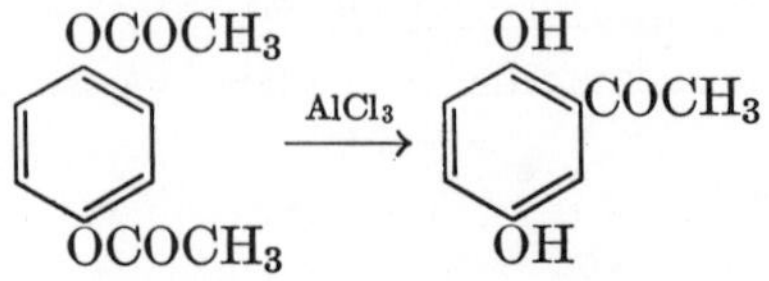

Submitted by G. C. Amin and N. M. Shah.
Checked by R. S. Schreiber and W. W. Prichard.

1. Procedure

A mixture of 50 g. (0.257 mole) of dry hydroquinone diacetate (p. 452) and 116 g. (0.87 mole) (Note 1) of anhydrous aluminum chloride is finely powdered in a mortar and introduced into a dry 500-ml. round-bottomed flask fitted with an air condenser protected by a calcium chloride tube and connected to a gas-absorption trap. The flask is placed in an oil bath (Note 2) which is heated slowly from room temperature so that at the end of about 30 minutes the temperature of the oil reaches 110–120°, at which point the evolution of hydrogen chloride begins. The temperature is then raised slowly to 160–165° and maintained at that point for about 3 hours (Note 3); at the end of about 2 hours the evolution of hydrogen chloride becomes very slow and the mass assumes a green color and becomes pasty in consistency (Note 4).

The flask is removed from the oil bath and allowed to cool to room temperature. The excess aluminum chloride is decomposed by treating the reaction mixture with 350 g. of crushed ice followed by 25 ml. of concentrated hydrochloric acid. The solid obtained is collected on a Büchner funnel and washed with two 100-ml. portions of cold water. The crude product weighs about 35 g. (89–90%). Recrystallization from 4 l. of water yields 25–30 g. (64–77%) of green, silky needles melting at 202–203° (Note 5).

2. Notes

1. Ordinary commercial aluminum chloride can be used. If the amount of this reagent is less than 3 moles per mole of the ester the yield diminishes. To compensate for any inert ingredients in the

commercial aluminum chloride, the reagent is employed in an excess of about 10% over 3 moles.

2. The flask should not touch the bottom of the oil bath; if it does, the lower portion of the mixture may char.

3. If the evolution of the hydrogen chloride becomes vigorous, the calcium chloride tube may be removed temporarily and the top of the condenser connected directly to the gas-absorption trap. When the gas evolution slackens and there is no longer any danger that the calcium chloride tube will be blown off, the guard tube is reinserted.

4. The reaction requires about 2 hours, but the heating is continued another hour to ensure its completion.

5. The product may be recrystallized from 250 ml. of 95% ethanol rather than from the much larger quantity of water.

3. Methods of Preparation

2,5-Dihydroxyacetophenone has been prepared in 54% yield[1] by heating hydroquinone diacetate, hydroquinone, and anhydrous aluminum chloride. It has also been prepared[2] by the reaction of hydroquinone with acetic acid in the presence of zinc chloride.

The procedure given has also been applied in the Fries rearrangement of hydroquinone dipropionate, with the production of 2,5-dihydroxypropiophenone in good yields.[3]

[1] Rosenmund and Lohfert, *Ber.*, **61**, 2606 (1928).
[2] Nencki and Schmid, *J. prakt. Chem.*, **23**, 546 (1881).
[3] Amin and Shah, *J. Indian Chem. Soc.*, **25**, 377 (1948).

2,6-DIHYDROXYACETOPHENONE

(Acetophenone, 2,6-dihydroxy-)

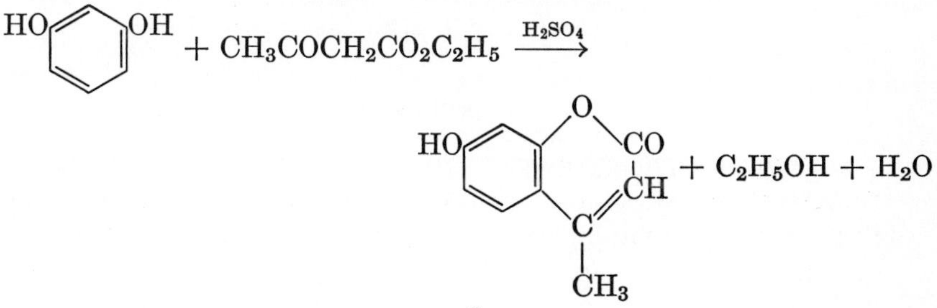

I

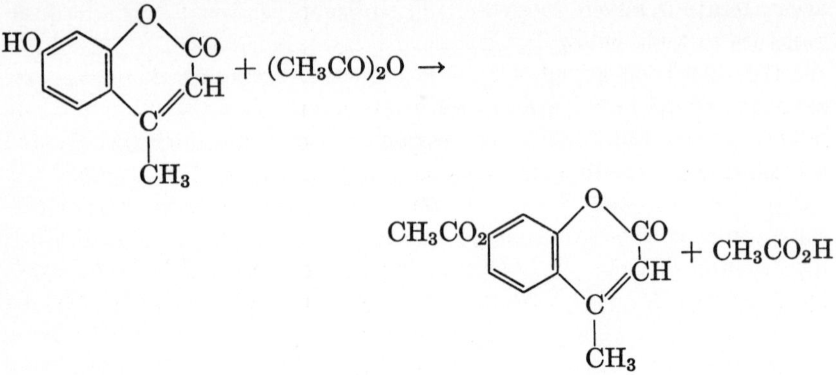

II

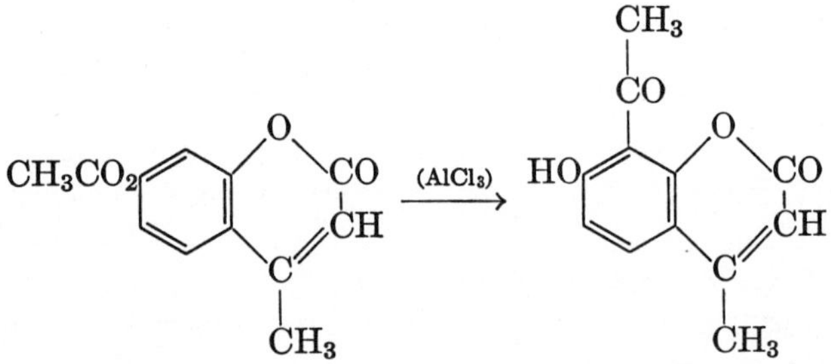

III

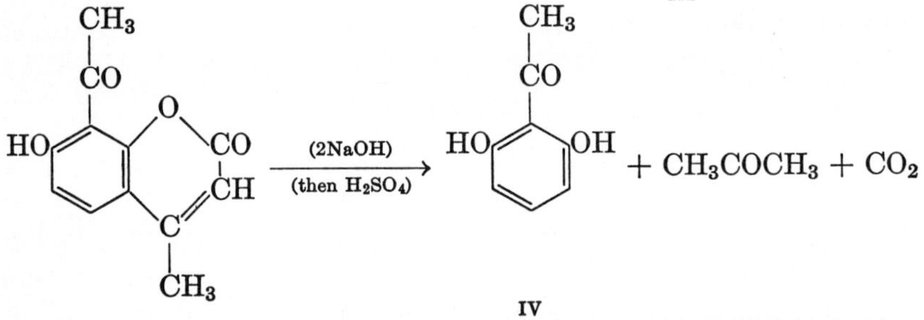

IV

Submitted by ALFRED RUSSELL and JOHN R. FRYE.
Checked by R. L. SHRINER and MICHAEL WITTE.

1. Procedure

A. 4-*Methyl-7-hydroxycoumarin* (I). In a 5-l. three-necked round-bottomed flask, fitted with a mechanical stirrer, a thermometer reaching to the bottom, and a dropping funnel, is placed 2 l. of concentrated sulfuric acid (sp. gr. 1.84). The flask is surrounded by an ice bath, and when the temperature falls below 10° a solution of 220 g. (2 moles) of resorcinol in 260 g. (2 moles) of freshly distilled ethyl acetoacetate is added dropwise. The mixture is stirred, and the temperature is kept below 10° by means of ice and salt. After all the solution has been added (about 2 hours) the reaction mixture is set aside for 12–24 hours without further cooling. The reaction mixture is now poured with vigorous stirring into a mixture of 4 kg. of ice and 6 l. of water. The precipitate is collected on a filter and washed with three 50-ml. portions of cold water. The crude product is then dissolved in 3 l. of 5% aqueous sodium hydroxide solution, the solution is filtered, and the substituted coumarin is reprecipitated from the filtrate by the slow addition of dilute (1:10) sulfuric acid until the solution is acid to litmus. About 1.1 l. of dilute sulfuric acid is required. During the neutralization, the reaction mixture must be well stirred. The product is collected on a Büchner funnel, washed with four 50-ml. portions of cold water, and dried. The yield of 4-methyl-7-hydroxycoumarin is 290–320 g. (82–90%). It is sufficiently pure for use in the next step but may be purified by recrystallization from 95% ethanol using about 15 ml. of ethanol for 5 g. of product. The recrystallized material forms stout almost colorless needles melting at 185°.

B. 4-*Methyl*-7-*acetoxycoumarin* (II). A mixture of 286 g. (1.6 moles) of crude, dry 4-methyl-7-hydroxycoumarin and 572 g. (5.6 moles) of acetic anhydride is placed in a 2-l. round-bottomed flask fitted to a reflux condenser by a ground-glass joint. The mixture is refluxed for 1.5 hours, cooled to about 50°, and poured with vigorous stirring into a mixture of 4 kg. of cracked ice and 4 l. of water. The precipitate is collected on a Büchner funnel, washed with five 50-ml. portions of cold water, and spread on absorbent paper to dry. The drying is completed by placing the product in a steam oven for 10 hours. The yield of crude 4-methyl-7-acetoxycoumarin is 320–340 g. (90–96%). It may be purified by recrystallization from 95% ethanol (5 g. of compound to 20 ml. of solvent) and forms fibrous needles melting at 150–151°. The crude oven-dried product is finely powdered and used in the next step.

C. 4-*Methyl*-7-*hydroxy*-8-*acetylcoumarin* (III). In a clean, dry, 5-l. round-bottomed flask are placed 200 g. (0.92 mole) of dry, powdered 4-methyl-7-acetoxycoumarin and 453 g. (3.4 moles) of technical anhydrous aluminum chloride. The flask is stoppered and shaken vigorously for 3–5 minutes in order to mix the ingredients thoroughly. The stopper is removed and the flask attached to a reflux condenser fitted with a gas-absorption tube. The flask is placed in an oil bath the temperature of which is raised quickly to 125° and then slowly over a period of 2 hours to 170°. At the end of this time the flask is removed from the oil bath, allowed to cool, and immersed in an ice bath. About 1 kg. of cracked ice is added, and then 2.4 l. of dilute (1:7) hydrochloric acid is added over a period of about 2 hours. The mixture is then heated on a steam bath for 30 minutes with vigorous stirring in order to effect complete decomposition. The mixture is filtered and the precipitate washed with four 50-ml. portions of cold water and sucked dry. This crude product is recrystallized by dissolving it in 4 l. of hot 95% ethanol, filtering the hot solution through a warm funnel, and chilling the filtrate. The crystals are collected on a funnel and air-dried. The product melts at 162–163°, and the yield is 145–155 g. (72.5–77.0%) (Note 1).

D. 2,6-*Dihydroxyacetophenone* (IV). A 5-l. three-necked round-bottomed flask is fitted with a reflux condenser, a dropping funnel, and a glass tube, extending to the bottom of the flask, connected to a cylinder of nitrogen (Note 2). In the flask are placed 148 g. (0.68 mole) of 4-methyl-7-hydroxy-8-acetylcoumarin and 500 ml. of distilled water. A rapid stream of nitrogen is bubbled through the water suspension until all the air in the apparatus is displaced, and then a slow stream of the gas is kept passing through the solution (Note 3). A solution of 129 g. (3.23 moles) of sodium hydroxide in 580 ml. of water is added through the dropping funnel, and the mixture is heated on a steam bath for 5 hours. The solution is then cooled and acidified by the addition of about 1 l. of dilute (1:3) hydrochloric acid. The stream of nitrogen gas is continued throughout the period of heating and while the solution is cooling. It may be stopped after the solution is acid. The crude 2,6-dihydroxyacetophenone which separates on acidification is collected on a filter, washed three times with 50-ml. portions of cold water, and air-dried. A yield of 90–95 g. (87–92% based on the 4-methyl-7-hydroxy-8-acetylcoumarin) of light-yellow solid is obtained.

The purification is accomplished by dissolving the crude product in 1 l. of 95% ethanol, adding 20 g. of Norit, and heating the mixture on a steam cone for 15 minutes with occasional shaking. After this time, 800 ml. of warm water is added, and the solution is heated 5 more minutes and filtered through a hot funnel. The greenish filtrate is chilled in an ice-salt bath, and the first crop (about 65 g.) of lemon-yellow needles of 2,6-dihydroxyacetophenone is removed by filtration. The filtrate is then concentrated under reduced pressure to a volume of 800 ml., again chilled, and the second crop of product (about 15 g.) collected on a filter. The total yield of purified 2,6-dihydroxyacetophenone, melting at 154–155°, is 75–85 g. (a recovery of 83–89%) (Note 4).

2. Notes

1. The mother liquor from this crystallization contains the isomeric acetyl derivative, 4-methyl-6-acetyl-7-hydroxycoumarin.

2. Hydrogen or illuminating gas may be used in place of nitrogen, provided that proper precautions are taken to conduct the gas from the condenser to a flue.

3. It is important to prevent oxygen from coming in contact with the alkaline solution of the 2,6-dihydroxyacetophenone since it causes the formation of oxidation products which materially lower the yield and cause difficulty in purification. The inert atmosphere must be maintained until after the mixture is acidified.

4. The first two steps in this preparation have been carried out using four times the quantities stated with no reduction in the yields. The third step, involving the Fries rearrangement, usually gives lower yields when larger amounts are used. The amounts of materials in the fourth step may be doubled.

3. Methods of Preparation

2,6-Dihydroxyacetophenone has been prepared by the action of methylmagnesium iodide on 2,6-dimethoxybenzonitrile [1] followed by cleavage of the ether linkages with aluminum chloride. The present method is based on the procedures described by Limaye [2] and by Baker.[3]

[1] Mauthner, *J. prakt. Chem.*, **139**, 290 (1934).

[2] Limaye, *Ber.*, **67**, 12 (1934).

[3] Baker, *J. Chem. Soc.*, **1934**, 1953.

2,5-DIHYDROXY-*p*-BENZENEDIACETIC ACID

(*p*-Benzenediacetic acid, 2,5-dihydroxy-)

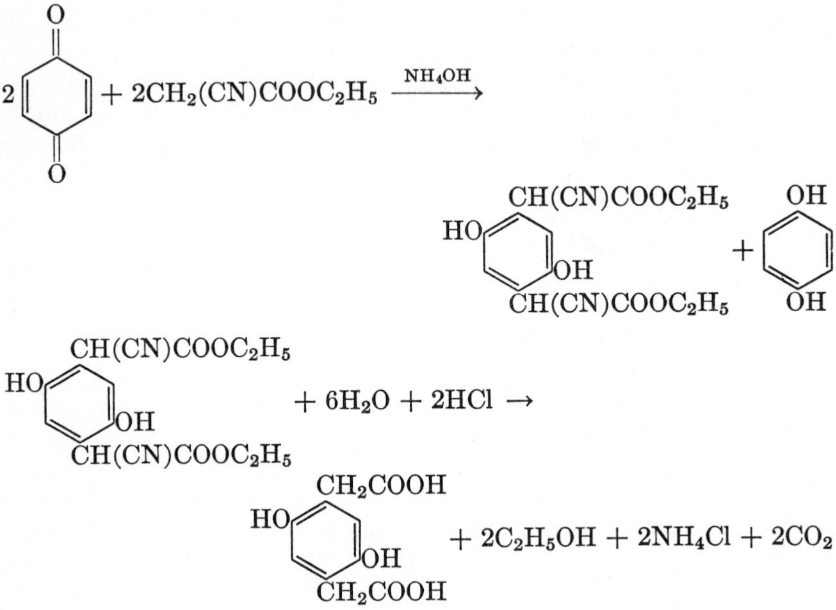

Submitted by J. H. Wood and Lucile Cox.
Checked by Lee Irvin Smith, Scott Searles, and R. T. Arnold.

1. Procedure

A 3-l. round-bottomed three-necked flask is equipped with a mechanical stirrer, a 500-ml. dropping funnel, and a 250-ml. dropping funnel. The flask is contained in a pneumatic trough through which water at room temperature is circulated. A mixture of 30 ml. of ethyl cyanoacetate and 100 ml. of ethanol is placed in the flask. Meanwhile 54 g. (0.5 mole) of *p*-benzoquinone in a 1-l. flask is heated to 40° with 500 ml. of 95% ethanol, and, when practically all the quinone is dissolved, 56 g. (0.5 mole) of ethyl cyanoacetate is added and the mixture is stirred until solution is complete (Note 1). Most of this solution is placed in the 500-ml. dropping funnel, the remainder being added when the funnel becomes sufficiently emptied after the reaction has started.

Twenty-five milliliters of concentrated ammonium hydroxide (Note 2) is introduced into the 3-l. flask. A solution of 100 ml. of concentrated ammonium hydroxide in 150 ml. of water is placed in the 250-ml. dropping funnel, which is loosely stoppered. Stirring is started, and the stopcocks of the two dropping funnels are adjusted so that the solutions will be delivered into the flask at a uniform rate in a period of about 45 minutes (Note 3). Stirring is continued for an additional 10 minutes. The mixture is then permitted to stand for 1 hour, after which the purplish red precipitate is filtered with suction. Any solid material adhering to the walls of the flask is washed onto the funnel with ethanol. The solid is then washed three times on the funnel with ethanol (Note 4). The filtrate and washings are discarded. The product is slightly impure diethyl α,α'-dicyano-2,5-dihydroxy-*p*-benzenediacetate and weighs 36–41 g. (43–49%) (Note 5).

For the hydrolysis, 36 g. (0.11 mole) of the residue is transferred to a 1-l. round-bottomed flask equipped with a reflux condenser, and 210 ml. of concentrated hydrochloric acid in 190 ml. of water is added. The mixture is refluxed gently at first, then vigorously (Note 6), until hydrolysis is essentially complete (about 20 hours). Then 180 ml. of water and 4 g. of Norit A are added and the mixture is stirred and boiled for 3 minutes, after which it is rapidly filtered with suction through two layers of hardened filter paper in a Büchner funnel (Note 7). The filtrate, upon cooling to 20° or below, deposits 18 g. (72%) of lightly colored 2,5-dihydroxy-*p*-benzenediacetic acid (Note 8). A snow-white product is obtained by dissolving 18 g. of this material in 375 ml. of boiling water, treating with 2 g. of Norit A, and filtering. Upon cooling, there is deposited 15 g. (61%) of the acid which melts at 233° (Note 9).

2. Notes

1. Precautions must be taken to prevent ammonia from coming in contact with the quinone before the desired time. The dropping funnel is loosely stoppered, and the flask containing the remainder of the solution is well stoppered.

2. Concentrated ammonium hydroxide (28% NH_3) must be used to obtain a good yield. The ammonium hydroxide may be introduced through either the central opening or the 250-ml. dropping funnel with the precaution given in Note 1.

3. Preferably, all the ammonium hydroxide solution should be added by the time 90% of the quinone solution has been added.

4. This product is insoluble in ethanol, and there is no loss in washing.

5. Slightly lower yields were obtained by the checkers at this point. The yields in three experiments were respectively 32.5, 32.0, and 31.5 g. (39, 38.5, and 38%).

6. In the early stages, considerable foaming usually occurs, and care must be exercised that the foam does not carry part of the insoluble product into the condenser. The insoluble material accumulating on the walls of the flask is occasionally returned to the liquid portion by whirling the flask.

7. Rapid filtration is essential to avoid crystallization in the funnel and to decrease the time the filter paper is in contact with the concentrated acid.

8. The checkers obtained, in two experiments, yields of 71 and 86% of material melting respectively at 232–238° (cor.) and 235–239° (cor.).

9. The checkers obtained, in two experiments, yields of 63 and 73% at this point. The product was white and melted at 233° (cor.).

3. Methods of Preparation

The method described above is a modification of that described by Wood, Colburn, Cox, and Garland.[1]

[1] Wood, Colburn, Cox, and Garland, *J. Am. Chem. Soc.*, **66**, 1540 (1944).

3,5-DIHYDROXYBENZOIC ACID

(α-Resorcylic acid)

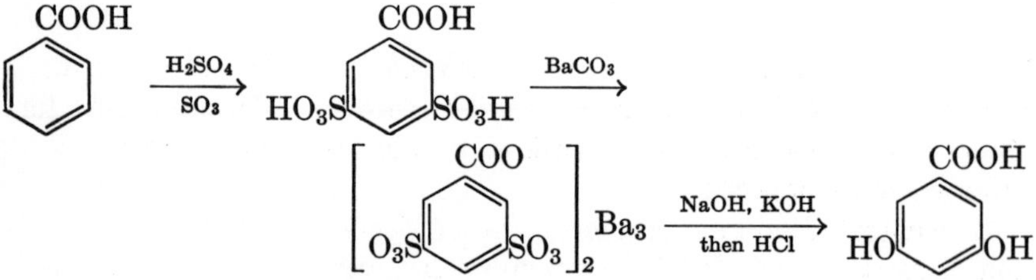

Submitted by Arthur W. Weston and C. M. Suter.
Checked by C. F. H. Allen and John W. Gates, Jr.

1. Procedure

To 200 g. (1.64 moles) of benzoic acid in a 1-l. Kjeldahl flask, in the top of which is inserted a loosely fitting cold finger (Note 1), is added

500 ml. of fuming sulfuric acid (Note 2). The mixture is heated in an oil bath for 5 hours at 240–250° (bath temperature).

After standing overnight the syrupy liquid is poured slowly with stirring into 3 kg. of ice in a 3-gal. crock (Note 3). The solution is then neutralized by adding barium carbonate in 100-g. portions, stirring occasionally until the gas evolution slackens before each addition: 2.4–2.5 kg. is required. The pasty mass is filtered by suction on a 30-cm. Büchner funnel, and the barium sulfate is washed with five 300-ml. portions of water. The combined filtrates are evaporated nearly to dryness on a steam bath and finally dried in an oven at 125–140° (Note 4). The yield of crude barium salt is 640–800 g.

The operator should wear goggles and long-sleeved gloves during the next two operations. The dried, pulverized barium salt is introduced in 200-g. portions into a melt of 600 g. each of sodium and potassium hydroxides contained in a 14 by 20 cm. copper beaker. The mixture is stirred with a copper stirrer and the temperature determined with a thermometer in a copper well (Note 5). Each portion is well stirred in before the next is added. The temperature is then slowly raised to 250–260°, at which point a vigorous reaction occurs with copious evolution of gas. After this has slackened (about 30 minutes) the temperature is raised to 280–310° and maintained there for 1 hour, then allowed to drop to 200°. The melt is ladled into 6 l. of water (Notes 6 and 7). The barium sulfite is filtered by suction, and the filtrate is acidified with concentrated hydrochloric acid (about 2.5 l. is required).

The resulting solution (about 9 l.) is divided into two portions and each is extracted three times, 600 ml. of ether being used each time (Note 8). The combined extracts are concentrated to about 1 l. and dried overnight over 150 g. of anhydrous sodium sulfate. After filtration and evaporation of the ether, there remains 137–160 g. (58–65%) of a slightly colored product, melting with decomposition at 227–229°. It is sufficiently pure for most purposes (Notes 9 and 10).

2. Notes

1. An ordinary glass funnel suspended in the neck serves less satisfactorily, whereas with a water condenser solidification of the sulfur trioxide causes difficulty.

2. This acid, approximately 30% sulfur trioxide, is made by mixing 500 g. each of fuming sulfuric acid containing 60% free sulfur trioxide and concentrated sulfuric acid (sp. gr. 1.84); 500 ml. of this mixture is used.

3. The reaction product may solidify at first, but it dissolves later.

4. The evaporation of the solution is most conveniently done overnight. The final drying is best done in flat metal trays; the length of time required depends upon the temperature. The checkers left trays on a steam coil over a week end.

5. An 18-cm. piece of 10- to 12-mm. copper tubing, pounded together at the lower end, is used. The stirrer can be made of the same tubing.

6. The addition of the hot melt causes some spattering.

7. Alternatively, the melt may be poured into shallow metal trays and allowed to cool and solidify. It is then pulverized and dissolved. Grinding is difficult and unpleasant. The melt should not be allowed to solidify in the reaction vessel.

8. A large automatic extraction apparatus may be used if available.

9. The color can be removed by recrystallization from hot acetic acid, using a decolorizing carbon. A solution of 16 g. of the crude acid in 85 ml. of hot acetic acid deposits 13.4 g. of white needles after filtration through a 2-mm. layer of Darco in a hot funnel and cooling. These are anhydrous after drying at 100° at 35 mm. for 2 hours; m.p. 234–235° with decomposition (cor.). The melting point varies with the rate of heating.

10. This material is partially hydrated.[1] The melting point is unchanged after recrystallization from water.

3. Methods of Preparation

3,5-Dihydroxybenzoic acid is most conveniently prepared by alkaline fusion of the disulfonic acid obtained by sulfonation of benzoic acid.[1–3] It has also been prepared by alkaline fusion of the 3,5-dihalobenzoic acids [4,5] or of 3-bromo-5-sulfobenzoic acid.[6] Benzoic acid has been sulfonated using various strengths of sulfuric acid with or without catalysts.[7,8]

[1] Barth and Senhoefer, *Ann.*, **159**, 222 (1871); **164**, 109 (1872).

[2] Graves and Adams, *J. Am. Chem. Soc.*, **45**, 2439 (1923).

[3] Hohenemser, *Ber.*, **35**, 2305 (1902).

[4] Ger. pat. 286,266 [*Frdl.*, **12**, 158 (1914–1916)].

[5] U. S. pat. 1,321,271 [*C. A.*, **14**, 186 (1920)].

[6] Böttinger, *Ber.*, **8**, 374 (1875); **9**, 180 (1876); Bülow and Riess, *Ber.*, **35**, 3901 (1902); Koller and Klein, *Monatsh.*, **64**, 85 (1935).

[7] Herzig and Epstein, *Monatsh.*, **29**, 661 (1901); Brunner, *Monatsh.*, **50**, 216 (1928); Lock and Nottes, *Monatsh.*, **68**, 51 (1936); Suter and Weston, *J. Am. Chem. Soc.*, **61**, 232 (1939).

[8] Birkenshaw and Bracken, *J. Chem. Soc.*, **1942**, 369.

DIISOVALERYLMETHANE

(4,6-Nonanedione, 2,8-dimethyl-)

$$CH_3COCH_2CH(CH_3)_2 + NaNH_2 \rightarrow Na^+[CH_2COCH_2CH(CH_3)_2]^- + NH_3$$

$$(CH_3)_2CHCH_2COOC_2H_5 + Na^+[CH_2COCH_2CH(CH_3)_2]^- \rightarrow C_2H_5OH + Na^+[(CH_3)_2CHCH_2COCHCOCH_2CH(CH_3)_2]^- \xrightarrow{HCl} (CH_3)_2CHCH_2COCH_2COCH_2CH(CH_3)_2$$

Submitted by C. R. Hauser, J. T. Adams, and R. Levine.
Checked by Arthur C. Cope and Frank S. Fawcett.

1. Procedure

Approximately 300 ml. of commercial anhydrous liquid ammonia is added to a 500-ml. three-necked round-bottomed flask equipped with a mercury-sealed stirrer and a reflux condenser connected to a soda-lime drying tube (Note 1). The drying tube is attached to a gas-absorption trap, or the apparatus is assembled in a well-ventilated hood. The third neck of the flask is closed with a stopper. Freshly cut sodium (13.8 g., 0.6 gram atom) is weighed under xylene or kerosene, and a small amount is added to the liquid ammonia, with stirring, until a permanent blue color is produced. A few small crystals of ferric nitrate are added to catalyze the conversion of sodium to sodium amide (Note 2), and when the blue color has disappeared the remainder of the sodium is added in small pieces. When the sodium is converted completely to sodium amide, as indicated by change of the blue solution to a gray suspension, the ammonia is evaporated by warming the flask on a steam bath. During this operation sufficient dry ether is added through a dropping funnel (attached to the third neck of the flask) so that the volume of the liquid remains approximately 300 ml. After practically all the ammonia has been evaporated, as indicated by refluxing of the ether, the sodium amide suspension is stirred and heated under reflux for a few minutes and then cooled to room temperature. The procedure to this point requires approximately 1 hour.

A solution of 30 g. (0.3 mole) of redistilled methyl isobutyl ketone in 50 ml. of absolute ether is added to the stirred suspension of sodium amide during 5–10 minutes. After an additional 5 minutes, a solution of 78 g. (0.6 mole) of redistilled ethyl isovalerate in 50 ml. of dry ether is added during about 15 minutes. Stirring is continued for 2 hours while the mixture is heated under reflux on the steam bath. The gelatinous suspension of the sodium salt of diisovalerylmethane is poured into 300 ml. of water, made neutral to litmus by dilute hydrochloric acid, and extracted with three 100-ml. portions of ether. The solvent is removed by distillation under reduced pressure, and the residue is dissolved in an equal volume of methanol. A solution prepared from 44 g. of cupric acetate monohydrate and 350 ml. of water is heated nearly to the boiling point, filtered, and added to the methanol solution. The resulting mixture is allowed to stand until it has cooled to room temperature. The blue copper salt of diisovalerylmethane is collected on a Büchner funnel, pressed as dry as possible, washed on the funnel with 100 ml. of petroleum ether (b.p. 30–60°), and again sucked dry. The yield of the copper salt after air drying is 44–51 g. (69–79%); it melts in the range 150–155° (Note 3).

The diisovalerylmethane is regenerated by shaking the copper salt vigorously with 500 ml. of 10% sulfuric acid and 200 ml. of ether until all the salt has dissolved. The aqueous acid layer is extracted with two 100-ml. portions of ether, and the combined ether solutions are dried over anhydrous sodium sulfate. The solvent is removed, and the residue is distilled under reduced pressure. The yield of diisovalerylmethane is 32–42 g. (58–76%, based on methyl isobutyl ketone), b.p. 115–116°/20 mm., n_D^{25} 1.4565 (Notes 4 and 5).

2. Notes

1. Apparatus fitted with standard-taper ground-glass joints is convenient for this preparation.

2. It may be helpful to provide additional catalysis of the conversion to sodium amide of the small amount of sodium added initially by bubbling dry air through the solution.[1]

3. After recrystallization from methanol or 95% ethanol the fluffy light blue copper salt melts at 157–158°.

4. The submitters state that the reaction may be conducted with equal success by using 0.6 mole each of sodium amide and methyl isobutyl ketone, and 0.3 mole of ethyl isovalerate (yield 75% based on the ester).[2,3]

5. Acetone may be acylated with ethyl laurate by either procedure (with excess ester or excess ketone). Lauroylacetone (m.p. 31.5–32°) is obtained in 75% yield[3] by either procedure.

3. Methods of Preparation

Diisovalerylmethane has been prepared by the method described [2,3] and by the use of sodium hydride [4] as the condensing agent.

[1] Leffler, *Org. Reactions,* **1,** 99, John Wiley & Sons, New York, 1942; Vaughn, Vogt, and Nieuwland, *J. Am. Chem. Soc.,* **56,** 2120 (1934).
[2] Adams and Hauser, *J. Am. Chem. Soc.,* **66,** 1220 (1944).
[3] Levine, Conroy, Adams, and Hauser, *J. Am. Chem. Soc.,* **67,** 1510 (1945).
[4] Swamer and Hauser, *J. Am. Chem. Soc.,* **72,** 1352 (1950).

2,6-DIMETHOXYBENZONITRILE

(Benzonitrile, 2,6-dimethoxy-)

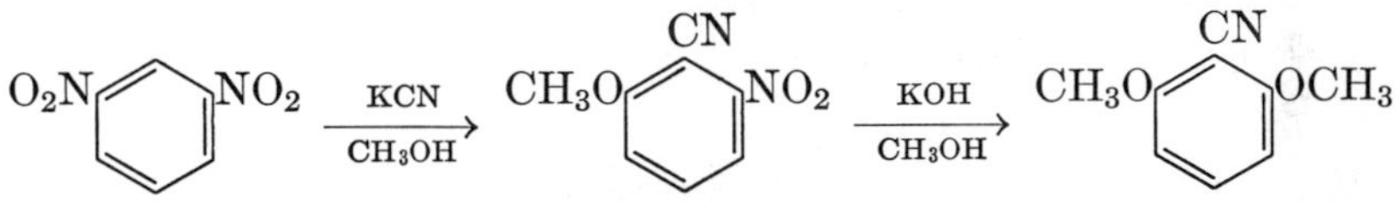

Submitted by Alfred Russell and W. G. Tebbens.
Checked by R. L. Shriner and Robert S. Voris.

1. Procedure

A. 2-*Nitro*-6-*methoxybenzonitrile.* Five hundred grams (2.97 moles) of technical *m*-dinitrobenzene (m.p. 88–89°) is dissolved in 7.5 l. of absolute methanol in a 12-l. round-bottomed flask fitted with an efficient mechanical stirrer. The temperature is raised to 40° by means of a water bath and maintained there while a solution of 230 g. of potassium cyanide in 400 ml. of water is added with stirring. The dark purple mixture is stirred for 2 hours and then is allowed to stand at room temperature for 2–3 days. The black precipitate is collected with suction on a Büchner funnel, pressed as dry as possible, and then spread out in the air to dry. It weighs about 185–188 g.

The filtrate is diluted with 60 l. of cold water (Note 1) and allowed to stand overnight. The brown sludge is filtered with suction (Note 2) and pressed as dry as possible on the funnel. After drying, it weighs about 164 g. The combined precipitates are refluxed for 30 minutes each with successive 650-, 500-, and 500-ml. portions of chloroform. The chloroform extracts are filtered while hot, and the bright red filtrates are combined and concentrated to a volume of 500 ml. by distillation. One liter of petroleum ether (b.p. 60–90°) is added, whereupon the crude 2-nitro-6-methoxybenzonitrile separates as a red powder. It is removed by filtration and air-dried. It weighs 120–125 g. (22–23%) and melts at 148–157° (Note 3). It is used without purification (Note 4).

B. 2,6-*Dimethoxybenzonitrile.* The crude nitrile is placed in a 3-l. flask and is refluxed for 2 hours with a solution of 75 g. of potassium hydroxide in 1.9 l. of methanol. The solution is concentrated by distillation to a volume of 400 ml. and then is poured into 4 l. of cold water. The brownish solid is filtered, washed thoroughly with cold water, and dried. The crude product (85–95 g.) is dissolved in 300 ml. of chloroform and refluxed with 8 g. of Darco for 30 minutes. The hot mixture is filtered, and 500 ml. of hot petroleum ether (b.p. 60–90°) is added to the filtrate. The solution is cooled, and the light-tan-colored needles of 2,6-dimethoxybenzonitrile are removed by filtration. The product weighs 75–86 g. (15–17%) and melts at 116–117° (Note 5).

2. Notes

1. The dilution is conveniently carried out by dividing the mixture among seven 3-gal. earthenware crocks.

2. The filtration can be performed by siphoning the mixture from the crocks through an 18.5 cm. Büchner funnel fitted to a 12-l. flask attached to a suction pump.

3. A fourth extraction with 500 ml. of choloroform gives only an additional 4 g. of product.

4. Dioxane is an excellent solvent for cleaning the flasks and crocks used in this first step.

5. This product is pure enough for most purposes. Pure white needles, melting at 117–118°, may be obtained by repeating the crystallization from the chloroform-petroleum ether mixture. The recovery is about 89–90%.

3. Methods of Preparation

The action of alcoholic potassium cyanide on *m*-dinitrobenzene was first studied by Lobry de Bruyn.[1] The present method is a modification of the procedure described by Mauthner.[2]

[1] Lobry de Bruyn, *Rec. trav. chim.*, **2**, 205 (1883).
[2] Mauthner, *J. prakt. Chem.*, (2) **121**, 259 (1929).

3,3′-DIMETHOXYBIPHENYL AND 3,3′-DIMETHYLBIPHENYL

(Biphenyl, 3,3′-dimethyl-, and biphenyl, 3,3′-dimethoxy-)

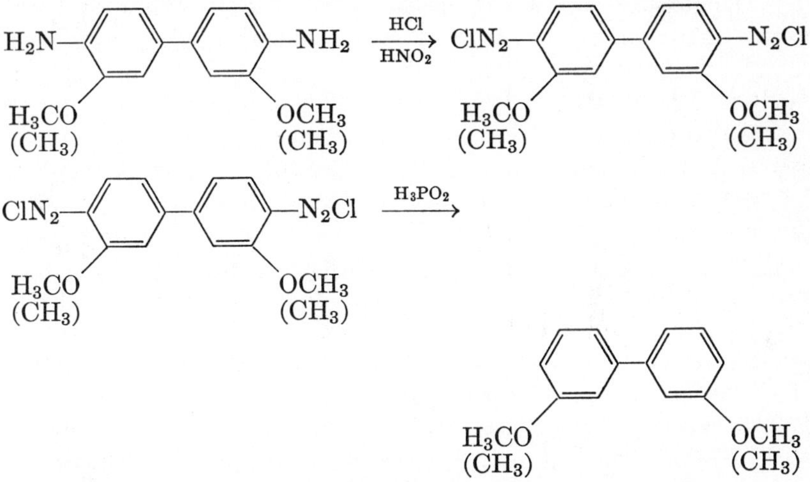

Submitted by Nathan Kornblum.
Checked by W. E. Bachmann and S. Kushner.

1. Procedure

A. 3,3′-*Dimethoxybiphenyl.* In a 1.5-l. beaker are placed 400 ml. of water and 31 ml. of concentrated hydrochloric acid (sp. gr. 1.19). This solution is heated to boiling, the flame is removed, and 40 g. (0.16 mole) of *o*-dianisidine (Note 1) is added. The hot mixture is stirred for about 3 minutes, until only a small amount of solid remains. The beaker is then placed in an ice-salt mixture and its contents are stirred mechanically until the temperature has dropped to about 15°. At this point 35 ml. more of concentrated hydrochloric acid is added. When the temperature of the mixture has fallen to 10–13°, a solution of 23.3 g. (0.33 mole) of 97% sodium nitrite (or an equivalent amount of sodium nitrite of higher or lower purity) in 50 ml. of water is added from a dropping funnel in the course of 10–15 minutes. The contents of the beaker are then stirred for 15–20 minutes at 5–10°, and finally filtered rapidly with suction from an appreciable amount of dark material (Note 2).

The cold, clear red filtrate is now poured rapidly into a 2-l. flask containing 325 ml. of ice-cold 30% hypophosphorous acid solution (Note 3). Immediate evolution of nitrogen occurs. The flask is stoppered loosely, placed in a refrigerator for 8–10 hours, and then allowed to stand at room temperature for another 8–10 hours (Note 4). The reaction product is transferred to a 2-l. separatory funnel, and the lower dark brown layer is separated from the aqueous phase. The aqueous layer is extracted with two 125-ml. portions of ether (Note 5). The brown oil combined with the ether extracts is washed in a separatory funnel with two 30-ml. portions of 20% sodium hydroxide solution (Note 6). Separation into two layers occurs rapidly, but occasionally the aqueous phase will remain turbid for some time. The loss by discarding the turbid sodium hydroxide washings is negligible. The ether solution is dried with a minimal amount of anhydrous potassium carbonate, 5 g. usually being sufficient (Note 7).

After filtration of the ether solution from the mixture of solid potassium carbonate and dark solids which have settled out, the ether is removed by distillation on a steam cone from a 50-ml. modified Claisen flask having a 15-cm. column. Usually suction must be applied to remove the last traces of ether. The residue is distilled at 4 mm. and a bath temperature of 200°. There is a small fore-run of 1–1.5 g. of a very pale yellow liquid which distils at 155–157°. The main fraction, boiling at 157–159°, is practically colorless and weighs 19–22 g. When the temperature of the thermometer in the flask starts to drop, the bath temperature is gradually raised to 240°. In this way, 3–4 g. more of light orange-yellow dianisyl is obtained. The yield based on the three fractions is 23–27.5 g. (66–78%). The fractions usually crystallize spontaneously; if they do not crystallize, they are cooled and scratched.

The main fraction melts at 41–43° (Note 8), and the other two fractions usually melt about a degree or two lower. The product is pure enough for most purposes; it may be recrystallized from the minimal volume of ethanol necessary for solution at 45–50°. The solution is cooled to about 35°; any oil that precipitates is brought back into solution by cautious addition of ethanol. The saturated solution is seeded, allowed to cool to room temperature, and then kept in a refrigerator for at least 4 hours. In this manner, 20–21 g. of flat, colorless needles is obtained; the product melts at 42–43.5°.

B. *3,3′-Dimethylbiphenyl.* Twenty-seven grams (0.13 mole) of *o*-tolidine (Note 9) is tetrazotized according to *Org. Syntheses* Coll. Vol. **2,** 145 (1943). It is not necessary to take the indicated precautions against a slight excess of nitrous acid. The clear orange tetrazonium solution is added to 290 ml. of 30% hypophosphorous acid (Note 10), and the mixture is allowed to stand, loosely stoppered, at room temperature for 16–18 hours (Note 11).

The reaction product is transferred to a 2-l. separatory funnel, and the red oily layer is separated from the aqueous phase. The aqueous layer is extracted once with 60 ml. of benzene. The combined red oil and benzene extract are dried with 1–5 g. of anhydrous sodium sulfate (Note 12). The benzene is removed by distillation from a 50-ml. modified Claisen flask having a 15-cm. column. The flask is heated in an oil bath to a final temperature of about 150° to ensure removal of the last traces of benzene. The residue is distilled at 3 mm. and a bath temperature of 155°. A practically colorless fore-run of 1–1.5 g. comes over from 109° to 114°. This is followed by 15.5–16 g. of a very pale lemon-yellow liquid which boils at 114–115°. When the temperature of the thermometer in the flask starts to fall the bath temperature is raised to 170°, and an additional 1–1.5 g. of light lemon-yellow material is obtained (Note 13). The total yield is 17.5–19 g. (76–82%) (Notes 14 and 15).

2. Notes

1. Eastman Kodak Company's technical grade, melting at 133–135°, was used.

2. Usually the pores of the filter paper become partially clogged after about half of the mixture has been filtered so that the rate of filtration is cut down considerably. In this event a fresh funnel and filter should be employed.

3. When less hypophosphorous acid solution is used, the yields are slightly lower, the product more highly colored than that obtained here, and the reaction time appreciably greater. When large amounts of a cheap amine are to be deaminated, it might be worth while to investigate the use of less hypophosphorous acid. In one run, 645 ml. of hypophosphorous acid solution was used for 40 g. of dianisidine. The yield was 29 g. (83%). With valuable amines it is probably best to use more hypophosphorous acid than is called for in these directions. Thus, in deaminating 33 g. of 2,2′-dimethyl-4,4′-diamino-5,5′-dimethoxybiphenyl, 1.2 l. of 30% hypophosphorous acid was used. There was obtained 24.5 g. (83%) of purified 2,2′-dimethyl-5,5′-dimethoxybiphenyl.

4. If the flask is permitted to stand at room temperature directly after the two solutions are mixed, the yield is a little lower than that obtained with these directions and the product is somewhat more highly colored. The entire reaction may be run at refrigerator temperature,

in which event 25–30 hours should elapse before the product is worked up.

When this reaction is applied to aniline derivatives, it appears advisable to conduct it entirely at refrigerator temperature owing to lowered thermal stability of the diazonium compounds as compared to the tetrazonium salts obtained from the benzidine type.

5. If the entire reaction is carried out at refrigerator temperatures the crude product is a light-brown solid, which is best worked up by filtering with suction, dissolving in ether, and proceeding as in the regular directions.

6. When this washing is omitted, the product contains a white, alkali-soluble solid and rapidly becomes orange-brown.

7. The potassium carbonate usually retains some dianisyl. It should be washed with a little of the ether which has been removed by distillation, and these washings should be redistilled.

8. In agreement with others, the checkers obtained a product melting at 31.5–33.5°. After recrystallization from ethanol, the 3,3′-dimethoxybiphenyl melted at 33.5–35°. This low-melting form changed rapidly to the form melting at 42–43.5° when a solution or melt of the former or even the solid came in contact with a crystal of the high-melting form (obtained from the submitter).

9. Eastman Kodak Company's *o*-tolidine, m.p. 128–129°, was used.

10. Use of 175 ml. of hypophosphorous acid gave a yield of 17.8 g. (77%). The first and last fractions were turbid.

11. The yield and quality of the product are identical with those obtained by conducting the reaction at refrigerator temperature.

12. Washing the benzene solution with aqueous alkali was found to be unnecessary.

13. In spite of the slight differences in color, the three fractions have almost identical refractive indices (n_D^{20} 1.5945).

14. One run was made using 27 g. of Eastman's practical *o*-tolidine, m.p. 121–125°. The procedure employed differed from the above only in that the benzene solution was washed once with 40 ml. of 10% aqueous alkali. There was obtained 18.9 g. (82%) of a light orange-yellow liquid possessing essentially the same refractive index as the product obtained from the better grade of *o*-tolidine.

15. This method of deaminating aromatic amines appears to be of general applicability, particularly to benzidine and its derivatives. Benzidine itself has been deaminated in 60% yield. The use of hypophosphorous acid in preference to ethanol for these deaminations arises from the fact that this procedure is much simpler, the yields are higher, and the products are of better quality.

This use of hypophosphorous acid for the deamination of aromatic amines appears to have originated with Mai.[1] It has also been used for this purpose by Raiford and Oberst.[2]

3. Methods of Preparation

3,3′-Dimethoxybiphenyl has been prepared by methylating 3,3′-dihydroxybiphenyl with methyl iodide[3] or dimethyl sulfate;[4] and by deaminating tetrazotized *o*-dianisidine with alcohol.[5–7] The present procedure is a slight modification of Mai's directions.

3,3′-Dimethylbiphenyl can be prepared from *m*-bromotoluene[8] or *m*-iodotoluene[9] and sodium; by the action of copper powder on *m*-iodotoluene;[10] by treating 4,4′-dihydroxy-3,3′-dimethylbiphenyl with zinc dust;[11] by heating 4,4′-dichloro-3,3′-dimethylbiphenyl with hydriodic acid and phosphorus;[11] by treating *o*-tolidine with nitrous acid in alcoholic solution;[12] by the decomposition of tetrazotized *o*-tolidine with methanol or ethanol in the presence of zinc dust.[13,14] It has been prepared by reduction of *m*-bromotoluene with hydrazine hydrate, using a palladium-calcium carbonate catalyst.[15] Upon treatment of lithium *m*-tolyl with oxygen, a small amount of 3,3′-dimethylbiphenyl is formed.[16]

3,3′-Dimethylbiphenyl, prepared by the general method described here, may be purified further.[17]

1 Mai, *Ber.*, **35**, 162 (1902).

2 Raiford and Oberst, *Am. J. Pharm.*, **107**, 242 (1935) [*C. A.*, **29**, 6216 (1935)].

3 Barth, *Ann.*, **156**, 98 (1870).

4 Schultz and Kohlhaus, *Ber.*, **39**, 3343 (1906).

5 Starke, *J. prakt. Chem.*, (2) **59**, 226 (1899).

6 Haeussermann and Teichmann, *Ber.*, **27**, 2108 (1894).

7 Mascarelli and Visintin, *Gazz. chim. ital.*, **62**, 358 (1932).

8 Perrier, *Bull. soc. chim. France*, (3) **7**, 182 (1892).

9 Schultz, Rohde, and Vicari, *Ber.*, **37**, 1401 (1904); *Ann.*, **352**, 112 (1907).

10 Ullmann and Meyer, *Ann.*, **332**, 43 (1904).

11 Stolle, *Ber.*, **21**, 1096 (1888).

12 Schultz, *Ber.*, **17**, 468 (1884).

13 Winston, *Am. Chem. J.*, **31**, 128 (1904).

14 Schlenk and Brauns, *Ber.*, **48**, 666 (1915).

15 Busch, Hahn, and Mathauser, *J. prakt. Chem.*, **146**, 22 (1936).

16 Müller and Töpel, *Ber.*, **72**, 273 (1939).

17 Mills, *Nature*, **167**, 726 (1951).

6,7-DIMETHOXY-3,4-DIHYDRO-2-NAPHTHOIC ACID

(2-Naphthoic acid, 3,4-dihydro-6,7-dimethoxy-)

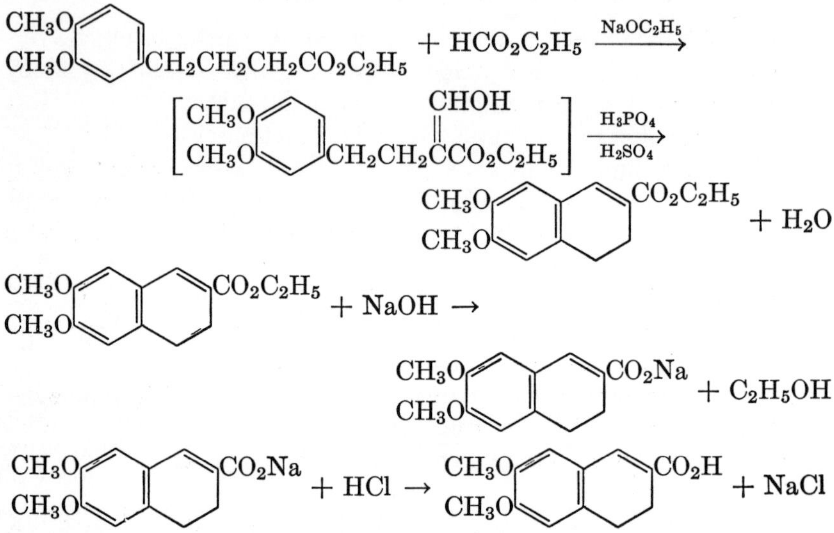

Submitted by H. L. Holmes and L. W. Trevoy.
Checked by Cliff S. Hamilton and Carol K. Ikeda.

1. Procedure

A. *Ester condensation.* A suspension of 9.40 g. (0.41 gram atom) of powdered sodium [1] in 100 ml. of absolute ether is placed in a 1-l. three-necked flask (Note 1) fitted with a reflux condenser, dropping funnel, and a calcium chloride tube. A solution of 23.8 ml. (0.41 mole) of absolute ethanol (Note 2) in 50 ml. of absolute ether is added through the dropping funnel, and the mixture is refluxed on a steam bath for 10–11 hours. The reflux condenser is replaced by a mercury-seal stirrer. A thermometer and a calcium chloride tube are fitted into the third neck of the flask. The suspension of sodium ethoxide in absolute ether is cooled to −10° to −15° in an ice-hydrochloric acid bath, and a solution of 47.85 g. (0.19 mole) of ethyl γ-veratrylbutyrate (Note 3) and 29.6 g. (0.40 mole) of ethyl formate in 100 ml. of absolute ether is added dropwise through the dropping funnel with vigorous stirring (Note 4). The mixture is kept at −10° for 4 hours, and then the thermometer and calcium chloride tube are removed and the stirrer is replaced by a reflux condenser equipped with a calcium chloride tube. The mixture is allowed to come to room temperature and stand for 72 hours. During this time a gas is evolved and a pale yellow solid separates.

On digestion of this solid mass with 1 l. of ice and water, the sodium salt of the enol dissolves in the water, and the unreacted ester is removed by extracting the aqueous layer with two 200-ml. portions of ether (Note 5). The formyl derivative settles out as an oil upon acidification of the aqueous layer with dilute sulfuric acid. The oil is extracted with three 200-ml. portions of ether, and the ethereal extract is washed several times with water and dried over anhydrous sodium sulfate. The ether is distilled, and, to remove traces of ethyl formate, the oil is heated on a steam bath under a pressure of 20–30 mm. for 1 hour. The remaining yellow formyl derivative weighs 27–29 g. (Note 6).

B. *Cyclization.* The above oil is poured dropwise into a well-stirred mixture of 110 ml. of 90% phosphoric acid (sp. gr. 1.75) and 23.4 ml. of sulfuric acid (sp. gr. 1.83) which is kept at −10°. The temperature is allowed to rise to 0–10°, and the stirring is continued for 2 hours. The viscous reaction mixture is poured into 500 ml. of ice and water, and the acid is partially neutralized with 300 ml. of 40% sodium hydroxide solution with efficient cooling. The viscous cream-colored oil is extracted with three 150-ml. portions of ether; the ether extract is washed well with water and sodium bicarbonate solution to remove the last traces of acid and then dried over anhydrous sodium sulfate. The cyclized ester, after removal of the ether, is a red oil which solidifies upon cooling. The yield is 23–24 g.

C. *Saponification.* The ester is saponified by refluxing for 5 hours with 75 ml. of a 10% sodium hydroxide solution containing 3 ml. of ethanol; it is then poured into 150 ml. of water and decolorized with Norit. Upon acidification of the alkaline solution with dilute hydrochloric acid, 18–19 g. (40–43%) (Notes 7 and 8) of acid is obtained, melting at 191–193° (cor.).

2. Notes

1. It is suggested that a flask with a ground-glass joint be employed for the condensation as it leads to less discoloration of the formylation product.

2. The absolute ethanol employed in the condensation was refluxed over calcium oxide for 20 hours and finally distilled from magnesium ethoxide.

3. The ethyl γ-veratrylbutyrate, b.p. 203–207°/20 mm., is obtained in 80% yield by esterification of γ-veratrylbutyric acid by the Fischer-

Speier method. The γ-veratrylbutyric acid is prepared by the method of E. L. Martin[2] from β-(3,4-dimethoxybenzoyl)propionic acid.

4. The ethereal solution of the two esters should be added at such a rate that the addition is complete in about 1 hour.

5. On distillation, after removal of the ether, 18–20 g. of the ethyl γ-veratrylbutyrate is recovered.

6. The formylation product is too sensitive to purify by distillation even at pressures of 1–2 mm. of mercury.

7. In a similar manner, 3,4-dihydro-2-naphthoic acid, m.p. 111–114° (cor.), and 7-methoxy-3,4-dihydro-2-naphthoic acid, m.p. 149–150.5° (cor.), have been obtained in 35% yields.

8. 6,7-Dimethoxy-3,4-dihydro-2-naphthoic acid can be dehydrogenated[3] to give 6,7-dimethoxy-2-naphthoic acid. The yield of the latter was 85% after distillation and crystallization from ethanol.

3. Methods of Preparation

This method of preparation is essentially that used by Robinson and Crowley[4] for the preparation of 6-methoxy-3,4-dihydro-2-naphthoic acid. They first used concentrated sulfuric acid but found that this reagent caused sulfonation.

[1] *Org. Syntheses,* **18,** 25 (1938); Coll. Vol. **2,** 195 (1943).

[2] E. L. Martin, *J. Am. Chem. Soc.,* **58,** 1440 (1936); *Org. Syntheses,* **17,** 97 (1937); Coll. Vol. **2,** 499 (1943).

[3] *Org. Syntheses,* **18,** 59 (1938); Coll. Vol. **2,** 423 (1943).

[4] Robinson and Crowley, *J. Chem. Soc.,* **1938,** 2003.

β,β-DIMETHYLACRYLIC ACID

(Senecioic acid)

$$(CH_3)_2C{=}CHCOCH_3 + 3KOCl \rightarrow (CH_3)_2C{=}CHCO_2K + CHCl_3 + 2KOH$$

$$2(CH_3)_2C{=}CHCO_2K + H_2SO_4 \rightarrow 2(CH_3)_2C{=}CHCO_2H + K_2SO_4$$

Submitted by Lee Irvin Smith, W. W. Prichard, and Leo J. Spillane.
Checked by Nathan L. Drake and Harry D. Anspon.

1. Procedure

A 5-l. round-bottomed three-necked flask is equipped with a Kyrides sealed stirrer and two long condensers, and the apparatus is so arranged that, if necessary, the exit water from the condensers may be quickly used to cool the flask.

In the flask are placed 200 ml. of dioxane, 100 g. (1.02 moles) of mesityl oxide, and a solution of 4.6 moles of potassium hypochlorite in 3 l. of water (Note 1), and the stirrer is started. The mixture becomes warm immediately, and within 5 minutes chloroform begins to reflux. When the reaction becomes very vigorous the stirrer is stopped and the flask is cooled with running water (Note 2). The stirring is resumed as soon as feasible and is continued for 3–4 hours, when the temperature of the mixture will have dropped to that of the room. Sodium bisulfite (about 5 g.) is then added to react with the excess hypochlorite (Note 3).

One of the condensers is then replaced by a dropping funnel, and enough 50% sulfuric acid (about 200 ml.) is added, with stirring and cooling, to make the solution acid to Congo red paper. When the solution has cooled, it is extracted with eight 400-ml. portions of ether (Note 4). The ether extract is *carefully* distilled on a steam bath until the ether and chloroform are removed.

The residue is then placed in a modified Claisen flask and distilled under reduced pressure. Dimethylacrylic acid distils at 100–106°/20 mm. The yield of white solid is 49–53 g. (49–53%). This product melts at 60–65°. It may be further purified by recrystallization from petroleum ether (b.p., 60–70°) or water (Note 5).

2. Notes

1. The hypochlorite is prepared according to *Org. Syntheses,* **17,** 66; Coll. Vol. **2,** 429 (1943), Note 2. It is cooled to 10° before it is placed in the flask.

2. The flask is cooled only as much as is necessary to keep the chloroform refluxing gently. After 30 minutes the reaction will have subsided enough so that no further cooling is necessary.

3. When a few milliliters of the solution no longer liberate iodine from a slightly acid potassium iodide solution, enough sodium bisulfite has been added.

4. It is important that the ether be *well shaken* with the solution during the extractions.

5. For recrystallization from water, 48–50 g. of the acid is dissolved in 450 ml. of hot water. The solution is cooled in ice for several hours, and the crystalline precipitate is filtered with suction and dried overnight in a desiccator. The yield of pure dimethylacrylic acid melting at 66–67.5° is 35–40 g., a recovery of 70–83%.

3. Methods of Preparation

This acid has been prepared from various α-haloisovaleric acid derivatives by elimination of the halogen together with one of the β-hydrogen atoms; [1–4] from β-hydroxyisovaleric acid derivatives by elimination of water; [5–7] by action of sodium isobutoxide on iodoform; [8] by the condensation of malonic acid with acetone; [9–11] by the reaction of acetone and ketene; [12] by condensation of acetoacetic ester with acetone and action of barium hydroxide on the product, isopropylidene acetoacetic ester; [13] by the action of alkali upon 2,4-dibromo-2-methylbutanone; [14] by the incomplete ozonolysis of phorone; [15] by the action of alcoholic potassium hydroxide upon 2,5,5,7-tetramethylocta-2,6-diene-4-one; [16] by the action of potassium hydroxide on isopropenylacetic acid; [17] by the action of aluminum chloride on α,α-dimethylsuccinic anhydride; [18] by metallating isobutylene with amylsodium, carbonating, and acidifying; [19] by the action of hypohalites on mesityl oxide; [20–22] and by the action of sulfuric acid on 3-methyl-3-butenoic acid.[23]

[1] Duvillier, *Compt. rend.*, **88**, 913 (1879); *Ann. chim. phys.*, (5) **19**, 428 (1880); *Bull. soc. chim. France*, (3) **3**, 507 (1890); (3) **5**, 848 (1891).

[2] Weinig, *Ann.*, **280**, 252 (1894).

[3] Perkin, *J. Chem. Soc.*, **69**, 1470 (1896).

[4] Sernow, *J. Russ. Phys. Chem. Soc.*, **32**, 804 (1900) [*Chem. Zentr.*, **1901**, I, 665].

[5] Semljanitzin and Saytzeff, *Ann.*, **197**, 72 (1879).

[6] v. Miller, *Ann.*, **200**, 261 (1880).

[7] Neubauer, *Ann.*, **106**, 63 (1858).

[8] Gorbow and Kessler, *Ber.*, **17**, Ref. 67 (1884); **20**, Ref. 776 (1887).

[9] Dutt, *J. Ind. Chem. Soc.*, **1**, 297 (1924).

[10] Masset, *Ber.*, **27**, 1225 (1894).

[11] Knoevenagel, *Chem. Zentr.*, **1905**, II, 726.

[12] U. S. pat. 2,382,464 [*C. A.*, **40**, 1867 (1946)].

[13] Pauly, *Ber.*, **30**, 481 (1897).

[14] Favorski and Wanscheidt, *J. prakt. Chem.*, (2) **88**, 665 (1913).

[15] Harries and Türk, *Ann.*, **374**, 347 (1910).

[16] Deux, *Compt. rend.*, **208**, 522 (1939).

[17] Morton, Brown, Holden, Letsinger, and Magat, *J. Am. Chem. Soc.*, **67**, 2224 (1945).

[18] Desfontaines, *Compt. rend.*, **134**, 293 (1902).

[19] U. S. pat. 2,454,082 [*C. A.*, **43**, 1795 (1949)].

[20] Kohn, *Montash.*, **24**, 770 (1903).

[21] Barbier and Lesser, *Bull. soc. chim. France*, (3) **33**, 815 (1905).

[22] Cuculescu, *Bull. Fac. Stiinte Cernauti*, **1**, 53 (1927); [*C. A.*, **26**, 1897 (1932)].

[23] Wagner, *J. Am. Chem. Soc.*, **71**, 3214 (1949).

β-DIMETHYLAMINOPROPIOPHENONE HYDROCHLORIDE

(Propiophenone, β-dimethylamino-)

$$C_6H_5COCH_3 + CH_2O + (CH_3)_2NH_2Cl \rightarrow C_6H_5COCH_2CH_2N(CH_3)_2 \cdot HCl + H_2O$$

Submitted by Charles E. Maxwell.
Checked by C. F. H. Allen and J. VanAllan.

1. Procedure

In a 500-ml. round-bottomed flask attached to a reflux condenser are placed 60 g. (58.5 ml., 0.5 mole) of acetophenone (Note 1), 52.7 g. (0.65 mole) of dimethylamine hydrochloride, and 19.8 g. (0.22 mole) of paraformaldehyde. After the addition of 1 ml. of concentrated hydrochloric acid (sp. gr. 1.19) in 80 ml. of 95% ethanol, the mixture is refluxed on a steam bath for 2 hours (Note 2). The yellowish solution is filtered, if it is not clear (Notes 3 and 4), and is transferred to a 1-l. wide-mouthed Erlenmeyer flask. While still warm, it is diluted by the addition of 400 ml. of acetone (Note 5), allowed to cool slowly to room temperature, and then chilled overnight in the refrigerator. The large crystals are filtered and washed with 25 ml. of acetone. After it has been dried for 2.5 hours at 40–50°, this crude product weighs 72–77 g. (68–72%) and melts at 138–141° (Notes 6 and 7); it is suitable for many reactions.

It may be recrystallized by dissolving it in 85–90 ml. of hot 95% ethanol and slowly adding 450 ml. of acetone to the solution. The recovery is about 90%. The purified material, dried at 70°, melts at 155–156° (Notes 8 and 9).

2. Notes

1. Acetophenone, m.p. 19–20°, and a practical grade of dimethylamine hydrochloride were used.

2. The reaction mixture, which at first forms two layers, soon becomes homogeneous, and the paraformaldehyde dissolves.

3. The filtration must be done rapidly, preferably through a preheated funnel. Any material that crystallizes in the receiver is brought into solution again by warming on the steam bath.

4. Alternatively, the reaction mixture may be cooled at once and the solid product removed. The filtrate is successively concentrated and chilled three times, each crop of crystals being rinsed with acetone. For example, in a run using 480 g. of acetophenone, the amounts obtained were 297, 92, 53, and 16 g. respectively, and 42 g. from the acetone washings, making a total of 500 g., or 66% of the theoretical amount.

5. The excess dimethylamine hydrochloride is held in solution by the acetone.

6. The material is somewhat hygroscopic and holds traces of water tenaciously. The melting point is lowered by the presence of moisture; a preliminary shrinking is usually observed.

7. After it has been dried for an additional 4 hours, the product melts at 152–153°. The product melts at this same temperature after it has been kept for 60 hours in a vacuum desiccator, except that then there is no preliminary shrinking.

8. These directions are applicable equally well for runs of larger size.

9. The diethylamino homolog results when diethylamine hydrochloride is used.

3. Method of Preparation

The procedure described is an example of a general reaction,[1,2] the Mannich reaction, a review of which, from the experimental point of view, has been published.[3]

[1] Mannich and Heilner, *Ber.*, **55**, 359 (1922).
[2] Blicke and Burckhalter, *J. Am. Chem. Soc.*, **64**, 453 (1942).
[3] Blicke, *Org. Reactions,* **1,** 303 (1942), New York, John Wiley & Sons.

3,4-DIMETHYLANILINE

(Xylidine, 3,4-)

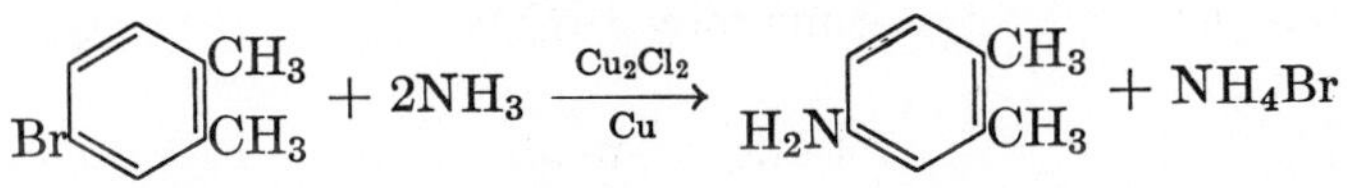

Submitted by W. A. WISANSKY and S. ANSBACHER.
Checked by N. L. DRAKE, WILKINS REEVE, and JOHN STERLING, JR.

1. Procedure

In a 1.1-l. steel reaction vessel (Note 1) are placed 200 g. (1.08 moles) of 4-bromo-*o*-xylene (p. 138), 14 g. copper wire, and 600 ml. (540 g., 9.0 moles) of 28% ammonia containing 12 g. of cuprous chloride. The steel reaction vessel is heated and rocked at 195° (thermostatic control) (Note 2) for 14 hours (Note 3); the pressure rises to 700–1000 lb. After cooling, the bomb is emptied and the two layers are separated; 40 ml. of 40% sodium hydroxide is added to the organic layer, and the mixture is steam-distilled. The amine distils and crystallizes when the distillate is cooled. It is separated from the water and dissolved in 500 ml. of 8% hydrochloric acid; the acid solution is extracted with two 100-ml. portions of ether (Note 4). The ether extracts are discarded, and the acid solution is made alkaline with 160 ml. of 40% sodium hydroxide. The resulting mixture is steam-distilled (Note 5). The distillate is cooled, most of the water is decanted, and the crystalline xylidine is dissolved by shaking with two 250-ml. portions of ether. The combined ether solutions are dried over calcium chloride and concentrated by distillation at a steam bath. The residue is distilled under reduced pressure from a Claisen flask. The yield of 3,4-dimethylaniline, boiling at 116–118°/22–25 mm., is 103 g. (79%). This crude product is recrystallized from 200 ml. of petroleum ether (b.p. 60–80°); the hot solution is placed in a refrigerator, and the crystallization is allowed to proceed overnight. The yield of recrystallized 3,4-dimethylaniline is 86 g. (66%) (Note 6); it melts at 47.3–49.2° (Note 7).

2. Notes

1. A hydrogenation vessel supplied by the American Instrument Company is satisfactory. A certain amount of copper plates out on

the walls of the bomb during the reaction, but most of it is removed when the apparatus is cleaned.

2. The heating and rocking are carried out in a hydrogenation assembly. The temperature must not be allowed to exceed 200°; at higher temperatures decomposition occurs and the yield suffers.

3. The optimum reaction time may vary slightly with apparatus of different types.

4. The crude product contains a small amount of unchanged 4-bromo-*o*-xylene which is removed by the ether extraction.

5. Caution must be exercised because a considerable amount of ether is present; it distils first and may be only partially condensed. A preliminary heating of the mixture on a steam bath does not remove the ether completely.

6. In an alternative method of purification the submitters collected the amine from the second steam distillation on a suction filter, pressed out the oily materials on the filter, and distilled the resulting crystalline product under diminished pressure; the distillate was not recrystallized. The checkers found it difficult to effect complete removal of oils by filtration; they preferred to omit this step and to recrystallize the distilled amine.

7. The reported melting point of 3,4-dimethylaniline varies from 47–48° [1] to 48.5–49°.[2]

3. Methods of Preparation

3,4-Dimethylaniline has been prepared by reduction of the corresponding nitro compound, either chemically [2,3] or catalytically using platinum,[4] Raney nickel,[5] or molybdenum [6,7] or tungsten sulfide [6] catalysts. It has been prepared from 3,4-dimethylphenol by heating with ammonia, ammonium bromide, and zinc bromide; [8] from *m*-toluidine hydrochloride by alkylation with methanol at high temperature,[1,9] from anhydro-4-amino-2-methylbenzyl alcohol by dry distillation from calcium hydroxide; [10] from 2-methyl-5-aminobenzyl alcohol by reduction with sodium; from 2-methyl-5-nitrobenzyl acetate by catalytic reduction; [11] from 2-methyl-5-nitrobenzyl or -benzal chloride (prepared from dichloromethyl ethyl ether [12] or *bis*-chloromethyl ether [13] and *p*-nitrotoluene) by catalytic,[12,13] chemical,[13] or electrochemical [12,14] reduction; from *o*-xylene by direct amination with hydroxylamine hydrochloride in the presence of aluminum chloride; [15] and from 3,4-dimethylacetophenone by Beckmann rearrangement of the oxime [16,17] or by reaction with hydrazoic acid in the presence of concentrated sulfuric acid.[18] The present method has been published.[19]

[1] Limpach, *Ber.,* **21,** 643 (1888).

[2] Bamberger and Blangey, *Ann.,* **384,** 318 (1911).

[3] Jacobsen, *Ber.,* **17,** 159 (1884).

[4] Karrer, Becker, Benz, Frei, Salomon, and Schöpp, *Helv. Chim. Acta,* **18,** 1435 (1935).

[5] Birch, Dean, Fidler, and Lowry, *J. Am. Chem. Soc.,* **71,** 1362 (1949).

[6] U. S. pat. 2,432,087 [*C. A.,* **42,** 2621 (1948)].

[7] Brown, Smith, and Scharmann, *Ind. Eng. Chem.,* **40,** 1538 (1948).

[8] Müller, *Ber.,* **20,** 1039 (1887).

[9] Cripps and Hey, *J. Chem. Soc.,* **1943,** 14.

[10] Barclay, Burawoy, and Thomson, *J. Chem. Soc.,* **1944,** 109.

[11] U. S. pat. 2,373,438 [*C. A.,* **39,** 3793 (1945)].

[12] Jap. pat. 162,726 [*C. A.,* **43,** 2637 (1949)].

[13] Sahashi, Akamatsu, and Genda, *Repts. Sci. Research Inst. Japan,* **24,** 72 (1948) [*C. A.,* **42,** 5458 (1948)].

[14] Sahashi, Akamatsu, and Inagaki, *Repts. Sci. Research Inst. Japan,* **24,** 80 (1948) [*C. A.,* **42,** 5459 (1948)].

[15] Graebe, *Ber.,* **34,** 1780 (1901).

[16] Zaugg, *J. Am. Chem. Soc.,* **67,** 1861 (1945).

[17] Berezovskii, Kurdyukova, and Preobrazhenskii, *Zhur. Priklad. Khim.,* **22,** 533 (1949); [*C. A.,* **44,** 2463 (1950)].

[18] Benson, Hartzel, and Savell, *J. Am. Chem. Soc.,* **71,** 1111 (1949).

[19] Wisansky and Ansbacher, *J. Am. Chem. Soc.,* **63,** 2532 (1941).

2,3-DIMETHYLANTHRAQUINONE

(Anthraquinone, 2,3-dimethyl-)

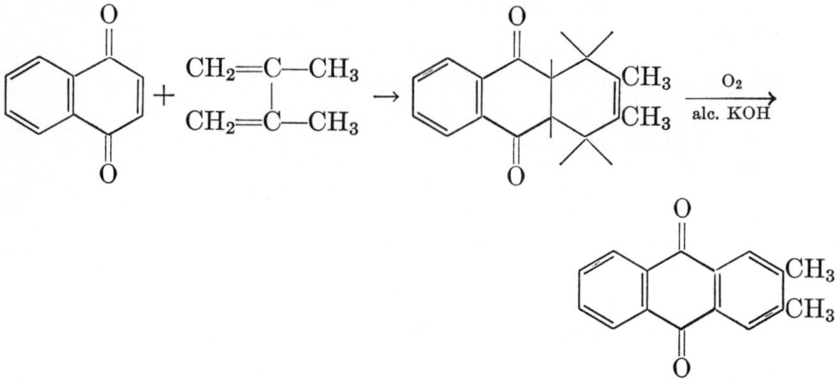

Submitted by C. F. H. ALLEN and ALAN BELL.
Checked by R. L. SHRINER and JOHN C. ROBINSON, JR.

1. Procedure

A solution of 80 g. (0.5 mole) of 1,4-naphthoquinone [*Org. Syntheses* Coll. Vol. **1,** 375 (1932), 383 (1941)] and 80 g. (1 mole) of 2,3-dimethylbutadiene-1,3 (p. 312) in 300 ml. of alcohol is refluxed for 5 hours, using a 1-l. round-bottomed flask and an efficient reflux condenser. The solution is cooled and placed in a refrigerator for 10–12 hours. The crystalline mass is then broken up with a spatula, and the addition product is filtered and washed with 50 ml. of cold ethanol. The product forms white, feathery crystals melting at 147–149° (Note 1). The yield is 116 g. (96% based on the 1,4-naphthoquinone).

For the dehydrogenation, 40 g. of the addition product is dissolved in 600 ml. of 5% ethanolic potassium hydroxide solution (Note 2) in a 1-l. three-necked flask equipped with a reflux condenser and inlet tube. A current of air is bubbled through the solution for 24 hours; considerable heat is generated, and the initial green color soon changes to yellow. The yellow quinone that has separated is then filtered with suction and is washed, first with 200 ml. of water, then with 100 ml. of ethanol, and finally with 50 ml. of ether. The yield of air-dried product (m.p. 209–210°) is 36.5–37.5 g. (94–96%) (Note 3). The over-all yield for both steps is 90% (Note 4).

2. Notes

1. The addition product is usually pure enough for the next step. It may be purified by recrystallization from acetone, ethanol, or methanol, and then it melts sharply at 150°. If the crude product is deeply colored, it should be recrystallized, using a decolorizing carbon.

2. This is prepared by dissolving 30 g. of potassium hydroxide in 570 g. of 95% ethanol.

3. The melting points given in the literature vary from 183° [1] to 208°.[2]

4. Essentially the same percentage yield has been obtained using three times the amounts given.

3. Methods of Preparation

2,3-Dimethylanthraquinone has been obtained by ring closure of the corresponding *o*-benzoylbenzoic acid; [1–3] by oxidation of the corresponding anthrone; [4] by decarboxylation of 2,3-dimethylanthraquinone-5-carboxylic acid; [5] from 2-chloro- and 2,3-dichloro-1,4-naphthoquinone and 2,3-dimethylbutadiene-1,3 by the action of sodium hydroxide; [6] and from 2-methyl-1,4-naphthoquinone and 2,3-dimethylbutadiene-1,3 with subsequent dehydrogenation by sulfur.[7] The addition product employed in the present procedure has been described.[8]

[1] Elbs and Emich, *Ber.*, **20**, 1361 (1887); *J. prakt. Chem.*, (2) **41**, 6 (1890).

[2] Heller, *Ber.*, **43**, 2891 (1910).

[3] Fairbourne, *J. Chem. Soc.*, **119**, 1573 (1921).

[4] Limpricht and Martens, *Ann.*, **312**, 103 (1900).

[5] Fieser and Newton, *J. Am. Chem. Soc.*, **64**, 917 (1942).

[6] Ger. pat. 500,160 [*Frdl.*, **17**, 1143 (1932); *C. A.*, **24**, 4790 (1930)].

[7] Fieser and Seligman, *J. Am. Chem. Soc.*, **56**, 2690 (1934).

[8] Fieser and Webber, *J. Am. Chem. Soc.*, **62**, 1362 (1940).

2,3-DIMETHYL-1,3-BUTADIENE

(1,3-Butadiene, 2,3-dimethyl-)

I. HYDROBROMIC ACID METHOD

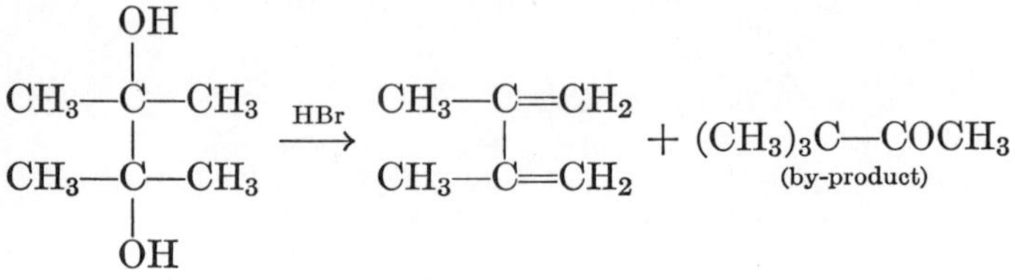

Submitted by C. F. H. ALLEN and ALAN BELL.
Checked by R. L. SHRINER and JOHN C. ROBINSON, JR.

1. Procedure

In a 2-l. round-bottomed flask, surmounted by a packed fractionating column (Note 1), is placed a mixture of 354 g. (3 moles) of pinacol (Note 2) and 10 ml. of commercial 48% hydrobromic acid. A few boiling chips are added, and the flask is then heated slowly with a colorless flame about 3 in. high (Note 3). The distillate is collected until the thermometer reads 95°; this requires about 2 hours when the rate of distillation is approximately 20–30 drops per minute. The upper, non-aqueous layer is washed twice with 100-ml. portions of water, 0.5 g. of hydroquinone is added, and the liquid is dried overnight with 15 g. of anhydrous calcium chloride. It is then fractionated as below, using the same column but a 1-l. flask (Note 4).

		Yield	
Fraction	*B.P.*	*Grams*	*Per Cent*
2,3-Dimethylbutadiene	69–70.5°	135–147	55–60
Intermediate	70.5–105°	10–15	
Pinacolone	105–106°	66–75	22–25
Residue		7–8	

2. Notes

1. The column described in *Organic Syntheses*, **20**, 96, filled with glass helices, and wrapped with asbestos paper, is satisfactory. Carborundum may also be used as the filling, but a longer time is required for the distillation. Cold water is circulated through the cold finger.

2. The pinacol used was the commercially available material, or that obtained by dehydration of pinacol hydrate. This dehydration may be accomplished by adding 2 l. of benzene to 1 kg. of pinacol hydrate [*Org. Syntheses* Coll. Vol. **1**, 448 (1932), 459 (1941)], and distilling the water-benzene mixture. The lower layer is separated, and the upper benzene layer is returned to the distilling flask. This is repeated until the benzene distillate is clear. The anhydrous pinacol is then distilled, and the fraction boiling from 168° to 173° is collected. Depending upon the quality of the material used, 1 kg. of pinacol hydrate yields about 500 g. of anhydrous pinacol.

3. If too large a flame is used the column floods.

4. 2,3-Dimethylbutadiene can be kept, without appreciable change, for a limited time in a refrigerator. If it is not to be used reasonably soon, a little hydroquinone should be added as an inhibitor.

II. ALUMINUM OXIDE METHOD

Submitted by L. W. Newton and E. R. Coburn.
Checked by Nathan L. Drake and Richard Tollefson.

1. Procedure

A Claisen flask, provided with the usual capillary inlet for air, is connected to a Pyrex tube (Note 1) which is drawn out at one end and packed with 8-mesh alumina (Note 2). The tube is inserted in an electric furnace (Note 3) capable of maintaining a temperature of 420–470°; the temperature is measured by a thermometer placed alongside the tube in the furnace. The drawn-out end of the tube is connected by a rubber stopper to an efficient Pyrex coil-condenser, which is in turn connected by a rubber stopper to the first of two receivers, arranged in series and connected by a short length of rubber tubing. Each receiver consists of a 500-ml. filter flask, which carries an inlet tube (Note 4) extending somewhat more than halfway to the bottom of the flask. The first receiver is immersed in an ice-salt mixture (Note 5); the second is immersed in a Dry Ice-methanol mixture contained in a Dewar flask. The exit tube of the second receiver is connected to a manometer (Note 6) and to a water pump.

Pinacol (Note 7) or pinacolone is placed in the Claisen flask and is distilled, under the vacuum of a water pump, through the Pyrex tube, which is maintained at a temperature of 420–470°. About 100 g. of pinacol is distilled in 15 minutes, and then the apparatus is swept out by maintaining the reduced pressure for 15 minutes longer. The water and resinous material collected in the ice-cooled receiver are discarded (Note 8). The second receiver is removed from the cooling bath, and the product is allowed to melt. Two layers are formed; these are allowed to separate, and then the receiver is replaced in the cooling bath until the lower water layer is frozen. The crude dimethylbutadiene is then decanted. The yellow product is dried over anhydrous calcium sulfate and is fractionated (Note 9) through a Widmer or other column; the portion boiling at 67–70° is collected. The yield from 100 g. of pinacol is 55–60 g. (79–86%) (Note 10). The diene is best stored over a small amount of hydroquinone in a refrigerator.

2. Notes

1. The Pyrex tube should have a diameter of 3 cm. and a length of 70 cm.

2. The grade of alumina is important. With an ordinary grade of desiccator alumina, the checkers were unable to obtain the yields stated. The alumina employed successfully was Alorco activated alumina, from the Aluminum Ore Company, East St. Louis, Illinois. The catalyst darkens in use, but it can be kept in a clean and active condition by the following treatment: after each run, or at least after two or three runs, the tube is heated to 420–470°, and a slow stream of air is drawn through it until the alumina is white (2–5 hours).

3. A very convenient furnace can be constructed by wrapping a 3-ft. length of 1½-in. iron pipe with asbestos paper and winding a suitable heating element over this paper (about 47.5 ft. of No. 18 B. & S. gauge Nichrome wire). The whole is enclosed in a length of steam-pipe insulation. The temperature is controlled by a properly chosen Variac (variable transformer).

4. The inlet tube should have a diameter of 10 mm.

5. This receiver serves to collect most of the water and polymer formed.

6. Another receiver, consisting of a 3 by 30 cm. side-arm test tube, may be used. It is immersed in a Dewar flask and cooled with Dry Ice-methanol. Only a small amount of material is collected in this receiver.

7. Anhydrous pinacol can be prepared from pinacol hydrate [*Org. Syntheses* Coll. Vol. **1**, 448 (1932), 459 (1941)] by distillation. The material that boils at 172–178° is collected, and the distillate should be refractionated. It is more convenient, however, to prepare anhy-

drous pinacol from the hydrate by the method described in Note 2 under part I (p. 313). Pinacol hydrate may be used as starting material; the yield will be slightly less than that obtained when the anhydrous material is used.

8. The contents of this receiver should be examined carefully for the presence of 2,3-dimethylbutadiene; the checkers usually found appreciable amounts present.

9. If the dehydration has been inefficient, some pinacolone (b.p. 107°) may be present in this product. The pinacolone may be distilled over the alumina again.

10. According to the submitters, the yield from 100 g. of pinacolone is 57–63 g. (70–77%).

3. Methods of Preparation

The most convenient method for preparation of 2,3-dimethylbutadiene involves the dehydration of pinacol. Many catalysts have been used, among them hydrobromic,[1] hydriodic,[1] and sulfuric acids; [2–4] sulfonic acids of the benzene [5] and naphthalene [6–8] series; acid potassium sulfate; [9] alum; [10] aniline hydrobromide; [1] iodine; [11] hot copper at 450–480°; [1] and alumina at 400°.[12–14] The diene has also been obtained by distillation of the product from the reaction between methylmagnesium iodide and ethyl α-methacrylate; [15] by treatment of tetramethylethylene dichloride with alcoholic potash; [4] from pinacol hydrochloride and sodium carbonate; [16] by the action of sodium or pyridine on α,α,β-trimethyl-βγ-dibromobutyric acid; [17] by passing 1,2,4-trimethyl-4-isopropenylcyclohexene over hot copper; [18] and by dehydration of dimethylisopropenylcarbinol with hydrochloric [19] or sulfuric acid.[17]

A patent reports the preparation of 2,3-dimethylbutadiene from trimethylethylene and formaldehyde in the presence of aqueous sulfuric acid.[20]

[1] Kyriakides, *J. Am. Chem. Soc.*, **36**, 985 (1914).

[2] Bayer and Company, Ger. pat. 253,081 [*Frdl.*, **10**, 1006 (1910–1912)].

[3] Couturier, *Ann. chim.*, (6) **26**, 485 (1892).

[4] Kondakow, *J. prakt. Chem.*, (2) **62**, 172 (1900).

[5] Ostromisslenski, *J. Russ. Phys. Chem. Soc.*, **47**, 1973 (1915) [*C. A.*, **10**, 1341 (1916)].

[6] Bayer and Company, Ger. pat. 249,030 [*Frdl.*, **10**, 1004 (1910–1912); *C. A.*, **6**, 3200 (1912)].

[7] Bayer and Company, Ger. pat. 253,082 [*Frdl.*, **10**, 1005 (1910–1912); *C. A.*, **7**, 1109 (1913)].

[8] Ruzicka and Schinz, *Helv. Chim. Acta*, **23**, 963 (1940).

[9] Bayer and Company, Ger. pat. 246,660 [*Frdl.*, **10**, 1003 (1910–1912); *C. A.*, **7**, 908 (1913)].

[10] Bayer and Company, Ger. pat. 250,086 [*Frdl.*, **10**, 1005 (1910–1912); *C. A.*, **6**, 3200 (1912)].

[11] Hibbert, *J. Am. Chem. Soc.*, **37**, 1754 (1915).

[12] Badische Anilin- u. Soda-Fabrik, Ger. pat. 235,311 [*Frdl.*, **10**, 1003 (1910–1912); *C. A.*, **5**, 3175 (1911)].

[13] Badische Anilin- u. Soda-Fabrik, Ger. pat. 256,717 [*Frdl.*, **11**, 796 (1912–1914); *C. A.*, **7**, 2488 (1913)].

[14] Fieser and Martin, *J. Am. Chem. Soc.*, **59**, 1019 (1937); Fieser and Seligman, *J. Am. Chem. Soc.*, **56**, 2694 (1934); Fieser, *Experiments in Organic Chem.*, 2nd Ed., p. 383, D. C. Heath and Co., Boston, 1941.

[15] Blaise and Courtot, *Compt. rend.*, **140**, 371 (1905).

[16] Harries, *Ann.*, **383**, 182 (1911).

[17] Courtot, *Bull. soc. chim. France,* (3) **35**, 972 (1906).

[18] Ostromisslenski, *J. Russ. Phys. Chem. Soc.*, **47**, 1949 (1915) [*Chem. Zentr.*, **1916**, I, 1133; *C. A.*, **10**, 1340 (1916)].

[19] Marinca, *J. Russ. Phys, Chem. Soc.*, **21**, 435 (1889) [*Chem. Zentr.*, **1890**, I, 520].

[20] U. S. pat. 2,350,485 [*C. A.*, **38**, 4957 (1944)].

3,5-DIMETHYL-4-CARBETHOXY-2-CYCLOHEXEN-1-ONE and 3,5-DIMETHYL-2-CYCLOHEXEN-1-ONE

(2-Cyclohexen-1-one, 4-carbethoxy-3,5-dimethyl-, and 2-cyclohexen-1-one, 3,5-dimethyl-)

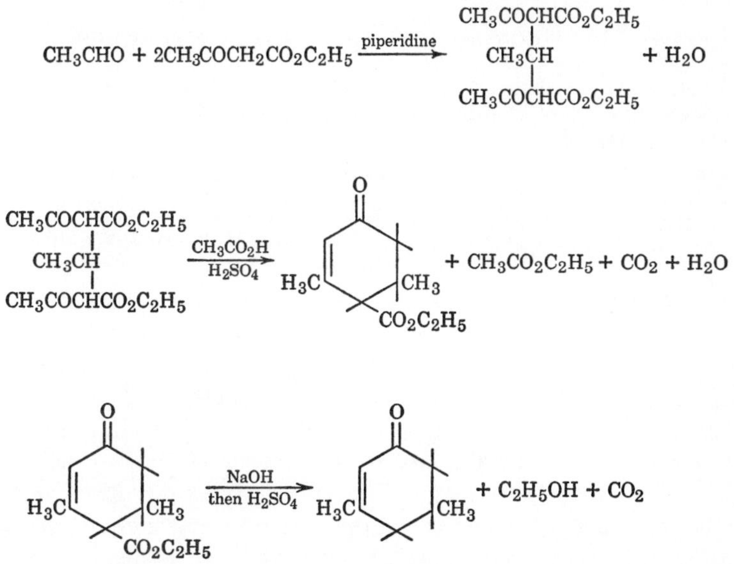

Submitted by E. C. Horning, M. O. Denekas, and R. E. Field.
Checked by Cliff S. Hamilton and Robert F. Coles.

1. Procedure

A. 3,5-*Dimethyl*-4-*carbethoxy*-2-*cyclohexen*-1-*one*. In each of three 500-ml. Erlenmeyer flasks (Note 1) is placed 210 ml. (210 g., 1.61 mole) of ethyl acetoacetate (Note 2). The flasks are placed in an ice-salt bath and chilled to 0°; to each flask there is then added 45 ml. (35.2 g., 0.78 mole) of acetaldehyde. When the contents of the flasks have cooled to −5° to 0° there is added to each flask, with shaking, a solution of 2 ml. of piperidine in 5 ml. of absolute ethanol. The flasks, the contents of which become cloudy in a short time because of the separation of water, are kept in the ice-salt bath for 6 hours. The reaction mixtures are then combined in a 1-l. flask and placed in an icebox. About 24 hours later, 3 ml. of piperidine in 5 ml. of absolute ethanol is added with shaking, and the flask is replaced in an icebox until the next day. The addition of 3 ml. of piperidine in 5 ml. of absolute ethanol is repeated once more, and the mixture is again returned to the icebox for 24 hours. The mixture is then allowed to stand for at least 1 day at room temperature. At some point during this reaction period the mixture should crystallize as a mass of yellow-white needles (Note 3). This product is crude ethyl ethylidenebisacetoacetate.

The crude bis ester is melted on a steam cone and poured into a 3-l. round-bottomed flask containing 600 ml. of glacial acetic acid, 40 ml. of concentrated sulfuric acid, and approximately 10 g. of small chips of porous plate (Note 4). The mixture is heated under reflux for 1 hour. There is a copious evolution of carbon dioxide which should be directed to a gas-absorption trap since it is accompanied by acetic acid vapors. The mixture is poured, with mechanical stirring, into 2 l. of ice water in a 4-l. beaker. Enough ether is added to allow separation of the layers (Note 5), and the organic layer is returned to the beaker with 1.2–1.3 l. of water. With the aid of good stirring the mixture is neutralized by slow addition of solid sodium carbonate until the effervescence ceases. The layers are separated, and the material so obtained may be used immediately for the preparation of 3,5-dimethyl-2-cyclohexen-1-one. To obtain 3,5-dimethyl-4-carbethoxy-2-cyclohexen-1-one, the ether solution is washed with 100 ml. of 5% sodium hydroxide solution and then with 100 ml. of water containing 2 ml. of acetic acid, is dried over anhydrous magnesium sulfate, and distilled through a short column under reduced pressure. After a very slight fore-run the main fraction is collected at 135–155°/10 mm. This is redistilled through a moderately good column, preferably of the Widmer or Vigreux type. The product is 3,5-dimethyl-4-carbethoxy-2-cyclohexen-1-one, b.p. 136–138°/9 mm. The yield is 220–234 g. (47–50%).

B. 3,5-*Dimethyl*-2-*cyclohexen*-1-*one*. The ether solution of crude 3,5-dimethyl-4-carbethoxy-2-cyclohexen-1-one obtained as described above is transferred to a 3-l. round-bottomed flask. The ether is removed on a steam cone, perferably with the aid of an aspirator, and there are added 1140 ml. of water, 60 ml. of ethanol (95%), and 130 g. of sodium hydroxide. The mixture is shaken continuously and heated on a steam cone until the alkali dissolves, and heating is continued on a steam cone with frequent shaking until the ester dissolves (Note 6). The solution is then refluxed for 15 minutes.

The flask is cooled with a stream of water while a solution of 100 ml. of concentrated sulfuric acid in 200 ml. of water is added slowly and cautiously (Note 7). The acidified mixture is heated under reflux for 15 minutes, allowed to cool, and the layers are separated. The crude product is diluted with 100 ml. of ether and washed successively with two 100-ml. portions of 5% sodium hydroxide solution and with 100 ml. of water containing 5 ml. of acetic acid. The ethereal solution, after drying over anhydrous magnesium sulfate, is distilled through a short column under reduced pressure. The product boiling at 84–86°/9 mm. is 3,5-dimethyl-2-cyclohexen-1-one; the yield is 155–165 g. (52–55%) (Note 8).

2. Notes

1. The checkers found it more convenient to use a 1-l. Erlenmeyer flask instead of three 500-ml. flasks.

2. The ethyl acetoacetate was obtained from the Carbide and Carbon Chemicals Company. The acetaldehyde was obtained from the Niacet Chemicals Corporation. The piperidine was the Practical grade of the Eastman Kodak Company.

3. Crystals may appear before the addition of all the piperidine has been completed. In this event the time of standing as described is followed, but no more catalyst need be added.

4. Ordinary boiling chips are convenient.

5. Usually 150–200 ml. of ether is ample. The layers may be separated in a separatory funnel, or the lower aqueous layer may be removed with a siphon.

6. The solution of the ester is an exothermic process. No attempt should be made to heat the mixture to reflux temperature until the ester has been dissolved.

7. It is important that this step be carried out slowly and with a hot solution. A satisfactory method is to place the flask in a sink while the contents are still hot and to cool with a stream of water directed over the entire flask. The sulfuric acid solution should be poured slowly down the wall of the flask, the rate of addition being regulated by the vigor of the decarboxylation. It is possible to carry out part of the decarboxylation in the alkaline solution by prolonging the reflux period, but this procedure offers no advantage over that described.

8. This general procedure can also be applied to compounds derived from other aliphatic aldehydes.[1]

3. Methods of Preparation

3,5-Dimethyl-4-carbethoxy-2-cyclohexen-1-one and 3,5-dimethyl-2-cyclohexen-1-one are usually prepared from acetaldehyde and acetoacetic ester through the Knoevenagel condensation.[2] The keto ester has been obtained previously by selective saponification and decarboxylation methods which have involved heating the crude condensation product with water at 140° [2,3] or with sodium ethoxide in ethanol.[3] The ketone has also been obtained from the same condensation product by prolonged refluxing in 20% sulfuric acid.[2,4,5] 3,5-Dimethyl-2-cyclohexen-1-one has also been obtained by the reduction of 2,4,6-trimethylpyridine with sodium and ethanol in liquid amomnia, followed by hydrolysis and cyclization.[6]

[1] Horning, Denekas, and Field, *J. Org. Chem.*, **9**, 548 (1944).

[2] Knoevenagel, *Ann.*, **281**, 104 (1894).

[3] Rabe, *Ann.*, **342**, 344 (1905).

[4] Gatterman, *Practical Methods of Organic Chemistry*, trans. by Schober and Babasinian, p. 202, The Macmillan Company, New York, 1921.

[5] Smith and Roualt, *J. Am. Chem. Soc.*, **65**, 634 (1943).

[6] Birch, *J. Chem. Soc.*, **1947**, 1270.

DIMETHYLETHYNYLCARBINOL

(3-Butyn-2-ol, 2-methyl-)

$$CH_3COCH_3 + NaNH_2 \rightarrow (CH_3COCH_2)Na + NH_3$$

$$(CH_3COCH_2)Na + C_2H_2 \rightarrow (CH_3)_2C(C{\equiv}CH)ONa$$

$$(CH_3)_2C(C{\equiv}CH)ONa + H_2SO_4 \rightarrow (CH_3)_2C(OH){-}C{\equiv}CH + NaHSO_4$$

Submitted by Donald D. Coffman.
Checked by C. F. H. Allen and Alan Bell.

1. Procedure

In a 2-l. round-bottomed flask fitted with a three-holed stopper bearing a mechanical stirrer, a separatory funnel, and a gas outlet tube leading to a hood (Note 1) are placed 1 l. of anhydrous ether (Note 2) and 156 g. (4 moles) of finely ground sodium amide (p. 778) (Note 3). The flask is surrounded by a well-packed ice-salt bath. To the vigorously stirred mixture 232 g. (4 moles) of dry acetone (Note 4) is added, dropwise, during a period of 3 hours. With the flask cooled to −10° (Note 5), a slow current of acetylene (Note 6) is passed through the reaction mixture for 2 hours to sweep out the ammonia. The three-holed stopper is then replaced by a two-holed stopper having a stopcock and an inlet tube reaching to the bottom of the flask and connected with a cylinder of acetylene. The stopper is wired in. The mixture is placed in an ice-salt mixture (Note 5), the whole being mounted on a shaking machine and agitated vigorously for 10 hours; the mixture is kept under a pressure of 10 lb. of acetylene. Every 30 minutes the pressure is released by means of the stopcock, to sweep out ammonia formed from small amounts of previously unreacted sodium amide.

The reaction mixture is poured cautiously into 800 g. of crushed ice and acidified in the cold by the addition of 400 ml. of 10 *N* sulfuric acid (Note 7). The ether layer is separated and the aqueous layer extracted twice with 100-ml. portions of ether. The combined ethereal solutions are dried over 100 g. of anhydrous potassium carbonate, and the filtered solution is fractionated (Note 8). The portion boiling at 103–107° is collected; any low-boiling fraction is dried and redistilled. The total yield is 135–155 g. (40–46%) of a colorless product that boils at 103–107° (Notes 9, 10, and 11).

2. Notes

1. In cold weather, it is convenient to carry out the reaction out-of-doors. This minimizes the attention needed to replace the ice. The outlet tube then opens to the air.

2. A commercial grade of anhydrous ether was dried over sodium.

3. The sodium amide, moistened by the heptane, was rapidly ground and the solvent allowed to evaporate.

4. An Eastman grade of acetone was used, after standing over anhydrous potassium carbonate.

5. It was found convenient to add Dry Ice to the freezing mixture, thus decreasing the frequency of packing. The temperature never rose above −10° and was usually considerably less.

6. Commercial acetylene used for welding was dried by passing over anhydrous calcium chloride.

7. This is prepared by adding 110 ml. of concentrated sulfuric acid to 290 ml. of water.

8. The checkers used a modified Widmer column.

9. There is a considerable quantity of high-boiling material; the quantity and boiling-point range are greater when the shaking is insufficient.

10. The reaction may be interrupted at several points. After the ammonia has been swept out by acetylene, it is usually convenient to place the mixture in a refrigerator overnight and start the shaking the next day. The shaking period need not be continuous; in this event the chilled mixture is placed in the icebox.

11. By the same procedure the submitter obtained methylethylethynylcarbinol, b.p. 119–123° (33% yield), using methyl ethyl ketone,[1,2] and 1-ethynylcyclohexanol-1, b.p. 53–55°/2 mm. (50% yield), using cyclohexanone and double the amount of ether.

3. Methods of Preparation

Dimethylethynylcarbinol has usually been prepared by the addition of acetylene to the sodium derivative of acetone,[3–8] but potassium metal [9] and sodium ethoxide [10] have also been used. The above method is based upon that described by Sung Wouseng.[3] Other methods use potassium hydroxide with calcium carbide,[11] or with acetylene and an immiscible alcohol such as butanol [12] or amyl alcohol.[13] Dimethylethynylcarbinol has also been prepared by the action of sodium acetylide on the bisulfite addition compound of acetone.[14]

A modified procedure, involving the addition of acetone to sodium acetylide in liquid ammonia, has been reported to give a 67% yield.[15]

The general procedure of Campbell, Campbell, and Eby [16] gives excellent yields of ethynylcarbinols.

[1] Ger. pat. 289,800 [*Frdl.*, **12**, 57 (1914–1916)].
[2] Ger. pat. 285,770 [*Frdl.*, **12**, 55 (1914–1916)].
[3] Sung Wouseng, *Ann. chim.*, (10) **1**, 359 (1924).
[4] Scheibler and Fischer, *Ber.*, **55**, 2903 (1922).
[5] Ger. pat. 280,226 [*Frdl.*, **12**, 51 (1914–1916)].
[6] U. S. pat. 2,106,180 [*C. A.*, **32**. 2547 (1938)].
[7] Ger. pat. 286,920 [*Frdl.*, **12**, 54 (1914–1916)].
[8] Froning and Hennion, *J. Am. Chem. Soc.*, **62**, 654 (1940).
[9] Ger. pat. 284,764 [*Frdl.*, **12**, 53 (1914–1916)].
[10] Ger. pat. 291,185 [*Frdl.*, **12**, 56 (1914–1916)].
[11] Kazarian, *J. Gen. Chem. U.S.S.R.*, **4**, 1347 (1934) [*C. A.*, **29**, 3978 (1935)].
[12] Brit. pat. 627,474 [*C. A.*, **44**, 3514 (1950)].
[13] U. S. pat. 2,488,082 [*C. A.*, **44**, 2544 (1950)].
[14] Cymerman and Wilks, *Nature*, **165**, 236 (1950); *J. Chem. Soc.*, **1950**, 1208.
[15] Newman, Fones, and Booth, *J. Am. Chem. Soc.*, **67**, 1053 (1945).
[16] Campbell, Campbell, and Eby, *J. Am. Chem. Soc.*, **60**, 2882 (1938).

5,5-DIMETHYLHYDANTOIN *

(Hydantoin, 5,5-dimethyl-)

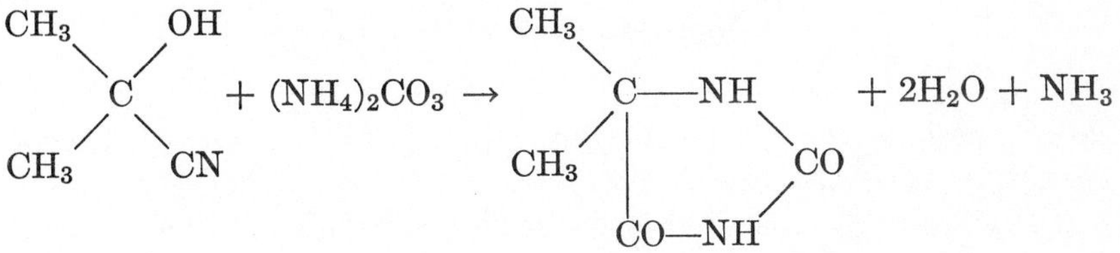

Submitted by E. C. WAGNER and MANUEL BAIZER.
Checked by C. F. H. ALLEN and G. F. FRAME.

1. Procedure

In a 600-ml. beaker are mixed 85 g. (1 mole) of acetone cyanohydrin (Note 1) and 150 g. (1.31 moles) of freshly powdered ammonium carbonate. The mixture is warmed on a steam bath, preferably in a hood (Note 2), and stirred with a thermometer (Note 3). Gentle action begins around 50° and continues during about 3 hours at 68–80°. To complete the reaction and to decompose excess ammonium carbonate, the temperature is finally raised to 90° and maintained at this point until the liquid mixture is quiescent (30 minutes). The residue is colorless or pale yellow (Note 1) and solidifies on cooling. It is dissolved in 100 ml. of hot water, digested with Norit, and filtered rapidly through a heated filter. The filtrate is evaporated on a hot plate until crystals appear at the surface of the liquid, which is then chilled in an ice bath. The white crystals are filtered with suction; the filter cake is pressed and sucked dry and then washed twice with small portions (5–7 ml.) of ether, each portion being well incorporated with the crystals and then drawn through with suction. The mother liquor is concentrated as before to a volume of 25 ml. or less and chilled, and a further crop of crystals is obtained by repetition of the operations outlined (Note 4). The yield is 65–72 g. (51–56%). The first crop is nearly pure and melts at 173°; the second crop melts at about 164°.

The dimethylhydantoin is dissolved in the least boiling water (about 65 ml.) and digested with charcoal, and the hot solution is filtered through a heated filter. The filtrate is chilled, and the separated crystals are filtered with suction and washed sparingly with cold water.

The recovery is about 80–85% of the crude weight. The recrystallized product melts at 174–175° (178° cor.). A further crop of less pure material (m.p. 171–172°) may be obtained by concentration of the mother liquor to small volume (Note 5).

2. Notes

1. Acetone cyanohydrin [*Org. Syntheses* Coll. Vol. **2,** 7 (1943)] is entirely satisfactory. For immediate use, a less pure cyanohydrin will serve; this is readily made as follows.[1]

A solution of 165 g. (pure basis) of sodium bisulfite in 300 ml. of cold water is transferred to a 1-l. flask, which is cooled in an ice bath, while 87 g. of acetone is dropped in slowly with rotation of the flask. A solution of 100 g. (pure basis) of potassium or 75 g. of sodium cyanide in 300 ml. of cold water is then added gradually. The cyanohydrin separates as an upper layer; when sodium cyanide is used, this separation is slower and is not complete until the mixture has come to room temperature. It is drawn off and dried for several hours over sodium sulfate in a stoppered flask kept in the dark. The yield is about 90 g. (70%); it may be somewhat increased by ether extraction of the aqueous liquid.

Acetone cyanohydrin so prepared is colorless, or nearly so, and if used promptly is satisfactory for the preparation of dimethylhydantoin. If it is kept more than a day or two, the cyanohydrin may become deep red and will then impart a red color to the dimethylhydantoin which is difficult to remove.

2. The reaction mixture evolves ammonia slowly in amounts which are unpleasant though tolerable in a well-ventilated room.

3. During most of the reaction, the mixture is partly solid or very viscous and cannot be stirred properly by a mechanically operated stirrer of the usual type. A Hershberg stirrer is unsatisfactory.

4. Dimethylhydantoin is highly soluble in hot water, and its solubility in cold water is considerable. Several crops may be removed by successive concentrations of the mother liquors, taken finally to very small volume. The conversion of acetone cyanohydrin to dimethylhydantoin is said to be practically quantitative.[2]

5. The method described serves for the preparation of various 5-substituted or 5,5-disubstituted hydantoins, using appropriate cyanohydrins. With methylethylketone cyanohydrin there was obtained a 75% yield of 5-methyl-5-ethylhydantoin, m.p. 141.5°.

3. Methods of Preparation

Hydantoins with one or two substituents in the 5 position have been prepared by heating cyanohydrins with urea and treating this reaction mixture with moderately concentrated hydrochloric acid; [3] by heating alanine sulfate with potassium cyanate; [4] by the action of phosgene or oxalyl chloride [5] or carbonic esters [6] on C-substituted aminoacetamides; by the fusion of amino acids with urea; [7] by the action of potassium cyanate on α-aminonitrile hydrochlorides, and heating the resulting ureido-nitriles with dilute hydrochloric acid; [8] by heating aldehydes or ketones with alkali cyanide and ammonium carbonate under pressure of several atmospheres of carbon dioxide; [9] by warming cyanohydrins with ammonium carbonate; [2] by the interaction of ketone or aldehyde and ammonium carbonate with hydrogen cyanide or alkali cyanide, in ligroin or in 50% ethanol, at room temperature or at 50–80°; [10] or by the interaction of ketone or aldehyde bisulfite compounds with cyanide and ammonium carbonate.[10] The procedure described [11] is that of Bucherer and Steiner.[2] In a more recent method, an equimolecular mixture of acetone and hydrogen cyanide is treated with ammonia and carbon dioxide.[12]

1 Houben, *Die Methoden der organischen Chemie,* 3rd Ed., Vol. III, p. 568, Verlag Georg Thieme, Leipzig, 1930.

2 Bucherer and Steiner, *J. prakt. Chem.,* **140,** 291 (1934).

3 Pinner, *Ber.,* **20,** 2355 (1887); **21,** 2320 (1888).

4 Urech, *Ann.,* **165,** 99 (1873).

5 Ger. pat. 310,427 [*Frdl.,* **13,** 803 (1923–1924)].

6 Read, *J. Am. Chem. Soc.,* **44,** 1749 (1922).

7 Griess, *Ber.,* **2,** 47 (1869); Halpern, *Monatsh.,* **17,** 243 (1896).

8 Herbst and Johnson, *J. Am. Chem. Soc.,* **54,** 2465 (1932).

9 Bergs, Ger. pat. 566,094 (1929) [*C. A.,* **27,** 1001 (1933)].

10 Bucherer and Lieb, *J. prakt. Chem.,* **141,** 5 (1934).

11 Wagner and Simons, *J. Chem. Education,* **13,** 266 (1936).

12 U. S. pat. 2,391,799 [*C. A.,* **40,** 1876 (1946)].

2,5-DIMETHYLMANDELIC ACID

(Mandelic acid, 2,5-dimethyl-)

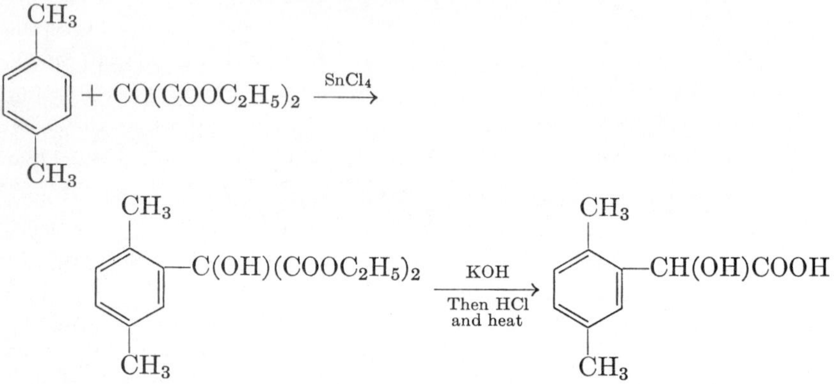

Submitted by J. L. Riebsomer and James Irvine.
Checked by H. R. Snyder and R. L. Rowland.

1. Procedure

A. *Ethyl 2,5-dimethylphenylhydroxymalonate.* One hundred and seventy-four grams (1 mole) of ethyl oxomalonate (Note 1) and 265 g. (2.5 moles) of *p*-xylene are mixed in a 1-l. three-necked flask equipped with a good stirrer, a dropping funnel, and a calcium chloride drying tube. Suitable precautions are taken to exclude moisture from the reaction mixture. While the mixture is cooled in an ice-water bath and stirred vigorously, 325 g. (1.25 moles) of anhydrous stannic chloride is added dropwise from the funnel. After the addition is complete, the cooling bath is removed and stirring is continued for 3 hours. The reaction mixture is poured with stirring into about 300 g. of cracked ice containing 50 ml. of concentrated hydrochloric acid. On standing this mixture separates into two layers. Two hundred milliliters of ether is added, and the ether layer is washed with 100-ml. portions of water until the wash water is free from chlorides. The ether solution is dried with anhydrous sodium sulfate and distilled. After removal of the ether and the excess of reagents (Note 2), the fraction boiling at 154–156°/5 mm. is collected. The yield of 2,5-dimethylphenylhydroxymalonic ester is 144.5–160 g. (51.5–57%).

B. *2,5-Dimethylmandelic acid.* A mixture prepared from 140 g. (0.5 mole) of 2,5-dimethylphenylhydroxymalonic ester and a cold solution of 140 g. of potassium hydroxide in 560 ml. of water in a 1-l. round-bottomed flask is warmed on a steam bath for 5 hours. The alkaline solution is cooled and extracted with one 100-ml. portion of ether to remove any material not soluble in alkali. The alkaline solution is acidified with 300 ml. of concentrated hydrochloric acid and then warmed on a steam bath and stirred for 2 hours (or until there is no further evidence that carbon dioxide is escaping). The mixture is cooled, the oily layer is extracted with ether, the ethereal solution is dried with anhydrous sodium sulfate, and the ether is distilled under partial vacuum (Note 3). The oily residue is crystallized from benzene. The yield of 2,5-dimethylmandelic acid melting at 116.5–117° is 55–63 g. (63–70%) (Note 4).

2. Notes

1. The ethyl oxomalonate must be of good quality. It has been found possible to use nitrogen peroxide (now available in cylinders from the du Pont Company) in the preparation of this reagent. The general conditions are the same as in the previous method in *Organic Syntheses* [1] except that the generating flask containing arsenious oxide and nitric acid is replaced with a cylinder of nitrogen peroxide. This method is much more satisfactory.

2. The excess *p*-xylene can be recovered easily.

3. It is important to avoid long heating after removal of the ether. The mandelic acid tends to react with itself.

4. This general method has been applied to the synthesis of a variety of alkyl-substituted mandelic acids.[2]

3. Methods of Preparation

2,5-Dimethylmandelic acid has been prepared by the procedure described above; [2] and by the reaction between 2,5-dimethyphenylmagnesium bromide and chloral, followed by alkaline hydrolysis.[3]

[1] *Org. Syntheses* Coll. Vol. **1**, 266 (1941).

[2] Ando, *J. Chem. Soc. Japan,* **56,** 745 (1935) [*C. A.,* **29,** 7960 (1935)]; Riebsomer, Irvine, and Andrews, *J. Am. Chem. Soc.,* **60,** 1015 (1938); *Proc. Indiana Acad. Sci.,* **47,** 139 (1938); Riebsomer, Baldwin, Buchanan, and Burkett, *J. Am. Chem. Soc.,* **60,** 2974 (1938); Riebsomer, Stauffer, Glick, and Lambert, *J. Am. Chem. Soc.,* **64,** 2080 (1942).

[3] Savarian, *Compt. rend.,* **146,** 297 (1908) [*C. A.,* **2,** 1443 (1908)].

1,5-DIMETHYL-2-PYRROLIDONE

(2-Pyrrolidone, 1,5-dimethyl-)

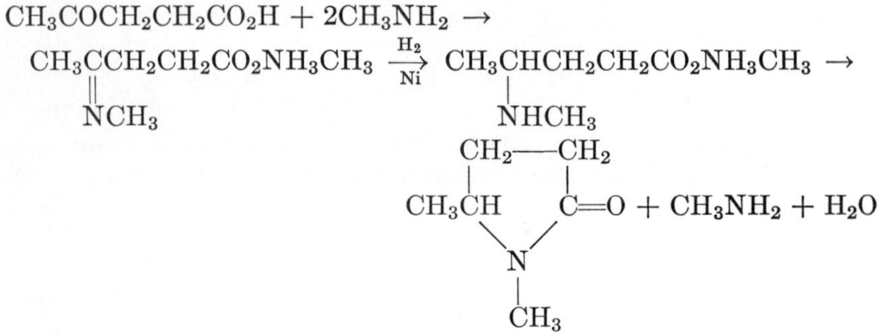

Submitted by Robert L. Frank, William R. Schmitz, and Blossom Zeidman.
Checked by Arthur C. Cope and W. H. Jones.

1. Procedure

One hundred and ninety-four grams (170 ml., 1.67 moles) of levulinic acid (Note 1) and 500 ml. of 35% aqueous methylamine (sp. gr., 0.89; 3.94 moles of methylamine) (Note 2) are placed in a 2-l. steel reaction vessel of a high-pressure hydrogenation apparatus. Ten grams of Raney nickel catalyst [1] is added, the vessel is closed, and hydrogen is admitted to a pressure of 1000–2000 lb. The bomb is then heated with continuous agitation to 140° and maintained at that temperature for 5 hours (Note 3). The contents are removed, and the bomb is washed with two 100-ml. portions of water. After removal of the catalyst by filtration (Note 4), the filtrate is distilled under reduced pressure using a good column. A low-boiling fore-run of water and methylamine distils first, followed by the product, a colorless liquid boiling at 84–86°/13 mm.; 102–104°/27 mm.; n_D^{25} 1.4611 (Note 5). The yield is 140–146 g. (74–77%).

2. Notes

1. A. E. Staley Manufacturing Company's Grade A levulinic acid was used.
2. Commercial 35% aqueous methylamine solution was used.

3. The course of the reaction may be followed by the drop in hydrogen pressure, the decrease depending upon the size of the bomb employed.

4. Because of its pyrophoric nature, the nickel catalyst should not be allowed to dry on the filter. A convenient alternative procedure for removing catalyst is to centrifuge the reaction mixture.

5. The product of some runs has a light yellow tint, but the color appears to cause no complications in subsequent reactions of the material.

3. Methods of Preparation

1,5-Dimethyl-2-pyrrolidone has been prepared by the reaction of 5-methyl-2-pyrrolidone with excess methyl iodide [2] and by the catalytic hydrogenation of a mixture of levulinic acid and methylamine.[3]

[1] *Org. Syntheses* Coll. Vol. **3**, 181 (1954).
[2] Senfter and Tagel, *Ber.*, **27**, 2313 (1894).
[3] Hoffman-LaRoche and Co., Ger. pat. 609,244 [*C. A.*, **29**, 3116 (1935)].

2,4-DIMETHYLQUINOLINE

(Quinoline, 2,4-dimethyl-)

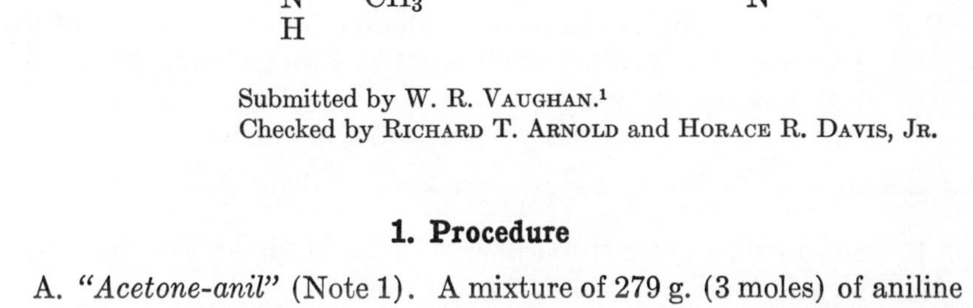

Submitted by W. R. Vaughan.[1]
Checked by Richard T. Arnold and Horace R. Davis, Jr.

1. Procedure

A. *"Acetone-anil"* (Note 1). A mixture of 279 g. (3 moles) of aniline and 9 g. of iodine is placed in a 1-l. three-necked round-bottomed flask. The flask is fitted with a dropping funnel (the delivery tube of which extends below the liquid surface), a mercury-sealed stirrer (Note 2), and an adapter carrying a thermometer which extends into the liquid and a condenser system such that most of the aniline will return and all the water produced (as well as excess acetone) will distil (Note 3). The flask is now heated in an oil bath (Notes 4 and 5), and the reaction mixture is maintained at 170–175° while acetone is passed in through the dropping funnel at such a rate that no more than 2 drops of liquid distils per second. During the addition the mixture is vigorously stirred. A total volume of 850 ml. (670 g., 11.6 moles) of acetone is added over a 4-hour period, and about 610 ml. of distillate is collected. At the end of this period the reaction mixture is cooled and distilled in vacuum, three fractions being collected (up to 136°/15 mm., 136–141°/15 mm., and 141–146°/15 mm.). The acetone-water distillate is now distilled through a simple column at atmospheric pressure until essentially all the acetone has come over. There remains a mixture of oil and water which is separated after being cooled. The oil is then used as the first fraction for a fractional distillation in vacuum of the reaction product, each of the previous fractions being added in succession. Three new fractions are collected: 78–82°/13 mm. (a little aniline); 82–133°/13 mm. (a small intermediate fraction); and 133–138°/13 mm. (acetone-anil). An appreciable quantity of tarry residue remains in the distillation flask. The yield based upon aniline actually consumed in the reaction is 61–68% (Note 5).

B. 2,4-*Dimethylquinoline.* A small quantity of copper powder (Note 6) is added to 4.6 g. (0.2 gram atom) of sodium metal (small pieces) and 56 g. (0.6 mole) of dry aniline (Note 7) contained in a 1-l. round-bottomed flask. The mixture is warmed carefully over a low flame until the evolution of hydrogen ceases, at which time the mixture is usually black. A few boiling chips and 346 g. (2 moles) of acetone-anil are then added. The resulting mixture is heated at the reflux temperature (220–230°) until the evolution of gas (methane) ceases (Note 8). At the end of the reaction the mixture is cooled and then distilled in vacuum. The fraction boiling at 135–140°/12 mm. weighs 252–283 g. (80–90%) (Note 9).

2. Notes

1. The so-called acetone-anil has been assigned the structure indicated in accordance with the work of Craig.[2]

2. A convenient stirrer may be made from tantalum wire as described by Hershberg.[3]

3. A Liebig condenser in reflux position is heated with steam and connected at the top to a water-cooled downward condenser.

4. For a reaction temperature of 170–175° the oil is best maintained at about 200° during the addition of the acetone.

5. The reaction may be carried out at any temperature from 70° to 175°. At lower temperatures it is much slower and is best carried out by refluxing the theoretical quantities of acetone and aniline with iodine catalyst for an extended time. Regardless of the time or reaction temperature, if due allowance is made for aniline reclaimed at the end of the reaction the yield falls between the limits indicated. At the higher temperature, equilibrium is attained more rapidly, as would be expected, and relatively little aniline is recovered. (A typical run may yield 40 g. of aniline and 297 g. of "anil," with 11 g. of middle fraction.)

6. The exact amount of copper powder used as a catalyst in the preparation of the sodium salt of aniline is not important. Approximately 0.2 g. has been found to be satisfactory in runs of the size described.

7. Aniline which has been recently distilled is satisfactory. An excess of aniline over the sodium is used to minimize decomposition of the 2,4-dimethylquinoline.[2]

8. The time required for the decomposition of the acetone-anil varies from 3 to 6 hours. The checkers, however, found that essentially all the methane is eliminated after 2 hours.

9. Fractionation of 245 g. of this product through a 14-plate Lecky-Ewell [4] column yielded 6 g. of aniline, 3 g. of an intermediate fraction, and 218 g. (89%) of 2,4-dimethylquinoline, b.p. 149–150°/20 mm. (controlled). If the material is to be nitrated, the crude product is quite satisfactory.

3. Methods of Preparation

The present procedure for the preparation of acetone-anil is described by Reddelien.[5] A reaction at lower temperature is reported by Craig,[2] who also describes the alkaline decomposition of the anil to 2,4-dimethylquinoline and methane and gives a method of purification of the final product. Other methods for the preparation of 2,4-dimethylquinoline involve the Beyer [6] synthesis from aniline hydrochloride and ethylidene acetone, a modification of this synthesis,[7] a synthesis using 2-chloro-2-pentenone-4,[8] and the Combes synthesis from acetylacetone and aniline.[9]

[1] Work done under contract with the Office of Scientific Research and Development.

[2] Craig, *J. Am. Chem. Soc.*, **60**, 1458 (1938).

[3] Hershberg, *Ind. Eng. Chem., Anal. Ed.,* **8,** 313 (1936).
[4] Lecky and Ewell, *Ind. Eng. Chem., Anal. Ed.,* **12,** 544 (1940).
[5] Reddelien and Thurm, *Ber.,* **65,** 1511 (1932).
[6] Beyer, *J. prakt. Chem.,* [2] **33,** 401, (1886).
[7] Mikeska, Stewart, and Wise, *Ind. Eng. Chem.,* **11,** 456 (1919).
[8] Julia, *Compt. rend.,* **228,** 1807 (1949).
[9] Combes, *Bull. soc. chim. France,* **49,** 89 (1888).

2,4-DIMETHYLTHIAZOLE

(Thiazole, 2,4-dimethyl-)

$$CH_3CONH_2 + P_2S_5 \rightarrow CH_3CSNH_2$$

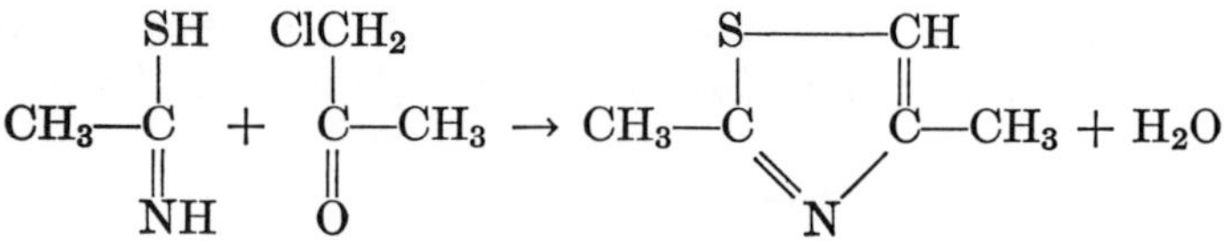

Submitted by George Schwarz.
Checked by C. S. Hamilton and Edward J. Cragoe, Jr.

1. Procedure

In a 2-l. round-bottomed flask provided with a reflux condenser is placed 200 ml. of dry benzene (Note 1). A mixture of 300 g. (5.08 moles) of finely divided acetamide and 200 g. (0.9 mole) of powdered phosphorus pentasulfide is prepared quickly and transferred immediately to the flask. To this is added 20 ml. of a mixture of 400 ml. (4.97 moles) of chloroacetone (Note 2) and 150 ml. of dry benzene. The exothermic reaction is started by careful heating in a water bath. The water bath is removed, and the remainder of the chloroacetone-benzene mixture is introduced gradually through the reflux condenser (Notes 3 and 4). When all the chloroacetone has been added and reaction is no longer apparent, the mixture is refluxed on the water bath for 30 minutes.

About 750 ml. of water is added to the mixture with shaking. After 30 minutes the mixture is poured into a separatory funnel, and the reddish upper layer containing the benzene with some impurities is discarded. The lower layer is made alkaline (Note 5) by the addition of 5 *N* sodium hydroxide or potassium hydroxide, and the crude thiazole, which separates as a black upper layer, is removed with ether, and the aqueous lower layer is extracted with five 120-ml. portions of ether. The combined ethereal extracts are dried over anhydrous sodium sulfate and filtered through glass wool. The ether is removed by distillation from a steam bath (Note 6), and the residual oil is fractionated at atmospheric pressure; the fraction boiling at 140–150° is collected and redistilled. The yield of 2,4-dimethylthiazole boiling at 143–145° (Note 7) is 210–230 g. (41–45% based on the phosphorus pentasulfide).

2. Notes

1. Commercial benzene is satisfactory after being dried over calcium chloride and distilled.

2. Commercial chloroacetone is distilled, and the fraction boiling at 116–122° is used.

3. The progress of the reaction is controlled by the portions of chloroacetone added. If too much is added at once the reaction may become too vigorous. Portions of 20 ml. are safe. Toward the end of the reaction, larger portions may be added.

4. As soon as the reaction has begun, the mixture becomes a gray-black oily liquid.

5. The alkalinity of the deeply colored solution can be tested with phenolphthalein paper. The edges of the wetted paper become visibly red. One should be sure to test the aqueous solution and not the separated thiazole floating on top of the liquid.

6. The thiazole is hygroscopic and should be protected from moisture.

7. Reported boiling points are 144–145.5°/719 mm.[1] and 143°.[2] No 2,4-dimethyloxazole (b.p. 108°) was obtained during the distillation.

3. Methods of Preparation

2,4-Dimethylthiazole has been prepared from chloroacetone and thioacetamide,[1] but forming the required thioacetamide in the reaction mixture is to be preferred since no additional manipulation is involved. The method here described is substantially that of E. Merck.[2] Other substituted thiazoles can be prepared by practically the same method.[2] 2,4-Dimethylthiazole has been obtained by dry distillation of 2-methylthiazyl-4-acetic acid,[3] and also by heating 2,4-dimethylthiazole-5-carboxylic acid with calcium oxide.[4]

[1] Hantzsch, *Ann.*, **250**, 265 (1889).

[2] E. Merck, Ger. pat. 670,131 [*C. A.*, **33**, 2909 (1939)].

[3] Steude, *Ann.*, **261**, 41 (1891).

[4] Rubleff, *Ann.*, **259**, 266 (1890).

2,5-DINITROBENZOIC ACID *

(Benzoic acid, 2,5-dinitro-)

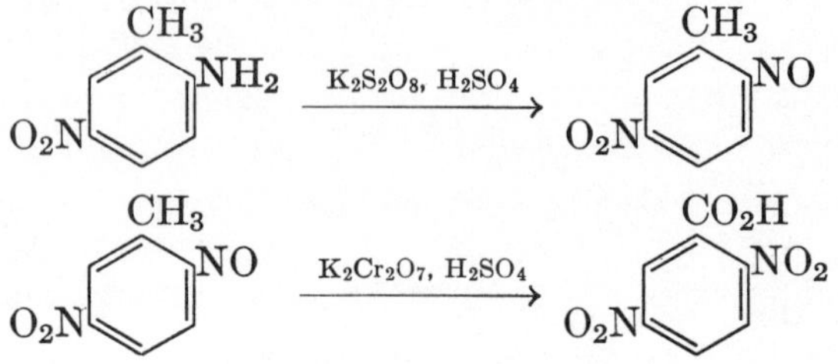

Submitted by WILSON D. LANGLEY.
Checked by W. E. BACHMANN and D. W. HOLMES.

1. Procedure

A. 2-*Nitroso-5-nitrotoluene.* In a 5-l. round-bottomed flask fitted with a mechanical stirrer is placed 50 g. (0.33 mole) of pulverized 5-nitro-2-aminotoluene (Note 1). To this is added an ice-cold solution of 200 ml. of concentrated sulfuric acid in 50 ml. of water. While the suspension is stirred at room temperature, a solution of Caro's acid is prepared, as follows: to 175 ml. of ice-cold sulfuric acid (sp. gr. 1.84) in a 2-l. beaker is added 300 g. (1.11 moles) of pulverized potassium persulfate. The mixture is thoroughly stirred with a glass rod, and to it is added 900 g. of crushed ice and 300 ml. of water.

The well-stirred solution of Caro's acid is poured into the suspension of nitroaminotoluene. The mixture is stirred and warmed, but as soon as the temperature reaches 40°, heating is discontinued. After the solution has been stirred for 2 hours longer, an additional 100 g. (0.37 mole) of powdered potassium persulfate is added in one portion. The heat of reaction is sufficient to maintain the temperature at 40°. Stirring is continued for 2 hours more, and the suspension is then diluted with water to 4 l. The solid is filtered with suction (Note 2) and washed with 400 ml. of water. The wet material is transferred to a 5-l. round-bottomed flask, about 700 ml. of water is added, and the mixture is steam-distilled. The 2-nitroso-5-nitrotoluene is filtered from the distillate (Note 3). The total yield of air-dried product is 30–39 g. (55–71%) (Note 4).

B. 2,5-*Dinitrobenzoic acid.* A suspension of 20 g. (0.12 mole) of the air-dried 2-nitroso-5-nitrotoluene in 100 ml. of water is prepared in a 500-ml. Erlenmeyer flask, and to it is added 50 g. (0.17 mole) of powdered potassium dichromate (Note 5). The flask is placed in an ice-salt bath, and the mixture is stirred vigorously by means of an efficient stirrer. When the temperature has dropped to 5°, 175 ml. of concentrated sulfuric acid is added in a thin stream (Note 6) while the temperature is not allowed to exceed 35°. After all the sulfuric acid has been added the mixture is stirred and heated to 50°. The source of heat is removed, and the temperature is maintained between 50° and 55° by cooling in an ice bath as the exothermic reaction takes place (Note 7). After 20 minutes the temperature is raised to 65° ± 3° and held there for 1 hour longer.

The solution is cooled to 20°, and 250 g. of ice is added. The mixture is stirred for a few minutes, then is filtered with suction (hardened filter paper), and the solid is washed with 35 ml. of ice water. The solid is suspended in 25 ml. of water in a 600-ml. beaker and is slowly dissolved by gradual addition of 55–65 ml. of a 10% solution of sodium carbonate. The solution is filtered (Note 8), and the filtrate is made strongly acidic to Congo red by addition of 1:1 hydrochloric acid. The mixture is chilled in ice for an hour and is then filtered. The product is washed with 12 ml. of ice water and air-dried. The acid melts at 174–176° and weighs 14–17 g. (55–66%). The acid may be recrystallized by dissolving 10 g. of it in 250 ml. of boiling 5% hydrochloric acid (Note 9). The solution, when chilled in ice, deposits 9.4 g. of 2,5-dinitrobenzoic acid as nearly colorless crystals which melt at 177–178° (Note 10).

2. Notes

1. Eastman Kodak Company's practical grade was used.

2. The filtrates, on standing overnight, deposit 5–5.6 g. of crude 2-nitroso-5-nitrotoluene, which is not included in the weights recorded. It may be added to the product obtained by steam distillation.

3. The submitters reported that successive 2.5-l. portions of distillate were filtered, and the following weights of product were obtained: 14.5, 10.7, 9.0, and 2 g., or 36.2 g. in 10 l. From one run the checkers obtained 19, 12, 6, and 2 g., or 39 g. In some runs it was necessary to collect 15–30 l. of distillate in order to obtain a 30-g. yield. The submitters report that unchanged amine, if present during the steam distillation, condenses with the nitroso compound, giving tarry products, and the yield of nitroso compound is reduced.

4. In favorable cases, the product is white and melts at 135–136°.

5. The procedures given in *Org. Syntheses* Coll. Vol. **1,** 385, 528 (1932), 392, 543 (1941), are not suitable for the oxidation of nitro-

sonitrotoluenes. Using the procedure given on p. 385 (1932) and p. 392 (1941) the yield of dinitro acid is low, and using the procedure given on p. 528 (1932) and p. 543 (1941) the dinitro acid is deeply colored.

6. This requires 20–25 minutes.

7. In a run twice this size the checkers observed that at this stage some of the semi-solid material on the top of the stirred solution burned. The burning ceased after a few minutes. The yield of final product was slightly lower. It must be emphasized that careful control of the temperature is necessary during the oxidation with dichromate in order to prevent the reaction from becoming too vigorous.

8. The undissolved residue is largely 2,5-dinitrotoluene, which may be used in subsequent oxidations. Larger yields of this compound may be obtained by shortening the time of stirring during the oxidation process.

9. Additional data on the recrystallization of 2,5-dinitrobenzoic acid are: a solution of 6.83 g. in 200 ml. of boiling toluene, when filtered and chilled, furnished 5.9 g.; a solution of 10 g. in 110 ml. of boiling water, when filtered and chilled, furnished 9.18 g.

10. When 3-nitro-4-aminotoluene was used, 3,4-dinitrobenzoic acid was obtained in approximately the same yields as those recorded for 2,5-dinitrobenzoic acid. The submitter reports: "3-nitro-4-aminotoluene dissolves readily in the sulfuric acid when the Caro's acid is added, and the 3-nitro-4-nitrosotoluene which separates is practically free of the amine. Therefore, the steam distillation may be omitted and the nitroso compound may be oxidized directly to the dinitro acid, which is obtained as a light-colored product. From 100 g. of 3-nitro-4-aminotoluene there was obtained 102 g. of yellow nitrosonitrotoluene which, when oxidized, gave 85 g. of 3,4-dinitrobenzoic acid melting at 158–162°. One crystallization gave the pure acid. A solution of 10 g. of the acid in 210 ml. of boiling water gave, after cooling, 9.1 g. of the acid. Similarly, a solution of 10 g. of the acid in 280 ml. of hot 5% hydrochloric acid was filtered quickly and then cooled. The filtrate deposited 9.4 g. of the acid."

3. Methods of Preparation

2,5-Dinitrobenzoic acid has been prepared by nitration of *o*-nitrobenzoic acid and fractional crystallization of the acids as the barium salts;[1] and by oxidation of 2,5-dinitrotoluene.[2]

[1] Griess, *Ber.*, **7**, 1223 (1874).

[2] Grell, *Ber.*, **28**, 2564 (1895).

3,5-DINITROBENZOIC ACID *

(Benzoic acid, 3,5-dinitro-)

$$C_6H_5CO_2H + 2HNO_3 \xrightarrow{(H_2SO_4)} (NO_2)_2C_6H_3CO_2H + 2H_2O$$

Submitted by (A) R. Q. BREWSTER and BILL WILLIAMS.
(B) ROSS PHILLIPS.
Checked by HOMER ADKINS and JAMES RUHOFF.

1. Procedure

A. In a 2-l. round-bottomed flask are placed 61 g. (0.5 mole) of benzoic acid and 300 ml. of concentrated sulfuric acid (sp. gr. 1.84). To this mixture is added 100 ml. of fuming nitric acid (sp. gr. 1.54) in portions of 2 or 3 ml. The temperature during the addition of the acid is kept between 70° and 90° by means of external cooling with cold water (Note 1), and the addition should be carried out in a good hood. The flask is covered with a watch glass and allowed to stand for an hour, or overnight, in the hood. The flask is then heated on a steam bath in the hood for 4 hours, during which time considerable amounts of brown fumes are evolved. The reaction mixture is then allowed to cool to room temperature (Note 2), whereupon yellow crystals separate from the solution. An additional 75 ml. of fuming nitric acid is added, and the mixture is heated on the steam bath for 3 hours and then in an oil bath at 135–145° for 3 hours (hood). Brown fumes are evolved continuously, especially during the heating in the oil bath. The color of the reaction mixture is light to reddish yellow.

The mixture is allowed to cool and is poured into 800 g. of ice and 800 ml. of water. After standing for 30 minutes the 3,5-dinitrobenzoic acid is filtered with suction and washed with water until free of sulfates. The crude product weighs 62–65 g. and melts at 200–202°. This product is recrystallized from 275 ml. of hot 50% ethanol. The purified 3,5-dinitrobenzoic acid weighs 57–61 g. (54–58%) and melts at 205–207° (Note 3).

B. Two hundred grams of benzoic acid (1.6 moles) is stirred into 1 l. of concentrated sulfuric acid contained in a 3-l. round-bottomed flask. The flask is surrounded by cold water, and the temperature is maintained below 45° while 80 ml. of fuming nitric acid is added. When the temperature of the reaction mixture has fallen to 30°, 240 ml. of fuming nitric acid is added, a beaker is inverted over the neck of the flask, and the mixture is allowed to stand for 6 weeks.

The flask is heated on a steam bath for 4 hours and then in an oil bath at 145° until all the crystals are dissolved. A mush of crystals separates when the mixture is cooled to room temperature. These are filtered on a Büchner funnel without any filtering medium. After the crystals are pressed dry, they are placed in 1.5 l. of cold water, washed thoroughly, filtered through paper, and dried in the air. The yield is 200–213 g. (60%), and the product melts at 205–207° (cor.) (Notes 4 and 5).

2. Notes

1. The temperature is so controlled that evolution of brown fumes, in other than small quantities, is avoided.
2. Mechanical stirring is advantageous but not necessary.
3. The melting points were taken with a short-stem total-immersion thermometer.
4. Procedure B has been carried out on ten times the scale specified, with yields a few per cent higher than those given above.
5. A purification procedure which gives material suitable for use in the determination of creatinine has been reported.[1]

3. Methods of Preparation

3,5-Dinitrobenzoic acid has been prepared by nitration of benzoic acid with sulfuric acid and fuming nitric acid;[2] by nitration of 3-nitrobenzoic acid;[3] and by oxidation of 3,5-dinitrotoluene.[2,4,5] It has been obtained as a product of the action of nitric acid on 1,5-dinitronaphthalene.[6]

[1] Jansen, Sombrock, and Noyons, *Chem. Weekblad,* **43,** 731 (1947).

[2] Cahours, *Jahresber. Fortsch. Chem.,* **1847–1848,** 533; Voit, *Ann.,* **99,** 104 (1856); Michler, *Ann.,* **175,** 152 (1875); Hübner, *Ann.,* **222,** 72 (1884); Saunders, Stacey, and Wilding, *Biochem. J.,* **36,** 368 (1942)

[3] Tiemann and Judson, *Ber.,* **3,** 224 (1870).

[4] Staedel, *Ann.,* **217,** 194 (1883).

[5] Goldstein and Voegeli, *Helv. Chim. Acta,* **26,** 475 (1943)

[6] Beilstein and Kuhlberg, *Ann.,* **202,** 220 (1880)

2,2′-DINITROBIPHENYL

(Biphenyl, 2,2′-dinitro-)

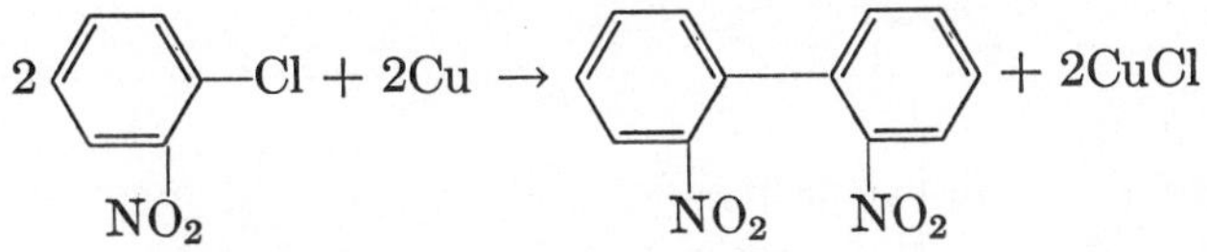

Submitted by REYNOLD C. FUSON and E. A. CLEVELAND.
Checked by W. E. BACHMANN, A. L. WILDS, and J. KORMAN.

1. Procedure

In a 1-l. flask equipped with a mechanical stirrer are placed 200 g. (1.27 moles) of *o*-chloronitrobenzene and 300 g. of clean dry sand. The mixture is heated in an oil bath to 215–225°, and then 200 g. of copper bronze (Note 1) is slowly added, the addition requiring about 1.2 hours (Note 2). The temperature is kept at 215–225° for 1.5 hours longer, the stirring being continued throughout. The mixture is poured while hot into a beaker containing 300–500 g. of sand and is then stirred until small clumps are formed; upon cooling, these are broken up in a mortar (Note 3). The mixture is boiled 10 minutes with two 1.5-l. portions of ethanol and removed by filtration each time. The filtrates are cooled in an ice bath, and the 2,2′-dinitrobiphenyl is collected on a filter (Note 4). A second crop is obtained by concentrating the filtrate. The product is dissolved in hot ethanol (Note 5), and the solution is treated with Norit, filtered, and cooled in an ice bath. The solid is recrystallized from hot ethanol and is obtained as pure, yellow crystals melting at 123.5–124.5° (cor). The yield is 80–95 g. (52–61%).

2. Notes

1. Ordinary copper bronze does not always give satisfactory results in the Ullmann reaction. More uniform results are obtained if the copper bronze is prepared as suggested by Kleiderer and Adams.[1] The copper bronze is treated with 2 l. of a 2% solution of iodine in acetone for 5–10 minutes. The product is then collected on a Büchner funnel, removed, washed by stirring into a slurry with 1 l. of a 1:1 solution of concentrated hydrochloric acid in acetone, and again filtered. The copper iodide dissolves, and the copper bronze remaining is separated

by filtration and washed with acetone. It is then dried in a vacuum desiccator. It should be used immediately.

2. The temperature of the mixture must not be allowed to rise much above 240° or reduction of the nitro groups will occur and carbazole will be formed.

3. The reaction mixture should not be allowed to cool in the flask, as it will set to a hard mass; it is almost impossible to remove this from the flask.

4. If the first two extractions with ethanol do not yield about 90 g. of crude product, a third extraction of the sand and residue is well worth while (the yield of impure product was never less than 90 g. when this procedure was carried out).

5. If the impure material is recrystallized from the minimum amount of ethanol, the suction funnel will become rapidly plugged during filtration with consequent loss of time and material. Two liters of alcohol per 100 g. of 2,2′-dinitrobiphenyl is preferable. All filtrates were reduced to a small volume, and the crude material obtained was recrystallized twice, using Norit. (This amounted to about 10% of the total yield.)

3. Methods of Preparation

2,2′-Dinitrobiphenyl has been prepared by the action of copper on *o*-chloronitrobenzene, on *o*-bromonitrobenzene,[2,3] and on diazotized *o*-nitroaniline.[4]

[1] Kleiderer and Adams, *J. Am. Chem. Soc.*, **55**, 4225 (1933).

[2] Ullman and Bielecki, *Ber.*, **34**, 2174 (1901).

[3] Mascarelli and Gatti, *Gazz. chim. ital.*, **67**, 807 (1937)

[4] Niementowski, *Ber.*, **34**, 3325 (1901).

1,4-DINITRONAPHTHALENE

(Naphthalene, 1,4-dinitro-)

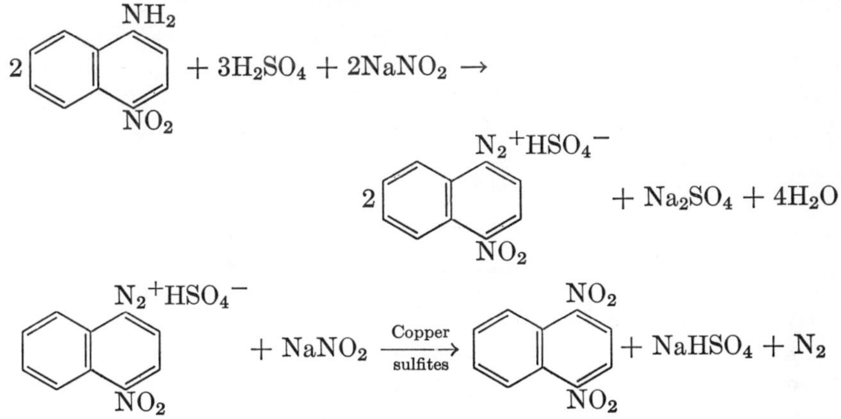

Submitted by H. H. HODGSON, A. P. MAHADEVAN, and E. R. WARD.
Checked by C. C. PRICE and SING-TUH VOONG.

1. Procedure

Ten grams (0.14 mole) of powdered sodium nitrite is dissolved in 50 ml. of concentrated sulfuric acid (sp. gr. 1.84) contained in a 1-l. beaker placed in an ice bath. A solution of 10 g. of 4-nitro-1-naphthylamine (0.053 mole) (p. 664) in 100 ml. of glacial acetic acid is prepared by heating, and the well-stirred solution is cooled below 20°. Some crystals separate. The resulting thin slurry is dropped slowly into the cold solution of nitrosylsulfuric acid with mechanical stirring. Throughout the addition, and for 30 minutes thereafter, the temperature is kept below 20°. Seven hundred milliliters of dry ether is added slowly with stirring, and the temperature of the mixture is kept at 0° for 1 hour. At the end of this period, the precipitation (aided by scratching) of the crystalline 4-nitronaphthalene-1-diazonium sulfate is complete (Note 1). This precipitate is collected, washed with ether and then with 95% ethanol until all the acid is removed, and finally dissolved in 100 ml. of iced water.

A saturated aqueous solution containing 50 g. of crystalline copper sulfate is treated with a similar solution of 50 g. of crystallized sodium sulfite. The greenish brown precipitate is collected, washed with water

(Note 2), and then stirred into a solution of 100 g. (1.45 moles) of sodium nitrite in 400 ml. of water contained in a 2-l. beaker provided with an efficient mechanical stirrer.

The cold aqueous solution of the diazonium salt is then added slowly to the decomposition mixture. Considerable frothing occurs, and 4–5 ml. of ether is added from time to time to break the foam. After stirring for 1 hour (Note 3), the mixture is filtered and the crude dark brown precipitate of the 1,4-dinitronaphthalene is washed several times with water, then with 2% aqueous sodium hydroxide, and again with water. The precipitate is dried and extracted three times with boiling 95% ethanol (450 ml. in all). The extract is concentrated to 75 ml.; most of the 1,4-dinitronaphthalene separates and is collected. Additional amounts can be obtained by further concentration. The resulting product melts at 130–132° and weighs 6.0–7.0 g. (52–60%). The product can be purified either by steam distillation (Note 4) or by recrystallization from aqueous ethanol. Pale yellow needles melting at 134° are obtained (Notes 5 and 6).

2. Notes

1. This precipitate is sometimes sticky. It can be made granular by treating it with a small amount of 95% ethanol (after the removal of the supernatant liquid). Alternatively, it may suffice to keep the ethereal diazotized solution cold and scratch the sides of the beaker with a glass rod.

2. The cupro-cupri sulfite of this variety is more efficient as a decomposition reagent than the red-violet precipitate obtained by treating a hot solution of copper sulfate with a solution of ammonium sulfite saturated with sulfur dioxide and subsequently heating the mixture for 10 minutes at 90°.

3. The decomposition appears to be immediate, and at the end of 1 hour most of the inorganic material has passed into solution.

4. Steam distillation of 1,4-dinitronaphthalene is very slow. However, the cupro-cupri sulfite method [1] is a general one for the replacement of the diazonium group by the nitro group, and steam distillation is preferable whenever the product is readily volatile.

5. The solution is decolorized with charcoal in the course of the recrystallization. The checkers obtained 5 g. (43%) of golden needles after three recrystallizations.

6. 1,2-Dinitronaphthalene may be obtained similarly from 2-nitro-1-naphthylamine, or less satisfactorily from 1-nitro-2-naphthylamine; 1,6- and 2,6-dinitronaphthalenes can be prepared from the 5-nitro- and

6-nitro-2-naphthylamines, respectively, by a modification of the process. Since the solubility of these amines in glacial acetic acid is slight, it is preferable to prepare the diazonium sulfate as follows: Ten grams of the amine is dissolved in 50 ml. of sulfuric acid (sp. gr. 1.84), and the solution is mixed with one of 10 g. of sodium nitrite in 50 ml. of sulfuric acid (sp. gr. 1.84). This mixture is stirred into 200 ml. of glacial acetic acid. The temperature is maintained below 20° throughout these operations. After 30 minutes, the diazonium sulfate is precipitated at 0° by the addition of 200–500 ml. of ether as previously described.

3. Methods of Preparation

1,4-Dinitronaphthalene has been prepared previously from diazotized 4-nitro-1-naphthylamine by a modified Sandmeyer procedure,[2,3] from 5,8-dinitrotetralin by dehydrogenation,[4] by the deamination of 1,4-dinitro-2-naphthylamine,[5] and by the decomposition of 4-nitro-1-naphthalenediazonium cobaltinitrite.[6] The method described above has been published.[1]

[1] Hodgson, Mahadevan, and Ward, *J. Chem. Soc.*, **1947**, 1392.

[2] Vesely and Dvorak, *Bull. soc. chim. France,* **33**, 319 (1923).

[3] Contardi and Mor, *Rend. ist. lombardo sci.*, **57**, 646 (1924) [*C. A.*, **19**, 827 (1925)].

[4] Chudozilov, *Collection Czechoslov. Chem. Commun.*, **1**, 302 (1929) [*C. A.*, **23**, 4212 (1929)].

[5] Hodgson and Hathway, *J. Chem. Soc.*, **1945**, 453.

[6] Hodgson and Ward, *J. Chem. Soc.*, **1947**, 127.

α,α-DIPHENYLACETONE

(2-Propanone, 1,1-diphenyl-)

$$C_6H_5CH_2COCH_3 + Br_2 \longrightarrow C_6H_5CH(Br)COCH_3 + HBr$$

$$C_6H_5CH(Br)COCH_3 + C_6H_6 \xrightarrow{AlCl_3} (C_6H_5)_2CHCOCH_3 + HBr$$

Submitted by EVERETT M. SCHULTZ and SALLY MICKEY.
Checked by EDWARD T. CLINE and R. S. SCHREIBER.

1. Procedure

A. *α-Bromo-α-phenylacetone.* A 1-l. three-necked flask is equipped with a sealed stirrer (Note 1), a dropping funnel, and a water-cooled reflux condenser provided with a short inverted U-tube on the open end; the equipment is assembled in the hood. In the flask are placed 200 ml. of dry, thiophene-free benzene (Note 2) and 37 g. (36.89 ml., 0.276 mole) of phenylacetone (Note 3). The stirrer is started, and 45 g. (14.42 ml., 0.28 mole) of reagent bromine is added dropwise during a period of 1 hour. The reaction mixture first becomes cloudy but changes to a clear orange-red solution by the time all the bromine is added. After the addition of bromine is complete, a rapid stream of dry nitrogen or carbon dioxide is passed through the solution by means of an inlet tube which now replaces the dropping funnel. The hydrogen bromide issuing from the inverted U-tube at the top of the condenser may be trapped in water in an Erlenmeyer flask or other container adjusted so that the water surface is about 1 cm. below the outlet of the tube. The reaction is complete when no more hydrogen bromide fumes can be detected in the gas issuing from the condenser. This operation requires 3–6 hours. During this period the reaction mixture becomes yellow to green. The benzene solution of α-bromo-α-phenylacetone is then transferred to a dry 500-ml. separatory funnel (Note 4).

B. *α,α-Diphenylacetone* (Note 5). The reaction flask used in A is again equipped as it was originally, and in it are placed 75 g. (0.56 mole) of anhydrous aluminum chloride (Note 6) and 150 ml. of dry benzene (Note 2). The stirrer is started, and the flask is heated by means of a steam bath so that the benzene boils gently. The benzene solution of α-bromo-α-phenylacetone (Part A) is added dropwise from the separatory funnel to the boiling mixture over a period of 1 hour. After the addition is complete, the almost black reaction mixture is heated to boiling for an additional hour, then cooled and poured with stirring onto 500 g. of crushed ice and 100 ml. of concentrated hydrochloric acid contained in a 2-l. beaker. The deeply colored benzene solution gradually becomes orange-yellow. When the ice has melted, the benzene layer is separated and the aqueous layer is extracted with three 50-ml. portions of ether. The combined ether and benzene solution is washed with 100 ml. of water and then with 100 ml. of saturated sodium bicarbonate solution. After the solution has been dried for at least 1 hour over 60 g. of anhydrous sodium sulfate, the solvents are evaporated on a steam bath until the solution no longer boils. The dark residue is transferred to a 250-ml. Claisen flask equipped for reduced-pressure distillation and having a receiver fused to the side arm (Note 7). The residual benzene is first distilled at 19–20 mm. while the flask is being heated on a steam bath. The steam bath is replaced by an oil or metal bath which is gradually heated to 190°. A small fore-run of yellow oil which darkens rapidly is collected. It is removed from the receiver by siponing and rinsing with acetone. The bath is allowed to cool to 100–120°, the system is evacuated with an oil pump (Note 8), and the diphenylacetone is distilled. The bulk of the material boils at 142–148°/2–3 mm. The distillate may solidify in the receiver (Note 9). The crude product is triturated with just sufficient petroleum ether (b.p. 35–60°) to moisten it, cooled in an ice-water bath, collected on a filter, and washed on the filter with small amounts of petroleum ether at 0–5° until nearly colorless. The amount of petroleum ether used up to this point should not exceed 50 ml. The product is then crystallized from petroleum ether using 8 ml. of solvent per gram of crude, dry solid. The hot solution is allowed to stand at room temperature until crystallization begins, and then at 0–5° for 16 hours (Note 10). The crystals are collected by filtration and dried in air at room temperature. The yield of nearly colorless product melting at 60–61° (Note 11) is 31–33 g. (53–57%).

2. Notes

1. A rubber tubing seal is satisfactory. The submitters used an all-glass seal made from a 5-ml. glass syringe.

2. The benzene was reagent grade and was dried by distilling until no more water collected in the condenser. The undistilled portion was used directly.

3. Phenylacetone was obtained from the Swope Oil Company, Philadelphia, Pennsylvania. On attempted distillation of this material as received, the checkers observed considerable water in the fore-run. The distillation was interrupted, the material was dried over anhydrous sodium sulfate and redistilled at atmospheric pressure, but small amounts of water still appeared in the condensate, indicating possible decomposition. The fraction boiling at 214–214.8° (uncor.)/756 mm. was used.

4. The transfer may be carried out easily by admitting dry nitrogen or air under pressure continuously to the flask and allowing the benzene solution to be blown out of the flask through the glass tube previously used for introducing nitrogen or carbon dioxide.

5. Once Part B is started, it should be continued without interruption until the Friedel-Crafts reaction product is decomposed.

6. Baker and Adamson reagent grade powdered aluminum chloride was used.

7. Diphenylacetone may solidify in the side arm, in which event it may become necessary to melt it by heating with a small flame.

8. It is necessary to protect the pump by means of a sodium hydroxide tower since hydrogen halide may be evolved during the distillation. The checkers, in addition, used a Dry Ice trap.

9. The checkers transferred the distillate to another container before solidification occurred.

10. The checkers found that the small additional amount (approximately 5 g.) of product obtained by cooling to −25° is more highly colored and considerably less pure than that obtained at 0–5°.

11. A labile form of α,α-diphenylacetone, melting at 46°, is described in the literature.[1,2]

3. Methods of Preparation

α,α-Diphenylacetone has been prepared by the oxidation of 1,1-diphenyl-2-propanol;[1] by the action of dilute mineral acid on 1,1-diphenyl-1,2-propanediol,[1,2] 1,1-diphenyl-1-hydroxy-2-ethoxypropane,[3] 1,1-diphenyl-1-hydroxy-2-aminopropane,[4] 1,1-diphenyl-2-ethoxy-1-propene,[5] and 2,2-diphenyl-3-methylethylenimine;[6] by the distillation at atmospheric pressure of 1,1-diphenyl-2-methylethylene oxide;[7] by the reaction of benzene with α-phenyl-α-chloro- (or bromo-) acetone;[8,9] by acetylation of potassium *t*-butyl diphenylacetate, followed by hydrolysis[10] and by the rearrangement of α-methyldesoxybenzoin in the presence of zinc chloride.[11] The present method is adapted from the procedure of Ruggli, Dahn, and Wegmann.[8]

[1] Tiffeneau and Dorlencourt, *Compt. rend.*, **143**, 127 (1906).

[2] Stoermer, *Ber.*, **39**, 2302 (1906); Tiffeneau and Levy, *Bull. soc. chim. France*, **33**, 776 (1923); Henze and Leslie, *J. Org. Chem.*, **15**, 901 (1950).

[3] Stoermer, *Ber.*, **39**, 2302 (1906).

[4] Thomas and Bettziechi, *Z. physiol. Chem.*, **140**, 265 (1924).

[5] Bardan, *Bull. soc. chim. France*, **49**, 1875 (1931).

[6] Campbell, McKenna, and Chaput, *J. Org. Chem.*, **8**, 107 (1943).

[7] Levy and Lagrave, *Compt. rend.*, **180**, 1032 (1925); *Ann. chim.*, (10) **8**, 365 (1927).

[8] Richard, *Compt. rend.*, **200**, 753 (1935); Ruggli, Dahn, and Wegmann, *Helv. Chim. Acta*, **29**, 113 (1946).

[9] Jilek and Protiva, *Chem. Listy*, **44**, 49 (1950).

[10] Yost and Hauser, *J. Am. Chem. Soc.*, **69**, 2325 (1947).

[11] Zalesskaya, *Zhur. Obscheĭ Khim.*, **18**, 1168 (1948) [*C. A.*, **43**, 1754 (1949)].

DIPHENYLACETONITRILE *

(Acetonitrile, diphenyl-)

$$C_6H_5CH_2CN + Br_2 \longrightarrow C_6H_5CH(Br)CN + HBr$$

$$C_6H_5CHBrCN + C_6H_6 \xrightarrow{AlCl_3} (C_6H_5)_2CHCN + HBr$$

Submitted by C. M. Robb and E. M. Schultz.
Checked by Richard T. Arnold and J. W. Britain.

1. Procedure

A. *α-Bromo-α-phenylacetonitrile.* In a well-ventilated hood (Note 1), a dry 500-ml. round-bottomed three-necked flask is equipped with a sealed stirrer, an air condenser (Note 2), and a cork carrying a dropping funnel and a thermometer. In the flask is placed 117 g. (1 mole) of benzyl cyanide (Note 3). The thermometer is adjusted so that the bulb is immersed in the benzyl cyanide. The flask is placed in a bath and heated to 105–110° (inside temperature). With good stirring, 176 g. (1.1 moles) of bromine (Note 4) is added over a period of 1 hour; during the addition and for 15 minutes thereafter the temperature of the liquid is maintained at 105–110°. At the end of the heating period the evolution of hydrogen bromide gas has practically ceased. The dropping funnel is replaced by a cork carrying a glass tube reaching to within 2–3 cm. of the surface of the reaction mixture. Dry nitrogen is then led through the apparatus for 30 minutes. The hot mixture is poured into the 500-ml. addition funnel fitted to the apparatus to be used in the next step (see part B). The reaction flask is rinsed with 100 g. (1.3 moles) of dry benzene (Note 5), and this is added to the bromonitrile. The benzene solution thus obtained is used immediately in the next step (Note 6).

B. *Diphenylacetonitrile.* A dry 2-l. round-bottomed three-necked flask, equipped with a sealed stirrer, a 500-ml. dropping funnel, and a dry reflux condenser (Note 2), is mounted on a steam bath. In the flask are placed 368 g. (4.7 moles) of dry benzene (Note 5) and 133.5 g. (1 mole) of powdered anhydrous aluminum chloride. The stirrer is started, and the benzene is heated to vigorous refluxing. The α-bromo-α-phenylacetonitrile solution is added to the boiling mixture over a period of 2 hours in small portions (Note 7). After the addition is complete, the reaction mixture is refluxed for an additional hour. The flask is cooled, and the mixture is poured into a stirred mixture of 1 kg. of crushed ice and 100 ml. of concentrated hydrochloric acid in a 3-l. beaker.

The benzene layer is separated. The aqueous layer is extracted with 500 ml. of ether in two equal portions. The ether and benzene solutions are combined and washed successively with 500 ml. of water, 250 ml. of saturated sodium bicarbonate solution, and 500 ml. of water. The organic layer is dried over 100 g. of anhydrous sodium sulfate. The drying agent is separated from the solution, and the solvents are removed by heating on a steam bath. The last traces of benzene are removed by vacuum distillation from a 250-ml. Claisen flask heated on a steam bath. The residue weighs about 190 g. A receiver is connected directly to the side arm (Note 8) of the Claisen flask, and the product is distilled under reduced pressure. It boils at 122–125°/1–2 mm. and crystallizes to a yellow solid which melts at 68–70°. The solid is recrystallized from isopropyl alochol (1 ml. per g.); the flask containing the filtered hot solution is placed on an asbestos mat, and the solution is allowed to cool only to room temperature without shaking or stirring. The crystals are then collected and washed on the funnel with chilled isopropyl alcohol (one-fifth the volume used in the recrystallization). The product is dried in a vacuum desiccator over concentrated sulfuric acid or phosphorus pentoxide. The yield of pure white product, melting at 74–75°, is 97–116 g. (50–60% based on benzyl cyanide) (Note 9).

2. Notes

1. α-Bromo-α-phenylacetonitrile is a lachrymator.

2. The mouth of the condenser is fitted with a cork carrying a short U-tube, to prevent condensed moisture from running back into the flask.

3. The benzyl cyanide used was a redistilled grade obtained from the Benzol Products Company.

4. Reagent grade bromine was used.

5. Reagent grade benzene was subjected to distillation until no droplets of water formed in the distillate. The residual benzene was used directly without further distillation or drying.

6. The checkers found that the work could be interrupted for as much as 1 day after the preparation of α-bromo-α-phenylacetonitrile without appreciable effect on the yield.

7. The reaction is vigorous, but it can be readily controlled by adding the bromonitrile frequently in small portions.

8. The flask must be of such a type that the side arm can be heated to prevent plugging by the solidified distillate.

9. A second crop, amounting to 10–15 g. and melting at 68–70°, can be obtained by chilling the mother liquor, but recrystallization of this fraction from isopropyl alcohol does not raise the melting point.

3. Methods of Preparation

Diphenylacetonitrile has been prepared by the dehydration of diphenylacetamide,[1–9] by the reaction of diphenylbromomethane and mercuric cyanide,[10] by the reaction of diphenylaectic acid and lead thiocyanate,[11] by the removal of ammonia from diphenylacetamidine,[12] by phenylation of potassium acetonitrile with chlorobenzene in liquid ammonia,[13] by the action of benzene on benzaldehyde cyanohydrin in the presence of aluminum chloride,[14] and from diphenylmethane and cyanogen chloride.[15] The present method is a modification of that of Hoch.[16,17]

[1] Neure, *Ann.*, **250,** 142 (1889).

[2] Zinsser, *Ber.*, **24,** 3556 (1891).

[3] Norris and Klemka, *J. Am. Chem. Soc.*, **62,** 1432 (1940).

[4] Rupe and Gisiger, *Helv. Chim. Acta,* **8,** 338 (1925).

[5] Trivedi, Phalnikar, and Nargund, *J. Univ. Bombay,* **10,** 135 (1942) [*C. A.*, **37,** 622 (1943)].

[6] Hellerman, Cohn, and Hoen, *J. Am. Chem. Soc.*, **50,** 1716 (1928).

[7] Smith, *Ber.*, **71,** 634 (1938).

[8] Wilson, *J. Chem. Soc.*, **1950,** 2173.

[9] Reid and Hunter, *J. Am. Chem. Soc.*, **70,** 3515 (1948).

[10] Anschutz, *Ann.*, **233,** 345 (1886).

[11] Wittig and Hopf, *Ber.*, **65,** 760 (1932).

[12] Lipp, *Ann.*, **449,** 15 (1926).

[13] Bergstrom and Agostinho, *J. Am. Chem. Soc.*, **67,** 2152 (1945).

[14] Homeyer and Splitter, U. S. pat. 2,443,246 [*C. A.*, **42,** 7338 (1948)].

[15] Dixon, U. S. pat. 2,553,404 [*C. A.*, **45,** 9081 (1951)]

[16] Hoch, *Compt. rend.*, **197,** 770 (1933)

[17] Shapiro, *J. Org. Chem.*, **14,** 839 (1949).

DIPHENYLACETYLENE

(Tolan)

$$C_6H_5CH{=}CHC_6H_5 + Br_2 \rightarrow C_6H_5CHBrCHBrC_6H_5$$

$$C_6H_5CHBrCHBrC_6H_5 + 2KOH \rightarrow C_6H_5C{\equiv}CC_6H_5 + 2KBr + 2H_2O$$

Submitted by Lee Irvin Smith and M. M. Falkof.
Checked by W. E. Bachmann and Charles E. Maxwell.

1. Procedure

A solution of 45 g. (0.25 mole) of *trans*-stilbene in 750 ml. of ether is prepared in a 1-l. three-necked round-bottomed flask fitted with a reflux condenser, an efficient mechanical stirrer, and a dropping funnel. To the well-stirred solution there is added 13.8 ml. (43 g.; 0.27 mole) of bromine, during the course of 10 minutes. A solid begins to separate in 5 minutes, but stirring is continued for 1 hour. The product is collected on a Büchner funnel and washed with ether until it is white. The yield of stilbene dibromide, melting at 235–237°, is 65.8–69.1 g. (77–81%).

A solution of 90 g. of potassium hydroxide in 150 ml. of absolute ethanol is prepared in a 500-ml. round-bottomed flask fitted with a reflux condenser (Note 1). The solution is cooled somewhat, and the stilbene dibromide is added in several portions (Note 2). The mixture is then refluxed for 24 hours in an oil bath (Note 3). The hot mixture is poured into 750 ml. of cold water, and the product is removed by filtration and washed with 50 ml. of water. The crude diphenylacetylene is dried over calcium chloride in a vacuum desiccator for 18 hours at room temperature. The chunky, pale yellow crystals melt at 58–60° and weight 33.7–37.8 g. By recrystallization of this product from 50 ml. of 95% ethanol there is obtained 29.2–30.5 g. (66–69% based on the stilbene) of pure diphenylacetylene, which forms white needles melting at 60–61° (Note 4).

2. Notes

1. Complete solution of the alkali in the alcohol is effected by placing the flask in an oil bath at 130–140°.

2. The addition of the stilbene dibromide causes an immediate, vigorous reaction, with the evolution of heat. It is necessary to re-

place the reflux condenser after each addition, until boiling has ceased.

3. An electrically heated oil bath is used to maintain a constant bath temperature of 130–140°.

4. The submitters obtained the same percentage yield on runs four times this size; the time employed was the same as that described here, except for the addition of bromine, which required 30 minutes.

3. Methods of Preparation

Diphenylacetylene has been prepared by the present method;[1] by the action of sodium amide on monochlorostilbene;[2] by the action of potassium amide on stilbene dichloride;[3] by heating 1,1-diphenyl-2-chloroethane with sodium ethoxide in a sealed tube;[4] and from benzil via the hydrazone.[5]

[1] Smith and Hoehn, *J. Am. Chem. Soc.*, **63**, 1180 (1941); Söderbäck, *Ann.*, **443**, 161 (1925).

[2] Paillard and Wieland, *Helv. Chim. Acta*, **21**, 1363 (1938).

[3] Coleman and Maxwell, *J. Am. Chem. Soc.*, **56**, 133 (1934).

[4] Buttenberg, *Ann.*, **279**, 327 (1894)

[5] Schlenk and Bergmann, *Ann.*, **463**, 76 (1928)

DIPHENYLDIAZOMETHANE

(Methane, diazodiphenyl-)

$$(C_6H_5)_2CO + H_2NNH_2 \rightarrow (C_6H_5)_2C{=}NNH_2 + H_2O$$

$$(C_6H_5)_2C{=}NNH_2 + HgO \rightarrow (C_6H_5)_2CN_2 + H_2O + Hg$$

Submitted by Lee Irvin Smith and Kenneth L. Howard.
Checked by H. R. Snyder, R. L. Rowland, and H. J. Sampson, Jr.

1. Procedure

In a pressure bottle are placed 19.6 g. (0.1 mole) of benzophenone hydrazone (Note 1), 22 g. (0.1 mole) of yellow oxide of mercury, and 100 ml. of petroleum ether (b.p. 30–60°). The bottle is closed, wrapped in a wet towel, and shaken mechanically at room temperature for 6 hours. The mixture is then filtered to remove mercury and any benzophenone azine (Note 2), and the filtrate is evaporated to dryness under reduced pressure at room temperature. The crystalline residue of diphenyldiazomethane melts when its temperature reaches that of the room (Note 3), but it is difficult to purify and this product is pure enough for all practical purposes. The material weighs 17.3–18.6 g. (89–96%). The product should be used immediately (Note 4).

2. Notes

1. Benzophenone hydrazone is most readily prepared from benzophenone (40 g., 0.22 mole) in 150 ml. of absolute ethanol with 41.2 g. (0.824 mole) of 100% hydrazine; the mixture is heated under reflux for 10 hours. On cooling in ice, colorless benzophenone hydrazone separates. The yield is 37.5 g. (87%), m.p. 97–98°.

If anhydrous hydrazine is not available, it may be prepared by the method described in *Org. Syntheses,* **24,** 53 (1944).

2. This ketazine is insoluble in petroleum ether.

3. The reported melting point of diphenyldiazomethane is 29–30°, but this melting point is shown only after the substance has been recrystallized from petroleum ether.

4. On standing, diphenyldiazomethane decomposes to yield benzophenone azine. In one of the checkers' runs the product was stored at room temperature; after 2 days, crystals of the azine were visible. The product at this stage was assayed by treatment with benzoic acid; addition of 6.8 g. of the diazo compound in a thin stream to a solution of 17 g. of benzoic acid in 90 ml. of ether, and, after 30 minutes, extraction of the excess benzoic acid with dilute sodium hydroxide followed by distillation of the ether gave 7.4 g. (75%) of crude benzohydryl benzoate melting at 83–85°. In the same procedure the freshly prepared diazo compound gave a quantitative yield of the crude ester.

3. Methods of Preparation

Diphenyldiazomethane has been prepared only by oxidation of benzophenone hydrazone.[1] The procedure given above is that of Staudinger, Anthes, and Pfenninger, with minor changes.

[1] Staudinger, Anthes, and Pfenninger, *Ber.,* **49,** 1932 (1916)

2,3-DIPHENYLINDONE (2,3-DIPHENYL-1-INDENONE)

(Indone, 2,3-diphenyl-)

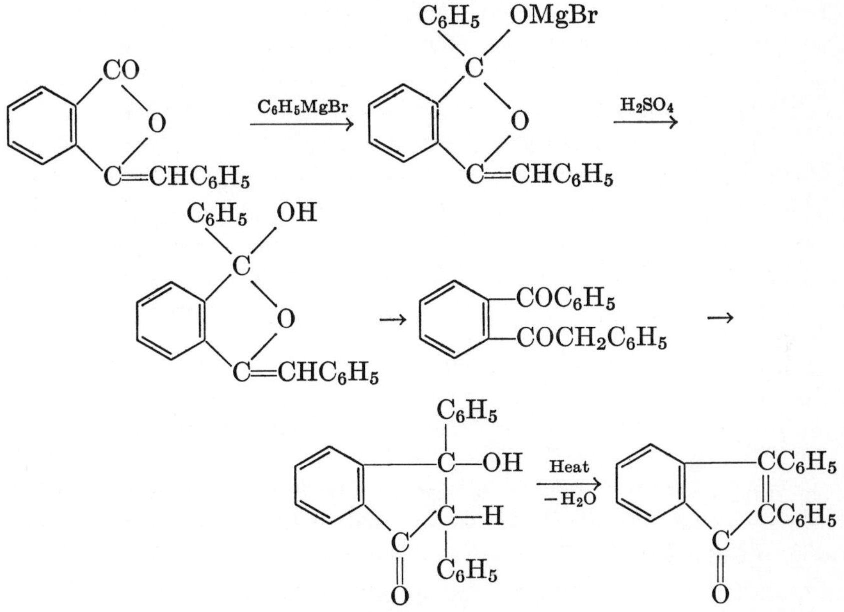

Submitted by C. F. H. ALLEN, J. W. GATES, JR., and J. A. VANALLAN.
Checked by R. L. SHRINER and H. W. JOHNSTON.

1. Procedure

A solution of phenylmagnesium bromide is prepared in the usual manner [1] using 12.2 g. (0.5 gram atom) of magnesium, 78.5 g. (52 ml., 0.5 mole) of bromobenzene, and 500 ml. of absolute ether in a 2-l. round-bottomed three-necked flask fitted with a reflux condenser, a mechanical stirrer, and a dropping funnel. No unreacted magnesium should remain; if any does, an additional 1–2 ml. of bromobenzene should be added.

To this vigorously stirred solution is added slowly a solution of 44.5 g. (0.20 mole) of benzalphthalide [2] in 400 ml. of dry benzene. After about half of this solution has been added, the magnesium complex begins to separate; it hinders the stirring somewhat. When all the phthalide solution has been admitted (about 1 hour), the reflux condenser is replaced by a still head carrying a thermometer and attached to a condenser set for downward distillation. The bulk of the solvent is then removed; this requires about 30 minutes, the temperature of the vapor remaining at 50° for about half this time and rising to about 65° toward the end. About 220–230 ml. of distillate is obtained. The flask and contents are then immersed in an ice bath, and the magnesium complex is decomposed by the slow addition of a cold solution of 15 ml. of concentrated sulfuric acid in 300 ml. of water (Note 1) with rapid stirring. The upper benzene layer is separated and transferred to a 1-l. Claisen flask, and the solvent is removed by distillation from a steam bath; this requires about 4 hours. The residual thick red syrup is transferred to a 125-ml. Claisen flask with a wide side arm set up for vacuum distillation. The residue is heated under a pressure of about 10 mm. to remove all low-boiling material (Note 2) and then is distilled under reduced pressure. The fraction boiling at 215–255°/6 mm. (195–220°/1 mm.) is collected, most of the distillate coming over at 235–240°/6 mm. The distillate is dissolved in 50 ml. of boiling benzene, 200 ml. of hot 95% ethanol is added, and the solution is chilled in an ice bath for 2 hours. The red, crystalline 2,3-diphenylindone is collected on a filter, washed with 50 ml. of cold 95% ethanol, and air-dried. The yield is 34–40 g. (60–71%) of red crystals melting at 149–151° (Notes 3 and 4).

2. Notes

1. The decomposition is vigorous, and the acid must be admitted slowly at first but may be added more rapidly toward the end.

2. During this heating any carbinol present is dehydrated. Low and erratic yields usually indicate incomplete dehydration.

3. The product may be recrystallized by dissolving it in 50 ml. of boiling benzene and diluting with 200 ml. of hot ethanol. The recovery of material, m.p. 150–151°, is about 90%.

4. This procedure is capable of considerable variation, by which other indenones may be secured. For example, the benzalphthalide may be replaced by other phthalides made from (*a*) other aldehydes, or (*b*) other anhydrides; the phenylmagnesium bromide can be replaced by other Grignard reagents.

3. Methods of Preparation

2,3-Diphenylindone has been prepared by the action of phenylmagnesium bromide upon benzalphthalide [3,4] and by ring closure from α,β-diphenylcinnamic acid [5,6] or from 2,3,3-triphenyl-3-hydroxypropionic acid.[7,8]

[1] *Org. Syntheses* Coll. Vol. **1**, 226 (1941).
[2] *Org. Syntheses* Coll. Vol. **2**, 61 (1943).
[3] Löwenbein and Ulich, *Ber.*, **58**, 2662 (1925).
[4] Weiss and Sauermann, *Ber.*, **58**, 2736 (1925).
[5] Meyer and Weil, *Ber.*, **30**, 1281 (1897).
[6] Weitz and Scheffer, *Ber.*, **54**, 2341 (1921).
[7] Ivanoff and Ivanoff, *Compt. rend.*, **226**, 1199 (1948).
[8] Ivanoff and Ivanoff, *Annuaire univ. Sofia, Faculté sci.*, **44**, (2), 121 (1947–1948) [*C. A.*, **44**, 3960 (1950)].

DIPHENYLIODONIUM IODIDE

(Iodonium compounds, diphenyl—iodide)

$$C_6H_5IO + C_6H_5IO_2 \xrightarrow{OH^-} [(C_6H_5)_2I]IO_3$$

$$[(C_6H_5)_2I]IO_3 + KI \longrightarrow [(C_6H_5)_2I]I + KIO_3$$

Submitted by H. J. Lucas and E. R. Kennedy.
Checked by John R. Johnson and M. W. Formo.

1. Procedure

A mixture of 22 g. (0.1 mole) of iodosobenzene (p. 483), 24 g. (0.1 mole) of iodoxybenzene (p. 485), and 200 ml. of 1 *N* sodium hydroxide (Note 1) is gently stirred for 24 hours. The resulting brown slurry is thoroughly stirred with 1 l. of cold water, and, after the mixture settles, the supernatant solution of diphenyliodonium iodate is decanted through a filter. The solid residue is extracted twice with 500-ml. portions of water, and the extracts are separated, by decantation through a filter, from the small amount of tarry residue. An aqueous solution of 20 g. (0.12 mole) of potassium iodide is added to the combined filtrates. After the bulky white precipitate of diphenyliodonium iodide has stood for an hour or two, with occasional shaking, it is filtered with suction, washed with water, and dried on porous tile at room temperature. The product weighs 29–30 g. (70–72%) and melts at 172–175° with vigorous decomposition.

2. Note

1. Since the reaction is catalyzed by hydroxide ions, the amount of base may be varied within wide limits. It is advantageous to grind the solid reactants with 50 ml. of water before the alkali is added.

3. Methods of Preparation

Diphenyliodonium iodide is obtained when sulfur dioxide or aqueous potassium iodide is added to a solution containing diphenyliodonium iodate.[1,2] Such solutions have been prepared by the action of moist silver oxide, or of aqueous sodium or potassium hydroxide, on an equimolar mixture of iodosobenzene and iodoxybenzene;[1,2] by shaking iodosobenzene with moist silver oxide;[3] by shaking iodoxybenzene with sodium hydroxide;[2] and by steam distillation of a mixture of iodosobenzene and iodoxybenzene.[2] Diphenyliodonium iodide has been prepared by heating iodoxybenzene with aqueous potassium iodide and barium hydroxide.[4]

[1] Hartmann and Meyer, *Ber.*, **27**, 504, 506 (1894).
[2] Hartmann and Meyer, *Ber.*, **27**, 1598 (1894).
[3] Hartmann and Meyer, *Ber.*, **27**, 503 (1894).
[4] Willgerodt, *Ber.*, **29**, 2009 (1896).

DIPHENYLKETENE

(Ketene, diphenyl-)

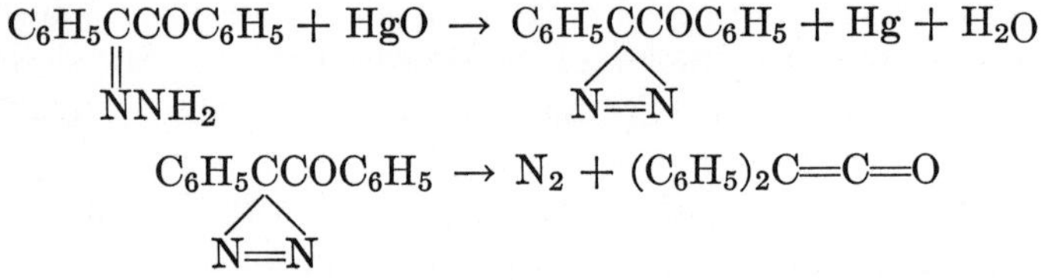

Submitted by LEE IRVIN SMITH and H. H. HOEHN.
Checked by NATHAN L. DRAKE and JONATHAN WILLIAMS.

1. Procedure

Fifty-six grams (0.25 mole) of benzil monohydrazone [*Org. Syntheses*, **15**, 62 (1935)] (Note 1) is mixed in a mortar with 81 g. (0.38 mole) of yellow mercuric oxide and 35 g. of anhydrous calcium sulfate (Note 2). The mixture is introduced into a 1-l. three-necked flask fitted with a stirrer, a condenser, and a thermometer. The flask is placed in a water bath, 200 ml. of dry thiophene-free benzene is added, and the suspension is stirred at 25–35° (thermometer in solution) for 4 hours (Notes 3 and 4). The reaction mixture is filtered through a fine-grained filter paper, with slight suction, and the residue is washed with dry benzene until the washings are colorless.

The benzene solution of the diazo compound is poured into a separatory funnel protected with a drying tube and connected to a 125-ml. Claisen distilling flask provided with a condenser set for downward distillation and arranged so that it can be heated in a bath of Wood's metal. The temperature of the metal bath being maintained at 100–110°, the benzene solution is dropped slowly into the hot flask. Under these conditions, the benzene is removed by distillation and the diazo compound is transformed into diphenylketene. The residue is distilled under reduced pressure in an atmosphere of nitrogen, and the fraction boiling at 115–125° at 3–4 mm. (Note 5) is collected. The yield is 31 g. (64%) of a product which, on redistillation, yields 28 g. of diphenylketene boiling at 119–121° at 3.5 mm. (58%).

Diphenylketene is best stored in an atmosphere of nitrogen; the addition of a small crystal of hydroquinone serves to inhibit polymerization (Note 6).

2. Notes

1. Benzil monohydrazone can also be obtained in practically quantitative yield using hydrazine hydrate, a method first suggested by Curtius and Thun.[1] Hydrazine hydrate (45 g., 0.75 mole, of an 85% solution of hydrazine hydrate in water) is slowly dropped into a hot solution of benzil (158 g., 0.75 mole) in alcohol (300 ml.) with stirring. The product begins to separate from the hot solution after three-fourths of the hydrazine hydrate has been added. The solution is heated under reflux for 5 minutes after all the hydrazine hydrate has been added. The flask is then cooled to 0°, and the hydrazone is filtered off and washed twice on the funnel with 100-ml. portions of cold ethanol. The product melts at 149–151° with decomposition.

2. Anhydrous calcium sulfate removes the water formed in the oxidation.

3. Considerable heat is generated at the beginning of the reaction, and ice must be used in the water bath to keep the temperature within the prescribed limits. After 10–15 minutes the ice is removed, and the temperature is maintained at 25–35° by the water bath.

4. Best results are obtained when the reaction mixture is stirred for 4 hours.

5. A viscous red residue always remains in the distilling flask, necessitating superheating to remove the last traces of diphenylketene.

6. According to the submitters, the preparation has been carried out using twice the quantities of material throughout with no loss in yield of diphenylketene.

3. Methods of Preparation

Diphenylketene has been prepared by action of tripropylamine on diphenylacetyl chloride,[2] by treating diphenylchloroacetyl chloride with granulated zinc,[3] and by the action of quinoline on diphenylacetyl chloride.[4] It is most conveniently prepared by heating phenylbenzoyldiazomethane—a method first described by Schroeter [5] and later used by Staudinger.[6]

[1] Curtius and Thun, *J. prakt. Chem.*, (2) **44**, 176 (1891).
[2] Staudinger, *Ber.*, **44**, 1619 (1911).
[3] Staudinger, *Ber.*, **38**, 1735 (1905).
[4] Staudinger, *Ber.*, **40**, 1148 (1907).
[5] Schroeter, *Ber.*, **42**, 2346 (1909).
[6] Staudinger, *Ber.*, **44**, 1623 (1911).

2,4-DIPHENYLPYRROLE

(Pyrrole, 2,4-diphenyl-)

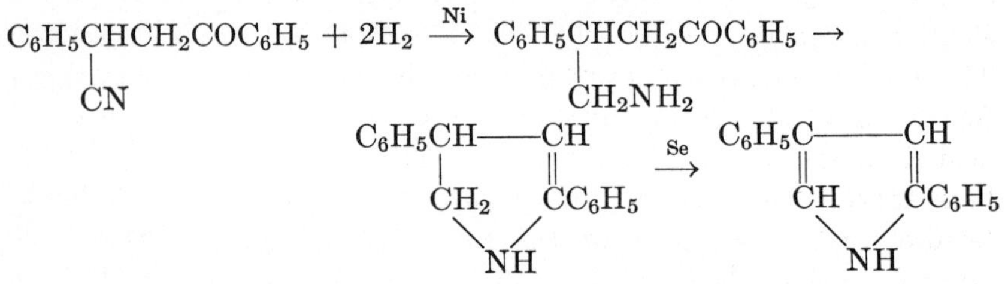

Submitted by C. F. H. Allen and C. V. Wilson.
Checked by Arthur C. Cope, Frank S. Fawcett, and Harold R. Nace.

1. Procedure

A suspension of 61 g. (0.26 mole) of α-phenyl-β-benzoylpropionitrile [1] (Note 1) in 150 ml. of methanol and 1 level teaspoonful of Raney nickel catalyst [2,3] (Note 2) are placed in a 400–500 ml. pressure bottle (Note 3) connected to a low-pressure reduction apparatus. The bottle is alternately evacuated and filled with hydrogen twice, and the reduction is conducted at 80–90° by shaking with hydrogen at an initial pressure of 50 lb. After 2 moles of hydrogen is absorbed the reduction is discontinued, the solution is filtered to separate the catalyst, and the solvent is removed from the green filtrate by distillation under reduced pressure. The crude 2,4-diphenyl-2-pyrroline (Note 4) is rinsed into a 300-ml. Kjeldahl flask with a little methanol, which is removed by distillation at reduced pressure (water pump) while the flask is immersed in a bath at 100°. Selenium (10 g.) is added, and the mixture is heated in a metal bath at 245–265° for 5 hours while a slow stream of nitrogen is passed over the surface of the liquid. This operation is conducted in a well-ventilated hood. The molten mass is poured from the reaction flask into a beaker (Note 5), and the product is extracted with 300 ml. of boiling toluene. The green toluene solution is filtered through a cotton plug to remove a small amount of suspended selenium, and the filtrate is cooled. The 2,4-diphenylpyrrole separates as light green crystals which melt at 174–176° and weigh 24–26 g. (42–46%) (Note 6). A second crop of 2–3 g. (4–5%), m.p. 160–170°, can be obtained by concentrating the filtrate and cooling.

2. Notes

1. The α-phenyl-β-benzoylpropionitrile [1] was used without recrystallization (m.p. 123–125°).

2. The time required for the hydrogenation was 2.5 hours with a very active nickel catalyst, and longer with a less active catalyst. Raney nickel prepared according to reference 2 is a particularly active catalyst and gives excellent results in this preparation.

3. A 500-ml. Pyrex centrifuge bottle is satisfactory.

4. The submitters state that distillation yields pure 2,4-diphenyl-2-pyrroline, b.p. 170–172°/3 mm., 203–205°/8 mm., which solidifies on cooling (yield 80–90%).

5. If the contents of the flask are not poured into a beaker while molten, it is difficult to remove the cake of selenium. If more than one preparation is to be made, the contents of the flask are allowed to cool and the product is extracted with hot toluene. The selenium can be reused.

6. The product is sufficiently pure for most purposes. Recrystallization from toluene gives a 90–93% recovery of a very light-green product, m.p. 179–180° (cor.). The submitters state that the product obtained in the first crystallization has this melting point if the 2,4-diphenyl-2-pyrroline is purified by distillation before dehydrogenation.

3. Methods of Preparation

2,4-Diphenylpyrrole has been prepared only by dehydrogenation of 2,4-diphenyl-2-pyrroline with selenium, Raney nickel, and nickel supported on pumice.[4]

[1] *Org. Syntheses* Coll. Vol. **2,** 498 (1943)
[2] Pavlic and Adkins, *J. Am. Chem. Soc.,* **68,** 1471 (1946).
[3] *Org. Syntheses* Coll. Vol. **3,** 176, 181 (1954).
[4] Rogers, *J. Chem. Soc.,* **1943,** 594.

DITHIZONE *

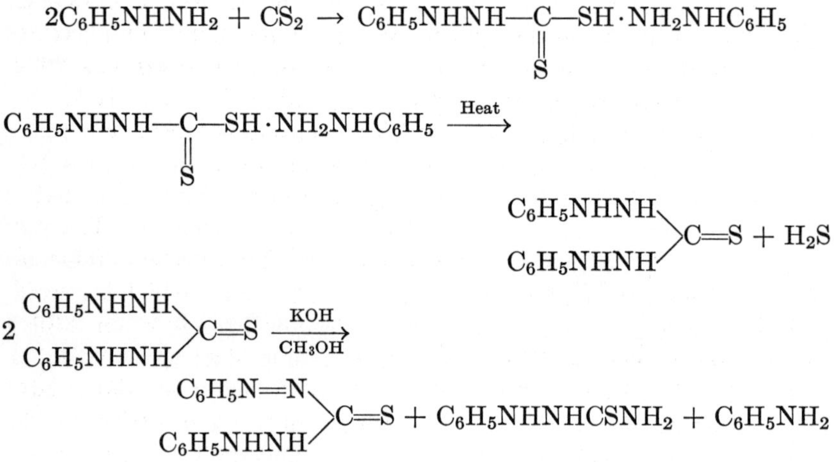

Submitted by John H. Billman and Elizabeth S. Cleland.
Checked by Lee Irvin Smith and Burris D. Tiffany.

1. Procedure

A. *Phenylhydrazine salt of β-phenyldithiocarbazic acid.* In a 1-l. three-necked flask, fitted with a mechanical stirrer, a condenser, and a dropping funnel, is placed a solution of 128 ml. (1.3 moles) of pure redistilled phenylhydrazine in 600 ml. of ordinary ether. To the vigorously stirred mixture, 52 ml. (0.86 mole) of carbon disulfide is added in the course of 30 minutes (Note 1). After the mixture has been stirred for an additional 30 minutes, the precipitate is filtered with suction, washed with 50 ml. of ether, and spread on filter paper for 15–20 minutes to allow evaporation of the ether. The yield of the salt is 181–185 g. (96–98%).

B. *Diphenylthiocarbazide.* The above salt is transferred to a 1-l. beaker, and, while it is continuously stirred by hand (Note 2), it is heated (hood) in a water bath maintained at 96–98° (Note 3). After about 10–15 minutes the material softens to a taffylike mass, becomes yellow, foams, and evolves hydrogen sulfide. After about 20–30 minutes ammonia is evolved. When a distinct odor of ammonia is *first detected* (Note 4), the beaker is removed from the bath, placed in a pan of cold water for 1 minute (Note 5), and then surrounded immediately by cracked ice. About 150 ml. of absolute ethanol is added, the mixture is warmed slightly to loosen the mass, and the taffylike material is stirred until it is transformed into a granular precipitate. After the mixture has stood at room temperature for 1 hour, the precipitate is collected on a Büchner funnel and washed with 50 ml. of absolute ethanol (Note 6). The yield of crude diphenylthiocarbazide is 100–125 g. (60–75% based on phenylhydrazine) (Note 7).

C. *Dithizone.* The crude diphenylthiocarbazide is added to a solution of 60 g. of potassium hydroxide in 600 ml. of methanol in a 1-l. round-bottomed flask. The flask is immersed in a boiling water bath, and the mixture is *refluxed* for exactly 5 minutes (Note 8). The red solution is cooled with ice water and filtered by gravity. Ice-cold 1 *N* sulfuric acid (900–1100 ml.) is added to the filtrate, which is stirred vigorously by means of a mechanical stirrer, until the solution is just acid to Congo red paper (Note 9). The blue-black precipitate is filtered with suction and washed with 50 ml. of cold water (Note 10). The crude carbazone is dissolved in 500 ml. of 5% sodium hydroxide solution, the mixture is filtered with suction, and the filtrate is cooled in an ice bath and acidified immediately with ice-cold 1 *N* sulfuric acid (about 650 ml. is required) until it is just acid to Congo red paper. The precipitate is filtered with suction and then washed by transferring it to a 2-l. beaker and stirring it thoroughly with 1.6–1.8 l. of cold water. The mixture is filtered, and the process of washing is repeated until there is no trace of sulfate in the washings (Note 11). After air has been drawn through the precipitate on the Büchner funnel for 20–30 minutes, the solid is dried in an oven at 40°. The product at this stage weighs 63–85 g. For purification, a portion of 5–10 g. of the carbazone is placed in the thimble of a Soxhlet extractor (Note 12), covered with ether, allowed to stand for 1 hour, and then extracted for 1.5 hours. The material in the thimble is transferred immediately to a beaker, stirred with 50 ml. of ether, and the mixture is filtered with suction until most, but not all, of the liquid is removed. The wet product is then dried by pressing it between filter papers. The yield of pure dithizone (diphenylthiocarbazone) which decomposes sharply at a temperature between 165° and 169° is 43–54.8 g. (50–64% based on phenylhydrazine) (Note 13).

2. Notes

1. A precipitate is formed immediately upon addition of the carbon disulfide; the mixture becomes warm, and the temperature soon approaches the boiling point. The temperature is maintained just below the boiling point; cooling with ice water may be necessary to keep it there. As the carbon disulfide is added, the mixture soon becomes pasty, then more fluid, and finally pasty again.

2. A heavy glass rod, bent or flattened at the bottom, serves as an efficient stirrer.

3. Care must be taken that the temperature of the bath does not exceed 98°. If the material remains above 98° for a long period of time and/or if it is not immediately cooled after removal from the water bath, the product decomposes violently after standing for 10–15 minutes.

4. Ammonia can be detected by means of litmus paper before the odor is noticeable, but the heating should be continued until there is a *distinct* odor of ammonia (but no longer). The yield depends upon the rapidity with which the heating is stopped after the first sign of ammonia is detected.

5. The mass may be olive-green or brown while hot, but it becomes light brown on cooling.

6. The precipitate is almost pure white. More diphenylthiocarbazide crystallizes from the red alcoholic filtrate and washings, if they are allowed to evaporate slowly. When the mixture is allowed to stand for some time before it is filtered, the amount of precipitate is increased and that of the material left in the filtrate is decreased.

7. Diphenylthiocarbazide does not have a sharp melting point. The compound is reported to become green at 130° and melt at 150° to a dark green liquid which decomposes on further heating.

8. The solution is heated until it is definitely boiling; then it is allowed to boil for 5 minutes only. If the solution is boiled for a longer time, the yield of product is decreased.

9. When the end point is reached, the mother liquor is no longer red but colorless.

10. Care should be exercised in handling this precipitate or suspensions of it. If any of it is spilled or splashed, it should be removed at once, for it dries to a light, fine powder which is readily scattered; it dyes the skin black and other material pink.

11. Four or five washings are usually necessary.

12. The Soxhlet extractor used by the submitters had a capacity of 85 ml. to the top of the siphon tube. The size of the thimble was 33 by 80 mm.

13. The pure compound is completely soluble in chloroform.

3. Methods of Preparation

This method of preparation is a modification of the method used by Emil Fischer.[1] Similar methods have been described by H. Fischer,[2] Wertheim,[3] and Grummitt and Stickle.[4]

[1] E. Fischer, *Ann.*, **190**, 118 (1878); **212**, 316 (1882).

[2] H. Fischer and Leopoldi, *Wiss. Veröffentl. Siemens-Konzern*, **12**, 44 (1933) [*C. A.*, **27**, 3418 (1933)].

[3] Wertheim, *Organic Chemistry Laboratory Guide*, 2nd ed., p. 504, P. Blakiston's Son & Co.

[4] Grummitt and Stickle, *Ind. Eng. Chem., Anal. Ed.*, **14**, 953 (1942).

n-DODECYL (LAURYL) MERCAPTAN *

$$n\text{-}C_{12}H_{25}Br + S{=}C(NH_2)_2 \rightarrow C_{12}H_{25}{-}S{-}C\begin{matrix}{\diagup}NH \\ {\diagdown}NH_2\end{matrix}\cdot HBr$$

$$2C_{12}H_{25}S{-}C\begin{matrix}{\diagup}NH \\ {\diagdown}NH_2\end{matrix}\cdot HBr + 2NaOH \rightarrow$$

$$2C_{12}H_{25}SH + H_2N{-}\overset{\overset{\displaystyle NH}{\|}}{C}{-}NHCN + 2NaBr + 2H_2O$$

Submitted by G. G. Urquhart, J. W. Gates, Jr., and Ralph Connor.
Checked by Lee Irvin Smith, R. T. Arnold, and Kenneth Stevenson.

1. Procedure

A mixture of 125 g. (0.5 mole) of *n*-dodecyl bromide (Note 1), 38 g. (0.5 mole) of thiourea, and 250 ml. of 95% ethanol is refluxed on a steam cone for 3 hours (Notes 2, 3, and 4). A solution of 30 g. (0.75

mole) of sodium hydroxide in 300 ml. of water is added, and the mixture is refluxed for 2 hours. During this period the mercaptan separates as a pink to red oil. The layers are separated, and the aqueous layer is acidified with dilute sulfuric acid (7 ml. of concentrated acid to 50 ml. of water) and then extracted with one 75-ml. portion of benzene (Note 5). The extract is added to the crude mercaptan layer, and the whole is washed twice with 200-ml. portions of water and then dried over 20 g. of anhydrous sodium sulfate (Note 6). The solvent is removed and the residual oil distilled from a modified Claisen flask. There is no appreciable fore-run (Note 6). The yield of *n*-dodecyl mercaptan, b.p. 165–169°/39 mm., is 80–84 g. (79–83%) (Notes 7 and 8).

2. Notes

1. The preparation of lauryl bromide has been described [*Org. Syntheses* Coll. Vol. **1,** 29 (1941)], but the following modification avoids the emulsion formed in washing lauryl bromide with sulfuric acid. The crude bromide from the reaction mixture is washed thoroughly with water and then with potassium carbonate solution. The washing with water must be thorough enough to remove most of the acid in order to prevent the formation of heavy foam by the carbonate. The bromide is then dried over calcium chloride and distilled. The distillate is washed with concentrated sulfuric acid and treated as described (*loc. cit.*). The checkers, however, found the method using anhydrous hydrogen bromide [*Org. Syntheses* Coll. Vol. **2,** 246 (1943)] to be preferable.

2. The intermediate laurylisothiourea hydrobromide may be obtained by cooling the ethanol solution in ice and filtering the precipitate. It may be further purified by washing with ether. An alternative method gives more satisfactory results. The ethanol solution is diluted with 150 ml. of water and heated until the solution is homogeneous, and then, after cooling slightly, 200 ml. of concentrated hydrochloric acid is added; the laurylisothiourea hydrochloride separates quantitatively as a waxy white solid.

3. The reflux period required for complete reaction varies considerably with the structure of the halide. Some alkyl bromides require more than 3 hours. Alkyl chlorides frequently require 16 hours and occasionally as much as 24 hours.

4. This procedure is useful for the preparation of other mercaptans. The yields are of the same order. The boiling points are as follows:

n-Heptyl		174–176°
n-Octyl	198–200°;	98–100°/22 mm.

n-Nonyl	220–222°;	98–100°/15 mm.
n-Decyl	88–91°/2 mm.;	96–97°/5 mm.
n-Undecyl		103–104°/3 mm.
n-Octadecyl		170–175°/4 mm.

5. Although benzene is the most satisfactory solvent for this extraction, emulsions are sometime produced when it is used for the extraction of other mercaptans. To obviate this difficulty the same amount of ether may be used, provided that the ethanol is first removed by distillation on a steam bath.

6. If the ether extracts are not thoroughly dry, the mercaptan appears slightly cloudy on subsequent distillation. The checkers found that this cloudiness was frequently due to colloidal sulfur. The sulfur can be removed by shaking an ether solution of the cloudy product with bone black, then filtering and distilling.

7. The odor of lauryl mercaptan is not disagreeable, and no unusual precautions need be taken in working with it. With lower-molecular-weight mercaptans a trap containing alkaline potassium permanganate solution will prevent the escape of unpleasant odors.

8. Other boiling points: 107–109°/2 mm.; 133–135°/7 mm.; 145–147°/15 mm.; 150–152°/20 mm.; 158–160°/28 mm.; 168–170°/38 mm.

3. Methods of Preparation

n-Dodecyl mercaptan has been prepared by the action of alkali on the salts of S-*n*-dodecyl thiourea [1] and by reduction of di-*n*-dodecyl disulfide.[2,3]

It has also been prepared from lauryl alcohol by the action of thiourea in the presence of hydrobromic acid, followed by alkaline hydrolysis.[4]

[1] Backer, Terpstra, and Dijkstra, *Rec. trav. chim.*, **51**, 1166 (1932).

[2] Noller and Gordon, *J. Am. Chem. Soc.*, **55**, 1090 (1933).

[3] Henkel et Cie., Fr. pat. 751,117; U. S. pat. 2,031,529 [*C. A.*, **30**, 2202 (1936)].

[4] Frank and Smith, *J. Am. Chem. Soc.*, **68**, 2104 (1946).

n-DODECYL (LAURYL) *p*-TOLUENESULFONATE

$$p\text{-}CH_3C_6H_4SO_2Cl + CH_3(CH_2)_{10}CH_2OH + C_5H_5N \rightarrow p\text{-}CH_3C_6H_4SO_3CH_2(CH_2)_{10}CH_3 + C_5H_5NH^+Cl^-$$

Submitted by C. S. MARVEL and V. C. SEKERA.
Checked by C. F. H. ALLEN and C. V. WILSON.

1. Procedure

In a 1-l. three-necked flask fitted with a stirrer and thermometer are placed 93 g. (0.5 mole) of dodecanol (Note 1) and 158 g. (2 moles) of pyridine. The flask is surrounded by a water bath sufficiently cold to lower the temperature of the mixture to 10°. At this temperature 105 g. (0.55 mole) of *p*-toluenesulfonyl chloride is added in portions over a 20- to 30-minute period, or at such a rate that the temperature does not exceed 20° at any time. The mixture is then stirred for 3 hours at a temperature below 20°, after which it is diluted with 300 ml. of hydrochloric acid (sp. gr. 1.19) in 1 l. of ice water. The ester which crystallizes is collected on a chilled Büchner funnel and sucked as dry as possible. The solid is transferred to a 600-ml. beaker, 250–300 ml. of methanol is added, and the mixture is warmed on the steam bath until the ester melts. It is then cooled in a freezing mixture while being stirred continuously; the ester separates in a fairly fine state. It is then collected on a chilled funnel and allowed to dry in the air, preferably at a temperature below 20°. The yield of ester is 152–156 g. (88–90% based upon the dodecanol used). It melts at 20–25° (Note 2) and is sufficiently pure for most purposes.

If a purer product is desired, it is recrystallized from petroleum ether (b.p. 30–60°), using 4 ml. per 3 g., and drying over anhydrous sodium sulfate. The solution is chilled to 0° and the ester filtered on a chilled funnel; the recovery is 90%, and the melting point is 28–30°. Evaporation of the solvent to a small volume deposits an additional amount (Note 3).

2. Notes

1. Dodecanol (lauryl alcohol), m.p. 20–22°; pyridine, b.p. 113–115°; and *p*-toluenesulfonyl chloride, m.p. 66–68°, are used.

2. The ester contains traces of water, which makes the melting point unreliable; the freezing point is 24–25°.

3. The following esters have been made in essentially the same yields: butyl *p*-toluenesulfonate [*Org. Syntheses* Coll. Vol. **1,** 145 (1941)]; *n*-tetradecyl *p*-toluenesulfonate, m.p. 35°; *n*-hexadecyl *p*-toluenesulfonate, m.p. 49°; *n*-octadecyl *p*-toluenesulfonate, m.p. 56°; *n*-decyl *p*-bromobenzenesulfonate, m.p. 43–44°; *n*-dodecyl *p*-bromobenzenesulfonate, m.p. 49°; *n*-tetradecyl *p*-bromobenzenesulfonate, m.p. 51.5°; *n*-hexadecyl *p*-bromobenzenesulfonate, m.p. 60°; *n*-octadecyl *p*-bromobenzenesulfonate, m.p. 64–65°.

3. Method of Preparation

n-Dodecyl *p*-toluenesulfonate has been prepared only by the action of *p*-toluenesulfonyl chloride on dodecanol-1 in the presence of pyridine [1] according to the general procedure developed by Patterson and Frew [2] for making esters of sulfonic acids.

[1] Sekera and Marvel, *J. Am. Chem. Soc.,* **55,** 345 (1933).
[2] Patterson and Frew, *J. Chem. Soc.,* **89,** 332 (1906).

DYPNONE *

$$2C_6H_5COCH_3 \xrightarrow{Al[OC(CH_3)_3]_3} C_6H_5(CH_3)C{=}CHCOC_6H_5 + H_2O$$

Submitted by Winston Wayne and Homer Adkins.
Checked by Nathan L. Drake, Wm. H. Souder, Jr., and Ralph Mozingo.

1. Procedure

In a 1-l. round-bottomed three-necked flask, equipped with a thermometer, an efficient mechanical stirrer (Note 1), and a 35-cm. Vigreux column fitted to a condenser and receiver protected by a calcium chloride tube, are placed 345 g. (400 ml.) of dry xylene, 120 g. (117 ml., 1 mole) of dry acetophenone, and 135 g. (0.55 mole) of aluminum *tert*-butoxide (p. 48) (Note 2). The stirrer is started and the flask heated in an oil bath so that the temperature of the reaction mixture is held between 133° and 137°. *tert*-Butyl alcohol slowly distils at a temperature in the vapor of 80–85°. The distillation of the alcohol can be accomplished by maintaining the temperature of the heating bath between 150° and 155° for 2 hours after distillation has commenced (Notes 3 and 4).

The reaction mixture is cooled to 100°, and 40 ml. of water is added cautiously in small portions with continued stirring. As the water is added the mixture sets to a gel, and then, upon the addition of the remainder of the water and gentle tapping of the reaction flask, the stiff mass breaks up and boiling begins. Refluxing is continued by heating the oil bath for another 15 minutes to ensure the complete hydrolysis of the aluminum *tert*-butoxide.

After cooling, the reaction mixture is transferred in nearly equal portions to four centrifuge bottles and the aluminum hydroxide is centrifuged out (Note 5). The upper liquid layer is removed, and the aluminum hydroxide is worked up to a smooth paste, by means of a spatula, with a total of approximately 250 ml. of ether for the four portions. The aluminum hydroxide is again centrifuged out and the ether extract poured off. The centrifuging is repeated three more times so that approximately a liter of ether is used to separate the product from the aluminum hydroxide (Note 6).

Ether and *tert*-butyl alcohol are removed by distillation at atmospheric pressure without a column, and the xylene is removed by distillation through a 35-cm. Vigreux column at 25–50 mm. The residue is transferred to a smaller flask, and the acetophenone is first distilled at about 80°/10 mm., and finally the dypnone is distilled at 150–155°/1 mm. (Notes 7 and 8). The yield is 85–91 g. (77–82%) of a yellow liquid.

2. Notes

1. Instead of the customary mercury seal, a simple rubber tube seal suggested by Dr. L. P. Kyrides is recommended. The upper end of an 8-cm. length of 10-mm. glass tubing, projecting through a stopper, is fitted with a 2-cm. piece of 6-mm. rubber tubing so that it projects 5 mm. beyond the end of the glass tube. This projecting portion fits snugly to form a seal round an 8-mm. stirrer shaft running through the glass tubing. Glycerol is applied at the point of contact of glass and rubber to act as a lubricant and sealing medium. This type of seal can be conveniently used with reduced pressure down to 10 mm.

2. Slightly more than a mole of aluminum *tert*-butoxide is used for each mole of water split out as the alkoxide loses its effectiveness after the replacement of one alkoxyl by an hydroxyl group.

3. The reaction may proceed so vigorously at first that the distillation temperature will rise to 100° or above. This is not detrimental and the temperature will soon fall.

4. As the reaction proceeds the color changes from yellow to deep orange and the mixture becomes viscous. After about 45 minutes the distillation temperature begins to fall and eventually reaches 70°

owing to the formation of butylene from the aluminum *tert*-butoxide which decomposes slightly at this temperature.

5. Ordinary undried ether is used in balancing the centrifuge bottles, in removing any aluminum hydroxide and product remaining in the flask, and in the subsequent extraction of the aluminum hydroxide.

6. This method eliminates the washing and drying of the customary extraction procedure. The small amount of water present distils off with the solvent. The extraction should be done carefully and thoroughly. If it has been carried out properly the aluminum hydroxide will be white or faintly yellow and at least 95% of the original acetophenone can be accounted for after the final distillation.

7. The acetophenone and dypnone are easily separated in any of the common types of fractionating columns. The column should be short (12–18 cm.) and electrically heated because of the high boiling point of the product. An undistillable residue of 5–10 g. remains.

8. The boiling point varies greatly with the rate of distillation and the amount of ebullition.

3. Methods of Preparation

Dypnone has been prepared by the action of sodium ethoxide,[1] aluminum bromide,[2] phosphorus pentachloride,[3] aluminum triphenyl,[4] zinc diethyl,[5] calcium hydroxide,[6] anhydrous hydrogen chloride,[5,7] anhydrous hydrogen bromide,[8] aluminum chloride,[4] aluminum *tert*-butoxide,[9] and hydrogen fluoride[10] on acetophenone. It has been obtained by the action of aniline hydrochloride on acetophenone anil followed by treatment with hydrochloric acid.[11] The preparation of aluminum *tert*-butoxide has been described.[9,12,13] The procedure described is a modification of that by Adkins and Cox.[9]

1 Eijkman, *Chem. Weekblad,* **1,** 349 (1904).

2 Konowalow and Finogejew, *J. Russ. Phys. Chem. Soc.,* **34,** 944 (1902) [*Chem. Zentr.,* **74,** I, 521 (1903)].

3 Taylor, *J. Chem. Soc.,* **1937,** 304.

4 Calloway and Green, *J. Am. Chem. Soc.,* **59,** 809 (1937).

5 Henrich and Wirth, *Monatsh.,* **25,** 423 (1904).

6 Porlezza and Gatti, *Gazz. chim. ital.,* **56,** 265 (1926).

7 Delacre, *Bull. classe sci. Acad. roy. Belg.,* (3) **20,** 467 (1898); Colonge, *Bull. soc. chim. France,* (4) **49,** 426 (1931); Kohler, *Am. Chem. J.,* **31,** 642 (1904).

8 Müller and Spinosa-Stöckel, *Österr. Chem. Ztg.,* **49,** 130 (1948).

9 Adkins and Cox, *J. Am. Chem. Soc.,* **60,** 1151 (1938).

10 Simons and Ramler, *J. Am. Chem. Soc.,* **65,** 1390 (1943).

11 Reddelien, *Ber.,* **46,** 2712 (1913).

12 Oppenauer, *Rec. trav. chim.,* **56,** 137 (1937).

13 *Org. Syntheses* Coll. Vol. **3,** 48 (1954).

β-ETHOXYETHYL BROMIDE

(Ethane, 1-bromo-2-ethoxy-)

$$3C_2H_5OCH_2CH_2OH + PBr_3 \rightarrow 3C_2H_5OCH_2CH_2Br + P(OH)_3$$

Submitted by GEORGE C. HARRISON and HARVEY DIEHL.
Checked by R. L. SHRINER and C. H. TILFORD.

1. Procedure

In a 2-l. three-necked flask, fitted with a mechanical stirrer (Note 1), a reflux condenser, and a dropping funnel, is placed 630 g. (670 ml., 7 moles) of β-ethoxyethyl alcohol (Note 2). The stirrer is started, and 600 g. (210 ml., 2.2 moles) (Note 3) of phosphorus tribromide is added from the dropping funnel over a period of 1.5–2 hours. The temperature is permitted to rise until the reaction mixture refluxes gently.

The mixture is then distilled, and the distillate boiling below 150° is collected in a 2-l. flask containing 1 l. of water. The lower layer of crude β-ethoxyethyl bromide is separated and dried over 10 g. of calcium chloride. The liquid is decanted and distilled through a 25-cm. fractionating column, and the fraction boiling at 125–127°/760 mm. is collected (Note 4). The yield of pure product is 660–670 g. (65–66% based on the phosphorus tribromide).

2. Notes

1. A rubber sleeve lubricated with a drop of oil provides an effective seal for the stirrer.

2. Technical β-ethoxyethyl alcohol is marketed commercially as Cellosolve; it should be dried over calcium oxide and distilled.

3. A slight excess of β-ethoxyethyl alcohol gives the best results.

4. The first fraction, boiling at 38–40°, is ethyl bromide, and it weighs 120–130 g. Lower yields (56–59%) of β-ethoxyethyl bromide are obtained by distillation without a fractionating column. The residue, however, should not be overheated, for at high temperatures phosphorous acid decomposes to give phosphine and perhaps even elementary phosphorus, and then, when air is admitted to the apparatus, an explosion may occur. A minor explosion in connection with this preparation has been reported (N. L. Drake, private communication),

and a similar situation has been observed in reactions with phosphorus trichloride.[1]

3. Methods of Preparation

β-Ethoxyethyl bromide has been prepared by the action of sodium ethoxide upon ethylene bromide;[2] by the action of bromine upon β-ethoxyethyl iodide;[3] and by the procedure adopted here, which was first used by Chalmers.[4] The action of sodium bromide and sulfuric acid on β-ethoxyethyl alcohol cleaves the ether linkage.

[1] Coghill, *J. Am. Chem. Soc.*, **60**, 488 (1938).
[2] Foran, *J. Soc. Chem. Ind.*, **44**, 173 (1925).
[3] Henry, *Jahresb.*, **1885**, 1163.
[4] Chalmers, *Can. J. Research*, **7**, 464 (1932).

β-ETHOXYPROPIONALDEHYDE ACETAL

(Propionaldehyde, β-ethoxy-, diethyl acetal)

$$CH_2{=}CH{-}CHO \xrightarrow[NH_4Cl]{C_2H_5OH} C_2H_5OCH_2CH_2CHO + C_2H_5OCH_2CH_2CH_2(OC_2H_5)_2$$

Submitted by C. G. Alberti and R. Sollazzo.
Checked by Miriam Sharp, V. L. Stromberg, and E. C. Horning.

1. Procedure

In a 1-l. three-necked round-bottomed flask fitted with a reflux condenser, a mechanical stirrer, and a thermometer dipping below the surface of the mixture are placed 158 g. (2.8 moles) of acrolein (Note 1), 500 ml. of absolute ethanol, and 10 g. of ammonium chloride (Note 2). The material is stirred for 1 hour without external heating; the temperature of the mixture should rise to about 30°. Over a 3-hour period the temperature is raised to 80°, and the mixture is maintained at this temperature for an additional hour. The flask is then cooled, and anhydrous sodium or magnesium sulfate is added as a drying agent.

After 24 hours the mixture is filtered and distilled under reduced pressure through a good column (Note 3). The following fractions are collected (Note 4):

Below 23°/20 mm.	Acrolein and alcohol
38–42°/16 mm.	β-Ethoxypropionaldehyde
75–78°/16 mm.	β-Ethoxypropionaldehyde acetal

The yield of β-ethoxypropionaldehyde is 22–24 g. (7–8%) and of β-ethoxypropionaldehyde acetal is 166–193 g. (31–39%), n_D^{20} 1.4067, d_4^{15} 0.898.

2. Notes

1. Commercial acrolein may be employed. It should be dried over anhydrous sodium sulfate.

2. Reagent grade ammonium chloride is used. It should be washed several times with absolute ethanol.

3. A 1-m. Fenske column packed with 3/32-in. glass helices was employed by the submitters.

4. Ammonium chloride separates during the distillation. It is advisable to decant the liquid into a clean flask for distillation of the last two fractions.

3. Methods of Preparation

β-Ethoxypropionaldehyde acetal is best prepared from acrolein by reaction with ethanol.[1,2]

[1] Hall and Stern, *Chemistry & Industry,* **1950,** 775.
[2] *Org. Syntheses,* **25,** 1 (1945).

β-ETHOXYPROPIONITRILE

(Propionitrile, β-ethoxy-)

$$C_2H_5OCH_2CH_2Br + NaCN \rightarrow C_2H_5OCH_2CH_2CN + NaBr$$

Submitted by George C. Harrison and Harvey Diehl.
Checked by R. L. Shriner and C. H. Tilford.

1. Procedure

In a 3-l. three-necked flask, equipped with a Hershberg stirrer, a reflux condenser, and a dropping funnel, are placed 175 g. (3.6 moles) of sodium cyanide and 125 ml. of distilled water (Note 1). The mixture is stirred until the sodium cyanide is dissolved, and then, while vigorous stirring is continued, 535 g. (360 ml., 3.5 moles) of β-ethoxyethyl bromide (p. 370) in 260 ml. of 95% ethanol is added over a period of 15 minutes. The separatory funnel is replaced by a 360° thermometer, and the mixture is stirred and gently refluxed for 10 hours.

The mixture is then fractionally distilled until the temperature reaches 140°. The first fraction boiling at 75–95° consists mostly of ethanol and water, and is discarded. The fraction boiling between 95° and 140° contains water and 25–30 g. of β-ethoxypropionitrile. This is extracted twice with 50-ml. portions of benzene, and the benzene extracts are added to the cooled residue in the distilling flask. The mixture is filtered (Note 2), the solid material is washed with 75 ml. of benzene, which is added to the filtrate, and the whole is distilled. The fraction boiling at 169–174° is collected; it weighs 178–200 g. (52–58%).

2. Notes

1. The reaction should be carried out under the hood.
2. Direct distillation without filtration may result in clogging the condenser by ammonium bromide.

3. Methods of Preparation

β-Ethoxypropionitrile has been prepared by the action of potassium cyanide on β-ethoxyethyl bromide.[1] The addition of ethanol to acrylonitrile with sodium ethoxide,[2,3] sodium hydroxide,[3] or benzyltrimethylammonium hydroxide [4] is reported to be a superior procedure. The action of sodium ethoxide on a mixture of β-hydroxypropionitrile and ethyl formate, ethyl benzoate, ethyl bromide, or ethyl sulfate also results in the formation of β-ethoxypropionitrile.[5]

[1] Henry, *Bull. soc. chim. France,* (2) **44,** 458 (1885).
[2] Koelsch, *J. Am. Chem. Soc.,* **65,** 437 (1943).
[3] MacGregor and Pugh, *J. Chem. Soc.,* **1945,** 535.
[4] Utermohlen, *J. Am. Chem. Soc.,* **67,** 1505 (1945).
[5] Chelintsev, Benevolenskaya, and Dubinin, *J. Gen. Chem. U.S.S.R.,* **17,** 269 (1947).

ETHYL β-ANILINOCROTONATE

(Crotonic acid, β-anilino-, ethyl ester)

$$C_6H_5NH_2 + CH_3COCH_2CO_2C_2H_5 \rightarrow \underset{\displaystyle NHC_6H_5}{CH_3\overset{}{C}}{=}CHCO_2C_2H_5 + H_2O$$

Submitted by George A. Reynolds and Charles R. Hauser.
Checked by Arthur C. Cope and William R. Armstrong.

1. Procedure

In a 1-l. round-bottomed flask attached to a Dean and Stark constant water separator (Note 1) which is connected to a reflux condenser are placed 46.5 g. (45.5 ml., 0.5 mole) of redistilled aniline, 65 g. (63.5 ml., 0.5 mole) of commercial ethyl acetoacetate, 100 ml. of benzene, and 1 ml. of glacial acetic acid. The flask is heated in an oil bath at about 125°, and the water which distils out of the mixture with the refluxing benzene is removed at intervals. Refluxing is continued until no more water separates (9 ml. collects in about 3 hours) and then for an additional 30 minutes. The benzene is then distilled under reduced pressure, and the residue is transferred to a 125-ml. modified Claisen flask with an insulated column. The flask is heated in an oil or metal bath maintained at a temperature not higher than 120° while the fore-run of aniline and ethyl acetoacetate is removed by distillation under reduced pressure, and at 140–160° during distillation of the product (Note 2). The yield of ethyl β-anilinocrotonate boiling at 128–130°/2 mm., n_D^{25} 1.5770, is 78–82 g. (76–80%).

2. Notes

1. A Dean and Stark separator made with a stopcock for removal of water [1] or any continuous water separator which will return the benzene to the reaction mixture may be used.

2. The submitters state that if the temperature of the bath rises much above 120° during distillation of the fore-run, or much above 160° during distillation of the product, the ethyl β-anilinocrotonate may be contaminated with diphenylurea, part of which may precipitate from the distillate on standing. Such a precipitate can be removed by adding an equal volume of 30–60° petroleum ether to the dis-

tillate and filtering. The petroleum ether is distilled from the filtrate, and the residue is placed in a vacuum desiccator over mineral oil or is redistilled.

3. Methods of Preparation

The procedure described is a modification of procedures employed previously. Ethyl β-anilinocrotonate is obtained when the reactants are allowed to stand at room temperature for several days without a catalyst,[2,3,4] or are heated on a steam bath,[4] or are allowed to stand for 12 hours in the presence of a catalytic amount of aniline hydrochloride.[5] Iodine also is a catalyst.[5] Properties of the ester are reported by Coffey, Thomson, and Wilson.[5]

[1] Cope, Hofmann, Wyckoff, and Hardenbergh, *J. Am. Chem. Soc.*, **63**, 3452 (1941).

[2] Knorr, *Ber.*, **16**, 2593 (1883).

[3] Limpach, *Ber.*, **64**, 969 (1931).

[4] Conrad and Limpach, *Ber.*, **20**, 944 (1887).

[5] Coffey, Thomson, and Wilson, *J. Chem. Soc.*, **1936**, 856.

ETHYL AZODICARBOXYLATE

(Formic acid, azodi-, diethyl ester)

$$H_2NNH_2 \cdot H_2O + 2ClCO_2C_2H_5 + Na_2CO_3 \rightarrow C_2H_5O_2CNHNHCO_2C_2H_5 + 2NaCl + CO_2 + 2H_2O$$

$$C_2H_5O_2CNHNHCO_2C_2H_5 + HOCl \rightarrow C_2H_5O_2CN{=}NCO_2C_2H_5 + HCl + H_2O$$

Submitted by Norman Rabjohn.
Checked by H. J. Sampson and R. S. Schreiber.

1. Procedure

A. *Ethyl hydrazodicarboxylate.* In a 2-l. three-necked flask, equipped with a mechanical stirrer, two 500-ml. dropping funnels, and a thermometer (Note 1), is placed a solution of 59 g. (1 mole) of 85% hydrazine hydrate in 500 ml. of 95% ethanol. The reaction flask is cooled by means of an ice bath. When the temperature of the solution has dropped to 10°, 217 g. (2 moles) of ethyl chloroformate is added dropwise with stirring at a rate sufficient to maintain the temperature between 15° and 20°. After one-half of the ethyl chloroformate has been introduced, a solution of 106 g. (1 mole) of sodium carbonate in 500 ml. of water is added dropwise simultaneously with the remaining ethyl chloroformate. The addition of these two reactants is regulated so that the temperature does not rise above 20° and so that the addition of the chloroformate is completed slightly in advance of the sodium carbonate in order to ensure a slight excess of ethyl chloroformate in the reaction mixture at all times.

After all the reactants have been added, the precipitate on the upper walls of the flask is washed down with 200 ml. of water and the reaction mixture is allowed to stir for an additional 30-minute period. The precipitate is then collected on a Büchner funnel, washed well with a total of 800 ml. of water, and dried in an oven at 80°. There is obtained 145–150 g. (82–85%) of ethyl hydrazodicarboxylate which melts at 131–133°. It is sufficiently pure (Note 2) for the preparation of ethyl azodicarboxylate.

B. *Ethyl azodicarboxylate.* A mixture of 100 g. (0.57 mole) of ethyl hydrazodicarboxylate, 500 ml. of benzene, and 500 ml. of water is placed in a 2-l. three-necked flask equipped with a mechanical stirrer and a gas inlet tube. The flask and contents are tared, the flask is placed in an ice bath, and a slow stream of chlorine is bubbled into the mixture with stirring. The temperature is maintained below 15°, and chlorine is introduced until the increase in weight amounts to 50–55 g. (Note 3). The flow of chlorine is stopped, and the reaction mixture is stirred until a clear, orange-colored benzene layer forms when the mixture is allowed to settle.

The layers are separated, and the water layer is extracted once with benzene. The benzene solutions are combined and washed twice with 100-ml. portions of water, then with 100-ml. portions of 10% sodium bicarbonate solution until neutral (usually four to six washes are required), and twice more with water, and then are dried over anhydrous sodium sulfate. The benzene is removed under reduced pressure on a steam bath, and the residue is distilled in vacuum through a short indented column. After a small fore-run, the main fraction is collected at 107–111°/15 mm. There is obtained 80–82 g. (81–83%) of ethyl azodicarboxylate.

2. Notes

1. The thermometer and one of the funnels are fitted to a two-necked adapter; the thermometer scale must be such that the range between 10° and 20° is easily visible, preferably outside the flask, when the bulb is inserted in the liquid.

2. Ethyl hydrazodicarboxylate may be purified by crystallization from dilute ethanol; m.p. 134–135°

3. A larger excess of chlorine causes the formation of higher-boiling materials and lowers the yield of ethyl azodicarboxylate.

3. Methods of Preparation

Ethyl hydrazodicarboxylate can be prepared by the reaction of ethyl chloroformate with hydrazine hydrate [1] or hydrazine sulfate in the presence of potassium hydroxide.[2] It can be prepared also by the treatment of symmetrical hydrazinedicarboxylic acid diazide with ethanol.[3]

Ethyl azodicarboxylate can be prepared by treating ethyl hydrazodicarboxylate with concentrated nitric acid [2] or a mixture of concentrated and fuming nitric acid.[4,5]

[1] Curtius and Heidenreich, *J. prakt. Chem.*, **52**, 476 (1895); Vogelesang, *Rec. trav. chim.*, **65**, 789 (1946).

[2] Ingold and Weaver, *J. Chem. Soc.*, **127**, 381 (1925).

[3] Stolle, *Ber.*, **43**, 2470 (1910).

[4] Curtius and Heidenreich, *Ber.*, **27**, 774 (1894).

[5] Diels and Fritzsche, *Ber.*, **44**, 3018 (1911).

ETHYL BENZALMALONATE

(Malonic acid, benzal-, diethyl ester)

$$C_6H_5CHO + CH_2(COOC_2H_5)_2 \xrightarrow[\text{benzoate}]{\text{Piperidine}} C_6H_5CH{=}C(COOC_2H_5)_2 + H_2O$$

Submitted by C. F. H. Allen and F. W. Spangler.
Checked by R. L. Shriner and Fred W. Neumann.

1. Procedure

In a 1-l. round-bottomed flask, which is fitted with a Clarke-Rahrs column [1] with a unit for removing water and surmounted by a reflux condenser, are placed 100 g. (0.63 mole) of ethyl malonate (Note 1), about 72–76 g. of commercial benzaldehyde (Note 2), 2–7 ml. of piperidine (Note 3), and 200 ml. of benzene. The mixture is refluxed vigorously in an oil bath at 130–140° until no more water (total, 12–13 ml.) is collected; this operation requires 11–18 hours. After the mixture has been cooled, 100 ml. of benzene is added and the solution is washed with two 100-ml. portions of water, with two 100-ml. portions of 1 *N* hydrochloric acid, and then with 100 ml. of a saturated solution of sodium bicarbonate. The aqueous wash solutions are shaken with a single 50-ml. portion of benzene, the benzene extract is added to the original organic layer, and the organic solution is dried with 30 g. of anhydrous sodium sulfate. After the benzene has been removed under reduced pressure on a steam bath, the residue is distilled under reduced pressure from a 250-ml. modified Claisen flask well wrapped with asbestos. The yield of colorless ethyl benzalmalonate boiling at 140–142°/4 mm. (Note 4) is 137–142 g. (89–91%) (Note 5).

2. Notes

1. B.p. 94–96°/16 mm. The checkers used redistilled ethyl malonate.

2. Commercial benzaldehyde which contains 2–8% of benzoic acid is eminently satisfactory. The acid content is determined (conveniently by titration of a sample in neutral ethanol with standard alkali), and a quantity of aldehyde containing 70 g. (0.66 mole) of benzaldehyde is used. The checkers used 76 g. of commercial benzaldehyde which contained 8% of benzoic acid. If pure benzaldehyde is employed, then 2 g. of benzoic acid should be added.

It has been shown [2] that piperidine salts and not the free base act as the catalyst in the Knoevenagel reaction. The checkers obtained only a 71% yield with benzaldehyde containing less than 0.2% of benzoic acid.

3. The amount of piperidine that is employed depends on the benzoic acid content of the aldehyde; it should be slightly in excess of that required to neutralize the benzoic acid. About 1.2 ml. of piperidine per gram of acid is satisfactory.

4. Another boiling point is 179–181°/10 mm. The boiling points reported for ethyl benzalmalonate vary widely. The temperature observed depends on the degree of superheating and the rapidity of distillation.

5. Methyl benzalmalonate (b.p. 143–146°/2 mm., 169–171°/10 mm.) can be prepared in the same manner and in the same apparatus in yields of 90–94%.

3. Methods of Preparation

Ethyl benzalmalonate has been prepared by the esterification of benzalmalonic acid in the presence of concentrated sulfuric acid; [3] by the action of benzal chloride on sodiomalonic ester; [4] by the condensation of benzaldehyde and malonic ester in the presence of hydrogen chloride,[5,6] acetic anhydride,[7] alcoholic piperidine or ammonia,[8] or aluminum chloride,[9] and by the action of phenylmagnesium bromide on hydroxymethylene malonic ester.[10]

[1] Clarke and Rahrs, *Ind. Eng. Chem.*, **18**, 1092 (1926).

[2] Kuhn, Badstubner, and Grundmann, *Ber.*, **69**, 98 (1936).

[3] Stuart, *J. Chem. Soc.*, **47**, 158 (1885).

[4] Avery and Bouton, *Am. Chem. J.*, **20**, 510 (1898).

[5] Claisen, *Ber.*, **14**, 348 (1891).

[6] Claisen and Crismer, *Ann.*, **218**, 131 (1883).

[7] Knoevenagel, *Ber.*, **31**, 2591 (1898); Gravel, *Nat. canadien*, **57**, 181 (1931) [*C. A.*, **28**, 169 (1934)].

[8] Ger. pat. 97,734 [*Frdl.*, **5**, 906 (1898–1900)].

[9] Breslow and Hauser, *J. Am. Chem. Soc.*, **62**, 2385 (1940).

[10] Reynolds, *Am. Chem. J.*, **44**, 314 (1910).

ETHYL BENZOYLACETATE *

(Acetic acid, benzoyl-, ethyl ester)

$$CH_3COCH_2CO_2C_2H_5 + C_2H_5ONa \rightarrow [CH_3COCHCO_2C_2H_5]Na + C_2H_5OH$$

$$[CH_3COCHCO_2C_2H_5]Na + C_6H_5CO_2C_2H_5 \rightarrow [C_6H_5COCHCO_2C_2H_5]Na + CH_3CO_2C_2H_5$$

$$[C_6H_5COCHCO_2C_2H_5]Na + H_2SO_4 \rightarrow C_6H_5COCH_2CO_2C_2H_5 + NaHSO_4$$

Submitted by S. M. McElvain and K. H. Weber.
Checked by R. L. Shriner and C. H. Tilford.

1. Procedure

A 2-l. three-necked flask is mounted on a steam bath and fitted with a reflux condenser, a separatory funnel, and an efficient sealed stirrer. In the flask is placed 600 ml. of absolute ethanol, and to this is added gradually 46 g. (2 gram atoms) of clean sodium cut into small pieces. The sodium ethoxide solution is stirred and cooled to room temperature, after which 267 g. (260 ml., 2.05 moles) of ethyl acetoacetate (Note 1) is added slowly through the separatory funnel. The reflux condenser is then replaced by a short still head, and the ethanol is removed by distillation at approximately room temperature and under the pressure of a water pump. When approximately half the ethanol has been removed, sufficient sodium enolate precipitates so that stirring has to be discontinued. When the residue appears dry (after about 2 hours) the last traces of ethanol are removed by heating for an hour on the steam bath under a pressure of 2 mm. The flask is allowed to cool to room temperature under reduced pressure.

To the cooled residue of sodium enolate is added 600 g. (570 ml., 4 moles) of ethyl benzoate. The steam bath is replaced by an oil bath, and the temperature of the bath is raised to 140–150° and maintained there for 6 hours. Then, over a period of another hour, the temperature of the bath is gradually raised to 180° (Note 2). The distillate that is collected during this period of heating amounts to 200–210 g. and consists chiefly of ethyl acetate and ethanol.

The reaction mixture is cooled, 250 ml. of water is added, and the mixture is made acid to litmus by addition of a cooled solution of 100 g. of concentrated sulfuric acid in 200 ml. of water. Chipped ice is added if necessary to keep the mixture cool. The upper ester layer is separated, and the aqueous layer is extracted with 200 ml. of ether. The combined ether and ester layers are shaken with 350 ml. of a saturated sodium bicarbonate solution until no more carbon dioxide is evolved, and then the organic layer is washed with 200 ml. of water. The water layer is combined with the sodium bicarbonate solution and extracted with 400 ml. of ether. The combined ether and ester layers are dried over sodium sulfate. The ether is removed by distillation on the steam bath, and the excess ethyl benzoate and acetoacetic ester (Note 3) are then removed by distillation under reduced pressure through a 15-cm. fractionating column. Finally, the ethyl benzoylactetate is distilled (Note 3) at 101–106°/1 mm. (130–135°/3 mm.). The yield of ester boiling over a 5° range is 190–210 g. (50–55% based on the ethyl acetoacetate).

2. Notes

1. Commercial ethyl acetoacetate was distilled, and the fraction boiling at 68–69°/11 mm. was used.

2. During the last hour of heating about 20 g. of the ethyl acetate-ethanol mixture distilled.

3. A mixture of ethyl acetoacetate and ethyl benzoate (100–150 g.) was collected at 75–90°/12 mm., after which 250–300 g. of pure ethyl benzoate, b.p. 90–93°/12 mm., was recovered. These products were removed through a 15-cm. fractionating column. The remaining ethyl

benzoylacetate was distilled through a short still head without a fractionating column.

3. Methods of Preparation

Methods of preparation are listed in an earlier volume.[1] In addition ethyl benzoylacetate has been prepared by the reaction of benzoyl chloride with the magnesium enolate of malonic ester [2] and subsequent decomposition with β-naphthalenesulfonic acid,[3] and by the condensation of acetophenone with ethyl carbonate in the presence of sodium hydride.[4]

[1] *Org. Syntheses* Coll. Vol. **2**, 266 (1943).
[2] *Org. Syntheses* Coll. Vol. **2**, 594 (1943).
[3] Riegel and Lilienfeld, *J. Am. Chem. Soc.*, **67**, 1273 (1945).
[4] Swamer and Hauser, *J. Am. Chem. Soc.*, **72**, 1352 (1950).

ETHYL BROMOACETATE *

(Acetic acid, bromo-, ethyl ester)

$$CH_3COOH + Br_2 \xrightarrow[(CH_3CO)_2O]{\text{Pyridine}} BrCH_2COOH + HBr$$

$$BrCH_2COOH + C_2H_5OH \xrightarrow{H_2SO_4} BrCH_2COOC_2H_5 + H_2O$$

Submitted by Samuel Natelson and Sidney Gottfried.
Checked by Nathan L. Drake and Stuart Haywood.

1. Procedure

A. *Bromoacetic acid.* (Note 1.) A mixture of 1 l. (17.5 moles, excess) of glacial acetic acid, 200 ml. of acetic anhydride, and 1 ml. of pyridine is placed in a 3-l. flask fitted with a dropping funnel and a reflux condenser, the end of which is protected with a drying tube (Note 2); the tip of the dropping funnel should reach below the level of the liquid. Some glass beads are added, and the mixture is heated to boiling. The flame is then removed, approximately 1 ml. of bromine is added, and the reaction is allowed to proceed until the liquid becomes colorless (Note 3). Then the remainder of 1124 g. (360 ml., 7.03 moles) of bromine (Note 4) is added as rapidly as it will *react* (Note 5); during this period (about 2.5 hours), the acid is kept boiling *gently* by means of a flame. After about half the bromine has been added, the liquid assumes a cherry color which is retained throughout the remainder of the bromination. After all the bromine has been added, the mixture is heated until it becomes colorless.

The mixture is allowed to cool, and 75 ml. of water is added slowly to destroy the acetic anhydride. Excess acetic acid and water are now removed on a boiling water bath under a pressure of approximately 35 mm. When the evaporation is complete, the residue will crystallize on cooling; this residue, which is almost pure bromoacetic acid, weighs 845–895 g. (Note 6).

B. *Ethyl bromoacetate.* For the esterification, an apparatus similar to that used in the preparation of anhydrous oxalic acid [*Org. Syntheses* Coll. Vol. **1,** 422 (1941)] may be used, but with the outlets from the trap reversed so that the lighter liquid returns to the mixture and the heavier liquid (water) is drawn off at the bottom. A somewhat simpler apparatus may be built using the water trap shown in Fig. 12 (Note 7). The crude bromoacetic acid is placed in a 3-l. flask, together with 610 ml. of ethanol (9.9 moles, excess) and 950 ml. of benzene. About 1.5 ml. of concentrated sulfuric acid is added to hasten the reaction (Note 8), and the mixture is refluxed on a boiling water bath while the water is separated and measured. Approximately 296 ml. of liquid (whose composition is approximately 50% ethanol and water) separates from the benzene; this includes all the water formed in the reaction, together with the excess ethanol. When no more water separates from the benzene, 75 ml. of ethanol is added to the reaction mixture and heating is continued for 30 minutes. If the reaction has been completed, there will not be a second phase in the distillate. The end of the reaction is also indicated when the benzene flowing through the side tube becomes clear and the rate of refluxing decreases considerably. At this point, 150 ml. of benzene is condensed and removed through the trap.

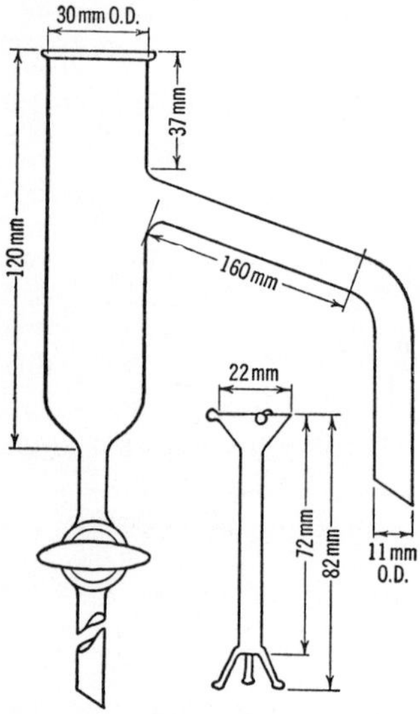

FIG. 12.

The mixture is transferred to a separatory funnel and washed once with 1.5 l. of water, once with 1.5 l. of 1% sodium bicarbonate solution, and finally with 1.5 l. of water. It is then dried over anhydrous sodium sulfate and fractionated at atmospheric pressure from an oil bath using a Vigreux column 1 ft. in length (Note 9). The fraction boiling at 154–155°/759 mm. is collected (Note 10). The yield is approximately 818 g. (65–70%).

2. Notes

1. The vapors of ethyl bromoacetate are extremely irritating to the eyes. Care should be taken to keep the material in closed containers and to manipulate it in open vessels only in a good hood.

2. An all-glass apparatus is advisable. If it is not available, one-holed asbestos stoppers may be made by soaking strips of asbestos in water, wrapping them around pieces of glass tubing of suitable size until the desired diameter has been reached, and then allowing them to dry at 110°.

3. At the beginning there is a lag of about 10 minutes before the reaction starts and the color of the bromine disappears.

4. If C.P. bromine is available it may be used directly. Technical bromine should be dried with concentrated sulfuric acid.

5. The bromine should not be added so rapidly that loss occurs through the condenser.

6. Pure bromoacetic acid may be obtained by distillation of this crude product from a Claisen flask immersed in an oil bath and fitted with an 8-in. insulated Vigreux column. The fraction boiling at 108–110°/30 mm. is collected. The yield is 775–825 g. (80–85%).

7. The trap shown in Fig. 12 is a modification of the moisture trap designed by Dean and Stark.[1] The dimensions may be varied to suit individual purposes, for the size is largely a matter of convenience. The trap may be used without the inner funnel, but with this funnel (C. F. Koelsch, private communication) the condensate separates into two layers rapidly and completely, and the liquid falling from the condenser does not agitate the two phases in the trap. The funnel tube must be of such a length that the top of the funnel is above the side arm of the trap. The tube of the trap may be graduated, but this is not necessary.

8. In the absence of a catalyst the reaction proceeds more slowly and smaller yields are obtained. Phosphoric acid may be substituted for sulfuric acid, but the use of sulfuric acid results in the best yield in the shortest time.

9. If a fractionating column is not used, as much as 15–20% of the product may be lost in the fore-run.

10. The fraction boiling over a 1-degree range is collected. The boiling point has been observed to range from 154–155° to 158–159° on different days.

3. Methods of Preparation

Bromoacetic acid has been prepared by direct bromination of acetic acid at elevated temperatures and pressures,[2–4] or with dry hydrogen chloride as a catalyst; [5] and with red phosphorus as a catalyst with the formation of bromoacetyl bromide.[6–10] Bromoacetic acid has also been prepared from chloroacetic acid and hydrogen bromide at elevated temperatures; [6] by oxidation of ethylene bromide with fuming nitric acid; [7] by oxidation of an ethanolic solution of bromoacetylene by air; [8] and from ethyl α,β-dibromovinyl ether by hydrolysis.[9] Acetic acid has been converted into bromoacetyl bromide by action of bromine in the presence of red phosphorus, and ethyl bromoacetate has been obtained by action of ethanol upon the acid bromide.[10–14] Ethyl bromoacetate has also been prepared by direct bromination of ethyl acetate at elevated temperatures; [15,16] by action of ethanol upon bromoacetic anhydride; [17] by action of phosphorus tribromide upon ethyl glycollate; [18] and by action of hydrogen bromide upon ethyl diazoacetate.[19] The method described above is based upon the procedure of Natelson and Gottfried.[20]

[1] Dean and Stark, *Ind. Eng. Chem.*, **12**, 486 (1920).
[2] Perkin and Duppa, *Ann.*, **108**, 106 (1858).
[3] Michael, *Am. Chem. J.*, **5**, 202 (1883).
[4] Hell and Muhlhauser, *Ber.*, **11**, 241 (1878); **12**, 735 (1879).
[5] Lapworth, *J. Chem. Soc.*, **85**, 41 (1904).
[6] Demole, *Ber.*, **9**, 561 (1876).
[7] Kachler, *Monatsh*, **2**, 559 (1881).
[8] Gloeckner, *Ann.*, Suppl. **7**, 115 (1870).
[9] Imbert and Konsort. Electrochem. Ind., Ger. pat. 216,716 [*Frdl.*, **9**, 28 (1908–10); *C. A.*, **4**, 952 (1910)].
[10] Ward, *J. Chem. Soc.*, **121**, 1161 (1922).
[11] Naumann, *Ann.*, **129**, 268 (1864).
[12] Auwers and Bernhardi, *Ber.*, **24**, 2218 (1891).
[13] Lassar-Cohn, *Ann.*, **251**, 341 (1889).
[14] Volhard, *Ann.*, **242**, 161 (1887).
[15] Crafts, *Compt. rend.*, **56**, 707 (1863); *Ann.*, **129**, 50 (1864).
[16] Schutzenberger, *Ber.*, **6**, 71 (1873).
[17] Gal, *Compt. rend.*, **71**, 274 (1870).
[18] Henry, *Ann.*, **156**, 176 (1870).
[19] Curtius, *J. prakt. Chem.*, (2) **38**, 430 (1888).
[20] Natelson and Gottfried, *J. Am. Chem. Soc.*, **61**, 970 (1939).

ETHYL *n*-BUTYLCYANOACETATE

(Caproic acid, α-cyano-, ethyl ester)

$$n\text{-}C_3H_7CHO + CH_2(CN)COOC_2H_5 + H_2 \xrightarrow{Pd} n\text{-}C_4H_9CH(CN)COOC_2H_5 + H_2O$$

Submitted by Elliot R. Alexander and Arthur C. Cope.
Checked by Cliff S. Hamilton and Robert F. Coles.

1. Procedure

A mixture of ethyl cyanoacetate (Note 1) (56.6 g., 0.5 mole), freshly distilled butyraldehyde (43.2 g., 0.6 mole), 1 g. of palladium on carbon (Note 2), and 80 ml. of glacial acetic acid is placed in a 500-ml. bottle suitable for attachment to a low-pressure reduction apparatus. A solution of piperidine (2.0 ml., 0.02 mole) in 20 ml. of glacial acetic acid is added, and the bottle is connected to the reduction apparatus.

The bottle is alternately evacuated and filled with hydrogen twice, and the mixture is reduced by shaking with hydrogen at 1 to 2 atm. (15–30 lb.) pressure. The reduction is rapid and exothermic. In 1–2 hours the theoretical amount of hydrogen (0.5 mole) is taken up and absorption ceases (Note 3).

The reaction mixture is filtered through a Hirsch or Büchner funnel, and the bottle is rinsed with 50 ml. of benzene, which is also poured through the funnel. The filtrate is washed with two 50-ml. portions of 10% sodium chloride solution and three 25-ml. portions of water (Note 4). The washings are extracted with three 10-ml. portions of benzene, and the combined benzene solutions are distilled under reduced pressure from a 250-ml. modified Claisen flask. The yield of ethyl *n*-butylcyanoacetate, b.p. 108–109°/9 mm., is 79–81 g. (94–96%, based on the ethyl cyanoacetate used) (Note 5).

2. Notes

1. Ethyl cyanoacetate was purchased from the Dow Chemical Company, Midland, Michigan, and redistilled before using.

2. The palladium on carbon catalyst was prepared by the following method, developed by Walter H. Hartung, School of Pharmacy, University of Maryland, Baltimore. Ten milliliters of a commercial palladium chloride solution containing 0.1 g. of palladium and approxi-

mately 0.05 g. of hydrogen chloride per milliliter (obtained from the J. Bishop Company, Malvern, Pennsylvania) is added to a solution of 27 g. of sodium acetate trihydrate in 100 ml. of water. Norit (9 g.) is added, and the mixture is hydrogenated until absorption ceases. The catalyst (10 g.) is filtered on a Büchner funnel, washed with water, dried by drawing air through the funnel for about 30 minutes, and stored in a desiccator over calcium chloride. The palladium catalysts, prepared as described elsewhere in this volume, are presumably also satisfactory for the reductive alkylation described above (p. 685).

3. It is advisable to start the reduction as soon as the reactants are mixed. The yield dropped to 87% when the reaction mixture was allowed to stand for 3 hours before hydrogenating.

4. If an emulsion is formed, it can be broken by adding a few drops of ethanol or several milliliters of ether.

5. Ethyl ethylcyanoacetate and ethyl propylcyanoacetate have been prepared by the same method in yields of 85 and 94%, respectively. Other aldehydes and ketones have been used under slightly different conditions to prepare other ethyl monoalkylcyanoacetates.[1]

3. Methods of Preparation

Ethyl *n*-butylcyanoacetate has been prepared by alkylation of the sodium enolate of ethyl cyanoacetate with butyl bromide [2] and by condensation of capronitrile with ethyl carbonate,[3] in addition to the method given above.[1]

[1] Alexander and Cope, *J. Am. Chem. Soc.*, **66**, 886 (1944).

[2] Hessler and Henderson, *J. Am. Chem. Soc.*, **43**, 674 (1921).

[3] Wallingford, Jones, and Homeyer, *J. Am. Chem. Soc.*, **64**, 577 (1942).

2-ETHYLCHROMONE

(Chromone, 2-ethyl-)

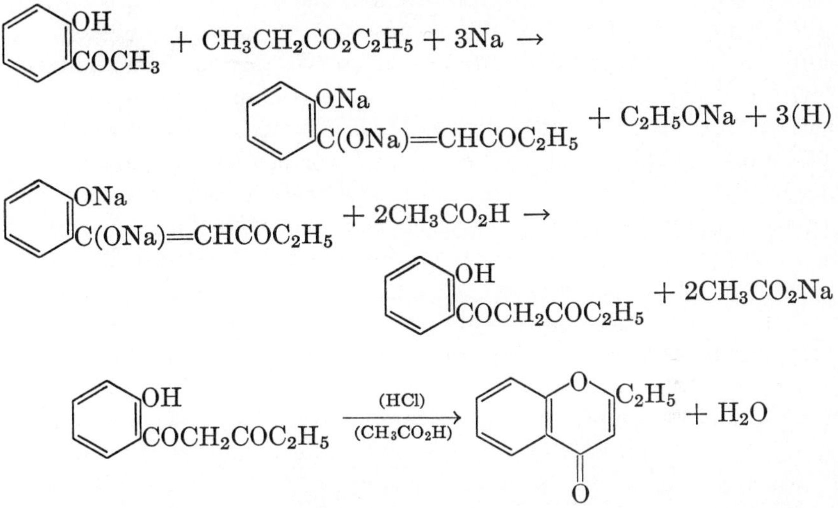

Submitted by Ralph Mozingo.
Checked by Homer Adkins and Robert J. Gander.

1. Procedure

In a 1-l. three-necked flask fitted (Note 1) with a reflux condenser, a dropping funnel, and a stirrer (Note 2) are placed 24 g. (1.05 gram atoms) of oxide-free sodium and 200 ml. of xylene (Note 3). The condenser is protected by a drying tube containing soda-lime. The flask is surrounded by an oil bath which is heated until the sodium is melted. The stirrer is started, and, after the sodium is powdered, the oil bath is removed. When the contents of the flask have cooled to room temperature, the stirrer is stopped, the xylene decanted, and the sodium washed with 100 ml. of dry ether to remove traces of xylene.

The flask is replaced and completely surrounded by an ice bath (Note 4). A current of nitrogen is passed through the flask for 5 minutes by means of a rubber tube attached to the dropping funnel. The nitrogen inlet tube is removed, and a mixture of 60 g. (0.44 mole) of *o*-hydroxyacetophenone (Note 5) and 125 g. (140 ml., 1.2 moles) of ethyl propionate (Note 3) is placed in the dropping funnel and protected by a soda-lime tube. The ester-ketone mixture is dropped very slowly onto the powdered sodium. After a small amount (2–5 ml.) of the mixture has been added, the ice bath is removed and the flask warmed with the hand to make sure that the reaction has started (Note 6). When the reaction has started, the funnel is adjusted so that one-half of the ester-ketone mixture is added over a period of 1.5–2 hours. After half of the mixture has been added, the rate of addition is again regulated so that the remainder is added in 30 minutes.

After all the ester-ketone mixture has been added and the reaction has subsided, the ice bath is removed and replaced by a steam bath. The crust on the mixture is broken up and the mixture heated on a steam bath until all the sodium has reacted (30–60 minutes). The reaction mixture is allowed to cool and is poured with stirring onto 300 g. of crushed ice. Ninety-five milliliters of glacial acetic acid in 350 ml. of water is added, and the mixture is stirred for 30 minutes. The organic layer which separates is removed by means of a separatory funnel, and the water layer is extracted three times with 200-ml. portions of ether. The extracts are combined with the original organic material, and the ether is distilled from a steam bath. To the residue are added 150 ml. of glacial acetic acid and 10 ml. of concentrated hydrochloric acid, and the solution is refluxed for 30 minutes, after which it is allowed to cool to room temperature (Note 7).

The entire reaction mixture is then subjected to fractional distillation through a Widmer column (Note 8). The material boiling up to 120° is removed at atmospheric pressure and discarded. The contents of the flask are cooled and the pressure reduced to 7 mm. The fraction up to 80°/7 mm. is removed and discarded. The material boiling at 80–90°/7 mm. is collected, and, after the contents of the flask have been cooled, the pressure is reduced to 2 mm. and all the material boiling up to 110°/2 mm. is collected and combined with the 80–90°/7 mm. fraction. The next fraction, boiling at 110–138°/2 mm., is collected separately. On refractionation of the lower fraction (b.p. 80°/7 mm.–110°/2 mm.) there is obtained 13–15 g. of *o*-hydroxyacetophenone (b.p. 87–88°/7 mm.). To the residue after recovery of the *o*-hydroxyacetophenone is added the fraction boiling at 110–138°/2 mm. The pressure is again reduced to 2 mm. and the fraction boiling at 124–126°/2 mm. collected. This fraction is pure 2-ethylchromone and weighs 42–45 g. (70–75% based on the ketone not recovered) (Note 9).

2. Notes

1. All the stoppers used should be rubber ones that have been boiled for 2 hours in 10% sodium hydroxide and thoroughly washed and dried before use.

2. A convenient stirrer may be made by fitting a ¼-in brass rod with a disk 1 in. in diameter. Four holes are drilled through the disk, and through each hole is placed a loop of 18-gauge Nichrome wire. The loops are each turned so that the twisted end is next to the stirrer shaft and each is twisted in the middle to give a double loop. The stirrer shaft is then passed through a 4-in. length of glass tubing which fits tightly about the shaft. A rubber seal is made by slipping a piece of rubber tubing over the brass rod and glass tubing. The rubber tubing extends about 3 mm. up onto the stirrer shaft, over which it fits rather tightly. A drop of glycerol placed on the stirrer shaft just above the rubber sleeve serves as a lubricant for the seal. The seal is similar to that described for the preparation of dypnone (p. 368, Note 1).

3. The xylene is dried over sodium and distilled from it. The ethyl propionate is dried over phosphorus pentoxide, more of the drying agent being added from time to time until some solid phosphorus pentoxide remains. The ester is then decanted into a flask containing fresh phosphorus pentoxide and fractionated. The fraction boiling at 98–99°/745 mm. is collected.

4. An ice bath is used to keep the reaction from becoming too violent. An ice-salt bath must not be used or the reaction may be delayed in starting.

5. The *o*-hydroxyacetophenone was made as described by Miller and Hartung [*Org. Syntheses* Coll. Vol. **2,** 543 (1943)] for *o*- and *p*-hydroxypropiophenone except that the procedure for isolating the product was modified as follows. The cold, partly solidified oil, separating after acidification, was diluted with benzene and the water layer extracted with benzene. After distillation of the benzene, the residue was distilled at 17 mm. until the solid *p*-isomer began to collect in the condenser. The distillate was then fractionated through a Widmer column, and the fraction boiling at 90–110°/17 mm. was saved. Refractionation gave *o*-hydroxyacetophenone boiling at 105–106°/20 mm. or 87–88°/7 mm. in a yield of 30% based upon the phenyl acetate.

6. It is *very important* that the operator assure himself that the reaction has started before more of the ester-ketone mixture is added, and that the first half of the mixture be added slowly, or the reaction may proceed with *explosive violence.* It is usually possible to tell when the reaction has started by the local evolution of heat when the ice bath is removed and by the appearance of the yellow salt of the diketone.

7. The preparation should be carried as far as this point without interruption.

8. The Widmer column had a spiral 15 cm. in length, and had an electrically heated jacket.[1]

9. The submitter carried out this preparation with twice the quantities recommended here. The yield was 68–77 g. (56–64%).

3. Methods of Preparation

This chromone has been prepared by the condensation of ethyl propionate with *o*-hydroxyacetophenone using sodium.[2,3]

[1] Smith and Adkins, *J. Am. Chem. Soc.*, **60**, 662 (1938).
[2] Heilbron, Hey, and Lowe, *J. Chem. Soc.*, **1934**, 1312.
[3] Mozingo and Adkins, *J. Am. Chem. Soc.*, **60**, 669 (1938).

ETHYL DIACETYLACETATE

$$CH_3COCH_2CO_2C_2H_5 + CH_3COCl \xrightarrow{Mg} (CH_3CO)_2CHCO_2C_2H_5 + (H_2 + MgCl_2)$$

Submitted by A. Spassow.
Checked by Nathan L. Drake, Leonard Smith, and Jonathan W. Williams.

1. Procedure

A mixture of 12 g. (0.50 gram atom) of magnesium turnings, 130 g. (1.0 mole) of ethyl acetoacetate, 200 g. of benzene (dried over sodium), and 120 g. (1.50 moles) of acetyl chloride is heated under reflux for 2 hours in a 1-l. round-bottomed flask provided with a condenser closed by a calcium chloride tube and supported in an oil bath (85–90°) (Note 1). The yellow reaction mixture is cooled in an ice bath, and the liquid portion is decanted into a separatory funnel. The residue in the flask is washed twice with 50-ml. portions of ether, and the ethereal solution is poured over ice. The ether-water mixture is then added to the benzene solution in the separatory funnel, and the mixture is shaken thoroughly (Note 2); the aqueous layer is drawn off and discarded. The benzene-ether solution is washed once with 500 ml. of 5% sodium bicarbonate solution and once with 50 ml. of water, and finally is dried over calcium chloride. The ether and most of the benzene are removed by distillation from a water bath, and the remainder of the benzene is driven off at 50°/50 mm. The ethyl diacetylacetate is then precipitated from the residue as the copper derivative by the addition of 1.2 l. of a saturated aqueous solution of copper acetate

(Note 3). After addition of the copper acetate solution, the contents of the flask are shaken vigorously now and then and allowed to stand for an hour to ensure complete precipitation of the copper derivative. The blue copper derivative is filtered on a Büchner funnel, washed with two 50-ml. portions of water, and transferred directly to a separatory funnel where it is mixed with 600 ml. of ether.

Four hundred milliliters of 25% sulfuric acid is added, and the contents of the funnel are shaken continually until the copper derivative has disappeared (5–10 minutes). After separation of the ethereal layer, the aqueous layer is extracted twice with 100-ml. portions of ether, and the combined ethereal extracts are dried over calcium chloride. The ether is removed on the steam bath and the residual ester distilled under diminished pressure. A few drops come over up to 90°, but the bulk of the material distils at 92–98°/12 mm. Redistillation yields pure ethyl diacetylacetate boiling at 95–97°/12 mm. The yield is 80–90 g. (46–52%).

2. Notes

1. Hydrogen and hydrogen chloride are evolved; this operation must be conducted in a hood.

2. The greater part of the magnesium remains unchanged. It should be removed by filtering the solution through a plug of *fine* glass wool in an ordinary funnel.

3. The saturated solution of copper acetate is prepared by dissolving 100 g. of finely pulverized copper acetate in 1.2 l. of boiling water. If the preparation contains basic salt, a few milliliters of acetic acid is added, and the solution is filtered. The solution is cooled to 35° before use.

3. Methods of Preparation

Ethyl diacetylacetate has been prepared by Claisen from the sodium derivative of acetylacetone and ethyl chloroformate.[1] It has also been prepared from the sodium derivative of ethyl acetoacetate and acetyl chloride,[2–4] from ethyl acetoacetate and acetyl chloride in the presence of magnesium,[5] and from ethyl acetoacetate and ketene.[6]

[1] Claisen, *Ann.*, **277**, 171 (1893).

[2] James, *Ann.*, **226**, 211 (1884).

[3] Elion, *Rec. trav. chim.*, **3**, 250 (1884).

[4] Michael, *Ber.*, **38**, 2088 (1905).

[5] Ogata, Nosaki, and Takagi, *J. Pharm. Soc. Japan*, **59**, 105 (1939) [*C. A.*, **33**, 4230 (1939)].

[6] U. S. pat. 2,417,381 [*C. A.*, **41**, 4169 (1947)].

ETHYL DIAZOACETATE

(Acetic acid, diazo-, ethyl ester)

$$CH_2(NH_3Cl)CO_2C_2H_5 + NaNO_2 \rightarrow N_2CHCO_2C_2H_5 + NaCl + 2H_2O$$

Submitted by Ennis B. Womack and A. B. Nelson.
Checked by R. L. Shriner and C. H. Tilford.

1. Procedure

A 1-l. three-necked round-bottomed flask is fitted with a 50-ml. separatory funnel and a mechanical stirrer sealed with a well-lubricated rubber collar. A stopper in the third neck of the flask carries a glass tube that reaches to the bottom of the flask, enters the top of a 1-l. separatory funnel, and extends down to the stopcock.

A solution of 140 g. (1 mole) of glycine ethyl ester hydrochloride and 3 g. of sodium acetate in 150 ml. of water is added to the flask and cooled to 2° by means of an ice-salt bath. A cold solution of 80 g. (1.15 moles) of sodium nitrite in 100 ml. of water is added, and the mixture is stirred until the temperature has fallen to 0°. The temperature is maintained below 2°, and stirring is continued throughout all the following operations. To the cold mixture are added 80 ml. of cold, ethanol-free ethyl ether (Note 1) and 3 ml. of cold 10% sulfuric acid. After 5 minutes, the reaction mixture is blown over into the 1-l. separatory funnel by application of air pressure. The lower aqueous layer is *quickly* sucked back into the reaction flask. The ether layer is removed and immediately washed with 50 ml. of cold 10% sodium carbonate solution. This ether solution should be neutral to moist litmus paper; if not, the washing with sodium carbonate is repeated. The ether solution is finally dried over 10 g. of anhydrous sodium sulfate.

A second portion of 80 ml. of ethanol-free ether is then added to the reaction mixture with stirring, followed by 15 ml. of cold 10% sulfuric acid over a period of 5 minutes. After 3 minutes' contact (Note 2), the ether layer is removed as before, washed immediately with 50 ml. of fresh 10% sodium carbonate solution, and dried over 10 g. of sodium sulfate. This procedure is repeated (about 6 or 7 times) until the ether layer is no longer yellow.

The combined ether solutions are then subjected to distillation at 20° or below under the vacuum obtainable from a water pump until all the ether is removed. Prolonged distillation results in decomposition of the diazo ester and in a decreased yield. The yellow residual oil is practically pure ethyl diazoacetate and is satisfactory for most synthetic purposes (Note 3). The yield is about 98 g. (85%) (Notes 4 and 5).

2. Notes

1. The ethanol is removed from 600 ml. of commercial ethyl ether by thorough washing with 75 ml. of a saturated calcium chloride solution.

2. The ether layer should be removed as rapidly as possible from the aqueous layer, because ethyl diazoacetate is rapidly decomposed by acid.

3. Ethyl diazoacetate may be purified, but with considerable loss, by steam distillation under reduced pressure.[1]

4. The submitters have carried out preparations using twice the amounts stated.

5. The product should be placed in dark brown bottles and kept in a cool place. It should be used as soon as possible. *Distillation, even under reduced pressure, is dangerous, for the substance is explosive.*

3. Methods of Preparation

Ethyl diazoacetate has been prepared by the action of sodium nitrite on glycine ethyl ester hydrochloride.[1–3] A modification of this method, which is reported to give better yields than those described here, has been published.[4]

[1] Gattermann, *Laboratory Methods of Organic Chemistry,* p. 277, The Macmillan Company, New York, 1948.

[2] Curtius, *J. prakt. Chem.,* (2) **38,** 396 (1888); Silberrad, *J. Chem. Soc.,* **81,** 600 (1902).

[3] U. S. pat. 2,490,714 [*C. A.,* **44,** 3519 (1950)].

[4] La Forge, Gersdorff, Green, and Schechter, *J. Org. Chem.,* **17,** 381 (1952).

ETHYLENE THIOUREA *

(2-Imidazolidinethione)

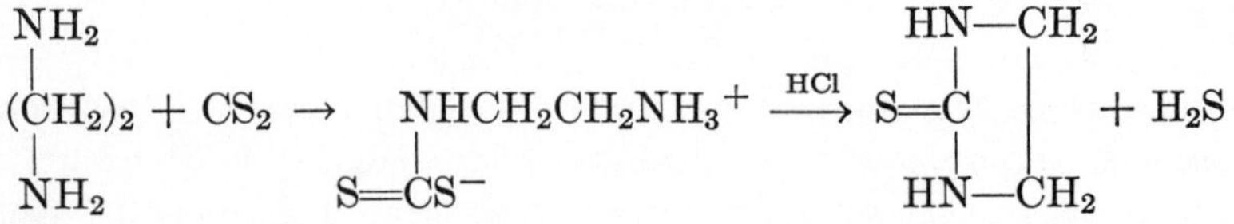

Submitted by C. F. H. ALLEN, C. O. EDENS, and JAMES VANALLAN.
Checked by R. L. SHRINER and CURTIS D. SNOW.

1. Procedure

In a 2-l. round-bottomed flask are placed 120 g. (1.83 moles) of 92% ethylenediamine (Note 1), 300 ml. of 95% ethanol, and 300 ml. of water. The flask is attached to an efficient reflux condenser, and 121 ml. of carbon disulfide is placed in a separatory funnel attached to the top of the condenser by means of a notched cork. About 15 to 20 ml. of the carbon disulfide is added, and the flask is shaken to mix the contents. A vigorous reaction takes place (Note 2), and it may be necessary to cool the flask. After the reaction has started, a water bath at 60° is placed under the flask and the balance of the carbon disulfide is added at such a rate that the vapors reflux about one-third the way up the condenser. About 2 hours is required for the addition of the carbon disulfide. At this time the bath temperature is raised to about 100°, and the mixture is allowed to reflux for 1 hour. Then 15 ml. of concentrated hydrochloric acid is added, and the mixture is refluxed under a good hood (bath at 100°) for 9–10 hours. The mixture is cooled in an ice bath, and the product is filtered by suction on a Büchner funnel and washed with 200–300 ml. of cold acetone (Note 3). A yield of 156–167 g. (83–89%) of white crystals is obtained melting at 197–198° (Notes 4 and 5).

2. Notes

1. If commercial ethylenediamine is used, it should be redistilled. The concentration of the ethylenediamine may be determined by titration with standard acid and the proper amount taken.

2. Care should be exercised to make certain that the reaction starts, before an additional quantity of carbon disulfide is added. In one ex-

periment in which the carbon disulfide was added all at once, a very violent reaction occurred.

3. Since all the likely contaminants are readily volatile, extensive washing is unnecessary.

4. This product is pure enough for most purposes. It gives no precipitate with copper sulfate solution, indicating the absence of the open-chain acid.[1]

5. It has been reported (J. VanAllan) that this procedure may also be used for the preparation of 2-thio-3,4,5,6-tetrahydropyrimidine; in this case 130 g. (150 ml., 1.76 moles) of propylenediamine is used in place of ethylenediamine. The yield is 160 g. (80%) of colorless product, m.p. 207–208°, after recrystallization from water.

3. Methods of Preparation

The only practical method for preparing alkylene thioureas is by the action of the diamines upon carbon disulfide in aqueous alcohol.[1–5] The final heating is essential to convert the thiocarbamic acid into the cyclic compound, the addition of hydrochloric acid being beneficial. Alkylene thioureas have also been prepared by heating N-formylalkylenediamines with sulfur.[6]

[1] Ruiz and Libenson, *Anales asoc. quím. argentina,* **18,** 37 (1930) [*C. A.,* **24,** 5726 (1930)].

[2] Hofmann, *Ber.,* **5,** 242 (1872).

[3] Johnson and Edens, *J. Am. Chem. Soc.,* **64,** 2707 (1942).

[4] Schackt, *Arch. Pharm.,* **235,** 442 (1897).

[5] Klut, *Arch. Pharm.,* **240,** 675 (1902).

[6] Zienty, *J. Am. Chem. Soc.,* **68,** 1388 (1946).

ETHYL ETHOXYMETHYLENEMALONATE

(Malonic acid, ethoxymethylene-, diethyl ester)

$$CH_2(CO_2C_2H_5)_2 + HC(OC_2H_5)_3 + 2(CH_3CO)_2O \rightarrow$$
$$C_2H_5OCH{=}C(CO_2C_2H_5)_2 + 2CH_3CO_2C_2H_5 + 2CH_3CO_2H$$

Submitted by W. E. Parham and L. J. Reed.[1]
Checked by J. R. Roland, C. W. Todd, and R. S. Schreiber.

1. Procedure

A mixture of 1000 g. (6.75 moles) of ethyl orthoformate, 1260 g. (12.3 moles) of acetic anhydride, 960 g. (6.0 moles) of ethyl malonate, and 0.5 g. of anhydrous zinc chloride is prepared in a 5-l. three-necked flask equipped with a thermometer, a gas inlet tube, and a 12-in. column packed with Berl Saddles (Note 1). The column is attached to a still head and condenser. The contents of the flask are well agitated for 5 minutes by a stream of dry air and then heated (Note 2) as follows: 102–115° for 2.5 hours, 115–127° for 7 hours [after the eighth hour of heating, 250 g. (2.45 moles) of acetic anhydride and 200 g. (1.35 moles) of ethyl orthoformate are added], 127–145° for 2 hours, and 145–155° for 2 hours (Note 3). At the end of 13.5 hours of heating the mixture is cooled to room temperature and filtered (Note 4). The filtrate is distilled under reduced pressure (15–20 mm.) until the temperature at the still head reaches 100° (Note 5). The distillation is then continued under lower pressure (0.25 mm.). The yield of ethyl ethoxymethylenemalonate, b.p. 108–110°/0.25 mm. (Note 6), is 650–780 g. (50–60% based on the ethyl malonate used).

2. Notes

1. The use of this column prevents loss of acetic anhydride during the heating process and permits volatile products formed during the reaction to be removed by distillation. The checkers used a 1 by 12 in. column packed with 6 by 6 mm. glass rings.

2. The zinc chloride dissolves, and a chlorine-free crystalline precipitate soon separates. The checkers observed only traces of this precipitate.

3. The final heating is necessary to convert unchanged ethyl diethoxymethylmalonate, an intermediate in the reaction, into ethyl ethoxymethylenemalonate. This intermediate is difficult to separate from ethyl ethoxymethylenemalonate by distillation, and it is important that it be converted as completely as possible during the heating process.

4. The mixture is filtered through a fluted filter paper to remove suspended zinc salts.

5. Unchanged ethyl malonate (as much as 15% of the amount used) can be obtained by redistillation of the fraction boiling at 70–100°/17 mm.

6. Since ethyl diethoxymethylmalonate is difficult to separate from ethyl ethoxymethylenemalonate by distillation, it is necessary to follow the course of the distillation by observation of the change in refractive index instead of the change in boiling point. After a low-boiling fraction is collected, there is obtained an intermediate fraction (n_D^{20} 1.4142–1.4580), the size of which depends upon the amount of ethyl diethoxymethylmalonate present. Under the best of conditions

it amounts to 15 g. In each of three runs, the checkers obtained about 300–400 g. of material having a refractive index in this range.

The ethyl ethoxymethylenemalonate was collected at n_D^{20} 1.4580–1.4623. The checkers used a 5-in. Vigreux column and observed that at the low pressures the boiling point of the distillate in several runs was about 10° lower than that reported by the submitters. For example, in one run at 0.25 mm. (gauge pressure) the boiling point of the distillate never rose above 97.2°; however, the refractive indices of twelve 50-ml. fractions taken in the course of this distillation were all in the proper range, n_D^{20} 1.4612–1.4623.

Interruption of the heating during the reaction period should be avoided. In an experiment in which the heating was discontinued after about 8 hours and resumed the next day the yield was about 50%.

3. Methods of Preparation

Ethyl ethoxymethylenemalonate has been prepared by heating ethyl orthoformate, ethyl malonate, and acetic anhydride in the presence of zinc chloride.[2,3] A higher yield of purer product is obtained by the method described above,[4] which is a modification of the Claisen procedure.

[1] Work done under contract with the Office of Scientific Research and Development.

[2] Claisen, *Ber.*, **26**, 2729 (1893); *Ann.*, **297**, 76 (1897).

[3] Wheeler and Johns, *Am. Chem. J.*, **40**, 237 (1908).

[4] Fuson, Parham, and Reed, *J. Org. Chem.*, **11**, 194 (1946).

ETHYL (1-ETHYLPROPENYL)METHYLCYANOACETATE

(3-Pentenoic acid, 2-cyano-3-ethyl-2-methyl-, ethyl ester)

$$(C_2H_5)_2C{=}C(CN)COOC_2H_5 \xrightarrow{NaOC_2H_5} [CH_3CH{=}C(C_2H_5)C(CN)COOC_2H_5]^-Na^+ \xrightarrow{CH_3I} CH_3CH{=}C(C_2H_5)C(CH_3)(CN)COOC_2H_5$$

Submitted by EVELYN M. HANCOCK and ARTHUR C. COPE.
Checked by H. R. SNYDER and J. H. SAUNDERS.

1. Procedure

A solution of sodium ethoxide is prepared from 9.2 g. (0.40 mole) of freshly cut sodium and 400 ml. of absolute ethanol (Note 1) in a 1-l.

three-necked flask fitted with a dropping funnel, a thermometer, a mercury-sealed stirrer, and a condenser protected by a drying tube (Note 2). To the stirred solution, which is kept at −5° (conveniently by partially immersing the flask in a Dry Ice bath), 72.4 g. (0.40 mole) of ethyl (1-ethylpropylidene) cyanoacetate (p. 399) is added dropwise from the funnel during 8–10 minutes. After the mixture has been stirred for an additional 20 minutes at −5°, 62.5 g. (0.44 mole) of methyl iodide is added from the dropping funnel as rapidly as possible. The flask is heated immediately with a strong flame which is withdrawn just as the solution reaches the boiling point. The alkylation is vigorous, but the flask is not cooled unless loss of material through the reflux condenser appears imminent (Note 3). After the spontaneous reaction has subsided, the solution is refluxed until a piece of red litmus paper dipped into the liquid and subsequently moistened shows a neutral reaction (15–30 minutes).

The solution is cooled and diluted with 1 l. of water, and the ester layer is separated. The aqueous layer is extracted with four 50-ml. portions of benzene, and the combined ester and benzene extracts are washed with two 25-ml. portions of water and then distilled from a 500-ml. modified Claisen flask. The fraction (above 75 g.) which is collected at 95–118°/10 mm. is shaken mechanically for 4 hours with 100 ml. of 20% sodium bisulfite solution (Note 4). The ester layer is separated, the aqueous layer is extracted with three 25-ml. portions of benzene, and the combined ester and benzene extracts are washed with 25 ml. of water. The product remaining after removal of the benzene is distilled under reduced pressure from a 250-ml. modified Claisen flask or through a Widmer column. The yield of ester boiling at 112–113°/8 mm. is 63.5–68 g. (81–87%).

2. Notes

1. Commercial absolute ethanol is dried with sodium and ethyl phthalate [1] and redistilled. The checkers obtained a considerably lower yield in a run employing ethanol dried over magnesium methoxide.

2. The various pieces of apparatus are dried in an oven and assembled rapidly in order to exclude moisture.

3. Excessive cleavage (alcoholysis) of the product is avoided by carrying out the alkylation rapidly. This procedure minimizes the time during which the product is in contact with sodium ethoxide.

4. This treatment removes ethyl (1-ethylpropylidene)cyanoacetate as a water-soluble sodium bisulfite addition product (p. 399).

3. Methods of Preparation

The above procedure illustrates a general method for preparing homologous ethyl (dialkylvinyl)alkylcyanoacetates by the alkylation of ethyl alkylidenecyanoacetates.[2]

[1] *Org. Syntheses* Coll. Vol. **2**, 155 (1943).
[2] Cope and Hancock, *J. Am. Chem. Soc.*, **60**, 2903 (1938).

ETHYL (1-ETHYLPROPYLIDENE)CYANOACETATE

(2-Pentenoic acid, 2-cyano-3-ethyl-, ethyl ester)

$$(C_2H_5)_2CO + CH_2(CN)COOC_2H_5 \xrightarrow[CH_3COOH]{CH_3COONH_4} (C_2H_5)_2C{=}C(CN)COOC_2H_5 + H_2O$$

Submitted by ARTHUR C. COPE and EVELYN M. HANCOCK.
Checked by H. R. SNYDER and J. H. SAUNDERS.

1. Procedure

In a 500-ml. round-bottomed flask attached to a modified Dean and Stark constant water separator[1] (Note 1) which is connected to a reflux condenser are placed 67.8 g. (0.60 mole) of ethyl cyanoacetate (Note 2), 56.8 g. (0.66 mole) of diethyl ketone (Note 3), 9.2 g. (0.12 mole) of ammonium acetate, 30 g. (0.48 mole) of glacial acetic acid, and 100 ml. of benzene. The flask is heated in an oil bath at 160–165°, and the water that distils out of the mixture with the refluxing benzene is removed from the separator at intervals. Refluxing is continued for 24 hours (several hours after the separation of water has ceased) (Note 4).

The solution is cooled and washed with three 25-ml. portions of 10% sodium chloride solution, after which the benzene is removed by distillation under reduced pressure. The residue is transferred to a 1-l. bottle, a solution of 78 g. (0.75 mole) of commercial sodium bisulfite in 310 ml. of water is added, and the mixture is shaken on a mechanical shaker for 2 hours. The turbid solution is diluted with 500 ml. of water and extracted with three 50-ml. portions of benzene. The extracts are discarded (Note 5). The bisulfite solution is then cooled in an ice bath, and an ice-cold solution of 32 g. (0.8 mole) of sodium hydroxide in 130 ml. of water is added dropwise with mechanical stirring. The ester which separates is extracted at once with four 25-ml. portions of benzene (Note 6). The benzene solution is washed with 50 ml. of 1% hydrochloric acid, dried for a short time over 20 g. of anhydrous sodium sulfate, filtered into a 250-ml. modified Claisen flask, and distilled under reduced pressure. The yield of ester boiling at 123–125°/12 mm. (Note 7) is 65.4–75 g. (60.5–68%) (Notes 8, 9, and 10).

2. Notes

1. Any continuous water separator that will return the benzene to the reaction mixture may be used.

2. Commercial ethyl cyanoacetate (Dow Chemical Company) was redistilled before use; b.p. 93–94°/12 mm.

3. Diethyl ketone was either purchased from the Eastman Kodak Company and redistilled, or prepared by passing propionic acid slowly over a mixture of manganous oxide and clay plate chips in a tube furnace at 420–440°; [2] the apparatus was similar to one described in *Organic Syntheses*.[3] When prepared by this method the ketone was distilled, dried over potassium carbonate, and redistilled; b.p. 100–101°.

4. The water layer (20–25 ml.) contains some acetic acid and acetamide, the acetamide being formed from the ammonium acetate catalyst.

5. Extraction of the aqueous solution removes ethyl cyanoacetate from the aqueous solution of the sodium bisulfite addition product of ethyl (1-ethylpropylidene)cyanoacetate.

6. The unsaturated ester is regenerated when the bisulfite is neutralized with sodium hydroxide. The solution is kept cold during neutralization and extraction, and but little excess sodium hydroxide is used in order to prevent hydrolytic cleavage of the ester to diethyl ketone and ethyl cyanoacetate.

7. Other boiling points are 135–137°/25 mm. and 88–89°/1 mm.

8. A number of ketones have been condensed with ethyl cyanoacetate by this procedure. Reactive ketones such as aliphatic methyl ketones and cyclohexanone condense with ethyl cyanoacetate much more rapidly and give better yields of alkylidene esters. It is advantageous with such ketones to use a lower ratio of ammonium acetate-acetic acid catalyst.[1]

9. The sodium bisulfite purification step may be omitted, and the alkylidene ester purified directly by distillation. Care must be taken to separate the product from ethyl cyanoacetate by fractionation through a moderately efficient column. Purification through the bisulfite addition compound is recommended for alkylidene cyanoacetic

esters derived from ketones containing four and five carbon atoms, but not for the higher homologs.

The checkers obtained a slightly higher yield by the method employing fractional distillation (64% vs. 60.5%) but the quality of the product appeared to be slightly inferior (n_D^{25} 1.4645 vs. 1.4649).

10. A trace of a water-soluble white solid may cause the distillate to be slightly turbid. It may be removed by washing the product with water and redistilling.

3. Methods of Preparation

The above procedure is a very slight modification of a general method[1] for condensing ketones with ethyl cyanoacetate. Ethyl (1-ethylpropylidene)cyanoacetate also has been prepared by condensing diethyl ketone with ethyl cyanoacetate in the presence of piperidine or acetic anhydride and zinc chloride,[4] or piperidine and anhydrous sodium sulfate in a pressure bottle at 100°.[5]

[1] Cope, Hofmann, Wyckoff, and Hardenbergh, *J. Am. Chem. Soc.*, **63**, 3452 (1941).
[2] Sabatier and Mailhe, *Compt. rend.*, **158**, 831 (1914).
[3] *Org. Syntheses* Coll. Vol. **2**, 389 (1943).
[4] Birch and Kon, *J. Chem. Soc.*, **1923**, 2448.
[5] Cowan and Vogel, *J. Chem. Soc.*, **1940**, 1528.

ETHYL 1,16-HEXADECANEDICARBOXYLATE

(Octadecanedioic acid, diethyl ester)

$$2C_2H_5O_2C(CH_2)_8CO_2K \xrightarrow{\text{electrolysis}} C_2H_5O_2C(CH_2)_{16}CO_2C_2H_5$$

Submitted by Sherlock Swann, Jr., René Oehler, and P. S. Pinkney.
Checked by Lee Irvin Smith, James W. Horner, Jr., and J. W. Clegg.

1. Procedure

To 86.5 g. (0.38 mole) of ethyl hydrogen sebacate [*Org. Syntheses* Coll. Vol. **2,** 276 (1943)] is added slowly and with cooling 125–130 ml. of approximately 3 *N* potassium hydroxide. The solution is then diluted to approximately 250 ml., yielding a 1.5 *N* solution of potassium ethyl sebacate.

The solution of potassium ethyl sebacate (Note 1) is poured into a 500-ml. tall beaker provided with a cooling coil (Note 2), a thermometer, a stirrer, a platinum sheet anode 45–55 sq. cm. in area (Note 3), and two platinum wire cathodes (Note 4). To the solution 10 g. of monoethyl sebacate (Note 5) is added.

The electrodes are connected to a suitable source of direct current, 10 amperes of which is allowed to pass through the cell. To the solution there is now added 40 g. more of monoethyl sebacate in portions of 10 g. each over a period of 10 minutes (Note 6). The temperature of the cell is held below 50° by running cold water through the cooling coil (Note 7).

The run is finished in 60–70 minutes. The reaction will have reached completion when a few drops of the electrolyte removed with a pipet show an alkaline reaction to phenolphthalein. The alkalinity should be tested every 10 minutes after the first 45.

When the electrolyte has become alkaline, the oily product floating on the top is removed by means of a pipet (Note 8). The oil is washed with an equal volume of 10% potassium carbonate (Note 9), then with an equal volume of 3.5% hydrochloric acid, and finally twice with half its volume of water (a little ether effectively breaks up any emulsions which may form at this point). It is then crystallized from methanol, filtered by suction, and washed twice while in the funnel with ice-cold methanol (Note 10). The product, snow-white and waxy in appearance, is dried in a desiccator over sulfuric acid. The yield is 16–22 g. (40–55%) of material melting at 41–42°. These figures for the percentage yields, however, are based only upon the 50 g. of ethyl hydrogen sebacate used in the second part and no account is taken of the 86.5 g. used in preparing the electrolyte solutions. This preparation becomes practicable, therefore, only when several consecutive runs are made (Note 8).

2. Notes

1. The reaction mixture may require slight warming on the water bath to make a homogeneous solution.

2. The checkers used a copper cooling coil. This was slightly attacked during the electrolysis: the solution became blue, and copper deposited on the cathodes.

3. Convenient dimensions for the anode are 3.75 by 6.0 cm. (the area of the anode [both sides] used by the checkers was 52 sq. cm.). Care should be taken to submerge it in the solution to a depth such that the oil that gathers on the top of the solution during the electrolysis will not insulate it, thus increasing the current density. The wire tab of the platinum sheet may be sealed into a glass tube so that only the sheet is exposed to the solution.

4. The cathodes should be equidistant from the anode to ensure uniform current density on both sides of the anode.

5. As the potassium salt of monoethyl sebacate is decomposed during the electrolysis, the free acid forms new salt. The alkalinity of the solution may, therefore, be used as the end point of the electrolysis.

6. If the 50 g. of monoethyl sebacate is added all at once, excessive foaming will occur.

7. Addition of more water (50 ml.) makes possible a much better control of the temperature.

8. If it is desired to prepare larger amounts of product, more free acid may be added to the solution without turning off the current after the product has been drawn off by the pipet. Between ten and fifteen 50-g. batches of monoethyl sebacate may be converted in this way. The length of time for a run will gradually increase with the number of runs. After ten to fifteen runs, the time required to convert a 50-g. batch is 3.5–4 hours. The electrolyte should then be renewed.

9. Unchanged mono ester may be recovered by acidification of the carbonate washings, followed by extraction with ether.

10. About the same amount of by-product as main product is in solution in the methanol. The first material that crystallizes is pure main product.

3. Methods of Preparation

Diethyl 1,16-hexadecanedicarboxylate has been prepared by the electrolysis of potassium ethyl sebacate [1] and by the electrolysis of a mixture of sodium and hydrogen sebacates.[2]

[1] Brown and Walker, *Proc. Roy. Soc. Edinburgh,* **17,** 297 (1890); *Trans. Roy. Soc. Edinburgh,* **36,** 222 (1890); *Ann.,* **261,** 125 (1891); Fairweather, *Proc. Roy. Soc. Edinburgh,* **45,** 283 (1925); Shiina, *J. Soc. Chem. Ind. Japan,* **40,** Suppl. binding 324 (1937) [*C. A.,* **32,** 499 (1938)].

[2] Drake, Carhart, and Mozingo, *J. Am. Chem. Soc.,* **63,** 617 (1941).

ETHYL HYDRAZINECARBOXYLATE

(Carbazic acid, ethyl ester)

AND DIAMINOBIURET

(Imidodicarboxylic acid dihydrazide)

$$2N(CO_2C_2H_5)_3 + 5NH_2NH_2 + H_2O \rightarrow$$
$$3NH_2NHCO_2C_2H_5 + NH(CONHNH_2)_2 + CO_2 + 3C_2H_5OH + NH_3$$

Submitted by C. F. H. Allen and Alan Bell.
Checked by Nathan L. Drake and Carl Blumenstein.

1. Procedure

In a 3-l. round-bottomed flask equipped with a reflux condenser are placed 582 g. (2.5 moles) of N-tricarboxylic ester (p. 415) and 800 g. (6.7 moles) of 42% hydrazine hydrate (Note 1). The flask is shaken by hand to mix the two layers. After a short time, reaction begins with considerable evolution of heat, and all the N-tricarboxylic ester goes into solution (Note 2). After the reaction subsides, the solution is heated for 1 hour on a steam bath and then evaporated under reduced pressure until the mixture becomes a thick slurry of diaminobiuret crystals (Note 3). The mixture is cooled, 2 l. of 95% ethanol is added, and the diaminobiuret which has crystallized is filtered, washed with 250 ml. of ethanol, and dried. The substance melts at 205° with decomposition. The yield is 115–125 g. (69–75%).

The filtrate is now evaporated at atmospheric pressure to remove the ethanol, and the residual oil is distilled using a Vigreux column. Ethyl hydrazinecarboxylate boils at 92–95°/13 mm. The yield is 350–370 g. (90–95%). After distillation in vacuum, the ester may crystallize; the crystals melt at 51–52°.

2. Notes

1. This reaction can be carried out successfully with these amounts, but, if larger quantities of starting materials were to be used, it would be advisable to dilute the ester with ethanol and run in the hydrazine slowly. Several hours of refluxing on a steam bath would serve to complete the reaction. In smaller runs no difficulty is encountered.

2. It is advisable to keep an ice bath at hand in case the reaction becomes too violent and it is necessary to cool the mixture rapidly.

3. If insufficient water is removed by evaporation, too much diaminobiuret will remain in solution and will interfere during the distillation of the ethyl hydrazinecarboxylate.

3. Methods of Preparation

Diaminobiuret has been prepared only from N-tricarboxylic ester and hydrazine hydrate.[1] Ethyl hydrazinecarboxylate has been prepared by reduction of nitrourethan electrolytically [2] or with zinc dust and acetic acid,[3] and by the action of hydrazine hydrate on diethyl carbonate,[4,5] ethyl chlorocarbonate,[6] and N-tricarboxylic ester.[1]

[1] Diels, *Ber.*, **36**, 736 (1903).
[2] Backer, *Rec. trav. chim.*, **31**, 20 (1912).
[3] Thiele and Lachmann, *Ann.*, **288**, 293 (1895).
[4] Diels, *Ber.*, **47**, 2186 (1914).
[5] Merck, Ger. pat. 285,800 [*Frdl.*, **12**, 94 (1914–1916)].
[6] Stollé and Benrath, *J. prakt. Chem.*, (2) **70**, 276 (1904).

ETHYL α-ISOPROPYLACETOACETATE

(Isovaleric acid, α-acetyl, ethyl ester)

$$CH_3COCH_2CO_2C_2H_5 + (CH_3)_2CHOH \xrightarrow{BF_3} CH_3COCH((CH_3)_2CH)CO_2C_2H_5 + H_2O$$

Submitted by Joe T. Adams, Robert Levine, and Charles R. Hauser.
Checked by H. R. Snyder and John Mirza.

1. Procedure

This reaction should be carried out under a well-ventilated hood (*Note* 1).

A mixture of 30.0 g. (0.5 mole) of anhydrous isopropyl alcohol and 65.0 g. (0.5 mole) of freshly distilled ethyl acetoacetate is placed in a 500-ml. round-bottomed three-necked flask equipped with a mercury-sealed stirrer, a gas inlet tube terminating about 1 cm. above the surface of the liquid, a gas outlet tube connected with a calcium chloride drying tube, and a thermometer (Note 2). The gas inlet tube is connected to a source of boron fluoride (Note 3), and an ice bath is applied to the reaction flask; when the temperature of the stirred mixture has fallen to approximately 0° the stream of boron fluoride is started and adjusted so that the temperature of the mixture does not exceed 7°. The addition of boron fluoride is continued until the mixture is saturated and for 15 minutes thereafter (Note 4). The reaction mixture then is stirred at 28° for 2.5 hours (Note 5), at the end of which period it is poured slowly into a stirred mixture of 130 g. of hydrated sodium acetate, 100 ml. of water, and 200 g. of crushed ice. The beaker containing the resulting mixture is allowed to stand in an ice bath for 2 hours with occasional stirring.

The mixture is poured into a 1-l. separatory funnel. The beaker is rinsed with 300 ml. of ether, and this portion of solvent is shaken with the mixture in the separatory funnel. The phases are separated, and the aqueous solution is extracted twice with 100-ml. portions of ether. The combined ether solutions are washed with saturated aqueous bicarbonate solution until carbon dioxide no longer forms. The solution is transferred to an Erlenmeyer flask and dried over about 25 g. of anhydrous sodium sulfate for 12 hours, after which it is decanted into another Erlenmeyer flask and dried over about 10 g. of Drierite for 6 hours.

A 125-ml. modified Claisen flask is arranged for distillation, but with a dropping funnel fitted in the neck intended for the ebullator tube. Portions of the dried ether solution are introduced into the flask through the funnel while the flask is heated on the steam bath for continuous removal of the solvent. After all the ether solution and the ether washings from the last drying agent have been concentrated, the dropping funnel is replaced by an ebullator tube and distillation under diminished pressure is begun cautiously. Fractions are collected at 60–96°/20 mm. and at 96–98°/20 mm. The first fraction is redistilled to give an additional quantity of material boiling at 96–98°/20 mm. (Note 6). The combined product weighs 52–58 g. (60–67%) (Note 7).

2. Notes

1. Although no definite data are available concerning the toxicity of boron fluoride, users should exercise caution and avoid breathing the fumes. The toxic effects of hydrogen fluoride and alkali fluorides are well known. Boron fluoride reacts with moisture in the air, forming white fumes of fluoboric acid and boric acid which cause a choking sensation when breathed.

2. The entire apparatus is dried in an oven at about 100° just before use.

3. Commercial boron fluoride from a tank is passed through a saturated solution of boric oxide in concentrated sulfuric acid.

4. The time required for saturation varies from 1 to 2 hours, depending upon the rate of addition. As the saturation point is approached, the fuming at the end of the exit tube increases rather sharply; the mixture is considered to be saturated when the fuming appears to have become constant.

5. The temperature and time at this point are critical and must be controlled carefully.

6. If the fractionation is conducted with a 30-cm. Vigreux column the redistillation of the fore-run is unnecessary.

7. By the same method ethyl α-cyclohexylacetoacetate (b.p. 146–148°/20 mm.) has been prepared in 34% yield from cyclohexanol and acetoacetic ester, and ethyl α-*tert*-butylacetoacetate (b.p. 101–102°/20 mm.) in 10–14% yield from *tert*-butyl alcohol and acetoacetic ester.

3. Methods of Preparation

The above procedure is based on that described by Adams, Abramovitch, and Hauser.[1] Ethyl α-isopropylacetoacetate has also been prepared by Hauser and Breslow[2] by the reaction of ethyl acetoacetate with isopropyl ether in the presence of boron trifluoride, by Bischoff[3] by the alkylation of the sodium derivative of ethyl acetoacetate with isopropyl bromide, and by Renfrow[4] by the alkylation of ethyl acetoacetate with isopropyl bromide using potassium *tert*-amyloxide as the condensing agent.

[1] Adams, Abramovitch, and Hauser, *J. Am. Chem. Soc.*, **65**, 552 (1943).

[2] Hauser and Breslow, *J. Am. Chem. Soc.*, **62**, 2389 (1940).

[3] Bischoff, *Ber.*, **28**, 2620 (1895).

[4] Renfrow, *J. Am. Chem. Soc.*, **66**, 144 (1944).

ETHYL β-PHENYL-β-HYDROXYPROPIONATE

(Hydracrylic acid, β-phenyl-, ethyl ester)

$$C_6H_5CHO + BrCH_2CO_2C_2H_5 + Zn \rightarrow C_6H_5\overset{\overset{\displaystyle OZnBr}{|}}{C}HCH_2CO_2C_2H_5$$

$$C_6H_5\overset{\overset{\displaystyle OZnBr}{|}}{C}HCH_2CO_2C_2H_5 + H_2SO_4 \rightarrow C_6H_5CHOHCH_2CO_2C_2H_5 + ZnSO_4 + HBr$$

Submitted by Charles R. Hauser and David S. Breslow.
Checked by R. L. Shriner, W. M. Hoehn, and Vera A. Patterson.

1. Procedure

In a clean, dry, 500-ml. three-necked flask, fitted with a mechanical stirrer, a 250-ml. separatory funnel, and a reflux condenser (Note 1), the upper end of which is protected by a calcium chloride drying tube, is placed 40 g. (0.62 gram atom) of powdered zinc (Note 2).

A solution of 83.5 g. (0.50 mole) of ethyl bromoacetate (Note 3) and 65 g. (0.61 mole) of benzaldehyde (Note 4) in 80 ml. of dry benzene and 20 ml. of absolute ether is placed in the separatory funnel. About 10 ml. of this solution is added to the zinc, and the flask is warmed until the reaction starts (Note 5). The mixture is then stirred and the rest of the solution added at such a rate that the reaction mixture refluxes, care being taken that the reaction does not become too vigorous. The addition should take about an hour. The reaction mixture is refluxed for 30 minutes on a water bath after the addition of the solution is complete.

The flask is then cooled in an ice bath and the reaction mixture hydrolyzed by the addition of 200 ml. of cold 10% sulfuric acid with vigorous stirring during the addition. The acid layer is drawn off and the benzene solution extracted twice with 50-ml. portions of 5% sulfuric acid. The benzene solution is washed once with 25 ml. of 10% sodium carbonate solution, then with 25 ml. of 5% sulfuric acid (Note 6), and finally with two 25-ml. portions of water. The combined acid solutions are extracted with two 50-ml. portions of ether, and the combined ether and benzene solutions are dried with 5 g. of magnesium sulfate or Drierite. The solution is filtered, the solvent removed by distillation at atmospheric pressure from a steam bath, and the residue fractionated under reduced pressure. The ester is collected at 151–154°/11–12 mm. (128–132°/5–7 mm.). The total yield is 59–62 g. (61–64%) (Note 7).

2. Notes

1. An efficient reflux condenser is necessary to prevent loss of solvent. The same precautions as to cleanliness of apparatus and absence of moisture should be observed in carrying out Reformatsky reactions as in reactions involving the Grignard reagent.

2. The zinc dust is purified by washing it rapidly with dilute sodium hydroxide solution, water, dilute acetic acid, water, ethanol, acetone, and ether. It is then dried in a vacuum oven at 100°.

3. Ethyl bromoacetate is a powerful lachrymator, and care should be exercised in handling it. A 10% solution of ammonium hydroxide should be kept available to neutralize any of the bromo ester which may be spilled.

4. The benzaldehyde should be washed with two 50-ml. portions of 10% sodium bicarbonate solution, dried, and distilled.

5. It is essential that the reaction start before any additional ethyl bromoacetate and benzaldehyde are added.

6. The acid extraction should be continued until no white precipitate of zinc hydroxide is formed on extraction with sodium carbonate. This step is important, since the zinc complex hydrolyzes much less readily than the corresponding magnesium complex. A small amount of unhydrolyzed complex forms at the interface of the benzene and acid solutions. It is most readily separated during the first alkaline extraction.

7. Ethyl α,α-dimethyl-β-phenyl-β-hydroxypropionate may be prepared in a similar manner. For instance, from 100 g. of ethyl α-bromoisobutyrate and 65 g. of benzaldehyde, 82.5 g. (73%) of ethyl α,α-dimethyl-β-phenyl-β-hydroxypropionate, boiling at 153–158°/11 mm., is obtained. It may be recrystallized by dissolving it at 30° in 100 ml. of ligroin (30–60°), chilling the solution for several days, and decanting the clear liquid from the precipitate; m.p. 38.5–39°.

3. Methods of Preparation

Ethyl β-phenyl-β-hydroxypropionate has been prepared by esterification of β-phenyl-β-hydroxypropionic acid [1,2] obtained in turn by the addition of hydrogen bromide to cinnamic acid and hydrolysis; by catalytic reduction of ethyl benzoylacetate in aqueous acetic acid or aqueous ethanol;[3] by catalytic reduction of ethyl α-anilino-β-phenyl-β-hydroxypropionate;[4] and by the procedure described.[5-7]

[1] Findlay and Hickmans, *J. Chem. Soc.*, **1909**, 1004.
[2] Fittig and Binder, *Ann.*, **195**, 131 (1877).
[3] Kindler and Blaas, *Ber.*, **76B**, 1211 (1943).
[4] Fourneau and Billeter, *Bull. soc. chim. France*, **7**, 593 (1940).
[5] Andrijewski, *J. Russ. Phys. Chem. Soc.*, **40**, 1635 (1908).
[6] Blaise and Herman, *Ann. chim. phys.*, [8] **23**, 532 (1911).
[7] Gaudini, *Gazz. chim. ital.*, **73**, 263 (1943).

4-ETHYLPYRIDINE

(Pyridine, 4-ethyl-)

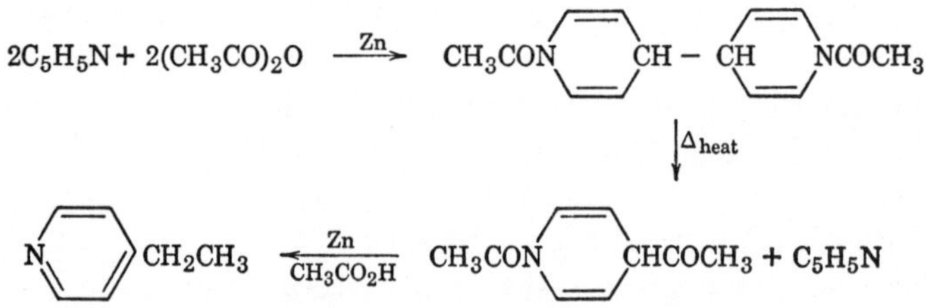

Submitted by Robert L. Frank and Paul V. Smith.[1]
Checked by Homer Adkins and Robert H. Jones.

1. Procedure

In a 3-l. three-necked round-bottomed flask fitted with a Hershberg stirrer and a thermometer are placed 1500 ml. of acetic anhydride and 300 g. (306 ml., 3.80 moles) of dry pyridine (Note 1). Three hundred grams (4.6 gram atoms) of zinc dust (Note 2), in amounts of 5–10 g., is added with stirring over a period of 3 hours. Heat is evolved almost immediately, and a cooling bath of water may be necessary. The reaction mixture becomes green after about 20 minutes. The temperature of the contents of the flask should be maintained between 25° and 30°. After the addition of the first 300-g. portion of zinc, 300 ml. of acetic acid is added to the reaction mixture and a reflux condenser is attached to the flask. Then 120 g. (1.83 gram atoms) of zinc dust is added in small portions. Heat is evolved during the addition, and the reaction may become rather violent. The mixture is refluxed with stirring for 30 minutes. A third portion of 180 g. (2.75 gram atoms) of zinc dust is added all at once, and refluxing is continued for an additional half hour. The solution is now orange-brown.

The flask is allowed to cool, and the contents are transferred to a 5-l. round-bottomed flask. The mixture is cautiously neutralized with 2 l. of a 40% aqueous solution of sodium hydroxide. The mixture is steam-distilled until 3 l. of distillate is collected, after which the residue is discarded. The distillate, which separates into two layers, is saturated with 1.5–1.8 kg. of solid potassium carbonate (Note 3). The organic layer is removed by decantation; the remaining water layer is divided into two portions, and each is extracted once with 150 ml. of chloroform. The chloroform extracts are combined with the organic layer.

The mixture is distilled using an efficient fractionating column (Note 4). There is a large fore-run of chloroform, pyridine, and water (Note 5), after which the temperature rises and 145–167 g. of material, b.p. 145–165°/760 mm., is collected. This is refractionated, and 135–155 g. (33–38%) of 4-ethylpyridine, b.p. 163–165°/760 mm., n_D^{20} 1.5010, is obtained (Note 6).

2. Notes

1. The pyridine is dried over calcium oxide and redistilled.
2. The zinc is activated before use by stirring 830 g. of zinc dust in 300 ml. of 10% hydrochloric acid for 2 minutes, filtering, and washing the zinc with 600 ml. of water, then with 200 ml. of acetone.
3. An equal weight of sodium chloride may be used instead of potassium carbonate.
4. A Fenske-type column 30 cm. in length and 18 mm. in diameter packed with glass helices gave satisfactory separations.
5. It does not appear feasible to recover the pyridine from the mixture of chloroform, water, and pyridine.
6. This procedure has been employed by Arens and Wibaut [2] to prepare other 4-alkyl derivatives of pyridine. The yields tend to decrease as the molecular weight of the anhydride increases. The boiling points are as given below: 4-propylpyridine 189°, 4-*n*-butylpyridine 207–209°, 4-isobutylpyridine 197–199°, 4-isoamylpyridine 222–223°, and 4-*n*-octylpyridine 265–268°.

3. Methods of Preparation

4-Ethylpyridine has been prepared by heating N-ethylpyridinium iodide in a sealed tube at 300°; [3] by heating pyridine with ethyl iodide; [4] from 4-ethylpyridinecarboxylic acid through distillation from lime;[5] and in small amounts by distilling brucine with potassium hydroxide.[6] Pyridine when heated with ferric chloride in an autoclave gives a mixture of alkylated pyridines from which 4-ethylpyridine can be isolated.[7] A general method for 4-alkylpyridines involves heating 5-(γ-pyridyl)-5-alkylbarbituric acids with alkali followed by an acid cleavage to remove the carbon dioxide.[8] 4-Ethylpyridine has been isolated from California petroleum.[9] The most useful method involves the treatment of pyridine with acetic anhydride and zinc.[10,11]

[1] Work done under contract with the Office of Rubber Reserve, Reconstruction Finance Corporation.

[2] Arens and Wibaut, *Rec. trav. chim.*, **61**, 59 (1942).

[3] Ladenburg, *Ber.*, **16**, 2059 (1883); **18**, 2961 (1885); *Ann.*, **247**, 1 (1888).

[4] Ladenburg, *Ber.*, **32**, 42 (1899).

[5] Gabriel and Colman, *Ber.*, **35**, 1358 (1902).

[6] Oechsner de Coninck, *Ann. chim.*, (5) **27**, 507 (1882).

[7] Morgan and Burstall, *J. Chem. Soc.*, **1932**, 20.

[8] Gebauer (Chemische Fabrik von Heyden, A.-G.), Ger. pat. 638,596 [*C. A.*, **31**, 3067 (1937)].

[9] Hackmann, Wibaut, and Gitsels, *Rec. trav. chim.*, **62**, 229 (1943).

[10] Dohrn and Horsters (Chemische Fabrik auf Actien vorm. E. Schering), Ger. pat. 390,333 [*Chem. Zentr.*, **1924**, II, 891].

[11] Wibaut and Arens, *Rec. trav. chim.*, **60**, 119 (1941).

ETHYL 2-PYRIDYLACETATE

(2-Pyridineacetic acid, ethyl ester)

$$C_6H_5Br + 2Li \rightarrow C_6H_5Li + LiBr$$

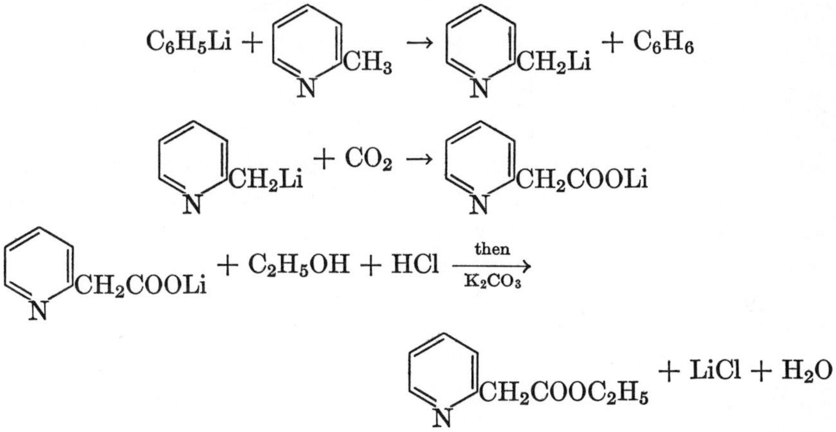

Submitted by R. B. WOODWARD and E. C. KORNFELD.
Checked by ARTHUR C. COPE and WILLIAM R. ARMSTRONG.

1. Procedure

A 2-l. round-bottomed three-necked flask is fitted with a reflux condenser, a dropping funnel, and an efficient mechanical stirrer. A calcium chloride tube is attached to the condenser to protect the apparatus from moisture. To the flask are added 800 ml. of absolute ether and 13.9 g. (2 gram atoms) of lithium chips or shavings (Note 1). The stirrer is started, and 105 ml. (157 g., 1 mole) of dry bromobenzene is placed in the dropping funnel. About 5–15 ml. of the bromobenzene is added to initiate the reaction; when the ether begins to reflux, the balance is added at such a rate that the solvent refluxes continuously (1 hour) (Note 2). The mixture is then stirred and refluxed until most of the lithium disappears (45–90 min.). While stirring is continued, 97 ml. (93.1 g., 1 mole) of α-picoline is added dropwise in about 5–10 minutes. The dark red-brown solution of picolyllithium is stirred for an additional 30 minutes and is then poured slowly and with shaking onto 500–750 g. of crushed Dry Ice contained in a 3-l. round-bottomed flask (Note 3). The mixture is stirred well until the dark color of the picolyllithium is discharged, and the excess of Dry Ice is allowed to evaporate. The ether is removed by distillation under reduced pressure at room temperature. The lumpy residue of lithium salts is broken up, and to it is added 750 ml. of commercial absolute ethanol. The solution is saturated with dry hydrogen chloride while cooling in an ice bath. The esterification mixture is allowed to stand overnight, after which the solvent is removed as completely as possible by distillation under reduced pressure on a steam bath. The syrupy residue is dissolved in 750 ml. of chloroform, and a paste prepared from 225 g. of potassium carbonate and 135 ml. of water is slowly added to the solution with mechanical stirring. After the paste has been added, the solution is stirred vigorously and is kept just below the boiling point for 1 hour. The chloroform solution is decanted from the inorganic salts, and the chloroform is removed by distillation. The residue is fractionated under reduced pressure from a modified Claisen flask with a fractionating side arm. About 40 g. of α-picoline is recovered in the fore-run, and the ethyl 2-pyridylacetate is obtained as a light yellow liquid, b.p. 135–137°/28 mm., 142–144°/40 mm., 109–112°/6 mm.; n_D^{25} 1.4979. The yield is 58–66 g. (35–40% based on lithium) (Notes 4 and 5).

2. Notes

1. The most convenient method of preparing the lithium chips is as follows. Pieces of lithium several grams each in size and slightly moist with paraffin oil are pounded with a hammer into thin sheets on a dry surface. The sheets are quickly cut into small chips by means of a pair of scissors and are added immediately to the absolute ether.

2. The use of a nitrogen atmosphere is not essential if the solution is kept protected from oxygen by an atmosphere of ether vapor. For this purpose the solution is kept at the reflux point throughout.

3. Rapid filtration of the picolyllithium solution onto the Dry Ice through a thin layer of glass wool is useful in removing unreacted lithium at this point.

4. Runs twice the size of the one described give comparable yields.

5. Methyl 2-pyridylacetate, b.p. 122–125°/21 mm., can be obtained in similar yield by use of methanol in the esterification.

3. Methods of Preparation

2-Pyridylacetic esters have been obtained by the alcoholysis of 2-pyridylacetanilide, in turn prepared by Beckmann rearrangement of the oxime of 2-phenacylpyridine,[1] and by the carbethoxylation of α-picoline in the presence of potassium amide.[2]

[1] Oparina and Smirnov, *Khim. Farm. Prom.*, **1934**, No. 4, 15 [*C. A.*, **29**, 1820 (1935)]; *J. Gen. Chem. U.S.S.R.*, **5**, 1699 (1935) [*C. A.*, **30**, 2567 (1936)].
[2] Weiss and Hauser, *J. Am. Chem. Soc.*, **71**, 2023 (1949).

ETHYL N-TRICARBOXYLATE

(N-Tricarboxylic ester)

$$NH_2CO_2C_2H_5 + 2ClCO_2C_2H_5 + 2Na \rightarrow N(CO_2C_2H_5)_3 + 2NaCl + H_2$$

Submitted by C. F. H. Allen and Alan Bell.
Checked by Nathan L. Drake and Carl Blumenstein.

1. Procedure

In a 12-l. round-bottomed flask equipped with stirrer, dropping funnel, and efficient reflux condenser are placed 450 g. (5 moles) of urethan (m.p. 46–48°), 7.5 l. of dry ether, and 218 g. of sodium wire (Notes 1 and 2). The mixture is warmed, and evolution of hydrogen accompanied by formation of sodium compound soon begins; after 2–3 hours the greater part of the metal will have been replaced by a gelatinous white precipitate. The stirrer is now started and the mixture warmed under reflux for an additional 2 hours. The flask is then cooled externally with running water, and 1050 g. (9.6 moles) of ethyl chlorocarbonate is added slowly over a period of 2 hours. The gelatinous precipitate becomes powdery, and the remainder of the sodium dissolves. After all the ester has been added, the mixture is stirred overnight at room temperature and then filtered (Note 3). The white residue is washed with 1 l. of ether, and the ether is removed from the combined solutions by evaporation on a steam bath (Note 4). The residual oil is distilled; after a fore-run containing some urethan, the product distils at 143–147°/12 mm. The yield is 575–640 g. (51–57%). On redistillation, all the product boils at 146–147°/12 mm.

2. Notes

1. Since water is usually present, it is advisable to distil the urethan before use, discarding the first 10%.

2. Sodium wire is not essential. The reaction will proceed just as well with sodium cut into pieces the size of a small pea. It is important that the sodium be cut into *small* pieces; otherwise a protective coating forms on the surface of the metal, preventing further reaction.

3. The sodium chloride is very finely divided and quickly clogs the pores of the filter paper. The mixture should be allowed to stand for a time before filtration to allow the salt to settle; the clear supernatant liquor can then be filtered rapidly.

4. Care should be taken in disposing of this residue; it invariably contains some unreacted sodium.

3. Methods of Preparation

Ethyl N-tricarboxylate has been prepared from urethan by reaction with sodium and chlorocarbonic ester[1] as well as from the potassium salt of ethyl imidodicarboxylate.[2]

[1] Diels, *Ber.*, **36**, 740 (1903).
[2] Diels, *Ber.*, **36**, 742 (1903).

1-ETHYNYLCYCLOHEXANOL

(Cyclohexanol, 1-ethynyl-)

$$2CH{\equiv}CH + 2Na \xrightarrow{NH_3} 2CH{\equiv}CNa + H_2$$

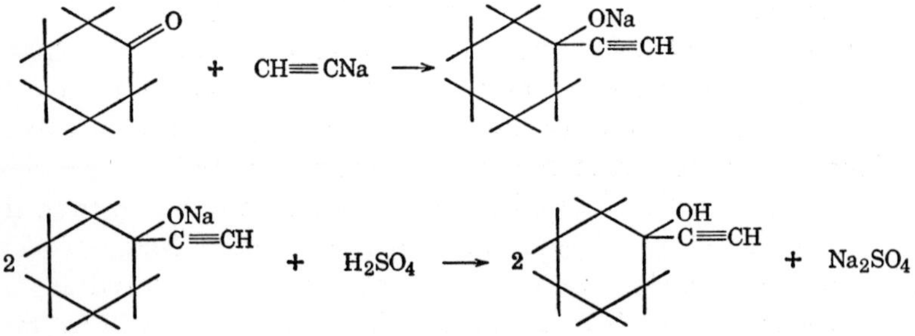

Submitted by J. H. Saunders.[1]
Checked by R. S. Schreiber and E. L. Jenner.

1. Procedure

A rapid stream of dry acetylene is passed into approximately 1 l. of liquid ammonia in a 2-l. three-necked flask equipped with a gas inlet tube and a mechanical stirrer while 23 g. (1 gram atom) of sodium is added over a period of 30 minutes (Notes 1 and 2). The flow of acetylene is then reduced (Note 3), and 98 g. (1 mole) of cyclohexanone is added dropwise. When this addition, which requires about an hour, is completed, the reaction mixture is allowed to stand for about 20 hours to permit the evaporation of nearly all the ammonia (Note 4).

The solid residue is decomposed by adding approximately 400 ml. of ice and water, and the resulting mixture is carefully acidified with 50% sulfuric acid (Note 5). The organic layer is dissolved in 100 ml. of ether and washed with 50 ml. of brine. The original aqueous phase and the brine wash are then extracted with two 50-ml. portions of ether. The combined ethereal solutions are dried over anhydrous magnesium sulfate and filtered, and the ether is distilled. The product is then distilled under reduced pressure through a good column (Note 6). The yield of 1-ethynylcyclohexanol is 81–93 g. (65–75%), b.p. 74°/14 mm.; n_D^{20} 1.4822 (Note 7).

2. Notes

1. The preparation of sodium acetylide is based on the procedure of Vaughn, Hennion, Vogt, and Nieuwland.[2]

2. The blue color of dissolved sodium is discharged so rapidly by the acetylene that it seldom spreads through the entire mixture.

3. The flow of acetylene may be terminated before the cyclohexanone is added. This operation, however, is alleged to increase the formation of glycol.[3]

4. If all the ammonia is allowed to evaporate and the residual solid is exposed to the air the yields may be decreased.

5. Upwards of 70 ml. is required. The amount depends upon the quantity of ammonia remaining.

6. The submitters and the checkers used a 15-cm. Vigreux column.

7. The product may solidify to a colorless solid, m.p. 30°.

3. Methods of Preparation

An attractive alternative procedure for the preparation of 1-ethynylcyclohexanol which gives yields of 80–90% employs the potassium salt of *tert*-amyl alcohol to effect the addition of acetylene to cyclohexanone.[4–6] This condensation has been brought about by a suspension of sodium amide in ether[7–11] and by potassium hydroxide in ether.[12] 1-Ethynylcyclohexanol has also been prepared by the action of acetylene on the sodium enolate of cyclohexanone[13] and by the action of sodium acetylide on cyclohexanone in liquid ammonia.[14] The procedure described here is essentially that of Campbell, Campbell, and Eby.[3]

[1] This investigation was carried out under the sponsorship of the Office of Rubber Reserve, Reconstruction Finance Corporation, in connection with the Government Synthetic Rubber Program.

[2] Vaughn, Hennion, Vogt, and Nieuwland, *J. Org. Chem.*, **2**, 1 (1937).

[3] Campbell, Campbell, and Eby, *J. Am. Chem. Soc.*, **60**, 2882 (1938).

[4] Pinkney, Nesty, Wiley, and Marvel, *J. Am. Chem. Soc.*, **58**, 972 (1936).

[5] Dimroth, *Ber.*, **71B**, 1333 (1938).

[6] Backer and van der Bij, *Rec. trav. chim.*, **62**, 561 (1943).

[7] Ger. pat. 289,800 [*Frdl.*, **12**, 55 (1914–1916).

[8] Locquin and Sung, *Bull. soc. chim. France*, **35**, 597 (1924).

[9] Rupe, Messner, and Kambli, *Helv. Chim. Acta*, **11**, 449 (1928).

[10] Levina and Vinogradova, *J. Applied Chem. U.S.S.R.*, **9**, 1299 (1936) [*C. A.*, **31**, 2587 (1937)].

[11] *Org. Syntheses*, **20**, 41 (1940), Note 11.

[12] Azerbaev, *J. Gen. Chem. U.S.S.R.*, **15**, 415 (1945) [*C. A.*, **40**, 4683 (1946)].

[13] Hurd and Jones, *J. Am. Chem. Soc.*, **56**, 1924 (1934).

[14] Milas, MacDonald, and Black, *J. Am. Chem. Soc.*, **70**, 1831 (1948).

o-EUGENOL *

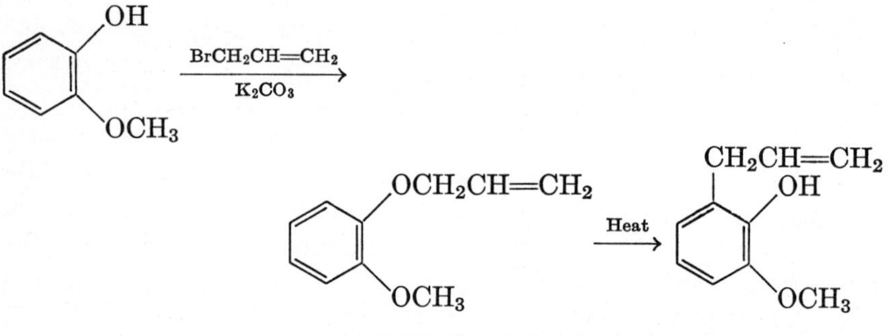

Submitted by C. F. H. Allen and J. W. Gates, Jr.
Checked by W. E. Bachmann and N. C. Deno.

1. Procedure

A. *Guaiacol allyl ether.* A mixture of 63 g. (0.5 mole) of guaiacol, 66 g. (0.55 mole) of allyl bromide, 70 g. of anhydrous potassium carbonate (0.5 mole), and 100 ml. of dry acetone in a 500-ml. round-bottomed flask is refluxed on a steam bath for 8 hours and cooled. The mixture is diluted with 200 ml. of water and extracted with two 100-ml. portions of ether. The combined extracts are washed with two

100-ml. portions of 10% sodium hydroxide (Note 1) and dried with 50 g. of anhydrous potassium carbonate. After removal of the solvent, the residual oil (Note 2) is distilled under reduced pressure. The yield of guaiacol allyl ether boiling at 110–113°/12 mm. is 66–75 g. (80–90%).

B. *o-Eugenol.* The allyl ether (70 g.) is cautiously (Note 3) brought to boiling in a 500-ml. round-bottomed flask, refluxed for 1 hour, and cooled. The oil is dissolved in 100 ml. of ether (Note 4), and the solution is extracted with three 100-ml. portions of 10% sodium hydroxide. The combined alkaline extracts are then acidified with 100 ml. of concentrated hydrochloric acid diluted with 100 ml. of water, and the mixture is extracted with three 100-ml. portions of ether. The combined ether extracts are dried with 50 g. of anhydrous sodium sulfate and evaporated, and the residual oil is distilled under reduced pressure. The yield of *o*-eugenol boiling at 120–122°/12 mm. (Note 5) is 56–63 g. (80–90%).

2. Notes

1. A small amount of guaiacol may be recovered by acidifying the alkaline wash and extracting with ether.
2. When this crude guaiacol allyl ether was rearranged without prior distillation, the yields of *o*-eugenol were about 10% lower than those obtained with the distilled ether.
3. This rearrangement is sometimes quite vigorous and needs little heat once it is started. Dimethylaniline is said to be a good solvent for use in this type of rearrangement.[1]
4. Benzene may be substituted for ether throughout.
5. Other boiling points are 250–251°/760 mm., 125°/14 mm., and 115°/9 mm.

3. Methods of Preparation

Guaiacol allyl ether has been prepared from guaiacol, ethanolic potassium hydroxide, and allyl iodide;[2] or from guaiacol, allyl bromide, and potassium carbonate in acetone.[3,4] *o*-Eugenol has been prepared by the rearrangement of guaiacol allyl ether;[3,4] from 3-methoxy-2-allyloxybenzaldehyde by heating to 210°;[5] and from 3-methoxy-2-allyloxybenzoic acid by heating above 110°.[6]

[1] Tarbell, *Org. Reactions,* **2,** 24 (1944).
[2] Marfori, *Annali di chimica e di farmacologia,* (5), **12,** 115; *Jahresb.,* **1890, 1196.**
[3] Claisen and Eisleb, *Ann.,* **401,** 52 (1913).

[4] Claisen, *Ber.*, **45**, 3161 (1912); Ger. pat. 268,099 [*Chem. Zentr.*, **1914**, I, 308; *Frdl.*, **11**, 181 (1912–1914)].
[5] Claisen and Eisleb, *Ann.*, **401**, 112, 114 (1913).
[6] Claisen, *Ann.*, **418**, 117 (1919).

FLUORENONE-2-CARBOXYLIC ACID

(2-Fluorenecarboxylic acid, 9-oxo-)

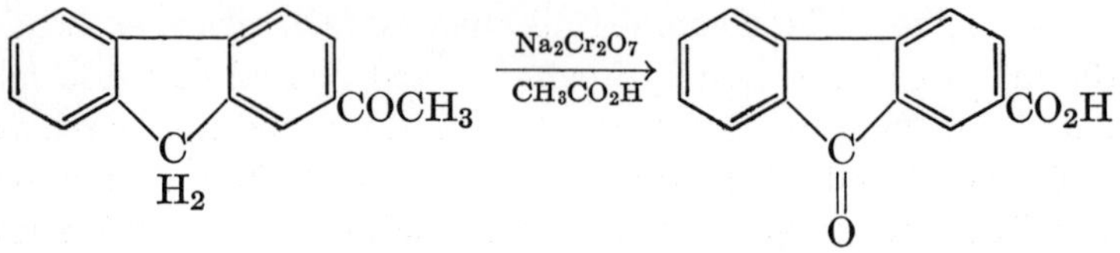

Submitted by George Rieveschl, Jr., and F. E. Ray.
Checked by R. L. Shriner and Arne Langsjoen.

1. Procedure

A 5-l. three-necked round-bottomed flask (Note 1) equipped with a mercury-sealed stirrer, a reflux condenser, and a dropping funnel is set up on a steam bath. In the flask are placed 50 g. (0.24 mole) of 2-acetylfluorene (p. 23) (Note 2) and 650 ml. of glacial acetic acid. After the ketone has been brought into solution by heating and stirring, 450 g. (1.5 moles) of sodium dichromate dihydrate, previously ground to a coarse powder, is added carefully in small portions (about 10 g. each) (Note 3). Heating and stirring are continued throughout the addition, which requires about 45 minutes. The reaction flask is then removed from the steam bath, and the solution is brought to gentle reflux over a flame. During a period of 1.5 hours, 200 ml. of acetic anhydride is added through the dropping funnel. The mixture is refluxed and stirred during this addition and for 8 hours longer (Note 4).

The hot mixture is stirred into 9 l. of hot water in a 5-gal. crock. The suspension is agitated for 15 minutes and is then filtered on a 24-cm. Büchner funnel. The filter cake is washed with four 400-ml. portions of 2% sulfuric acid. The wet filter cake is transferred to a 2-l. beaker containing 700 ml. of a 5% solution of potassium hydroxide (Note 5). The mixture is stirred and heated to about 80°, and the hot solution is filtered. The alkali-insoluble material (Note 6) is washed with 50 ml. of hot 5% potassium hydroxide solution, and the combined alkaline filtrates are placed in a beaker, stirred with 5 g. of decolorizing carbon for 20 minutes, and again filtered. The solution of the potassium salt is heated with vigorous mechanical stirring to 70°, and 200 ml. of 18% hydrochloric acid is added dropwise, the heating being continued so that the temperature rises to about 85°. The fluorenone-2-carboxylic acid separates as a thick yellow mass. It is allowed to digest for 10 minutes with stirring, and the hot mixture is filtered on a 24-cm. Büchner funnel. The acid is washed free of potassium chloride by five or six 200-ml. portions of hot water. The product is pressed as dry as possible and then dried for 3 hours in an oven at 150°. It is a bright canary-yellow, weighs 36–40 g. (67–74%), and melts on an aluminum block at 339–341° (uncor.) with partial sublimation (Note 7).

2. Notes

1. A relatively large flask is used so that the sodium dichromate may be added without the escape of much acetic acid vapor.

2. 2-Acetylfluorene melting at 124–126° has been found to give as good results as the pure compound of melting point 128–129°. The submitters report that the over-all yield from fluorene to fluorenone-2-carboxylic acid can be improved by the use of the crude product of the acylation (m.p. 113–117°) without purification (p. 23).

3. The acetic acid solution of 2-acetylfluorene is conveniently and safely heated to the maximum temperature attainable on the steam bath, and then small portions of sodium dichromate are added as the speed of the reaction permits. To prevent a violent reaction, not more than 10 g. of dichromate is added until the oxidation has begun (as indicated by the appearance of a green color).

4. Small yellow crystals usually appear after the mixture has refluxed for about 7 hours; their quantity does not increase much during the remaining heating period.

5. Sodium hydroxide should not be substituted here because of the lower solubility of the sodium salt.

6. The alkali-insoluble material (3–9 g.) can be washed with water, dried, and recrystallized from 95% ethanol to yield from 1 to 5 g. of pure 2-acetylfluorenone melting at 154–155°. Alternatively, it can be added to the next lot of 2-acetylfluorene if another oxidation is to be carried out.

7. This melting point corresponds to that given in the literature.[1,2] The submitters report that the product can be recrystallized from acetic anhydride or Cellosolve.

3. Methods of Preparation

The method described is a modification of the method of Dziewonski and Schnayder[1] and is described in a patent by the submitters.[2] Fluorenone-2-carboxylic acid has also been obtained by the oxidation of 2-fluorenealdehyde[3] and by the reaction of 2-acetylfluorene with potassium hypochlorite.[4]

[1] Dziewonski and Schnayder, *Bull. intern. acad. polon. sci.*, **1930A,** 529 [*C. A.*, **25,** 5416 (1931)].

[2] Rieveschl and Ray, U. S. pat. 2,377,040 [*C. A.*, **39,** 3305 (1945)].

[3] Hinkel, Ayling, and Beynon, *J. Chem. Soc.*, **1936,** 345.

[4] Schiessler and Eldred, *J. Am. Chem. Soc.*, **70,** 3958 (1948).

FUMARYL CHLORIDE *

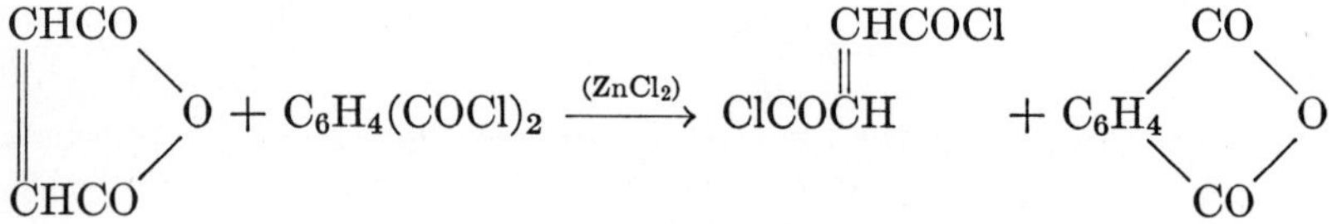

Submitted by L. P. Kyrides.
Checked by C. F. H. Allen and F. P. Pingert.

1. Procedure

A mixture of 98 g. (1 mole) of maleic anhydride, m.p. 52–54°, 230 g. of commercial phthaloyl chloride (Note 1), and 2 g. of anhydrous zinc chloride is placed in a 500-ml. three-necked round-bottomed flask. The flask is provided with a thermometer, the bulb of which extends into the liquid nearly to the bottom, and an efficient fractionating column (Notes 2, 3, and 4). A 500-ml. water-cooled distilling flask is connected to the side arm of the column to serve as a receiving vessel.

The reaction mixture is heated by means of an oil bath (inside temperature 130–135°) for 2 hours, care being taken to avoid overheating (Note 5), and then allowed to cool to 90–95°. The fumaryl chloride is distilled as rapidly as possible (20 minutes), and the portion boiling over a 25° range (60–85°/13–14 mm.) is collected. It is then redistilled slowly (1 hour), and the portion boiling over a 2° range (62–64°/13 mm.) is collected (Note 6). The yield is 125–143 g. (82–95%) (Notes 7 and 8.

2. Notes

1. Commercial phthaloyl chloride contains about 94% of halide and some phthalic anhydride. The amount of chloride specified corresponds to a slight molar excess.

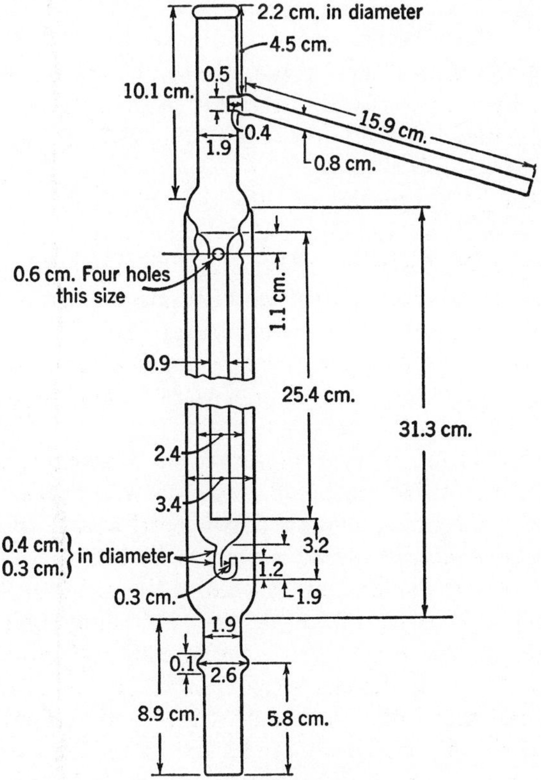

FIG. 13.

2. The checkers employed a modified Widmer column [1] (Fig. 13) that has been used in many organic laboratories, but not officially described. They also used a Vigreux column (50 cm. effective length, 2.7 cm. inside diameter); the first distillation then required 1.5 hours, and the yield was 82–83%. The final temperature of the reaction mixture and observed boiling point will depend upon the type of apparatus used.

3. This procedure has also been checked using a fractionating column 30 cm. in length and 1.5 cm. in inside diameter packed with glass Wilson rings [2] and provided with the usual jackets for electrical heating.

The distillation requires 3 hours. If this column is used, a single distillation gives a product pure enough for most purposes. An unpacked, indented column of about the same dimensions was unsatisfactory (checked by N. L. Drake).

4. Rubber stoppers are used throughout. Tightly fitting ground-glass connections are convenient but unnecessary. The third neck of the flask is used when acid chlorides are prepared from the corresponding acids (Note 9).

5. Above 135° the reaction is likely to get out of control; the ensuing decomposition seriously reduces the yield. For this reason the flask should be immersed only slightly in the oil bath (about one-third of the depth of the liquid layer).

6. With some lots of phthaloyl chloride, owing to the presence of an unknown impurity, the first few drops of the distillate have a reddish color. If the distillation is interrupted and air admitted to the system, the same phenomenon is observed on resuming the distillation.

7. The yield obtained is usually nearer the higher figure. The checkers carried out this preparation using five times these amounts and a 3-l. flask. The distillation times were 1 hour for the first and 3 hours for the final distillation.

8. Fumaryl chloride is best preserved in sealed glass containers. Bottles, closed by rubber stoppers free from sulfur, can be used for short periods. Ground-glass-stoppered bottles are unsuitable, the joints readily becoming "frozen," owing to hydrolysis of the chloride.

9. According to the submitters, yields of the order of 95% of other acid chlorides can be obtained by the use of phthaloyl chloride (1 mole of chloride to 1 mole of a monobasic acid, 2 moles of chloride to 1 mole of a dibasic acid). Zinc chloride, as catalyst, is not necessary in the reaction of most acids and their anhydrides with phthaloyl chloride. When acids are used, it is best to add one of the components slowly, in order to avoid a too violent evolution of hydrogen chloride on warming.

3. Methods of Preparation

Fumaryl chloride has been prepared from fumaric acid and phthaloyl chloride,[3] from maleic acid by the action of thionyl chloride in the presence of zinc chloride, and from maleic anhydride by the use of phthaloyl chloride in the presence of zinc chloride.[4]

[1] Widmer, *Helv. Chim. Acta,* **7**, 59 (1927).
[2] Wilson, Parker, and Laughlin, *J. Am. Chem. Soc.*, **55**, 2795 (1933).
[3] Van Dorp and Van Dorp, *Rec. trav. chim.*, **25**, 96 (1906).
[4] Kyrides, *J. Am. Chem. Soc.*, **59**, 207 (1937).

FURYLACRYLIC ACID

(2-Furanacrylic acid)

I. METHOD A

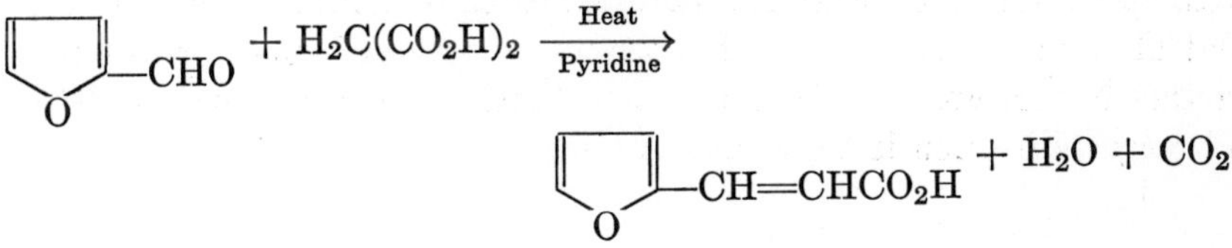

Submitted by S. Rajagopalan and P. V. A. Raman.
Checked by C. S. Hamilton and R. A. Alberty.

1. Procedure

In a 1-l. round-bottomed flask fitted with a reflux condenser are placed 192 g. (166 ml., 2 moles) of freshly distilled furfural (Note 1), 208 g. (2 moles) of malonic acid (Note 2), and 96 ml. (1.2 moles) of pyridine (Note 3). The flask is heated on a boiling water bath for 2 hours, and the reaction mixture is cooled and diluted with 200 ml. of water. The acid is dissolved by the addition of concentrated aqueous ammonia, the solution is filtered through a fluted filter paper, and the paper is washed with three 80-ml. portions of water. The combined filtrates are acidified with an excess of diluted (1:1) hydrochloric acid with stirring. The mixture is cooled by running water and then allowed to stand in an ice bath for at least 1 hour. The furylacrylic acid is filtered, washed with four 100-ml portions of water, and dried. The yield of practically colorless needles melting at 141° is 252–254 g. (91–92%). If a purer product is desired, recrystallization is best effected from dilute alcohol (Note 4). On slow cooling of the solution, needles melting at 141° separate.

2. Notes

1. Commercial furfural is subjected to a single distillation; b.p. 160–161°.

2. Commercial malonic acid is dried at 100° for 2 hours and stored in a stoppered bottle.

3. The pyridine, which may be a commercial grade, is dried over sticks of potassium hydroxide for a few hours and filtered.

4. A convenient procedure is to dissolve the acid in a slight excess of 50% ethanol, reflux the solution with charcoal (5 g. per 100 g. of acid) for 5–10 minutes, and filter through a fluted filter in a preheated funnel. Any solid material that separates toward the later stages is redissolved by the addition of a few drops of ethanol. The residue is thoroughly washed with boiling water. The hot solution is then diluted with cold water until crystals separate, heated to boiling, cooled slowly, and allowed to stand in a refrigerator for several hours. When the mother liquors are used for subsequent batches, the usual loss (2–3%) by recrystallization is more than halved.

II. METHOD B

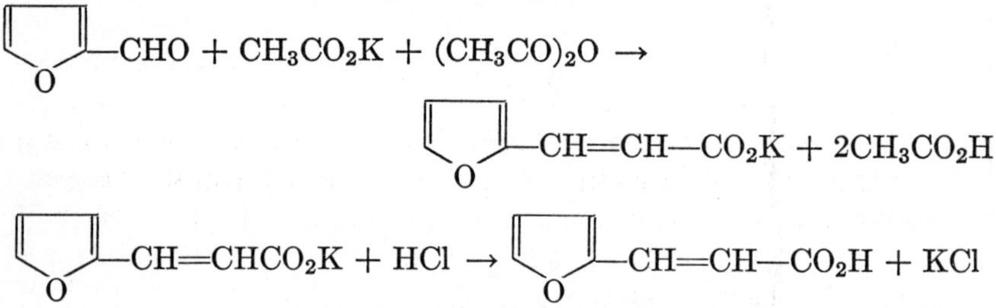

Submitted by John R. Johnson.
Checked by R. L. Shriner and C. M. Stevens.

1. Procedure

In a 3-l. round-bottomed flask provided with a mechanical stirrer and a 90-cm. air-cooled condenser are placed 288 g. (3 moles) of freshly distilled furfural, 460 g. (425 ml., 4.5 moles) of acetic anhydride, and 294 g. (3 moles) of dry, pulverized, freshly fused potassium acetate (Note 1). The ingredients are mixed thoroughly, stirring is started, and the flask is heated in an oil bath at 150° (bath temperature) for 4 hours, without interruption (Note 2). It is well to make provision for acetic acid vapor which escapes through the air condenser.

After cooling slightly, the reaction mixture is transferred to a large flask and treated with 3.5 l. of water. Part of this is used to rinse out the reaction flask. The mixture is boiled with 30 g. of Norit for about 10 minutes and is filtered while still *hot* with suction, using a preheated Büchner funnel. Furylacrylic acid tends to separate quickly and sometimes offers trouble in clogging the funnel. The filtrate is acidified to Congo red by the addition of a 1:1 solution of concentrated hydrochloric acid in water. After being cooled to 20° or below, preferably with stirring, and being allowed to stand for at least 1 hour, the acid is filtered with suction and washed with a small quantity of ice water. The yield is 270–290 g. (65–70%). The light tan crude acid melts at 138–139° (Note 3)

2. Notes

1. If fused sodium acetate is used the reaction is slower and 6–8 hours' heating is required.

2. When the temperature of the bath approaches 145–150° a rapid exothermic reaction sets in. This must be controlled (by application of cold, wet towels to the flask) to avoid too vigorous ebullition.

3. Furylacrylic acid melting at 138–139° is sufficiently pure for most purposes. The acid is perfectly white when pure, but many recrystallizations are required to attain that state. It may be recrystallized from benzene or ligroin (90–100°) with the addition of Norit. The loss is about 20–25%, and the product melts at 139–140° (sometimes 140–141°).

3. Methods of Preparation

This acid is usually prepared by the condensation of furfural with malonic acid in the presence of pyridine.[1] It may also be prepared from furfural by the Perkin reaction.[2] The use of potassium acetate is advantageous since it allows the reaction to proceed rapidly at relatively lower temperatures. A preparation from furfuralacetone by oxidation has also been reported.[3]

[1] Dutt, *J. Indian Chem. Soc.*, **1**, 297 (1925); Kurien, Pandya, and Surange; *J. Indian Chem. Soc.*, **11**, 824 (1934) [*C. A.*, **29**, 3325 (1935)].

[2] Marckwald, *Ber.*, **10**, 357 (1877); Gibson and Kahnweiler, *Am. Chem. J.*, **12**, 314 (1890).

[3] Hurd and Thomas, *J. Am. Chem. Soc.*, **55**, 1646 (1933).

β-GENTIOBIOSE OCTAACETATE

(Gentiobiose, β-octaacetyl-)

$$2C_6H_{12}O_6 \xrightarrow{\text{Emulsin}} C_{12}H_{22}O_{11} + H_2O$$

$$C_{12}H_{22}O_{11} + 8(CH_3CO)_2O \rightarrow C_{12}H_{14}O_3(OCOCH_3)_8 + 8CH_3CO_2H$$

Submitted by B. HELFERICH and J. F. LEETE.
Checked by HOMER ADKINS and E. E. BOWDEN.

1. Procedure

A solution of 1.65 kg. of crystallized glucose or 1.5 kg. of anhydrous glucose (8.3 moles) is prepared by heating the solid on a steam bath with 1.35 l. of distilled water. The solution is cooled and placed in a 2.5-l. glass-stoppered bottle. Fifteen grams of emulsin (Note 1) is added, then 20 ml. of toluene, and the flask is closed and allowed to stand at room temperature, with occasional shaking, for 5 weeks. The solution is then boiled, diluted with 8.5 l. of water, and filtered. To the filtrate is added 56 g. of baker's yeast in 650 ml. of water, and the temperature of the mixture is maintained at 28–32° for 12–14 days (Note 2). The mixture is then boiled for 30 minutes with an excess of powdered calcium carbonate and filtered.

The filtrate is evaporated under a pressure of 20–30 mm. to as thick a syrup as possible. For this operation the solution is placed in a 12-l. flask set in a steam bath and connected by a goose-neck to a condenser, which in turn is connected to a suction or distilling flask. A very fine capillary inlet tube into the 12-l. flask prevents any violent ebullition (Notes 3 and 4).

To the thick syrup (170–185 g.) are added 100 g. of anhydrous sodium acetate and 1.125 l. of acetic anhydride. The mixture is carefully heated to the boiling point with provision for cooling the flask with wet cloths should the acetylation become too violent. When the reaction is complete (about 20 minutes is required) the solution is poured into 10 l. of ice water. The water is decanted and renewed twice at 24-hour intervals in order to render the acetylated sugar filterable.

The dark brown product is filtered and, after drying in the air, is extracted with ether in a Soxhlet apparatus. The ether is removed by evaporation, and the light-colored residue (170 g.) is recrystallized from 1.65 l. of hot methanol. The crystals are washed on a Büchner funnel with methanol until free from colored impurities. The yield is 77–87 g. of material which melts at 187–190°. A second recrystallization from methanol (1.25 ml. per g. of product) gives pure β-octaacetylgentiobiose, with about 10% loss. The pure substance melts at 196° (cor.).

2. Notes

1. A commercial sample of emulsin was purchased from Merck and Company. If freshly prepared emulsin of good quality is available, the quantity may be reduced to one-tenth or less of the specified amount.

2. The yeast, by fermentation, removes the glucose which has not been transformed into gentiobiose. Fleischmann's and "Red Star" yeast have been used. The flask should be stoppered and protected with the conventional trap for the escape of carbon dioxide.

3. The submitters suggest a more complicated apparatus (Note 4) for this operation. However, the checkers have had no trouble with foaming or bumping when using the simple apparatus described above. Very little water condenses in the receiver, which serves as a safety trap in case foaming or bumping should occur.

4. The submitters used quantities four times as large as those stated above. They give the following directions for the distillation:

"The solution is drawn slowly through a glass tube provided with a stopcock into a 3-l. distillation flask standing in a bath of rapidly boiling water. The small portion which foams over is collected and evaporated in a second 3-l. distillation flask, likewise standing in rapidly boiling water. The distillate is passed through an efficient condenser and is collected in a thick-walled bottle or flask to which the vacuum pump or water pump is connected. A high vacuum is essential. The tube connecting the condenser and the receiver is provided with a stopcock so that the distillate may be poured out without interrupting the vacuum to any extent."

3. Methods of Preparation

The chief methods for the preparation of gentiobiose and its octaacetate are discussed in a paper by Reynolds and Evans.[1,2] These methods involve isolation of the substance from gentian root;[3,4] the action of emulsin on glucose;[5,6] the catalytic hydrogenolysis of amygdalin;[7] separation from the mother liquors ("Hydrol") obtained in the manufacture of glucose;[8] and the condensation of acetobromoglucose with *β-d*-glucose-1,2,3,4-tetraacetate.[1,9,10] Reynolds and Evans recommend the last-mentioned method.

The method given above is a modification of those originally described by Bourguelot[3] and Zemplén.[11]

[1] Reynolds and Evans, *J. Am. Chem. Soc.*, **60**, 2559 (1938).
[2] Haworth and Wylam, *J. Chem. Soc.*, **1923**, 123.
[3] Bourquelot and Hérissey, *Compt. rend.*, **132**, 571 (1901).
[4] Bourquelot and Hérissey, *Bull. soc. chim. France*, (3) **29**, 363 (1903).
[5] Bourquelot and Hérissey, *Compt. rend.*, **157**, 732 (1913).
[6] J. F. Leete, Ph.D. dissertation, Univ. of Greifswald, 1929.
[7] Bergmann and Freudenberg, *Ber.*, **62**, 2783 (1929).
[8] Berlin, *J. Am. Chem. Soc.*, **48**, 2627 (1926).
[9] Helferich and Klein, *Ann.*, **450**, 219 (1926).
[10] Gilbert, Smith, and Stacey, *J. Chem. Soc.*, **1946**, 622.
[11] Zemplén, *Ber.*, **48**, 232 (1915).

d-GLUCOSAMINE HYDROCHLORIDE *

Crab shells → Chitin →
$CH_2OH—CHOH—CHOH—CHOH—CHNH_2 \cdot HCl—CHO$

Submitted by EARL R. PURCHASE and CHARLES E. BRAUN.
Checked by H. R. SNYDER and NELSON R. EASTON.

1. Procedure

Two hundred grams of cleaned and dried crab shells (Note 1) ground to a fine powder is placed in a 2-l. beaker, and an excess of dilute (approximately 6 *N*) commercial hydrochloric acid is added slowly to the powdered material until no further action is evident. Much frothing occurs during the addition of the acid, and care must be exercised to avoid loss of material due to foaming over the sides of the beaker. After the reaction has subsided, the reaction mixture is allowed to stand from 4 to 6 hours to ensure complete removal of calcium carbonate. The residue is then filtered, washed with water until neutral to litmus, and dried in an oven at 50–60°. The weight of dried chitin is usually about 70 g., but with some lots of crab shells it may be as low as 40 g.

To 40 g. of dry chitin in a 500-ml. beaker is added 200 ml. of concentrated hydrochloric acid (C.P., sp. gr. 1.18), and the mixture is heated on a boiling water bath for 2.5 hours with continuous mechanical agitation. At the end of this time solution is complete, and 200 ml. of water and 4 g. of Norit are added. The beaker is transferred to a hot plate, and the solution is maintained at a temperature of about 60° and is stirred continuously during the process of decolorization. After an hour the solution is filtered through a layer of a filter aid such as Filter-Cel. The filtrate is usually a pale straw color; however, if an excessive color persists, the decolorization may be repeated until the solution becomes almost colorless. The filtrate is concentrated under diminished pressure at 50° until the volume of the solution is 10–15 ml. The white crystals of glucosamine hydrochloride are washed onto a sintered-glass filter with 95% ethanol. The white crystalline product, after being washed with 95% ethanol and dried, weighs 24–28 g. (60–70% of the weight of the chitin used). The optical rotation of a stable solution of the product containing the α- and β-isomers at equilibrium, $[\alpha]_D^{25.5}$, varies from +68.8° to +70.1° (c = 4.75 in water).

The product is pure enough for most uses. If a purer product is desired, the crystals may be dissolved in the minimum amount of boiling water and treated with Norit. The resulting solution is filtered and added to a large excess of 95% ethanol and stirred vigorously for several hours. The product is collected after 4–6 hours. An appreciable amount of the β-form of the amino sugar remains dissolved in the alcohol and may be precipitated by adding ether.

2. Note

1. The cleaned and dried crab shells were obtained from Carter and Lanhardt Company, Eleventh and Maine Avenue, S.W., Washington, D. C.

3. Methods of Preparation

The preparation of glucosamine hydrochloride from lobster shells and crab shells by essentially this method has been reported by Irvine, McNicoll, and Hynd [1] and Hudson and Dale.[2] Other methods involving the use of cicad larvae shells [3] and shrimp shells [4,5] also have been reported.

[1] Irvine, McNicoll, and Hynd, *J. Chem. Soc.*, **99**, 256 (1911).

[2] Hudson and Dale, *J. Am. Chem. Soc.*, **38**, 1434 (1916).

[3] Komori, *J. Biochem. Japan*, **6**, 1–20 (1926) [*C. A.*, **21**, 372 (1927)].

[4] van Alphen, *Chem. Weekblad*, **26**, 602 (1929) [*C. A.*, **24**, 2113 (1930)].

[5] Rigby, U. S. pat. 2,040,879 [*C. A.*, **30**, 4598 (1936)].

β-*d*-GLUCOSE-1,2,3,4-TETRAACETATE

(D-Glucose, β-1,2,3,4-tetraacetyl-)

$$HOCH_2\underbrace{CH(CHOH)_3CH}_{O}OH + (C_6H_5)_3CCl + C_5H_5N \rightarrow$$

$$(C_6H_5)_3COCH_2\underbrace{CH(CHOH)_3CH}_{O}OH + C_5H_6NCl$$

$$(C_6H_5)_3COCH_2\underbrace{CH(CHOH)_3CH}_{O}OH + 4(CH_3CO)_2O + 4C_5H_5N \rightarrow$$

$$(C_6H_5)_3COCH_2\underbrace{CH(CHOCOCH_3)_3CH}_{O}OCOCH_3$$

$$+ 4C_5H_5N \cdot CH_3COOH$$

$$(C_6H_5)_3COCH_2\underbrace{CH(CHOCOCH_3)_3CH}_{O}OCOCH_3 + HBr \rightarrow$$

$$(C_6H_5)_3CBr + HOCH_2\underbrace{CH(CHOCOCH_3)_3CH}_{O}OCOCH_3$$

Submitted by DELBERT D. REYNOLDS and WILLIAM LLOYD EVANS.
Checked by LEE IRVIN SMITH, R. T. ARNOLD, NEWMAN BORTNICK, AARON LERNER, and EVERETT SCHULTZ.

1. Procedure

A. 6-*Trityl-β-d-glucose*-1,2,3,4-*tetraacetate.* A mixture containing 120 g. (0.67 mole) of anhydrous glucose, 193.2 g. (0.7 mole) of trityl chloride, and 500 ml. of anhydrous pyridine [*Org. Syntheses* Coll. Vol. **1,** 100 (1941)], is heated on the steam cone until solution is complete (Note 1). Without cooling, 360 ml. of acetic anhydride is added in one portion (Note 2). After standing for 12 hours, the reaction mixture is poured slowly into 10 l. of ice water, to which 500 ml. of acetic acid has been added, and the resulting mixture is vigorously stirred mechanically for 2 hours (Note 3). The precipitate is filtered and is immediately stirred for a short time with 10 l. of ice water. The white, granular precipitate is filtered, washed well with cold water, and then air-dried (Note 4). The dried solid is digested with 500 ml. of ether (Note 5). The insoluble portion is dissolved in hot 95% ethanol (approximately 3 l.), and the solution is decolorized and filtered while hot. The filtrate, upon cooling, deposits fine needles of 6-trityl-*β-d*-glucose-1,2,3,4-tetraacetate of sufficient purity for further use. The yield at this point is about 169 g. (43%). Recrystallization from 95% ethanol gives the pure compound which melts at 166–166.5°. In pyridine, $[\alpha]_D^{19}$ is +44.8°; $[\alpha]_D^{28}$, +45.3°. The yield of purified material is 137 g. (35%) (Note 6).

B. *β-d-Glucose*-1,2,3,4-*tetraacetate*. A solution of 46 g. (0.078 mole) of 6-trityl-β-*d*-glucose-1,2,3,4-tetraacetate in 200 ml. of acetic acid is prepared by warming on the steam bath. The solution is then cooled to approximately 10°, 18 ml. of a saturated solution of dry hydrogen bromide in acetic acid is added, and the reaction mixture is shaken for about 45 seconds. The trityl bromide formed during the reaction is removed *at once* by filtration, and the filtrate is poured *immediately* into 1 l. of cold water. The tetraacetate is extracted with 250 ml. of chloroform; the chloroform extract is washed four times with ice water and dried over anhydrous sodium sulfate. The drying agent is removed, and the chloroform is evaporated, under reduced pressure at room temperature. The remaining syrup is covered with 100 ml. of anhydrous ether and is rubbed with a glass rod. Crystallization takes place immediately. The product is removed and is purified by dissolving it in the minimum amount of chloroform and adding anhydrous ether until crystallization begins. The purified product melts at 128–129°. In chloroform, $[\alpha]_D^{20}$ is +12.1°. The yield is 15 g. (55%).

2. Notes

1. The materials and apparatus used for this reaction must be strictly anhydrous in order to prevent hydrolysis of the trityl chloride.
2. Higher temperatures favor formation of the β-isomer.
3. The stirring must be unusually rapid, and the solution must be added to the water slowly and in a fine stream. If this is not done, the precipitate will not be granular and will be extremely difficult to filter.
4. If the material warms up to room temperature during the time required for drying, it becomes exceedingly sticky, and mechanical difficulties in manipulation result. This behavior appears to be due to the presence of traces of pyridine. The checkers found that the pure product, m.p., 166–166.5°, when recrystallized from pyridine, gave a sticky material.
5. The α-isomer is soluble in ether, whereas the β-isomer is insoluble. In some cases the entire product dissolves, but the β-isomer separates when the solution is allowed to stand.

6. The submitters state that the yield of purified product varies from 148 g. to 180 g.

3. Methods of Preparation

These directions are modifications of the methods used by Helferich and Klein for the original preparation of the substance,[1] although it had apparently previously been obtained by Oldham by hydrolysis of tetraacetylglucose-6-mononitrate.[2]

[1] Helferich and Klein, *Ann.*, **450**, 219 (1926); Helferich, Moog, and Junger, *Ber.*, **58**, 877 (1925).

[2] Oldham, *J. Chem. Soc.*, **127**, 2840 (1925).

β-*d*-GLUCOSE-2,3,4,6-TETRAACETATE

(D-Glucose, β-2,3,4,6-tetraacetyl-)

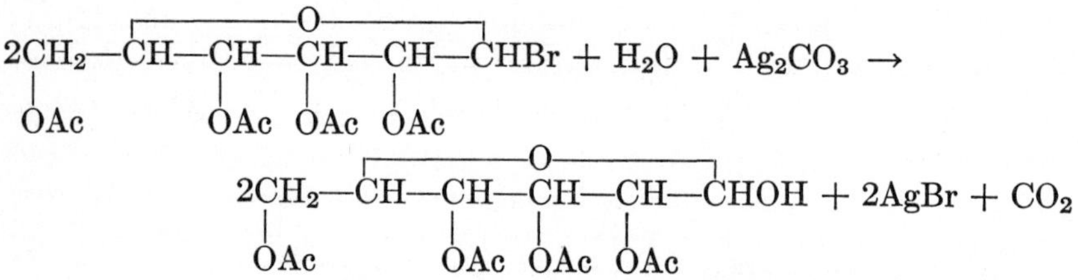

Submitted by Chester M. McCloskey and George H. Coleman.
Checked by C. S. Hamilton, Robert Angier, and Ivan Baumgart.

1. Procedure

A solution of 82.2 g. (0.2 mole) of acetobromoglucose (p. 11) (Note 1) in 125 ml. of dry acetone (Note 2) in a 250-ml. flask is cooled to 0° in an ice bath. To the cold solution is added 2.3 ml. of water and then 46.5 g. (0.17 mole) of silver carbonate (Note 3) in small portions in the course of 15 minutes. The mixture is shaken well during the addition and for 30 minutes longer (Note 4). The mixture is then warmed to 50–60° and filtered. The mass of silver salts is washed with 65 ml. of dry acetone (Note 5), removed from the funnel, warmed in a flask with 65 ml. more of acetone, filtered, and washed again on the funnel.

The combined filtrates are concentrated under reduced pressure in a 500-ml. filter flask (Note 6) until most of the solution is filled with crystals. The mixture is warmed to dissolve the crystals, the solution is poured into a 600-ml. beaker, and an equal volume of absolute ether and a similar volume of ligroin are added. The resulting solution is cooled in a freezing mixture with gentle stirring. The crystals of the tetraacetate form quickly and after about 10 minutes are filtered and air-dried. The crystals so obtained melt at 132–134° (Note 7). If a purer product is required, the product is dissolved in acetone, and ether and ligroin are added to the solution in the manner described. The yield of once-crystallized product melting at 132–134° is 52–56 g. (75–80%).

2. Notes

1. The acetobromoglucose that was used had a melting point of 87–88°.
2. The acetone was dried over calcium chloride.
3. The silver carbonate should be freshly prepared and finely ground. Silver carbonate can be prepared by the addition of a solution of sodium carbonate (53 g. in 600 ml. of water) to one of silver nitrate (172 g. in 2 l. of water). This is a very slight excess of the silver nitrate. The sodium carbonate solution is added slowly (10 minutes), and the reaction mixture is stirred vigorously with a mechanical stirrer. The silver carbonate is filtered, washed with a little acetone to facilitate drying, and then air-dried. All operations are carried out in dim light.
4. At the end of this time, evolution of carbon dioxide should no longer be appreciable. The time required for the reaction depends largely on the agitation of the silver carbonate. In large runs mechanical stirring is required.
5. Anhydrous chemicals are used throughout as the presence of any appreciable amount of water interferes with the crystallization of the tetraacetate.
6. The solution is not heated during the concentration, and thus the temperature is maintained below that of the room by the evaporation of the solvent. If a capillary is used it should be equipped with a drying tube.
7. β-*d*-Glucose-2,3,4,6-tetraacetate (2,3,4,6-tetraacetyl-β-*d*-glucose) decomposes slightly on prolonged standing and after 1–2 months has a melting range of 5–8 degrees. Such material can be purified by recrystallization.

3. Methods of Preparation

β-d-Glucose-2,3,4,6-tetraacetate has been prepared usually by the hydrolysis of acetobromoglucose.[1]

[1] Fischer and Delbruck, *Ber.*, **42**, 2776 (1909); Georg, *Helv. Chim. Acta*, **15**, 924 (1932); Hendricks, Wulf, and Liddel, *J. Am. Chem. Soc.*, **58**, 1998 (1936); McCloskey, Pyle, and Coleman, *J. Am. Chem. Soc.*, **66**, 349 (1944).

GLYCOLONITRILE *

$$CH_2O + KCN + H_2O \rightarrow HOCH_2CN + KOH$$

Submitted by Roger Gaudry.
Checked by C. F. H. Allen and J. A. VanAllan.

1. Procedure

This preparation should be carried out under a good hood since poisonous hydrogen cyanide may be evolved.

In a 1-l. three-necked flask, fitted with a stirrer, thermometer for reading low temperatures, and a dropping funnel, and surrounded by an ice-salt bath, is placed a solution of 130 g. (2.0 moles) of potassium cyanide in 250 ml. of water. With stirring, a solution of 170 ml. (2.0 moles) of commercial 37% formaldehyde solution [1] and 130 ml. of water is admitted slowly from the dropping funnel at such a rate that the temperature never rises above 10° (about 40 minutes is required). After 10 minutes' standing, 230 ml. of dilute sulfuric acid (57 ml. of concentrated sulfuric acid, sp. gr. 1.84, in 173 ml. of water) is added with stirring, the same low temperature being maintained. A copious precipitate of potassium sulfate is formed. The *p*H of the solution is then about 1.9. A 5% potassium hydroxide solution is then added, dropwise, and with cooling, until the *p*H is about 3.0 (determined either by means of a *p*H meter or tropaeolin 00 paper); about 4 ml. of the solution is required. The flask is then removed from the cooling bath, 30 ml. of ether is added, and the mixture is well shaken. The salt is removed by filtration, using a 14-cm. Büchner funnel, and washed with 30 ml. of ether. The filtrate is poured into a 1-l. continuous ether extractor [2] and extracted for 48 hours with 300 ml. of ether (Note 1). The ether extract is dried for 3–4 hours over 15 g. of anhydrous calcium sulfate (Drierite) (Note 2) and filtered. Ten milliliters of absolute ethanol is added to the filtrate, and the ether is removed on a steam bath (Note 3). The residue is distilled under reduced pressure using a flask having a Vigreux side arm. After a small (2–3 ml.) fore-run, the glycolonitrile distils smoothly at 86–88°/8 mm. (102–104°/16 mm.). The yield of pure glycolonitrile (Note 4) amounts to 86.5–91 g. (76–80%).

2. Notes

1. It is impractical to extract more than 40–45% of the nitrile without using a continuous ether extractor. A slightly lower yield is obtained if the extraction is continued for only 24 hours. The reaction mixture may be extracted in portions if the available flask is smaller than that specified.

2. Anhydrous sodium sulfate can be used equally well, but its drying action is slower, at least 24 hours being advisable.

3. If the ethanol is omitted, the nitrile shows a strong tendency to polymerize during the removal of the ether, especially when most of the ether has distilled.

4. The ethanol serves as a preservative before and after the distillation. Glycolonitrile obtained without the use of ethanol usually cannot be kept more than a few days; it sometimes turns brown within 24 hours. Some samples of ethanol-stabilized glycolonitrile have been preserved in sealed bottles for 2 years, whereas other samples polymerized in a few months.

3. Methods of Preparation

Glycolonitrile has usually been prepared by the interaction of formaldehyde and an alkali cyanide in aqueous solution [3] of which the procedure outlined is a modification. A more recent development is the cyanohydrin interchange method.[4,5] Glycolonitrile has also been prepared by a catalytic oxidation of methanol and ammonia in the vapor phase.[6]

[1] *Org. Syntheses* Coll. Vol. **1**, 378, Note 1 (1941).
[2] *Org. Syntheses* Coll. Vol. **2**, 378 (1943).
[3] Polstorff and Meyer, *Ber.*, **45**, 1911 (1912).
[4] Kung, U. S. pat. 2,259,167 [*C. A.*, **36**, 494 (1942)].
[5] Mowry, *J. Am. Chem. Soc.*, **66**, 372 (1944).
[6] U. S. pat. 2,405,963 [*C. A.*, **40**, 7231 (1946)].

GLYOXAL BISULFITE

$$(CH_3CHO)_3 + 3H_2SeO_3 \rightarrow 3OHCCHO + 3Se + 6H_2O$$

$$OHCCHO + 2NaHSO_3 + H_2O \rightarrow OHCCHO \cdot 2NaHSO_3 \cdot H_2O$$

Submitted by ANTHONY R. RONZIO and T. D. WAUGH.
Checked by C. F. H. ALLEN and J. VANALLAN.

1. Procedure

In a 2-l. round-bottomed flask, attached to an efficient reflux condenser (Note 1) and set in a hot water bath, are placed 222 g. (1.72 moles) of selenious acid (Note 2), 270 ml. of paraldehyde (Note 3), 540 ml. of dioxane, and 40 ml. of 50% acetic acid (Note 4), and the mixture is refluxed for 6 hours (Note 5). The solution is then decanted from the inorganic material (Note 6), which is washed with two 150-ml. portions of water. The combined solutions are steam-distilled through a still head (p. 65) until the paraldehyde and dioxane have been removed; this requires about 3.5 hours (Note 7). The mixture is decanted from a little selenium (Note 6), and to the solution, without filtration, is added a slight excess of 25% lead acetate solution (Notes 8 and 9). The lead selenite is removed by filtration, and the filtrate is saturated with hydrogen sulfide in a hood (Note 8). Then 20 g. of Norit is added; the whole is warmed to 40° in a hood and filtered with suction. The water-clear solution is concentrated on a hot water bath under reduced pressure to about 150 ml. in the usual apparatus (Note 10).

This concentrate is added to a previously prepared and filtered solution of sodium bisulfite in 40% ethanol (Note 11) contained in a 4-l. beaker provided with a mechanical stirrer (Note 12). The mixture is stirred for 3 hours, and the addition product then is filtered with suction on an 18-cm. Büchner funnel and washed, first with two 150-ml. portions of ethanol and then with 150 ml. of ether. The yield of air-dried product is 350–360 g. (72–74%, based on the selenious acid used) (Note 13).

2. Notes

1. The loss of large quantities of acetaldehyde is avoided by use of a spiral condenser, with sufficient heating so that a vapor lock is formed—the entrapped liquid should fill about two-thirds of the spiral.

In cold weather, the tap water is usually cold enough so that any efficient long condenser, or two in series, is sufficient.

2. The selenious acid does not need to be freshly prepared. A larger amount does not increase the yield. The submitters specified selenium dioxide as the oxidizing agent, but the checkers prepared their material by evaporating an aqueous solution to dryness on the water bath. They, therefore, have considered the oxidizing agent to be selenious acid and have calculated the yield on this basis. If this product is in reality selenium dioxide, then the yield is 62–64%, and 222 g. is 2.0 moles.

3. This amount of paraldehyde represents a considerable excess over the theoretical but was found to be most satisfactory.

4. Acetic acid appears to function both as an accelerator for the oxidation and an inhibitor of the rearrangement to glycolic acid.

5. This is regulated by the temperature of the water bath, 65–80° being the required range.

6. This material, impure selenium, may be reoxidized,[1] and then it is suitable for a subsequent preparation. About 130 g. is recovered at this point, and 8–10 g. after the concentration.

7. Alternatively, direct distillation may be employed; this requires only 2.5 hours but demands more attention. The volume is reduced to 200–300 ml., and then 800 ml. of water is added.

8. A test sample is filtered, and the clear filtrate is treated with more of the reagent to determine the end point.

9. Lead acetate is more satisfactory than sulfur dioxide for the removal of selenious acid, provided that the solution is kept cool and a large excess is avoided.

10. This volume of solution is most easily handled. Should glyoxal itself be desired, the solution may be evaporated to dryness in a desiccator. The product thus obtained is identical with that sold as "polyglyoxal."

11. The solution is prepared by dissolving 312 g. of technical sodium bisulfite in 2.1 l. of warm (about 40°) water, and adding 1.4 l. of 95% ethanol.

12. Alternatively, this may be done in a flask, which is shaken by hand frequently to prevent formation of a solid cake. The use of a stirrer results in a granular product. The mother liquor retains about 7 g. of glyoxal bisulfite per liter.

13. This product is pure enough for most purposes. It can be recrystallized by dissolving it in water and adding enough alcohol to make a 40% solution. The recovery is 90–92%.

3. Methods of Preparation

Glyoxal has been obtained by several methods, only a few of which are of preparative value. The most feasible are the oxidation of acetaldehyde by nitric[2–4] or selenious[5] acid; the hydrolysis of dichlorodioxane;[6] and the hydrolysis of the product resulting from the action of fuming sulfuric acid upon tetrahaloethanes.[7]

1 *Org. Syntheses* Coll. Vol. **2**, 510, Note 2 (1943).
2 Lubawin, *Ber.*, **8**, 768 (1875).
3 Wyss, *Ber.*, **10**, 1366 (1877).
4 Behrend and Kölln, *Ann.*, **416**, 230 (1918).
5 Riley, Morley, and Friend, *J. Chem. Soc.*, **1932**, 1881.
6 Butler and Cretcher, *J. Am. Chem. Soc.*, **54**, 2988 (1932).
7 Ott, Ger. pat. 362,743 [*C. A.*, **18**, 991 (1924)].

GUANIDOACETIC ACID

(Glycocyamine)

$$NH_2CSNH_2 + C_2H_5Br \rightarrow HN{=}C(SC_2H_5)NH_2 \cdot HBr$$

$$HN{=}C(SC_2H_5)NH_2 + NH_2CH_2COOH \rightarrow \underset{\displaystyle \overset{\|}{NH}}{NH_2C}NHCH_2COOH + C_2H_5SH$$

Submitted by E. Brand and F. C. Brand.
Checked by C. F. H. Allen and John W. Gates, Jr.

1. Procedure

A. *S-Ethylthiourea hydrobromide.* A mixture of 150 g. of powdered thiourea (1.97 moles) (Note 1), 250 g. of ethyl bromide (2.29 moles), and 200 ml. of absolute ethanol is placed in a 1-l. round-bottomed flask equipped with an efficient condenser. The mixture is warmed on a water bath (bath temperature 55–65°) for 3 hours, with occasional shaking. During this time all the thiourea dissolves. The reflux condenser is replaced by one set for downward distillation, and the ethanol and excess ethyl bromide are removed under the vacuum of a water pump. During the distillation, the temperature of the bath is slowly raised to the boiling point (Note 2). The residual oil is poured into a 500-ml. beaker and allowed to crystallize (Note 3). The solid is pulverized and dried in a desiccator (Notes 4 and 5). The yield is 340–360 g. (93–99%).

B. *Guanidoacetic acid.* This reaction should be carried out in a well-ventilated hood, as considerable amounts of ethyl mercaptan are evolved. In a 1-l. Erlenmeyer flask is placed 92.5 g. (0.50) mole of S-ethylthiourea hydrobromide. The flask is immersed in an ice bath, and 252 ml. of 2 N sodium hydroxide solution is added. A hot (80+°) solution of 41 g. of glycine in 90 ml. of water is added rapidly. When the temperature reaches 25° (Note 6), the flask is removed from the ice water. After about 30 minutes crystallization begins; then approximately 100 ml. of ether is added, and the mixture is left in the hood overnight (Note 7). The mixture is then chilled for 2 hours in an ice bath, the ether layer is decanted, and the solid is filtered with suction. The crystals are washed on the funnel successively with two 20-ml. portions of ice water (Note 8), two 150-ml. portions of 95% ethanol, and two 150-ml. portions of ether. The yield of air-dried guanidoacetic acid is 47–53 g. (80–90%). This product is pure enough for most purposes (Note 9); it melts with decomposition at 280–284°.

2. Notes

1. Commercial thiourea is usually sufficiently finely divided so that it may be used directly.

2. The last traces are more quickly removed if the vacuum line is attached directly to the flask.

3. If the product is inoculated, the liquid solidifies at once.

4. The crude product is sufficiently pure for the subsequent reaction. If kept in a cool place in the absence of air, it is stable for several months.

5. This is a general method for preparing S-alkylthiourea hydrobromides and hydriodides. The yields are always over 90%. The hydrochlorides are not so readily prepared; it is necessary to determine, by experiment, the optimum conditions for each hydrochloride.

6. The temperature may rise or fall to 25°, depending upon the temperatures of the component solutions.

7. The yield is slightly lower (45 g.) if the mixture is filtered after standing for only 3 hours.

8. The product is appreciably soluble in water.

9. Further purification of this product may be accomplished by (*a*) recrystallizing it from hot water (125 ml. per 5 g.), or (*b*) dissolving it in slightly more than the calculated amount of 2 N hydro-

chloric acid and reprecipitating by adding an equivalent quantity of 2 *N* sodium hydroxide. Analytical values for nitrogen (Dumas) given by the crude and purified products were as follows: Calculated: 35.9%. Found: acid as prepared, 35.4%; once recrystallized, 35.9%; reprecipitated, 35.7%.

3. Methods of Preparation

S-Ethylthiourea has been prepared as the hydrobromide [1-3] and hydriodide.[2,4] Guanidoacetic acid has previously been made from S-ethylthiourea hydriodide [2] and from S-methylthiourea.[5]

1 Claus, *Ber.*, **7**, 236 (1874).
2 Wheeler and Merriam, *Am. Chem. J.*, **29**, 483 (1903).
3 Schotte, Priewe, and Roescheisen, *Z. physiol. Chem.*, **174**, 119 (1928).
4 Claus, *Ber.*, **8**, 41 (1875).
5 Mourgue, *Bull. soc. chim. France,* **1948**, 181.

HEMIN *

$C_{34}H_{32}O_4N_4FeCl$

Submitted by Hans Fischer.
Checked by C. R. Noller and G. A. Smith.

1. Procedure

In a 5-l. round-bottomed flask equipped with a thermometer, reflux condenser, and dropping funnel are placed 4 l. of glacial acetic acid and 1 g. of sodium chloride. The acid is heated to boiling on a sand bath until the sodium chloride is in solution, and then 1 l. of defibrinated blood (Note 1) is added in a thin stream from the dropping funnel over a period of about 30 minutes. The blood should not touch the sides of the flask. During this time the temperature is kept at 100–105°, and heating is continued for 10 minutes after all the blood has been added. The flame is then removed and the mixture allowed to cool and stand overnight.

The precipitated hemin is removed by centrifuging (Note 2). If the centrifuging is carried out in 100-ml. tubes, each lot of tubes is centrifuged 10 minutes, the supernatant liquid is decanted, more of the mixture added, and the centrifuging repeated. The hemin is allowed to accumulate in the tubes until all the mixture has been centrifuged, after which it is stirred with a glass rod and washed from the several tubes into one with 75 ml. of 50% aqueous acetic acid.

After centrifuging and decanting, the hemin is washed successively in the same manner with two 75-ml. portions of distilled water, one 50-ml. portion of 95% ethanol, and one 50-ml. portion of ether. After the ether has been decanted the hemin is transferred to a watch glass by means of a rubber policeman and about 5 ml. of ether. After evaporation to dryness 3.5–4.5 g. of crude product is obtained.

For recrystallization, 5 g. of the crude hemin is placed in a 100-ml. Erlenmeyer flask, 25 ml. of pyridine is added, and the flask is shaken until the hemin has dissolved. Forty milliliters of chloroform is added, and the flask is stoppered with a cork and shaken for 15 minutes; the cork is carefully removed from time to time to release the pressure. The solution is then filtered with slight suction through a small Büchner funnel, and the Erlenmeyer flask and filter are washed with 15 ml. of chloroform.

During the shaking 350 ml. of glacial acetic acid is heated to boiling in a 600-ml. beaker under a hood, and 5 ml. of a saturated sodium chloride solution and 4 ml. of concentrated hydrochloric acid are added. The flame is extinguished, and the filtered hemin solution poured in a steady stream with stirring into the hot mixture; the suction flask is rinsed with 15 ml. of chloroform. After the mixture has stood for 12 hours, the crystals are filtered with suction on a small Büchner funnel and washed with 50 ml. of 50% aqueous acetic acid, 100 ml. of distilled water, 25 ml. of ethanol, and 25 ml. of ether. Suction is continued until the crystals are dry, when they can be readily removed. The recovery is 75–85%.

2. Notes

1. Fresh blood obtained from a slaughter house is defibrinated by whipping it with a stiff vegetable-fiber brush followed by filtration with suction through a large Büchner funnel. The blood is stirred during the filtration to prevent settling of the erythrocytes. Beef blood was used for checking this preparation.

2. The hemin may be removed by filtration but it is usually so finely divided that centrifuging is easier and less loss results.

3. Methods of Preparation

Hemin has been synthesized,[1] but it is always prepared from blood.[2]

[1] Fischer and Zeile, *Ann.*, **468**, 98 (1929).

[2] Nencki and Zaleski, *Z. physiol. Chem.*, **30**, 390 (1900); Piloty, *Ann.*, **377**, 358 (1910).

o-n-HEPTYLPHENOL

(Phenol, *o-n*-heptyl-)

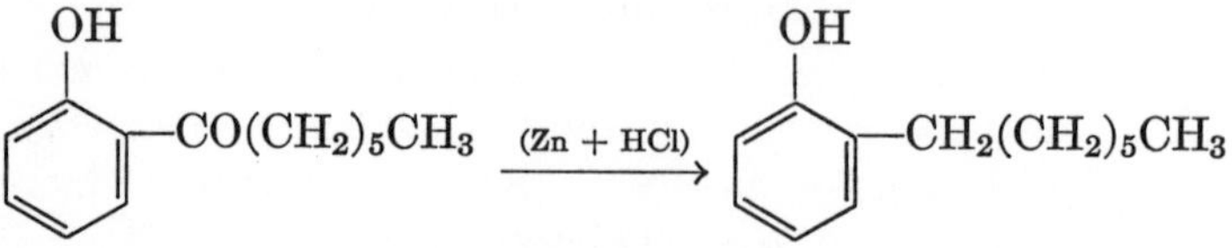

Submitted by R. R. READ and JOHN WOOD, JR.
Checked by W. E. BACHMANN and M. C. KLOETZEL.

1. Procedure

In a 1-l. three-necked flask fitted with a stirrer (Note 1) and a reflux condenser is placed 200 g. of amalgamated mossy zinc (Note 2). A mixture of 200 ml. of water and 200 ml. of concentrated hydrochloric acid is added and then a solution of 60 g. of *o*-heptanoyl phenol (Note 3) in 100 ml. of ethanol. The mixture is agitated vigorously and refluxed until reduction is complete (Note 4).

To the mixture is added 120 ml. of toluene, stirring being continued for a few minutes. The toluene solution is separated from the aqueous solution and washed three times with water. The solution is filtered from suspended matter, and the toluene is distilled from a Claisen flask until a thermometer in the liquid reads 170° (Note 5). The residue is then distilled under reduced pressure, the portion boiling at 118–123° at 1 mm. being collected (Note 6). The yield of colorless *o-n*-heptylphenol is 45–47 g. (81–86%) (Note 7).

2. Notes

1. The stirrer should be as large and substantial as the flask will accommodate, since the rate of reduction depends greatly on complete emulsification of the oil.

2. The zinc is amalgamated in the reaction flask by covering it with a solution of 4 g. of mercuric chloride in 300 ml. of water. Occasional agitation during 30 minutes is sufficient for amalgamation. The solution is poured off, and the zinc is rinsed once with water.

3. *o*-Heptanoyl phenol may be prepared by the method of Miller and Hartung [*Org. Syntheses* Coll. Vol. **2,** 543 (1943)]. The checkers found that, by keeping the mixture of *ortho* and *para* isomers in a cool place overnight, most of the *para* isomer crystallized and could be

separated by filtration. The *ortho* isomer is then obtained by fractional vacuum distillation, repeated two or three times. *o*-Heptanoyl phenol boils at 135–140°/3 mm.; *p*-heptanoyl phenol boils at 200–207°/4 mm.

4. The reduction requires at least 8 hours and may take twice that long. For testing completeness of reduction, 0.1 ml. of the oil is withdrawn and dissolved in 2 ml. of ethanol; to the solution is added 2–4 drops of a 10% solution of ferric chloride in ethanol. A deep red or reddish brown color is produced by the ketone; a light brownish yellow color indicates completion of the reduction. A standard solution containing 0.5 g. of the ketone per liter of ethanol is used for comparison, 2 ml. of this solution being treated with a few drops of the ferric chloride solution. Since the *para* acylphenols usually do not give a pronounced color with ferric chloride, at least 8 hours should be allowed for their reduction.

5. Drying of the toluene extract is unnecessary, since the water is carried over during the removal of the toluene. If necessary, the toluene may be returned once to the flask to effect complete removal of the water.

6. The residue from the distillation is usually less than 5 g. and may be discarded.

7. Other ketones may be reduced by this same procedure. The submitters report that the following have been reduced with yields of 70–90%: *o*- and *p*-butyryl phenols, *o*- and *p*-valeryl phenols, *o*- and *p*-caproyl phenols, *o*- and *p*-heptanoyl phenols, *o*- and *p*-octanoyl phenols, *o*- and *p*-pelargonyl phenols, *o*- and *p*-undecylyl phenols. The same procedure applies to acyl resorcinols and acyl chlororesorcinols. Caproyl and octanoyl resorcinols reduce to the corresponding alkyl derivatives in yields of 70–80%. Butyryl, valeryl, caproyl, heptanoyl, and octanoyl chlororesorcinols reduce to the corresponding alkyl chlororesorcinols in yields of 60–75%.

3. Methods of Preparation

Primary alkyl phenols have been prepared by the reduction of acyl phenols;[1] by the demethylation of the corresponding ethers;[1,2] by the diazotization of the corresponding amines;[3] and by the alkali fusion of sulfonates.[4] Alkyl resorcinols have been prepared by the reduction of acyl resorcinols.[1,5,6] Alkyl chlororesorcinols have been prepared from the corresponding acyl chlororesorcinols by reduction.[7]

[1] Johnson and Hodge, *J. Am. Chem. Soc.*, **35**, 1014 (1913); Coulthard, Marshall, and Pyman, *J. Chem. Soc.*, **1930**, 280.

[2] Klages, *Ber.*, **32**, 1438 (1899).

[3] *Org. Syntheses* Coll. Vol. **1**, 128, 407 (1941).
[4] Ullmann, *Enzyklopädie der technischen Chemie,* Vol. 9, p. 35, Urban and Schwarzenberg, Berlin, 1921.
[5] Johnson and Lane, *J. Am. Chem. Soc.,* **43,** 348 (1921).
[6] Dohme, Cox, and Miller, *J. Am. Chem. Soc.,* **48,** 1688 (1926).
[7] Read, Reddish, and Burlingame, *J. Am. Chem. Soc.,* **56,** 1377 (1934).

HEXAMETHYLENE CHLOROHYDRIN

(1-Hexanol, 6-chloro-)

$$HOCH_2CH_2CH_2CH_2CH_2CH_2OH + HCl \rightarrow HOCH_2CH_2CH_2CH_2CH_2CH_2Cl + H_2O$$

Submitted by Kenneth N. Campbell and Armiger H. Sommers.[1]
Checked by Richard T. Arnold and Roger Amidon.

1. Procedure

The apparatus shown in Fig. 14 is constructed from a 1-l. distilling flask (*A*) and a 500-ml. distilling flask (*B*) the side arm of which has been replaced by 10-mm. Pyrex tubing sealed at an upward angle to the side arm of the larger flask (*A*). Charred cork stoppers are used, and a hydrogen chloride trap [2] is connected to the water-cooled reflux condenser. The bent side arm (*C*) is insulated by a wrapping of cloth. Reaction flask *A* is charged with 105 g. (0.89 mole) of hexamethylene glycol,[3] 785 ml. (9.5 moles) of concentrated hydrochloric acid, 130 ml. of water (Note 1), and 55 ml. of toluene (Notes 2 and 3). In flask *B* are put 350 ml. of toluene and a few boiling chips. The apparatus is assembled, and the flasks are heated in oil baths. The bath covering about two-thirds of flask *A* is kept at 95°; the one surrounding most of flask *B* is kept between 160° and 165° (Note 4). As the reaction proceeds and the products are removed by the continuous flow of toluene through the mixture, the organic-aqueous interface in *A* falls steadily. After 9 hours the heating baths and condensers are removed, and water is added to *A* to force the

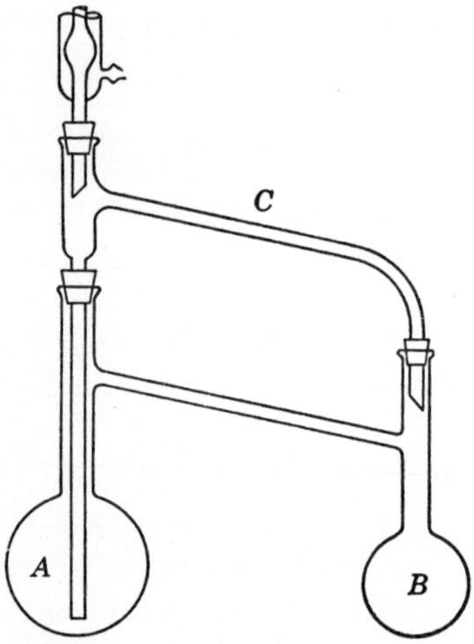

Fig. 14.

upper layer of toluene into *B*. The aqueous solution is siphoned out of *A*, and then the toluene extract in *B* is poured into a dropping funnel inserted through a stopper in the neck of a 500-ml. Claisen flask. Some of the extract is run into the flask, and most of the toluene is removed by distillation at atmospheric pressure while the remainder of the solution is slowly run into the flask. To remove the last of the solvent, the pressure is lowered to 65 mm. while the bath temperature is kept at 100°.

The material obtained from two such runs is combined and fractionated through a Fenske-Whitmore column [4] packed with glass helices. Distillation is carried out at a pressure of 8–12 mm., with a reflux ratio of 5:1. After a fore-run of 5–15 g., about 45 g. (16%) of hexamethylene dichloride distils at 80–84°/9 mm., n_D^{20} 1.4585–1.4565. An intermediate fraction of 10–20 g. distils at 84–100°/9 mm., and then 108–122 g. (45–50%) of hexamethylene chlorohydrin boiling at 100–104°/9 mm., n_D^{20} 1.4551–1.4557, is collected. If several runs are made the residues may be distilled from a Claisen flask and the distillate combined with the intermediate fractions for refractionation. In this way the average yield may be raised to 50–55%.

2. Notes

1. The use of concentrated hydrochloric acid without dilution results in violent bubbling during the reaction.

2. In the original work of Bennett and Turner [5] a petroleum fraction of b.p. 90–120° was used. When the submitters used Skellysolve L, b.p. 90–100°, the hexamethylene chlorohydrin was not sufficiently soluble in it to be removed from the reaction mixture. A study of the solubility of hexamethylene chlorohydrin in various commonly available solvents was made, and toluene was found to be the most satisfactory.

3. The exact volume of free space below the side arm of *A* will vary somewhat according to the exact dimensions of the flask and the adapter tube, and it may be desirable to make corresponding changes in the volumes of hydrochloric acid and water used. In the apparatus constructed by the checkers the aqueous phase expanded during the heating to such an extent that only about 5 ml. of toluene remained in the neck below the side arm; this apparatus operated satisfactorily when the volumes of hydrochloric acid and water were reduced to 720 ml. and 115 ml., respectively.

4. The bath temperature may be controlled by heating with an immersed loop of Nichrome resistance coil, such as the heating element

from an electrical appliance. The coil is connected to a 110-volt line in series with a 27-ohm 5-ampere variable resistor and a 600-watt heater element, which may be cut out of the circuit to lower the fixed resistance. An ammeter in the circuit helps to determine the proper adjustment of the sliding contact of the resistor. To avoid an excessive load on the coil, the bath about *B* should be heated in addition by a hot plate adjusted to give a temperature somewhat below 160°.

3. Methods of Preparation

Hexamethylene chlorohydrin has been prepared by the reaction of hydrochloric acid with hexamethylene glycol, without a catalyst[5–7] or in the presence of cuprous chloride.[8]

[1] Work done under contract with the Office of Scientific Research and Development.

[2] *Org. Syntheses* Coll. Vol. **2**, 4 (1943).

[3] *Org. Syntheses* Coll. Vol. **2**, 325 (1943).

[4] *Org. Syntheses*, **25**, 2 (1945).

[5] Bennett and Turner, *J. Chem. Soc.*, **1938**, 814.

[6] Müller and Vanc, *Monatsh.*, **77**, 259 (1947).

[7] Campbell, Sommers, Kerwin, and Campbell, *J. Am. Chem. Soc.*, **68**, 1556 (1946).

[8] Coleman and Bywater, *J. Am. Chem. Soc.*, **66**, 1821 (1944).

HOMOPHTHALIC ACID AND ANHYDRIDE

(Phenylacetic acid, *o*-carboxy-, and 1,3-Isochromandione)

A.
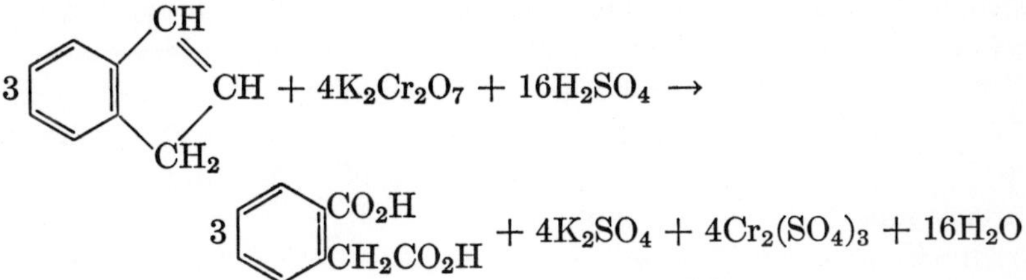

B.
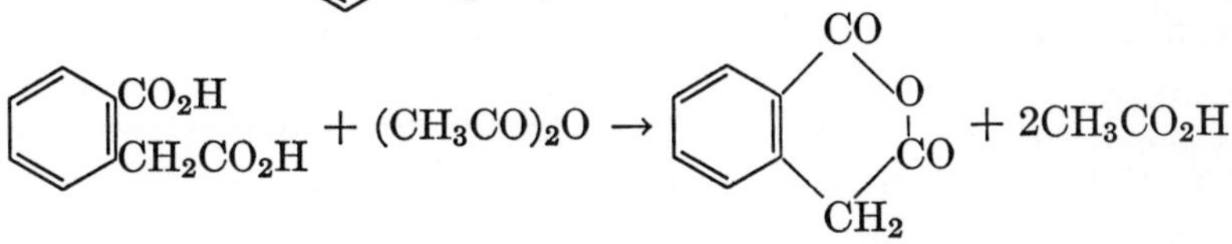

Submitted by Oliver Grummitt, Richard Egan, and Allen Buck.
Checked by R. L. Shriner and Walter R. Knox.

1. Procedure

A. *Homophthalic acid.* A solution of 243 g. (0.83 mole) of technical potassium dichromate in 3.6 l. of water and 1330 g. (725 ml., 13 moles) of concentrated sulfuric acid is prepared in a 5-l. three-necked flask fitted with a sealed mechanical stirrer, a thermometer, a 100-ml. dropping funnel, and a reflux condenser. The mixture is warmed to 65°, and 72 g. (72 ml., 0.56 mole) of technical 90% indene (Note 1) is added dropwise from the dropping funnel. The temperature must be kept at 65 ± 2°, and cooling by a water bath is necessary during the addition. After the addition is complete the mixture is stirred for 2 hours at 65 ± 2° (Note 2). At the end of this period the mixture is cooled with stirring to 20–25° and then further cooled in an ice-salt bath for 5 hours at 0°. The homophthalic acid which separates is collected on a 10-cm. Büchner funnel with the aid of suction, then washed with two 75-ml. portions of ice-cold 1% sulfuric acid and once with 75 ml. of ice water (Note 3). The precipitate is then dissolved in 215 ml. of 10% sodium hydroxide solution, and the resulting solution is extracted with two 50-ml. portions of benzene (Note 4), which are discarded. The aqueous solution is added to 160 ml. of 33% sulfuric acid with vigorous stirring, and the mixture is chilled in an ice-salt bath for 2–3 hours. The homophthalic acid is collected on a 10-cm. Büchner funnel with the aid of suction, washed with three 25-ml. portions of ice water, and pressed and sucked as dry as possible. The acid is transferred to a 500-ml. distilling flask, 300 ml. of benzene is added, and the mixture is distilled from a steam bath until about 250 ml. of distillate (benzene and water) has been collected (Note 5). The slurry of acid and benzene is filtered with the aid of suction through a 10-cm. Büchner funnel, and the product is spread out on a porous plate to allow the last traces of benzene to evaporate. The yield of white crystals of homophthalic acid, melting at 180–181° (Note 6), amounts to 67–77 g. (66–77% calculated on the basis of 90% indene content of the commercial indene) (Note 7).

B. *Homophthalic anhydride.* A mixture of 60 g. (0.33 mole) of dry homophthalic acid and 33.7 g. (31 ml., 0.33 mole) of acetic anhydride in a 200-ml. round-bottomed flask fitted to a reflux condenser by a ground-glass joint is refluxed for 2 hours. The mixture is cooled to about 10° for 30 minutes, and the solid anhydride is collected on a Büchner funnel with the aid of suction. It is washed with 10 ml. of glacial acetic acid and pressed, and as much of the solvent as possible is removed by suction. The product is spread out on a porous plate for several hours (Note 8); it amounts to 46–47.5 g (85–88%) of white crystals melting at 140–141° (Note 9).

2. Notes

1. Indene of approximately 90% purity may be obtained from the Koppers Company, Pittsburgh, Pennsylvania, or United Gas Improvement Company, Philadelphia, Pennsylvania. The practical grade of the same purity from the Eastman Kodak Company may also be used. If the material is dark colored, it should be redistilled and the fraction boiling from 180° to 182° should be used.

2. It is important to control the temperature of this oxidation. If the oxidation is carried out at the reflux temperature, the yield of purified acid drops to about 52–54%.

3. Homophthalic acid possesses an appreciable solubility in water (about 1.6 g. per 100 ml. at 20°); hence it is necessary to cool the solutions and wash liquids to obtain good yields.

4. This extraction removes about 7–8 g. of oily alkali-insoluble products.

5. This azeotropic distillation is the quickest method for drying homophthalic acid. The acid turns dark colored if dried in an oven at 110°. It may be dried over anhydrous calcium chloride in a vacuum desiccator for 24–36 hours; it melts at 180–181°.

6. The melting point depends on the rate of heating. When the melting-point tube was placed in the bath preheated to 170°, the acid melted at 180–181°. When the heating was started at room temperature, the observed melting point was 172–174°.

7. If phthalide is available, an alternative procedure via *o*-carboxyphenylacetonitrile may be preferred for the preparation of homophthalic acid [*Org. Syntheses,* **22,** 61 (1942)].

8. Drying in an oven causes some loss because the anhydride sublimes.

9. If all the acetic acid is not removed the product melts lower, *ca.* 138–139°.

3. Methods of Preparation

Homophthalic acid has been prepared from naphthalene [1] or tetralin [2] via phthalonic acid; by hydrolysis of *o*-cyanobenzylcyanide; [3] by oxidation of β-indanone with alkaline permanganate solution; [4] from α-indanone by chromic acid oxidation,[5] by nitrosation, Beckmann rearrangement, and hydrolysis,[6] or by nitrosation, hydrolysis, and hydrogen peroxide oxidation; [7] from *o*-toluic acid by bromination of the acid chloride followed by treatment with ethanol and sodium cyanide and hydrolysis with 50% sulfuric acid; [8] from indene by oxidation with chromic and sulfuric acid; [9] and from *o*-acetylbenzoic acid by Willgerodt reaction with morpholine and sulfur followed by hydrolysis of the dimorpholide.[10] The oxidation of indene by alkaline permanganate produces homophthalic acid [11] and phthalic acid; [12] oxidation with chromic acid leads to a purer product.[13] The hydrolysis of *o*-carboxyphenylacetonitrile yields homophthalic acid.[14]

The procedure described here is a modification of that described by Meyer and Vittenet [9] and Whitmore and Cooney.[15]

Homophthalic anhydride has been obtained by heating the acid alone,[1] and by refluxing with acetyl chloride [14] or with acetic anhydride.[16]

1 Graebe and Trümpy, *Ber.,* **31,** 375 (1898).

2 Davies and Poole, *J. Chem. Soc.,* **1928,** 1616.

3 Gabriel and Otto, *Ber.,* **20,** 2224, 2502 (1887).

4 Benedikt, *Ann.,* **275,** 354 (1893).

5 Ingold and Piggott, *J. Chem. Soc.,* **123,** 1497 (1923).

6 Perkin and Robinson, *J. Chem. Soc.,* **91,** 1082 (1907).

7 Perkin, Roberts, and Robinson, *J. Chem. Soc.,* **101,** 232 (1912).

8 Price, Lewis, and Meister, *J. Am. Chem. Soc.,* **61,** 2760 (1939).

9 Meyer and Vittenet, *Compt. rend.,* **194,** 1250 (1932); *Ann. chim.,* (10) **17,** 274 (1932).

[10] Schwenk and Papa, *J. Org. Chem.*, **11**, 798 (1946).
[11] Hensler and Schieffer, *Ber.*, **32**, 29 (1899).
[12] Cooney, PhD. thesis, Polytech. Inst. Brooklyn, 1943.
[13] Fieser and Pechet, *J. Am. Chem. Soc.*, **68**, 2577 (1946).
[14] Wislicenus, *Ann.*, **233**, 102 (1886); *Org. Syntheses*, **22**, 61 (1942).
[15] Whitmore and Cooney, *J. Am. Chem. Soc.*, **66**, 1237 (1944).
[16] Dieckmann, *Ber.*, **47**, 1432 (1914).

HYDROQUINONE DIACETATE *

(Hydroquinone, diacetate)

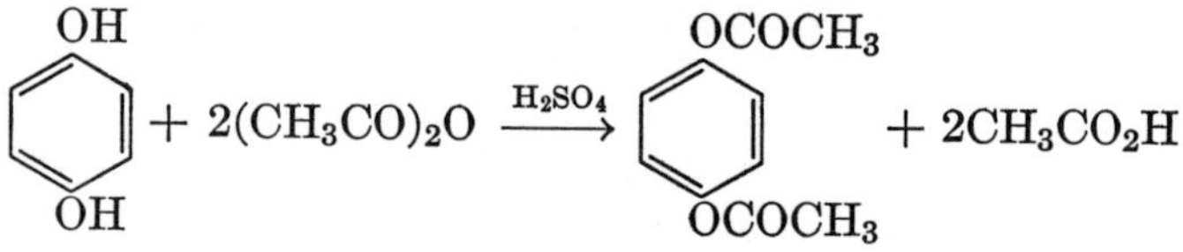

Submitted by W. W. PRICHARD.
Checked by HOMER ADKINS and MAYNETTE VERNSTEN.

1. Procedure

One drop of concentrated sulfuric acid is added to a mixture of 110 g. (1.0 mole) of hydroquinone and 206 g. (190.3 ml., 2.02 moles) of acetic anhydride (Note 1) in a 1-l. Erlenmeyer flask. The mixture is stirred gently by hand; it warms up very rapidly, and the hydroquinone dissolves. After 5 minutes the clear solution is poured onto about 800 ml. of crushed ice. The white crystalline solid which separates is collected on a Büchner filter and washed with 1 l. of water. The filter cake is pressed occasionally to facilitate the removal of water; the solid is dried to constant weight over phosphorus pentoxide in a vacuum desiccator. The nearly pure product weighs 186–190 g. (96–98%) and melts at 121–122° (Note 2); it can be recrystallized from dilute ethanol (Note 3).

2. Notes

1. The use of commercial acetic anhydride in this preparation sometimes results in appreciably lower yields. The checkers used the freshly redistilled reagent.

2. The melting point is recorded in the literature [1,2] as 121° and as 123–124°.

3. Recrystallization from 50% ethanol (by weight) permits a 93–94% recovery of material melting at 121.5–122.5°; about 365 g. (400 ml.) of the solvent is required for 100 g. of the crude product.

3. Methods of Preparation

Hydroquinone diacetate has been prepared by the treatment of hydroquinone with acetic anhydride, both in the presence [1,3,4] and in the absence [5–7] of strong acid catalysts, by the treatment of the sodium salt of hydroquinone with acetic anhydride, and by the reaction of hydroquinone with acetic anhydride in the presence of sodium acetate.[8,9] It has also been prepared from hydroquinone and acetyl chloride; [10] the acetylation with acetyl chloride is reported to be improved by the addition of metallic magnesium.[11]

1 Ciusa and Sollazzo, *Chem. Zentr.*, **114**, II, 615 (1943).
2 Chattaway, *J. Chem. Soc.*, **1931**, 2495.
3 Henle, *Ann.*, **350**, 344 (1906).
4 Reychler, *Chem. Zentr.*, **79**, I, 1042 (1908).
5 Shaw, *J. Chem. Soc.*, **99**, 1610 (1911).
6 Hesse, *Ann.*, **200**, 244 (1880).
7 Kaufmann, *Ber.*, **42**, 3482 (1909).
8 Sarauw, *Ann.*, **209**, 128 (1881).
9 Buchka, *Ber.*, **14**, 1327 (1881).
10 Nietzki, *Ber.*, **11**, 470 (1878).
11 Spasov, *Ann. univ. Sofia II, Faculté phys.-math.*, livre 2, **35**, 289 (1938–1939) [*C. A.*, **34**, 2343 (1940)].

m-HYDROXYBENZALDEHYDE *

(Benzaldehyde, *m*-hydroxy-)

$$m\text{-}NO_2C_6H_4CHO \xrightarrow[HCl]{SnCl_2} m\text{-}NH_2C_6H_4CHO\cdot(SnCl_4)_x \xrightarrow[HCl]{NaNO_2}$$
$$(m\text{-}ClN_2C_6H_4CHO)_2\cdot SnCl_4 \xrightarrow[Heat]{H_2O} m\text{-}HOC_6H_4CHO$$

Submitted by R. B. Woodward.
Checked by C. S. Hamilton and Raymond J. Andres.

1. Procedure

A solution of 450 g. (2 moles) of powdered stannous chloride dihydrate (Note 1) in 600 ml. of concentrated hydrochloric acid in a 3-l. beaker provided with an efficient mechanical stirrer is cooled in an ice-salt bath. When the temperature of the solution has fallen to 5°, the ice bath is removed, and 100 g. (0.66 mole) of *m*-nitrobenzaldehyde (Note 2) is added in one portion. The temperature rises slowly at first, reaching 25–30° in about 5 minutes. As it then rises rapidly to about 100° the ice-salt bath is again placed around the beaker. Stirring must be vigorous or the reaction mixture may be forced out of the beaker (Note 3). During the reaction the *m*-nitrobenzaldehyde dissolves, and an almost clear red solution is obtained. The solution is cooled with very slow stirring (Note 4) in an ice-salt mixture for 2.5 hours. The orange-red paste of the stannichloride of *m*-aminobenzaldehyde (Note 5) is filtered on a sintered-glass funnel (Note 6).

A suspension of the material in 600 ml. of concentrated hydrochloric acid is stirred mechanically in a 3-l. beaker which is set in an ice-salt bath. A solution of 46 g. of sodium nitrite in 150 ml. of water in a 250-ml. dropping funnel fixed with its stem below the surface of the suspension is added slowly while the temperature of the mixture is kept at 4–5° (Note 7). After the addition is completed (in approximately 80 minutes), stirring in the ice-salt bath is continued for 1 hour to crystallize completely the stannichloride of the diazonium salt. The reddish brown salt is filtered on a sintered glass funnel (Note 6).

The damp salt is added cautiously in small portions to 1.7 l. of boiling water in a 4-l. beaker in the course of 40 minutes. During the reaction, the water lost by vaporization is replaced (Note 8). The solution is treated with 4 g. of Norit, boiled a few minutes, and filtered hot. The red filtrate is kept in an icebox for 12–16 hours; during this time scratching is employed occasionally to induce crystallization (Note 9). The yield of orange crystals melting at 99–101° is 48–52 g. (59–64%). A solution of this product in about 1 l. of boiling benzene is treated with Norit, filtered, and concentrated to 300 ml. On cooling, 41–45 g. (51–56%) of light-tan crystals melting at 101–102° is obtained (Note 10).

2. Notes

1. A chemically pure grade of stannous chloride ($SnCl_2 \cdot 2H_2O$) should be used. Lower yields are obtained with technical stannous chloride.

2. A practical grade of *m*-nitrobenzaldehyde (m.p. 52–55°) was used by the submitter. The checkers used Eastman's *m*-nitrobenzaldehyde (m.p. 57–58°).

3. Less satisfactory yields are obtained when the reaction is moderated by the addition of the *m*-nitrobenzaldehyde in several portions.

4. Stirring must be very slow since gradual cooling is necessary in order to obtain particles large enough to permit fairly rapid filtration.

5. The exact formula of the *m*-aminobenzaldehyde stannichloride is not known.

6. As a substitute for the sintered-glass funnel, a large Büchner funnel provided with an asbestos fiber mat under the filter paper will serve. The asbestos mat can be washed and used repeatedly.

7. Best results are obtained by keeping the temperature during diazotization at 4–5°. Below 0° the speed of diazotization is greatly decreased, and above 5° the diazonium salt begins to decompose.

8. Concentrating the solution below 1.7 l. results in markedly decreased purity.

9. If scratching does not succeed in inducing crystallization, a small portion of the solution may be removed and evaporated for seed.

10. According to the submitter, further purification can be effected by sublimation: the material is placed in the bottom of a Pyrex desiccator which is then evacuated and placed on a steam bath overnight. The mat of interwoven needles can be removed easily. One crystallization from benzene gives 37–40 g. (47–49%) of needles melting at 103–104°.

3. Methods of Preparation

m-Hydroxybenzaldehyde has been made by reduction of *m*-hydroxybenzoic acid with sodium amalgam in weak acid solution;[1] by oxidation of the arylsulfonic ester $ArSO_2OC_6H_4CH_3$ with manganese dioxide and sulfuric acid to the corresponding derivative of *m*-hydroxybenzaldehyde $ArSO_2OC_6H_4CHO$ and hydrolysis of the latter with concentrated sulfuric acid;[2] from *m*-dichloromethylphenyl benzoate by heating with water and calcium carbonate under a pressure of 4–5 atm.[3] The present method is derived from that of Tiemann and Ludwig[4] and is similar to the procedure for the preparation of *m*-chlorobenzaldehyde.[5]

[1] Sandmann, *Ber.*, **14**, 969 (1881).

[2] Ger. pat. 162,322 (*Chem. Zentr.*, **1905** II, 726).

[3] Raschig, Ger. pat. 233,631 [*Chem. Zentr.*, **1911** I, 1388; *Frdl.*, **10**, 163 (1910–1912)].

[4] Tiemann and Ludwig, *Ber.*, **15**, 2045 (1882); see also Rieche, *Ber.*, **22**, 2348 (1889); Ger. pat. 18,016 [*Frdl.*, **1**, 586 (1888)]; Subak, *Monatsh.*, **24**, 167 (1903).

[5] *Org. Syntheses* Coll. Vol. **2**, 130 (1943).

2-HYDROXYCINCHONINIC ACID

(Cinchoninic acid, 2-hydroxy-)

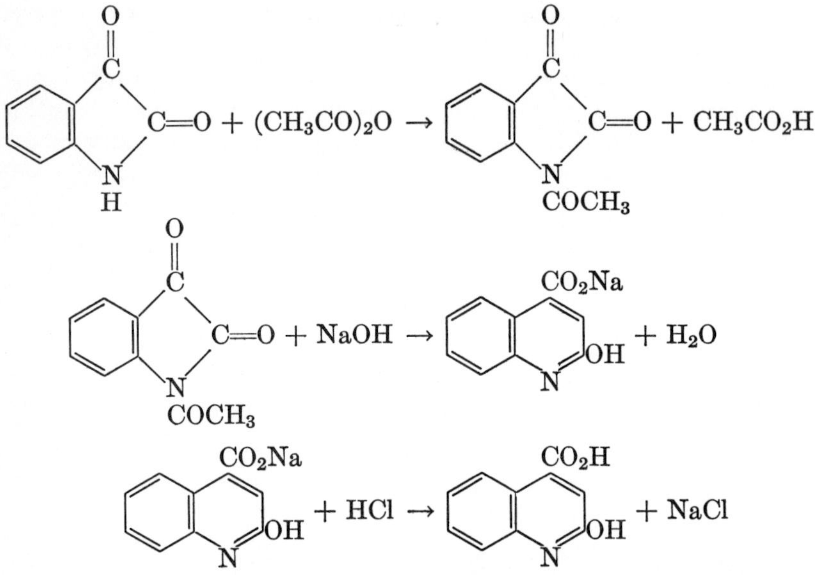

Submitted by Thomas L. Jacobs, S. Winstein, Gustave B. Linden, Jean H. Robson, Edward F. Levy, and Dexter Seymour.[1]
Checked by R. T. Arnold and F. M. Robinson.

1. Procedure

A. *N-Acetylisatin.* In a 3-l. three-necked flask equipped with a mercury-sealed stirrer and reflux condenser are placed 600 g. (4.1 moles) of isatin (Note 1) and 1390 ml. (14.7 moles) of acetic anhydride (Note 2). This mixture is refluxed for 4 hours over a free flame with constant stirring. The solution is allowed to cool slowly to room temperature with stirring and is then cooled to about 10° (Note 3). The slurry of crystals is collected on a 12-in. Büchner funnel and washed with five 175-ml. portions of ether or until the washings are light red. The yield of crude air-dried N-acetylisatin, m.p. 141–142° (cor.), is 635 to 643 g. (82–83%) (Note 4).

B. 2-*Hydroxycinchoninic acid.* To a solution of 160 g. (about 3.8 moles) of U.S.P. sodium hydroxide in 9.6 l. of water in a 12-l. flask is added 310 g. (1.64 moles) of crude N-acetylisatin. The mixture is stirred mechanically and heated to boiling; the flame is adjusted to maintain gentle boiling. After the mixture has boiled 1 hour (Note 5) the flame is removed, 50 g. of decolorizing carbon is added cautiously (vigorous boiling and frothing may occur if the solution is too hot), and the suspension is stirred for 30 minutes. The charcoal is allowed to settle for a few minutes, and the solution is siphoned into a filter funnel containing a layer of Supercel. The filtrate is allowed to cool overnight in a 5-gal. crock. The solution is stirred with a hook-type stirrer driven by a powerful motor (Note 6), and 6 *N* hydrochloric acid is added until the mixture is neutral to Congo red. The suspension is filtered immediately (Note 7) through a 12-in. Büchner funnel, and the solid is washed with 2 l. of water. The filtrate is made distinctly acid to Congo red and set aside for a few hours. The solid acid is dried at 90° for 2 days; the anhydrous acid, m.p. 346–347° (cor.), so obtained has the correct neutral equivalent (Note 8) and weighs 200 to 205 g. (70–73% over-all yield, based on isatin consumed). From the acidified filtrate, after precipitation is complete, about 60 g. of isatin is recovered.

2. Notes

1. Commercial isatin, m.p. 199.0–200.5° (cor.), was obtained from the National Aniline and Chemical Company.

2. Acetic anhydride obtained from the Carbide and Carbon Chemicals Corporation was employed. Redistillation of this reagent did not increase the yield.

3. If the solution is not stirred during the cooling, the product crystallizes as a solid cake which is difficult to remove from the flask.

4. Recrystallization from benzene, unnecessary for the present purpose, gives a yellow crystalline product of m.p. 142.0–143.5° (cor.).

5. It is convenient to make two of these runs simultaneously. A reflux condenser is unnecessary, but if none is used the stopper holding the stirrer must be provided with a vent to permit steam to escape.

6. The ordinary small laboratory stirring motor is insufficiently powerful to maintain rapid agitation after most of the cinchoninic acid has precipitated.

7. If the mixture is not filtered immediately after acidification, or if the first end point is overrun, isatin is precipitated along with the cinchoninic acid and the product has a deep red color. Such material can be freed of isatin by dissolving it in sodium bicarbonate solution, filtering, and reprecipitating.

8. This crude acid, which is satisfactory for most purposes, can be recrystallized from glacial acetic acid to yield a product melting at 352–353° (cor.).

3. Methods of Preparation

2-Hydroxycinchoninic acid has been prepared by the rearrangement of N-acetylisatin in alkali,[2–4] by the treatment of isatin with malonic acid in glacial acetic acid,[5] by heating cinchoninic acid with concentrated potassium hydroxide,[6] and by the treatment of ethyl 2-chlorocinchoninate with 30% sodium hydroxide.[7]

[1] Work done under contract with the Office of Scientific Research and Development.

[2] Camps, *Arch. Pharm.*, **237**, 659 (1899).

[3] Ainley and King, *Proc. Roy. Soc. London,* **125B**, 60 (1938).

[4] Campbell and Kerwin, *J. Am. Chem. Soc.*, **68**, 1837 (1946).

[5] Borsche and Jacobs, *Ber.*, **47**, 354 (1914).

[6] Koenigs and Koerner, *Ber.*, **16**, 2152 (1883).

[7] Thielpape, *Ber.*, **71B**, 387 (1938); Ger. pat. 670,582 [*C. A.*, **33**, 6348 (1939)].

β-(2-HYDROXYETHYLMERCAPTO)PROPIONITRILE

(Propionitrile, β-(2-hydroxyethylmercapto)-)

$$HOCH_2CH_2SH + CH_2{=}CHCN \rightarrow HOCH_2CH_2SCH_2CH_2CN$$

Submitted by Leon L. Gershbein and Charles D. Hurd.
Checked by Cliff S. Hamilton and John A. Stephens.

1. Procedure

In a 500-ml. three-necked round-bottomed flask equipped with a sealed stirrer, a reflux condenser, a dropping funnel, and a thermometer is placed 78 g. (70 ml., 1 mole) of 2-mercaptoethanol (Note 1). Into the dropping funnel is poured 67 ml. (54.3 g., 1 mole) of acrylonitrile (Note 2), and after the addition of about 3 ml. of the nitrile, with stirring, the contents are warmed with a water bath to about 35–40° for 5 minutes. The remainder of the acrylonitrile is then added dropwise during 10 minutes. The temperature soon mounts to about 65° and is kept between 55° and 60° by intermittent short cooling with water until it only slowly increases or remains stationary at 55–60° (Note 3). Forty milliliters of acrylonitrile is then added all at once, cooling being applied if necessary, and the contents are stirred for 16 hours at room temperature. The product is distilled from a 250-ml. Claisen flask after removal of excess acrylonitrile under reduced pressure. The yield of nitrile distilling at 178–180°/14 mm., n_D^{25} 1.5101, as a colorless viscous liquid is 121–123 g. (92–94%) (Note 4).

2. Notes

1. The 2-mercaptoethanol was obtained from Carbide and Carbon Chemicals Corporation.

2. Commercial acrylonitrile may be used without further purification.

3. This requires about 30 minutes. As an inhibition period generally occurs, care must be taken in the initiation of the reaction and subsequent moderation of the heat evolved, but this operation can easily be controlled.

4. In the presence of alcoholic sodium hydroxide, either 2-mercaptoethanol or β-(2-hydroxyethylmercapto)propionitrile is converted to the dicyanoethylated product, 4-oxa-7-thiadecanedinitrile, $NCCH_2CH_2OCH_2CH_2SCH_2CH_2CN$. This basic agent can also be applied to the general reaction of thiophenols or mercaptans with acrylonitrile.

3. Methods of Preparation

This method is a modification of the directions of Hurd and Gershbein.[1] The compound has been made also [2] with piperidine as the basic catalyst.

[1] Hurd and Gershbein, *J. Am. Chem. Soc.*, **69**, 2331 (1947).

[2] Gribbins, Miller, and O'Leary, U. S. pat. 2,397,960 [*C. A.*, **40**, 3542 (1946)].

4(5)-HYDROXYMETHYLIMIDAZOLE HYDROCHLORIDE

(4-Imidazolemethanol, hydrochloride)

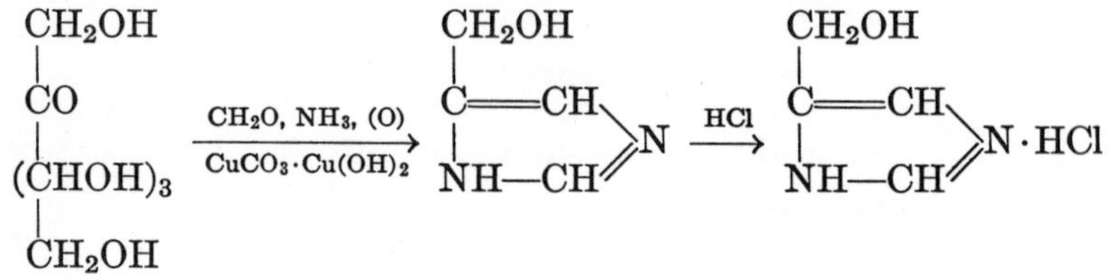

Submitted by JOHN R. TOTTER and WILLIAM J. DARBY.
Checked by H. R. SNYDER and R. L. ROWLAND.

1. Procedure

A. *Hydroxymethylimidazole picrate.* To 222 g. (1 mole) of basic cupric carbonate (Note 1) in a 5-l. flask are added 1.5 l. of distilled water and 720 g. (800 ml., 12 moles) of 28% ammonia. The bulk of the copper carbonate is brought into solution by swirling; 112 g. (100 ml., 1.3–1.4 moles) of 37–40% formaldehyde and 90 g. (0.475 mole) of commercial 95% fructose are added. The solution is well mixed and placed on a steam bath under a hood. After 30 minutes of heating with occasional shaking, a moderate current of air is bubbled through the solution (500–600 ml. per minute) (Note 2), with continued heating, for 2 hours longer (Note 3).

The mixture is then chilled in an ice bath for at least 3 hours, and the olive-brown precipitate of the sparingly soluble copper complex of imidazole derivatives is filtered. The product is washed with about 500 ml. of cold water, suspended while moist (Note 4) in 1 l. of water, and rendered just acid to litmus by the addition of concentrated hydrochloric acid (about 40 ml.). Hydrogen sulfide is then passed into the suspension, with frequent shaking, until precipitation of the copper is complete (2–3 hours). The precipitate is filtered and extracted with 500 ml. of hot water in two or three portions. The clear, light-brown to reddish brown filtrate and washings are boiled for 15 minutes, and then 60 g. (0.26 mole) of picric acid is added with stirring; heating is continued until solution is complete.

The greenish yellow plates, which separate as the solution is cooled to room temperature, are filtered, washed 3 times with 150- to 200-ml. portions of water, and air-dried. The filtrate and first washings are combined and heated, 10 g. of picric acid is added, and the mixture is cooled and filtered. This process is repeated, using 10-g. portions of picric acid, until the air-dried picrate fraction so obtained melts below 195° (Note 5).

All fractions melting above 200° (Note 5) are combined and recrystallized from water by adding 700 ml. of water for each 30 g. of crystals, heating the mixture in a covered beaker until solution occurs, treating with charcoal, and filtering through a warm funnel. The crystals deposit upon the slightest cooling. After cooling, the yellow needles (occasionally plates) are filtered, washed, and air-dried; their melting point is 204° or higher, with decomposition (Note 5). The fractions melting at 195–200° (Note 5) are recrystallized in like manner until the melting point is raised to 203°. The yield of crude picrate is 95–100 g. (61–64%); of recrystallized picrate, 84–94 g. (54–60%) (Note 6). The melting point varies slightly with the rate of heating between 203.5° and 206° with decomposition (Note 5).

B. *Hydroxymethylimidazole hydrochloride.* One hundred and twenty grams (100 ml., 1.2 moles) of 37% hydrochloric acid, 250 ml. of water, and 500 ml. of benzene are placed in a 2-l. round-bottomed flask, which is then immersed in a water bath maintained at 80°. One hundred grams of the pure picrate (0.306 mole) is added to this mixture, and the flask is shaken thoroughly until the picrate dissolves. The benzene layer is decanted. The aqueous layer is then extracted 5 times with 330-ml. portions of benzene, treated with about 3 g. of Norit, and filtered through a wet filter paper. The clear, pale yellow filtrate is evaporated to dryness at 60–70° under reduced pressure. The resulting pale yellow to slightly brown crystals are taken up in the minimum quantity (30–35 ml.) of hot absolute ethanol. Colorless needles deposit on cooling. Three to four volumes of ethyl ether are added, and the mixture is kept in the refrigerator overnight. The almost colorless needles are filtered and, after being washed with a small amount of ether, are dried in a vacuum desiccator. The yield amounts to 37–39 g. (90–95%) of a product which melts at 107–109° after sintering a few degrees lower (Note 6).

2. Notes

1. Equally good results are obtained with technical or pure grades of basic copper carbonate, $CuCO_3 \cdot Cu(OH)_2$. Satisfactory results are obtained also when the reagent is prepared by adding an equivalent amount of a solution of sodium carbonate to a solution of copper sul-

fate and washing the resulting precipitate until it is nearly free of sulfate ion. When the reagent is prepared in this way, 4 moles of copper sulfate are used for each mole of fructose, and the wet basic cupric carbonate is used without drying or weighing.

2. If the aeration is omitted, the yield is about 10% lower.

3. No attempt is made to recover ammonia which escapes during the aeration.

4. The copper precipitate can be dried unchanged but is then difficult to decompose with hydrogen sulfide.

5. One to three such additions usually are necessary. The melting points given are those of the submitters and were observed by immersing the capillary melting-point tubes containing the samples in a bath preheated to a temperature 20° to 50° below the expected melting point and heating rather rapidly. The checkers determined the melting points in the ordinary manner, starting with the bath at room temperature, and observed melting points about 5° lower than those of the submitters. In one of the checkers' preparations the main fraction of crude picrate weighed 86 g. and melted at 193–194°; the second fraction weighed 12 g. and melted at 191–193°; the third fraction, which melted at 186–188°, was rejected. One recrystallization raised the melting point of the first fraction to 199° (75 g. recovered) and that of the second fraction to 197° (9.5 g. recovered). The once-recrystallized products were combined and employed in the preparation of the hydrochloride, which was obtained in the yield described and with a melting point of 107–108°.

6. Equally good yields are obtainable with double the quantities indicated.

3. Methods of Preparation

4(5)-Hydroxymethylimidazole has been synthesized in a long series of steps from citric acid.[1] Weidenhagen and Herrmann prepared it from dihydroxyacetone.[2] The reaction which provides the basis of the procedure here described[3] was discovered by Parrod[4] and developed by Weidenhagen, Herrmann, and Wegner.[5] The method developed by these last-named investigators, however, provides considerably smaller yields than that described above.

[1] Pyman, *J. Chem. Soc.*, **99**, 668 (1911); Koessler and Hanke, *J. Am. Chem. Soc.*, **40**, 1716 (1918).

[2] Weidenhagen and Herrmann, *Ber.*, **68**, 1953 (1935).

[3] Darby, Lewis, and Totter, *J. Am. Chem. Soc.*, **64**, 463 (1942).

[4] Parrod, *Bull. soc. chim. France*, (4) **51**, 1424 (1932).

[5] Weidenhagen, Herrmann, and Wegner, *Ber.*, **70**, 570 (1937).

2-HYDROXY-1-NAPHTHALDEHYDE

(1-Naphthaldehyde, 2-hydroxy-)

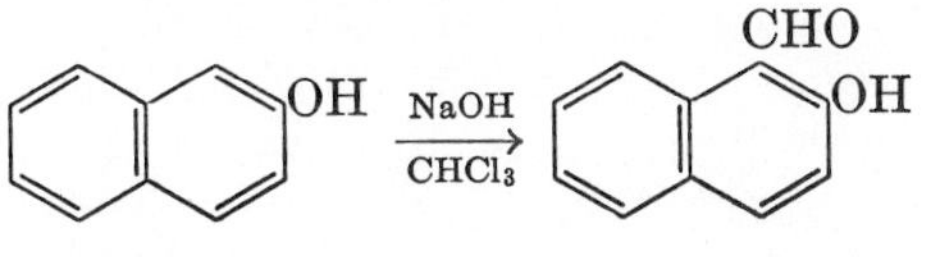

Submitted by ALFRED RUSSELL and LUTHER B. LOCKHART.
Checked by W. E. BACHMANN and CHARLES E. MAXWELL.

1. Procedure

In a 2-l. three-necked round-bottomed flask (Note 1) fitted with a 40-in. reflux condenser, a mercury-sealed stirrer, and a dropping funnel are placed 100 g. (0.69 mole) of β-naphthol and 300 g. of 95% ethanol. The stirrer is started, and a solution of 200 g. (5 moles) of sodium hydroxide in 415 g. of water is rapidly added.

The resulting solution is heated to 70–80° on a steam bath, and the dropwise addition of chloroform is started. After the reaction begins (Note 2), further heating is unnecessary. A total of 131 g. (1.1 moles) of chloroform is added at such a rate that gentle refluxing is maintained (Note 3). Near the end of the addition the sodium salt of the phenolic aldehyde separates. Stirring is continued for 1 hour after all the chloroform has been added.

The ethanol and excess chloroform are removed by distillation from a steam bath. Hydrochloric acid (sp. gr. 1.18) is added dropwise to the residue, with good stirring, until the contents of the flask are acid to Congo red paper (Note 4). The dark oil that separates is mixed with a considerable amount of sodium chloride. Sufficient water to dissolve the salt is added, and the oil is separated and washed several times with hot water. By distillation of the oil under reduced pressure there is obtained 87.5–93.7 g. of a slightly colored distillate which boils at 163–166° at 8 mm. (Note 5) and which solidifies on cooling. Recrystallization of the solid from 75 ml. of ethanol yields 45.2–57.5 g. (38–48%) of pure 2-hydroxy-1-naphthaldehyde, melting at 79–80°.

2. Notes

1. The submitters employed a 5-l. flask for a run four times this size. They obtained the same percentage yield of product.

2. The beginning of the reaction is indicated by the formation of a deep blue color.

3. The chloroform is added over a period of 1 to 1.5 hours.

4. About 175 ml. of hydrochloric acid is required to neutralize the excess sodium hydroxide and to liberate the phenolic aldehyde from its sodium salt.

5. The boiling range of the aldehyde is 177–180°/20 mm. The lower pressure is preferable to avoid decomposition. The checkers found a boiling point of 139–142°/4 mm. The color of the distillate varies in different runs; it may be green, pink, or amber.

3. Methods of Preparation

2-Hydroxy-1-naphthaldehyde has been prepared from β-naphthol, zinc chloride, and hydrogen cyanide; [1] from β-naphthol, zinc cyanide, and anhydrous hydrogen chloride; [2] from β-naphthol, chloroform, and sodium hydroxide (Reimer-Tiemann reaction); [3,4] from β-naphthol and hexamethylene tetramine in glycerin-boric acid, followed by hydrolysis; [5] and by hydrolysis of 2-hydroxynaphthaldehyde anil,[6] prepared in turn from 2-hydroxydithionaphthoic acid.[7]

[1] Gattermann and von Horlacher, *Ber.*, **32**, 285 (1899).

[2] Adams and Levine, *J. Am. Chem. Soc.*, **45**, 2373 (1923).

[3] Fosse, *Bull. soc. chim. France,* (3) **25**, 373 (1901).

[4] Kolesnikov, Korshak, and Krakovnaya, *Zhur. Obshcheĭ Khim.*, **21**, 397 (1951) [*C. A.*, **45**, 7553 (1951)].

[5] Duff, *J. Chem. Soc.*, **1941**, 547.

[6] Brit. pat. 563,747 [*C. A.*, **40**, 3141 (1946)].

[7] Brit. pat. 563,942 [*C. A.*, **40**, 3141 (1946)].

2-HYDROXY-1,4-NAPHTHOQUINONE

(1,4-Naphthoquinone, 2-hydroxy-)

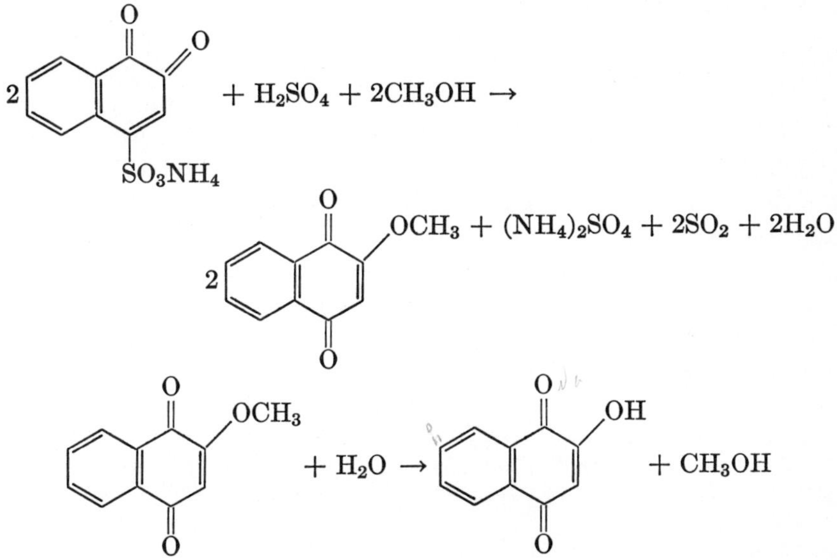

Submitted by L. F. Fieser and E. L. Martin.
Checked by Reynold C. Fuson and E. A. Cleveland.

1. Procedure

One liter of absolute methanol is cooled in a 3-l. round-bottomed flask to 0° in an ice-salt bath, and 80 ml. of concentrated sulfuric acid is slowly added, with good shaking, the temperature being kept at 0°. The flask is removed from the freezing mixture, and 255 g. (1 mole) of ammonium 1,2-naphthoquinone-4-sulfonate (p. 633) is added and made into an even paste by thorough shaking. After standing for 30 minutes, during which time the temperature rises to 15–20°, the flask is heated gradually on the steam bath with continuous shaking and rotating so that the solution reaches its boiling point in about 15 minutes. The solution becomes red, sulfur dioxide is evolved, and methoxynaphthoquinone begins to separate. The mixture is kept boiling very gently, with continued shaking, for 15 minutes, when the paste of separated material becomes very stiff. Two hundred and fifty milliliters of methanol is added, and the heating and rotating continued for an additional 15–20 minutes. The reaction mixture is cooled to 20–25°, water and ice are added until the flask is nearly filled, and the methoxynaphthoquinone is collected on a 15-cm. Büchner funnel and washed with cold water until the filtrate is nearly colorless; about 2–2.5 l. of water is required (Note 1).

The moist material is washed into a solution of 30 g. of sodium hydroxide in 1.5 l. of water, and the mixture is heated rapidly nearly to the boiling point. In about 10 minutes all the ether is hydrolyzed and a deep red solution results (Note 2). The hot solution is filtered by suction from a trace of residue, transferred to a 2-l. beaker, and acidified while still hot by adding 130 ml. of 6 N hydrochloric acid slowly, and with good stirring. The yellow suspension of hydroxynaphthoquinone thus obtained is cooled to 0° and allowed to stand for 2 hours (Note 3). The hydroxynaphthoquinone is collected, washed with 2 l. of cold water, dried overnight at room temperature, and finally to constant weight at 60–80°. The yield is 101–112 g. (58–65% based on ammonium 1,2-naphthoquinone-4-sulfonate; 99%, based on methoxynaphthoquinone) (Note 4). The hydroxynaphthoquinone thus obtained is bright yellow, is granular, and melts, with decomposition, at about 188–189° (Note 5). It is of high quality and for ordinary uses requires no further purification (Note 6).

2. Notes

1. The methoxynaphthoquinone weighs 111–122 g. (59–65%). It melts at 181–182° and can be further purified by crystallization from ethanol. The pure substance forms pale yellow needles, m.p. 183.5°.

2. Should the sodium salt separate during the heating or filtration, it is brought into solution by adding about 1 l. of water and heating.

3. By allowing the precipitate to stand for the indicated period, the final product is granular, and the filtration is rapid.

4. Numerous modifications have been tried without improving the yield. The loss is probably due to a partial reduction of the quinone sulfonate by the sulfur dioxide liberated, but this was not prevented by adding manganese dioxide to the reaction mixture, and no pure product could be obtained from the mother liquor.

5. The temperature of decomposition varies with the rate of heating and with the nature of the glass capillary.

6. Crystallization from ethanol containing a trace of acetic acid gives glistening yellow needles, melting with decomposition at about 191–192°. The red samples of hydroxynaphthoquinone often men-

tioned in the literature are not completely pure. Such material, or crude material of any kind, is best purified through either the sodium salt or the methyl ether.

3. Methods of Preparation

Of the many reactions by which hydroxynaphthoquinone has been obtained,[1,2] two have been developed into practical preparative methods, and both utilize the inexpensive β-naphthol as the primary starting material. In the first method, this is converted into β-naphthoquinone [*Org. Syntheses* Coll. Vol. **2,** 430 (1943)], which reacts with acetic anhydride-sulfuric acid to give 1,2,4-trihydroxynaphthalene triacetate, which is then hydrolyzed and oxidized to the desired product. The yield of the acetylation reaction is about 75%; that in the final step can be brought to 93% of the theoretical amount by hydrolyzing with ethanolic alkali in an atmosphere of nitrogen and with a trace of sodium hydrosulfite present, then diluting with water, acidifying, and adding ferric chloride. The overall yield from β-naphthol can thus be brought to 54%. The method is a good one, and it can be used to advantage for the preparation of many similar hydroxyquinones. With ordinary laboratory equipment, however, one is limited to 0.5-mole runs, and not more than about 50 g. of hydroxynaphthoquinone can be prepared at a time.

The second method is that described above: β-naphthol is converted through the nitroso derivative and 1-amino-2-naphthol-4-sulfonic acid into naphthoquinone sulfonate, and this is subjected to acid hydrolysis. The sulfonate can be converted directly into hydroxynaphthoquinone by the action of concentrated sulfuric acid,[3,4] but the process is not so easily controlled as when the quinone is etherified as it is formed and the ether subsequently hydrolyzed.[4] The overall yield from β-naphthol is 46% of the theoretical amount, but all the reagents are inexpensive, and with ordinary apparatus, 150 g. of hydroxynaphthoquinone can be made conveniently in one run (from 300 g. of β-naphthol).

The compound has also been prepared in good yield by air oxidation of 1,3-dihydroxynaphthalene,[5] which in turn was obtained from ethyl γ-phenylacetoacetate.[6]

[1] Beilstein-Prager-Jacobson, VIII, 300 (1925).
[2] Thiele and Winter, *Ann.*, **311,** 347 (1900).
[3] Akt.-Ges. Anilinf., Ger. pat. 100,703 [*Chem. Zentr.*, **70,** 766 (1899)].
[4] Fieser, *J. Am. Chem. Soc.*, **48,** 2929 (1926).
[5] Soliman and Latif, *J. Chem. Soc.*, **1944,** 55.
[6] Soliman and West, *J. Chem. Soc.*, **1944,** 53.

2-HYDROXY-5-NITROBENZYL CHLORIDE

(Toluene, α-chloro-2-hydroxy-5-nitro-)

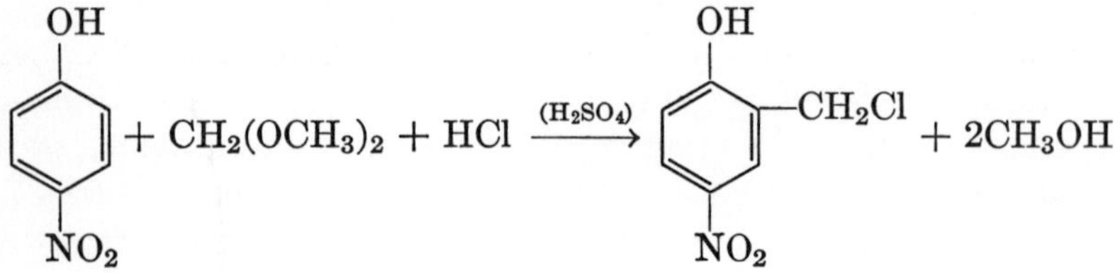

Submitted by C. A. Buehler, Fred K. Kirchner, and George F. Deebel.
Checked by C. F. H. Allen and Alan Bell.

1. Procedure

In a 1-l. three-necked round-bottomed flask equipped with a mechanical stirrer, short reflux condenser, and bent glass tube reaching below the surface of the liquid for the introduction of hydrogen chloride, are placed 50 g. (0.36 mole) of *p*-nitrophenol (Note 1), 650 ml. of concentrated hydrochloric acid, 5 ml. of concentrated sulfuric acid (Note 2), and 76 g. (1 mole) of methylal (Note 3). The mixture is stirred while the temperature is maintained at 70 ± 2° for 4–5 hours by means of a water bath (Note 4). During this time hydrogen chloride is bubbled into the reaction mixture through the bent glass tube, and the excess gas is carried away through the reflux condenser to a hood or gas-absorption trap (Note 5).

The 2-hydroxy-5-nitrobenzyl chloride begins to separate as a solid about 1 hour after the beginning of the reaction. At the end the mixture is cooled in ice for 1 hour whereby more crystals separate, after which the acid liquors are either filtered or decanted from the crystals (Note 6). The 2-hydroxy-5-nitrobenzyl chloride is purified by recrystallization from 125 ml. of hot benzene (Note 7). The yield is 46 g. (69% based on *p*-nitrophenol) of a white product melting at 129–130°.

2. Notes

1. Best results are obtained by using *p*-nitrophenol of a grade melting above 112°.

2. The reaction will proceed in the absence of sulfuric acid, but in its presence a greater reaction velocity results.

3. Methylal, used in excess, is prepared according to the method given by Houben.[1] Twelve hundred grams of methanol is added to 800 g. of anhydrous calcium chloride in a 5-l. round-bottomed flask equipped with a reflux condenser. Twenty-four grams of concentrated hydrochloric acid is added, and then ,with cooling, 800 g. of technical 35–40% formaldehyde is slowly dropped in through a dropping funnel. The reaction is strongly exothermic, requiring about 2 hours for complete addition of the formaldehyde. When all the formaldehyde has been added, the mixture is heated with a Bunsen flame for a few minutes until the liquid boils vigorously. The methylal forms quickly as an upper layer and, after 6 hours' standing, is fractionally distilled, preferably using a Clarke and Rahrs column.[2] The 42–46° fraction is sufficiently pure for use.

4. Temperatures higher than 72° may result in the formation of oily-resinous material with a corresponding decrease in chloride. The reaction is nearly complete at the end of 4 hours, but additional time will result in a slightly increased yield.

5. Rubber joints should be used sparingly as the hydrogen chloride causes an internal swelling of the rubber with subsequent blocking of the lines.

6. The high acid concentration makes ordinary filtration difficult. By allowing the reaction mixture to stand in an ice bath for 1 hour, the solid clings together, and filtration with gentle suction or decantation can be used with ease.

7. The solid material should be air-dried for several hours or, better, overnight to remove all water before recrystallizing from hot benzene. The chlorine atom is very labile and hydrolyzes easily to form 2-hydroxy-5-nitrobenzyl alcohol.

3. Methods of Preparation

2-Hydroxy-5-nitrobenzyl chloride can also be prepared from 2-hydroxy-5-nitrobenzyl alcohol by passing hydrogen chloride into an ethanolic solution at 70 ± 2° for 2 hours. In this case the yield of chloride is almost quantitative. The foregoing detailed procedure has been arrived at from that given in German Patent 132,475.[3]

[1] Houben, *Die Methoden der organischen Chemie,* 3rd Ed., Vol. III, p. 193, Verlag Georg Thieme, Leipzig, 1930.

[2] Clarke and Rahrs, *Ind. Eng. Chem.,* **18,** 1092 (1926).

[3] *Frdl.,* **6,** 142 (1904).

5-HYDROXYPENTANAL

(Valeraldehyde, δ-hydroxy-)

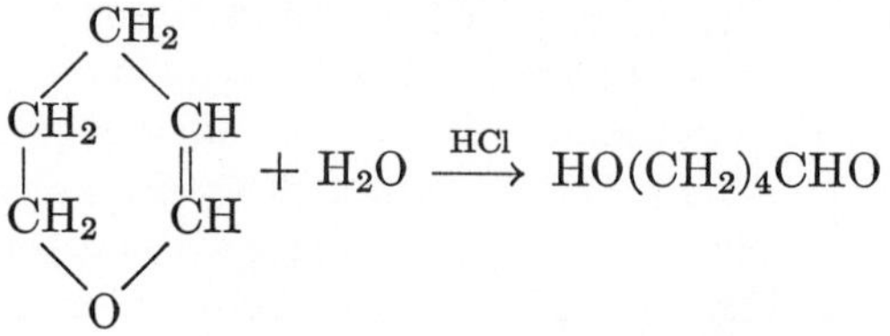

Submitted by G. Forrest Woods, Jr.
Checked by Cliff S. Hamilton and Wm. V. Ruyle.

1. Procedure

In a 1-l. three-necked flask provided with a Hershberg stirrer are mixed 300 ml. of water, 25 ml. of concentrated hydrochloric acid (Note 1), and 100 g. of 2,3-dihydropyran (p. 276). The mixture is stirred until the solution has become homogeneous and then for an additional 20 minutes (Note 2). After the addition of a few drops of phenolphthalein indicator to the mixture, the acid is neutralized with 20% sodium hydroxide; enough alkali is added so that a faint pink color just persists.

The solution is then transferred to a continuous extractor and extracted with ether for about 16 hours. The ether extract is added in convenient portions to a 250-ml. modified Claisen flask fitted with a condenser and a suitable fraction-cutter for distillation under reduced pressure. The ether is removed by distillation under the diminished pressure of a water pump, and the residue is then distilled at about 10 mm. pressure. After a small fore-run which weighs 2–5 g., the product distils as a clear, colorless, viscous oil at 62–66°/9–10 mm.; n_D^{25} 1.4513. The yield is 90–95 g. (74–79%) (Notes 3 and 4).

2. Notes

1. The quantity of acid is arbitrary. This amount was chosen to minimize the time of hydration. The order of addition of the reactants should be that stated above. Less acid may be used with increased reaction time; more should not be used.

2. About 5–10 minutes is required before the solution becomes homogeneous. Some heat is evolved during the hydration. The sub-

mitter has found that the amount of 2,3-dihydropyran may be increased up to 300 g. without using more water or acid; however, if larger quantities of the pyran are used, the pyran must be added dropwise to the acid solution with cooling.

3. A convenient apparatus for the distillation is a modified Claisen flask whose side arm is provided with a short water-cooled condenser. A fraction-cutter of the "pig" type is satisfactory.

4. According to the submitter the product can be converted smoothly to pentamethylene glycol by hydrogenation over Raney nickel at 90° and 2000 lb. pressure. It will also undergo reductive alkylamination by a procedure similar to that described for 2-isopropylaminoethanol (p. 501).

3. Methods of Preparation

5-Hydroxypentanal has been prepared only by the method of Paul,[1] of which the above is an adaptation.

[1] Paul, *Bull. soc. chim. France,* (5), **1**, 976 (1934).

IMIDAZOLE *

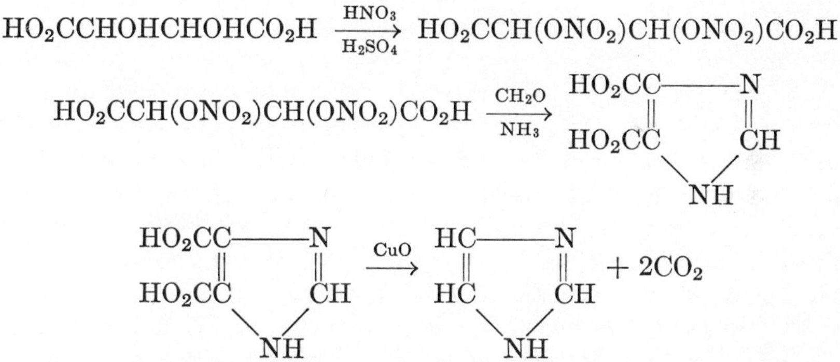

Submitted by H. R. Snyder, R. G. Handrick, and L. A. Brooks.
Checked by C. F. H. Allen, C. J. Kibler, and James VanAllan.

1. Procedure

In a 2-l. three-necked round-bottomed flask fitted with a thermometer, mechanical stirrer, and 1-l. dropping funnel is placed 200 g. (1.33 moles) of powdered *d*-tartaric acid (Note 1). To this are added, successively, 432 ml. of nitric acid (sp. gr. 1.42) and 432 ml. of fuming nitric acid (sp. gr. 1.50). The mixture is stirred until the tartaric acid is all, or nearly all, in solution (5–10 minutes) (Note 2). Then 800 ml. of concentrated sulfuric acid (sp. gr. 1.84) is slowly added from the dropping funnel. As soon as the temperature of the reaction mixture reaches 38°, the flask is surrounded by a vessel of ice water, and the rate of addition of acid is regulated so that the temperature is maintained at 38–43° (Note 3). Near the end of the addition (about 15 minutes), tartaric acid dinitrate sometimes begins to crystallize. After all the sulfuric acid has been added the mixture is allowed to stand in a cool (20–25°) place for 3 hours.

The crystalline mass is broken up with a glass rod and collected on a glass filter cloth (Note 4) in an 18-cm. Büchner funnel. The resulting cake is pressed nearly dry by means of a large flat glass stopper or the bottom of a 125-ml. Erlenmeyer flask. After most of the mother liquor has been removed (Note 5), the solid is transferred in portions to a 4-l. beaker containing about 3 l. of *finely cracked* ice. The ice is stirred to effect solution of each portion of the solid as it is added; a wooden stirring rod with a flat end is advisable. Not more than 10 minutes should be required.

The cold solution is poured *immediately* into a previously assembled apparatus, consisting of a 5-l. three-necked flask, immersed in a 20-in. tub of ice-salt mixture (Note 6), and provided with a stirrer, dropping funnel, and thermometer for reading low temperatures. The solution is neutralized by the addition of 600–700 ml. of concentrated ammonium hydroxide (sp. gr. 0.90) at such a rate that the temperature *never* exceeds −5° (Note 7); the tip of the dropping funnel should be placed directly over the vortex created by the stirrer. After 500 ml. of ammonium hydroxide has been added, the solution is tested with Congo red paper at intervals corresponding to the addition of 50 ml. of ammonium hydroxide; 3–4 hours is required for the neutralization. An additional 100 ml. of ammonium hydroxide is then added.

While the neutralization is in progress, a solution of hexamethylenetetramine (Note 8) is prepared by the cautious addition of 520 ml. of formalin (sp. gr. 1.08; approximately 7 moles) to 500 ml. of ammonium hydroxide (sp. gr. 0.90; approximately 7.5 moles). The temperature must be maintained below 20° by external cooling with ice water. The solution is finally chilled to 0° and added dropwise to the cold ammoniacal solution of tartaric acid dinitrate; the temperature should not exceed 2°. The addition requires 30 minutes. Stirring is then discontinued, and the mixture is allowed to stand overnight. During this period the cooling bath and the reaction mixture come to room temperature.

The solution is now filtered, and 100 ml. of ethanol is added to the filtrate, which is then acidified (hood) to Congo red paper by the slow (30 minutes) addition of about 400 ml. of concentrated hydrochloric acid (sp. gr. 1.19). The acidified mixture is cooled in an icebox for 4–5 hours, and the imidazole-4,5-dicarboxylic acid is collected on an 18-cm. Büchner funnel. The solid is transferred to a 1-l. beaker, stirred with 400–500 ml. of water, and again filtered. It is washed on the funnel successively with three 150-ml. portions of water, two 75-ml. portions of methanol, and finally with 75 ml. of ether. After drying in the air, it weighs 90–100 g. (43–48%) and melts with decomposition at about 280°.

The imidazole-4,5-dicarboxylic acid is divided into two portions. Each portion is intimately mixed with about 0.5 g. of copper-chromium oxide catalyst [*Org. Syntheses* Coll. Vol. **2**, 142 (1943)] or powdered copper oxide (Note 9), and the resulting mixture is transferred to a 250-ml. Claisen flask having a modified side arm; the receiving flask is loosely placed over the side arm. The flask is heated gently with a free flame. After a small fore-run at 95–100°, the temperature rises sharply to 260° (Note 10), and the imidazole distils at 262–264°. The product is purified by dissolving it in 60–70 ml. of benzene, boiling the solution for a few minutes with 2–3 g. of decolorizing carbon (Note 11), filtering the mixture through a preheated Büchner funnel, and cooling the filtrate to 10° for 2 hours. The yield is 13–14.5 g. (68–76%) of a pure white product, m.p. 88–90° (Note 12).

2. Notes

1. The submitters used U.S.P. tartaric acid. The checkers used the Eastman Kodak Company acid, m.p. 169–171°.

2. It is possible to carry out the preparation without mechanical stirring, merely using a single-necked flask and shaking by hand. Temperature control is not so satisfactory under these conditions.

3. At lower temperatures tartaric acid dinitrate separates during the addition of the sulfuric acid. Under these conditions, it forms very fine crystals that are not easily filtered.

4. Vinyon fabric serves equally well. If a filter cloth is not available, a layer of glass wool is prepared on the funnel; the mixed wool and solid are then added to the cracked ice.

5. A fairly dry cake is obtained in about 45 minutes. Since the substance decomposes in the air, it is advisable not to leave it for a longer period.

6. Efficient cooling is extremely important; a small ice bath will not suffice.

7. The decomposition temperature is about 0°. A centrifugal-type stirrer is advisable.

8. No product was obtained when solid hexamethylenetetramine was substituted at this point.

9. Copper oxide gives a slightly lower yield.

10. If an ordinary 250-ml. distilling flask is used, the fraction boiling from 200° to 270° is collected as crude imidazole. The fore-run contains about 1 g. of imidazole, which can be recovered by boiling this fraction of the distillate with benzene, until the water is removed, and then evaporating to crystallization.

11. If a flask and fractionating column are used, the crude product is white; otherwise it is colored.

12. The benzene filtrate from the recrystallization contains about 0.5 g. of imidazole.

3. Methods of Preparation

Imidazole has been prepared from glyoxal and ammonia, with [1] or without [2] the addition of formaldehyde; from imidazolthione-2 and nitric acid; [3] and by decarboxylation of imidazole-4,5-dicarboxylic acid.[4–6] The procedure described is essentially that of Fargher and Pyman.[6]

[1] Wallach, *Ber.*, **15**, 645 (1882).
[2] Behrend and Schmitz, *Ann.*, **277**, 338 (1893).
[3] Marckwald, *Ber.*, **25**, 2361 (1892).
[4] Maquenne, *Ann. chim.*, (6) **24**, 525 (1891).
[5] Dedichen, *Ber.*, **39**, 1835 (1906).
[6] Fargher and Pyman, *J. Chem. Soc.*, **115**, 227 (1919).

INDAZOLE

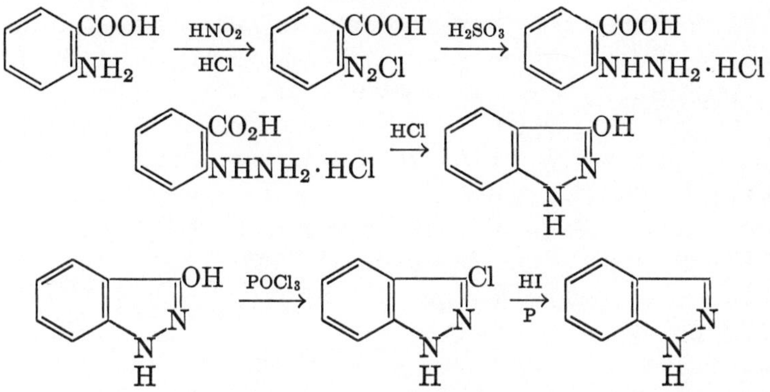

Submitted by Emily F. M. Stephenson.
Checked by C. F. H. Allen, C. V. Wilson, and Jean V. Crawford.

1. Procedure

A. *o-Hydrazinobenzoic acid hydrochloride.* In a 2-l. beaker, provided with a stirrer and a low-temperature thermometer, and cooled in an ice-salt bath, are placed 42 g. (0.31 mole) of anthranilic acid and 300 ml. of water. The stirrer is started, and 340 ml. of concentrated hydrochloric acid (sp. gr. 1.18) is added in one portion; the anthranilic acid dissolves, and its hydrochloride begins to separate almost immediately. After the mixture has been cooled to 0°, a solution of 21.6 g. (0.31 mole) of technical sodium nitrite in 210 ml. of water is added from a dropping funnel, the tip of which extends below the surface of the suspension, at such a rate that the temperature never rises above 3°. The addition requires about 30 minutes; stirring is continued for 15 minutes longer, and at the end of this period a positive test with starch-iodide paper should be obtained (Note 1). The clear brown solution is then diluted with 150 ml. of ice water.

In a 12-l. flask, equipped with a low-temperature thermometer and surrounded by an ice-salt bath, a solution of sulfurous acid is prepared by saturating 2.4 l. of water at 0–5° with sulfur dioxide from a cylinder. A brisk stream of the gas is continued (Note 2) while the cold diazonium salt solution is added in about 150-ml. portions over a 30-minute period and the temperature is maintained at 5–10°; the reaction mixture assumes a dark orange color (Note 3). The cooling bath is removed, but sulfur dioxide is passed into the mixture for an additional 30 minutes. After the mixture has been allowed to stand for 12 hours at room temperature, 3 l. of concentrated hydrochloric acid (sp. gr. 1.18) is added; the *o*-hydrazinobenzoic acid hydrochloride separates at once. The mixture is chilled to 0–5° and filtered through a precooled Büchner funnel; the product is washed with two 50-ml. portions of ice-cold dilute (1:1) hydrochloric acid. The yield is 50–51 g. (86–88%); the salt melts at 194–195° with decomposition (Note 4) and is suitable for the next step without further purification (Note 5).

B. *Indazolone.* In a 2-l. round-bottomed flask to which a reflux condenser is attached are placed 47.1 g. (0.25 mole) of *o*-hydrazinobenzoic acid hydrochloride, 1.25 l. of water, and 12.5 ml. of concentrated hydrochloric acid (sp. gr. 1.18). The mixture is refluxed for 30 minutes. The resulting pale yellow solution is transferred in two portions to a 23-cm. evaporating dish and concentrated on the steam bath to about one-fourth its original volume. The indazolone separates at an early stage of the evaporation but redissolves as the concentration of acid increases. Sodium carbonate is added to the warm solution in small portions until the acid is neutralized (Note 6), and the suspension is allowed to stand for 2 hours. The nearly colorless indazolone is removed by filtration, washed with two 25-ml. portions of cold water, and air-dried. The yield of product, m.p. 246–249°, is 30–33 g. (90–98%) (Note 7).

C. 3-*Chloroindazole.* In a 200-ml. flask connected by a glass joint to an air condenser protected by a drying tube are placed 26.8 g. (0.2 mole) of dry indazolone and 15.8 g. (16 ml., 0.2 mole) of dry pyridine (Note 8); 46.1 g. (27.6 ml., 0.3 mole) of phosphorus oxychloride is then added, with shaking, over a 10-minute period. Heat is evolved, and acid fumes are generated. The mixture is heated with occasional shaking in an oil bath, which is maintained at 128–130° for 1 hour and at 130–140° for 4 hours. The clear brown solution is then cooled to 70° and poured, with hand stirring, upon 500 g. of cracked ice. This mixture is allowed to stand for 24 hours. The pale buff solid is removed, washed on the filter, first with 100 ml. of 0.5 N hydrochloric acid and then with 40 ml. of cold water, and air-dried (Note 9). The 3-chloroindazole is crystallized from 3 l. of 20% ethanol. The yield is 21–22.5 g. (68–74%) of material melting at 148–148.5° (Note 10).

D. *Indazole.* In a 300-ml. flask are placed 15.3 g. (0.1 mole) of 3-chloroindazole, 18.6 g. (0.15 mole) of red phosphorus, and 100 ml. of constant-boiling hydriodic acid (sp. gr. 1.7) (Note 11). The mixture is refluxed for 24 hours (Note 12), cooled, and filtered through a sintered-glass funnel (Note 13) to remove the phosphorus; the flask and the solid are washed with two 20-ml. portions of water. The clear filtrate is transferred to a 300-ml. Claisen flask and concentrated to about 40 ml. by heating in a water bath at a reduced pressure. The residue is washed into a 250-ml. beaker with 70–80 ml. of hot water, and the clear solution is cooled in an ice bath and made strongly alkaline with concentrated ammonium hydroxide (about 80 ml. is required). The next day, the indazole is collected and dried; the white solid melts at 143–145° (Note 14).

The product is added to 75 ml. of benzene, and the suspension is boiled until the frothing has ceased, the benzene lost being replaced (Note 15); the resulting suspension is filtered to remove the insoluble material. The clear filtrate is heated to boiling, diluted with 25 ml. of petroleum ether (b.p. 70–90°), and allowed to cool slowly. The yield of white product, m.p. 145–146.5°, is 9.7–10.2 g. (82–86%). The over-all yield from anthranilic acid is 43–55%.

2. Notes

1. If the starch-iodide test is negative at this point a little solid sodium nitrite may be added.

2. This operation should be carried out in a hood or out-of-doors.

3. Small amounts of a red crystalline solid were obtained at this point by the checkers in several runs. This substance can be converted to *o*-hydrazinobenzoic acid by the addition of 5 ml. of concentrated hydrochloric acid to a suspension of 1 g. of the solid in 25 ml. of dilute (1:1) hydrochloric acid. The red solid changes to the white *o*-hydrazinobenzoic acid hydrochloride without apparent solution.

4. The melting point varies with the rate of heating. The values given were obtained with a bath preheated to 180°.

5. The free acid may be obtained by treatment of a solution of the hydrochloride with a concentrated aqueous solution of sodium acetate. The powdered hydrochloride (18.9 g., 0.1 mole) is dissolved in 567 ml. of water, and sodium acetate solution (8.2 g. [0.1 mole] of anhydrous sodium acetate in 30 ml. of water) is added. *o*-Hydrazinobenzoic acid separates at once; the mixture is chilled, and the light-tan acid is removed by filtration, washed with two 25-ml. portions of water, and air-dried. The yield is 13.1 g. (86%); m.p. 248–250°. If a purer acid is required, the crude material may be recrystallized from ethanol (50 ml. per g.); the pale tan product then melts at 250–251.5°.

6. About 20 g. of sodium carbonate is required.

7. The indazolone may be purified further by recrystallization from methanol (24 ml. per g.), with filtration of the hot solution through a layer of Norit. It separates as white needles, m.p. 250–252°; the recovery is about 50%. An additional 10% of material (m.p. 246–248°) may be obtained by dilution of the filtrate with 2 volumes of water.

The submitter reports that the described method of purification gives a better product than is obtained by solution in dilute sodium hydroxide and reprecipitation with acid.

8. The submitter reports that dimethylaniline can be used but that it is less desirable because a small amount of a green by-product is formed.

9. The crude chloroindazole, m.p. 143–145°, is difficult to dry. Small quantities may be crystallized satisfactorily from water (250 ml. per g.). The submitter reports that a good product can be obtained by steam distillation but that even with superheated steam the distillation is very slow.

10. In a run 2.5 times this size, the checkers dissolved the crude product in 190 ml. of ethanol and diluted the hot filtrate with 260 ml. of water; the chloroindazole was obtained in 80% yield.

11. It is essential to use acid of this concentration.

12. This reaction time ensures complete conversion of the chloroindazole.

13. As an alternative procedure, the mixture may be diluted with 70 ml. of water and filtered through S & S No. 596 filter paper.

14. The crude indazole is so difficult to dry that the weight at this stage is not significant.

15. This operation is carried out in an open flask in the hood and at a point remote from flames; the indazole is dried by the steam distillation of the water with the benzene.

3. Methods of Preparation

The preparations of *o*-hydrazinobenzoic acid hydrochloride and indazolone are essentially those given by Pfannstiel and Janecke.[1] The procedure for the conversion of indazolone to indazole is a modification of that of Fischer and Seuffert.[2] A procedure involving the decarboxylation of indazole-3-carboxylic acid is described by Schad.[3]

Indazole has been obtained in a variety of ways which are of no preparative value. The elimination of the amino group from aminoindazoles, first utilized by Witt,[4] by the action of ethanol or sodium stannite on the diazonium compounds appears to be the only other useful procedure.

[1] Pfannstiel and Janecke, *Ber.*, **75**, 1104 (1942).
[2] Fischer and Seuffert, *Ber.*, **34**, 796 (1901).
[3] Schad, *Ber.*, **26**, 217 (1893).
[4] Witt, Nölting, and Grandmougin, *Ber.*, **23**, 3642 (1890).

INDOLE *

$$o\text{-}CH_3C_6H_4NHCHO + (CH_3)_3COK \rightarrow o\text{-}CH_3C_6H_4N(CHO)K + C_4H_9OH$$

$$o\text{-}CH_3C_6H_4N(CHO)K \rightarrow o\text{-}CH_3C_6H_4NHK + CO$$

$$o\text{-}CH_3C_6H_4NHK + o\text{-}CH_3C_6H_4N(CHO)K \rightarrow o\text{-}CH_3C_6H_4NH_2 + KOH +$$

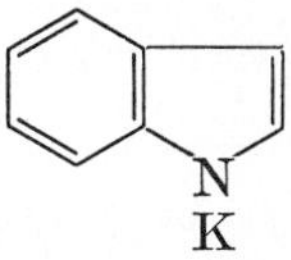

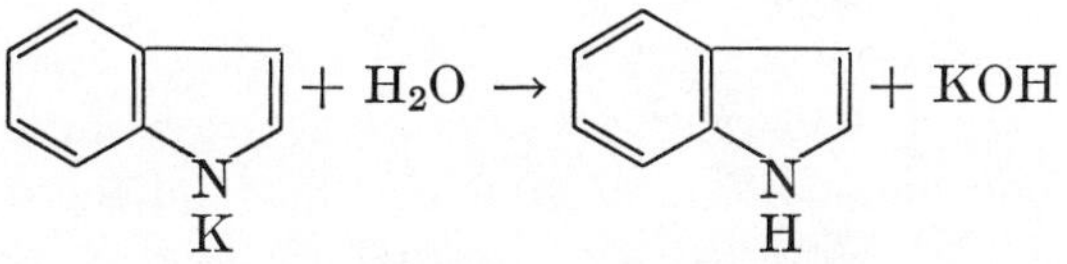

Submitted by F. T. Tyson.
Checked by R. L. Shriner and C. H. Tilford.

1. Procedure

A 2-l. three-necked round-bottomed flask is fitted with a reflux condenser and a glass inlet tube connected to a cylinder of nitrogen. The third opening of the flask is closed by a stopper. The top of the condenser is connected to an air trap which consists of two 500-ml. suction flasks joined in series. The first suction flask is empty; the second contains 100 ml. of paraffin oil, and the inlet tube of this flask extends slightly below the surface of the oil.

In the reaction flask is placed 600 ml. of commercial *tert*-butyl alcohol (Note 1), and the air in the flask is displaced by dry nitrogen gas. Then 29 g. (0.75 gram atom) of metallic potassium is added, in portions, to the alcohol. The mixture is heated on a water bath until all the potassium has dissolved, and then 68 g. (0.5 mole) of *o*-formotoluide (Note 2) is added and brought into solution. The condenser is set for distillation with a filter flask as the receiver; this flask is protected from the air by connecting it to the trap used in the initial operation. The reaction flask is surrounded by a metal bath, and the excess alcohol is removed by distillation. The residue is heated to 350–360° for about 20 minutes (Note 3) and then is allowed to cool in a stream of nitrogen. The residue is decomposed by addition of 300 ml. of water, and the mixture is steam-distilled to remove the indole. The distillate is extracted successively with 300 ml. and 100 ml. of ether, and the combined ether extracts are shaken with cold dilute 5% hydrochloric acid to remove small amounts of *o*-toluidine. The ether extract is washed with 100 ml. of water, followed by 100 ml. of 5% sodium carbonate solution, and is dried over 20 g. of sodium sulfate. The ether is removed by distillation, and the residue is distilled under reduced pressure. Indole distils at 142–144°/27 mm. (128°/10 mm.; 121°/5 mm.) as a pale yellow oil which solidifies and then melts at 52–53°. The yield is 23 g. (79%) (Note 4).

2. Notes

1. Alcohols other than *tert*-butyl alcohol, such as methyl, ethyl, butyl, or isobutyl alcohol, may be used, but with a decrease in yield. If methyl or ethyl alcohol is substituted for *tert*-butyl alcohol, the potassium should be added in smaller portions and the more vigorous reaction must be controlled by external cooling. Furthermore, if methyl alcohol is used the amount of potassium should be decreased from 0.75 gram atom to 0.5 gram atom.

2. *o*-Formotoluide can be prepared by heating a mixture of 856 g. (8 moles) of *o*-toluidine and 403 g. (8.4 moles) of 90% formic acid on a steam bath for about 3 hours and allowing the reaction mixture to stand overnight. The mixture is fractionated under reduced pressure; there is obtained 920–963 g. (85–89%) of *o*-formotoluide, b.p. 173–5°/25 mm., m.p. 55–58°. This product is a pale yellow solid which contains traces of toluidine and possesses an odor indicating the presence of traces of an isocyanide. However, the material is sufficiently pure for conversion to indole. By use of 99% formic acid, a quantitative yield may be obtained. If a purer product is desired, the original reaction mixture is mixed with about a liter of water, the crude formotoluide is filtered, then washed with 1% hydrochloric acid and with water. After drying, the crude formotoluide is recrystallized from benzene-petroleum ether. The yield is practically the same as that obtained by direct distillation, and the product melts at 61°. In order to assure freedom from moisture, the recrystallized product may be

distilled under reduced pressure; the loss in this distillation is negligible.

3. During this interval combustible gases, chiefly carbon monoxide and hydrogen, are evolved, and a liquid, which is principally *o*-toluidine, distils.

4. The product is pale yellow. The color may be removed by crystallizing the material from a mixture of 100 ml. of petroleum ether and about 10 ml. of ethyl ether. The recovery is 21 g. (91%).

3. Methods of Preparation

Indole has been obtained through many syntheses which have little value as methods of preparation. In addition to these syntheses it has been prepared by heating 2-amino-ω-chlorostyrene with sodium methoxide;[1] by pyrolysis of 2,2′-diaminostilbene hydrochloride under reduced pressure;[2] by reduction of indoxyl with zinc dust and alkali;[3] by reduction of 2-aminoindole with sodium and alcohol;[4] by reduction of methyl *o*-nitrostyrylcarbamate,[5] by heating *o*-formylphenylglycine with acetic anhydride and sodium acetate;[6] by passing a mixture of acetylene and aniline through an iron tube at 700°;[7] and by reduction of ω,2-dinitrostyrene with iron and acetic acid.[8] The method described above is related to the Madelung synthesis of 2-alkylindoles[9] and has been published.[10] The methods for preparation of indole have been examined experimentally.[11] The authors recommend the method of Reissert,[12] which involves condensation of *o*-nitrotoluene with ethyl oxalate, followed by reduction to indole-2-carboxylic acid and decarboxylation of the latter. A patent reports the preparation of indole by catalytic dehydrogenation of *o*-ethylaniline at 650° in the presence of titania gel.[13] The cyclization of N-formyl-*o*-toluidine with sodium in methanol has been described and is recommended as a safer and less expensive method than that involving the use of potassium metal.[14]

The reaction of carbon monoxide with *o*-toluidine, as the sodium salt, under pressure, at an elevated temperature, has been patented.[15]

[1] Lipp, *Ber.*, **17**, 1072 (1884).
[2] Thiele and Dimroth, *Ber.*, **28**, 1411 (1895).
[3] Vorländer and Apelt, *Ber.*, **37**, 1134 (1904).
[4] Pschorr and Hoppe, *Ber.*, **43**, 2550 (1910).
[5] Weerman, *Ann.*, **401**, 14 (1913).
[6] Gluud, *Ber.*, **48**, 420 (1915).
[7] Majima, Unno, and Ono, *Ber.*, **55**, 3854 (1922).
[8] Nenitzescu, *Ber.*, **58**, 1063 (1925).
[9] Madelung, *Ber.*, **45**, 1128 (1912); cf. Verley, *Bull. soc. chim. France,* (4) **35**, 1039 (1924).

[10] Tyson, *J. Am. Chem. Soc.*, **63**, 2024 (1941).

[11] Shorygin and Polyakova, *Khim. Referat. Zhur.*, **1940**, No. 4, 114 [*C. A.*, **36**, 3802 (1942)].

[12] Reissert, *Ber.*, **30**, 1045 (1897).

[13] U. S. pat. 2,409,676 [*C. A.*, **41**, 998 (1947)].

[14] Galat and Friedman, *J. Am. Chem. Soc.*, **70**, 1280 (1948).

[15] Tyson, *J. Am. Chem. Soc.*, **72**, 2801 (1950); U. S. pat. 2,537,609 [*C. A.*, **45**, 5190 (1951)].

IODOBENZENE DICHLORIDE

(Benzene, iodo-, dichloride-)

$$C_6H_5I + Cl_2 \rightarrow C_6H_5ICl_2$$

Submitted by H. J. Lucas and E. R. Kennedy.
Checked by John R. Johnson and M. W. Formo.

1. Procedure

In a 1-l. three-necked flask, protected from the light and equipped with a mechanical stirrer, an inlet tube for the introduction of chlorine (Note 1), and an exit tube carrying a calcium chloride drying tube, are placed 150 ml. of dry chloroform (Note 2) and 102 g. (0.5 mole) of iodobenzene. The flask is cooled in an ice-salt mixture, and dry chlorine (Note 3) is introduced, as rapidly as the solution will absorb it, until an excess is present (usually about 3 hours is required). The yellow, crystalline iodobenzene dichloride is filtered with suction, washed sparingly with chloroform, and dried in the air on filter paper. The yield is 120–134 g. (87–94%) (Notes 4 and 5). The product is quite pure and may be used directly for the preparation of iodosobenzene and iodoxybenzene. Since iodobenzene dichloride decomposes slowly on standing, it should not be stored indefinitely.

2. Notes

1. The delivery tube should be at least 10 mm. in diameter and should terminate about 5 mm. above the surface of the liquid.

2. Chloroform was dried and rendered free of ethanol by allowing it to stand over anhydrous calcium chloride for 24 hours. It was then decanted through a filter and distilled through dry apparatus. The initial low-boiling fraction was rejected.

3. Chlorine was dried by passing it through at least two wash bottles containing sulfuric acid. Spray was removed by passing the gas through a plug of glass wool. The yields are reduced appreciably if the reactants are not dry.

4. In some runs an additional quantity of the product may be obtained by evaporating the chloroform filtrate to a small volume under reduced pressure. The solvent must not be evaporated completely if a pure product is desired.

5. The submitters report that the three isomeric iodotoluene dichlorides may be prepared in good yields by a similar procedure, but in these cases the chloroform solution must be concentrated by evaporation under reduced pressure (Note 4), since the tolyl homologs are more soluble in chloroform.

3. Methods of Preparation

The best preparative method for iodobenzene dichloride is the direct combination of iodobenzene and chlorine in chloroform.[1]

[1] Willgerodt, *J. prakt. Chem.*, (2) **33**, 155 (1886).

IODOSOBENZENE

(Benzene, iodoso-)

$$C_6H_5ICl_2 + 2NaOH \rightarrow C_6H_5IO + 2NaCl + H_2O$$

Submitted by H. J. Lucas, E. R. Kennedy, and M. W. Formo.
Checked by Lee Irvin Smith, R. T. Arnold, and R. A. Matthews.

1. Procedure

In a large mortar chilled in an ice bath are placed 55 g. (0.2 mole) of iodobenzene dichloride (p. 482), 50 g. of anhydrous sodium carbonate, and 100 g. of finely crushed ice. The mixture is ground thoroughly (Note 1) until all the ice has melted and a thick paste results. To this suspension 140 ml. of 5 *N* sodium hydroxide is added, in 20-ml. portions, with repeated trituration after each addition. Finally, 100 ml. of water is added to render the mixture more fluid and the material is allowed to stand overnight. The product is collected with suction, pressed thoroughly on the filter, transferred to a beaker, and washed thoroughly with 300 ml. of water (Note 2). The material is filtered with suction, washed again in a beaker with 300 ml. of water, collected with suction, and washed with about 250 ml. of water on the filter. After thorough drying in the air, the product is stirred to a thin mush with a little chloroform (Note 3), freed of solvent by suction, and spread on filter paper to dry in the air. The yield is 26–27 g. (60–62%) of a product having a purity of about 99%, as determined by titration (Notes 4 and 5).

2. Notes

1. The solid forms a caked mass, which is disintegrated by trituration.

2. The filtrate contains some diphenyliodonium salts, which may be recovered in the form of the sparingly soluble iodide by the addition of potassium iodide (p. 355). Usually 7–9 g. of diphenyliodonium iodide is obtained.

3. The chloroform removes iodobenzene, which may be recovered.

4. The following procedure is used in the analysis of iodoso and iodoxy compounds. In a 200-ml. iodine flask are placed 100 ml. of water, 10 ml. of 6 *N* sulfuric acid, 2 g. of iodate-free potassium iodide, 10 ml. of chloroform, and finally the sample, about 0.25 g. The flask is shaken for 15 minutes (or longer, if the reaction is not complete), and then the mixture is titrated with 0.1 *N* sodium thiosulfate. If the sample is pure the change of color in the chloroform layer may be taken as the end point, but if impurities are present starch must be used, for the impurities impart a brownish color to the chloroform. This solvent is desirable, as it facilitates the reaction with potassium iodide by dissolving the reaction products. Iodosobenzene may be differentiated from iodoxybenzene, for the former reduces iodide ion in a saturated sodium borate solution, whereas the latter does not.[1] The reactions involved are:

$$C_6H_5IO + 2HI \rightarrow C_6H_5I + H_2O + I_2$$
$$C_6H_5IO_2 + 4HI \rightarrow C_6H_5I + 2H_2O + 2I_2$$

5. For use in the preparation of iodoxybenzene by the disproportionation method (p. 485) it is superfluous to dry the crude product and to wash it with chloroform to remove iodobenzene. The crude wet iodosobenzene may also be used directly for the preparation of diphenyliodonium iodide (p. 355), but it is desirable to assay the wet product by titration to determine the quantity of iodoxybenzene needed.

3. Methods of Preparation

Iodosobenzene has been prepared by the action of aqueous sodium or potassium hydroxide upon iodobenzene dichloride;[2] and by repeated additions of water to iodobenzene dichloride.[3]

[1] Masson, Race, and Pounder, *J. Chem. Soc.*, **1935**, 1678.

[2] Willgerodt, *Ber.*, **25**, 3495 (1892); **26**, 357, 1807 (1893); Askenasy and Meyer, *Ber.*, **26**, 1356 (1893); Hartmann and Meyer, *Ber.*, **27**, 505 (1894).

[3] Willgerodt, *Ber.*, **26**, 357 (1893); Ortoleva, *Chem. Zentr.*, **1900, I,** 722.

IODOXYBENZENE

(Benzene, iodoxy-)

A. DISPROPORTIONATION OF IODOSOBENZENE

$$2C_6H_5IO \xrightarrow[\text{distil}]{\text{Steam-}} C_6H_5IO_2 + C_6H_5I$$

Submitted by H. J. LUCAS and E. R. KENNEDY.
Checked by JOHN R. JOHNSON and M. W. FORMO.

1. Procedure

In a 5-l. flask (Note 1), 110 g. (0.5 mole) of iodosobenzene (p. 483) is made into a thin paste with water, and the mixture is rapidly steam-distilled (Note 2) until almost all the iodobenzene (Note 3) is removed. The distillation should not be continued any longer than necessary, and the contents of the flask should be cooled at once. The white solid is filtered with suction and dried in the air at room temperature. It is then washed with chloroform, again dried, and the resulting cake is ground lightly in a mortar to facilitate removal of moisture in a final drying. The yield is 54–56 g. (92–95%). The product has a purity of about 99%, as determined by iodimetry (Note 4). The solid may be crystallized from hot water (Note 5).

2. Notes

1. Because of a tendency to froth, a large flask should be used.

2. Direct heating must be avoided, as the solid reactant and product may decompose with explosive violence on heating, especially when dry.

3. The recovered iodobenzene is quite pure and may be used for the preparation of iodobenzene dichloride. The recovery is about 46 g. (90%).

4. The iodimetric method described under iodosobenzene (p. 484) is applicable here. The main impurity in the iodoxybenzene is iodosobenzene.

5. The solubility of iodoxybenzene, per liter of water, is 2.8 g. at 12°, and about 12 g. at 100°.

B. HYPOCHLORITE OXIDATION OF IODOBENZENE DICHLORIDE

$$C_6H_5ICl_2 + NaOCl + H_2O \rightarrow C_6H_5IO_2 + NaCl + 2HCl$$

Submitted by M. W. Formo and John R. Johnson.
Checked by Lee Irvin Smith, R. T. Arnold, and Louis E. DeMytt.

1. Procedure

In a 2-l. round-bottomed flask equipped with an efficient mechanical stirrer (Note 1) are placed 110 g. (0.4 mole) of freshly prepared, pulverized iodobenzene dichloride (p. 482) (Note 2), 1.0 mole of sodium hypochlorite solution (Note 3), and 2 ml. of glacial acetic acid. The vigorously stirred mixture is heated on a water bath maintained at 65–75°. After 10–15 minutes the heated mixture becomes frothy and the yellow color of iodobenzene dichloride is displaced by the white color of iodoxybenzene. The stirring is stopped after 1 hour, and the flask is cooled in an ice bath. The product is filtered with suction, pressed with a glass stopper, and transferred to an 800-ml. beaker. The material is stirred thoroughly with 300 ml. of water, filtered with suction, washed on the filter with 100 ml. of water, pressed, and dried in the air. The crude product weighs 85–89 g. (90–94%) and has a purity, as determined iodimetrically, of 97–99% (Note 4). The air-dried material may be washed with 50–60 ml. of chloroform, air-dried, and finally dried in a vacuum desiccator. This product weighs 82–87 g. (87–92%) and has a purity of 99.0–99.9% as determined by iodimetry (Note 5).

2. Notes

1. A Hershberg stirrer is well suited for stirring the pasty suspension.

2. Any lumps of iodobenzene dichloride should be broken up by pressing with a spatula; otherwise the reaction may be incomplete.

3. Stabilized hypochlorite solutions, commercially available, are satisfactory. The submitters used 1.15 l. of "Clorox" solution, containing 5.25% sodium hypochlorite by weight.

4. The iodimetric method as described under iodosobenzene (p. 484) is applicable here. The crude product is satisfactory for the preparation of diphenyliodonium iodide (p. 355).

5. The checkers obtained the yields claimed, but the purity of the product was 90% before, and 95.7% after, washing with chloroform.

3. Methods of Preparation

Iodoxybenzene has been prepared by oxidizing iodobenzene with Caro's acid;[1,2] by treating iodobenzene with hypochlorous acid or with aqueous sodium hydroxide and bromine;[3] by action of chlorine upon iodobenzene dissolved in pyridine;[4] by oxidation of iodosobenzene with hypochlorous acid or bleaching powder;[5] by heating iodosobenzene;[6] by steam distillation of iodosobenzene;[7] by heating iodobenzene dichloride with aqueous sodium hypochlorite containing some acetic acid;[8] and by oxidation of iodobenzene with concentrated chloric acid solution.[9]

[1] Bamberger and Hill, *Ber.*, **33**, 534 (1900).

[2] Masson, Race, and Pounder, *J. Chem. Soc.*, **1935**, 1678.

[3] Willgerodt, *Ber.*, **29**, 1571, 1572 (1896).

[4] Ortoleva, *Chem. Zentr.*, **1900, I,** 723.

[5] Willgerodt, *Ber.*, **29**, 1568, 1569 (1896).

[6] Willgerodt, *Ber.*, **25**, 3500 (1892); **26**, 1806 (1893); Askenasy and Meyer, *Ber.*, **26**, 1356 (1893).

[7] Willgerodt, *Ber.*, **26**, 358, 1307 (1893).

[8] Willgerodt and Wiegand, *Ber.*, **42**, 3765 (1909).

[9] Datta and Choudhury, *J. Am. Chem. Soc.*, **38**, 1085 (1916).

ISATOIC ANHYDRIDE

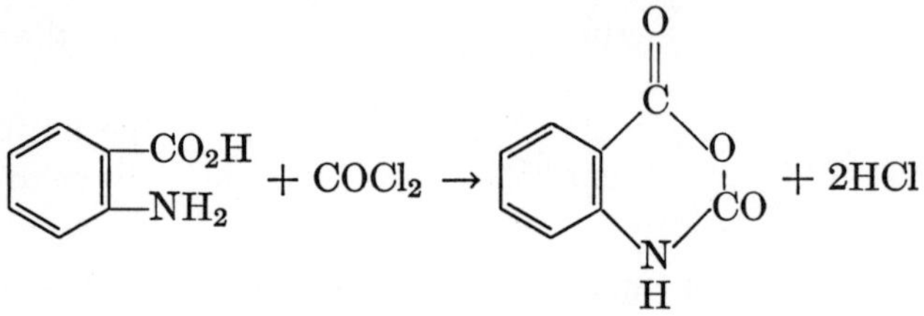

Submitted by E. C. Wagner and Marion F. Fegley.
Checked by R. L. Shriner and R. M. Hedrick.

1. Procedure

Phosgene is toxic. The apparatus should be set up in a good hood.

One hundred and thirty-seven grams (1 mole) of anthranilic acid is dissolved, with the aid of gentle warming, in a mixture of 1 l. of water and 126 ml. of concentrated hydrochloric acid (sp. gr. 1.19). The solution is filtered into a 2-l. three-necked flask fitted with a gas-tight, mechanically driven Hershberg stirrer. Through one of the side necks extends an inlet tube which ends in a coarse sintered-glass gas-dispersing tip extending well into the liquid in the flask. The inlet tube is connected with a cylinder of phosgene through an empty safety flask. In the third neck of the flask is mounted an addition tube, through the vertical arm of which a thermometer (Note 1) is mounted so that the bulb is immersed in the reaction liquid. The outlet tube is attached to the lateral arm of the addition tube and is connected to an empty safety flask which in turn is connected to a Drechsel bottle charged with ammonium hydroxide.

With the stirrer in rapid motion (Note 2) phosgene is passed into the solution of anthranilic acid at such a rate that bubbles of gas escape slowly into the ammonia scrubber (about two bubbles per second). Isatoic anhydride appears as a precipitate soon after the stream of phosgene is started. The temperature rises but is prevented from exceeding 50° (Note 3) by regulation of the rate at which phosgene is introduced. The stream of phosgene is continued for 2–4 hours, or until the rate of absorption is clearly much decreased (Note 4). The flask is disconnected, and residual phosgene is blown out by passing a current of air through the mixture. The product is collected on a Büchner funnel and is washed with three 500-ml. portions of cold water. The first crop amounts to 54–56 g.

The mother liquor is returned to the reaction flask, the apparatus reassembled, and the passage of phosgene resumed (Note 5). When the rate of absorption has noticeably decreased (1–1.5 hours) the precipitated isatoic anhydride is collected on a filter and washed. The second crop amounts to 34–54 g. A third passage of phosgene at a considerably reduced rate will often yield a small additional crop (10–24 g.) of isatoic anhydride (Note 6).

The product is dried in air and then at 100°. The total yield is 118–123 g. (72–75%) of a white or nearly white product which decomposes at 237–240° cor.; this material is pure enough for most purposes. It may be recrystallized from 95% ethanol (about 30 ml. per gram) or from dioxane (about 10 ml. per gram). The former solvent permits the higher recovery (89–90%) and, except for the large volume required, is to be preferred. The mother liquor may be used for recrystallization of several successive lots of isatoic anhydride. The purified compound decomposes at 243° cor. (Note 7).

2. Notes

1. A thermometer with the graduation marks on the upper half of the stem is convenient.

2. The rate of absorption of phosgene is dependent upon the speed and efficiency of the stirring. This also determines the amount of product obtained in each treatment with phosgene.

3. Operation below room temperature is without advantage. At 60° or above the yield of isatoic anhydride is decreased, or the process may yield precipitated material from which little or no isatoic anhydride can be obtained. Cooling of the mixture might become advisable during operation on a scale larger than specified.

4. It is advisable to precipitate the isatoic anhydride in several successive crops as directed, rather than to attempt to complete the reaction in one step, because the accumulation of precipitated isatoic anhydride slows the rate of absorption of phosgene to such an extent that prolonged passage of the gas at a decreasing and eventually very low rate would be necessary.

5. It is important to clean the glass gas-dispersing tip with hot dioxane before reassembling the apparatus.

6. A small (usually trifling) final crop of isatoic anhydride can be obtained by addition of 40 g. (1 mole) of sodium hydroxide to the liquid and by introducing phosgene slowly.

7. The method is capable of extension to other *o*-aminocarboxylic acids,[1] e.g., to 3-amino-2-naphthoic acid, 4,4′-diaminobiphenyl-3,3′-dicarboxylic acid, and 2-amino-*p*-toluic acid. With some acids other than anthranilic, difficulty may be encountered owing to the readiness with which their hydrochlorides are salted out of solution by hydrochloric acid.

3. Methods of Preparation

Isatoic anhydride has been prepared by prolonged refluxing of a mixture of anthranilic acid and ethyl chlorocarbonate,[2] a reaction usually accompanied by formation of considerable monoethyl or/and diethyl isatoate; or by action of phosgene upon anthranilic acid in a solution the acidity of which is moderated by occasional addition of sodium carbonate.[2] The method described is based upon a patented procedure[1] in which, under conditions not fully specified, phosgene is passed into a solution of anthranilic acid hydrochloride with no subsequent adjustment of the acidity.

[1] Ger. pat. 500,916 [*Frdl.*, **17**, 500 (1930)]; Clark and Wagner, *J. Org. Chem.*, **9**, 60 (1944).

[2] Erdmann, *Ber.*, **32**, 2159 (1899).

ISOBUTYRAMIDE *

$$(CH_3)_2CHCOOH \xrightarrow{SOCl_2} (CH_3)_2CHCOCl \xrightarrow{NH_3} (CH_3)_2CHCONH_2$$

Submitted by R. E. Kent and S. M. McElvain.
Checked by C. R. Noller and D. Frazier.

1. Procedure

A. *Isobutyryl chloride.* A 1-l. three-necked flask is equipped with a 250-ml. dropping funnel, an efficient stirrer sealed with a glycerol-lubricated rubber tube, and an efficient condenser (Note 1). The water supplied to the condenser is cooled to 0°, and the flask is cooled in a large water bath. The apparatus is set up in a hood, and a gas-absorption trap is attached to the top of the condenser.

In the flask is placed 542 g. (4.55 moles) of thonyl chloride; to this is added dropwise, and with rapid stirring, 352 g. (4 moles) of isobutyric acid (Note 2). A vigorous evolution of hydrogen chloride and sulfur dioxide takes place. When all the acid has been added, the water bath is heated to 80° and is kept at this temperature for 30 minutes. Stirring is continued throughout the heating. The reaction mixture is then distilled through a 30-cm. Vigreux column by means of an oil bath heated on a hot plate (Note 3). The fore-run boiling up to 89° weighs 44 g. The isobutyryl chloride which is collected at 89–93° weighs 351 g. The residue weighs 49 g. On combining the fore-run and residue and redistilling slowly, an additional 33 g. of isobutyryl chloride is obtained; the total yield is 384 g. (90%). Redistillation gives a faintly yellow product boiling at 90–92° (Notes 4 and 5).

B. *Isobutyramide.* In a 3-l. flask, equipped with an efficient stirrer and a 500-ml. dropping funnel, and surrounded by an ice-salt freezing mixture, is placed 1.25 l. of cold concentrated aqueous ammonia (about 28%). To this 318 g. (3 moles) of isobutyryl chloride is added dropwise with rapid stirring at such a rate that the temperature of the reaction mixture does not rise above 15°, and the evolution of white fumes (mostly ammonium chloride) does not become vigorous. Stirring is continued for 1 hour after the addition of the acid chloride is finished.

The flask is heated by steam in a large can, and the reaction mixture is evaporated to dryness under reduced pressure (Note 6). The dry residue of ammonium chloride and isobutyramide is boiled 10 minutes with 2 l. of dry ethyl acetate, and the boiling solution is filtered quickly through a fluted filter paper on a large hot funnel. The residue on the filter is extracted in the same way with two 1-l. portions of ethyl acetate. The combined ethyl acetate extracts are cooled to 0°, and the crystalline amide which separates is removed by filtration. The filtrate is concentrated to about 300 ml. and chilled, and a second crop of amide is collected (Notes 7 and 8). The two crops of isobutyramide are combined and dried, first in an oven at 70° for 3 hours and then in a vacuum desiccator. The yield of glistening white needles melting at 127–129° is 203–215 g. (78–83%) (Note 9). This material is suitable for the preparation of isobutyronitrile.

2. Notes

1. A Friedrichs condenser is recommended. This efficient condenser has an inner cooling coil around which the vapors pass and condense. The submitters used rubber stoppers throughout.

2. Eastman's technical grade of isobutyric acid was distilled, and the fraction boiling at 151–153° was used.

3. The submitters used an electrically heated oil bath.

4. When 5 moles of thionyl chloride and 4 moles of isobutyric acid were used, the yield on the first distillation was 83%, and redistillation

of the fore-run gave an additional 7%. There was practically no high-boiling material.

5. The submitters obtained the same percentage yields in runs four times as large.

6. The steam can should be large enough to contain the entire reaction flask; otherwise the evaporation is very slow.

7. If the mixture of amide and ammonium chloride is not thoroughly dry, the ethyl acetate removed at this point will contain water and must be dried and redistilled before further use.

8. The filtrate from the second crop of amide may be evaporated to dryness and the residue crystallized from a mixture of ethyl acetate and ligroin (60–68°). It is profitable to work up this third crop of amide only when the mother liquors from several runs are combined.

9. The submitters obtained the same percentage yield in runs twice as large.

3. Methods of Preparation

Isobutyramide has been prepared by the action of concentrated aqueous ammonia on isobutyryl chloride [1] or methyl isobutyrate; [2] by the action of liquid ammonia on ethyl isobutyrate; [3] by the Willgerodt reaction with aqueous ammonium polysulfide; [4] by distillation of ammonium isobutyrate [5] or a mixture of isobutyric acid and potassium thiocyanate; [6] by hydrolysis of isobutyronitrile; [7] and from the ozonolysis of 2,3-dimethyl-6-isopropylpyridine.[8]

[1] Aschan, *Ber.*, **31**, 2348 (1898).

[2] Meyer, *Monatsh.*, **27**, 43 (1906).

[3] Hauser, Levine, and Kibler, *J. Am. Chem. Soc.*, **68**, 26 (1946).

[4] King and McMillan, *J. Am. Chem. Soc.*, **68**, 1369 (1946).

[5] Hofmann, *Ber.*, **15**, 982 (1882).

[6] Letts, *Ber.*, **5**, 671 (1872).

[7] Hoffmann and Barbier, *Bull. soc. chim. Belg.*, **45**, 565 (1936) [*C. A.*, **31**, 919 (1937)].

[8] Lochte, Barton, Roberts, and Bailey, *J. Am. Chem. Soc.*, **72**, 3007 (1950).

ISOBUTYRONITRILE

$$(CH_3)_2CHCONH_2 \xrightarrow{P_2O_5} (CH_3)_2CHCN + H_2O$$

Submitted by R. E. Kent and S. M. McElvain.
Checked by C. R. Noller and D. Frazier.

1. Procedure

To 308 g. (2.1 moles) of phosphorus pentoxide (reagent grade) in a 3-l. round-bottomed flask is added 174 g. (2 moles) of finely powdered, dry isobutyramide (Note 1). The flask is tightly stoppered, and the two dry solids are thoroughly mixed by shaking. The flask then is attached to a water-cooled condenser set for downward distillation; a 500-ml. suction flask connected to the condenser by a rubber stopper is a convenient receiver. A calcium chloride tube is attached to the side arm of the receiver, and the receiver is surrounded by crushed ice. The reaction flask is heated for 8–10 hours in an electrically heated oil bath maintained at 200–220°. The nitrile starts to distil almost at once. The reaction mixture becomes a thick, brown syrup which foams considerably toward the end of the distillation. The time of reaction may be cut to 1–2 hours by connecting the distillation system to an aspirator and intermittently removing the nitrile under reduced pressure (Note 2). Very little additional nitrile can be obtained by further heating of the reaction mixture.

The contents of the receiver are transferred to a 500-ml. modified Claisen flask, 10–15 g. of phosphorus pentoxide is added (Note 3), and the product is distilled from an oil bath held at 145–155°. After only a few drops of fore-run, the main fraction distils at 99–102°/740 mm. The yield is 96–120 g. (69–86%). The product is colorless when first distilled but acquires a yellow tint after standing a few days. If it is again distilled from phosphorus pentoxide or Drierite, it boils at 101–103° and remains colorless; n_D^{25} 1.3713.

2. Notes

1. The amide must be thoroughly dry or immediate reaction will take place with considerable generation of heat, and the yield will be lowered owing to the failure to obtain proper mixing of the reactants. If the amide is not finely powdered the yield of nitrile falls off sharply.

2. The yield is not altered by this procedure if care is taken not to evaporate the nitrile in the receiver. Keeping the flask surrounded by crushed ice and using a Dry Ice-acetone safety trap between the receiver and aspirator will guard against such loss.

3. When phosphorus pentoxide is added to isobutyronitrile, the liquid sets to a semisolid gel which is difficult to transfer from a flask. Hence, the addition of one to the other should be made only in the flask from which the distillation is to be made.

3. Methods of Preparation

Isobutyronitrile has been prepared by a number of catalytic vapor-phase reactions at elevated temperatures: isobutylamine over copper [1] or nickel,[2] isobutyramide over alumina,[3] a mixture of ammonia and isobutyraldehyde over thorium dioxide [4] or other catalysts,[5] a mixture of ammonia and isobutyl alcohol over copper,[6] and a mixture of isobutylene oxide and ammonia over alumina and copper.[7] It has also been prepared by decarboxylation of 2-methyl-2-cyanopropanoic acid,[8] and by the reaction of isobutyric acid with potassium thiocyanate.[9] The above procedure is a modification of the method used by Walter and McElvain.[10]

[1] Mailhe and de Godon, *J. pharm. chim.*, (7) **16**, 225 (1917) (*Chem. Zentr.*, **1918**, I, 819).

[2] Mailhe and de Godon, *Bull. soc. chim. France*, (4) **21**, 288 (1917).

[3] Boehmer and Andrews, *J. Am. Chem. Soc.*, **38**, 2503 (1916).

[4] Mailhe, *Compt. rend.*, **166**, 215 (1918).

[5] U. S. pat. 2,535,818 [*C. A.*, **45**, 2015 (1951)].

[6] Hara and Komatsu, *Mem. Coll. Sci. Kyoto Imp. Univ.*, **8A**, 241 (1925) [*C. A.*, **19**, 3248 (1925)].

[7] U. S. pat. 2,500,256 [*C. A.*, **44**, 7342 (1950)].

[8] Hoffmann and Barbier, *Bull. soc. chim. Belg.*, **45**, 565 (1936) [*C. A.*, **31**, 919 (1937)].

[9] Letts, *Ber.*, **5**, 671 (1872).

[10] Walter and McElvain, *J. Am. Chem. Soc.*, **56**, 1614 (1934).

dl-ISOLEUCINE *

(α-Amino-β-methylvaleric acid)

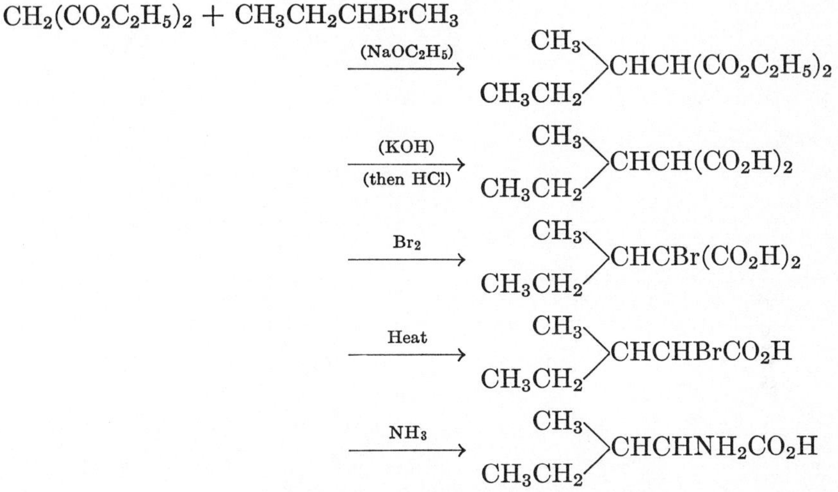

Submitted by C. S. MARVEL.[1]
Checked by W. E. BACHMANN and D. W. HOLMES.

1. Procedure

A. *Diethyl sec.-butylmalonate.* To 700 ml. of absolute ethanol in a 2-l. three-necked round-bottomed flask equipped with a long, wide-bore reflux condenser is added 35 g. (1.52 gram atoms) of sodium cut in pieces of suitable size. When all the sodium has reacted, the flask is placed on a steam cone and fitted with a mercury-sealed stirrer, a dropping funnel, and a reflux condenser bearing a calcium chloride tube (Note 1). The flask is heated, and 250 g. (1.56 moles) of diethyl malonate is added in a steady stream with stirring. After the ester addition, 210 g. (1.53 moles) of *sec*-butyl bromide is added at such a rate that the heat of reaction causes refluxing. The mixture is then stirred and refluxed for 48 hours. At the end of this time, the reflux condenser is exchanged for a downward condenser and the ethanol is removed by distillation (Note 2). The residue is treated with 200 ml. of water, shaken, and allowed to stand until the ester layer separates. The ester layer is separated from the aqueous layer and distilled from a 500-ml. two-necked flask fitted with a well-wrapped 18-in. Vigreux column. The fraction boiling at 110–120°/18–20 mm. is diethyl *sec*-butylmalonate, and it amounts to 274–278 g. (83–84%).

B. *α-Bromo-β-methylvaleric acid.* In a 2-l. three-necked round-bottomed flask equipped with a stirrer and dropping funnel and placed on a steam cone, 250 g. of technical potassium hydroxide is dissolved in 200 ml. of water. To the hot solution, 250 g. (1.16 moles) of diethyl *sec*-butylmalonate is added in a steady stream with vigorous stirring (Note 3). A tube connected to a vacuum line assists in the removal of ethanol. The mixture is stirred and heated for 5 hours (Note 4), and then the contents of the flask are transferred to a beaker fitted with a stirrer and surrounded by an ice bath. The cooling is hastened by the addition of 50 g. of ice, and, when the temperature reaches 15°, technical hydrochloric acid is added at such a rate that the temperature does not rise above 20°. After the addition of about 250 ml. of acid, the monopotassium salt separates, necessitating stirring by hand until solution again occurs. When the solution is acid to Congo red (Note 5) it is transferred to a separatory funnel (Note 6).

The *sec*-butylmalonic acid is extracted with three 200-ml. portions of ether, and the combined extracts are dried over calcium chloride overnight. The ether solution is then decanted into a 2-l. three-necked flask fitted with a mercury-sealed stirrer, reflux condenser, and dropping funnel. Five milliliters of bromine is added at one time, and the solution is stirred until decolorized (Note 7). Then 50 ml. more bromine is added dropwise at such a rate that the ether refluxes gently. When all the bromine has been added, 200 ml. of water is added through the dropping funnel dropwise so as to produce no foaming or violent reaction.

The ether layer containing the bromomalonic acid is separated from the aqueous layer, and the ether is removed by distillation from a steam cone. The residual liquid is decarboxylated by refluxing for 5 hours in a 500-ml. round-bottomed flask on an oil bath heated to 130°. The bromo acid is then separated from the small amount of water and distilled. The material distilling at 125–140°/18–20 mm. is α-bromo-β-methylvaleric acid (Note 8). The yield is 150.5 g. (66.7%).

C. *dl-Isoleucine.* One hundred and fifty grams (0.77 mole) of α-bromo-β-methylvaleric acid is added to 645 ml. of technical ammonium hydroxide (sp. gr. 0.90) in a 1.5-l. round-bottomed flask. A stopper is wired in, and the flask is allowed to stand at room temperature for a week (Note 9). The stopper is removed and the mixture heated on a steam cone overnight to remove ammonia. The aqueous solution is concentrated under reduced pressure until bumping becomes violent (about 300 ml.). The mixture is then cooled to 15° and the crystals are collected on a filter. The crystals are washed with 40 ml. of ethanol and dried. The filtrate is again concentrated to about 150 ml., and a second crop of crystals is obtained. This second crop is washed with 25 ml. of water and then with 25 ml. of 95% ethanol. The yield of crude product is 65 g.

The isoleucine is recrystallized by dissolving it in 850 ml. of water heated to 95° on a steam cone. The solution is decolorized by treatment with a gram of Norit for 30 minutes and is then filtered hot. To the hot solution is added 425 ml. of 95% ethanol, and the flask is placed in an ice chest overnight. The yield of pure product is 38 g. An additional crop of 12 g. may be obtained by concentrating the mother liquors from the recrystallization to about 100 ml. and adding an equal volume of ethanol. This second crop is washed with 10 ml. of cold water and 10 ml. of cold ethanol. The total yield is 50 g. (49%). The product decomposes at 278–280° in a sealed evacuated capillary (Note 10).

2. Notes

1. The submitters carried out the preparation on a run ten times this size, using a 12-l. round-bottomed flask. After the sodium had reacted, the flask was fitted with a stopper containing the stirrer and two angle tubes connected respectively to a reflux condenser and a dropping funnel. The time allowed for the various reactions to take place was the same as for the smaller run. The percentage yields of the various products were practically identical.

2. On a run ten times this size, the submitters distilled the ethanol into another 12-l. flask connected by means of an adapter and fitted with the wide-bore reflux condenser originally used. The sodium necessary for a second run was added to the second flask as the ethanol distilled into it; on the large run this took 4–6 hours.

3. The ester is added quite slowly at first, until the reaction gets under way, and then more rapidly.

4. Water is added if necessary to keep the mass from solidifying.

5. About 400 ml. of acid is required.

6. For a run ten times this size, the solution is transferred to a 12-l. round-bottomed flask fitted with a stopper containing a large stopcock which barely pierces the stopper and a glass tube which reaches to the bottom of the flask. A flask so fitted can be used as a large separatory funnel. The stopper is wired when the flask is upturned.

7. It is important that the first 5-ml. portion react completely; otherwise the bromination will not go smoothly and the solution may foam out of the condenser.

8. If the decarboxylation was not complete it will be finished here. Occasionally it is some time before a good water pump will maintain constant pressure.

9. On a larger run, the submitters allowed 350 g. of the bromoacid and 1.5 l. of ammonium hydroxide in a 3-l. flask to stand for a week. The contents of four such flasks, including any solid material, were combined in a 12-l. round-bottomed flask and heated on a steam cone overnight.

10. The product obtained in this manner has the calculated amount of amino nitrogen.

3. Methods of Preparation

dl-Isoleucine has been prepared by the reduction and subsequent hydrolysis of ethyl α-oximino-β-methyl-*n*-valerate;[2] by the action of aqueous ammonia on α-bromo-β-methyl-*n*-valeric acid;[3] and from ethyl *sec*-butylbromomalonate by saponification, decarboxylation, and amination.[4] The procedure described above is a combination of these last two methods.

[1] These directions are the result of the efforts of many men who have worked on the preparation of isoleucine at the University of Illinois.

[2] Bouveault and Locquin, *Compt. rend.*, **141**, 115 (1905); *Bull. soc. chim. France*, (3), **35**, 965 (1906).

[3] Ehrlich, *Ber.*, **41**, 1453 (1908); Brasch and Friedmann, *Beitr. Chem. Physiol. Path.*, (2) 376 (1908).

[4] Romburgh, *Rec. trav. chim.*, **6**, 150 (1887).

ISOPRENE CYCLIC SULFONE

(Thiophene, 2,5-dihydro-3-methyl-, 1-dioxide)

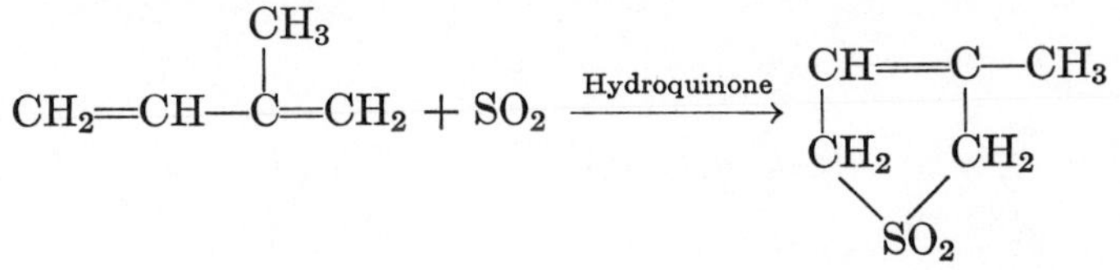

Submitted by Robert L. Frank and Raymond P. Seven.[1]
Checked by A. E. Barkdoll and R. S. Schreiber.

1. Procedure

Caution! If peroxides are present in the isoprene, they should be destroyed by washing with 10% acidified ferrous ammonium sulfate before distillation.

A 600-ml. steel reaction vessel (Note 1) is precooled before loading by filling it between one-fourth and one-half full of methanol and Dry Ice. After removal of the methanol and Dry Ice, the autoclave is charged with 120 g. (176 ml., 1.76 moles) of isoprene (Note 2), 113 g. (80 ml., 1.76 moles) of liquid sulfur dioxide, 88 ml. of methanol, and 4 g. of hydroquinone. The vessel is sealed, heated slowly to 85°, and maintained at that temperature for 4 hours. It is then cooled, the sulfone removed, the bomb rinsed with methanol, and the combined product and washings are treated hot with 5 g. of Norite. The filtered solution is concentrated to a volume of 250–300 ml., and the sulfone is allowed to crystallize. The material is filtered and washed with 50 ml. of cold methanol. Recrystallization from methanol (20 ml. per 25 g. of sulfone) yields 140–150 g. of thick, colorless plates. Concentration of the mother liquors raises the total yield to 182–191 g. (78–82%) (Note 3), melting at 63.5–64° (Note 4).

2. Notes

1. A steel reaction vessel of the type used for high-pressure catalytic hydrogenations is satisfactory. The pressure generated is less than 200 lb. For smaller quantities a heavy glass tube can also be used with proper precautions.

2. Commercial isoprene, obtained from Phillips Petroleum Company, should be freshly distilled before use, in order to eliminate isoprene dimers and polymers which are likely to accumulate in storage. If peroxides are present, they should be destroyed before distillation of the isoprene by washing with 10% acidified ferrous ammonium sulfate. As an added precaution, the distillation flask should not be permitted to go dry.

3. The yield of the cyclic sulfone depends upon the purity of isoprene. In one experiment, the checkers obtained the sulfone in 86% yield, using freshly distilled isoprene of 99 mole per cent purity, while the yield fell to 77% with commercial isoprene which had been distilled and stored at 5° for 1 week before use.

4. The checkers consistently obtained a slightly higher melting point (uncor.) in the range 64.4–65.4°. The purified cyclic sulfone serves as an ideal intermediate for the preparation of extremely pure isoprene, since isoprene can be regenerated nearly quantitatively at 135–140°.[2] Other sulfones that can be prepared by this method and that are useful in the purification of dienes are those of butadiene, m.p. 65.5°,[3] and 2,3-dimethylbutadiene, m.p. 135°.[3] The sulfone of piperylene is a liquid.[4]

3. Methods of Preparation

Isoprene cyclic sulfone has been prepared only from isoprene and sulfur dioxide.[2,3,5]

[1] Work done under contract with the Office of Rubber Reserve, Reconstruction Finance Corporation.

[2] Frank, Adams, Blegen, Deanin, and Smith, *Ind. Eng. Chem.*, **39**, 887 (1947).

[3] Staudinger and Ritzenthaler, *Ber.*, **68**, 455 (1935).

[4] Craig, *J. Am. Chem. Soc.*, **65**, 1006 (1943); Frank, Emmick, and Johnson, *J. Am. Chem. Soc.*, **69**, 2313 (1947).

[5] Eigenberger, *J. prakt. Chem.*, (2) **127**, 307 (1930).

2-ISOPROPYLAMINOETHANOL

(Ethanol, 2-isopropylamino-)

$$(CH_3)_2CO + H_2NCH_2CH_2OH + H_2 \xrightarrow{Pt} (CH_3)_2CHNHCH_2CH_2OH + H_2O$$

Submitted by Evelyn M. Hancock and Arthur C. Cope.
Checked by Nathan L. Drake and Larry Green.

1. Procedure

The reaction is carried out in a catalytic hydrogenation apparatus similar to the one described by Adams and Voorhees.[1] In a 1-l. reduction bottle are placed 0.5 g. of platinum oxide catalyst [1] and 50 ml. of commercial absolute ethanol. The bottle is connected to a calibrated low-pressure hydrogen tank and alternately evacuated and filled with hydrogen twice. Hydrogen is then admitted to the system until the pressure is 1–2 atm. (15–30 lb.), and the bottle is shaken for 20–30 minutes to reduce the platinum oxide (Note 1). The shaker is stopped, air is admitted to the bottle, and a solution of 61.0 g. (1.0 mole) of ethanolamine (Note 2), 75.4 g. (94 ml., 1.3 moles) of acetone, and 100 ml. of absolute ethanol is rinsed into the reduction bottle with 50 ml. of absolute ethanol. The bottle is alternately evacuated and filled with hydrogen twice. Hydrogen is admitted to the system until the pressure is approximately 25 lb., and the bottle is shaken until the pressure drop indicates that the theoretical amount (1 mole) of hydrogen has been taken up and absorption ceases (6–10 hours). Air is admitted to the bottle, and the catalyst is removed by filtration through a Hirsch funnel with a filter plate of small diameter (Note 3). The bottle is rinsed with a total of 75 ml. of benzene, which is also poured through the funnel. The filtrate is rinsed into a 500-ml. modified Claisen flask with 25 ml. of benzene, and most of the solvent is distilled at atmospheric pressure. Distillation of the residue under reduced pressure yields 97–98 g. of 2-isopropylaminoethanol, b.p. 86–87°/23 mm. (94–95%, based on the ethanolamine used) (Note 4).

2. Notes

1. If hydrogenation of the reaction mixture is begun in the presence of platinum oxide, a long induction period or lag occurs before the catalyst is reduced.

2. Commercial ethanolamine (Carbide and Carbon Chemicals Corporation) was dried by distillation with a small amount of benzene and redistilled; b.p. 70–72°/12 mm.

3. The usual precaution was observed of keeping the catalyst wet with the solution being filtered to prevent ignition of the filter paper.

4. Similar procedures have been used in preparing other 2-alkylaminoethanols [2] and N-alkyl derivatives of 1-amino-2-propanol, 2-amino-1-propanol, 3-amino-1-propanol, 2-amino-1-butanol, and 1-amino-2-methyl-2-propanol.[3]

3. Methods of Preparation

2-Isopropylaminoethanol has been prepared by the reaction of isopropylamine with ethylene oxide,[4,5] and by the method given above.[2]

[1] *Org. Syntheses* Coll. Vol. **1**, 61, 463 (1941).

[2] Cope and Hancock, *J. Am. Chem. Soc.*, **64**, 1503 (1942).

[3] Cope and Hancock, *J. Am. Chem. Soc.*, **66**, 1453 (1944); Hancock and Cope, *J. Am. Chem. Soc.*, **66**, 1738 (1944); Hancock, Hardy, Heyl, Wright, and Cope, *J. Am. Chem. Soc.*, **66**, 1747 (1944).

[4] Matthes, *Ann.*, **315**, 104 (1901).

[5] Biel, *J. Am. Chem. Soc.*, **71**, 1306 (1949).

dl-ISOPROPYLIDENEGLYCEROL

(Glycerol, isopropylidene-; also 1,3-dioxolane-4-methanol, 2,2-dimethyl-)

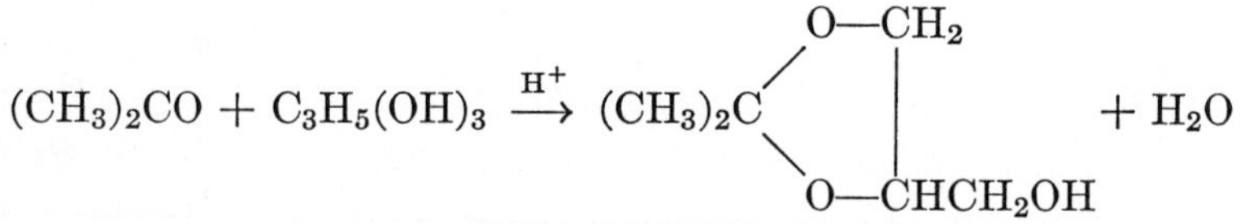

Submitted by MARY RENOLL and MELVIN S. NEWMAN.
Checked by R. L. SHRINER and ARNE LANGSJOEN.

1. Procedure

In a 1-l. three-necked flask, fitted with a sealed mechanical stirrer and a fractionating column (about 2 by 45 cm., packed with glass helices) attached to a total reflux phase-separating head (Fig. 15) (Note 1), are placed 237 g. (300 ml., 4.09 moles) of acetone (Note 2), 100 g. (1.09 moles) of glycerol (Note 3), 300 ml. of low-boiling petroleum ether (Note 4), and 3.0 g. of *p*-toluenesulfonic acid monohydrate. The third neck is closed with a cork or a ground-glass stopper, and the mixture is heated (Note 5) with stirring so that the petroleum ether refluxes as rapidly as the column permits. The stirring and refluxing are continued until no more water collects in the trap of the separating head; the time required varies between 21 and 36 hours (Note 6).

The mixture is cooled to room temperature, and 3.0 g. of powdered, freshly fused sodium acetate is added. Stirring is continued for 30 minutes; the mixture is then filtered, and the petroleum ether and excess acetone are removed by distillation under reduced pressure (water pump). The residual liquid is distilled from a modified Claisen flask. The fraction boiling at 80–81°/11 mm. is collected. The yield of colorless isopropylideneglycerol (n_D^{25} 1.4339, d_4^{25} 1.062) is 125–129 g. (87–90%).

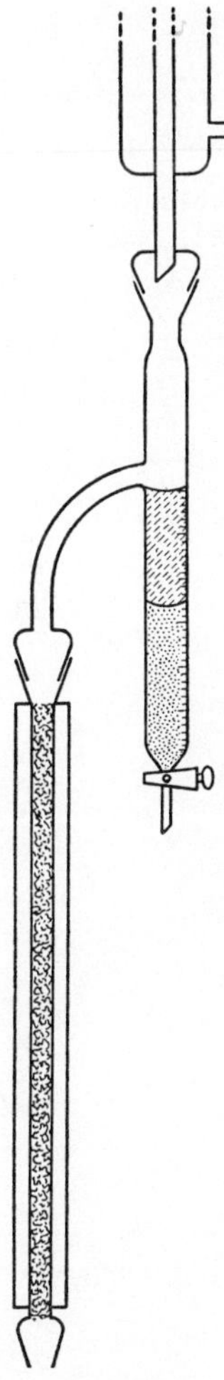

FIG. 15.

2. Notes

1. During operation, the apparatus shown in Fig. 15 requires no attention beyond occasional draining of the water trap. It is suitable for a number of preparations in which water is removed by distillation with an immiscible solvent; it functions only when the condensate separates into two phases, of which water is the more dense.

2. The acetone was the 99.5% grade obtained from the Carbide and Carbon Chemicals Company.

3. The glycerol should be of U.S.P. grade; if it has absorbed moisture, it may be dehydrated by heating at 170° in an open dish under a hood for 3 hours.[1]

4. Skellysolve F, b.p. 35–55°, obtainable from the Skelly Oil Company, is suitable.

5. The mixture may be heated by means of a steam or water bath, but in view of the long reflux period it is better to use a hemispherical electric heating mantle controlled by a variable transformer.

6. The period of refluxing need not be continuous. A longer reflux time, up to 70 hours, does not increase the yield. The volume of the aqueous phase collected in the separating head varies from 32 to 42 ml., depending on the quality of the glycerol.

3. Methods of Preparation

Isopropylideneglycerol has been prepared from acetone and glycerol in the presence of the following acidic catalyst: hydrogen chloride,[2,3] hydrogen chloride and anhydrous sodium sulfate,[4] phosphorus pentoxide,[5] and anhydrous copper sulfate.[6] It has also been prepared from acetone and glycerol in the presence of calcium carbide and a neutral surface-active agent.[7] The two optically active isomers of isopropylideneglycerol have been prepared from 1,2,5,6-diacetone-D-mannitol and 1,2,5,6-diacetone-L-mannitol.[8] The procedure given is based on the method of Newman and Renoll.[9]

1 *Org. Syntheses* Coll. Vol. **1,** 17 (1941).
2 Fischer, *Ber.,* **28,** 1167 (1895).
3 Irvine, Macdonald, and Soutar, *J. Chem. Soc.,* **107,** 343 (1915).
4 Fischer and Pfähler, *Ber.,* **53,** 1607 (1920).
5 Smith and Lindberg, *Ber.,* **64,** 510 (1931).
6 Hibbert and Morazain, *Can. J. Research,* **2,** 38 (1930).
7 Maglio and Burger, *J. Am. Chem. Soc.,* **68,** 529 (1946).
8 Baer and Fischer, *J. Biol. Chem.,* **128,** 468 (1939); *J. Am. Chem. Soc.,* **67,** 2035 (1945).
9 Newman and Renoll, *J. Am. Chem. Soc.,* **67,** 1621 (1945).

JULOLIDINE

(Benzo[*ij*]quinolizine, 1,2,3,5,6,7-hexahydro-)

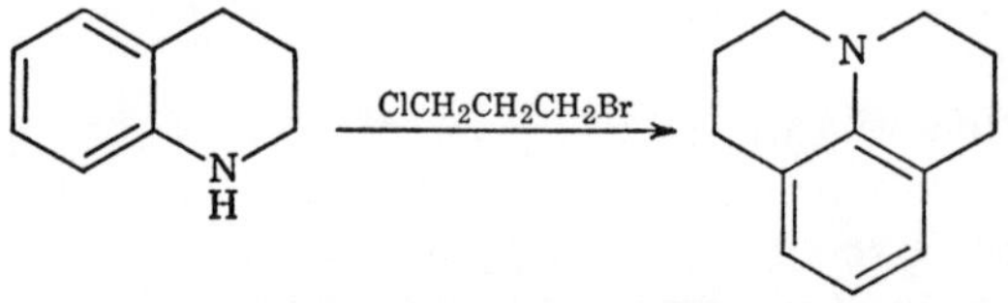

Submitted by D. B. Glass and A. Weissberger.
Checked by Cliff S. Hamilton and Carol K. Ikeda.

1. Procedure

A mixture of 66.5 g (0.5 mole) of tetrahydroquinoline and 400 g. of trimethylene chlorobromide (Note 1) is placed in a 1-l. round-bottomed flask attached to a reflux condenser, and heated in an oil bath held at 150–160° for 20 hours (Note 2). The reaction mixture is cooled, a solution of 50 ml. of concentrated hydrochloric acid in 500 ml.

of water is added, and the excess trimethylene chlorobromide is removed by distillation with steam (Note 3). The acid residue from the steam distillation is made alkaline with a 40% solution of sodium hydroxide (about 75 ml.), and the julolidine is extracted with two 150-ml. portions of ether. The ethereal solution is washed with 150 ml. of water and dried over sodium hydroxide pellets. The ether is evaporated and the residue distilled under reduced pressure. The portion that boils at 105–110°/1 mm. is collected (Notes 4 and 5). The yield is 67–70 g. (77–81%).

2. Notes

1. The tetrahydroquinoline and trimethylene chlorobromide were Eastman grade materials of the Eastman Kodak Company.

2. The heating should be carried out in a hood, or a gas trap should be used to remove the hydrogen halides that are evolved.

3. The trimethylene chlorobromide recovered may be dried over calcium chloride and used in a subsequent run.

4. The julolidine solidifies in the receiver, m.p. 39–40°.

5. After standing in contact with air for several weeks, the julolidine may become colored by the formation of a red compound. This red compound may be removed by distillation or by dissolving the amine in 2 or 3 volumes of hexane, treating the hexane solution with Norit or Darco, and filtering. The julolidine is crystallized from the hexane solution by cooling in an acetone-Dry Ice bath. The resulting product melts at 39–40° and amounts to 85–90% of the original.

3. Methods of Preparation

Julolidine has been prepared by the reaction of trimethylene chlorobromide with formanilide,[1] aniline,[1] methylaniline,[1] and tetrahydroquinoline;[1,2] by the reduction of 8,10-diketojulolidine;[1] by the intramolecular condensation of N-(γ-bromopropyl)tetrahydroquinoline;[3] and by dehydration of N-(γ-hydroxypropyl)tetrahydroquinoline or di-(γ-hydroxypropyl)aniline with phosphorus pentoxide.[4]

[1] Pinkus, *Ber.*, **25**, 2802 (1892).

[2] von Braun, Heider, and Wyczatkowska, *Ber.*, **51**, 1219 (1918).

[3] Jones and Dunlop, *J. Chem. Soc.*, **101**, 1752 (1912).

[4] Rindfusz and Harnack, *J. Am. Chem. Soc.*, **42**, 1724 (1920).

KETENE DIETHYLACETAL

(Ketene, diethyl ketal)

$$BrCH_2CH(OC_2H_5)_2 + (CH_3)_3COK \rightarrow CH_2{=}C(OC_2H_5)_2 + KBr + (CH_3)_3COH$$

Submitted by S. M. McElvain and D. Kundiger.
Checked by R. L. Shriner and C. H. Tilford.

1. Procedure

In a 2-l. round-bottomed flask, preferably fitted with an interchangeable ground-glass joint, are placed 650 g. (820 ml.) of absolute *tert*-butyl alcohol (Note 1) and 39.1 g. (1 gram atom) of potassium (Note 2). A reflux condenser is attached to the flask and the mixture is refluxed until all the potassium is dissolved (about 8 hours). The solution is allowed to cool slightly, and 198 g. (1 mole) of bromoacetal, together with some boiling chips (Note 3), is quickly added. A cream-colored precipitate of potassium bromide begins to deposit immediately. The flask is attached at once to a closely indented 46-cm. Vigreux column, equipped with a glass insulating jacket and a total-reflux partial take-off still head [1] (Note 4), and the *tert*-butyl alcohol is distilled from an oil bath (about 120–130°) at the rate of 25 drops per minute with a reflux ratio at the still head of about 6:1. This operation requires 16–18 hours (Note 5), and at the end of this time the temperature of the oil bath is raised to 160° and maintained there until no more alcohol comes over. The bath is then lowered and allowed to cool while the pressure within the fractionating system is very gradually reduced to 200 mm. and held there by a barostat (Note 6).

A small amount of the alcohol comes over at 51–52°/200 mm., and then, after the heating bath is replaced, 4–6 ml. of an intermediate fraction distils. This is followed by a fraction which boils at 83–86°/200 mm. and which is collected as pure ketene acetal. A total of 78–87 g. is obtained (67–75%). The major portion is collected while the temperature of the bath is 120–140°, and the remainder is obtained by raising the bath temperature to 170–180°.

Ketene acetal is best stored in a bottle made of alkaline glass which is preferably new and dusted with sodium *tert*-butoxide (Note 7). The glass stopper should be very well greased. Even with these precautions a small amount of a voluminous precipitate of the white polymer will develop.

2. Notes

1. The *tert*-butyl alcohol is refluxed over quicklime, distilled, and then redistilled over 1 g. of potassium per 100 g. of the alcohol. Improved yields of ketene acetal are obtained from *tert*-butyl alcohol that has been recovered from a previous preparation of the acetal.

2. The potassium should be cut into pieces sufficiently small to pass through the neck of the flask. Sodium in *tert*-butyl alcohol can be used, but it is necessary to carry out the subsequent reaction at 125° in sealed tubes. The amount of *tert*-butyl alcohol specified is sufficient to provide for complete solution of the potassium as the *tert*-butoxide.

3. An ebullator tube through which dry nitrogen was drawn has been used for the subsequent distillation under reduced pressure, but it is far more advantageous to use about six boiling chips. Because of its rapid reaction with water, ketene acetal must be protected from moisture of the air.

4. The still head described by Whitmore and Lux[1] is most satisfactory for controlling this distillation. The tube leading from the take-off of the column is attached to the receiver through a fraction cutter protected from moisture by a large tube of calcium chloride. The checkers replaced the Vigreux column by a 50-cm. column filled with glass helices and surrounded by a heating jacket. With this column the removal of the *tert*-butyl alcohol was complete in 5–6 hours.

5. A slower rate of fractionation does not result in an increased yield, but interrupted fractionation results in a decreased yield.

6. A good barostat is necessary. Control of the reduced pressure by adjusting a "leak" in the system is entirely unsatisfactory, for, as a result of a small increase in pressure, the liquid ceases to boil, the column drains, and the boiling chips are rendered ineffective. The barostat used by the submitters is essentially that described by Ellis[2] in which the relay is replaced by the thermionic relay described by Waddle and Saeman.[3]

7. The column and apparatus should not be washed with acid cleaning solution because the glass surface is left acidic and it then catalyzes the polymerization of ketene acetal.[4] A thin coating of the polymer on the walls of the apparatus is not detrimental. If polymer must be removed, it is best done by dissolving it in a 10% solution of hydrochloric acid in acetone; a deep red solution results.

3. Method of Preparation

Ketene acetal may be prepared by the action of potassium *tert*-butoxide on iodoacetal[5] or bromoacetal.[4]

[1] Whitmore and Lux, *J. Am. Chem. Soc.*, **54**, 3451 (1932).

[2] Ellis, *Ind. Eng. Chem., Anal. Ed.*, **4**, 318 (1932).

[3] Waddle and Saeman, *Ind. Eng. Chem., Anal. Ed.*, **12**, 225 (1940); Ferry, *Ind. Eng. Chem., Anal. Ed.*, **10**, 647 (1938).

[4] Johnson, Barnes, and McElvain, *J. Am. Chem. Soc.*, **62**, 964 (1940).

[5] Beyerstedt and McElvain, *J. Am. Chem. Soc.*, **58**, 529 (1936).

KETENE DIMER

(Acetylketene)

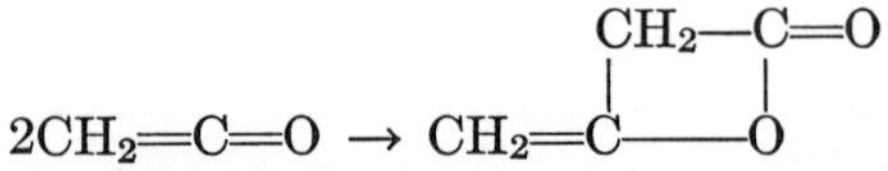

Submitted by Jonathan W. Williams and John A. Krynitsky.
Checked by Nathan L. Drake and Joseph Lann.

1. Procedure

Three 300-ml. gas-washing cylinders (Note 1) are connected in series, and the second and third cylinders are charged with 150 ml. each of dry acetone. Each of the three cylinders is immersed, in a thermos bottle, in sufficient Dry Ice-acetone cooling mixture to cover half of the cylinder. Ketene gas, prepared by the pyrolysis of acetone (Note 2), is passed through the system (Note 3) until a quantity of 2 moles has been introduced. This process requires 4–4.5 hours. During this time, after the ketene has been passing through the system for 1.5 hours, the Dry Ice-acetone cooling mixture is removed from the thermos bottle around the first cylinder. The cold thermos bottle is then replaced around the cylinder. Two hours after the completion of the ketene passage, the cooling mixture is removed from the second thermos bottle, and 6 hours later the third thermos bottle is emptied, both bottles being immediately restored to position. The entire system should be at room temperature 24 hours after the beginning of the run.

The liquids from all three cylinders are combined and fractionally distilled (Note 4). Most of the acetone is removed at room temperature under a pressure of 20 mm.; the last small portion is removed under atmospheric pressure. When the distillation temperature reaches 120°, the system is evacuated to a pressure of 80–100 mm. (Note 5), and the ketene dimer is collected within the boiling range 67–69°/92 mm. The yield of pure product is 42–46 g. (50–55%) (Notes 6 and 7).

2. Notes

1. The gas-washing cylinders are preferably without flanges. Those used by the submitters were prepared from 45-mm. Pyrex tubing, measured 28 cm. in length, and were fitted with 29/42 standard taper ground-glass joints. The inlet tubes extended two-thirds of the way into the cylinders.

2. Ketene may be generated conveniently from acetone by means of a "ketene lamp."[1] This apparatus was used by submitters and checkers. Other apparatus [*Org. Syntheses* Coll. Vol. **1,** 331 (1941)] might also be used.

3. The effluent gases from the third cylinder should be conducted to an efficient hood, or passed through a washing bottle containing a 10% aqueous solution of sodium hydroxide.

4. The submitters and checkers used a column of the Whitmore-Lux type.

5. Pressure between these limits is the optimum for this distillation. At higher pressure too much polymerization to dehydroacetic acid occurs, and at a lower pressure special cooling methods are necessary to prevent loss of distillate by evaporation.

6. A viscous dark red residue remains in the distillation flask. This material is mostly dehydroacetic acid.

7. The ketene dimer may be kept in a tightly stoppered bottle in the dark without appreciable further polymerization for about a week.[2c]

3. Method of Preparation

The procedure described is a modification of the method of Chick and Wilsmore,[3] which has been studied by several other workers.[2]

Homologous diketenes have been prepared by the action of tertiary amines on the corresponding acid chlorides.[4]

[1] Williams and Hurd, *J. Org. Chem.,* **5,** 122 (1940).

[2] (*a*) Hurd, Sweet, and Thomas, *J. Am. Chem. Soc.,* **55,** 335 (1933); (*b*) Hurd and Williams, *J. Am. Chem. Soc.,* **58,** 962 (1936); (*c*) Boese, *Ind. Eng. Chem.,* **32,** 16 (1940).

[3] Chick and Wilsmore, *J. Chem. Soc.,* **1908,** 946; **1910,** 1978.

[4] Sauer, *J. Am. Chem. Soc.,* **69,** 2444 (1947).

α-KETOGLUTARIC ACID *

(Glutaric acid, α-oxo-)

$$\begin{array}{l} CH_2CO_2C_2H_5 \\ | \\ CH_2CO_2C_2H_5 \end{array} + \begin{array}{l} CO_2C_2H_5 \\ | \\ CO_2C_2H_5 \end{array} \xrightarrow{(C_2H_5OK)} \begin{array}{l} COCO_2C_2H_5 \\ | \\ CHCO_2C_2H_5 \\ | \\ CH_2CO_2C_2H_5 \end{array}$$

$$\begin{array}{l} COCO_2C_2H_5 \\ | \\ CHCO_2C_2H_5 \\ | \\ CH_2CO_2C_2H_5 \end{array} + 3H_2O \xrightarrow{(HCl)} \begin{array}{l} COCO_2H \\ | \\ CH_2 \\ | \\ CH_2CO_2H \end{array} + 3C_2H_5OH + CO_2$$

Submitted by Lester Friedman and Edward Kosower.
Checked by Reynold C. Fuson and Elliott N. Marvell.

1. Procedure

A. *Ethyl oxalylsuccinate.* Potassium (39.5 g., 1 gram atom) is cut into pieces under xylene (Note 1) in a wide evaporating dish. The xylene is poured off, and the metal is washed with three 50-ml. portions of absolute ether. The potassium is then transferred quickly to a 2-l. three-necked flask containing 650 ml. of anhydrous ether and fitted with a reflux condenser, a mercury-sealed mechanical stirrer (Note 2), and a dropping funnel containing 150 ml. of anhydrous ethanol. The ethanol is added over a period of about 1.5 hours; stirring is unnecessary. After most of the ethanol has been added, the flask is heated on a water bath to ensure complete solution of the potassium. This usually takes from 3 to 4 hours. After the reaction is completed, the flask is cooled to room temperature, and 146 g. (1 mole) of ethyl oxalate is added rapidly through the dropping funnel, with stirring, to the solution of potassium ethoxide in ether. A yellow color develops at this point. The stirring is continued for an additional 10 minutes. Then 174 g. (1 mole) of ethyl succinate is added rapidly, with vigorous stirring (Note 3). After a few minutes, the potassium salt crystallizes, making further stirring impracticable. It is collected on a filter and washed with ether until the salt is colorless.

The salt is dissolved in 270 ml. of water, and 100 ml. of concentrated hydrochloric acid is added. The ethyl oxalylsuccinate separates as an oil and rises to the surface. It is removed by extracting the mixture with 100-ml. portions of ether until the aqueous solution is almost colorless. The extracts are dried over sodium sulfate, and the ether is distilled under reduced pressure. The ethyl oxalylsuccinate remains in the flask as a yellow oil. The yield is 225–227 g. (82–83%) (Note 4).

B. *α-Ketoglutaric acid.* The ester obtained by the foregoing procedure is mixed with 600 ml. of concentrated hydrochloric acid and left overnight. The mixture is concentrated by distillation (Note 5) until the temperature of the liquid reaches 140°. It is poured into an evaporating dish and allowed to cool. The solid mass, weighing 110–112 g., is then pulverized. The yield of α-ketoglutaric acid is 92–93% of the theoretical for the last step, or 75–77% based upon diethyl succinate. The light-tan product, obtained as described above, is suitable for most purposes, but a purer acid, m.p. 109–110° (cor.), may be obtained by recrystallization from an acetone-benzene mixture.

2. Notes

1. All containers must be absolutely dry, and anhydrous xylene must be used. Important! Destroy with anhydrous ethanol all potassium remaining in the xylene and waste ether.
2. The stirrer is not necessary until the ethyl oxalate is added. The submitters found the use of a nitrogen atmosphere to be unnecessary.
3. The ethyl oxalate was redistilled, and the fraction boiling at 106–107°/25 mm. was used. The ethyl succinate was C.P. material obtained from Eimer and Amend and was used without further purification.
4. For further purification, the product can be distilled at about 115°/1 mm.
5. This should be done in a hood, or a trap should be used to remove the hydrochloric acid.

The submitters wish to thank Mr. Andrew Streitweiser for his invaluable assistance.

3. Methods of Preparation

Ethyl oxalylsuccinate has been prepared by the condensation of ethyl oxalate with ethyl succinate in the presence of sodium ethoxide [1] or of potassium ethoxide.[2] The method described above is somewhat more convenient, and has given a higher yield of a better product, than one based upon sodium ethoxide, submitted by A. E. Martell and R. M. Herbst.

α-Ketoglutaric acid has been prepared by the hydrolysis of ethyl oxalylsuccinate with concentrated hydrochloric acid;[3] by the distillation of ethyl oxalylsuccinate with concentrated hydrochloric acid;[4] by treating α,β-dibromoglutaric acid with 2 *N* sodium carbonate solution;[5] by treatment of ethyl α-bromoglutaconate with alkalies;[6] and by treating ethyl α,α′-dibromoglutarate with alcoholic potash.[7]

[1] Wislicenus, *Ber.*, **22,** 885 (1889); *Ann.*, **285,** 1 (1895).

[2] (*a*) Wislicenus, *Ber.*, **44,** 1567 (1911); (*b*) Kögl, Halberstadt, and Barendregt, *Rec. trav. chim.*, **68,** 387 (1949).

[3] Blaise and Gault, *Compt. rend.*, **147,** 199 (1908); Gabriel, *Ber.*, **42,** 655 (1909).

[4] Blaise and Gault, *Bull. soc. chim. France,* **9,** 455 (1911).

[5] Ingold, *J. Chem. Soc.*, **119,** 2014 (1921).

[6] Ingold, *J. Chem. Soc.*, **119,** 2019 (1921).

[7] Ingold, *J. Chem. Soc.*, **119,** 326 (1921).

KRYPTOPYRROLE

(Pyrrole, 2,4-dimethyl-3-ethyl)

$$CH_3COCH_2CO_2C_2H_5 + HNO_2 \rightarrow CH_3COC(NOH)CO_2C_2H_5 + H_2O$$

$$CH_3COC(NOH)CO_2C_2H_5 + 4(H) \xrightarrow{(Zn\ +\ CH_3CO_2H)} CH_3COCH(NH_2)CO_2C_2H_5 + H_2O$$

$$CH_3COCH(NH_2)CO_2C_2H_5 + CH_3COCH_2COCH_3 \rightarrow$$

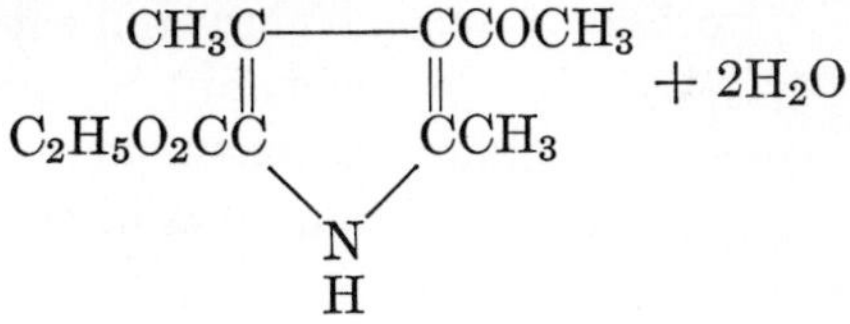

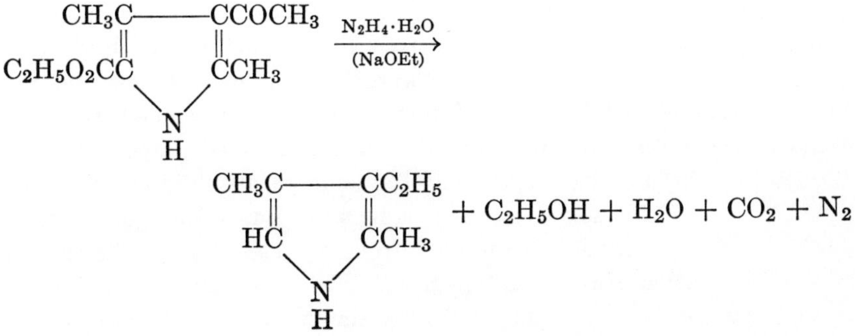

Submitted by HANS FISCHER.
Checked by HOMER ADKINS and IVAN A. WOLFF.

1. Procedure

In a 3-l. three-necked flask provided with a stirrer and surrounded by an ice bath are placed 402 g. (3.09 moles) of ethyl acetoacetate (Note 1) and 1.2 l. of glacial acetic acid. To this solution is then added dropwise with stirring a solution of 246 g. (3.55 moles) of sodium nitrite in 400 ml. of water. The rate of addition is controlled so that the temperature does not rise above 12°. After the sodium nitrite solution has been added, the mixture is stirred an additional 2–3 hours. It is then allowed to warm up to room temperature and stand about 12 hours, after which 348 g. (3.48 moles) of acetylacetone is added at one time.

To the reaction mixture 450 g. of zinc dust (Note 2) is added in portions of about 10 g. with vigorous stirring. The rate of addition is regulated so that the temperature never rises above 60°. After the addition is complete (Note 3), the mixture is refluxed for 2–3 hours on a hot plate until the unreacted zinc dust collects in balls. The hot solution is then poured through a fine copper sieve, with stirring, into 30 l. of ice water. The crude product which separates is contaminated with zinc (Note 4). On recrystallization from 1.5 l. of 95% ethanol, 360–390 g. of 2,4-dimethyl-3-acetyl-5-carbethoxypyrrole (m.p. 143–144°) is obtained (55–60% based on the ethyl acetoacetate used) (Note 5). A second recrystallization may be necessary to secure a perfectly white product, but the product of the first recrystallization is sufficiently pure for conversion to kryptopyrrole.

Thirty grams of sodium is dissolved in 430 ml. of absolute ethanol, the last portions by heating under reflux. The hot ethanolic solution is poured into an autoclave (Note 6), 75 g. (0.36 mole) of 2,4-dimethyl-3-acetyl-5-carbethoxypyrrole is stirred in, and 36 ml. of hydrazine hydrate (Note 7) is added. The autoclave is then heated and the mixture kept at 165–170° for 12 hours. After being cooled, the contents of the autoclave are emptied into a 2-l. round-bottomed flask. The autoclave is rinsed with a small amount of absolute ethanol, and these washings are added to the 2-l. flask. Then 50 ml. of water is added, and the ethanol is distilled from a steam bath, followed by steam distillation of the kryptopyrrole. The ethanolic distillate is collected separately. A glass condenser should be used for the steam distillation, as in some runs a white solid which melts around room temperature begins to appear after about 1 l. has distilled. The condenser water is turned off periodically to allow the solid in the condenser to melt. Steam distillation is continued until the drops of distillate are no longer cloudy, that is, after about 1.7 l. has distilled. A slow stream of nitrogen is passed over the surface of the distillate during the steam distillation (Note 8).

The ethanolic distillate is diluted to 2 l. with distilled water and extracted with a 500-ml. and then a 300-ml. portion of ether. The steam distillate is extracted twice with this ether extract, and twice more with 250-ml. portions of fresh ether. The ether extracts are combined and dried with 150 g. of anhydrous sodium sulfate. The space above the solution is filled with nitrogen. The ether solution is decanted, and the sodium sulfate is washed three times with distilled ether. The ether is then distilled, the temperature of the bath about the flask being raised finally to 130°. The kryptopyrrole is then fractionated under reduced pressure (Note 9). The yield of water-white product boiling within a 1.5° range (85.5–87°/12.5 mm.; 92.5–94°/18 mm.) is 22–25.5 g. (50–58%).

2. Notes

1. The ester used was a commercial product and was not further purified.

2. The zinc dust should be at least 80% pure.

3. Before the reaction mixture is refluxed, enough time should be allowed for the zinc dust to react completely; otherwise considerable trouble with foaming may be encountered.

4. The crude pyrrole darkens on exposure to light, especially when exposed to direct sunlight. The recrystallized product is unaffected by light.

5. The preparation can be carried out in larger or smaller quantities with proportionate amounts of materials and volumes of containers without affecting the yield. The amounts specified here are 60% of those used by the submitter.

6. The checkers used a steel hydrogenation bomb with a void of 650 ml. from which the inside bent steel tube leading to the gauge assembly was removed. A steel plug was inserted in the opening which usually carries a gauge. The temperature was controlled by an automatic regulator, and the reaction mixture was not shaken or stirred.

7. Hydrazine hydrate may be prepared by the ammonolysis of hydrazine sulfate,[1] followed by the addition of water.

Liquid ammonia is placed in a 1-pint thermos bottle, and 60 g. of Eastman's hydrazine sulfate is added in small portions while the liquid is stirred mechanically. After the addition is complete, stirring is continued for 30 minutes. The mixture is filtered into another thermos bottle through a fluted filter paper, and the remaining solid is washed twice with liquid ammonia by transferring it back to the original flask, stirring, and again filtering. On evaporation of the liquid ammonia 7.5–10 g. of colorless liquid is left, 51–68% calculated as anhydrous hydrazine. Five milliliters of water is then added. Three to four such runs are required to obtain the hydrazine hydrate needed in this preparation.

8. Kryptopyrrole is very sensitive to oxidation and should be handled with a minimum amount of exposure to the air. It is best stored in sealed glass tubes. If a capillary ebullition tube is used during distillation under reduced pressure, the tube should be connected to a supply of nitrogen.

9. A Vigreux, a modified Widmer, or other column may be used for this separation.

3. Methods of Preparation

Kryptopyrrole has been obtained by the degradation of bilirubin, hemin, rhodoporphyrin, etc.[2] The synthesis given above is based upon the work of Knorr and Hess.[3]

[1] Browne and Welsh, *J. Am. Chem. Soc.*, **33**, 1728 (1911).
[2] Fischer-Orth, *Die Chemie des Pyrrols*, p. 53, Akademische Verlagsgesellschaft, Leipzig, 1934.
[3] Knorr and Hess, *Ber.*, **44**, 2765 (1911).

LACTAMIDE

$$CH_3CH(OH)CO_2C_2H_5 + NH_3 \rightarrow CH_3CH(OH)CONH_2 + C_2H_5OH$$

Submitted by J. Kleinberg and L. F. Audrieth.
Checked by Homer Adkins and William H. Bateman.

1. Procedure

One hundred and twenty-five grams (1.06 moles) of ethyl lactate is placed in a suitable Pyrex container which is subsequently cooled in a Dry Ice-acetone bath. When the ester has been cooled below the boiling point of ammonia (Note 1), 125 ml. of liquid ammonia (Note 2) is added. The mixture is then placed in a specially constructed steel pressure apparatus (Note 3) and permitted to come to room temperature. After 24 hours (Note 4) the excess of ammonia is allowed to escape slowly through the gas outlet of the bomb. The last traces of ammonia are removed under reduced pressure. The reaction product is stirred with 200 ml. of absolute ether to dissolve unchanged ester and ethanol. The residue is filtered, washed with ether, and air-dried. The yield of lactamide melting at 74–75° amounts to 65–70 g. (70–74%) (Note 5).

2. Notes

1. Care must be taken to cool the ester below the boiling point of ammonia before addition of the ammonia, to avoid loss of ester by spattering.

2. For manipulative procedures employing liquid ammonia see Franklin, *The Nitrogen System of Compounds*, A.C.S. Monograph 68, Appendix, Reinhold Publishing Corporation, New York, 1935; also, Fernelius and Johnson, *J. Chem. Education*, **6**, 441 (1929).

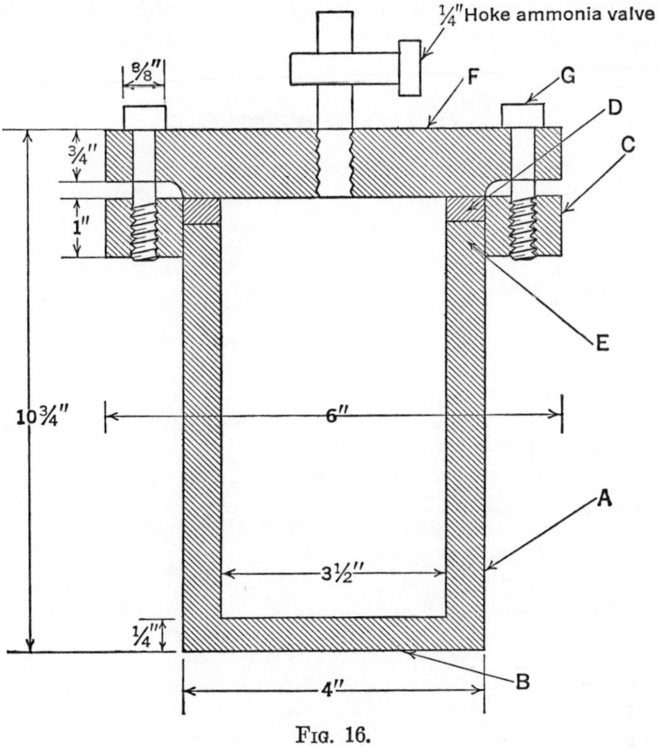

FIG. 16.

3. The steel bomb in which the reaction is carried out is depicted in Fig. 16. It consists of a cylindrical tube (*A*) of ordinary steel to which a steel bottom (*B*) has been welded. A 1-in. flange (*C*) is welded to the top of the container extending about $\frac{3}{16}$ in. above the top of the bomb. The lead gasket (*D*) is pressed into the groove (*E*) when the cover of stainless steel is tightened by means of six steel bolts (*G*). The top (*F*) is machined to make a tight seal on the gasket (*D*). A ¼-in. steel Hoke ammonia valve serves as the gas outlet. The cross-sectional dimensions are noted in the diagram. The checkers carried out the reaction in a glass beaker which was set in a steel reaction vessel such as is used for hydrogenations at pressures of 50–500 atm.[1]

4. Increase in reaction time causes no appreciable increase in yield of amide.

5. This method has been used for the preparation of numerous amides.[2] However, with many esters it is necessary to heat the reaction mixture to 200–250° for a few hours. Ethyl mandelate is like ethyl lactate in that it gives a good yield (75–80%) of mandelamide at room temperatures.

3. Methods of Preparation

Lactamide has been prepared by the action of ammonia on ethyl lactate,[3] methyl lactate,[4] lactic anhydride,[5] lactide,[6] and the condensation product of lactic acid with acetone.[7] In general, amides have been prepared by the reaction of liquid ammonia with esters at temperatures varying from −33° to 250°.[2,8,9]

[1] Adkins, *Reactions of Hydrogen with Organic Compounds over Copper-Chromium Oxide and Nickel Catalysts*, p. 31, University of Wisconsin Press, Madison, Wisconsin, 1937; *Ind. Eng. Chem., Anal. Ed.*, **4**, 342 (1932).

[2] Wojcik and Adkins, *J. Am. Chem. Soc.*, **56**, 2421 (1934), Paden and Adkins, *J. Am. Chem. Soc.*, **58**, 2497 (1936).

[3] Brüning, *Ann.*, **104**, 197 (1857).

[4] Ratchford, *J. Org. Chem.*, **15**, 326 (1950).

[5] Wurtz and Friedel, *Ann. chim. phys.*, (3) **63**, 108 (1861).

[6] Wislicenus, *Ann.*, **133**, 259 (1865).

[7] Oeda, *Bull. Chem. Soc. Japan*, **11**, 385 (1936).

[8] Glattfeld and MacMillan, *J. Am. Chem. Soc.*, **58**, 898 (1936).

[9] Audrieth and Kleinberg, *J. Org. Chem.*, **3**, 312 (1938).

LEPIDINE *

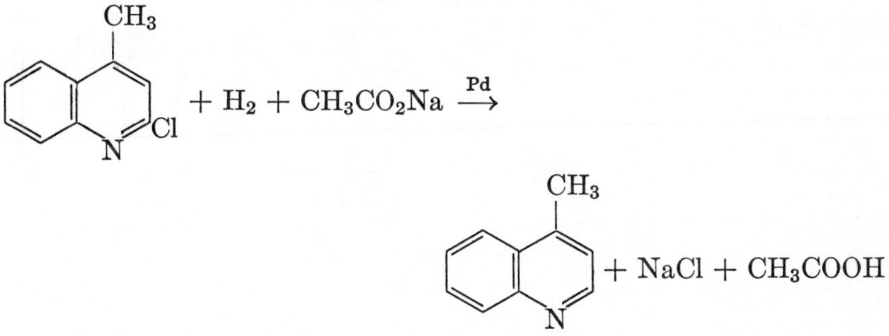

Submitted by FRED W. NEUMANN, NOLAN B. SOMMER, C. E. KASLOW, and R. L. SHRINER.
Checked by CLIFF S. HAMILTON and ROBERT F. COLES.

1. Procedure

In a 500-ml. Erlenmeyer flask are placed 20 g. (0.11 mole) of pure 2-chlorolepidine [1] (Note 1), 9.3 g. (0.11 mole) of powdered anhydrous sodium acetate, and 200 ml. of glacial acetic acid. The mixture is heated to about 70° and shaken until solution is complete. The solution is transferred to a pressure bottle of an apparatus for catalytic reduction,[2] equipped with a heating element and a variable resistance. The flask is rinsed with two 10-ml. portions of hot glacial acetic acid. Then 3 g. of palladium on carbon is added (Note 2), the bottle is attached to the shaking machine, and the variable resistance is adjusted until the temperature of the liquid is between 55° and 70° (Note 3). The bottle is swept out with hydrogen, an initial pressure of about 1.8–2.2 atm. (26–33 lb.) is applied, and the shaking is started. Hydrogen absorption is rapid during the first 15 minutes and then gradually slackens; the theoretical amount is absorbed in 1.5–2 hours. To ensure complete reduction, shaking is continued an additional 30 minutes. The warm acid solution is separated from the catalyst by filtration through a 1- to 2-mm. layer of Norit on a Büchner funnel. The bottle and funnel are washed with three 30-ml. portions of glacial acetic acid. The acetic acid is removed from the combined filtrates by heating to 70° under reduced pressure (water pump, 25 mm.). The residue is dissolved in 50 ml. of water and transferred to a 500-ml. separatory funnel, an additional 25 ml. of water being used for washing. The water solution is made basic to litmus with 30% sodium hydroxide (about 40–100 ml.) and extracted with one 100-ml. portion of ether and then with two 50-ml. portions. The ether extracts are combined and dried overnight with about 30 g. of solid potassium hydroxide. The ether is removed by distillation from a 250-ml. flask, and the residue is transferred to a modified 50-ml. distilling flask (Note 4), three 5-ml. portions of anhydrous ether being used to ensure complete transference. After the ether is removed, the residue distils at 126–127°/14–15 mm. The product is colorless, water clear, and weighs 13–14 g. (81–87%) (Notes 5 and 6).

2. Notes

1. Pure 2-chlorolepidine, m.p. 58–59°, should be used.

2. The catalyst is previously prepared in an apparatus for catalytic hydrogenation,[2] in which are placed 0.5 g. of palladous chloride, 3.0 g. of Norit, and 20 ml. of distilled water. The bottle is swept out with hydrogen and then shaken with hydrogen for 2–3 hours at 2–3 atm. (40 lb.) pressure. The palladium on carbon is collected on a Büchner funnel, washed with five 50-ml. portions of distilled water, then with five 50-ml. portions of 95% ethanol, and finally twice with ether. Upon drying, about 3 g. of the catalyst is obtained. It is stored in a vacuum desiccator over solid sodium hydroxide. If the reduction of the chlorolepidine does not proceed normally, the used catalyst should be removed by suction filtration and a fresh 3-g. portion of catalyst added. Failure of the reduction step is usually due to an inactive catalyst or to impurities in the acetic acid or chlorolepidine. The palladium catalysts, prepared as described elsewhere in this volume, are also satisfactory for the reduction of 2-chlorolepidine.

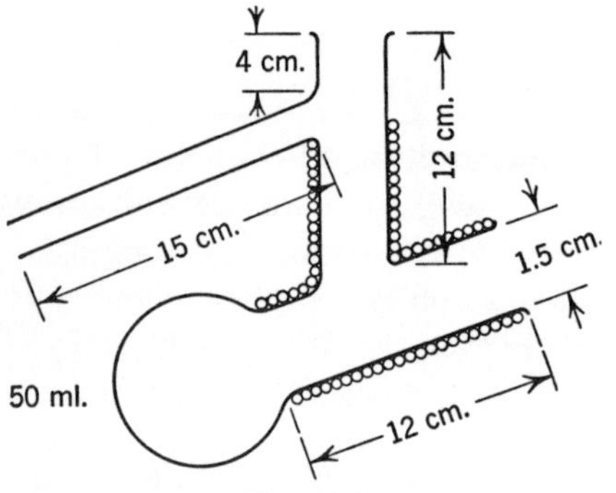

Fig. 17.

3. The reduction does not proceed smoothly at room temperature with the palladium catalyst. Raney nickel may be used as a catalyst with ethanol containing potassium hydroxide at room temperature, but about 15 hours is required for reduction.

4. The submitters used a special flask having the shape and dimensions shown in Fig. 17. The two necks were wrapped with asbestos cord. The checkers used an ordinary Claisen flask (50 ml.).

5. By distillation of the crude product from four runs, a yield of 92% was obtained.

6. The submitters have followed the same procedure in preparing the compounds listed below from the corresponding 2-chloro derivatives. The ether extractions and distillation steps were omitted when solid products were obtained.

	Product		
Compound	*% Crude*	*% Purified*	*B.P. or M.P.*
6-Methyllepidine	100	87	B.p. 137°/12 mm.
8-Methyllepidine	94	90	M.p. 54–55°
5,8-Dimethyllepidine	96	86	B.p. 154–56°/13 mm.
6,8-Dimethyllepidine	100	91	M.p. 55–56°
6-Methoxylepidine monohydrate	100	80	M.p. 50–52°
5,8-Dimethoxylepidine	98	90	M.p. 94–95°
2-Methyl-6-methoxyquinoline *	96	..	M.p. 62–65°

* From 2-methyl-4-chloro-6-methoxyquinoline.

3. Methods of Preparation

The process described above is essentially that of Ainley and King,[3] who prepared 6-methoxylepidine. Lepidine has also been prepared by the reduction of 2-chlorolepidine with hydrogen and Raney nickel,[4] with tin [5–7] or zinc [7a] and hydrochloric acid, and with concentrated hydriodic acid and red phosphorus; [8] by the reduction of 2-iodolepidine with iron and dilute sulfuric acid; [9] by the zinc dust distillation of 2-hydroxy-4-methylquinoline under reduced pressure [10] or of 2-hydroxy-3-cyano-4-methylquinoline; [11] by the reduction of 2-hydroxy-4-methylquinoline with concentrated hydriodic acid and red phosphorus; [12] by the distillation of 1,2,3,4-tetrahydroquinoline-4-carboxylic acid with zinc dust in a stream of hydrogen; [13] by decarboxylation of 4-methylquinoline-2-carboxylic acid; [14] by leading vapors of aniline and crotonaldehyde over a contact catalyst at above 500°; [15] by heating aniline and vinyl methyl ketone with sulfuric acid and nitrobenzene; [16] by heating aniline and β-hydroxyethyl methyl ketone in the presence of concentrated sulfuric acid and nitrobenzene [17] or aniline hydrochloride and ethanol; [18] by heating aniline and β-chloroethyl methyl ketone in the presence of concentrated hydrochloric acid and nitrobenzene or arsenic acid [19,20] or in the presence of aniline hydrochloride, ethanol, and nitrobenzene; [21] by heating a mixture of acetone, formaldehyde, and aniline hydrochloride; [22,23] by passing vapors of acetylene and aniline over aluminum oxide at 360–420°; [24] and by passing vapors of aniline and acetaldehyde or paraldehyde over aluminum oxide in a copper tube at 480°.[25] Campbell and Schaffner[26] have described the preparation of lepidine in 70–73% yields by the reaction of aniline hydrochloride with methyl vinyl ketone, 1,3,3,-trimethoxybutane, or 1-methoxybutanone-3 in ethanol in the presence of ferric and zinc chloride.

[1] *Org. Syntheses* Coll. Vol. **3**, 194 (1954).

[2] *Org. Syntheses* Coll. Vol. **1**, 61–63 (1941).

[3] Ainley and King, *Proc. Roy. Soc.*, **B125**, 84 (1938).

[4] Krahler and Burger, *J. Am. Chem. Soc.*, **63**, 2369 (1941).

[5] Mikhailov, *J. Gen. Chem. U.S.S.R.*, **6**, 511 (1936) [*C. A.*, **30**, 6372 (1936)]; Mikhailov, Russ. pat. 39,104 [*C. A.*, **30**, 3446 (1946)].

[6] Ainley and King, *Proc. Roy. Soc.*, **B125**, 72 (1938).

[7] (*a*) Manske, Marion, and Leger, *Can. J. Research*, **20**, 149 (1942); (*b*) Yoshihisa, *J. Pharm. Soc. Japan*, **69**, 126 (1949).

[8] Knorr, *Ann.*, **236**, 99 (1886).

[9] Byvanck, *Ber.*, **31**, 2153 (1898).

[10] Knorr, *Ann.*, **236**, 94 (1886); *Ber.*, **16**, 2596 (1883).

[11] Guareschi, *Atti accad. sci. Torino*, **1893**, 28 [*Ber.*, **26**, Ref. 944 (1893)].

[12] Hammick and Thewlis, *J. Chem. Soc.*, **1948**, 1457.

[13] Weidel, *Monatsh.*, **3**, 75 (1882) (Beilstein, *Handbuch der organischen Chemie*, 4th ed., **20**, 395, J. Springer, Berlin, 1935).

[14] Koenigs and Mengel, *Ber.*, **37**, 1328 (1904).

[15] Huntenberg, Ger. pat. 661,902 [*C. A.*, **32**, 8443 (1938)].

[16] Prill and Walter, U. S. pat. 1,806,564 [*C. A.*, **25**, 3668 (1931)].

[17] Prill and Walter, U. S. pat. 1,806,563 [*C. A.*, **25**, 3668 (1931)]; Ger. pat. 505,320 [*C. A.*, **26**, 479 (1932)].

[18] Tseou Heou-Feo, *Bull. soc. chim. France*, (5) **2**, 94 (1935).

[19] Zollner, U. S. pat. 1,804,045 [*C. A.*, **25**, 3668 (1931)]; Ger. pat. 518,291 [*C. A.*, **25**, 2442 (1931)].

[20] Brit. pat. 283,577 [*C. A.*, **22**, 4132 (1928)].

[21] Kenner and Statham, *Ber.*, **69B**, 17 (1936).

[22] Mikeska, *J. Am. Chem. Soc.*, **42**, 2396 (1920).

[23] Beyer, *J. prakt. Chem.*, (2) **33**, 418 (1885).

[24] Tschitschibabin, *J. Russ. Phys. Chem. Soc.*, **47**, 713 (1915) [*Chem. Zentr.*, **1916**, I, 920]; (Beilstein, *Handbuch der organischen Chemie*, 4th ed., Suppl. I, **20**, 150, J. Springer, Berlin, 1935).

[25] Tschitschibabin, Ger. pat. 468,303 [*C. A.*, **23**, 607 (1929)].

[26] Campbell and Schaffner, *J. Am. Chem. Soc.*, **67**, 86 (1945).

dl-LEUCINE *

(Isocaproic acid, α-amino)

$$(CH_3)_2CHCH_2CH_2CO_2H + Br_2 \xrightarrow{(PCl_3)} (CH_3)_2CHCH_2CHBrCO_2H + HBr$$

$$(CH_3)_2CHCH_2CHBrCO_2H + 2NH_3 \rightarrow (CH_3)_2CHCH_2CHNH_2CO_2H + NH_4Br$$

Submitted by C. S. Marvel.[1]
Checked by Homer Adkins and Robert Gander.

1. Procedure

A. *α-Bromoisocaproic acid.* Five hundred grams (4.3 moles) of commercial isocaproic acid is mixed with 250 ml. of benzene in a 2-l. round-bottomed flask, and the water and benzene are removed by distillation through a short column until the temperature of the vapors reaches 100°. The temperature rises rapidly as soon as the last of the benzene is removed. The residual acid is cooled to room temperature, 743 g. (4.65 moles, 243 ml.) of dry bromine (Note 1) is added, and the flask is fitted with a long condenser and placed in an oil bath. The top of the condenser is connected to an empty 500-ml. Erlenmeyer flask which acts as a safety flask, and this in turn leads to a gas-absorption trap (Note 2). Ten milliliters of phosphorus trichloride is added to the mixture through the top of the condenser, and the flask is heated to 80–85°. The bromination proceeds smoothly at this temperature and is allowed to continue for 8–15 hours until the dark red color of bromine disappears from the condenser. When it has, the temperature is raised to 100–105° and kept there 2 hours. The contents of the flask are transferred to a 1-l. modified Claisen flask or a flask attached to a Widmer column and distilled. The fraction boiling at 125–131°/12 mm. is collected. The yield amounts to 530–550 g. (63–66%). The low-boiling fraction is mainly isocaproic acid (Note 3).

B. *dl-Leucine.* To 1.5 l. of technical ammonium hydroxide (sp. gr. 0.90) in a 3-l. round-bottomed flask is added 300 g. (1.56 moles) of α-bromoisocaproic acid. A rubber stopper is wired in, and the flask is allowed to stand for a week at room temperature. The crude leucine from four such flasks is collected on a filter and washed with 400 ml. of ethanol. This crop amounts to about 300 g. The ammonia is removed from the filtrate by heating the solution in a 12-l. flask on a steam cone overnight. The solution is concentrated under reduced pressure until vigorous bumping occurs (about 2.5 l.). The mixture is then cooled to about 15° and filtered. The precipitate is washed with 250 ml. of cold water and 250 ml. of 95% ethanol. The total yield of crude leucine in the two fractions is 440–460 g.

The amino acid is recrystallized by dissolving all the crude material in 12.5 l. of water heated to 95° on a steam cone. The hot solution is treated with 20 g. of Norit for 30 minutes and filtered hot. An equal volume of 95% ethanol is added immediately, and the flask is placed in the ice chest overnight. The crystalline material is collected on a filter and washed with 200 ml. of 95% ethanol. The yield of pure leucine in this fraction is 290–300 g. An additional crop is obtained by evaporating the mother liquors under reduced pressure until considerable solid separates (liquid volume about 1 l.), adding an equal volume of ethanol, and cooling. This crop is washed with 100 ml. of cold water and then with 200 ml. of ethanol; it amounts to 60–65 g. The total yield of pure leucine is 350–365 g. (43–45%). It decomposes at 290–292° (uncor.) in a sealed capillary (Note 4).

2. Notes

1. The bromine is dried by shaking with 500 ml. of c.p. concentrated sulfuric acid.

2. The hydrogen bromide may be collected in water and distilled to give constant-boiling hydrobromic acid. See *Org. Syntheses* Coll. Vol. **1**, 26 (1941).

3. The low-boiling fractions (105–115 g.) may be combined with the next portion of acid to be brominated, or several such fractions may be collected and brominated together. If this last is done only two-thirds as much bromine is used as in the original run.

4. The amino nitrogen content of leucine prepared in this way checks with the theoretical value.

3. Methods of Preparation

dl-Leucine has been prepared by the hydrolysis of isobutylhydantoin with barium hydroxide;[2] by reduction and hydrolysis of α-oximinoisocaproate;[3] by racemization of *l*-leucine;[4] by the action of ammonia and hydrogen cyanide on isovaleraldehyde followed by hydrolysis;[5] by the action of heat on isobutylmalonylazidic acid followed by hydrolysis;[6] by the action of ammonia on α-bromoisocaproic acid;[7] by the condensation of isobutyl halides and sodio aminomalonic ester [8] or sodio benzoylaminomalonic ester [9] followed by hydrolysis; by the condensation of isobutyraldehyde and hippuric acid followed by reduction and hydrolysis,[10] by reduction of α-ketoisocaproic acid phenylhydrazone; [11] by the action of sodium hypobromite on ethyl isobutylmalonamate; [12] by the action of hexamethylene tetramine on α-chloro- and α-bromoisocaproic acids; [13] and by hydrogenation and subsequent hydrolysis of methallyl acetamidomalonic ester.[14] The method described above is essentially that of Fischer [7] and is the cheapest and best procedure for the synthesis of large amounts of this amino acid.

[1] These directions are the results of the efforts of many men who have worked on the preparation of leucine at the University of Illinois.

[2] Pinner and Spilker, *Ber.*, **22**, 696 (1889).

[3] Bouveault and Locquin, *Bull. soc. chim. France,* (3) **31,** 1181 (1904).

[4] Schulze and Bosshard, *Ber.*, **18,** 389 (1885); *Z. physiol. Chem.*, **10,** 135 (1886); Fischer, *Ber.,* **33,** 2372 (1900).

[5] Limpricht, *Ann.*, **94,** 243 (1855); Hüfner, *J. prakt. Chem.*, (2) **1,** 10 (1870); Schulze and Likiernik, *Z. physiol. Chem.*, **17,** 516 (1893); Abderhalden and Wybert, *Ber.*, **49,** 2455 (1916).

[6] Curtius, *J. prakt. Chem.*, (2) **125,** 211 (1930).

[7] Fischer, *Ber.*, **37,** 2486 (1904).

[8] Locquin and Cherchez, *Compt. rend.*, **186,** 1360 (1928).

[9] Redemann-Schmidt, *Chemistry of the Amino Acids and Proteins,* p. 50, Thomas, Baltimore, Md., 1938.

[10] Erlenmeyer and Kunlin, *Ann.*, **316,** 145 (1901).

[11] Feofilaktov, *Bull. acad. sci. U.R.S.S.*, **1941,** 521.

[12] Huang, Lin, and Li, *J. Chinese Chem. Soc.*, **15,** 31 (1947).

[13] Hillmann and Hillmann, *Z. physiol. Chem.*, **1948,** 71.

[14] Albertson and Archer, *J. Am. Chem. Soc.*, **67,** 308 (1945).

LINOLEIC ACID

$$CH_3(CH_2)_4CH{=}CHCH_2CH{=}CH(CH_2)_7CO_2H + 2Br_2 \rightarrow CH_3(CH_2)_4CHBrCHBrCH_2CHBrCHBr(CH_2)_7CO_2H$$

$$CH_3(CH_2)_4CHBrCHBrCH_2CHBrCHBr(CH_2)_7CO_2H + 2Zn \xrightarrow[HCl]{C_2H_5OH} CH_3(CH_2)_4CH{=}CHCH_2CH{=}CH(CH_2)_7CO_2C_2H_5 + 2ZnBr_2$$

$$CH_3(CH_2)_4CH{=}CHCH_2CH{=}CH(CH_2)_7CO_2C_2H_5{+}NaOH \xrightarrow{H_2SO_4} CH_3(CH_2)_4CH{=}CHCH_2CH{=}CH(CH_2)_7CO_2H{+}NaHSO_4{+}C_2H_5OH$$

Submitted by J. W. McCutcheon.
Checked by R. L. Shriner and S. P. Rowland.

1. Procedure

A. *Fatty acids.* In a 1-l. Erlenmeyer flask are placed 250 ml. of dynamite-grade glycerin (Note 1) and 40 g. (0.71 mole) of potassium hydroxide. The mixture is heated to 120–140° and is shaken by hand until the alkali is dissolved. To the hot solution there is added, in one portion (Note 2), 110 ml. (100 g., 0.11 mole) of sunflower-seed oil (Notes 3 and 4) which has been preheated to 110–115°. The hot solution is swirled vigorously until saponification is complete (Note 5). This is indicated by the formation of a permanent lather. After the mixture has cooled somewhat, 150 ml. of sulfuric acid (25% by volume) is added cautiously (Note 6) while the flask is swirled. Then 200 ml. of hot water is added, and, if necessary, the mixture is heated until the layer of fatty acid is clear. The water layer is removed, the acid layer is washed with two 500-ml. portions of hot water, and then it is filtered with the aid of suction through a large Hirsch funnel which is heated by a steam or hot-water jacket. The acids (90–95 g.) (Note 7) are dried thoroughly by heating them rapidly to 130°, with stirring.

B. *Tetrabromostearic acid.* In a 4-l. beaker equipped with a mechanical stirrer the above fatty acids are dissolved in 2 l. of petroleum ether (Note 8), and the solution is chilled to 0–10°. After 20–25 minutes, the solution deposits about 10 g. of solid saturated fatty acids, which are removed by rapid filtration with gentle suction. The filtrate is transferred to the 4-l. beaker and cooled to 0–10°. Then, with stirring, 30.7 ml. (90 g., 0.56 mole) of bromine (Note 9) is slowly introduced from a dropping funnel (Note 10). The bromine is added at such a rate that the temperature remains between 10° and 15°; about 20 minutes is usually required. Tetrabromostearic acid begins to separate toward the end of the addition. The solution, which should contain a slight excess of bromine (reddish color), is allowed to remain in the ice bath for 15–20 minutes (Note 11). The precipitate settles rapidly, and a large portion of the liquid is removed by decantation, after which the precipitate is finally collected with suction on a 12.5-cm. Büchner funnel and is washed with 200 ml. of petroleum ether. The solid is then thoroughly stirred with 300 ml. of petroleum ether (Note 12) in a 600-ml. beaker and is filtered as before. The gray-white tetrabromide, after being dried at 50°, is transferred to a 2-l. beaker and dissolved in 90–100 ml. of hot ethylene dichloride (Note 13). The solution is filtered through filter paper and allowed to crystallize at room temperature. The solid is removed by filtration and is washed on the funnel with 200 ml. of fresh petroleum ether. The tetrabromide is transferred to a *glass plate* and is worked to a powder (Note 14) with a spatula. The needle-shaped crystals are snow-white, with a silvery sheen, and have a melting point of 114.7–115.2°. The yield (Note 15) is 35–40 g. (32–36%) (Note 16). The product should be stored in brown screw-top bottles and kept away from sunlight.

C. *Ethyl linoleate.* In a 300-ml. Erlenmeyer flask are placed 26–30 g. of recrystallized tetrabromostearic acid, 85 ml. of absolute ethanol, and 30 g. of granulated zinc (20 mesh). This mixture is warmed gently until the debromination reaction begins. Since the reaction is exothermic, it is usually necessary to moderate it from time to time by dipping the flask into a basin of cold water. The vigorous reaction subsides in about 5 minutes, after which the flask is fitted with a reflux condenser and the mixture is refluxed for 30 minutes. In order to esterify the linoleic acid, 10 ml. of a 4 N solution of hydrochloric acid in ethanol (Note 17) is poured into the refluxing mixture through the top of the condenser, and successive 5-ml. portions are then added every 30 minutes for 2 hours. At the end of this time the solution is removed from the unchanged zinc by decantation into another flask. The zinc is washed with 15 ml. of absolute ethanol to complete the transfer of the solution. A 10-ml. portion of 4 N ethanolic hydrochloric acid is added and the solution is refluxed for an hour, with addition of a second 10-ml. portion of ethanolic acid after 30 minutes. The mixture is poured into 300 ml. of hot, saturated brine solution in a 500-ml. separatory funnel, and the crude ester is allowed to settle for 10–20 minutes. The brine layer is removed, and the ester is washed (Note 18) at room temperature with 300 ml. of 0.5% sodium carbonate solution. The rather stable emulsion is broken by centrifuging the solution, in 100-ml. tubes, for 5 minutes at 3300 r.p.m. The ester and any remaining emulsion are washed with 80 ml. of warm water, and the emulsion is again broken by centrifuging. Washing is continued (Note 19) until the wash water is neutral to methyl orange. The neutral ester is then transferred to a 50-ml. distilling flask and is distilled under reduced pressure; boiling point, 175°/2.5 mm., 193°/6 mm. The product is water-white (Note 20), gives a negative Beilstein test, has an iodine value (Wijs, 30-minute exposure) of 162.3-162.5, an n_D^{48} of 1.4489, and a true specific gravity of 0.8846 at 15.5/4°. The yield (Note 15) is 12–15 g.

D. *Linoleic acid.* The ester is dissolved in 200 ml. of a 5% ethanolic (Note 21) solution of sodium hydroxide in a 400-ml. beaker and is allowed to saponify overnight at room temperature. The resulting jelly is dissolved in 200 ml. of warm water, and a slow stream of carbon dioxide is introduced, beneath the surface of the liquid, while it is acidified with 50 ml. of dilute sulfuric acid (1:1 by volume). The stream of carbon dioxide is maintained throughout the subsequent operations. The linoleic acid rises to the surface as a clear layer. The water layer is siphoned off; the acid is washed once with hot water and then dried over anhydrous sodium sulfate and preserved under carbon dioxide. The yield is 10–12 g. of material having a melting point of −8° to −9° (Note 22).

2. Notes

1. Dynamite-grade glycerin is specified, because of its high glycerol content (99 plus per cent). Water interferes with the ease of saponification.

2. The oil must not be allowed to run down the side of the flask, as this interferes with complete saponification.

3. The following oils are recommended in decreasing order of preference, the numbers in parentheses indicating the approximate linoleic acid content: sunflower-seed oil (60), poppy-seed oil (60), cotton-seed oil (45).

4. The sample used must have a negative hexabromide test, indicating the absence of acids more highly unsaturated than linoleic. The test is carried out as follows: 2 ml. of the oil and 25 ml. of a 4:1 ethyl ether-glacial acetic acid mixture are chilled to 0° for 15–20 minutes; the mixture is filtered if it is not clear. To this solution is added sufficient bromine to give a deep red color, and the whole is allowed to stand for 15 minutes. The absence of a precipitate constitutes a negative test.

5. Usually 4–6 minutes of vigorous swirling is sufficient, but it may be necessary to reheat the mixture to 150° if difficulty is encountered. Saponification, once started, proceeds to completion within a few seconds. The emulsion thickens to a viscous paste which is transformed into a clear, limpid solution on continued swirling.

6. By adding the acid to the *hot* solution, with vigorous agitation, delay in clearing the fatty acid layer is avoided.

7. When large quantities of the fatty acids are desired, the following method is more economical: In a 5-l. round-bottomed flask, fitted with a mechanical stirrer and a reflux condenser, are placed 1.5 l. of methanol and 350 g. of potassium hydroxide. After solution is complete, 1.1 l. (1 kg.) of cotton-seed oil is added. The solution is stirred and refluxed for 1–2 hours, and then the methanol is removed by distillation. To the residue are added 1 l. of water and (slowly) 1.5 l. of cold 20% sulfuric acid. The fatty acid layer is separated from the water layer, washed with two 1.5-l. portions of hot water, and then filtered with the aid of a hot-water funnel. The acids are dried as described above. The yield is 950–975 g. of crude acids.

8. The petroleum ether used throughout had a boiling range of 40–60°.

9. The weight of bromine is independent of the amount (per cent) of linoleic acid in the sample, but it varies with the iodine value.[1] For the purpose of this preparation, the amount of bromine (grams) may be calculated as 0.7 times the iodine value of the oil. This allows an excess of 10–11%.

10. The tip of the dropping funnel should be just above the fatty acid solution.

11. Tetrabromostearic acid has a decided tendency to form a supersaturated solution in petroleum ether.

12. The solubility of the tetrabromide[2] is about 2–3 g. per l. of petroleum ether at room temperature, hence there is little danger of losing appreciable quantities by solution during the washing.

13. As an alternative but less satisfactory method of purification, the crude tetrabromide may be dissolved in 800 ml. of ethyl ether, filtered, and reprecipitated by stirring the solution into 800 ml. of petroleum ether. This solution is cooled below 20° and allowed to stand overnight. Ninety per cent of the precipitate will be deposited in 2 hours. The solid is removed by filtration and is washed on the funnel as described above.

14. The ease with which the product dries and powders is an indication of its purity. Usually the material falls apart to a fine white

powder at the mere touch of the spatula. A glass spatula should be used in the transfer of the crude material before the recrystallization.

15. These figures are based on the use of sunflower-seed oil containing 57% linoleic acid. With other oils, the yield will be proportionally larger or smaller, depending on the analysis of the oil. Approximate values for three oils may be calculated from the values given in Note 3 above. The yield from cotton-seed oil is 26–30 g. The checkers have brominated about nine times the amounts stated in these directions and have obtained proportionate yields.

16. Two isomers, a solid and a liquid, are formed during the bromination.

17. The 4 *N* ethanolic-hydrochloric acid is prepared by passing dry hydrogen chloride gas into slightly warmed absolute ethanol until the calculated increase in weight is obtained.

18. The ester contains about 0.5% of free fatty acid, and for some purposes this is not objectionable, particularly if the acid itself is desired. It is recommended, therefore, that this step be omitted where possible since the yields, particularly from the small batches, may be as low as 80% of the crude ester used. In the absence of an alkaline wash, centrifuging is not necessary, as water alone causes no emulsions to form.

19. Usually three washings are sufficient.

20. A small amount of decolorizing carbon may be added before distillation, but it is usually unnecessary and is to be avoided if possible, for it is likely to cause priming in the still.

21. Ordinary 95% denatured ethanol is satisfactory here.

22. The ester is much more stable toward oxidation than the acid; it is recommended that the material be stored as the ester, and that the acid be prepared only for immediate use. The acid oxidizes to some extent under the best of conditions and, unlike the ester, cannot be distilled without some decomposition. It is for this reason that the constants are determined on the ester rather than on the acid.

3. Methods of Preparation

Ethyl linoleate is prepared by debromination of the tetrabromide by action of zinc, or nascent hydrogen from zinc and glacial acetic acid;[3] by zinc and ethanolic-hydrochloric acid;[2,4] and by zinc and ethanolic-sulfuric acid.[5] The pure acid can be obtained by saponification of the ester, by hydrolysis of the hydroxamic acid,[6] and directly by the action of zinc and pyridine (or quinoline, aniline, piperidine) on tetrabromostearic acid.[7] It has also been prepared from 1-iodohexadecadiene-7,10 by condensation with malonic ester and subsequent hydrolysis.[8]

[1] *Official and Tentative Methods of the American Oil Chemists Society,* revised to Jan. 1, 1941, p. 31, Gillette Publishing Co., 330 S. Wells St., Chicago, Illinois.

[2] McCutcheon, *Can. J. Research,* **B16,** 158–175 (1938).

[3] Erdmann and Bedford, *Ber.,* **42,** 1324 (1909); Erdmann, Bedford, and Raspe, *Ber.,* **42,** 1334 (1909).

[4] Rollett, *Z. physiol. Chem.,* **63,** 410–421 (1909).

[5] Kimura, *Fettchem. Umschau,* **4,** 78 (1935).

[6] Inoue and Yukawa, *J. Agr. Chem. Soc. Japan,* **17,** 771; *Bull. Agr. Chem. Soc. Japan,* **17,** 89 (1941) [*C. A.,* **36,** 4803 (1942)].

[7] Kaufmann and Mestern, *Ber.,* **69,** 2684 (1936).

[8] Baudart, *Peintures, pigments, vernis,* **22,** 375 (1946); [*C. A.,* **41,** 2694 (1947)].

LINOLENIC ACID

$$CH_3CH_2CH{=}CHCH_2CH{=}CHCH_2CH{=}CH(CH_2)_7CO_2H + 3Br_2 \rightarrow CH_3CH_2CHBrCHBrCH_2CHBrCHBrCH_2CHBrCHBr(CH_2)_7CO_2H$$

$$CH_3CH_2CHBrCHBrCH_2CHBrCHBrCH_2CHBrCHBr(CH_2)_7CO_2H + 3Zn \xrightarrow[HCl]{C_2H_5OH} CH_3CH_2CH{=}CHCH_2CH{=}CHCH_2CH{=}CH(CH_2)_7CO_2C_2H_5 + 3ZnBr_2$$

$$CH_3CH_2CH{=}CHCH_2CH{=}CHCH_2CH{=}CH(CH_2)_7CO_2C_2H_5 + NaOH \xrightarrow{H_2SO_4} CH_3CH_2CH{=}CHCH_2CH{=}CHCH_2CH{=}CH(CH_2)_7CO_2H + NaHSO_4 + C_2H_5OH$$

Submitted by J. W. McCutcheon.
Checked by R. L. Shriner and S. P. Rowland.

1. Procedure

A. *Fatty acids.* The procedure outlined for linoleic acid (p. 526) is followed, omitting Notes 3 and 4. Linseed oil is used as the raw material.

B. *Hexabromostearic acid.* In a 4-l. beaker equipped with a mechanical stirrer (Note 1), 90–95 g. of the fatty acids is dissolved in 2.5 l. of ethyl ether, and the solution is chilled to 0–10°. Then, with stirring, 35 ml. of bromine (Notes 2 and 3) is introduced slowly, from a dropping funnel (Note 4), at such a rate that the temperature does not exceed 20°; about 50 minutes is usually required. The solution, which must contain an excess of bromine (deep red color), is allowed to stand in an ice bath at 0–10° overnight. The excess bromine is removed by addition of a small amount of amylene, and then the white precipitate is collected with suction on a 12.5-cm. Büchner funnel and is washed with 200 ml. of ethyl ether. The hexabromide is thoroughly stirred with 300 ml. of ethyl ether in a 600-ml. beaker and is filtered as before. The yield is 20–22 g. of gray-white hexabromostearic acid (Note 5). It is recommended that 100 g. of the crude hexabromide (Note 6) be accumulated before proceeding to the next step.

In a 1-l. beaker 100 g. of crude hexabromostearic acid is heated to 70–80° with 600 ml. of dioxane (Note 7). The mixture is filtered, and the filtrate is set aside at 15–20° for several hours or preferably overnight. The precipitate is collected on a Büchner funnel with gentle suction and is washed thoroughly on the funnel with several 300-ml. portions of ethyl ether. The hexabromide is then transferred to a glass plate and spread out with a glass spatula to dry in the air. The fine snow-white crystals weigh 80 g. and melt at 181.5–181.9° (Note 8).

C. *Ethyl linolenate.* In a 1-l. Erlenmeyer flask 80 g. of the pure hexabromide is dissolved in 190 ml. of absolute ethanol, and 40 g. of granulated zinc (20 mesh) is added. The mixture is refluxed for 1 hour, at the end of which time the solution should be clear. About 10 g. of zinc dust is added, refluxing is continued for a short time, and then 10 ml. of 4 N ethanol-hydrochloric acid (p. 530) is added. Refluxing is continued for 6 hours, and during this time 5 ml. of the ethanolic hydrochloric acid is added every 30 minutes. At the end of 6 hours, the solution is removed from the unchanged zinc by decantation into another flask, and the zinc is washed with 15 ml. of absolute ethanol to complete the transfer of the solution. A 10-ml. portion of ethanolic hydrochloric acid is added, and the solution is refluxed for a period of 3 hours, with addition of 5-ml. portions of ethanolic hydrochloric acid at 30-minute intervals. The mixture is poured into 500 ml. of hot saturated brine solution in a 1-l. separatory funnel, and the crude ester is allowed to settle for 10–20 minutes. The brine is removed, and the ester is washed (Note 9) at room temperature with 500 ml. of 0.5% sodium carbonate solution. The rather stable emulsion is broken by centrifuging the mixture for 5 minutes at 3300 r.p.m. The ester and any remaining emulsion are washed with 150 ml. of warm water, and the emulsion is broken by centrifuging. Washing (Note 10) is continued until the wash water is neutral to methyl orange. The neutral ester is distilled under reduced pressure. It boils at 174°/2.5 mm. and 198°/6.5 mm.

D. *Linolenic acid.* The ester is dissolved in 400 ml. of a 5% ethanolic solution of sodium hydroxide in a 1-l. beaker, and the solution is allowed to stand overnight at room temperature. The resulting jelly is dissolved in 400 ml. of warm water, and a slow stream of carbon dioxide is introduced beneath the surface of the liquid while it is acidified with 50 ml. of dilute sulfuric acid (1:1 by volume). The stream of carbon dioxide is maintained throughout the subsequent operations. The linolenic acid rises to the surface as a clear layer, which is washed once with hot water. The acid is dried over anhydrous sodium sulfate and is preserved under carbon dioxide. The yield of acid is 22–24 g., and the product has a melting point of −17° to −16° (Note 11).

2. Notes

1. In order to guard against fire hazard it is advisable to use hand stirring when proper safety equipment is not available.

2. Care must be observed that the stopcock does not slip out of place.

3. The weight of bromine is independent of the amount (per cent) of linolenic acid in the sample, but it varies with the iodine value. For the purposes of this experiment the amount of bromine (grams) may be calculated as 0.7 times the iodine value of the oil. This allows an excess of 10–11%.

4. The tip of the dropping funnel should be just above the fatty acid solution in order to avoid plugging the exit.

5. The yield is based on a linseed oil containing approximately 45% of linolenic acid. Because of the formation of stereoisomers, only 25% of the bromoacid is precipitated in the solid form. By using water-white distilled linseed fatty acids, obtained from Archer-Daniels-Midland Company, Minneapolis, Minnesota, the yield of solid bromoacid may be increased to 48–53 g.

6. The theoretical yield of the ethyl ester from 20 g. of pure hexabromide is 8 g. Losses through recrystallization, distillation, etc., amount to approximately 50%; hence the actual yield is about 4 g. of ester.

7. The by-product, insoluble in dioxane, is heptabromostearic acid. If the solution is yellow at this point, a small amount of decolorizing charcoal should be added.

8. The melting point varies with the rate of heating. Any uniform method of taking the melting point may be adopted. It has been found convenient to insert the capillary tube in the bath when the

temperature is 5° below the expected melting point, and then to raise the temperature 1° per minute.

9. The ester contains about 0.5% of free fatty acid, and for some purposes this is not objectionable, particularly if the acid itself is desired. It is recommended, therefore, that this step be omitted where possible, since the yields, particularly from small batches, may be as low as 80% of the crude ester used. In the absence of an alkaline wash centrifuging is not necessary, as water alone causes no emulsions to form.

10. Usually three washings are sufficient.

11. The ester is much more stable toward oxidation than the acid; it is recommended that the material be stored as the ester, and that the acid be prepared only for immediate use. The acid oxidizes to some extent under the best of conditions and, unlike the ester, cannot be distilled without some decomposition. It is for this reason that the constants are determined on the ester rather than on the acid.

3. Methods of Preparation

Linolenic acid is always obtained from natural sources, chiefly from the oils of various seeds, such as hemp seed,[1] walnut,[2] poppy seed,[2] cotton seed,[2] and, best of all, linseed.[3,4] The crude acid has been purified via the solid hexabromide, either directly,[5–7] or by the hydrolysis of the methyl or ethyl ester, obtained by simultaneous debromination and esterification of hexabromostearic acid.[5,8] The method described above is a modification [9,10] of the procedure of Rollett.[8] The acid has also been obtained from soy oil and has been purified via the hydroxamic acid.[11]

[1] Hazura, *Monatsh.*, **8,** 268 (1887).
[2] Hazura and Grüssner, *Monatsh.*, **9,** 204 (1888).
[3] Hazura, *Monatsh.*, **8,** 158 (1887); Erdmann and Bedford, *Ber.*, **42,** 1328 (1909).
[4] Erdmann, *Z. physiol. Chem.*, **74,** 180 (1911).
[5] Erdmann and Bedford, *Ber.*, **42,** 1330 (1909).
[6] Hazura, *Monatsh.*, **8,** 267 (1887).
[7] Kaufmann and Mestern, *Ber.*, **69,** 2684 (1936).
[8] Rollett, *Z. physiol. Chem.*, **62,** 422 (1909).
[9] McCutcheon, *Can. J. Research,* **B16,** 158 (1938).
[10] McCutcheon, *Can. J. Research,* **B18,** 231 (1940).
[11] Inoue and Yukawa, *J. Agr. Chem. Soc. Japan,* **17,** 771; *Bull. Agr. Chem. Soc. Japan,* **17,** 89 (1941) [*C. A.*, **36,** 4803 (1942)].

MALONONITRILE *

$$2NCCH_2CONH_2 + POCl_3 \rightarrow 2CH_2(CN)_2 + HPO_3 + 3HCl$$

Submitted by Alexander R. Surrey.
Checked by C. F. H. Allen and J. VanAllan.

1. Procedure

In a 12-l. three-necked round-bottomed flask, fitted with a powerful stirrer (Note 1) and a reflux condenser, are placed 1260 g. (15 moles) of cyanoacetamide, 1 kg. of salt (Note 2), and 5 l. of ethylene dichloride. After the mixture has been stirred rapidly for 15 minutes, 800 ml. (8.75 moles) of phosphorus oxychloride is added and the mixture is refluxed for 8 hours in an oil bath (Note 3). After the mixture has been cooled to room temperature, it is filtered and the solid is washed with 500 ml. of ethylene dichloride. The solvent is distilled from the combined filtrates in a 12-l. flask, and the residual crude nitrile is decanted into a 1-l. flask (Note 4) from any solid that may have separated. A fractionating column, condenser, and fractionating receiver are attached, and the malononitrile is distilled under reduced pressure. The fraction boiling at 113–118°/25 mm. weighs 570–654 g. (57–66%) (Note 5). This material may be freed from a small amount of phosphorus oxychloride which is present by redistillation. After a fore-run of about 5 ml., the malononitrile distils smoothly at 92–94°/8 mm.

2. Notes

1. A heavy motor-driven stirrer is advisable, because frequently the solid that separates contains some viscous material that makes stirring difficult.

2. The addition of salt gives a lighter-colored, granular solid that can be easily removed by filtration and washed.

3. The reaction can be done in a hood, or the hydrogen chloride that is evolved can be absorbed in a gas trap.

4. Ground-glass equipment is advisable. The fractionating column is the standard modified Claisen type.

5. The submitter obtained a 70–72% yield at this point in a run of this size and 80% yields in small runs. He also obtained yields of the same order by using a mixture of phosphorus oxychloride and a small amount of phosphorus pentachloride.

3. Methods of Preparation

In addition to the methods cited previously,[1] malononitrile has been prepared by the dehydration of cyanoacetamide by phosphorus oxychloride either with salt,[2] as in the above procedure, or with sodium metabisulfite.[3] The vapor-phase reaction of cyanogen chloride and acetonitrile has been patented.[4]

[1] *Org. Syntheses* Coll. Vol. **2**, 379 (1943).

[2] Surrey, *J. Am. Chem. Soc.*, **65**, 2471 (1943); U. S. pat. 2,389,217 [*C. A.*, **40**, 900 (1946)].

[3] U. S. pat. 2,459,128 [*C. A.*, **43**, 3470 (1949)].

[4] U. S. pat. 2,553,406 [*C. A.*, **45**, 9081 (1951)].

MANDELAMIDE

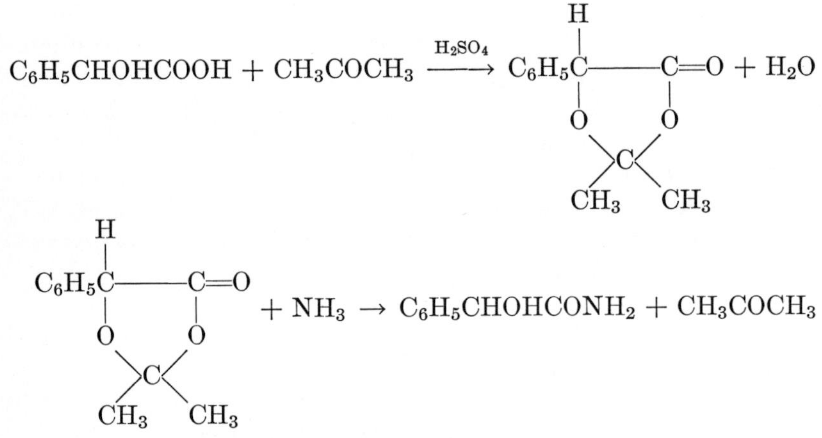

Submitted by L. F. Audrieth and M. Sveda.
Checked by Homer Adkins and William H. Bateman.

1. Procedure

One hundred and forty-six grams (0.96 mole) of mandelic acid is dissolved in 440 ml. (6.2 moles) of acetone, and the resulting solution is placed in a 2-l. three-necked flask fitted with an efficient stirrer, a dropping funnel, and a thermometer. The reaction flask is placed in an ice-salt bath, and 98 g. of concentrated sulfuric acid (sp. gr. 1.84) is added through the dropping funnel at such a rate that the temperature does not exceed −10°. The reaction mixture is then poured into an ice-cold solution of sodium carbonate containing 200 g. of the anhydrous salt in 1800 ml. of water. The mandelic acid-acetone condensation product precipitates from the solution. The curdy product is washed by grinding with ice water (200 ml.) and is then filtered and dried over calcium chloride under reduced pressure. The crude product weighs 181 g. (Note 1).

The mandelic acid-acetone condensation product is added in small portions to about 1.8 l. of liquid ammonia (Note 2) contained in two silvered 1-l. Dewar flasks. Each flask is fitted with a stopper containing a capillary tube to serve as an ammonia outlet. The ammonolysis is allowed to proceed overnight, after which the contents of the flasks are poured into open beakers to facilitate rapid evaporation of the liquid ammonia.

After the ammonia has been removed to a point where a pulverulent mass remains, the product is treated with 475 ml. of hot absolute ethanol and the resulting solution is filtered through a hot funnel to remove insoluble impurities. The filtrate is cooled in an ice bath to give about 90 g. of glistening white crystals of mandelamide melting at 132° (62% based upon the mandelic acid) (Notes 3 and 4).

2. Notes

1. The crude product contains varying quantities of sodium carbonate and sodium sulfate, which are difficult to remove. These impurities are insoluble in liquid ammonia; consequently the crude compound can be ammonolyzed without further purification. The mandelic acid-acetone condensation product may be purified by recrystallization from absolute ethanol; it then melts at 45°.

2. The solubility of the mandelic acid-acetone condensation compound in liquid ammonia at its boiling point is approximately 10 g. per 100 ml.

3. Additional quantities of mandelamide may be obtained by concentrating the ethanolic mother liquor carefully.

4. This method is generally applicable to the preparation of the amides of α-hydroxy acids.

3. Methods of Preparation

Mandelamide has been prepared by treating the ethyl ester with concentrated aqueous ammonia,[1,2] and a saturated ethanolic solution of ammonia has been used to effect ammonolysis of the methyl ester.[3] Esters of mandelic acid were treated with liquid ammonia at its boiling point; [4] this procedure was improved by use of ammonia at superatmospheric pressures and higher temperatures.[5] The procedure described in this synthesis was first used by Ôeda.[6]

1 Beyer, *J, prakt. Chem.*, (2) **31,** 390 (1885).
2 Einhorn and Feibelmann, *Ann.*, **361,** 145 (1908).
3 McKenzie and Wren, *J. Chem. Soc.*, **93,** 311 (1908).
4 Glattfeld and MacMillan, *J. Am. Chem. Soc.*, **58,** 898 (1936).
5 Kleinberg and Audrieth, *J. Org. Chem.*, **3,** 312 (1938).
6 Ôeda, *Bull. Chem. Soc. Japan,* **11,** 385 (1936).

MANDELIC ACID *

$$C_6H_5COCH_3 + 2Cl_2 \rightarrow C_6H_5COCHCl_2 + 2HCl$$

$$C_6H_5COCHCl_2 + 3NaOH \rightarrow C_6H_5CHOHCO_2Na + 2NaCl + H_2O$$

$$C_6H_5CHOHCO_2Na + HCl \rightarrow C_6H_5CHOHCO_2H + NaCl$$

Submitted by J. G. Aston, J. D. Newkirk, D. M. Jenkins, and Julian Dorsky.
Checked by R. L. Shriner and C. H. Tilford.

1. Procedure

A. *Dichloroacetophenone.* A 3-l. round-bottomed flask is fitted with a three-holed rubber stopper through which are passed an inlet tube extending to the bottom of the flask, an outlet tube, and a thermometer. The inlet tube is connected to a cylinder of chlorine through a bubble counter consisting of a 500-ml. wash bottle which contains about 200 ml. of concentrated sulfuric acid. The outlet tube is connected with a gas trap in which the evolved hydrochloric acid is absorbed by running water. It is best to set up the apparatus under a good hood.

In the flask are placed 240 g. (2 moles) of acetophenone and 1 l. of glacial acetic acid. The thermometer is adjusted so that it extends considerably below the surface of the solution, and chlorine is admitted at such a rate that the temperature does not exceed 60° (Note 1). Chlorination is continued until an excess of the halogen has been absorbed. This requires about 5 hours; completion of the reaction is indicated by the development of a yellow color. The reaction mixture is poured over 6 l. of crushed ice in a 2-gal. jar. The mixture is stirred several times (Note 2) and allowed to stand until the ice has melted. The dichloroacetophenone, which separates as a heavy lachrymatory oil, is removed. The yield is 340–370 g. (90–97%). This product, containing only a few per cent of water and acetic acid, is pure enough for the preparation of mandelic acid. It may be purified by adding about 100 ml. of benzene, removing the water and benzene by distillation, and fractionally distilling the residual oil under reduced pressure. There is obtained 302–356 g. (80–94%) of a colorless oil boiling at 132–134°/13 mm. (142–144°/25 mm.).

B. *Mandelic acid.* In a 2-l. three-necked round-bottomed flask, fitted with an efficient mechanical stirrer, a dropping funnel, and a thermometer, is placed 156 g. (3.9 moles) of sodium hydroxide dissolved in 1.4 l. of water. The solution is warmed to 60° (Note 3), vigorous stirring is begun, and 200 g. (1.06 moles) of dichloroacetophenone (either crude or distilled) is added from the dropping funnel. The dichloroacetophenone is added slowly at first so that the temperature does not exceed 65°. The addition requires about 2 hours (Note 4). Stirring is continued for 1 hour longer while the temperature is maintained at 65° by means of a water or steam bath. After addition of 170 ml. of 12 *M* hydrochloric acid (Note 5), the solution is extracted with ether. The continuous extractor shown in Fig. 18 is very useful for this purpose. About 250–300 ml. of ether is used, and the extraction is continued until no more material is obtained. With this apparatus, about 130 g. of crude mandelic acid is recovered after 24 hours, and 150 g. after 48 hours. The volume of liquid in the larger flask must be great enough so that, when 250–300 ml. of ether is used for the extraction, there will be a continuous overflow from the larger to the smaller flask; the latter serves as the "boiler."

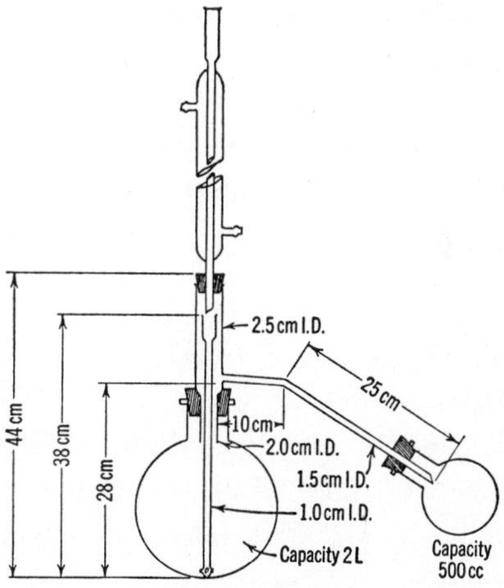

Fig. 18.

The ether extracts are transferred to a 1-l. round-bottomed flask, the ether is removed by distillation, and the residue of crude mandelic acid is dried by warming it on a steam bath under the vacuum of a water pump. About 400 ml. of benzene is added, and the mixture is distilled until 100 ml. of distillate is collected. The acid in the residual mixture is brought completely into solution by the addition of 6–10 ml. of ethanol. The hot solution is then filtered through a warm Büchner funnel, and the filtrate is cooled overnight at 6°. The first crop of pure mandelic acid weighs 100–120 g. A second crop is obtained by evaporation of the mother liquor to about one-fourth its volume; this weighs 20–40 g. The total yield is 136–144 g. (85–90% based on the dichloroacetophenone, or 76–87% based on the acetophenone). The white crystalline product melts at 115–117° (Note 6).

2. Notes

1. According to Beilstein (**VII,** 283) trichloroacetophenone is obtained by chlorination of acetophenone at elevated temperatures.[2] However, at 60° less than 1% of trichloroacetophenone is formed.

2. Stirring results in a better separation of acetic acid from the oil and prevents the formation of an emulsion.

3. At lower temperatures, hydrolysis is slow.

4. If the temperature becomes much higher, side reactions occur and loss in yield and purity results. The heat of reaction is sufficient to maintain the required temperature.

5. If the reaction mixture is cooled at this point, some of the mandelic acid may crystallize. If this happens, the precipitate should not be filtered, as it is contaminated with sodium chloride.

6. The submitters report that when 1 kg. of acetophenone is used yields of 0.98–1.06 kg. of mandelic acid are obtained.

3. Methods of Preparation

Dichloroacetophenone has been prepared by chlorination of acetophenone with and without aluminum chloride;[1] by action of dichloroacetyl chloride upon benzene and aluminum chloride;[1] by action of hypochlorous acid upon phenylacetylene;[2] by heating trichloromethylphenylcarbinol;[3] and by chlorination of phenylacetylene in alcohol.[4]

Mandelic acid has been prepared by hydrolysis of mandelonitrile (prepared in turn from benzaldehyde and hydrogen cyanide or from benzaldehyde, sodium bisulfite, and sodium cyanide);[5] by action of water at 180° upon trichloromethylphenylcarbinol;[6] by action of potassium carbonate upon a heated mixture of benzaldehyde and chloroform;[7] by action of warm, dilute alkali upon dibromoacetophenone;[8] by the action of warm, dilute sodium hydroxide upon phenylglyoxal;[9] by the hydrolysis of α-bromophenylacetic acid[10] or dimethyl α-cyano-α-hydroxybenzylphosphonate;[11] and by the hydrolysis of ethyl mandelate, prepared in turn by the catalytic reduction of phenylglyoxalate.[12]

α-C^{14}-Mandelic acid has been prepared from α-C^{14}-phenylglyoxal, and from α-C^{14}-α,α-dibromoacetophenone.[13] α-C^{13}-Mandelic acid has been prepared from α-C^{13}-α,α-dibromoacetophenone.[14]

[1] Gautier, *Ann. chim. phys.*, (6) **14,** 348, 379, 385, 396 (1888).

[2] Wittorf, *J. Russ. Phys. Chem. Soc.*, **32,** 88 (1900) [*Chem. Zentr.*, **1900,** II, 29].

[3] Kötz, *J. prakt. Chem.*, (2) **90,** 299, 304 (1914).

[4] Jackson, *J. Am. Chem. Soc.*, **56,** 977 (1934).

[5] *Org. Syntheses* Coll. Vol. **1,** 336 (1941); Spiegel, *Ber.*, **14,** 239 (1881); Ultee, *Rec. trav. chim.*, **28,** 254 (1909); Müller, *Ber.*, **4,** 980 (1871).

[6] Jocicz, *J. Russ. Phys. Chem. Soc.*, **29,** 100 (1897) [*Chem. Zentr.*, **1897,** I, 1013].

[7] Savariau, *Compt. rend.*, **146,** 297 (1908).

[8] Engler and Wöhrle, *Ber.*, **20,** 2202 (1887).

[9] Pechmann, *Ber.*, **20,** 2905 (1887).

[10] Fredga, *Arkiv Kemi, Mineral. Geol.*, **24B** #15 (1947).

[11] Kabachnik, Rossiisskaya, and Shepeleva, *Bull. acad. sci. U.R.S.S., Classe sci. chim.*, **1947,** 163 [*C. A.*, **42,** 4133 (1948)].

[12] Kindler, Metzendorf, and Dschi-yin-Kwok, *Ber.*, **76B,** 308 (1943).

[13] Neville, *J. Am. Chem. Soc.*, **70,** 3499 (1948).

[14] Doering, Taylor, and Schoenewaldt, *J. Am. Chem. Soc.*, **70,** 455 (1948).

D-MANNOSE *

$$C_6H_{11}O_5OCH_3 + H_2O \rightarrow C_6H_{12}O_6 + CH_3OH$$

Submitted by J. T. SHEEHAN and W. FREUDENBERG.
Checked by HOMER ADKINS, WINSTON WAYNE, and RICHARD JUDAY.

1. Procedure

A solution of 200 g. (1.03 moles) of α-methyl-*d*-mannoside [*Org. Syntheses* Coll. Vol. **1,** 362 (1932), 371 (1941)] in 3.2 l. of 2 *N* sulfuric acid (178 ml. concentrated sulfuric acid diluted to 3.2 l.) is steam-distilled from a 5-l. flask for 1 hour. The flask is heated externally during this time so that the volume of the solution remains constant. The mixture is transferred to a 2- or 3-gal. crock, 10 g. of decolorizing charcoal (Norit) is added, and the solution is rapidly stirred while it is neutralized (litmus) by the addition of barium carbonate (Note 1).

The mixture is filtered with suction through an asbestos mat supported on a filter paper in a 25-cm. Büchner funnel, and the barium sulfate is washed with two 100-ml. portions of water. The filtrate and washings are combined, and the clear colorless solution of the sugar is concentrated to about 300 ml. under reduced pressure (water pump) on a water bath whose temperature does not exceed 60°. The mixture is filtered, and the solid is washed on the filter with a little water. The combined filtrate and washings are then concentrated to a volume of 100 ml. under the conditions described above. The warm, slightly tan-colored syrup is poured into a 500-ml. Erlenmeyer flask. The transfer is completed by rinsing the distilling flask five times with 40-ml. portions of glacial acetic acid, each portion of the acetic acid being warmed to 50° in the bath before it is transferred. The acetic acid solution of the sugar is seeded with crystalline D-mannose (Note 2), and, after standing overnight at room temperature, the solution is placed in a refrigerator for a day or two.

The crystalline D-mannose is filtered and washed successively with two 20-ml. portions of cold glacial acetic acid, two 20-ml. portions of cold dry ethanol, and three 30-ml. portions of cold, dry ether (Note 3). After drying in the air for 3 days, the product weighs 112 to 115 g. (60%) (Note 4) and melts at 126.5–127.5° (Note 5).

2. Notes

1. The calculated quantity of pure barium carbonate required to neutralize the sulfuric acid is 630 g. However, the checkers found that the mixture was still acid after stirring it for 2–3 hours with 780 g. of carbonate. An additional 100 g. of carbonate was required to bring the mixture to the neutral point. The checkers found that it was preferable to add 975 g. of barium hydroxide [$Ba(OH)_2 \cdot 8H_2O$], stir for 30 minutes, and then to complete the neutralization by stirring with 75 g. of barium carbonate.

2. Crystallization may be induced by scratching with a glass rod if crystalline D-mannose is not available.

3. The sugar should be ground with each portion of the wash liquid, if the maximum purification with the minimum amount of solvent is to be obtained.

4. The submitters obtained 18–19 g. of crystalline D-mannose from the mother liquors, which increased the yield to about 70% of the theoretical amount.

5. The sugar so prepared is said to be predominately α-D-mannose. If a purer product is desired, the substance may be recrystallized from 80% ethanol. This procedure gives a product which melts at 131–132°; α_D^{23} +15°. The recovery in this crystallization is about 80%.

3. Methods of Preparation

D-Mannose has been prepared by hydrolysis of the mannan in vegetable ivory,[1-3] but the crystalline sugar is apparently best obtained from mannan through the isolation of α-methyl-*d*-mannoside.[4] The method described above is similar to that of Hudson and Jackson [4] except that sulfuric acid, rather than hydrochloric acid, is used for hydrolysis of the α-methyl-*d*-mannoside. Levene [5] has described what is probably the best method for the conversion of mannan to D-mannose without the isolation of α-methyl mannoside. The D-mannose obtained by Levene was predominately the β-isomer.

The electrolytic reduction of D-mannonic lactone to D-mannose has been described.[6]

[1] Hudson and Sawyer, *J. Am. Chem. Soc.*, **39**, 470 (1917).
[2] Narayanan, *Indian J. Med. Research,* **29**, 1 (1941).
[3] Isbell, *J. Research Natl. Bur. Standards,* **26**, 47 (1941).
[4] Hudson and Jackson, *J. Am. Chem. Soc.,* **56**, 958 (1934).
[5] Levene, *J. Biol. Chem.,* **108**, 419 (1935).
[6] Sato, *J. Chem. Soc. Japan, Pure Chem. Sect.,* **71**, 194 (1950).

l-MENTHOXYACETIC ACID

(Acetic acid, (menthyloxy)-, *l*-)

$$2C_{10}H_{19}OH + 2Na \rightarrow 2C_{10}H_{19}ONa + H_2$$

$$2C_{10}H_{19}ONa + ClCH_2CO_2H \rightarrow C_{10}H_{19}OCH_2CO_2Na + C_{10}H_{19}OH + NaCl$$

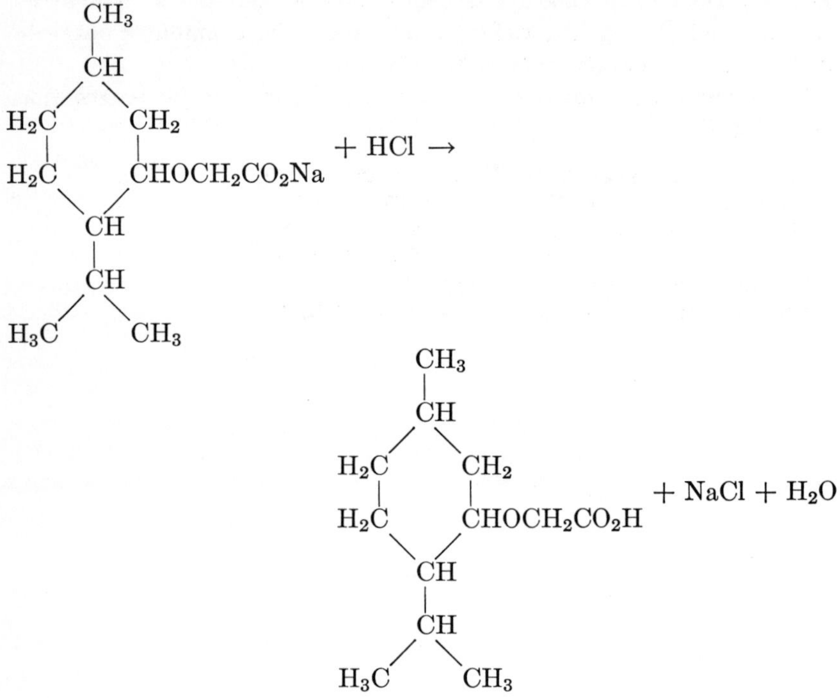

Submitted by M. T. LEFFLER and A. E. CALKINS.
Checked by R. L. SHRINER and C. H. TILFORD.

1. Procedure

A 5-l. three-necked round-bottomed flask is fitted with a mechanical stirrer (Note 1) and a reflux condenser bearing a calcium chloride tube. A solution of 400 g. (2.56 moles) of *l*-menthol (crystals, m.p. 41–42°) in 1 l. of dry toluene (Note 2) is placed in the flask, and to it is added 70 g. (3.04 gram atoms) of clean sodium. The third neck of the flask is then tightly closed with a cork stopper, and the flask is heated in an oil bath until the toluene refluxes gently. As soon as the sodium has melted, stirring is begun and is maintained at such a rate that the sodium is broken into fine globules.

After the reaction mixture has been refluxed for 15 hours, the stirrer is stopped, the reaction mixture is allowed to cool, and the excess sodium (Note 3) is carefully removed. The apparatus is then assembled as before (Note 1), but with a 1-l. separatory funnel fitted into the third neck of the flask (Note 4). The temperature of the oil bath is raised to 85–90°, and with continued stirring, a solution of 95 g. (1.01 moles) of monochloroacetic acid (Note 5) in 800 ml. of warm dry toluene is added from the separatory funnel at such a rate that refluxing is not too vigorous. A heavy precipitate of sodium chloroacetate forms immediately. After all the chloroacetic acid has been added, the mixture is refluxed and stirred for 48 hours. During this period, the stirring must be as thorough as possible; it is necessary to add 1–1.5 l. of dry toluene, and the stirrer must be stopped at frequent intervals while the solid material is removed from the side of the flask.

When the reaction is complete, the flask is removed from the oil bath and the cooled reaction mixture is transferred to a 5-l. separatory funnel and extracted with three 1-l. portions of water (Note 6). The water extract is carefully acidified with 20% hydrochloric acid, and the crude menthoxyacetic acid, which collects on top as a brown oil, is extracted with three 200-ml. portions of benzene. The benzene extracts are combined, and the solvent is removed by distillation on a steam cone. The residue is then fractionally distilled under reduced pressure. The fraction boiling below 100°/20 mm. is mainly water and toluene. The second fraction boiling at 100–115°/8–10 mm. is impure *l*-menthol and may be saved for redistillation (Note 6). The yield of *l*-menthoxyacetic acid, boiling at 134–137°/2 mm. (150–155°/4 mm.), $[\alpha]^{2}_{D}$ −92.4°, amounts to 166–180 g. (78–84%).

2. Notes

1. The success of the reaction depends largely on the type of agitation used. A stainless-steel stirrer of the anchor type with a gas-tight rubber or metal-graphite bearing serves well because of its strength. It should be operated at high speed in the first part of the reaction in order to powder the sodium; in the second stage vigorous stirring is not so essential, but the solid and liquid phases should be mixed efficiently.

2. The toluene may be dried by refluxing it over metallic sodium for several hours. It is distilled from the sodium into a receiver protected from the air by a calcium chloride tube.

3. The excess sodium should amount to about 11 g. If the sodium remains divided, it is removed by filtering the hot mixture through glass wool.

4. It is not advisable to leave the separatory funnel attached to the flask during the long period of stirring which follows, as the constant, heavy vibration tends to loosen the connection. As soon as it has been used, the funnel should be replaced by a tightly fitting cork stopper.

5. The commercial grade (m.p. 61–63°) of monochloroacetic acid should be ground and thoroughly dried over concentrated sulfuric acid in a vacuum desiccator for 2 days. The yield was lowered by about 10% when the chloroacetic acid was used without previous drying.

6. The toluene layer, containing the menthol formed in the reaction, is saved for the recovery of both the toluene and the menthol, which are separated by distillation under atmospheric pressure. The *l*-menthol collected at 210–212° (cor.) amounts to 150–225 g.

3. Method of Preparation

l-Menthoxyacetic acid has been prepared only by the action of sodium menthoxide upon monochloroacetic acid.[1]

[1] Frankland and O'Sullivan, *J. Chem. Soc.*, **99**, 2325 (1911); Rule and Tod, *J. Chem. Soc.*, **1931**, 1929.

l-MENTHOXYACETYL CHLORIDE

(Acetyl chloride, (menthyloxy)-, *l*-)

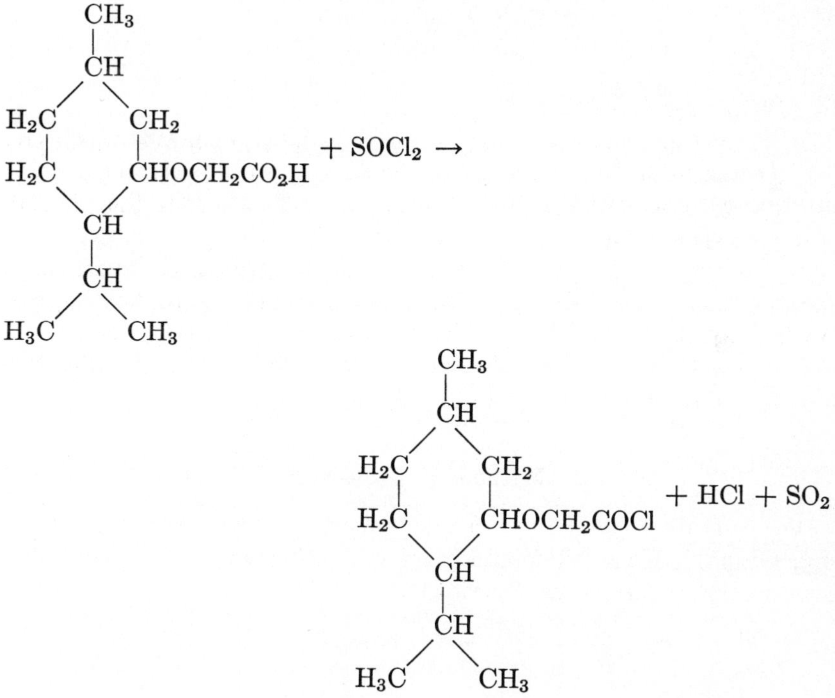

Submitted by M. T. LEFFLER and A. E. CALKINS.
Checked by R. L. SHRINER and C. H. TILFORD.

1. Procedure

A 1-l. three-necked round-bottomed flask is mounted on a steam cone and is fitted with a 250-ml. separatory funnel and a reflux condenser connected to a trap (Note 1) for absorbing gases. In the flask is placed 325 g. (198 ml., 2.73 moles) of thionyl chloride (Note 2), and to it is added, during the course of 1 hour, 125 g. (0.58 mole) of *l*-menthoxyacetic acid (p. 544). The flask is shaken frequently during the addition of the acid and, if necessary, is warmed to start the reaction. When all the acid has been added (Note 3), the reaction mixture is refluxed gently for 5 hours. After the reaction is complete, the excess of thionyl chloride is removed by distillation on the steam bath

(Note 4) and the residue is distilled under reduced pressure. The yield of *l*-menthoxyacetyl chloride boiling at 117–120°/3 mm. (120–125°/5 mm.), $[\alpha]_D^{25}$ −89.6°, amounts to 115–118 g. (Note 5) (85–87%). The product turns dark on standing; it should be stored in a glass-stoppered amber bottle.

2. Notes

1. The gas-absorption trap shown in *Org. Syntheses* Coll. Vol. **1**, 97 (1941) may be used.

2. The commercial grade (b.p. 74–78°) of thionyl chloride was used.

3. Owing to the high viscosity of the acid, it is desirable to rinse the separatory funnel with a little thionyl chloride which is then added to the reaction mixture.

4. The recovered thionyl chloride may be redistilled for future runs. It is best to remove the last traces of thionyl chloride by heating the crude product to about 140° under the vacuum of the water pump.

5. The submitters report the same yields (per cent) when twice the amounts of materials are used.

3. Method of Preparation

The procedure given above is adapted from that described by Read and Grubb.[1] No other methods for the preparation of *l*-menthoxyacetyl chloride have been described.

[1] Read and Grubb, *J. Soc. Chem. Ind.*, **51**, 330T (1932).

MESITALDEHYDE

(Benzaldehyde, 2,4,6-trimethyl-)

I. METHOD A

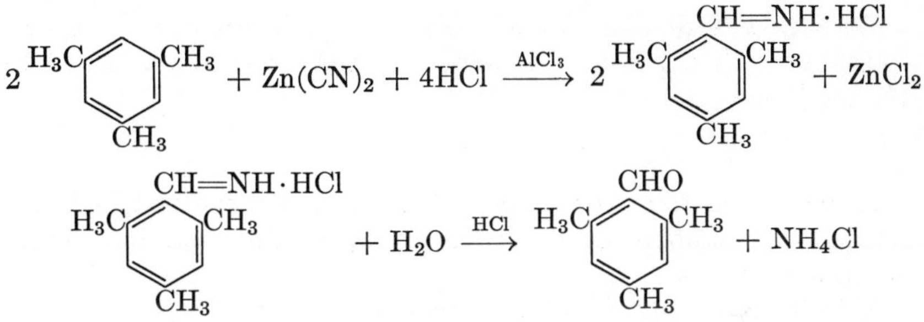

Submitted by R. C. FUSON, E. C. HORNING, S. P. ROWLAND, and M. L. WARD.
Checked by C. F. H. ALLEN and J. VANALLAN.

1. Procedure

In a 1-l. three-necked round-bottomed flask, fitted with an efficient stirrer, a reflux condenser, an inlet tube, and a thermometer (Notes 1 and 2), are placed 102 g. (118 ml., 0.85 mole) of mesitylene, 147 g. (1.25 moles) of zinc cyanide (Note 3), and 400 ml. of tetrachloroethane (Note 4). The inlet tube is connected to a source of hydrogen chloride (Note 5), and the mixture is stirred at room temperature while a rapid stream of dry hydrogen chloride is passed through it. This is continued until the zinc cyanide is decomposed; usually about 3 hours is required. The flask is then immersed in an ice bath, the inlet tube is removed, and 293 g. (2.2 moles) of finely ground anhydrous aluminum chloride is added to the mixture (Notes 6 and 7), with *very vigorous* stirring. The ice bath is then removed, and the passage of hydrogen chloride is resumed for the remainder of the reaction period. The heat of reaction is sufficient to warm the mixture slowly, and a temperature of about 70° is reached at the end of an hour. A temperature of 67–72° is maintained for an additional 2.5 hours. The cooled mixture is decomposed by pouring it cautiously, with stirring by hand, into a 4-l. container about half full of crushed ice, to which has been added 100 ml. of concentrated hydrochloric acid. After the mixture has stood overnight, it is transferred to a 3-l. round-bottomed flask and refluxed for 3 hours. The organic layer is then separated, and the aqueous layer is extracted once with 50 ml. of tetrachloroethane. The combined tetrachloroethane solutions are washed with 150 ml. of a 10% solution of sodium carbonate and distilled with steam. The first 800–900 ml. of distillate is set aside for recovery of the solvent (Notes 8 and 9), and the second portion is collected as long as oily drops are observed (Note 10). This distillate is extracted with 500 ml. of benzene, the solvent is removed on the steam bath, and the residue is distilled from a 250-ml. modified Claisen flask. After a small fore-run, the mesitaldehyde distils at 118–121°/16 mm. The yield is 95–102 g. (75–81%) (Notes 11 and 12).

2. Notes

1. Because of the toxic nature of tetrachloroethane and hydrogen cyanide, *all operations* as far as the final distillation should be carried out in a *good* hood.

2. It is convenient to place the thermometer in the gas inlet tube. The bulb should be immersed in the liquid, but the inlet tube need extend only just below the liquid surface.

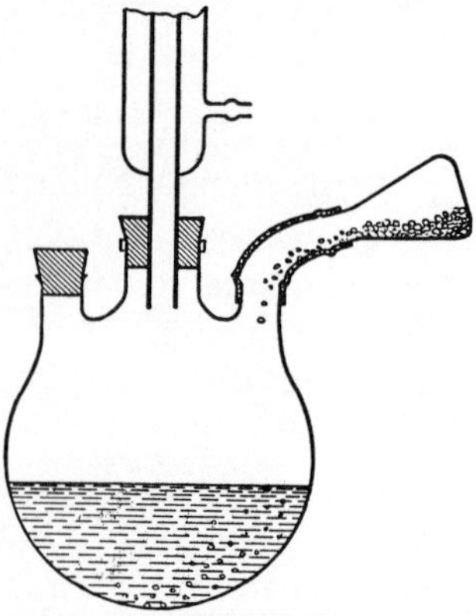

FIG. 19.

3. The zinc cyanide may be commercial material, or it may be prepared as directed by Adams and Levine.[1] However, the zinc cyanide does not react well if it is too carefully purified.[2]

4. Tetrachloroethane is a toxic substance; it should be handled with due care.

5. Commercial cylinders of hydrogen chloride, now available, are most convenient.

6. Although it is more convenient to add the zinc cyanide and aluminum chloride together, the procedure results in lower yields.

7. The apparatus shown in Fig. 19 is most convenient. A 250-ml. Erlenmeyer flask is connected to the side neck of the flask by a 13-cm. length of 20-mm. thin-walled rubber tubing.[3]

8. If difficulty is experienced in separating the organic layer, the entire solution may be subjected to steam distillation.

9. The first portion of the distillate consists almost entirely of tetrachloroethane and water. The solvent may be recovered by separating the organic layer, drying it with calcium chloride, and distilling.

10. About 9 l. of water is obtained; the time required is about 4 hours.

11. If a smaller yield of mesitaldehyde is acceptable and time is of importance, the preparation may be carried out without a solvent, and with other changes as follows: The zinc cyanide and aluminum chloride are mixed by shaking, the mesitylene is added, and the flask is immersed in an oil bath at 100°. The stirrer is started, and a fairly rapid current of dry hydrogen chloride is passed into the mixture, below the surface of the liquid, for 4 hours; at the end of this time, the current of gas is discontinued, but stirring and heating are maintained for 2 hours longer. The reaction mixture is decomposed and processed as already described. The fraction which boils at 110–120°/9–10 mm. is taken as mesitaldehyde; the yield is 73% (private communication, D. B. Glass).

12. 2,4,6-Triethylbenzaldehyde and 2,4,6-triisopropylbenzaldehyde may be prepared in yields of 69 and 65%, respectively, by the procedure described above, with a reaction time of 8 hours and the following modifications:

	Triethylbenzaldehyde	*Triisopropylbenzaldehyde*
Hydrocarbon	100 g. (0.62 mole)	100 g. (0.49 mole)
Zinc cyanide	115 g. (0.97 mole)	102 g. (0.74 mole)
Aluminum chloride	215 g. (1.60 moles)	134 g. (1.0 mole)
Boiling point	146–149°/21 mm.	123–126°/4 mm.

[1] Adams and Levine, *J. Am. Chem. Soc.*, **45**, 2375 (1923).

[2] Arnold and Sprung, *J. Am. Chem. Soc.*, **60**, 1699 (1938).

[3] Fieser, *Experiments in Organic Chemistry*, p. 311, New York, 1941. Reprinted by special permission of D. C. Heath and Company.

II. METHOD B

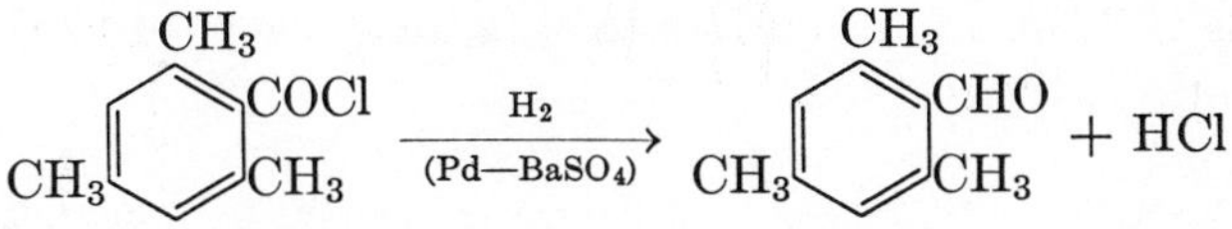

Submitted by R. P. Barnes.
Checked by Nathan L. Drake, Harry D. Anspon, and Ralph Mozingo.

1. Procedure

A 1-l. three-necked flask *with ground-glass joints* is fitted with a mercury-sealed stirrer (Notes 1 and 2), a glass tube of 6-mm. internal diameter which runs to the bottom of the flask, and a Friedrich condenser, protected by a drying tube. A solution of 90 g. (0.49 mole) of mesitoyl chloride (p. 555) in 270 g. of carefully dried xylene is placed in the flask together with 20 g. of palladium-barium sulfate (Note 3). The contents of the flask are refluxed while a stream of hydrogen (Note 4), which has been freed from oxygen by passage through Fieser's solution,[1] and dried by passage through concentrated sulfuric acid followed by a drying tube of Drierite (Note 5), is bubbled through the suspension until hydrogen chloride ceases to be evolved (Note 2). The catalyst is then removed by filtration and the xylene distilled. The residual liquid is transferred to a 125-ml. modified Claisen flask and distilled. The product boils at 96–98°/6 mm., and weighs 53–60 g. (70–80%) (Note 6).

2. Notes

1. The stirrer is not essential, but without it the time required for the reduction is increased about threefold.

2. The course of the reaction may be followed conveniently by passing the exit gases into water and titrating with approximately 1 *N* sodium hydroxide. The time for complete reduction is about 6–7 hours with stirring, and about 18 hours with no stirrer.

3. The palladium-barium sulfate is prepared by the method described by Houben.[2] The barium sulfate should be freshly precipitated material.

4. The checkers found that electrolytic hydrogen directly from the cylinder without purification gives only a slightly decreased yield.

5. Dehydrite or Anhydrone should not be used because of the danger of sulfuric acid spraying into it. Drierite is much more efficient than calcium chloride.

6. This preparation illustrates Rosenmund's method of synthesizing aldehydes without use of a poisoned catalyst. See p. 629 for the more general form of the method.

3. Methods of Preparation

Mesitaldehyde has been prepared from mesitylglyoxylic acid;[3,4] from mesitylmagnesium bromide and ethoxymethyleneaniline,[5] ethyl orthoformate,[5] or N-methylformanilide;[6] from mesitylene, hydrogen cyanide, and hydrogen chloride[7] or mesitylene, carbon monoxide, and hydrogen chloride[8] in the presence of aluminum and cuprous chlorides; and from mesitylene, zinc cyanide, and hydrogen chloride in the presence of aluminum chloride.[9–11]

The procedures described here are based on the method applied to the preparation of other aldehydes by Rosenmund and Zetzsche;[12] and by Nencki, as adapted for mesitaldehyde.[9-11]

[1] Fieser, *J. Am. Chem. Soc.*, **46**, 2639 (1924).

[2] Houben, *Die Methoden der organischen Chemie,* 3rd ed., Vol. II, p. 500, Verlag Georg Thieme, Leipzig, 1930; Schmidt, *Ber.*, **52**, 409 (1919).

[3] Bouveault, *Compt. rend.*, **124**, 156 (1897).

[4] Feith, *Ber.*, **24**, 3542 (1891).

[5] Smith and Nichols, *J. Org. Chem.*, **6**, 489 (1941).

[6] Smith and Bayliss, *J. Org. Chem.*, **6**, 437 (1941).

[7] German pat. 99,568 [*Chem. Zentr.*, **70**, I, 461 (1899)].

[8] German pat. 98,706 [*Chem. Zentr.*, **69**, II, 952 (1898)].

[9] Fuson, Horning, Ward, Rowland, and Marsh, *J. Am. Chem. Soc.*, **64**, 30 (1942).

[10] Hinkel, Ayling, and Morgan, *J. Chem. Soc.*, **1932**, 2793.

[11] Hinkel, Ayling, and Beynon, *J. Chem. Soc.*, **1936**, 339.

[12] Rosenmund, *Ber.*, **51**, 585 (1918); Rosenmund and Zetzsche, *Ber.*, **54**, 425 (1921).

MESITOIC ACID

(β-Isodurylic acid)

I. METHOD A

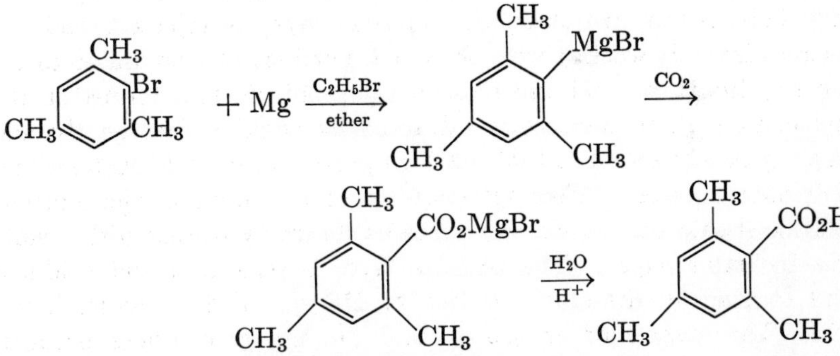

Submitted by Douglas M. Bowen.
Checked by H. R. Snyder and John R. Demuth.

1. Procedure

In a dry 2-l. three-necked flask, equipped with a sealed wire stirrer, a condenser protected by a drying tube, and a dropping funnel, are placed 85.0 g. (3.5 gram atoms) of magnesium turnings and 150 ml. of dry ether (Note 1). A solution of 199 g. (1 mole) of carefully fractionated bromomesitylene and 218 g. (2 moles) of ethyl bromide (Note 2) in 1 l. of dry ether (Note 1) is placed in the funnel. Stirring is commenced, and about 25 ml. of the ether solution is added; the reaction begins almost at once. The rest of the solution of halides is added during the course of 1.25–1.5 hours to the vigorously refluxing mixture; moderate cooling is necessary to permit addition within the specified period. After completion of the addition, refluxing is maintained by external heating for 30 minutes. The reaction mixture is then cooled, and the solution of alkyl magnesium bromides is decanted slowly from the excess magnesium onto 600 g. of Dry Ice which is stirred manually in a 4-l. beaker. The Dry Ice should be in the form of small lumps, and the addition must be slow enough to avoid spattering. The flask is rinsed with two 200-ml. portions of dry ether, which are added to the carbonation mixture. When most of the Dry Ice has evaporated, an additional 200-g. portion is added along with 250 ml. of dry ether. The viscous mixture is stirred until it becomes largely granular.

When the bulk of the Dry Ice has evaporated, 800 ml. of 20% hydrochloric acid and enough ice to keep the mixture cold are added with stirring. After most of the solid has dissolved, the mixture is transferred to a separatory funnel with the addition of ordinary ether if the volume of the organic layer is much less than 1 l. After agitation until both layers are clear, the aqueous layer is rejected, and the ethereal layer is washed with three 1-l. portions of cold water to remove hydrochloric acid and most of the propionic acid formed in the carbonation. The product is extracted by shaking first gently and then vigorously with a 540-ml. (600-g.) portion of *ice-cold* 10% sodium hydroxide solution. After agitation for several minutes the aqueous solution should still be strongly basic as shown by testing with a suitable indicator paper. The aqueous layer is separated and acidified with stirring by the slow addition of 250 ml. of 20% hydrochloric acid. The suspension is cooled, and the nearly colorless product, consisting of small granules, is collected and washed well with water. The crude acid, amounting to 141–143 g. (86–87%) (Note 3), melting at 152–154° (cor.), is satisfactory for most purposes. A purer product may be obtained as large, nearly colorless crystals, m.p. 153.4–154.4° (cor.), by crystallization from a solution, saturated at the boiling point, in 45% methanol. The yield of the recrystallized acid is 138–141.5 g. (84–86%) (Note 4).

2. Notes

1. The checkers used commercial anhydrous ether which had been stored over sodium wire, and they dried the various pieces of apparatus in an oven before assembly. If the redrying of the ether over sodium was omitted, the yield was lowered appreciably.

2. When 1 mole of ethyl bromide is used with 1 mole of bromomesitylene, the yield of crude acid is 81–82%. No appreciable improvement in yield is effected by use of more than 2 equivalents of ethyl bromide. If the ethyl bromide is omitted entirely, bromomesitylene reacts with magnesium very slowly and the Grignard reagent obtained affords mesitoic acid in 61–66% overall yield. Except in the absence of ethyl bromide there is no advantage in the use of an atmosphere of nitrogen.

3. This procedure may be conducted on one-fifth the scale without alteration in yield.

4. The checkers employed 20 ml. of a solvent mixture containing 45% of methanol by weight for each 10 g. of the crude acid. To avoid the loss of methanol from the hot solution the operation was carried out under a reflux condenser, and the clear, nearly colorless solution was cooled rapidly without transferral. If filtration of the hot solution should be desirable a higher concentration of methanol would be more convenient.

II. METHOD B

MESITOIC ACID AND MESITOYL CHLORIDE

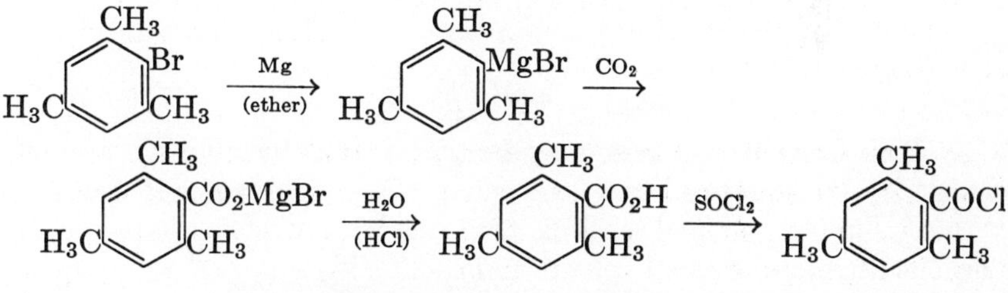

Submitted by R. P. Barnes.
Checked by Nathan L. Drake and Ralph Mozingo.

1. Procedure

In a 2-l. three-necked flask, fitted with a condenser protected by a drying tube, a dropping funnel, and a sealed stirrer, are placed 24.3 g. (1 gram atom) of magnesium turnings, a small crystal of iodine, enough absolute ether to cover the magnesium, and 10 g. (0.05 mole) of bromomesitylene. The bottom of the flask is warmed with the hand or a warm cloth until the reaction begins. The mixture is then stirred gently during the gradual addition of 190 g. (0.95 mole) of bromomesitylene dissolved in 500 g. (700 ml.) of absolute ether. When all the ethereal solution has been added, the reaction mixture is refluxed for about 2 hours, or until all the magnesium has dissolved. A large excess of solid carbon dioxide is now added slowly in small pieces with rapid stirring (Note 1). The resulting tough addition product is decomposed by pouring, with stirring, into a large volume of finely crushed ice to which has been added 100 ml. (1.2 moles) of concentrated hydrochloric acid. The ether is removed by evaporation, and the resulting oily solid is filtered, dissolved in 200–400 ml. of hot methanol, filtered, and thrown out by dilution with 1 l. of ice water. The crude mesitoic acid melts between 135° and 148° and weighs 110–120 g. The acid is recrystallized from petroleum ether (b.p. 90–100°), about 10 ml. of petroleum ether being used per gram of the crude acid. The product, melting at 150–152° after one recrystallization, weighs 90–100 g. (55–61%) (Note 2).

A mixture of 90 g. (0.55 mole) of mesitoic acid and 100 g. (63 ml., 0.84 mole) of thionyl chloride in a 500-ml. large-mouthed Claisen flask, with the side arm and adjacent neck closed, and with the other neck fitted with a condenser protected by a drying tube, is refluxed gently until the evolution of sulfur dioxide and hydrogen chloride ceases (Note 3). The excess thionyl chloride is removed by distillation at atmospheric pressure, and the residual acid chloride is distilled at 143–146°/60 mm. (Note 4). The yield is 90–97 g. (90–97%).

2. Notes

1. The Grignard reagent may be carbonated by pouring it slowly over Dry Ice contained in a 2-l. beaker. The yield is unchanged.

2. A small amount of mesitoic acid can be recovered by concentrating the filtrate to a small volume and cooling it.

3. This requires from 1 to 2 hours. The reaction mixture may also be allowed to stand overnight at room temperature, in which case no heating is necessary.

4. The boiling point varies with the rate of distillation.

3. Methods of Preparation

Mesitoic acid has been obtained by hydrolysis of its amide which was prepared from mesitylene, carbamyl chloride, and aluminum chloride in carbon disulfide.[1] It has been prepared by heating isodurene with dilute nitric acid,[2,3] in small yields by the distillation of 2,4,6-trimethylmandelic acid,[4] by dry distillation of 2,4,6-trimethylphenylglyoxylic acid,[4,5] by oxidation of 2,4,6-trimethylphenylglyoxylic acid with potassium permanganate,[6] and by treating 2,4,6-trimethylphenylglyoxylic acid with concentrated sulfuric acid either with heating [7] or in the cold.[8] The preparation of the acid from 2,4,6-trimethylphenylmagnesium bromide and a stream of carbon dioxide has been described.[9]

The method for the preparation of the chloride is the general method for preparing aromatic acid chlorides with thionyl chloride,[10] which has been applied to the preparation of mesitoyl chloride.[9]

1 Michael and Oechslin, *Ber.*, **42**, 317 (1909).
2 Jannasch and Weiler, *Ber.*, **27**, 3441 (1894).
3 Jacobsen, *Ber.*, **15**, 1853 (1882).
4 Meyer and Molz, *Ber.*, **30**, 1270 (1897).
5 Feith, *Ber.*, **24**, 3542 (1891).
6 Claus, *J. prakt. Chem.*, (2) **41**, 506 (1890).
7 van Scherpenzeel, *Rec. trav. chim.*, **19**, 377 (1900).
8 Hoogewerff and van Dorp, *Rec. trav. chim.*, **21**, 349 (1902).
9 Kohler and Baltzly, *J. Am. Chem. Soc.*, **54**, 4015 (1932).
10 Meyer, *Monatsh.*, **22**, 415 (1901).

MESITYLACETIC ACID

(Acetic acid, mesityl-)

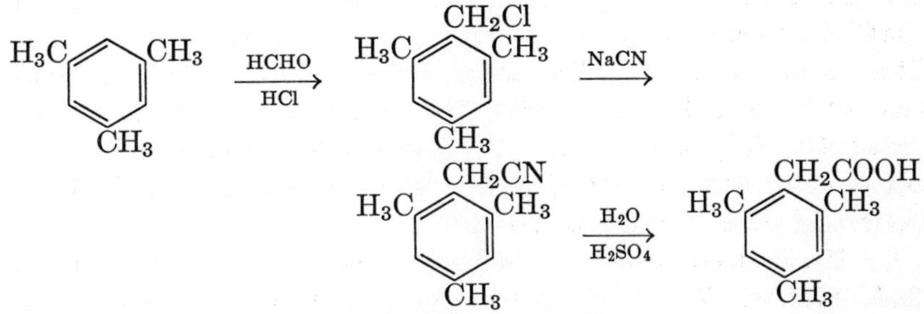

Submitted by REYNOLD C. FUSON and NORMAN RABJOHN.
Checked by W. E. BACHMANN, E. L. JENNER, and G. DANA JOHNSON.

1. Procedure

A. *α²-Chloroisodurene.* In a 2-l. round-bottomed flask, equipped with a sealed mechanical stirrer (Note 1), a gas inlet tube, and a reflux condenser, are placed 200 g. (1.66 moles) of mesitylene, 1 l. of concentrated hydrochloric acid, and 63 ml. (0.84 mole) of formaldehyde solution (concentration, 37%) (Note 2). Hydrogen chloride is introduced below the surface of the mixture (Note 3), which is stirred vigorously and heated in a water bath kept at 55°. These conditions are maintained throughout the reaction, which requires a total of 5.5 hours. At the halfway point, an additional 63 ml. (0.84 mole) of formaldehyde solution is added (Note 4).

After the mixture has been cooled to room temperature (Note 5), it is extracted with three 300-ml. portions of benzene. The combined benzene extracts are washed successively with water, dilute sodium hydroxide, and water, dried over calcium chloride, and filtered. The mixture is distilled through a still head under reduced pressure; the pressure is reduced only slightly while the benzene is distilling. The yield of α^2-chloroisodurene boiling at 130–131°/22 mm. is 155–170 g. (55–61%) (Note 6); the distillate solidifies to a crystalline mass which melts at 37°.

B. *Mesitylacetonitrile.* In a 1-l. three-necked flask, fitted with a mechanical stirrer, a reflux condenser, and a dropping funnel, are placed 77 g. (1.57 moles) of sodium cyanide, 110 ml. of water, and 160 ml. of ethanol. The flask is heated in a boiling water bath, and the contents are stirred until all the sodium cyanide is dissolved. Then 152 g. (0.90 mole) of α^2-chloroisodurene is added slowly, and stirring and heating are continued for 3 hours.

The reaction mixture is allowed to cool to about 40° (Note 7) and extracted with three 300-ml. portions of benzene. The benzene solution is washed well with water, dried over calcium chloride, filtered, and distilled under slightly reduced pressure to remove all the benzene. The residue is then distilled under reduced pressure from a Claisen flask with a wide-bore side arm. The yield of mesitylacetonitrile boiling at 160–165°/22 mm. is 128–133 g. (89–93%) (Note 8). This product is sufficiently pure for the next step. When recrystallized from petroleum ether, it melts at 79–80°.

C. *Mesitylacetic acid.* To 900 ml. of water in a 3-l. three-necked flask is added 750 ml. of concentrated sulfuric acid. When the mixture has cooled to about 50°, 127 g. (0.80 mole) of mesitylacetonitrile is added (Note 9), and the mixture is refluxed and stirred mechanically for 6 hours. At the end of this period, a large amount of mesitylacetic acid has precipitated from the solution. The contents of the flask are cooled and poured into 3 l. of ice water. The acid is collected on a Büchner funnel and washed well with water. A solution of the acid in dilute alkali is boiled with Norit, and the acid is precipitated from the filtered solution by acidifying with dilute hydrochloric acid. The mesitylacetic acid is collected on a filter, washed well with water, and dried in an oven at about 80°. The yield of mesitylacetic acid melting at 163–166° is 123 g. (87%). After recrystallization from dilute alcohol or ligroin, the acid melts at 167–168°.

2. Notes

1. The checkers used an Ace Tru-bore stirrer. This convenient sealed stirrer, which owes its seal to the snug fit between a ground-glass section of the stirrer shaft and a ground-glass bearing, can be obtained from the Ace Glass Company, Vineland, New Jersey.

2. Commercial formalin was used. This aqueous-methanolic solution contains 37% formaldehyde by weight. It is sometimes called "40% formalin" because 100 ml. of the solution contains 40 g. of formaldehyde.

3. The hydrogen chloride is introduced at such a rate that the bubbles form a little faster than they can be counted.

4. The yield of monochloromethyl compound seems to be improved slightly by adding the formalin in two portions instead of introducing the entire amount at the beginning of the reaction.

5. If the reaction mixture is cooled to too low a temperature, the chloromethyl derivatives will solidify. It is much easier to carry out the extractions while the mixture is still liquid.

6. The residue consists chiefly of α^2,α^4-dichloropentamethylbenzene. The fraction boiling at 131–200°/22 mm. is collected and recrystallized from petroleum ether; m.p. 105°.

7. If the reaction mixture is allowed to cool much below this temperature, it will solidify in the flask.

8. This method has been applied successfully to the preparation of 2,4,6-triethylphenylacetonitrile (b.p. 127°/3–4 mm.; yield, 72%); durylacetonitrile (m.p. 80–81°; yield, 75%); and isodurylacetonitrile (m.p. 74–75°; yield, 74%).

9. If the mesitylacetonitrile has not been purified by distillation, or if it is added to a boiling solution of sulfuric acid, the reaction mixture has a tendency to become very dark.

3. Methods of Preparation

α^2-Chloroisodurene can be prepared by the action of chloromethyl ether on mesitylene in the presence of stannic chloride [1] or acetic acid.[2] The procedure described is based on that of Nauta and Dienske.[3]

Mesitylacetonitrile can be prepared by the action of potassium cyanide on α^2-chloroisodurene [4] or by treating α^2-chloroisodurene with cuprous cyanide in the presence of pyridine.[5] The procedure described is based upon the method for the preparation of benzyl cyanide from benzyl chloride.

Mesitylacetic acid has been prepared from 2,4,6-trimethylacetophenone by treatment with yellow ammonium sulfide and hydrolyzing the resulting amide with alkali,[6] by the dry distillation of 2,4,6-trimethylmandelic acid,[7] by heating 2,4,6-trimethylphenylglyoxylic acid with hydriodic acid and red phosphorus [8] or with hydrazine hydrate,[9] by the action of boiling water on mesitoyldiazomethane [10] (prepared in turn from mesitylglyoxal monohydrazone), and by treating mesitylacetonitrile with potassium hydroxide.[4]

[1] Sommelet, *Compt. rend.*, **157**, 1443 (1913).
[2] Vavon and Bolle, *Compt. rend.*, **204**, 1826 (1937).
[3] Nauta and Dienske, *Rec. trav. chim.*, **55**, 1000 (1936).
[4] Hoch, *Compt. rend.*, **192**, 1464 (1931).
[5] Newman, *J. Am. Chem. Soc.*, **59**, 2472 (1937).
[6] Claus, *J. prakt. Chem.*, (2) **41**, 508 (1890); Willgerodt, *J. prakt. Chem.*, (2) **80**, 185 (1909); Willgerodt and Merk, *J. prakt. Chem.*, (2) **80**, 193 (1909).
[7] Meyer and Molz, *Ber.*, **30**, 1274 (1897).
[8] Dittrich and Meyer, *Ann.*, **264**, 140 (1891).
[9] Lock, *Oesterr. Chem. Ztg.*, **51**, 77 (1950) [*C. A.*, **44**, 8338 (1950)].
[10] Fuson, Armstrong, and Shenk, *J. Am. Chem. Soc.*, **66**, 964 (1944).

METHACRYLAMIDE *

$$(CH_3)_2C(OH)CN \xrightarrow{H_2SO_4} CH_2{=}C(CH_3){-}CONH_2$$

Submitted by Richard H. Wiley and Walter E. Waddey.
Checked by R. L. Shriner and Arne Langsjoen.

1. Procedure

In a 1-l. three-necked round-bottomed flask fitted with an efficient stirrer, a dropping funnel, and a thermometer, are placed 150 g. (82 ml., 1.5 moles) of 98% sulfuric acid (Note 1) and 1 g. of flowers of sulfur. To this is added, with rapid stirring, 85 g. (91 ml., 1 mole) of acetone cyanohydrin [1] (Note 2) over a period of 25 minutes, the temperature of the contents of the flask being kept at 75–80° by cooling in a water bath. At the end of this period, the water bath is replaced with an oil bath preheated to 160°. With continued stirring, the temperature of the reaction mixture is raised to 150° within 5 minutes and maintained at 150° for 15 minutes (Note 3). The reaction mixture is quickly cooled to room temperature by replacing the oil bath with an ice bath and is then poured into 300 ml. of cold water, and the flask is rinsed with 75 ml. of water. The diluted mixture is filtered (Note 4) through a 10-cm. Büchner funnel with the aid of suction. The filtrate is placed in a 1-l. beaker, cooled with an ice bath and 180–190 g. of anhydrous sodium carbonate sifted in with vigorous stirring (*Caution! foaming*) (Note 5). The precipitate is collected on a 20-cm. Büchner funnel and pressed and sucked as dry as possible. The crude product is dried in a vacuum desiccator over anhydrous calcium chloride for 36–48 hours (Note 6). The light cream- or tan-colored crude product weighs 300–400 g.

The crude dried solid is crushed to break up lumps and is placed in a 2-l. flask and heated and stirred mechanically with 500 ml. of boiling benzene. The solution is decanted, and the extraction is repeated four times using 200-ml. portions of benzene. The combined benzene solutions are heated to boiling, treated with 2–4 g. of Norit, and filtered. On cooling, 48–52 g. of methacrylamide separates; m.p. 105–107° (Note 7). An additional 5–8 g., m.p. 103–105°, is obtained when the mother liquor is concentrated to 150 ml. and cooled. The yield of methacrylamide is 52–60 g. (61–70%) after storing in a vacuum desiccator over paraffin wax and anhydrous calcium chloride.

2. Notes

1. Acid of 98% strength may be prepared by the addition of 33.5 ml. of fuming sulfuric acid (15% SO_3) to 48 ml. of concentrated sulfuric acid of sp. gr. 1.84.

2. Acetone cyanohydrin of 98% purity may be purchased from the Rohm and Haas Company, Philadelphia, Pennsylvania.

3. This heating converts the intermediate compounds into methacrylamide sulfate. Longer periods of heating decrease the yield.

4. A small amount of dark polymeric material, which may form, is separated at this point.

5. Care is taken to crush lumps of sodium carbonate formed during the addition. The final reaction mixture should be slightly alkaline to litmus paper, and the temperature of the mixture should not rise above 25–30°.

6. The precipitated methacrylamide contains varying amounts of sodium sulfate. It is essential to obtain a dry product for the benzene

extraction. The desiccator should be evacuated with an oil pump to about 5 mm. several times during 24–36 hours in order to obtain a sufficiently dry material.

7. The methacrylamide tends to retain some solvent. By placing the product in a vacuum desiccator containing paraffin wax shavings and anhydrous calcium chloride for 36–48 hours material with a melting point of 109–110° may be obtained.

3. Methods of Preparation

Methacrylamide has been prepared by the reaction of acetone cyanohydrin with concentrated sulfuric acid [2–7] and sulfuric acid with ammonium sulfate,[8] and by the hydrolysis of methacrylonitrile.[9] The method described is an adaptation of that of Crawford and McGrath.[4]

[1] *Org. Syntheses* Coll. Vol. **2**, 7 (1943).

[2] Verhulst, *Bull. soc. chim. Belg.*, **39**, 563 (1930); **40**, 475 (1931).

[3] Crawford, *J. Soc. Chem. Ind.*, **64**, 231 (1945).

[4] Crawford and McGrath, U. S. pat. 2,140,469 [*C. A.*, **33**, 2536 (1939)]; Brit. pat. 440,967 [*C. A.*, **30**, 4180 (1936)].

[5] Crawford and Grigor, U. S. pat. 2,101,822 [*C. A.*, **32**, 952 (1938)]; Brit. pat. 456,533 [*C. A.*, **31**, 2230 (1937)].

[6] I. G. Farbenind. A.-G., Fr. pat. 813,844 [*C. A.*, **32**, 953 (1938)].

[7] Rohm und Haas A.-G., Fr. pat. 815,908 [*C. A.*, **32**, 1816 (1938)].

[8] U. S. pat. 2,431,468 [*C. A.*, **42**, 3429 (1948)].

[9] Bruylants and Castille, *Bull. sci. acad. roy. Belg.*, **13**, 767 (1928) [*C. A.*, **27**, 2366 (1928)].

ω-METHOXYACETOPHENONE

(Acetophenone, ω-methoxy-)

$$C_6H_5MgBr + CH_3OCH_2CN \rightarrow CH_3OCH_2C(=NMgBr)C_6H_5$$

$$2CH_3OC(=NMgBr)C_6H_5 + H_2O + 2H_2SO_4 \rightarrow 2CH_3OCH_2COC_6H_5 + (NH_4)_2SO_4 + MgSO_4 + MgBr_2$$

Submitted by R. B. Moffett and R. L. Shriner.
Checked by W. E. Bachmann and W. S. Struve.

1. Procedure

A solution of phenylmagnesium bromide is prepared in a 2-l. three-necked flask, fitted with a separatory funnel, reflux condenser, and a mercury-sealed stirrer, from 8.8 g. (0.36 gram atom) of magnesium, 56.5 g. (38 ml., 0.36 mole) of bromobenzene, and a total of 350 ml. of dry ether by the procedure described in *Org. Syntheses* Coll. Vol. **1,** 226 (1941).

To the solution of the Grignard reagent, cooled by an ice-salt bath, a mixture of 21.3 g. (0.3 mole) of methoxyacetonitrile [*Org. Syntheses* Coll. Vol. **2,** 387 (1943)] and 50 ml. of dry ether is slowly added with stirring. The colorless addition product separates at once. After standing at room temperature for 2 hours, the mixture is again cooled and then decomposed by adding, with stirring, 500 ml. of water and cracked ice, and then 100 ml. of cold dilute sulfuric acid (Note 1). When the decomposition is complete (Note 2), the ether layer is separated and the aqueous layer is extracted with a little ether. This ether extract is combined with the ether layer, and the whole is washed with 5% aqueous sodium carbonate solution and then with water. The solution is dried with anhydrous sodium sulfate.

The ether is removed by distillation from a steam bath, and the residue is distilled under diminished pressure. ω-Methoxyacetophenone is a colorless liquid which boils at 118–120°/15 mm. or 228–230°/760 mm. (Note 3). The yield is 32–35 g. (71–78% based on the methoxyacetonitrile).

2. Notes

1. One volume of concentrated sulfuric acid is added to 2 volumes of water, and the mixture is cooled in an ice-salt bath.

2. The two layers should be light yellow in color with only a small amount of solid or tarry material present.

3. The checkers observed a boiling point of 110–112°/9 mm.

3. Method of Preparation

The method described is essentially that of Pratt and Robinson.[1] ω-Methoxyacetophenone has also been prepared by chromic acid oxidation of α-phenyl-β-methoxyethanol, which in turn was prepared from styrene oxide.[2]

[1] Pratt and Robinson, *J. Chem. Soc.*, **1923,** 748.

[2] Kaelin, *Helv. Chim. Acta,* **30,** 2132 (1947).

m-METHOXYBENZALDEHYDE

(Benzaldehyde, *m*-methoxy-)

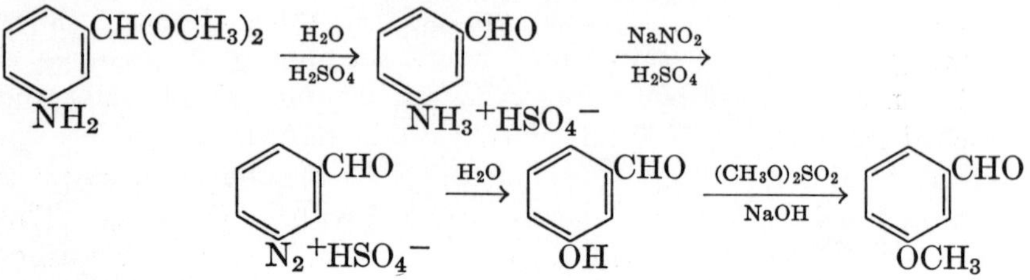

Submitted by Roland N. Icke, C. E. Redemann, Burnett B. Wisegarver, and Gordon A. Alles.
Checked by H. R. Snyder and Frank X. Werber.

1. Procedure

A. *m-Hydroxybenzaldehyde.* In a 2-l. three-necked flask, equipped with a mechanical stirrer, a thermometer, and a 250-ml. dropping funnel, 575 ml. of 6 *N* sulfuric acid is cooled to 0° by means of a salt-ice bath. The acid is stirred and maintained at 0° or below while 167 g. (1 mole) of *m*-aminobenzaldehyde dimethylacetal (p. 59) is added dropwise. The solution becomes deep orange or red. When the addition of the amino compound is complete, a solution of 71 g. (1 mole) of 97% sodium nitrite in about 175 ml. of water is introduced slowly while the temperature of the acid solution is maintained at 5°. Stirring at 5° is continued for 1 hour to complete the reaction.

In each of two 4-l. beakers are placed 450 ml. of water and 50 ml. of concentrated sulfuric acid, and the solutions are heated to boiling with large burners. The cold diazonium solution is divided into two approximately equal portions which are placed in 500-ml. separatory funnels suspended above the beakers containing the boiling acid. The two portions of the diazonium solution are run dropwise into the strongly heated acid at such a rate that boiling continues. The solutions are boiled for 5 minutes after the additions are complete. They are then allowed to cool to room temperature and are finally stored overnight in a refrigerator. The crude product separates as a dark oil which crystallizes (Note 1) and becomes lighter in color upon standing. It is collected on a Büchner funnel and used in part B without purification (Note 2).

Methyl sulfate is quite toxic.
Caution! The methylation should be carried out in a good hood.

B. *m-Methoxybenzaldehyde.* The crude *m*-hydroxybenzaldehyde is dissolved in about 550 ml. of 2 *N* sodium hydroxide in a 2-l. three-necked flask equipped with a mechanical stirrer, a thermometer, and a 125-ml. dropping funnel. The dark-colored solution is stirred while 126 g. (95 ml., 1 mole) of methyl sulfate (Note 3) is added dropwise and the temperature is maintained at 40–45°. When the addition is complete the mixture is stirred for 5 minutes. A 275–ml. portion of 2 *N* sodium hydroxide (Note 4) is added in one lot, and then 63 g. (47.5 ml.) of methyl sulfate is added as before, except that the temperature is allowed to rise to 50°. Stirring at 50° is continued for 30 minutes, the mixture is cooled, and the organic layer is extracted with ether (Note 5). The ether solution is dried over anhydrous sodium sulfate for 8 hours, then filtered and concentrated by distillation. The residue is distilled under reduced pressure. The yield of *m*-methoxybenzaldehyde, a pale yellow liquid boiling at 88–90°/3 mm., is 86–98 g. (63–72%) (Note 6).

2. Notes

1. Seeding the mixture helps to initiate crystallization.

2. If *m*-hydroxybenzaldehyde is desired, the crude product may be purified as described elsewhere (p. 453).

3. A good technical grade of methyl sulfate was used.

4. The optimum amount of sodium hydroxide solution apparently varies according to the amount of acid remaining in the crude, wet hydroxybenzaldehyde employed in the methylation. The checkers found it advisable to increase the amount added at this point to 345 ml. It is wise to test the reaction mixture with litmus paper occasionally during the final heating period and to add alkali as necessary to keep the solution from becoming acid.

5. If the methylation is not complete, some *m*-hydroxybenzaldehyde will remain dissolved in the aqueous phase. This may be recovered by acidifying the alkaline solution and collecting any crystalline solid which separates.

6. As with other aromatic aldehydes, *m*-methoxybenzaldehyde is susceptible to air oxidation and should be stored in a bottle which will just hold the product, so that air space above the liquid is minimized.

3. Methods of Preparation

m-Methoxybenzaldehyde has been prepared by the reduction of *m*-methoxybenzoic acid,[1] by the reaction of diazotized *m*-aminobenzaldehyde with methanol,[2] by an acid hydrolysis of the phenylhydrazone which was obtained by oxidation of the hydrazine analog,[3] and by the methylation of *m*-hydroxybenzaldehyde, with methyl iodide,[4–7] and with methyl sulfate.[2,7–9]

[1] Asano and Huziwara, *J. Pharm. Soc. Japan,* **50,** 141 (1939).
[2] Noelting, *Ann. chim.,* (8) **19,** 541 (1910).
[3] Grammaticakis, *Compt. rend.,* **210,** 303 (1940).
[4] Tiemann and Ludwig, *Ber.,* **15,** 2043 (1882).
[5] Pschorr and Jaeckel, *Ber.,* **33,** 1826 (1900).
[6] Staudinger and Kon, *Ann.,* **384,** 90 (1911).
[7] Späth, *Monatsh.,* **34,** 1998 (1913).
[8] Posner, *J. prakt. Chem.,* (2) **82,** 431 (1910).
[9] Livshits, Bazilevskaya, Bainova, Dobrovinskaya, and Preobrazhenskii, *J. Gen. Chem. U.S.S.R.,* **17,** 1671 (1947) [*C. A.,* **42,** 2606 (1948)].

2-METHOXYDIPHENYL ETHER

(Benzene, 1-methoxy-2-phenoxy-)

Submitted by H. E. Ungnade and E. F. Orwoll.
Checked by Homer Adkins and E. E. Burgoyne.

1. Procedure

Powdered potassium hydroxide (29.4 g., 0.43 mole) is placed in a 500-ml. round-bottomed flask. Guaiacol (75 g., 0.60 mole) is added, and the mixture is allowed to react exothermically. After the reaction is complete, the mixture is stirred with a glass rod and then heated under reduced pressure for 3 hours at 150° in an oil bath (Note 1).

To the dry salt is added 0.3 g. of copper powder (Note 2), 81 g. (0.51 mole) of bromobenzene, and a few drops of guaiacol (Note 3).

The mixture is stirred thoroughly with a glass rod; the flask is fitted with an air condenser and heated in a metal bath (Note 4). A reaction becomes evident at a bath temperature of 160–180°, liquefaction occurs, and the color of the mixture changes to red or purple. The temperature is gradually raised to 200° and maintained at 200° for 2 hours.

After cooling, the products are extracted from the reaction mixture with successive portions of water and ether. Extraction is facilitated by breaking up the solid material with a glass rod. The total amounts of solvents required are approximately 750 ml. of water and 150 ml. of ether. The combined ether and water solutions are transferred to a 3-l. round-bottomed flask and steam-distilled with superheated steam maintained at 180–200°. After removal of the ether, 300 ml. of distillate is collected. This distillate contains the unreacted starting materials. Continued distillation gives 64–69 g. (62–67%) of crude solid 2-methoxydiphenyl ether in 14 l. of distillate. The product is filtered with suction and dried. Crystallization from a mixture of 600 ml. of low-boiling petroleum ether (b.p. 30–60°) and 435 ml. of higher-boiling petroleum ether (b.p. 60–70°) yields 54–61 g. of 2-methoxydiphenyl ether melting at 77–78°. Other 2-methoxydiphenyl ethers have been prepared by this procedure (Note 5).

2. Notes

1. The salt of guaiacol is heated under reduced pressure, in order to remove water, which is a negative catalyst in the Ullmann reaction.[1]

2. The copper catalyst may be prepared by the method of Brewster and Groening.[2]

3. An excess of guaiacol is essential. Weston and Adkins[1] have found that the phenol, copper, and air form the active catalyst in the Ullmann reaction.

4. The checkers used an electrically heated oil bath.

5. Yields of 54% of 2-methoxy-4′-methyldiphenyl ether from *p*-bromotoluene and guaiacol, and 60% of 2-methoxy-5-methyldiphenyl ether from 3-bromo-4-methoxytoluene and phenol, have been obtained by the same method in the laboratory of the submitters.

3. Methods of Preparation

The procedure above is a modification of the method of Ullmann and Stein[3] for the same compound. Sartoretto and Sowa[4] used the same general method. The need for a catalyst can be avoided by heating a mixture of guaiacol potassium, guaiacol, and chlorobenzene at 200° under pressure.[5] Ullmann and Stein [6] have prepared the compound by using phenol, *o*-bromoanisole, copper powder, and potassium hydroxide. 2-Hydroxydiphenyl ether can be converted to the methoxy derivative by treating it with methanol, methyl iodide, and potassium hydroxide.[7]

[1] Weston and Adkins, *J. Am. Chem. Soc.*, **50**, 859 (1928).

[2] Brewster and Groening, *Org. Syntheses*, **14**, 66 (1934); Coll. Vol. **2**, 446 (1943).

[3] Ullmann and Stein, *Ber.*, **38**, 2212 (1905); **39**, 623 (1906).

[4] Sartoretto and Sowa, *J. Am. Chem. Soc.*, **59**, 603 (1937).

[5] Fritzsche and Co., Ger. pat. 269,543 (1914) (*Chem. Zentr.*, **1914**, I, 591).

[6] Ullmann and Stein, *Ber.*, **39**, 623 (1906).

[7] Norris, MacIntire, and Corse, *Am. Chem. J.*, **29**, 127 (1903).

6-METHOXY-8-NITROQUINOLINE *

(Quinoline, 6-methoxy-8-nitro-)

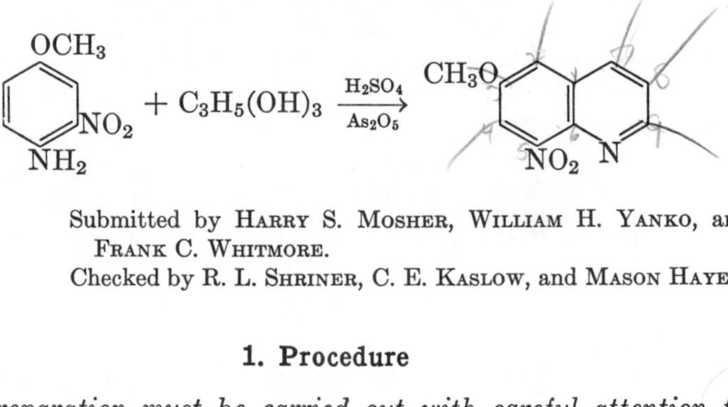

Submitted by Harry S. Mosher, William H. Yanko, and Frank C. Whitmore.
Checked by R. L. Shriner, C. E. Kaslow, and Mason Hayek.

1. Procedure

This preparation must be carried out with careful attention to the times and temperatures specified, since Skraup reactions are likely to become violent. The operator should wear goggles, and there should be a safety shower close at hand.

In a 5-l. three-necked round-bottomed flask, a homogeneous slurry of the following compounds is made by mixing in the order given (Note 1): 588 g. (2.45 moles) of powdered arsenic oxide (Note 2), 588 g. (3.5 moles) of 3-nitro-4-aminoanisole (Note 3), and 1.2 kg. (950 ml., 13 moles) of U.S.P. glycerol. The flask is fitted with an efficient mechanical stirrer and a 500-ml. dropping funnel in which is placed 315 ml. (579 g., 5.9 moles) of concentrated sulfuric acid (sp. gr. 1.84). With good mechanical stirring the sulfuric acid is dropped into the orange reaction mixture over a period of 30–45 minutes. During this addition, the temperature spontaneously rises to 65–70°.

The stirrer and dropping funnel are removed, and a thermometer is inserted in one neck of the flask by means of a stopper so that the bulb is well below the surface of the reaction mixture. A 10-mm. bent glass tube is attached to the second neck by a rubber stopper and attached through a trap to a water aspirator (Note 4). The third neck is closed with a stopper, and the flask and its contents are weighed. The flask, clamped in place in an oil bath which rests on a hot plate (Note 5), is evacuated, and heat is carefully applied at such a rate that the internal temperature slowly rises to 105° (Note 6). The mixture is kept between 105° and 110° until the loss in weight amounts to 235–285 g. (Note 7), which requires approximately 2–3 hours, depending on the efficiency of the aspirator. If the temperature shows any tendency to rise above 110°, the oil bath should be lowered and the mixture cooled.

When the removal of the water is complete, the suction tube is removed and the stirrer and dropping funnel are replaced. The internal temperature is raised with extreme care to 118° and is held rigidly between 117° and 119° during the subsequent addition of 438 g. (236 ml.) of concentrated sulfuric acid (sp. gr. 1.84) from the dropping funnel. The sulfuric acid must be added dropwise over a period of 2.5–3.5 hours, and the temperature *must not* vary from 117–119° (Note 8). After the addition is complete the temperature is maintained at 120° for 4 hours (Note 9) and finally at 123° for 3 hours. The reaction mixture is cooled below 100°, diluted with 1.5 l. of water, and allowed to cool overnight, preferably with stirring (Note 10). The diluted reaction mixture is poured with stirring into a mixture of 1.8 l. (1580 g.) of concentrated ammonium hydroxide (sp. gr. 0.9) and 3.5 kg. of ice in a 12-l. enameled pail. The resultant thick slurry is filtered through a large Büchner funnel (24–30 cm.), and the filtrate is discarded. The earth-colored precipitate is washed with four 700-ml. portions of water and then transferred to a 3-l. beaker and stirred with 1 l. of methanol for 15 minutes. The slurry is filtered and this process repeated (Note 11). The crude product at this point is a light chocolate brown and weighs about 800 g. when dried. It is purified by boiling for 30 minutes with 4.5 l. of chloroform (Note 12) to which has been added 30 g. of decolorizing carbon. The carbon and other humus-like material are removed by filtration from the hot chloroform solution by means of a 24-cm. warm Büchner funnel. The insoluble material is boiled with 500 ml. of chloroform, the resultant mixture filtered, and the combined filtrates concentrated by distillation from a steam bath to a volume of 1.5–2.5 l., at which point crystals of the 6-methoxy-8-nitroquinoline separate. The solution is then cooled to 5°, and the first crop of crystals is obtained by collection on a 24-cm. Büchner funnel. The crystals are transferred to a 1-l. beaker, stirred with 400 ml. of methanol for 15 minutes, and again collected on a Büchner funnel and washed with 200 ml. of methanol. The product consists of light-tan crystals which weigh 435–500 g. and melt at 158–160° (Note 13). By concentrating the filtrate to 400–500 ml., a second crop of crystals weighing 25–65 g. is obtained which, after washing with methanol in the same way as above, melts at 158–159°. The total yield is 460–540 g. (65–76%) (Notes 14, 15).

2. Notes

1. It is desirable to carry out the reaction without interruption. In order to do this, it is best to set up the apparatus and mix the arsenic oxide, 3-nitro-4-aminoanisole, and glycerol on one day and start the sulfuric acid addition the first thing in the morning of the following day. If the reaction must be interrupted, the mixture can be allowed to cool after it has been heated under reduced pressure. The yield is not affected, but great care must be exercised in reheating the reaction mixture since it sets to a glass, and it is very easy to superheat the outer portions while the center is still cold. If this happens, vigorous decomposition is very likely to occur.

2. Although powdered arsenic oxide was used, probably an equivalent amount of any form of the oxide, such as syrupy arsenic acid, would be suitable since the excess water would be removed when heated under reduced pressure.

3. The 3-nitro-4-aminoanisole was the technical commercial product, m.p. 124–126°, obtained from E. I. du Pont de Nemours and Company.

4. It is desirable to have a manometer in the system. The pressure should drop to approximately 30 mm. as the temperature approaches 105°. If, during the initial evacuation, foaming occurs, it will be necessary to loosen the stopper on the reaction flask or trap and let in air until foaming subsides and steady bubbling commences.

5. The oil bath can be heated by a hot plate which is connected in series to a variable resistance for temperature control. It may be heated with a Bunsen burner, but the temperature of the reaction must then be watched continually. The flask should be clamped in place in such a manner that the source of heat can be removed quickly if necessary.

6. The external temperature may be kept at about 110–115° but should not be raised above this, particularly after the internal temperature has reached 105°. If the temperature rises uncontrolled at this point, complete decomposition of the reaction mixture will result.

7. The loss in weight is dependent upon the quality of the arsenic oxide and glycerol used. If these are dry, the weight loss should be as indicated. The lower figure is equivalent to 1 mole of water for each mole of glycerol.

8. The success of the reaction depends upon the temperature control at this and subsequent points. Since the temperature is dependent upon at least four variables—the rate of addition of sulfuric acid, the rate of stirring, the heat of reaction which decreases as the reaction proceeds, and the heat applied—the temperature must be watched very closely. Too rapid addition of the sulfuric acid during the initial stages of the addition will result in an uncontrolled rise in temperature, terminating in the complete oxidation of the reaction mixture by the sulfuric acid. A large volume of sulfur dioxide is evolved, and the reaction mixture is converted into a voluminous, porous, carbon mass. *The progress of the reaction must be watched throughout its complete course and cannot be left unattended.*

9. The reaction is complete when a drop of the mixture on a piece of wet filter paper does not give an orange ring due to the presence of unreacted 3-nitro-4-aminoanisole. If the test is still positive at the end of the heating period, the temperature should be cautiously raised to 125° and heating continued until the test is negative. If the reaction does not go to completion, it is very difficult to remove this unused starting material from the product and it must be repeatedly decolorized and recrystallized.

10. On cooling, the sulfate salt of the 6-methoxy-8-nitroquinoline partially crystallizes from the reaction mixture. By stirring the mixture, the product comes down as a fine tan, microcrystalline precipitate. This can be isolated and crystallized, but it is more convenient to work it up in the form of the base.

11. These methanol washings remove much of the impurities, especially small amounts of unreacted nitroanisidine, with a minimum loss of product. The solubility of 6-methoxy-8-nitroquinoline in methanol at room temperature is 0.8 g. per 100 g. of solvent and at the boiling point is 4.1 g. per 100 g. of solvent. The solubility in chloroform, on the other hand, is 3.9 g. per 100 g. of solvent at room temperature and 14.2 g. at the boiling point. It is possible to obtain about 12 g. of product by working up these washings, but it is usually not worth the trouble.

12. The 4.5 l. of chloroform used is more than is necessary to dissolve the product, but this amount prevents the difficulties arising from crystallization in the Büchner funnel during filtration.

13. If, during the reaction, the temperature has not been properly regulated, the crystals may require another treatment with decolorizing carbon and crystallization from chloroform.

14. The product may also be recrystallized from hot ethylene dichloride using 300 ml. of solvent for each 100 g. The recovery is 80–90% of 6-methoxy-8-nitroquinoline melting at 160–161°.

15. By the same procedure the submitters have prepared 6-ethoxy-8-nitroquinoline in 70% yield, 8-methoxy-6-nitroquinoline in 68% yield, 6-chloro-8-nitroquinoline in 75% yield, and 6-methoxy-5-bromo-8-nitroquinoline in 69% yield from the properly substituted aromatic amines.

3. Methods of Preparation

6-Methoxy-8-nitroquinoline was first obtained by Schulemann and coworkers [1] by a modification of the Skraup reaction. Various other modifications of this reaction have been used.[2–9] The modification given here was the best of several tried and is taken from the procedure of Strukov.[10] It has been reported that 85% phosphoric acid may be used in place of sulfuric acid in certain Skraup reactions; thus with 3-nitro-4-aminoanisole, acrolein, and arsenic trioxide, with 85% phosphoric acid, a 60% yield of 6-methoxy-8-nitroquinoline was obtained.[11] A preparation from α-bromoacrolein has been reported.[12]

[1] Schulemann, Schuenhöfer, and Wingler, Ger. pat. 486,079 [*C. A.*, **24**, 1937 (1930)]; F. Bayer and Co., Brit. pat. 267,457 [*C. A.*, **22**, 1216 (1928)].

[2] Magidson and Strukov, *Arch. Pharm.*, **271**, 359 (1933).

[3] Fourneau, Trefouel, Bovet, and Benoit, *Ann. inst. Pasteur*, **46**, 514 (1931).

[4] Altman, *Rec. trav. chim.*, **57**, 941 (1938).

[5] Brahmachari and Bhattachryee, *J. Indian Chem. Soc.*, **8**, 571 (1931).

[6] Misani and Bogert, *J. Org. Chem.*, **10**, 347 (1945).

[7] Morgan and Tipson, *J. Am. Chem. Soc.*, **68**, 1569 (1946).

[8] Yale, *J. Am. Chem. Soc.*, **69**, 1230 (1947).

[9] Haskelberg, *J. Org. Chem.*, **12**, 434 (1947).

[10] Strukov, *Org. Chem. Ind. U.S.S.R.*, **4**, 523 (1937) [*C. A.*, **32**, 4987 (1938)].

[11] Yale and Bernstein, *J. Am. Chem. Soc.*, **70**, 254 (1948).

[12] Baker, Tinsley, Butler, and Riegel, *J. Am. Chem. Soc.*, **72**, 393 (1950).

1-METHYLAMINOANTHRAQUINONE *

(Anthraquinone, 1-methylamino-)

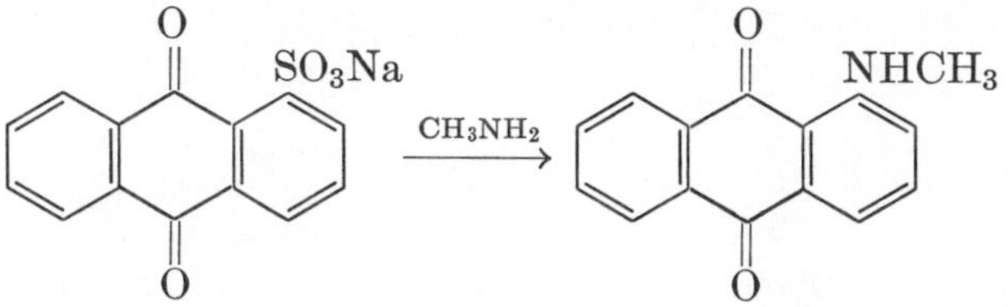

Submitted by C. V. Wilson, J. B. Dickey, and C. F. H. Allen.
Checked by George L. Evans and R. S. Schreiber.

1. Procedure

A 1-gal. autoclave (p. 80) (Note 1) is charged with 399 g. (1.29 moles) of technical sodium anthraquinone-α-sulfonate (Note 2), 45 g. (0.43 mole) of sodium chlorate, 780 g. (6.25 moles) of a 25% aqueous solution of methylamine, and 1.2 l. of water. The mixture is heated, with stirring, for 12 hours at 130–135° (Note 3). The heat is then shut off, but stirring is continued so that the product separates in an easily removable form. When cold, the autoclave is opened and the contents are removed; the material adhering to the walls is removed by water. The solid is filtered on a 13-cm. Büchner funnel. The red product is washed with two 500-ml. portions of hot water (70°) and dried in the air. The yield of 1-methylaminoanthraquinone, melting at 166–171° (Note 4), is 180–199 g. (59–65%) (Notes 5 and 6).

2. Notes

1. A shaking autoclave employed for high-pressure hydrogenations may be used equally well, but the quantities taken must be reduced to 155 g. (0.5 mole) of sodium anthraquinone-α-sulfonate, 17.5 g. of sodium chlorate, 300 g. of a 25% aqueous methylamine solution, and 600 ml. of water. The checkers used a 2-gal. stirred autoclave (stainless steel).

2. A corresponding quantity of the potassium salt [1] can be utilized.

3. According to the checkers the heating time may be decreased to 8 hours if desirable.

4. The melting point varies slightly with the method of heating; if the bath is preheated to 160° before the sample is inserted, the melt-

ing point is 168–169.5° When taken in the ordinary way, the melting point is 166–171°. This product is sufficiently pure for most purposes. One recrystallization from toluene raises the melting point 1°.

5. The yield depends upon the purity of the sodium anthraquinone-α-sulfonate. Apparent yields of as high as 87% have been obtained. The checkers employed technical-grade material, which apparently resulted in an appreciable decrease in the yields of 76–80% reported by the submitters.

6. α-Chloroanthraquinone [2] can be used as a starting material. In this case, 433 g. (1.79 moles) is taken, together with 1.5 l. of pyridine, 600 ml. of 25% aqueous methylamine, and 2.5 g. of a copper salt. The product is washed with dilute (2%) hydrochloric acid. The yield is 380–400 g. (90–95%).

3. Methods of Preparation

1-Methylaminoanthraquinone has been prepared from 1-chloro-, 1-bromo-, and 1-nitroanthraquinone by treatment with alcoholic methylamine under pressure,[3] from 1-methoxy- and 1-phenoxyanthraquinone with methylamine in pyridine solution at 150°; [4] from potassium anthraquinone-1-sulfonate with aqueous methylamine at 150–160°; [5,6] from 1-aminoanthraquinone by treatment with formaldehyde,[7] or methanol [8] in sulfuric acid or oleum; and by hydrolysis of *p*-toluenesulfonylmethylaminoanthraquinone with sulfuric acid.[9]

[1] *Org. Syntheses* Coll. Vol. **2,** 539 (1943).
[2] *Org. Syntheses* Coll. Vol. **2,** 128 (1943).
[3] Ger. pat. 144,634 [*Frdl.,* **7,** 201 (1902–1904)].
[4] Ger. pat. 165,728 [*Frdl.,* **8,** 289 (1905–1907)].
[5] Ger. pat. 175,024 [*Frdl.,* **8,** 283 (1905–1907)].
[6] Ger. pat. 256,515 [*Frdl.,* **11,** 551 (1912–1914)].
[7] Ger. pat. 156,056 [*Frdl.,* **8,** 288 (1905–1907)].
[8] Ger. pat. 288,825 [*Frdl.,* **12,** 414 (1914–1916)].
[9] Ullmann and Fodor, *Ann.,* **380,** 320 (1911).

1-METHYLAMINO-4-BROMOANTHRAQUINONE

(Anthraquinone, 1-methylamino-4-bromo-)

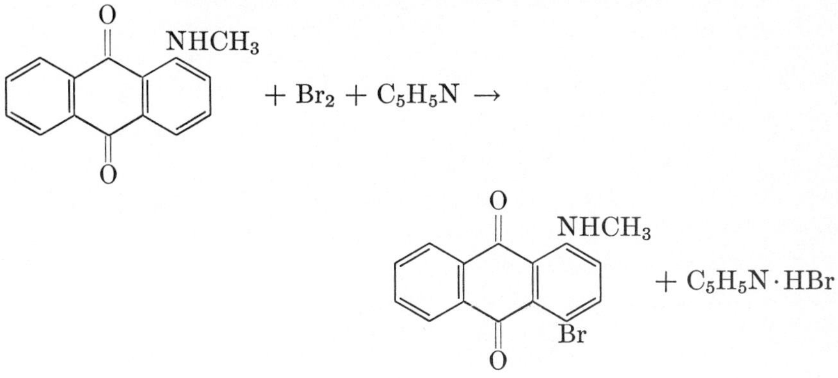

Submitted by C. V. Wilson.
Checked by G. L. Evans and R. S. Schreiber.

1. Procedure

In a 2-l. three-necked flask having ground-glass joints and equipped with a mechanical stirrer, a condenser (Note 1), and a dropping funnel are placed 119 g. (0.5 mole) of 1-methylaminoanthraquinone (p. 573) and 600 ml. of pyridine (Notes 2 and 3). The stirrer is started, and 90 g. (29 ml., 0.56 mole) of bromine is added over a period of 9–10 minutes. The flask and contents are now heated on the steam bath for 6 hours with continuous stirring. At the end of this period the hot mixture is transferred from the flask to a beaker and allowed to cool. The solid that separates (Note 4) is collected on a Büchner funnel and is washed thoroughly with hot water to remove a considerable portion of pyridine hydrobromide which is precipitated along with the desired product. The resulting deep red 1-methylamino-4-bromoanthraquinone, after thorough drying, weighs 111–117 g. (70–74%). It melts at 193–195° and is pure enough for most purposes (Note 5).

2. Notes

1. An air condenser of any type is sufficient.

2. A good grade of pyridine is essential. Very poor results are obtained with the practical or technical grades.

3. Larger volumes of pyridine have been used, but the yield drops progressively with increasing amounts.

4. Some of the solid separates during the heating on the steam bath.

5. It may be recrystallized, if desired, from pyridine, by the use of 3.5 ml. per g.; the melting point is raised to 195–196°.

3. Methods of Preparation

4-Bromo-1-methylaminoanthraquinone has been prepared from 4-bromo-1-nitroanthraquinone and methylamine at 60°,[1] and from 1-methylaminoanthraquinone in pyridine solution by treatment with two moles of bromine.[2] This procedure is based on the latter method.

[1] Ger. pat. 144,634 [*Frdl.*, **7**, 201 (1902–1904)].

[2] Ger. pat. 164,791 [*Frdl.*, **8**, 280 (1905–1907)].

METHYL β-BROMOPROPIONATE *

(Propionic acid, β-bromo-, methyl ester)

$$CH_2{=}CHCOOCH_3 + HBr \rightarrow BrCH_2CH_2COOCH_3$$

Submitted by RALPH MOZINGO and L. A. PATTERSON.
Checked by NATHAN L. DRAKE and HOMER CARHART.

1. Procedure

Six hundred and fifty grams of a 60% solution of methyl acrylate in methanol to which has been added 4 g. of hydroquinone (Note 1) is washed, successively, with 800-ml., 400-ml., and 200-ml. portions of a 7% sodium sulfate solution. The methyl acrylate layer is dried by shaking with 45 g. of anhydrous sodium sulfate for 20–30 minutes. The ester is then removed from the sodium sulfate by filtration and used without distillation. The yield is 280–325 g.

A solution of 258 g. (3 moles) of washed and dried methyl acrylate in 500 ml. of anhydrous ether is placed in a 1-l. round-bottomed flask. The flask is fitted with a rubber stopper carrying a drying tube and an 8-mm. glass inlet tube for hydrogen bromide. The inlet tube, which extends almost to the bottom of the flask, is connected through a 1-l. safety trap to a hydrogen bromide generator (Note 2). The flask with its contents is placed in an ice bath, and 245 g. (3.03 moles) of anhydrous hydrogen bromide is passed into the solution (Note 3).

After the hydrogen bromide has been added, the flask is stoppered and allowed to stand for about 20 hours at room temperature.

The ether is removed by distillation (Note 4) from a hot-water bath. At the end of the distillation, the water bath is heated to 80–85°, and when no more liquid comes over at this temperature the residue is transferred to a 500-ml. modified Claisen distilling flask and distilled under reduced pressure. The methyl β-bromopropionate distils at 64–66°/18 mm. and weighs 410–428 g. (80–84%) (Notes 5 and 6).

2. Notes

1. Methyl acrylate in methanol is available from Rohm and Haas Company, Philadelphia, Pennsylvania. Since the ester polymerizes in the presence of peroxides, it is necessary to add some hydroquinone as an inhibitor. The ester should not be stored for long periods of time, even when it contains hydroquinone. Storage should be in a refrigerator.

2. The hydrogen bromide [*Org. Syntheses* Coll. Vol. **2,** 338 (1943)] may be completely freed from bromine by bubbling it through a solution of phenol in carbon tetrachloride.

3. The addition may be as rapid as is convenient without the loss of ether due to the exothermic reaction.

4. The ether may be removed under anhydrous conditions and used for a subsequent preparation of the ester without further treatment.

5. Ethyl β-bromopropionate may be prepared in the same manner in about 90% yield. The boiling point of the ethyl ester is 77–79°/19 mm.

6. The residue consists largely of β-bromopropionic acid which may be recovered by distillation, b.p. 115–120°/18 mm., followed by recrystallization from carbon tetrachloride. The yield of this acid has never been more than 5% of the theoretical amount.

3. Methods of Preparation

Methyl β-bromopropionate has been prepared by the esterification of β-bromopropionic acid with methyl alcohol alone [1] and through the use of hydrogen bromide as a catalyst,[2] and by the direct addition of hydrogen bromide to methyl acrylate.[2]

[1] Le Mer and Kamner, *J. Am. Chem. Soc.*, **53**, 2833 (1931).

[2] Moureu, Murat, and Tampier, *Ann. chim.*, **15**, 221 (1921).

METHYL 5-BROMOVALERATE

(Valeric acid, δ-bromo-, methyl ester)

$$CH_3O_2C(CH_2)_4CO_2H + AgNO_3 \xrightarrow{KOH} CH_3O_2C(CH_2)_4CO_2Ag + KNO_3 + H_2O$$

$$CH_3O_2C(CH_2)_4CO_2Ag + Br_2 \xrightarrow{CCl_4} CH_3O_2C(CH_2)_3CH_2Br + AgBr + CO_2$$

Submitted by C. F. H. Allen and C. V. Wilson.
Checked by Cliff S. Hamilton and Noboru Tosaya.

1. Procedure

To a solution of 33 g. (0.5 mole) of potassium hydroxide (Note 1) in 1.5 l. of distilled water in a 5-l. flask or other appropriate container fitted with a mechanical stirrer is added 80 g. (0.5 mole) of methyl hydrogen adipate (Note 2). With continuous stirring a solution of 85 g. (0.5 mole) of silver nitrate in 1 l. of distilled water is added rapidly (about 15 minutes). The precipitated methyl silver adipate is collected on a Büchner funnel, washed with methanol, and dried in an oven at 50–60°. For the next step the dried silver salt is finely powdered and sieved through a 40-mesh screen. The combined yield from two such runs is 213 g. (80%).

The 213 g. (0.8 mole) of finely powdered silver salt is placed in a 1-l. three-necked flask (Note 3); two necks of the flask are stoppered, and the third is connected to a water pump through a U-tube or flask containing Drierite. The flask is then placed in an oil bath and evacuated to a pressure of about 15 mm. The temperature of the oil bath is maintained at 100–110° for 36 hours (Notes 4 and 5).

The pressure in the flask is restored to that of the atmosphere. The flask is removed from the oil bath and equipped with a dropping funnel, condenser, and mechanical stirrer (Note 6). To the salt is added 350 ml. of dry carbon tetrachloride (Note 7); the stirrer is started, and 117 g. (40 ml., 0.73 mole) of dry bromine (p. 790, Note 2) is added through the dropping funnel over a 30- to 40-minute period. Occasional cooling may be necessary as the reaction is quite vigorous at first. When all the bromine has been added, the mixture is heated for 1 hour on a steam bath. It is then filtered and the silver bromide washed thoroughly on the filter with 100 ml. of warm carbon tetrachloride. The filtrate is washed once with 100 ml. of 10% sodium carbonate solution and dried over 30–40 g. of Drierite. The solvent is removed and the residue distilled under reduced pressure. The yield of product boiling at 75–80°/4 mm. is 101–106 g. The yield is 65–68% based on the weight of the methyl silver adipate before drying under reduced pressure, or 52–54% based upon methyl hydrogen adipate.

2. Notes

1. Reagent grade potassium hydroxide containing 85% potassium hydroxide is used.

2. The methyl hydrogen adipate was Eastman Kodak grade.

3. Drying is carried out in the flask in which the final reaction is to be run in order to avoid a transfer. The success of this preparation depends upon the exclusion of moisture. The silver salt retains traces of water tenaciously.

4. If the indicated pressure is maintained the water pump may be disconnected, but owing to leaks it will usually be found necessary to re-evacuate several times over the 36-hour period.

5. A good vacuum oven would serve for drying just as well, but the temperature of the salt should not exceed 110°.

6. For best results all equipment should be thoroughly dried.

7. The carbon tetrachloride is dried over phosphorus pentoxide or some other drying agent.

3. Methods of Preparation

This method with some slight modifications is applied in the synthesis of ω-bromo esters from C_5 to C_{17}.[1] Methyl 5-bromovalerate has been prepared by treating the silver salt of methyl hydrogen adipate with bromine.[1] The ethyl ester has been prepared from the acid by esterification [2,3] or through the acid chloride.[3]

[1] Hunsdiecker and Hunsdiecker, *Ber.*, **75**, 296 (1942).

[2] Cloves, *Ann.*, **319**, 367 (1901).

[3] Merchant, Wickert, and Marvel, *J. Am. Chem. Soc.*, **49**, 1829 (1927).

4-METHYLCARBOSTYRIL

(Carbostyril, 4-methyl-)

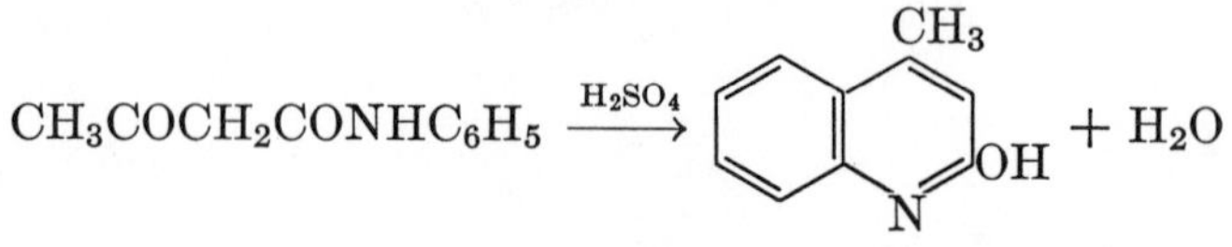

Submitted by W. M. LAUER and C. E. KASLOW.
Checked by C. F. H. ALLEN and H. W. J. CRESSMAN.

1. Procedure

One hundred seventy-seven grams (1 mole) of acetoacetanilide is added in small portions by means of a spatula to 185 ml. of concentrated sulfuric acid which has been heated previously to 75° (Note 1) in a 1-l. three-necked flask provided with a mechanical stirrer and a thermometer which extends into the liquid. The temperature of the mixture is maintained at 70–75° by intermittent cooling until nearly all the acetoacetanilide has been added. The last 10–15 g. is added without cooling, and the temperature rises to 95°; the addition requires 20–30 minutes. The heat of reaction maintains the temperature at 95° for about 15 minutes; the reaction mixture is then kept an additional 15 minutes at 95° by external heating. After the solution has cooled to 65°, it is poured into 5 l. of water with vigorous stirring.

After cooling, the product is filtered by suction, washed with four 500-ml. portions of water and two 250-ml. portions of methanol, and air-dried. The yield of 4-methylcarbostyril is 138–144 g. (86–91%). This material, which melts at 219–221°, is suitable for preparing 2-chlorolepidine (p. 194). It may be purified further by recrystallization from 95% ethanol. For recrystallization 39 g. is dissolved in 650 ml. of solvent; the recovery is 33–33.5 g., and the melting point of the product is 222–224°.

2. Note

1. The reaction flask must be so situated that it can be cooled rapidly. The submitter reports that the yield was reduced to 72% in one run in which the temperature reached 120°.

3. Methods of Preparation

The only useful method for preparing 4-methylcarbostyril is that of Knorr,[1] described by Mikhaĭlov.[2] Two modifications include the use of aniline and ethyl acetoacetate, without isolation of acetoacetanilide,[3] and the use of boron trifluoride as a cyclization agent.[4]

[1] Knorr, *Ann.*, **236**, 83 (1886).
[2] Mikhaĭlov, *J. Gen. Chem. U.S.S.R.*, **6**, 511 (1936) [*C. A.*, **30**, 6372 (1936)].
[3] Hauser and Reynolds, *J. Am. Chem. Soc.*, **70**, 2402 (1948).
[4] Killelea, *J. Am. Chem. Soc.*, **70**, 1971 (1948).

4-METHYLCOUMARIN

(Coumarin, 4-methyl-)

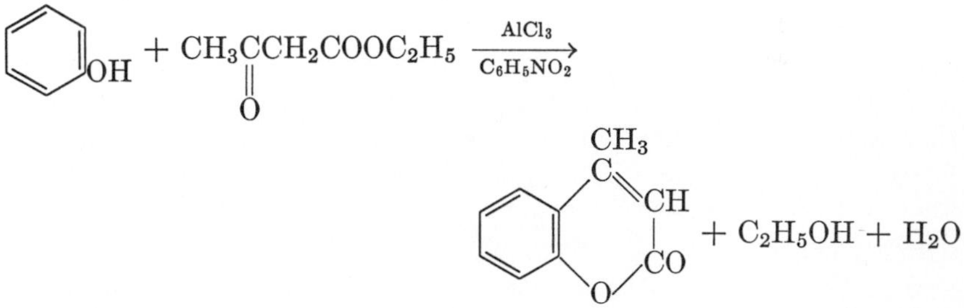

Submitted by Eugene H. Woodruff.
Checked by Nathan L. Drake and Carl Blumenstein.

1. Procedure

In a 5-l. three-necked round-bottomed flask, fitted with a sealed stirrer, dropping funnel, and air condenser, the open end of which is connected to a gas-absorption trap, are placed 188 g. (2 moles) of phenol and 268 g. (2 moles) of ethyl acetoacetate in 300 ml. of dry nitrobenzene (Notes 1 and 2). The mixture is heated to 100° by means of an oil bath, stirring is started, and 532 g. (4 moles) of anhydrous aluminum chloride dissolved in 2.1 l. of dry nitrobenzene is added over a period of 30–45 minutes (Note 3). The dropping funnel is then replaced by a thermometer, and the temperature of the solution is raised to 130° and held there 3 hours (Note 4). By this time the evolution of hydrogen chloride will have practically ceased.

The solution is cooled until its temperature is approximately that of the room, and 500 ml. of a mixture of equal parts of concentrated hydrochloric acid and water is added with stirring. The flask is then arranged for steam distillation and heated with a free flame while steam is passed into the reaction mixture until about 200 ml. of distillate is collected (Note 5). While hot, the mixture is placed in a large separatory funnel and the lower aqueous layer removed; the nitrobenzene layer is filtered through a Büchner funnel (Notes 6 and 7).

The nitrobenzene is then removed by vacuum distillation using a 2-l. Claisen flask, and the residue in the flask is distilled under diminished pressure; the fraction boiling at 180–195°/15 mm. is collected (Notes 8 and 9). The red-yellow oil, which solidifies on cooling, amounts to 128–176 g. (40–55%) and is sufficiently pure for some purposes (Note 10). A nearly white product can be obtained by dissolving the crude product in ether, shaking the ether solution with portions of 5% sodium hydroxide solution until no color appears in the aqueous layer, evaporating the ether, and recrystallizing the residue from a mixture of petroleum ether (b.p. 60–90°, Skellysolve B) and benzene (Note 11). The recrystallized product melts at 83–84°; the recovery is 80–85%.

2. Notes

1. Methyl acetoacetate has been used but insufficient data have been obtained as to the yield produced.

2. Commercial nitrobenzene is dried by distillation under reduced pressure until the distillate is clear. The residue is suitable for use without other treatment.

3. The aluminum chloride is added in 25- to 50-g. portions to the 2.1 l. of dry nitrobenzene contained in a 5-l. round-bottomed flask with stirring after each addition. The temperature may rise to 80–90° during the addition; occasional cooling of the flask under running water is necessary. After all the aluminum chloride has been added, the solution is cooled to room temperature. A small amount of aluminum chloride may settle to the bottom.

In one instance when the final cooling was omitted the temperature continued to rise and a large amount of hydrogen chloride was evolved. The residue in the flask was a thick paste. Addition of a portion of the paste to water caused nitrobenzene to separate. However, this batch was not suitable for condensation.

4. Below 100° the condensation does not take place; at 150° tar is formed and no coumarin can be isolated.

5. Some unchanged ester distils before the nitrobenzene; its removal aids in the subsequent separation of the two layers.

6. If a sufficiently large funnel is not available the separation may be done by dividing the solution in portions to suit the funnel available. If too much steam has condensed during the distillation, the aqueous layer may appear on top.

7. The tarry by-products clog filter paper immediately; a small quantity of filter aid may be used to promote the filtration. If the condensation has been satisfactory only a small amount of material will be retained, and this step may be omitted. Occasionally a considerable amount of lumpy, tarry material is removed; filtration is then more difficult, and the yield is lower. The tar, on distillation, yields no coumarin.

8. The recovered nitrobenzene, after being washed with water and dried as in Note 2, may be used again. The loss is from 400 to 700 ml.

9. Considerable tar remains in the distillation flask. The tar is best poured immediately into a flat pan of water or a sink, where it solidifies to a hard brittle mass; that adhering to the walls of the flask is easily removed by adding 100–200 ml. of nitrobenzene to the flask and heating the solvent to boiling under atmospheric pressure.

10. According to the submitter, when quantities 3 times this size are used, the yield of distilled material is 38–50%.

11. The ratio of petroleum ether to benzene should be 8 to 2.

3. Methods of Preparation

4-Methylcoumarin is prepared by the condensation of phenol and acetoacetic ester. Concentrated sulfuric acid [1,2] and 73% sulfuric acid have been used.[3,4] The method given here was mentioned by Sethna, Shah, and Shah,[5] and the procedure is adapted from that of the same authors [6] for another coumarin derivative.

[1] Von Pechmann and Duisberg, *Ber.*, **16**, 2127 (1883).

[2] Von Pechmann and Kraft, *Ber.*, **34**, 421 (1901).

[3] Peters and Simonis, *Ber.*, **41**, 831 (1908).

[4] Fries and Volk, *Ann.*, **379**, 94, footnote 1 (1911).

[5] Sethna, Shah, and Shah, *Current Sci.*, **6**, 93 (1937) [*C. A.*, **32**, 549 (1938)].

[6] Sethna, Shah, and Shah, *J. Chem. Soc.*, **1938**, 228.

METHYL ω-CYANOPELARGONATE

(Pelargonic acid, θ-cyano-, methyl ester)

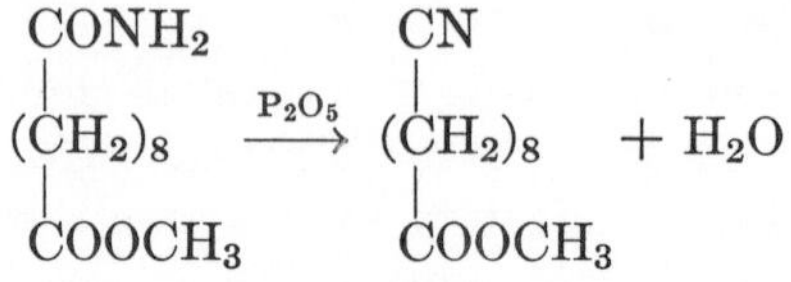

Submitted by W. S. Bishop.
Checked by C. F. H. Allen and J. VanAllan.

1. Procedure

Caution! Tetrachloroethane is toxic; the operations should be conducted in a hood.

In a 1.5-l. Erlenmeyer flask to which is attached a reflux condenser, 190 g. (0.88 mole) of methyl sebacamate (p. 613) is dissolved in 200 ml. of boiling tetrachloroethane. The solution is allowed to cool to 40–50° (Note 1), 95 g. (0.67 mole) of phosphorus pentoxide is added (Note 2), and the mixture is stirred well by means of a glass rod. The mixture is heated in an oil bath to 120° (thermometer in oil), and a second 95-g. portion of phosphorus pentoxide is added. After the mixture has been heated at 145° for 30 minutes with occasional hand stirring, the liquid is decanted. The residue is heated at 145° with 200 ml. of tetrachloroethane for 30 minutes with occasional stirring, and the liquid is decanted. This process is repeated once. The combined extracts are placed in a 1-l. flask, and most of the solvent is distilled under the reduced pressure of a water pump. The residue is transferred to a 300-ml. flask, and the remainder of the solvent is removed (Note 3). When no more distillate comes over, the receiver is changed, the water pump is replaced by an oil pump, and the residue is fractionated (Note 4). The yield of methyl ω-cyanopelargonate boiling at 121–124°/1 mm. (Note 5) is 119–124 g. (69–71%).

2. Notes

1. The slush that results on cooling is easily mixed with the phosphorus pentoxide.

2. The phosphorus pentoxide is weighed rapidly on a piece of paper, from which it can be slid quickly into the flask.

3. About 550–560 ml. of tetrachloroethane is recovered. Ground-glass equipment is preferable for the distillations.

4. There is no fore-run. At 1 mm., the thermometer reads about 118° when the first drop appears at the end of the side tube of the distilling flask. There is a 5- to 6-g. fraction, b.p. 124–135°/1 mm., and some residue.

5. Other boiling points are 154°/5 mm. and 170°/14 mm.

3. Methods of Preparation

Methyl ω-cyanopelargonate has also been prepared by esterification of ω-cyanopelargonic acid with methyl sulfate [1] or methanol,[2] and by dehydration of methyl sebacamate with phosphorus pentoxide [3] or thionyl chloride.[4] The procedure described appears in the literature.[1,3]

[1] Biggs and Bishop, *J. Am. Chem. Soc.*, **63**, 944 (1941).

[2] Biggs, U. S. pat. 2,339,672 [*C. A.*, **38**, 3990 (1944)].

[3] Bishop, U. S. pat. 2,277,033 [*C. A.*, **36**, 4636 (1942)].

[4] Soffer, Strauss, Trail, and Sherk, *J. Am. Chem. Soc.*, **69**, 1884 (1947).

N-METHYL-3,4-DIHYDROXYPHENYLALANINE

(Alanine, β-(3,4-dihydroxyphenyl)-N-methyl-)

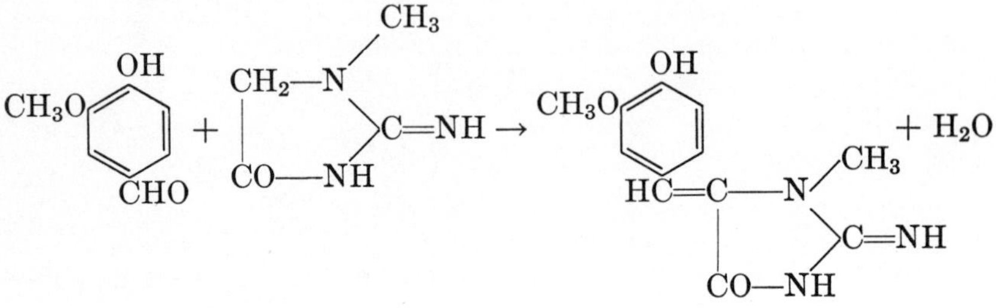

I

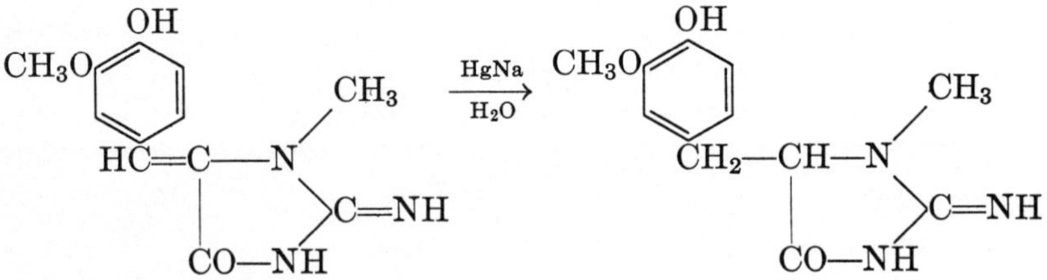

II

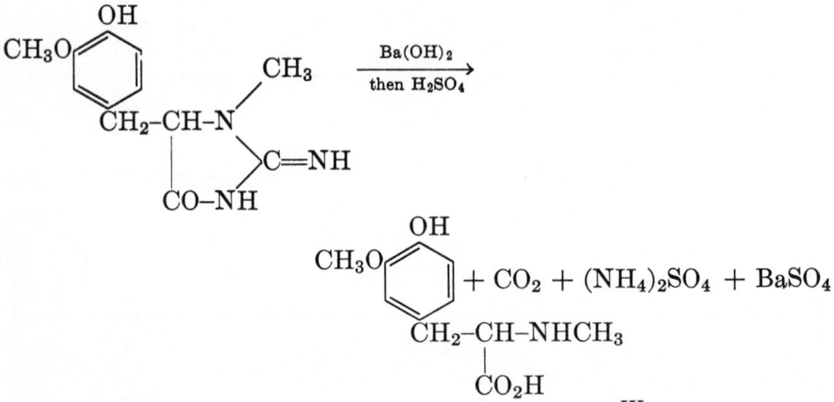

III

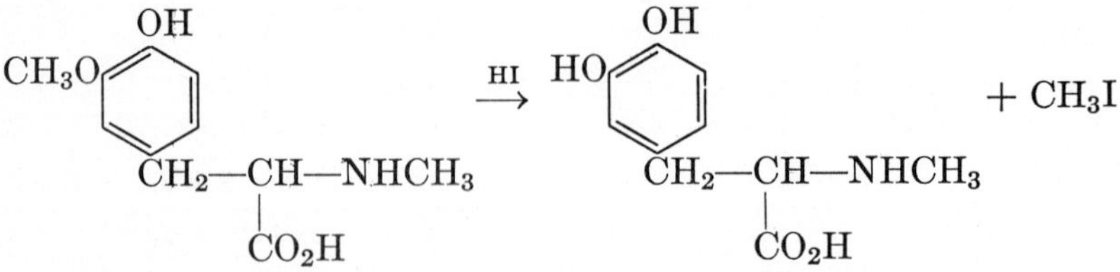

IV

Submitted by V. DEULOFEU and T. J. GUERRERO.
Checked by HOMER ADKINS, HAROLD H. GELLER, and EVERETT E. BOWDEN.

1. Procedure

A. 5-(3-*Methoxy*-4-*hydroxybenzal*)*creatinine* (I). In a large Pyrex test tube (51 mm. outside diameter and 200 mm. long) is placed an intimate mixture of 11.3 g. (0.1 mole) of creatinine [*Org. Syntheses* Coll. Vol. **1,** 172 (1941)] and 24 g. (0.16 mole) of vanillin. The tube is placed in an oil bath, which is then heated to 170°, and the mixture is melted while it is constantly agitated (Note 1). The temperature of the mixture reaches 155° in about 10 minutes; reaction then begins, and water is evolved. After 3–5 minutes longer, evolution of water ceases, and the mixture solidifies. The tube is heated for 3 minutes more, and then it is removed from the bath and allowed to cool.

When the temperature has fallen to 50–60°, 50 ml. of ethanol is added and the mixture is heated gently by occasional immersion in the warm oil bath. The solid partially disintegrates and forms a suspension. The suspension is filtered, and the solid remaining in the tube is warmed with a second 50-ml. portion of ethanol. This operation is repeated until all the orange-colored condensation product has been transferred to the filter. The material on the filter is then washed with three successive 30-ml. portions of water at 60°.

After drying, the crude product weighs 24 g. (95%), melts at 261–263°, and is suitable for use in the next step. A pure product, which melts at 273°, may be obtained by recrystallizing the crude material from acetic acid.

B. 5-(3-*Methoxy*-4-*hydroxybenzyl*)*creatinine* (II). To a suspension of 24 g. (0.1 mole) of the crude condensation product in 150 ml. of water, contained in a 500-ml. beaker, there is added, with continuous agitation, 180 g. of 3% sodium amalgam (Note 2). The amalgam is added in six portions, at intervals of 5 minutes. The solid dissolves, and the initial orange-red color of the solution slowly fades as the reduction proceeds. With good agitation, decolorization is complete in 45–60 minutes, if the starting material is pure. When the crude condensation product is used, the color of the solution fades to a faint, but permanent, yellow tint, which should mark the end point of the reduction.

The solution is decanted from the mercury and filtered from suspended impurities. The filtrate is stirred and is acidified with hydrochloric acid to *p*H 6.6, phenol red being used as the indicator (Note 3). After standing for 2 hours at 0°, the mixture is filtered and the solid is washed with a little cold water and dried. The product (free base) is microcrystalline, weighs 17.5–18 g. (72–74%), and usually melts at 167–169° (Note 4). After solidification from fusion, the substance melts at 226–228°. When recrystallized from water, the substance melts at 231–233°. The crude product may be used for the subsequent hydrolysis.

C. *N-Methyl-(3-methoxy-4-hydroxyphenyl)alanine* (III). In a 2-l. round-bottomed flask (Note 5), 18 g. (0.07 mole) of the crude reduction product is refluxed for 12 hours with a solution of 180 g. of crystalline barium hydroxide in 270 ml. of water. The hot solution is diluted with 1.2 l. of water, and the barium is precipitated by addition of 250–270 ml. of 6 *N* sulfuric acid (Note 6). The precipitated barium sulfate is separated by centrifuging and washed with two 100-ml. portions of water; the combined water solutions are evaporated under reduced pressure at 50° to a volume of about 50 ml. The acid solution is made alkaline to litmus by addition of about 10 ml. of a 12% solution of ammonium hydroxide in water. After standing for 24 hours at 0°, the mixture is filtered, and the solid is washed with cold water and dried. The yield is 12 g. (74%) (Note 7). On rapid heating, the solid melts at 273–275°. When recrystallized from water, the substance melts at 276–278°. The crude product may be used for the next step.

D. *N-Methyl-3,4-dihydroxyphenylalanine* (IV). In a carbon dioxide atmosphere, 12 g. (0.05 mole) of the methoxy compound is boiled gently for 3 hours with 24 g. of red phosphorus and a mixture of 60 ml. of acetic anhydride and 60 ml. of hydriodic acid (sp. gr. 1.7). The phosphorus is then removed by filtration and washed with 25 ml. of 50% acetic acid. The filtrates are combined and, in a current of carbon dioxide, are evaporated to a syrup at 50° under 35 mm. pressure. A 60-ml. portion of warm water is then added, and the solution is evaporated as before. The residue is dissolved in 100 ml. of water, and dilute ammonia (10% by volume) is added until the solution does not change Congo red paper to blue (Note 8). The mixture is allowed to stand for 2 hours at 0°, and then the white crystalline precipitate is filtered. The precipitate is washed on the funnel with a little water containing sulfur dioxide, and is dried by washing with ethanol and ether. The product weighs 9.5 g. (82%). When slowly heated it becomes slightly brown at 230° and melts at 282–283°; when rapidly heated it becomes slightly brown at 255–260° and melts at 290–292°.

This material may be purified by boiling 1 g. of it with 50 ml. of water containing sulfur dioxide, filtering the solution, and keeping the filtrate at 0° for 24 hours. The purified product (0.74 g.), when slowly heated, becomes slightly brown at 245° and melts at 287°; when rapidly heated it becomes slightly brown at 260° and melts at 298–300°.

2. Notes

1. Sometimes the evolution of water is too rapid, and there is excessive foaming. The reaction may be moderated by removing the tube from the oil bath.

2. The amalgam may be made by adding 5-mm. cubes of sodium (5.4 g.) to 175 g. of mercury, warmed to 30–40°, and contained in a mortar or Erlenmeyer flask. The mortar is covered with an asbestos board having a small hole in the center. The cubes of sodium are fixed on the end of a pointed glass rod and are quickly pushed through this hole beneath the surface of the mercury. A more convenient method is described in Fieser, *Experiments in Organic Chemistry*, 2nd Ed., 1941, D. C. Heath and Company, Boston, p. 419.

3. The solution becomes dark in color if allowed to stand overnight.

4. In some cases the melting point may be as low as 98°; but after melting and solidifying, the substance melts at 226–228°.

5. A large flask should be used, since the mixture foams during the hydrolysis.

6. An excess of acid must be added to make certain that all the amino acid is in solution.

7. The yield can be increased slightly by evaporation of the mother liquor.

8. If the solution is made too alkaline, colored impurities precipitate. Neutralization must be performed carefully since the end point to Congo red is reached just before the colored impurities precipitate.

3. Methods of Preparation

5-(3-Methoxy-4-hydroxybenzal)creatinine was first prepared by a similar method by Richardson, Welch, and Calvert.[1]

N-Methyl-3,4-dihydroxyphenylalanine has been prepared by methylation of α-acetamino-3,4-methylenedioxycinnamic acid, followed by reduction and hydrolysis of the product;[2] and by a method similar to that outlined above.[3]

N-Methylaminoaromatic acids have been prepared by a variety of methods: by the reaction between methylamine and an α-bromo acid;[4] by condensing methylhydantoin with aromatic aldehydes;[5] by condensation of creatinine[6] or benzoyl sarcosine with aromatic aldehydes;[7] by methylation of the toluenesulfonyl derivative of the amino acid;[8] and by substituting methylamine for ammonia in the Strecker synthesis.[9]

[1] Richardson, Welch, and Calvert, *J. Am. Chem. Soc.*, **51**, 3074 (1929).

[2] Heard, *Biochem. J.*, **27**, 54 (1933).

[3] Guerrero and Deulofeu, *Ber.*, **70**, 947 (1937).

[4] Friedmann and Gutmann, *Biochem. Z.*, **27**, 491 (1910).

[5] Johnson and Nicolet, *Am. Chem. J.*, **47**, 459 (1912).

[6] Nicolet and Campbell, *J. Am. Chem. Soc.*, **50**, 1155 (1928); Deulofeu and Mendivelzua, *Ber.*, **68**, 783 (1935).

[7] Deulofeu, *Ber.*, **67**, 1542 (1934).

[8] Fischer and Lipschitz, *Ber.*, **48**, 360 (1915).

[9] Kanewskaja, *J. prakt. Chem.*, (2) **124**, 48 (1929).

N-METHYLFORMANILIDE *

(Formanilide, N-methyl-)

$$C_6H_5NHCH_3 + HCOOH \rightarrow C_6H_5N(CH_3)CHO + H_2O$$

Submitted by L. F. Fieser and J. E. Jones.
Checked by C. F. H. Allen and J. VanAllan.

1. Procedure

In a 3-l. round-bottomed flask fitted with a 3-ft. indented column to which is attached a condenser set for downward distillation are placed 321 g. (3 moles) of methylaniline, 300 g. of formic acid (85–90%), and 1.8 l. of toluene (Note 1). The solution is distilled slowly. As long as the azeotrope containing water is present, the temperature of the vapor is 87–88°; when the water has been removed, the temperature rises to 108–110° (Note 2). The distillation is continued until approximately 1.5 l. of toluene has been collected (5–6 hours). The residue is then transferred to a modified Claisen flask and distilled under reduced pressure, the portion boiling at 114–121°/8 mm. being collected. This has a freezing point of 13.6–13.7°; n_D^{29} 1.553–1.555. The yield is 380–393 g. (93–97%). This product is satisfactory for the preparation of aldehydes. Upon redistillation it boils at 117–121°/8 mm., 130–132°/22 mm. The freezing point and refractive index are unchanged (Note 3).

2. Notes

1. The toluene serves to remove the water and minimize side reactions.

2. The water layer of the distillate is separated; it amounts to 140–150 ml.

3. If a water separator with a stopcock is employed, in such a way that the organic solvent is returned to the reaction flask, it is possible to use 200 ml. of benzene in place of the 1.8 l. of toluene (J. Meek, private communication).

3. Methods of Preparation

N-Methylformanilide has been obtained in a yield of 67.5% by heating methylaniline with formamide in glacial acetic acid solution.[1] The method above is a modification of that of Morgan and Grist,[2] who heated the amine and formic acid in the absence of a solvent or water-carrier (the present authors obtained a yield of only 40–50% by that procedure).

[1] Hirst and Cohen, *J. Chem. Soc.*, **67**, 830 (1895).
[2] Morgan and Grist, *J. Chem. Soc.*, **113**, 690 (1918).

β-METHYLGLUTARIC ACID

(Glutaric acid, β-methyl-)

$$2CNCH_2CONH_2 + CH_3CHO \rightarrow CH_3CH[CH(CN)CONH_2]_2 + H_2O$$

$$CH_3CH[CH(CN)CONH_2]_2 + 6H_2O + 4HCl \rightarrow HO_2CCH_2CH(CH_3)CH_2CO_2H + 4NH_4Cl + 2CO_2$$

Submitted by Robert E. Kent and S. M. McElvain.
Checked by Lee Irvin Smith and G. A. Boyack.

1. Procedure

A. *α,α′-Dicyano-β-methylglutaramide.* In a 6-l. flask, 520 g. (6.2 moles) of recrystallized cyanoacetamide (Note 1) is dissolved in 3.4 l. of water, and the solution is cooled to 10° and filtered if it is not clear (Note 2). While the flask is shaken constantly, 137.5 g. (3.1 moles) of freshly distilled acetaldehyde and 20 ml. of piperidine are added *successively* to the solution. After the mixture has stood at room temperature for 2 hours, the flask is transferred to an ice-salt bath and the mixture is partially frozen. During this operation, the flask should be shaken frequently. After 30 minutes, α,α′-dicyano-β-methylglutaramide begins to deposit, and when the precipitation is complete (about 1 hour), the mixture is allowed to come to room temperature in order to melt the ice that is present. The precipitate is then filtered with suction and washed thoroughly with cold distilled water. The yield is 420–425 g. (71%) of a white, powdery solid which melts at 152–157° (Note 3).

B. *β-Methylglutaric acid.* In a 5-l. flask are placed 400 g. of the amide and 1 l. of concentrated hydrochloric acid; the mixture is warmed on a steam bath until solution is complete, after which it is diluted with 1 l. of water and refluxed for 8 hours. The amber-colored solution is saturated with sodium chloride and extracted with five 600-ml. portions of ether. The combined ether extracts are dried for 1 hour over phosphorus pentoxide, and the solvent is removed by distillation. The residue, crude β-methylglutaric acid, weighs 238–240 g. (80%) and melts at 79–82° with previous softening. This product is recrystallized from about 250 ml. of 10% hydrochloric acid. The recovery is about 90% (Note 4), and the purified product melts at 85–86°.

2. Notes

1. Ammonia exerts a hindering effect on this condensation, and the yield is greatly reduced when crude cyanoacetamide is used.

2. At 10°, cyanoacetamide sometimes crystallizes. The mixture should not be filtered until it is certain that the precipitate contains no cyanoacetamide. Any cyanoacetamide that separates redissolves quickly after the acetaldehyde and piperidine have been added.

3. This amide is insoluble in the usual solvents, but it may be further purified, if desired, by trituration with dilute hydrochloric acid, followed by washing with hot absolute ethanol. It then melts sharply at 160–161°.

4. The checkers used one-tenth of the specified amounts of reagents. They obtained, in two runs, the following average yields of products: crude α,α′-dicyano-β-methylglutaramide, m.p. 152–157°, 75%; crude β-methylglutaric acid, m.p. 79–82°, 78%; purified β-methylglutaric acid, m.p. 84–85°, 80% (recovery). In the second run, however, the yields (except for the recovery in the recrystallization) were slightly higher than those given for the larger runs.

3. Methods of Preparation

The above method is adapted from the procedure of Day and Thorpe.[1] β-Methylglutaric acid has been prepared by hydrolysis of β-methylglutaronitrile;[2] by condensation of crotonic ester with ethyl sodiocyanoacetate,[3] and with sodiomalonic ester; [4,5] and by condensation of acetaldehyde with malonic ester.[6]

[1] Day and Thorpe, *J. Chem. Soc.*, **117**, 1465 (1920).
[2] Blaise and Gault, *Bull. soc. chim. France*, (4) **1**, 88 (1907).
[3] Howles, Thorpe, and Udall, *J. Chem. Soc.*, **77**, 948 (1900); Darbishire and Thorpe, *J. Chem. Soc.*, **87**, 1716 (1905).
[4] Hunsdiecker, *Ber.*, **75B**, 1199 (1942).
[5] Auwers, Kobner, and Meyenberg, *Ber.*, **24**, 2887 (1891).
[6] Knoevenagel, *Ber.*, **31**, 2585 (1899).

2-METHYL-4-HYDROXYQUINOLINE

(4-Quinolinol, 2-methyl-)

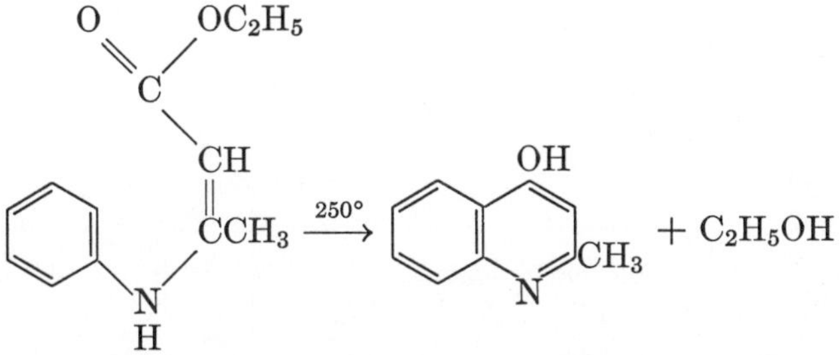

Submitted by George A. Reynolds and Charles R. Hauser.
Checked by Arthur C. Cope and William R. Armstrong.

1. Procedure

In a 500-ml. three-necked round-bottomed flask equipped with a dropping funnel, a sealed mechanical stirrer, and an air condenser is placed 150 ml. of Dowtherm (Note 1). The Dowtherm is stirred and heated at the reflux temperature while 65 g. (0.32 mole) of ethyl β-anilinocrotonate (p. 374) is added rapidly through the dropping funnel. Stirring and refluxing are continued for 10–15 minutes after the addition is completed. The ethanol formed in the condensation reaction may be allowed to escape from the condenser through a tube leading to a drain, or it may be collected by attaching a water-cooled condenser set for distillation to the top of the air condenser. The mixture is allowed to cool to room temperature, at which stage a yellow solid separates. Approximately 200 ml. of petroleum ether (b.p. 60–70°) is added; the solid is collected on a Büchner funnel and washed with 100 ml. of petroleum ether (b.p. 60–70°). After air drying, the crude product is treated with 10 g. of Darco or Norit in 1 l. of boiling water (Note 2). The hot solution is filtered and allowed to cool. The white, hairlike needles of 2-methyl-4-hydroxyquinoline are separated by filtration. The yield of product, melting at 235–236° (cor.), is 43–46 g. (85–90%).

2. Notes

1. Dowtherm is a mixture of diphenyl and diphenyl ether, obtainable from the Dow Chemical Company.

2. If the treatment with decolorizing carbon is omitted, the product has a low melting point after several crystallizations.

3. Methods of Preparation

The present procedure is a modification of preparations described previously.[1–3] The cyclization has been effected by allowing a solution of ethyl β-anilinocrotonate in concentrated sulfuric acid to stand at room temperature for several hours; by rapidly heating the ester to 240–250° without a solvent;[2] or by adding the ester to paraffin oil at 250°[3] or to refluxing diphenyl ether.[4] The procedure described has been applied to the preparation of other 2-substituted 4-hydroxyquinolines.[5]

[1] Knorr, *Ber.,* **16,** 2593 (1883).
[2] Conrad and Limpach, *Ber.,* **20,** 944 (1887).
[3] Cavalito and Haskell, *J. Am. Chem. Soc.,* **66,** 1166 (1944).
[4] Kaslow and Marsh, *J. Org. Chem.,* **12,** 456 (1947).
[5] Hauser and Reynolds, *J. Am. Chem. Soc.,* **70,** 2402 (1948).

1-METHYL-2-IMINO-β-NAPHTHOTHIAZOLINE

(Naphtho[1,2]thiazole, 1-methyl-1,2-dihydro-2-imino-)

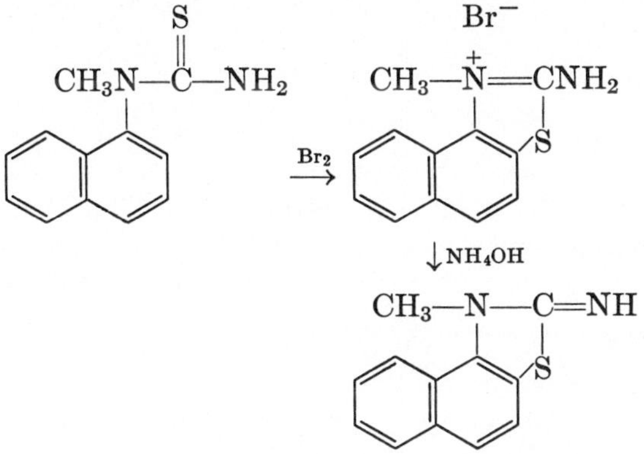

Submitted by HOMER W. J. CRESSMAN.
Checked by NATHAN L. DRAKE and WERNER R. BOEHME.

1. Procedure

In a 1-l. three-necked flask fitted with a thermometer, a dropping funnel, and a mechanical stirrer are placed 65 g. (0.3 mole) of 1-methyl-1-(1-naphthyl)-2-thiourea (p. 609) (Note 1) and 400 ml. of glacial acetic acid (Note 2). To the mechanically stirred suspension, maintained at 18–20° by cooling in a water bath, is added dropwise over a 30-minute period 48 g. (0.3 mole) of bromine in 50 ml. of glacial acetic acid. The light-yellow addition product is stirred an additional 15 minutes at 18–20°. After the thermometer is replaced by an outlet tube (Note 3), the mixture is heated in a water bath maintained at 80–85° (Note 4) for 3 hours. Hydrogen bromide is evolved copiously with the formation of a white, crystalline hydrobromide. When the mixture is cool, it is filtered and the precipitate is washed on the filter with 50 ml. of acetone and then with two 250- to 300-ml. portions of dry ether. One hundred milliliters of concentrated ammonia is then added with stirring to the salt suspended in 700 ml. of warm water (60–65°). The imine base first separates as an oil (Note 5); after the mixture has been stirred and warmed on the steam bath for 10 minutes, it is extracted with 500 ml. of chloroform and filtered by suction through a Norit filter pad 5–6 mm. thick. The bottom layer is separated, washed with 350 ml. of water, and dried by stirring with 40 g. of potassium carbonate. The residual oil, after removal of the solvent on the steam bath under reduced pressure, is poured into an evaporating dish and stirred while cooling. The brownish-colored crystals (Note 6) are dried at 80–85°; they melt at 97–99° and weigh 58–62 g. (90–97%). If desired, the naphthothiazoline (Note 7) can be crystallized from petroleum ether (60–90°) using 75 ml. per gram of product. The recovery of the almost colorless (cream) crystals is 85%. The recrystallized thiazoline melts at 97–99°.

2. Notes

1. The thiazoline was obtained in the best yields and with the lightest color from freshly prepared 1-methyl-1-(1-naphthyl)-2-thiourea. The air-dried thiourea loses hydrogen sulfide and ammonia after several days' storage.

2. The use of glacial acetic acid for cyclization is essential; in chloroform, the bromo-addition product does not lose hydrogen bromide.

3. A gas-absorption trap can be used to absorb the hydrogen bromide evolved.

4. The temperature of the water bath is easily maintained at 80–85° by heating it on a steam bath.

5. The oil solidifies on cooling. The crystalline product can be separated by filtration and recrystallized from 50% aqueous ethanol or petroleum ether (60–90°).

6. This product is satisfactory for synthetic purposes.

7. This type of reaction can be used successfully for the preparation of other naphthothiazolines and naphthoselenazolines. The submitter has prepared 2-imino-1-methylnaphtho[1,2]selenazoline, m.p. 94–95°, and 2-imino-1-ethylnaphtho[1,2]selenazoline, m.p. 82–84°, by similar methods.

3. Methods of Preparation

The preparation of this compound is not described in the literature.

2-METHYLINDOLE *

(Indole, 2-methyl-)

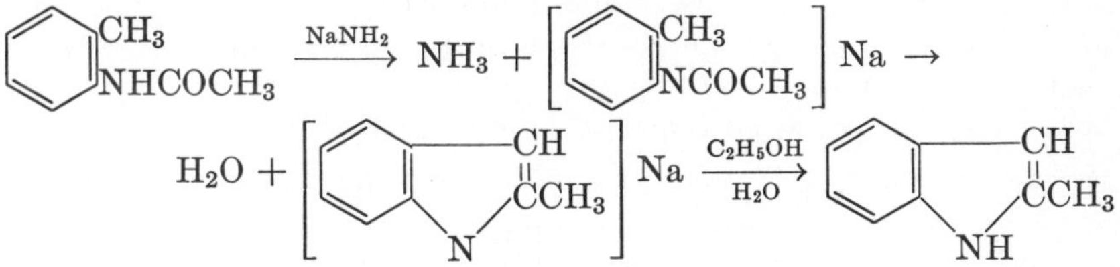

Submitted by C. F. H. ALLEN and JAMES VANALLAN.
Checked by NATHAN L. DRAKE and RICHARD TOLLEFSON.

1. Procedure

In a 1-l. Claisen flask is placed a mixture of 64 g. of finely divided sodium amide (Note 1) and 100 g. of acetyl-*o*-toluidine (Note 2). About 50 ml. of dry ether is added (Note 3), and the apparatus is swept out with dry nitrogen. Then, with a slow current of nitrogen passing through the mixture, the reaction flask (Note 4) is heated in a metal bath (Note 5). The temperature is raised to 240–260° over a 30-minute period and is maintained in this range for 10 minutes. A vigorous evolution of gas occurs, the cessation of which indicates that the reaction is complete (Note 6). The metal bath is removed, the flask is allowed to cool, and 50 ml. of 95% ethanol and 250 ml. of warm (about 50°) water are added, successively, to the reaction mixture. The decomposition of the sodium derivative of 2-methylindole, and of any excess sodium amide, is completed by warming the mixture gently with a Bunsen burner. The cooled reaction mixture is extracted with two 200-ml. portions of ether (Note 7). The combined ether extracts are filtered, and the filtrate is concentrated to about 125 ml. The solution is then transferred to a 250-ml. modified Claisen flask and distilled. The 2-methylindole distils at 119–126°/3–4 mm. as a water-white liquid, which rapidly solidifies in the receiver to a white crystalline mass. This product melts at 56–57°. The yield is 70–72 g. (80–83%) (Note 8).

The product may be further purified by dissolving it in 100 ml. of methanol, adding 30 ml. of water, and allowing the solution to stand in the ice chest for 5 hours. The pure white plates (52 g.) melt at 59°.

An additional 10 g. may be recovered by cooling the filtrate after it has been diluted with about 20 ml. of water.

2. Notes

1. The sodium amide was ground in an open mortar, and at no time was difficulty experienced. As a precautionary measure, the grinding could be carried out under ether.

2. Acetyl-*o*-toluidine, m.p. 110–111°, obtained from the Eastman Kodak Company, was used. After the reactants are introduced into the flask, they should be mixed thoroughly with a long spatula.

3. The ether is added to facilitate the formation of the sodium salt of the amide.

4. It is advisable to cover the bottom of the reaction flask with soot, to prevent the metal from adhering to the glass.

5. Sand and salt baths are not satisfactory.

6. Reaction begins when the bath temperature has risen to approximately 200°. Frothing occurs, and the froth solidifies. The checkers found it necessary to stir the solidified froth into the reaction mixture, so that heating of the whole mass could be uniform. This stirring is necessary throughout—i.e., from the beginning of vigorous gas evolution until completion of the period of heating.

7. The indole may be isolated, less conveniently, by steam distillation; the crystalline product (m.p. about 56–57°) can be filtered from the cold distillate.

8. The method described here is of general application to substituted acetyl- and benzoyl-*o*-toluidines.

3. Methods of Preparation

The method described here is a modification of Verley's procedure.[1] 2-Methylindole may also be prepared by the treatment of acetone phenylhydrazone with zinc chloride at 180°.[2]

The cyclization of acetyl-*o*-toluidine has been reported to occur in low yield on distillation with zinc dust.[3] Better yields have been obtained by heating the toluidide with sodium ethoxide or with barium oxide in a current of hydrogen at 360–380°.[4] A vapor-phase cyclization over a silica-alumina catalyst has also been reported.[5]

[1] Verley, *Bull. soc. chim. France,* (4) **35,** 1039 (1924).

[2] Fischer, *Ann.,* **236,** 116 (1886); *Ber.,* **19,** 1563 (1886); Marion and Oldfield, *Can. J. Research,* **25B,** 1 (1947).

[3] Mauthner and Suida, *Monatsh.,* **7,** 230, 237 (1886).

[4] Madelung, Ger. pat. 262,327 [*Frdl.*, **11**, 278 (1912–1914)]; *Ber.*, **45**, 1128 (1912).
[5] Ger. pat. 458,383 [*Frdl.*, **16**, 709 (1931)].

METHYL ISOTHIOCYANATE *

(Isothiocyanic acid, methyl ester)

$$CH_3NH_2 + CS_2 + NaOH \rightarrow CH_3NHCS_2Na + H_2O$$
$$CH_3NHCS_2Na + ClCO_2C_2H_5 \rightarrow CH_3NHCS_2CO_2C_2H_5 + NaCl$$
$$CH_3NHCS_2CO_2C_2H_5 \rightarrow CH_3NCS + COS + C_2H_5OH$$

Submitted by MAURICE L. MOORE and FRANK S. CROSSLEY.
Checked by NATHAN L. DRAKE and RALPH MOZINGO.

1. Procedure

In a 1-l. round-bottomed three-necked flask, surrounded by an ice bath and fitted with a mechanical stirrer, a reflux condenser, thermometer, and a 250-ml. dropping funnel, are placed 137 g. (110 ml., 1.8 moles) of carbon disulfide and a cold solution of 72 g. (1.8 moles) of sodium hydroxide in 160 ml. of water. To this mixture, cooled to 10–15°, is added, with stirring, 180 ml. (56 g., 1.8 moles of methylamine) of 35% aqueous methylamine solution (Notes 1 and 2) over a period of 30 minutes. Stirring is continued, and the mixture is warmed gently on a steam bath for 1–2 hours to ensure complete reaction (Note 3). The bright red solution is cooled to 35–40°, and to it is added over a period of 1 hour, with stirring, 196 g. (175 ml., 1.8 moles) of ethyl chlorocarbonate (Note 4). The stirring is continued for 30 minutes after all the ethyl chlorocarbonate has been added, at which time the temperature should have fallen to 30–40°. The methyl isothiocyanate, which separates on top, is removed from the reaction mixture and weighs 170–190 g.

The product is dried over 10 g. of sodium sulfate and distilled under atmospheric pressure through a short Vigreux column; the fraction which boils at 115–121° is collected. The yield is 85–100 g. (65–76%) (Notes 5 and 6). The product may be further purified by refractionation. The portion which boils at 117–119° is collected.

2. Notes

1. The monomethylamine used in this preparation was a "Commercial Special 35% Solution" obtained from Rohm and Haas Company.

2. Methylamine hydrochloride can be used in place of the commercial aqueous methylamine solution by a slight modification of the above procedure. The carbon disulfide and a solution of 122 g. (1.8 moles) of methylamine hydrochloride in 200 ml. of water are mixed together in the flask, and a cold solution of 144 g. (3.6 moles) of sodium hydroxide in 320 ml. of water is added, with stirring, over a period of 30 minutes. Two equivalents of sodium hydroxide must be used in this case. The remainder of the procedure is the same as with the free base.

3. The temperature gradually rises from 25°, which is that noted at the end of the addition of the methylamine solution, to 75–85°.

4. The temperature may rise rather rapidly during this addition; it is advisable to maintain the rate of addition constant so that the reaction does not become too vigorous.

5. Larger runs, up to 5.4 moles of carbon disulfide, have been made with only a slight reduction in yield.

6. This reaction is general for the preparation of alkyl isothiocyanates in good yields; thus, according to the submitters, ethyl isothiocyanate is obtained in yields of 60–70% from ethylamine hydrochloride.

3. Methods of Preparation

Methyl isothiocyanate has been prepared from methyl thiocyanate by rearrangement with heat [1] and from N,N′-dimethylthiuramdisulfide by the action of iodine [2] or by heating with water or methanol.[3] The most useful method of preparation has been the reaction of methylamine with carbon disulfide to form methyldithiocarbamic acid which is decomposed by steam distillation of the lead salt,[4] or by reaction with ethyl chlorocarbonate.[5]

[1] Hofmann, *Ber.,* **13,** 1349 (1880).

[2] v. Braun, *Ber.,* **35,** 817 (1902).

[3] Freund and Asbrand, *Ann.,* **285,** 166 (1895).

[4] Delépine, *Bull. soc. chim. France,* (4) **3,** 641 (1908); Delépine, *Compt. rend.,* **144,** 1125 (1907); Worrall, *J. Am. Chem. Soc.,* **50,** 1456 (1928).

[5] Slotta and Dressler, *Ber.,* **63,** 888 (1930).

METHYL 4-KETO-7-METHYLOCTANOATE

(Caprylic acid, ζ-methyl-γ-oxo-, methyl ester)

$$(CH_3)_2CHCH_2CH_2Br + Mg \xrightarrow{(C_2H_5)_2O} (CH_3)_2CHCH_2CH_2MgBr$$

$$2(CH_3)_2CHCH_2CH_2MgBr + CdCl_2 \xrightarrow{(C_2H_5)_2O} [(CH_3)_2CHCH_2CH_2]_2Cd + MgCl_2 + MgBr_2$$

$$[(CH_3)_2CHCH_2CH_2]_2Cd + 2ClCOCH_2CH_2CO_2CH_3 \xrightarrow{C_6H_6} 2(CH_3)_2CHCH_2CH_2COCH_2CH_2CO_2CH_3 + CdCl_2$$

Submitted by James Cason and Franklin S. Prout.
Checked by Homer Adkins and Robert Turner.

1. Procedure

A 2-l. three-necked flask (Note 1) is fitted with a reflux condenser having a take-off attachment,[1] a mercury-sealed Hershberg stirrer (Note 2), and a 500-ml. dropping funnel (Note 1). A nitrogen inlet tube is connected to the top of the condenser, and a branch of the tube, connected by means of a T-tube, is connected to a mercury valve[1] consisting of a U-tube, the bend of which is just filled with mercury. Unless the dropping funnel is equipped with a pressure-equalizing side tube,[1] a second branch of the nitrogen line is arranged for connection to its mouth. In the flask is placed 24.3 g. (1.0 gram atom) of magnesium turnings, and the entire flask is warmed with a soft flame while a slow stream of nitrogen is passed through and permitted to escape by way of the dropping funnel. The flask is allowed to cool, the dropping funnel is closed, and the nitrogen flow is reduced until the gas bubbles very slowly through the mercury valve. The magnesium is covered with 150 ml. of dry ether, introduced from the dropping funnel, and a solution of 151 g. (1.0 mole) of pure isoamyl bromide (Note 3) in 350 ml. of dry ether is placed in the dropping funnel. A few milliliters of the bromide solution is added to the flask, and the stirrer is started. The flask is warmed gently if the reaction does not start spontaneously. The remainder of the bromide is added during 1–2 hours, and the mixture is refluxed for 15 minutes longer. The flask is then cooled in an ice bath, the dropping funnel is removed, and 98 g. (0.535 mole) of anhydrous C.P. cadmium chloride (Note 4) is added over a period of 5–10 minutes. After all the cadmium chloride has been added, the ice bath is removed and the mixture is stirred for 5 minutes and then heated under reflux with stirring for an additional 45 minutes (Note 5).

Ether is now rapidly distilled from the reaction mixture (Note 6) by heating on a steam bath. Distillation is continued with stirring until it becomes very slow and a dark viscous residue remains. The distillate amounts to 250–325 ml. At this point, 350 ml. of dry thiophene-free benzene is added to the flask and the distillation is continued until an additional 100 ml. of distillate has been collected. A second 350-ml. portion of dry benzene is added to the flask, and the mixture is refluxed with vigorous stirring for a few minutes in order to break up the cake in the flask and disperse it through the mixture. The heating bath is then removed, and 120 g. (0.8 mole) (Note 7) of β-carbomethoxypropionyl chloride (p. 169) in 150 ml. of dry benzene is added from the dropping funnel. This addition, which causes vigorous refluxing, requires 10–20 minutes. During this time the heavy precipitate changes in appearance and stirring becomes more difficult. After the addition of the acid chloride is complete and spontaneous refluxing has stopped, the mixture is stirred and heated under reflux for an additional hour.

The reaction mixture is cooled in an ice bath and decomposed in the usual way by the careful addition of about 600 g. of ice and water, followed by sufficient 20% sulfuric acid to give two clear phases (Note 2). The aqueous phase is separated in a 2-l. separatory funnel and extracted with two 100-ml. portions of benzene. The two benzene extracts are placed in two 500-ml. separatory funnels. The original benzene layer and each extract are washed successively (Note 8) with 200 ml. of water, 200 ml. of 5% sodium carbonate solution, 200 ml. of water, and 100 ml. of saturated sodium chloride solution, and then each is filtered through a layer of anhydrous sodium sulfate. A little fresh solvent is used to rinse the separatory funnels and the filter.

A fractionating column (Note 9) is connected to a 250-ml. flask carrying a dropping funnel and heated in an oil bath at 150–160°. The combined benzene solutions are run into the flask from the dropping funnel so that the benzene is removed by flash distillation at atmospheric pressure. The pressure is then lowered, and after the distillation of the last of the solvent and a little 2,7-dimethyloctane (from coupling of the Grignard reagent) the vapor temperature rises to about 100°/20 mm. About 5 g. of methyl ethyl succinate (Notes 10 and 11) is collected at 100–106°/20 mm., and after an intermediate fraction (about 3 g.) the keto ester is collected at 136–137°/20 mm. (Notes 11 and 12). The yield of methyl 4-keto-7-methyloctanoate is

108.5–111.5 g. (73–75%, based on β-carbomethoxypropionyl chloride) (Notes 13 and 14).

2. Notes

1. Connections may be made with rubber stoppers, but ground-glass joints are preferable.

2. The heavy sludges encountered in this preparation make a strong and efficient stirrer essential. A stirrer of the Hershberg type of tantalum wire is preferable, but new, unetched Nichrome is quite satisfactory. If a Nichrome stirrer is used it should be removed before acid is added to complete the decomposition of the organometallic complex.

3. Isoamyl bromide, b.p. 120–120.5°, was prepared as previously described.[2] The product was fractionated through a packed column, such as that referred to in Note 9 below, to remove small quantities of *tert*-amyl bromide arising from *sec*-butyl carbinol present in the isoamyl alcohol.

4. The cadmium chloride is dried to constant weight at 110° in an oven, ground thoroughly, and stored in a desiccator. If stored in a screw-cap bottle in the laboratory it may absorb moisture slowly.

5. It is best to continue stirring at this point until the mixture gives a negative Gilman test for the Grignard reagent.[3] One-half milliliter of the reaction mixture is added to an equal volume of a 1% solution of Michler's ketone in dry benzene. After this mixture is shaken briefly, 1 ml. of water is added, followed by a few drops of a 0.2% solution of iodine in glacial acetic acid and 0.5 ml. of glacial acetic acid. If any Grignard reagent is still present, a greenish blue color is observed.

6. If the reaction with the acid chloride is carried out in ether solution larger amounts of methyl ethyl succinate are formed, and the yield of keto ester is only 42–59%.

7. The use of β-carbomethoxypropionyl chloride of poor quality frequently resulted in poor yields of the desired keto ester. It is most important that the methyl hydrogen succinate, m.p. 56–57°, from which the acyl chloride is prepared, be of high quality and that it be ground thoroughly before drying. The melting point of a sample of methyl hydrogen succinate is not a sufficient criterion of purity, so that the neutral equivalent of the sample of ester should be determined. If the neutral equivalent is not in the range 130–134, the reagent should be purified by recrystallization.

8. The indicated portion of each wash solution is used to wash each of the benzene solutions in turn. If three separatory funnels are used this process is not laborious and results in very thorough extraction.

9. An efficient column packed with glass helices and provided with a heating jacket and a suitable head [4] should be used. Columns of dimensions 50 cm. by 14 mm. i.d. and 43 cm. by 12 mm. i.d. have been found satisfactory.

10. The methyl ethyl succinate apparently results from the reaction of the acid chloride with ether.

11. The apparent boiling points of the fractions may vary somewhat, depending on the fractionating column.

12. Boiling points of the keto ester at other pressures are as follows: 117°/8 mm., 120–122°/11 mm., 125°/14 mm., 139.5°/22 mm., and 145°/27 mm.

13. In an alternative procedure the fractionating column is removed after the intermediate fraction has been collected, and the residual ester is distilled through a short still head.

14. The procedure described has been used by the submitters for the preparation of the following keto esters:

Keto Ester	*B.P.*	*% Yield*
Ethyl 10-keto-13-methyltetradecanoate	180–182°/3 mm.	85
Ethyl 10-keto-16-methyloctadecanoate	192–195°/2 mm.	76.5
Ethyl 6-ketoheptanoate	143–145°/33 mm.	76
Ethyl 10-ketohendecanoate	137–138°/2 mm.	83
Ethyl 10-keto-14-methyltetracosanoate	242–245°/1 mm.	77

The ester acid chlorides used in these preparations, ω-carbethoxyvaleryl chloride and ω-carbethoxynonanoyl chloride, were obtained from the corresponding half esters in yields of 90–95%. The method described for β-naphthoyl chloride (p. 629, Note 2) was used, but since with these acids the reaction is much more vigorous the phosphorus pentachloride was added in small portions and the mixture was not heated until after the exothermic reaction had subsided.

The procedure described failed completely when a secondary bromide, 2-bromopentane, was used; however, when the cadmium derivative was prepared at −5° to −7° and allowed to react in ether at this temperature with β-carbomethoxypropionyl chloride, methyl 4-keto-5-methyloctanoate was obtained in a yield of 19.3%, b.p. 130.5–130.7°/21 mm.

3. Methods of Preparation

Methyl 4-keto-7-methyloctanoate has been prepared by the procedure given, which is based on the method of Cason and Prout.[5] The use of cadmium derivatives for the preparation of simple ketones was introduced by Gilman and Nelson.[6]

A preparation from isobutyl-2-furylcarbinol (obtained by a Grignard reaction from furfural) has also been described.[7]

[1] Fieser, *Experiments in Organic Chemistry,* 2nd Ed., p. 323, Fig. 49; p. 404, Fig. 75, D. C. Heath and Company, Boston, 1941.

[2] *Org. Syntheses* Coll. Vol. **1,** 27 (1941).

[3] Gilman and Schulze, *J. Am. Chem. Soc.,* **47,** 2002 (1925); Gilman and Heck, *J. Am. Chem. Soc.,* **52,** 4949 (1930).

[4] Whitmore and Lux, *J. Am. Chem. Soc.,* **54,** 3448 (1932).

[5] Cason and Prout, *J. Am. Chem. Soc.,* **66,** 46 (1944); Cason, *J. Am. Chem. Soc.,* **68,** 2078 (1946); *Chem. Revs.,* **40,** 15 (1947).

[6] Gilman and Nelson, *Rec. trav. chim.,* **55,** 518 (1936).

[7] Kucherov, *Zhur. Obshcheĭ Khim.,* **20,** 1885 (1950) [*C. A.,* **45,** 2928 (1951)].

METHYL MYRISTATE * AND METHYL PALMITATE * AND THE CORRESPONDING ACIDS *

$$\text{Bayberry wax} + CH_3OH \xrightarrow{(H_2SO_4)} C_{13}H_{27}CO_2CH_3 + C_{15}H_{31}CO_2CH_3$$

Submitted by J. C. Sauer, B. E. Hain, and P. W. Boutwell.
Checked by Homer Adkins and Robert E. Burks, Jr.

1. Procedure

In a 3-l. round-bottomed flask are placed 500 g. of bayberry wax (Note 1), 600 g. of methanol, and 31 g. of sulfuric acid. Boiling chips are added, and the mixture is refluxed for 48 hours on the steam bath. About half of the methanol is removed by distillation, and 500 ml. of water is added to the remaining mixture. The ester layer is separated and washed several times with water. About 500 g. of crude methyl esters is obtained.

The esters are fractionally distilled through a suitable fractionating column (Note 2). A second fractionation is then made so that each successive fraction is added to the distilling flask after the temperature has reached the upper limit for the fraction being collected. There is obtained 170–190 g. of methyl myristate, b.p. 91–94°/0.5 mm., 112–116°/1 mm., or 157–162°/10 mm., and 170–190 g. of methyl palmitate, b.p. 115–118°/0.5 mm., 129–133°/1 mm., or 180–183°/10 mm. (Notes 3, 4, 5, and 6).

2. Notes

1. Bayberry wax, also known as myrtle or laurel wax, is obtained from the berries of various species of *Myrica.* The commercial wax,

prepared from the berries of *Myrica cerifera,* may be obtained from chemical supply houses for about $0.60 per pound. The relative proportion of myristin and palmitin in bayberry wax varies somewhat, but these two compounds constitute about 95% of the wax.

2. Several columns are suitable for this separation. The submitters used a three-ball Snyder column 26 cm. in length (similar to No. 28575 of E. H. Sargent, Chicago), sealed to a 500- or 200-ml. modified Claisen flask. The method of operation was essentially that of Simons and Wagner.[1] The column was not electrically heated but was snugly wrapped with asbestos paper. The rate of distillation was about 20 drops per minute, and the reflux ratio 2 or 3 to 1. Flooding of the column must be carefully avoided. The checkers used a modified Widmer column which carried a spiral, 14 cm. long with eleven turns of the helix, contained in a glass tube 12 mm. in internal diameter. A 500-ml. flask was employed for the first fractionation and a 250-ml. one for the refractionation. The column and the oil bath were heated electrically. A ground-glass joint was used for connecting the flask and column, the design and method of operation being that described by Smith and Adkins.[2] The temperature of the bath and column heater was so adjusted that the rate of distillation was 3–4 drops per minute, the spiral being well covered with returning liquid.

3. Suitable ranges for taking off fractions at various pressures on the first fractionation are as follows: 6 g., 152–157°/10 mm.; 176 g., 157–160°/10 mm.; 119 g., 160–180°/10 mm.; 126 g., 180–183°/10 mm.; 163 g., 115–123°/1 mm.; 115 g., 123–130°/1 mm.; 141 g., 130–133°/1 mm.; 29 g., below 91.5°/0.5 mm.; 116 g., 91.5–94.5°/0.5 mm.; 71 g., 94.5–115°/0.5 mm.; 170 g., 115–118°/0.5 mm.; 28 g., 118–125°/0.5 mm. Upon refractionation cuts were made as follows: 6 g., 152–157°/10 mm.; 161 g., 157–162°/10 mm.; 81 g., 162–171°/10 mm.; 64 g., 171–180°/10 mm.; 120 g., 180–183°/10 mm.; 198 g., 112–116°/1 mm.; 72 g., 116–129°/1 mm.; 147 g., 129–133°/1 mm.; 8 g., below 91.5°/0.5 mm.; 172 g., 91.5–94.5°/0.5 mm.; 10 g., 94.5–115.2°/0.5 mm.; 200 g., 115.2–118.2°/0.5 mm.; 25 g., 118.2–125°/0.5 mm. The temperature at which the esters distil varies with the rate of distillation and with the details of construction and operation of the column. The figures given above are intended to suggest the approximate ranges for cutting fractions.

4. This procedure has been used successfully for many years in the preparation of ethyl laurate, caprylate, and myristate by the alcoholysis of cocoanut oil (1 kg.) in ethanol (1.9 kg.) with hydrogen chloride (50 g.) as a catalyst.[3] The method differs slightly from the one described above. The alcoholysis is complete after 15 or 20 hours,

and the solution is then neutralized to methyl orange with barium carbonate. The mixture is added to an equal volume of a saturated sodium chloride solution, whereupon 1100–1300 g. of the mixture of crude ethyl esters separates. This mixture of esters is washed with water and fractionated as described above. The yields are approximately 50 g. of ethyl caprylate, 350 g. of ethyl laurate, and 60 g. of ethyl myristate from 1 kg. of cocoanut oil.

5. The methyl myristate obtained (n_D^{25} 1.4353) showed a melting point of 19° when at equilibrium with the liquid. The methyl palmitate (n_D^{25} 1.4391 when supercooled) gave a melting point of 29.5°.

6. Myristic and palmitic acids can be obtained from their esters by the procedure in *Org. Syntheses* Coll. Vol. **1,** 379 (1941). From 100 g. of methyl myristate is obtained 85–89 g. (90–95%) of colorless myristic acid melting at 52–53°. From 100 g. of methyl palmitate is obtained 84–88 g. (90–95%) of colorless palmitic acid melting at 62–63°.

3. Methods of Preparation

Of the early references to the preparation of methyl myristate and methyl palmitate, few are of preparative value. Methyl myristate can be prepared by the fractional distillation of the methyl esters from ucuhuba fat [4] and from cocoanut oil.[5,6] Methyl palmitate can be prepared in a similar manner from cocoanut oil [6] and from bayberry wax.[7,8]

Myristic and related acids may be prepared synthetically by a mixed Kolbe reaction.[9]

[1] Simons and Wagner, *J. Chem. Education,* **9,** 122 (1932).

[2] Smith and Adkins, *J. Am. Chem. Soc.,* **60,** 662 (1938).

[3] *Organic Chemical Reagents,* Vol. 3, p. 62, University of Illinois, Urbana, Illinois, 1921.

[4] Verkade and Coops, *Rec. trav. chim.,* **46,** 528 (1927).

[5] Taylor and Clarke, *J. Am. Chem. Soc.,* **49,** 2829 (1927).

[6] Lepkovsky, Feskov, and Evans, *J. Am. Chem. Soc.,* **58,** 978 (1936).

[7] Wenzel, *Ind. Eng. Chem. Anal. Ed.,* **6,** 1 (1934).

[8] McKay, *J. Org. Chem.,* **13,** 86 (1948).

[9] Greaves, Linstead, Shephard, Thomas, and Weedon, *J. Chem. Soc.,* **1950,** 3326.

N-METHYL-1-NAPHTHYLCYANAMIDE

(Cyanamide, methyl-(1-naphthyl)-)

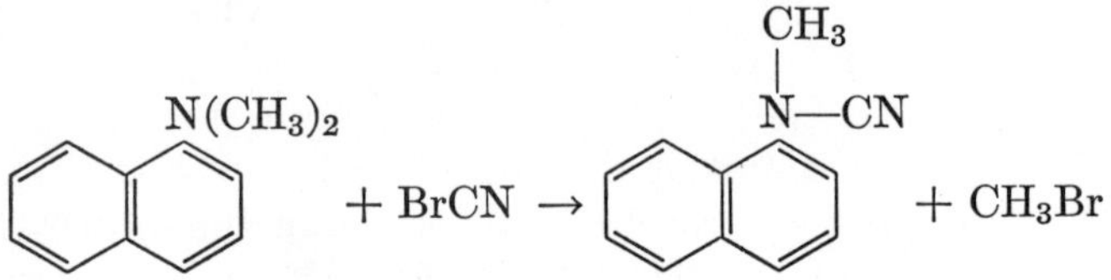

Submitted by HOMER W. J. CRESSMAN.
Checked by NATHAN L. DRAKE and WERNER R. BOEHME.

1. Procedure

Caution! Cyanogen bromide is highly toxic. This preparation should be carried out under a good hood.

A mixture of 171 g. (1 mole) of dimethyl-α-naphthylamine (Note 1) and 125 g. (1.2 moles) of cyanogen bromide [1] in a 1-l. flask (Note 2) is heated under reflux on the steam bath for 16 hours (Note 3). The cooled reaction mixture is added to 2.5 l. of dry ether, and the insoluble quaternary salt (Note 4) is filtered. The ether filtrate is extracted with four 800-ml. portions of approximately 15% hydrochloric acid solution (Note 5) and washed with five 500-ml. portions of water. The ether solution is dried with 30–35 g. of anhydrous calcium sulfate (Drierite). After filtration, the solvent is removed by distillation at atmospheric pressure on the steam bath; the residue is fractionated under reduced pressure. The yield of pale yellow oil, boiling at 170–171°/1 mm. (185–187°/3 mm.), amounts to 115–122 g. (63–67%) (Note 6).

2. Notes

1. Better yields are obtained by dealkylating the tertiary amine in this manner than by starting with the secondary amine. In a similar manner, N-ethyl-1-naphthylcyanamide, b.p. 165–168°/2 mm., is obtained in a 48% yield from diethyl-α-naphthylamine; the yield from the secondary amine is only 25%.
2. Ground-glass equipment is preferable.
3. In order to avoid loss of cyanogen bromide an efficient condenser must be used and the heating should be very gradual at first.

4. About 10 g. of α-naphthyltrimethylammonium bromide, m.p. 160°, is generally obtained.

5. About 25 g. of crystalline dimethyl-α-naphthylamine hydrochloride may be recovered by efficient chilling of the hydrochloric acid extracts.

6. The N-methyl-1-naphthylcyanamide turns dark green upon standing exposed to the air for several days. It may be preserved in sealed, evacuated ampoules in the absence of light.

3. Methods of Preparation

The preparation described is based on the method of von Braun and co-workers.[2]

[1] *Org. Syntheses* Coll. Vol. **2**, 150 (1943).
[2] von Braun, Heider, and Muller, *Ber.*, **51**, 281 (1918).

1-METHYL-1-(1-NAPHTHYL)-2-THIOUREA

(Urea, 1-methyl-1-(1-naphthyl)-2-thio-)

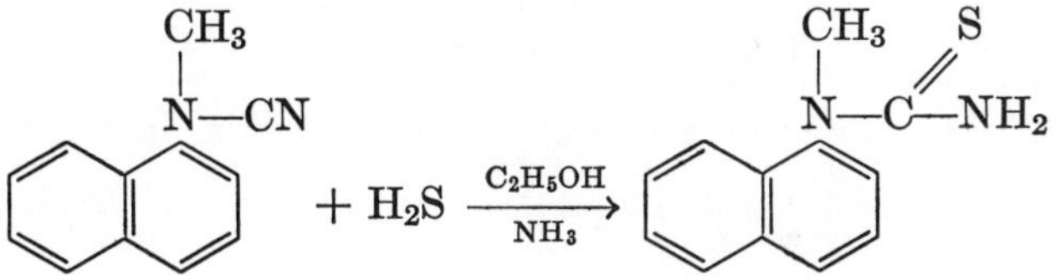

Submitted by Homer W. J. Cressman.
Checked by Nathan L. Drake and Werner R. Boehme.

1. Procedure

The apparatus, consisting of a 1-l. three-necked flask, fitted with a glycerol-sealed mechanical stirrer, two inlet tubes (10-mm. bore) extending nearly to the blades of the stirrer, and an outlet tube (Note 1), is set up in a good hood. Into a solution of 91 g. (0.5 mole) of N-methyl-1-naphthylcyanamide (p. 608) in 350 ml. of absolute ethanol, mechanically stirred and cooled by immersion of the flask in a water bath maintained at 20–25°, ammonia is bubbled at a moderate rate. After 5 minutes, hydrogen sulfide (Note 2) is passed into the solution at about the same rate. The introduction of ammonia is continued for 1.5 hours; hydrogen sulfide is passed into the mixture an additional 30 minutes. A white solid begins to separate soon after the first addition of hydrogen sulfide. After the mixture is chilled in an ice-salt bath, it is filtered. The solid is suspended in 350 ml. of cold water, again separated by filtration, and finally washed with two 100-ml. portions of water, then with two 100-ml. portions of methanol, and dried. The yield of 1-methyl-1-(1-naphthyl)-2-thiourea, m.p. 168–170° (Note 3), amounts to 89–97 g. (82–90%); this material is suitable for most purposes.

2. Notes

1. The use of an alkali trap [1] is recommended.

2. This general reaction can be used for preparing other unsymmetrical thioureas when the desired cyanamides can be obtained. The submitter has also prepared 1-methyl-1-(1-naphthyl)-2-selenourea, m.p. 174–175° with decomposition, and 1-ethyl-1-(1-naphthyl)-2-selenourea, m.p. 168–170° with decomposition, by substitution of hydrogen selenide for hydrogen sulfide.

3. The thiourea can be further purified by recrystallization from ethanol. The recrystallized product melts at 170–171°.

3. Methods of Preparation

The preparation of this compound is not reported in the literature.

[1] *Org. Syntheses* Coll. Vol. **1**, 266 (1941).

METHYL PYRUVATE

(Pyruvic acid, methyl ester)

$$CH_3COCO_2H + CH_3OH \xrightarrow{CH_3C_6H_4SO_3H} CH_3COCO_2CH_3 + H_2O$$

Submitted by A. Weissberger and C. J. Kibler.
Checked by C. S. Hamilton and R. F. Coles.

1. Procedure

A solution of 88 g. (1 mole) of freshly distilled pyruvic acid,[1] 128 g. (4 moles) of absolute methanol, 350 ml. of benzene, and 0.2 g. of *p*-toluenesulfonic acid is placed in a 1-l. round-bottomed flask connected through a ground-glass joint (Note 1) to a methyl ester column

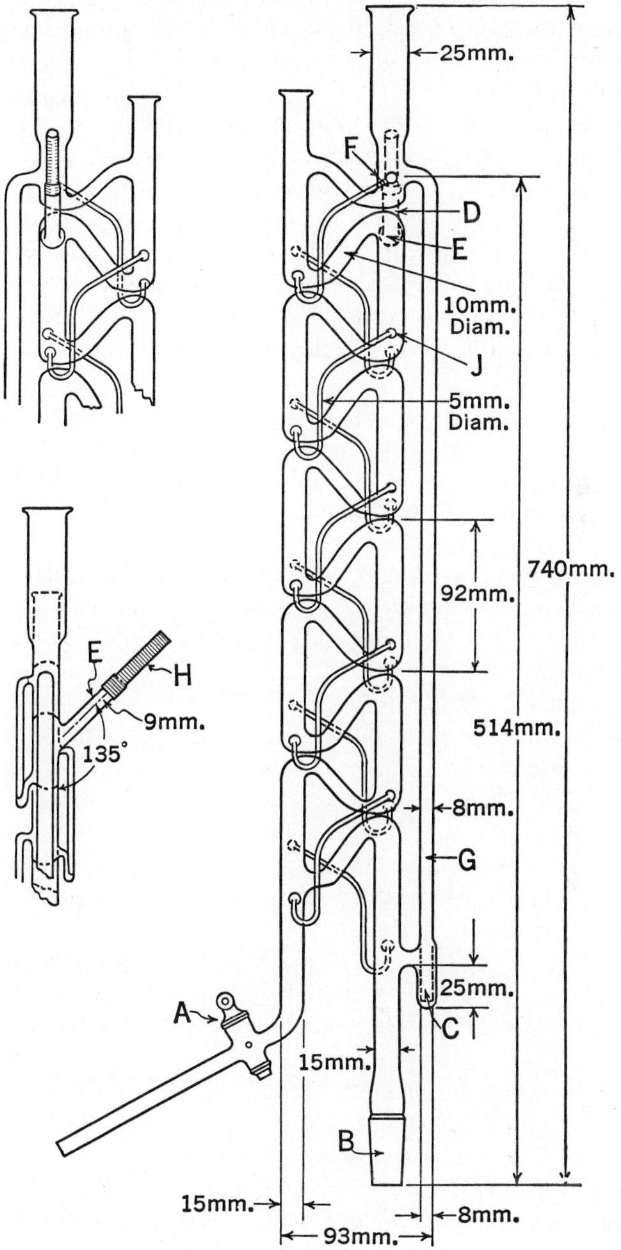
25mm.
F
D
E
10mm.
Diam.
J
5mm.
Diam.
740mm.
92mm.
514mm.
8mm.
G
25mm.
C
A
15mm.
B
15mm.
93mm.
8mm.
E
H
9mm.
135°

FIG. 20.

shown in Fig. 20 [2] (Note 2). The column is fitted at the top with a cold finger, a thermometer, and an efficient condenser. The solution is refluxed vigorously using an oil bath maintained at 150–155°. The temperature at the top of the column is 59–60°. After refluxing overnight, the liquid in the lower bubbler becomes cloudy and separates into two layers. The lower layer is removed as fast as it is formed (i.e., every 20–30 minutes) throughout the day. Refluxing is continued overnight, and the next morning the lower layer is again removed until cloudiness persists in the upper bubblers (Note 3). The ester is then isolated by fractional distillation of the remaining liquid. The fraction boiling at 136–140° at atmospheric pressure is collected (Note 4). It weighs 66–73 g. (65–71%) (Notes 5 and 6).

2. Notes

1. A ground-glass joint is advisable on account of the long reflux period. Benzene attacks a rubber stopper, and pyruvic acid destroys cork.

2. The Clarke-Rahrs methyl ester column [2] is illustrated in Fig. 20. *A* is the stopcock, above which the aqueous phase collects and is drawn off as necessary. *B* is a standard-taper ground joint; *C* is a trap whose outside diameter is 12 mm. At *D* the space between the consecutive bubblers is shown. *E* is the thermometer tube, set in at an angle of about 45°; it carries a piece of rubber tubing *H* which holds the thermometer. *F* is a solid spot in the top return tube *only*, where that tube has been sealed off; the apparatus will not function without this seal. It should be pointed out that the upper ends of all the return tubes should terminate just above the bends, as shown at *J;* otherwise there will be too great a pressure due to the height of the liquid. The overall length as given is not critical.

3. A total of about 300 ml. of liquid will have separated when cloudiness persists in the fourth and fifth bubblers. The liquid collected separates into two layers on cooling. It contains a trace of pyruvic acid.

4. A tarry residue of 10–17 g. is obtained. If an efficient column is not used in the distillation, the fore-runs will contain 5–10 g. of recoverable ester. The yield given is the total yield.

5. The difficulty in the preparation of methyl pyruvate is caused by the fact that this ester is very easily hydrolyzed and that the ester equilibrium is far on the side of the hydrolysis products.

6. Other methyl esters can be made by this procedure.

3. Methods of Preparation

Methyl pyruvate has been prepared from the silver salt of pyruvic acid and methyl iodide;[3] from the free acid, by the ethanol-vapor method without a catalyst,[4] by azeotropic removal of the water produced by the reaction of methanol in the presence of *p*-toluenesulfonic acid (the present method), and by refluxing with methanol in ethylene dichloride using ethanesulfonic acid as a catalyst (73% yield).[5] Pyruvic esters have also been prepared by the catalytic dehydrogenation of lactic acid esters[6] and by the oxidation of ethyl lactate with potassium permanganate.[7]

[1] *Org. Syntheses* Coll. Vol. **1,** 475 (1943).

[2] Rahrs, *Synthetic Organic Chemicals,* Vol. XI, No. 1, The Eastman Kodak Company, February, 1938.

[3] Oppenheim, *Ber.,* **5,** 1051 (1872).

[4] Baker and Laufer, *J. Chem. Soc.,* **1937,** 1345.

[5] Clinton and Laskowski, *J. Am. Chem. Soc.,* **70,** 3135 (1948).

[6] U. S. pat. 1,614,195 [*C. A.,* **21,** 746 (1927)].

[7] *Org. Syntheses,* **31,** 59 (1951).

METHYL SEBACAMATE

(Sebacamic acid, methyl ester)

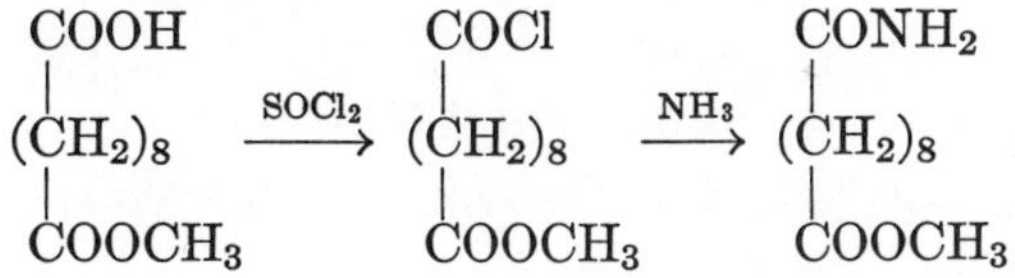

Submitted by W. S. Bishop.
Checked by C. F. H. Allen and J. VanAllan.

1. Procedure

A. *ω-Carbomethoxypelargonyl chloride.* A mixture of 216 g. (1 mole) of methyl hydrogen sebacate (Note 1) and 147 g. (1.24 moles) of thionyl chloride in a 2-l. round-bottomed flask attached by a standard ground-glass joint to an upright condenser is refluxed on a steam bath for 5 hours. The condenser is then replaced by a still head having a modified side arm, and the unused thionyl chloride is removed. The residual ω-carbomethoxypelargonyl chloride is suitable for the next step. If a purer product is desired it is distilled under reduced pressure. The yield of ester chloride boiling at 158–160°/10 mm. is 194–201 g. (83–86%).

B. *Methyl sebacamate.* Two and one-half liters of concentrated aqueous ammonia (about 28%) in a 4-l. beaker or enameled pot (Note 2) is stirred vigorously with an off-center stirrer and chilled to 8° in a cooling bath. The crude chloride from part A is added slowly from a dropping funnel to the solution, which is kept below 8° throughout the addition. A vigorous reaction takes place, and methyl sebacamate precipitates immediately. After the addition has been completed, the product is filtered by suction and washed with 200 ml. of cold water. After 3 days' drying in a vacuum desiccator, the methyl sebacamate weighs 200–204 g. (93–95%) and melts at 72–74° (Note 3).

2. Notes

1. Methyl hydrogen sebacate (b.p. 168–170°/3 mm.) was prepared by the checkers in yields of 52–61% from sebacic acid and methanol by the procedure described for the ethyl analog.[1] The ethyl ester can be used to obtain ethyl sebacamate. The preparation of the acid chloride from ethyl hydrogen sebacate is mentioned in Note 12, p. 171.
2. A variety of enameled-steel pots known as bain-marie is available from the American Specialty Company, Rochester, New York.
3. The yield is no higher with the distilled chloride.

3. Methods of Preparation

These compounds have been described in the literature.[2,3]

[1] *Org. Syntheses* Coll. Vol. **2**, 276 (1943).
[2] Biggs and Bishop, *J. Am. Chem. Soc.*, **63**, 944 (1941).
[3] Bishop, U. S. pat. 2,277,033 [*C. A.*, **36**, 4636 (1942)].

METHYLSUCCINIC ACID

(Pyrotartaric acid)

$$CH_3CH{=}CHCOOC_2H_5 + NaCN + H_2O \rightarrow CH_3CH(CN)CH_2COONa + C_2H_5OH$$

$$CH_3CH(CN)CH_2COOH + 2H_2O \xrightarrow{[Ba(OH)_2]} CH_3CH(COOH)CH_2COOH + NH_3$$

Submitted by George Bosworth Brown.
Checked by Lee Irvin Smith and Vincent J. Webers.

1. Procedure

The procedure should be carried out under a hood since the poisonous hydrogen cyanide may be evolved.

In a 3-l. flask provided with a reflux condenser are placed 114 g. (1.0 mole) of ethyl crotonate (Note 1), 460 ml. of 95% ethanol, and a solution of 54 g. (1.06 moles) of 95% sodium cyanide in 128 ml. of water. The solution is refluxed on a steam bath for 5 hours, during which time some ammonia is evolved and the solution becomes dark yellow. A suspension of 150 g. of barium hydroxide octahydrate in 286 ml. of hot water is added to the solution, and the mixture is concentrated under reduced pressure (about 30 mm.) to a volume of about 400 ml. It is again refluxed on the steam bath for 4 hours or until the evolution of ammonia almost ceases (Note 2). The solution is then concentrated to a thick paste under reduced pressure.

The residue is cooled and dissolved in 171 ml. of nitric acid (sp. gr. 1.4) (Note 3), and the solution is warmed for 30 minutes on the steam bath. It is immediately concentrated to complete dryness under reduced pressure (Note 4). The flask is cooled, 300 ml. of benzene is added, and the mixture is refluxed for a short time to render the cake friable. The benzene is removed by decantation, and the cake is pulverized and extracted six times by refluxing it briefly with 300-ml. portions of ether. The combined benzene and ether extracts are filtered and concentrated to a volume of about 225 ml. In the meantime the residual salts are extracted twice by refluxing them vigorously for a short time with 300-ml. portions of benzene. The benzene solutions are separated by decantation and added to the ether concentrate. The distillation is then continued until about two-thirds of the benzene has been removed, when the benzene solution is poured into a beaker and allowed to cool. The methylsuccinic acid is collected on a filter and is washed by shaking a suspension of it in 150 ml. of chloroform (Note 5). The yield of air-dried product, melting at 110–111°, amounts to 87–93 g. (66–70%) (Note 6).

2. Notes

1. Ethyl crotonate may be prepared readily from technical crotonic acid by action of sulfuric acid and ethanol. The checkers obtained 72% yield both by the ordinary procedure and by the method of azeotropic distillation.

2. The checkers refluxed the mixture for 8 hours at this point; traces of ammonia were still present. Most of the ammonia was evolved in 4 hours, however.

3. The excess nitric acid is used in order to oxidize unchanged crotonic acid. Since hydrocyanic acid may be evolved the operation should be carried out under a hood.

4. The residue must be dry because methylsuccinic acid is extremely soluble in water.

5. If a series of runs is to be made, the chloroform may be used repeatedly.

6. The submitter states that the same percentage yield is obtained when four times the above quantities of reagents are used.

3. Methods of Preparation

Methylsuccinic acid has been prepared by the pyrolysis of tartaric acid;[1] from 1,2-dibromopropane or allyl halides by the action of potassium cyanide followed by hydrolysis;[2] by reduction of itaconic, citraconic, and mesaconic acids;[3] by hydrolysis of ketovalerolactonecarboxylic acid;[4] by decarboxylation of 1,1,2-propanetricarboxylic acid;[5] by oxidation of β-methylcyclohexanone;[6] by fusion of gamboge with alkali;[7] by hydrogenation and condensation of sodium lactate over nickel oxide;[8] from acetoacetic ester by successive alkylation with a methyl halide and a monohaloacetic ester;[9] by hydrolysis of α-methyl-α′-oxalosuccinic ester[10] or α-methyl-α′-acetosuccinic ester;[11] by action of hot, concentrated potassium hydroxide upon methylsuccinaldehyde dioxime;[12] from the ammonium salt of α-methylbutyric acid by oxidation with hydrogen peroxide;[13] from β-methyllevulinic acid by oxidation with dilute nitric acid[14] or hypobromite;[15] from

β-methyladipic acid;[16] from the decomposition products of glyceric acid[17] and pyruvic acid;[18] and by the action of methylmagnesium bromide on ethylene tetracarboxylic ester, followed by hydrolysis and decarboxylation.[19] The method described above is a modification of that of Higginbotham and Lapworth.[20]

1 Fourcroy and Vauquelin, *Ann. chim.*, (1), **35**, 164 (1799); (1), **64**, 42 (1807); de Clermont, *Ber.*, **6**, 72 (1873); Wolff, *Ann.*, **317**, 22 (1901).

2 Simpson, *Ann.*, **121**, 160 (1862); Claus, *Ann.*, **191**, 37 (1878); Pinner, *Ber.*, **12**, 2054 (1879); Euler, *Ber.*, **28**, 2953 (1895).

3 Kekulé, *Ann.*, Supl. Bd. I, 342 (1861); II, 100 (1862); Swarts, *Zeit. für Chem.*, **9**, 723 (1866); Eijkman, *Chem. Zentr.*, **1907**, I, 1617.

4 Wolff, *Ann.*, **317**, 25 (1901); de Jong, *Rec. trav. chim.*, **21**, 199 (1902).

5 Bischoff and Guthzeit, *Ber.*, **14**, 615 (1881); Higson and Thorpe, *J. Chem. Soc.*, **89**, 1455 (1906); Bone and Sprankling, *J. Chem. Soc.*, **75**, 842 (1899).

6 Markownikoff, *Ann.*, **336**, 299 (1904).

7 Hlasiwetz and Barth, *Ann.*, **138**, 73 (1866).

8 Ipatieff and Rasuvajev, *Ber.*, **59**, 2031 (1926).

9 Kressner, *Ann.*, **192**, 135 (1878).

10 Blaise and Gault, *Bull. soc. chim. France*, (4), **9**, 458 (1911).

11 Conrad, *Ann.*, **188**, 226 (1877).

12 Oddo and Mameli, *Gazz. chim. ital.*, **44**, II, 167 (1914).

13 Raper, *Biochem. J.*, **8**, 320 (1914).

14 Bischoff, *Ann.*, **206**, 331 (1880).

15 Pauly, Gilmour, and Will, *Ann.*, **403**, 123 (1914).

16 v. Braun and Jostes, *Ber.*, **59**, 1444 (1926).

17 Moldenhauer, *Ann.*, **131**, 340 (1864).

18 Urion, *Ann. chim.*, (11), **1**, 84 (1934).

19 Hsing and Li, *J. Am. Chem. Soc.*, **71**, 774 (1949).

20 Higginbotham and Lapworth, *J. Chem. Soc.*, **121**, 49 (1922).

METHYLTHIOUREA

(Urea, 1-methyl-2-thio-)

$$CH_3NCS + NH_3 \rightarrow CH_3NHCSNH_2$$

Submitted by Maurice L. Moore and Frank S. Crossley.
Checked by Nathan L. Drake and Ralph Mozingo.

1. Procedure

In a 500-ml. three-necked flask, equipped with a stirrer, a reflux condenser, and a dropping funnel, is placed 140 ml. (34 g.; 2 moles of ammonia) of concentrated ammonium hydroxide solution, and 95 g.

(1.3 moles) of methyl isothiocyanate (p. 599), b.p. 115–121°, is added, with stirring, over a period of 1 hour (Note 1). After the addition has been completed, the condenser is removed and the solution is heated on a water bath for 30 minutes to remove excess ammonia. The solution is then boiled with 2 g. of Norit and filtered, and the filtrate is chilled in an ice bath. The methylthiourea crystallizes as a colorless, compact, solid mass, which is collected on a filter, washed three times with 25-ml. portions of ice water, and dried. The first crop of crystals weighs 65–75 g. A second crop amounting to 15–20 g. is obtained by concentrating the mother liquor and washings to a volume of 75 ml. and again chilling in the ice bath. The total yield of methylthiourea, m.p. 119–120.5° (Note 2), is 85–95 g. (74–81%) (Notes 3 and 4).

2. Notes

1. The addition of the methyl isothiocyanate should be maintained at a constant rate since the reaction is slow to start; when the mixture has warmed up, the reaction becomes very vigorous and hard to control.

2. Further purification by crystallizing from boiling anhydrous ethanol yields a product melting at 120.5–121°.

3. The yield is only slightly higher if purer methyl isothiocyanate is used.

4. This is a general method for the preparation of alkyl thioureas. Ethylthiourea, m.p. 103–106°, has been prepared from ethyl isothiocyanate in the same manner. Di- and trialkyl thioureas may be prepared from alkyl isothiocyanates in a similar manner by substituting an equivalent amount of an amine solution in place of the ammonium hydroxide. Thus, *sym*-dimethylthiourea is prepared from methyl isothiocyanate and methylamine solution. A solution of dimethylamine and methyl isothiocyanate gives trimethylthiourea.

3. Methods of Preparation

Methylthiourea has been prepared from methyl isothiocyanate and ammonia in ethanolic [1] or aqueous [2] solution, from methylammonium thiocyanate,[3] and by heating the methyl or ethyl ester of N-methyldithiocarbamic acid with alcoholic ammonia.[4]

[1] Näf, *Ann.*, **265**, 108 (1891).

[2] Andreasch, *Monatsh.*, **2**, 276 (1881).

[3] Salkowski, *Ber.*, **26**, 2497 (1893).

[4] Delépine, *Compt. rend.*, **134**, 1222 (1902); *Bull. soc. chim. France*, (3) **27**, 812 (1902).

MONOPERPHTHALIC ACID

(Phthalic monoperacid)

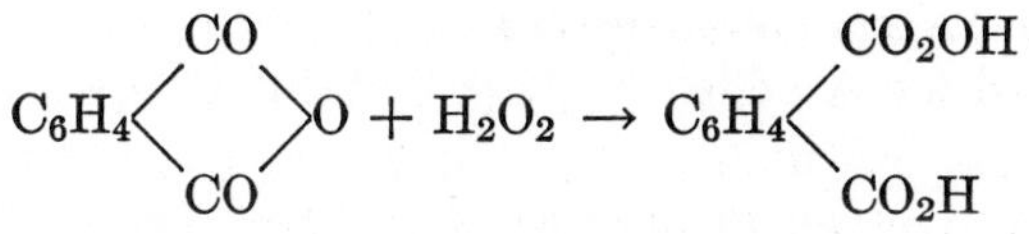

Submitted by Horst Böhme.
Checked by Homer Adkins and E. Leon Foreman.

1. Procedure

In a 1-l. round-bottomed flask, equipped with a mechanical stirrer and cooled in an ice-salt bath, is placed 275 g. (250 ml., approximately 1 mole) of 15% sodium hydroxide solution. This is cooled to −10° (Note 1), and 115 g. (105 ml., approximately 1 mole) of 30% hydrogen peroxide which has been similarly cooled is added in one portion. The heat of reaction causes the temperature to rise markedly. When the temperature has again dropped to −10°, 75 g. (0.5 mole) of phthalic anhydride which has been pulverized to pass a 40-mesh sieve is added as quickly as possible while the contents are stirred vigorously in the freezing mixture (Notes 2 and 3). As soon as all the anhydride has dissolved, 250 ml. (0.5 mole) of 20% sulfuric acid which has been previously cooled to −10°, but not frozen (Note 4), is added.

The acid solution is filtered without suction through glass wool into a 2-l. separatory funnel and extracted once with 500 ml. of ether, then three times with 250-ml. portions of the same solvent. The combined ether extracts are shaken out with three 150-ml. portions of 40% ammonium sulfate solution and dried for 24 hours, preferably in a refrigerator, over 50 g. of anhydrous sodium sulfate.

If the ether is evaporated under reduced pressure (Note 5), crystalline monoperphthalic acid is obtained. It is more convenient, however, to use the ether solution directly (Note 6). Its peracid content is determined by adding to 2 ml. of the solution 30 ml. of 20% potassium iodide solution and titrating the iodine after 10 minutes with 0.05 *N* thiosulfate solution. The yield is 60–65 g. (65–70% based on the phthalic anhydride) (Note 7).

2. Notes

1. If the solutions are cooled to −10°, little oxygen is evolved and the yields of peracid are good. If the reaction is carried out at 0°, a large amount of oxygen is evolved and the yields are poor.

2. Commercial phthalic anhydride may be used directly. If excessive decomposition occurs, however, the anhydride should be purified by distillation under reduced pressure.

3. The anhydride is added in large portions or, better, in one portion.

4. The decisive factor in the success of this preparation is the time interval between the addition of the anhydride and the acidification of the reaction mixture. All the anhydride should dissolve, but prolonged stirring results in excessive oxygen evolution. The quicker the anhydride dissolves, and the smaller the oxygen evolution, the better the yield of the peracid. Hence, stirring must be vigorous.

5. If crystalline monoperphthalic acid is desired, it may be prepared conveniently as follows: The dried ether solution is placed in a distilling flask equipped with a capillary tube connected with a drying tube, and the flask is connected with the water pump. The ether is evaporated at the pressure thus obtained without the application of heat (ice will form on the flask) to a thin syrup (approximately 150 ml.). The syrup is transferred to an evaporating dish and the flask rinsed with a small amount of dry ether, the washings being added to the syrup. The remainder of the ether is then evaporated in a vacuum desiccator over concentrated sulfuric acid. For good results in this preparation the drying must be very thorough, for only 1% of water in the ether solution will be more than sufficient to destroy the entire amount of peracid.

6. If ether is not suitable for the oxidation reactions in which the peracid is to be used, the material can be dissolved in another solvent after removal of the ether. An excellent solvent for monoperphthalic acid oxidations is dioxane, and a solution of the peracid in dioxane is readily prepared by adding dioxane to the dried ether extract and then removing the ether under reduced pressure at 15°. The dioxane must be peroxide-free.[1]

7. As originally submitted, this preparation was on one-fifth the scale indicated here. However, the checkers have had no difficulty with the larger-scale preparation.

It has been reported[2] that an 86% yield of monoperphthalic acid results when a single ether extraction is employed. In the modified procedure, 40% alkali was employed, and crushed ice was added directly for cooling.

3. Methods of Preparation

Monoperphthalic acid has been prepared by hydrolysis of phthalyl peroxide with sodium hydroxide[3] and by shaking phthalic anhydride with excess alkaline peroxide solution.[3] The method described here is a modification of the latter process.[4]

[1] Eigenberger, *J. prakt. Chem.*, (2) **130**, 75 (1931).
[2] Bachman and Cooper, *J. Org. Chem.*, **9**, 307 (1944).
[3] Baeyer and Villiger, *Ber.*, **34**, 764 (1901).
[4] Böhme, *Ber.*, **70**, 379 (1937).

MUCOBROMIC ACID

(Acrylic acid, α,β-dibromo-β-formyl-)

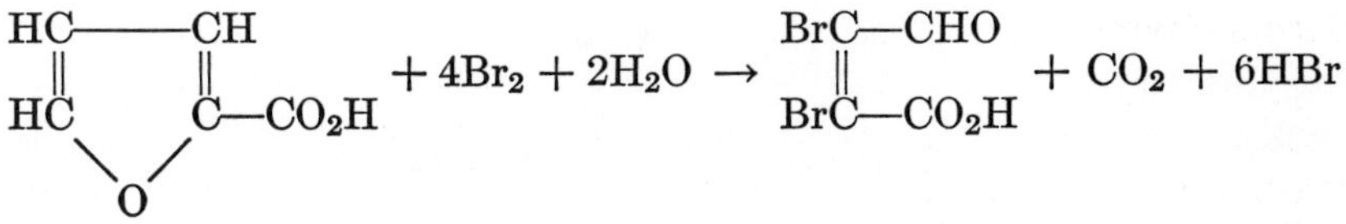

Submitted by C. F. H. ALLEN and F. W. SPANGLER.
Checked by R. L. SHRINER and NICHOLAS J. KARTINOS.

1. Procedure

In a 2-l. three-necked round-bottomed flask, equipped with a reflux condenser attached to a gas trap, a dropping funnel, and a mechanical stirrer, are placed 100 g. (0.9 mole) of furoic acid (Note 1) and 440 ml. of water. The flask is immersed in a pan of crushed ice, and 686 g. (220 ml., 4.3 moles) of bromine is placed in the dropping funnel. The stirrer is started, and the bromine is added over a period of about 1 hour with constant stirring and cooling (Note 2). The bromine is decolorized almost immediately at first and then more slowly as the reaction proceeds. The mixture is then heated to boiling and refluxed for 30 minutes. The condenser is removed and the boiling continued for an additional 30 minutes with one neck of the flask open. The mixture is cooled and chilled thoroughly, whereupon the mucobromic acid separates and is collected on a filter. The filter cake is removed and thoroughly triturated with a solution of 5 g. of sodium bisulfite in 150 ml. of water (Note 3), and the product is again removed by filtration. After drying in the air the crude product weighs 155–170 g. (67–73%) and melts at 120–122° (cor.).

The mucobromic acid may be recrystallized by solution in 250 ml. of boiling water with the addition of 2 g. of Darco. The hot solution is filtered, and the filtrate is thoroughly chilled in an ice bath. After filtration and drying there is obtained 148–155 g. (64–67%) of white crystals melting at 123–124° (cor.).

2. Notes

1. The furoic acid may be obtained from furfural by means of the Cannizzaro reaction [1] or by oxidation with alkaline potassium permanganate.[2] It is important to purify the furoic acid to a melting point of 131–132°.

2. It is essential that the flask and its contents be well cooled; otherwise the yield is materially decreased.

3. The sodium bisulfite solution removes the color produced by the excess bromine.

3. Methods of Preparation

The only practical methods of preparation are the action of bromine on furfuraldehyde [3] or furoic acid [4–6] in aqueous solution.

[1] *Org. Syntheses* Coll. Vol. **1,** 276 (1941).
[2] Wagner and Simons, *J. Chem. Education,* **13,** 270 (1936).
[3] Simonis, *Ber.,* **32,** 2085 (1899).
[4] Jackson and Hill, *Am. Chem. J.,* **3,** 41 (1881–2); *Ber.,* **11,** 1671 (1878).
[5] Schmelz and Beilstein, *Ann. Suppl.,* **3,** 276 (1864).
[6] Limpricht, *Ann.,* **165,** 293 (1873).

MUCONIC ACID

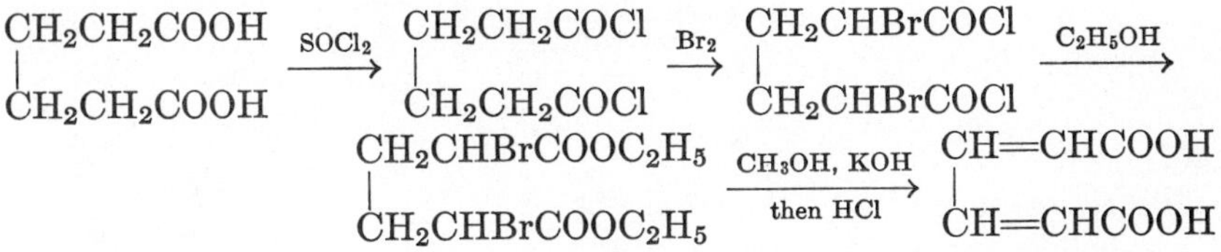

Submitted by P. C. GUHA and D. K. SANKARAN.
Checked by C. F. H. ALLEN and H. W. J. CRESSMAN.

1. Procedure

A. *Diethyl α,δ-dibromoadipate.* In a 3-l. three-necked flask (Note 1) fitted with a reflux condenser, a dropping funnel, and a mechanical stirrer are placed 1 kg. (6.85 moles) of adipic acid and 2 kg. (1220 ml., 16.8 moles) of thionyl chloride (Notes 2 and 3). The mixture is stirred and heated gently on the steam bath until solution is effected, and the evolution of hydrogen chloride (Note 4) ceases after about 3 hours. The excess thionyl chloride is distilled by heating on a steam bath, the last portion of the reagent being taken off under reduced pressure.

To the acid chloride, mechanically stirred and heated on the steam bath, is added 2.5 kg. (805 ml., 15.6 moles) of dry bromine (p. 124) as rapidly as it will react (Note 5). The addition requires about 12 hours. The contents of the flask are stirred and heated an additional 2 hours, transferred to a dropping funnel (Note 6), and added in a thin stream to 5 l. of absolute ethanol, which has previously been placed in a 12-l. flask provided with a stopper carrying an efficient reflux condenser, a separatory funnel, and a mechanical stirrer. The resulting vigorous reaction is controlled by external cooling. After the dibromoacid chloride has been added, the reaction mixture is allowed to stand at room temperature overnight and is then poured into 5 l. of cold water. The top ethanolic aqueous layer is decanted and extracted once with 8 l. of ether. The oily bottom layer is dissolved in the ether extract and washed first with 1 l. of a 2% sodium bisulfite solution, then with two 1-l. portions of 3% sodium carbonate solution, and finally with several portions of water. The ether solution is dried over 175 g. of potassium carbonate; the solvent is distilled on the steam bath. The yield of residual ester (Note 7) amounts to 2260–2400 g. (91–97%).

B. *Muconic acid.* To a solution of 3 kg. of potassium hydroxide and 5 l. of methanol (Note 8) in a 12-l. flask, equipped as described above, is added in a thin stream from a separatory funnel (Note 6), and with stirring, 1130 g. (3.14 moles) of diethyl α,δ-dibromoadipate. The ester is heated on the steam bath to nearly 100° before it is added. The addition is so regulated as to permit rather vigorous refluxing and requires 45 minutes. Heating on the steam bath and stirring are continued an additional 2 hours. The mixture is allowed to stand at room temperature overnight and is cooled. The potassium muconate and potassium bromide are filtered (Note 9), washed on the funnel with two 350-ml. portions of methanol and 400 ml. of ether, transferred without further drying to an enameled 11- to 12-l. pail, and dissolved in 8 l. of hot water. The aqueous solution, to which is added 30 g. of Norit, is heated to boiling by introduction of steam, filtered through a Norit filter pad by suction, and cooled in an ice-salt bath. The muconic acid is precipitated by the addition of a large excess (1.5 l.) (sp. gr. 1.18) of concentrated hydrochloric acid to the cold solution. The acid is added in a thin stream to the well-stirred solution. After 2 hours the muconic acid is filtered, washed first with two 400-ml. portions of cold water and then with 200 ml. of methanol, and dried at 85°. The yield of nearly colorless product melting at 296–298°, with decomposition, amounts to 165–195 g. (37–43%).

2. Notes

1. Apparatus with ground-glass joints is preferable.

2. The yield obtained in molar-sized runs is also about 40% of the theoretical.

3. The submitters used 1 mole of adipic acid and added 2 moles of phosphorus pentachloride in 20-g. portions over an interval of 30 minutes.

4. The considerable quantities of hydrogen chloride and hydrogen bromide evolved are best handled by means of a gas-absorption trap. The insertion of a calcium chloride tube between the trap and the reflux condenser is recommended.

5. The bromine is added as rapidly as it will react but not so rapidly that loss occurs through the condenser. A photoflood lamp accelerates the rate of bromination.

6. The dibromoacid chloride solidifies at room temperature. It is advisable to heat the funnel by means of a coil of copper tubing through which steam is passed.

7. Two isomers, a solid and a liquid, are formed during the bromination and subsequent esterification.

8. A commercial grade of methanol was used; absolute methanol did not result in an increased yield.

9. A pad of Filter-Cel aids in the slow filtration. About 100 g. is needed with a 30-cm. funnel.

3. Methods of Preparation

Muconic acid has been obtained in a variety of ways. The procedures that seem most important from a preparative point of view are by treatment of ethyl α,δ-dibromoadipate with ethanolic potassium hydroxide,[1,2] by condensation of glyoxal (as the sodium bisulfite addition product) with malonic acid,[3,4] by heating ethyl 1-acetoxy-1,4-dihydromuconate (obtained by condensing ethyl oxalate and ethyl crotonate, acetylating, and reducing),[5,6] by oxidation of phenol[7] or catechol[8] with peracetic acid, and by hydrolysis of the half-ester obtained by the condensation of ethyl fumaraldehydate with malonic acid.[9]

Three steroisomeric muconic acids are known.[10]

1 Ingold, *J. Chem. Soc.*, **119**, 967 (1921).

2 Guha and Sankaran, *Ber.*, **70**, 2110 (1937).

3 Doebner, *Ber.*, **35**, 1147 (1902).

4 Behrend and Koolman, *Ann.*, **394**, 244 (1912).

5 Kuhn and Grundmann, *Ber.*, **69**, 1759 (1936).

6 Brit. pat. 477,774 [*C. A.*, **32**, 4174 (1938)].

7 Boeseken and Engelbert, *Proc. Acad. Sci. Amsterdam*, **34**, 1292 (1931) [*C. A.*, **26**, 2970 (1932)].

8 U. S. pat. 2,534,212 [*C. A.*, **45**, 3415 (1951)].

9 Funke and Karrer, *Helv. Chim. Acta*, **32**, 1016 (1949).

10 Elvidge, Linstead, Sims, and Orkin, *J. Chem. Soc.*, **1950**, 2235.

β-NAPHTHALDEHYDE

(2-Naphthaldehyde)

I. METHOD A

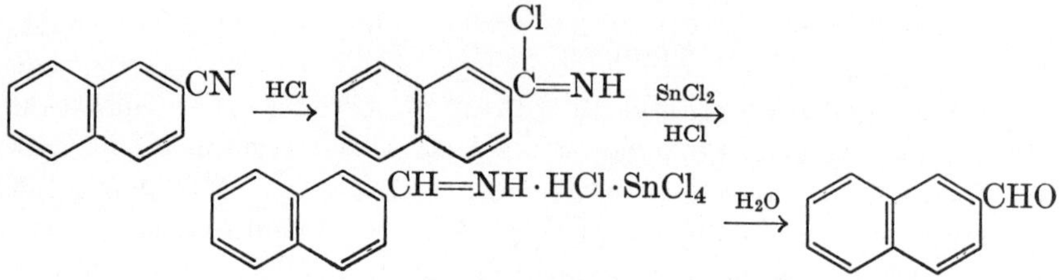

Submitted by Jonathan W. Williams.
Checked by C. F. H. Allen and J. VanAllan.

1. Procedure

In a 2-l. three-necked round-bottomed flask, provided with a mechanical stirrer, a reflux condenser carrying a drying tube, and an inlet tube reaching nearly to the bottom of the flask, are placed 76 g. (0.4 mole) of anhydrous stannous chloride (Note 1) and 400 ml. of anhydrous ether. The mixture is then saturated with dry hydrogen chloride, while it is slowly stirred; this requires 2.5–3 hours, during which time the stannous chloride forms a viscous lower layer.

The inlet tube is replaced by a dropping funnel, and a solution of 30.6 g. (0.2 mole) of β-naphthonitrile, m.p. 60–62° (Note 2), in 200 ml. of dry ether is added rapidly. Hydrogen chloride is again passed into the mixture until it is saturated, and the mixture is then stirred rapidly for 1 hour and allowed to stand overnight while the yellow aldimine-stannichloride separates completely.

The ethereal solution is decanted, and the solid is rinsed with two 100-ml. portions of ether. The solid is transferred to a 5-l. flask fitted for steam distillation and immersed in an oil bath, the temperature of which is maintained at 110–120° (Note 3). Dry steam is passed through the mixture (Note 4) until the aldehyde is completely removed; this requires 8–10 hours, and 8–10 l. of distillate is collected.

The white solid is filtered and allowed to dry in the air; it amounts to 23–25 g. (73–80%) and melts at 53–54°. For further purification, it is distilled under reduced pressure (Note 5); the water-clear distillate (b.p. 156–158°/15 mm.) is poured into a mortar while hot and is pulverized when cool. The recovery is 93–95%, and the melting point is 57–58°.

2. Notes

1. The success of this type of reaction depends on the quality of the catalyst. The most active and dependable form of anhydrous stannous chloride[1] is prepared as follows: In a 600-ml. beaker is placed 204 g. (189 ml., 2 moles) of acetic anhydride (99–100%) and, while the liquid is stirred by hand, 226 g. (1 mole) of commercial c.p. crystalline stannous chloride dihydrate is added. This operation should be performed in a hood, for the heat of the reaction is sufficient to cause the acetic anhydride to boil. After about 1.5 hours, the anhydrous stannous chloride is filtered on a large Büchner funnel, rinsed with two 50-ml. portions of dry ether, and dried overnight in a vacuum desiccator. The yield is quantitative (189 g.). The product may be kept in a tightly closed bottle until it is wanted. The product secured by dehydrating crystalline stannous chloride in an oil bath at 195–200° is satisfactory in many instances but is not dependable.

2. β-Naphthonitrile is prepared by the procedure described under *o*-Tolunitrile, in *Org. Syntheses* Coll. Vol. **1**, 514 (1941).

3. The use of dry, slightly superheated steam reduces the time of distillation but is not essential.

4. A superheater obtained from the Fisher Scientific Company was used. It was preceded by the usual steam trap to remove the condensed water. The thermometer in the superheater recorded 260°.

5. It is convenient to combine the material from several runs.

II. METHOD B

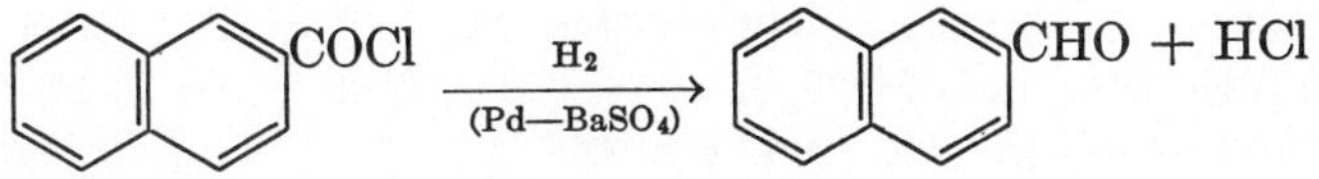

Submitted by E. B. Hershberg and James Cason.
Checked by Nathan L. Drake, Harry D. Anspon, and Ralph Mozingo.

1. Procedure

A 500-ml. three-necked flask, *equipped with ground joints,* is fitted with a mercury-sealed stirrer (Note 1), a reflux condenser, and a gas-inlet tube extending to a point just above the stirrer. In the flask are placed 57 g. (0.30 mole) of β-naphthoyl chloride (Note 2), 200 ml. of xylene (Note 3), 6 g. of palladium-barium sulfate catalyst (p. 685), and 0.6 ml. of stock poison solution (Note 4). The top of the condenser is connected by a rubber tube to a 6-mm. glass tube extending to the bottom of a 500-ml. Erlenmeyer flask containing 400 ml. of distilled water and a few drops of phenolphthalein indicator solution. A buret containing approximately 5 *N* sodium hydroxide solution (prepared by dissolving 20.5 g. of analytical reagent sodium hydroxide in water and diluting to 100 ml.) is arranged for delivery into this flask, which for safety should be placed at least 2 ft. away from any flame. Commercial electrolytic hydrogen is passed from a cylinder directly into the reaction flask at such a rate that 100–300 bubbles per minute emerge in the Erlenmeyer flask.

After the air in the reaction flask has been displaced by hydrogen, the flask is heated in an oil bath at 140–150°, the stirrer is started (Note 5), and 1 ml. of alkali is run into the Erlenmeyer flask. The course of the reaction is followed by the rate of hydrogen chloride evolution. The first 5 ml. of alkali should be neutralized in 12–15 minutes, and the reaction should be complete in approximately 3 hours. About 92% of the theoretical amount of hydrogen chloride (equivalent to 55 ml. of 5 *N* sodium hydroxide solution) is recovered. The end of the reaction is evidenced by a rather abrupt cessation of hydrogen chloride evolution, and the reaction is discontinued at this point.

The flask is cooled, 1–2 g. of Norit added with stirring, and the solution filtered with suction through a hardened filter paper (Note 6). The xylene is removed from the nearly colorless filtrate by flash distillation under diminished pressure. For this purpose, a 125-ml. modified Claisen flask is arranged for vacuum distillation, the usual capillary being replaced by a separatory funnel whose stem extends to the bottom of the flask. The flask is heated in an oil bath at 90–100° and the solution added from the funnel as rapidly as possible without causing accumulation of xylene in the distilling flask. After all the solution has been added, the separatory funnel is replaced by a capillary and the bath temperature is raised. After a small fore-run consisting mostly of naphthalene, the β-naphthaldehyde distils at 147–149°/11 mm. (bath temperature 170–180°), leaving a small non-volatile residue. In this way, 34.5–38 g. (74–81%) of white aldehyde, m.p. 59–60°, is obtained (Note 7).

2. Notes

1. A Hershberg tantalum or platinum wire stirrer whose shaft runs in a ball bearing is convenient, but an ordinary all-glass stirrer may

be used. The stirrer must be capable of running at a high speed, for the rate of reaction is dependent to a high degree on the speed of stirring. It is also extremely important that the stirrer be carefully lined up so that there is a minimum of splashing of mercury in the seal. If mercury works down into the flask, the reaction will not proceed properly (Note 5).

2. β-Naphthoyl chloride is conveniently prepared from β-naphthoic acid and phosphorus pentachloride. A mixture of 57.4 g. (0.33 mole) of acid and 69 g. (0.33 mole) of phosphorus pentachloride in a 250-ml. modified Claisen distilling flask is warmed on a steam bath in a hood. As soon as the vigorous reaction sets in, the flask is removed from the steam bath until the rapid evolution of hydrogen chloride has moderated, then warmed on the steam bath for 30 minutes. After removal of the phosphorus oxychloride at diminished pressure, using a water pump, the acid chloride is distilled. The fraction boiling at 160–162°/11 mm. (bath temperature 170–180°) weighs 57-60 g. (90–95%) and melts at 51–52°. The distillation should be carefully conducted, and a quite colorless product should result.

3. One liter of technical xylene is refluxed overnight with 2 g. of sodium, distilled, and stored over sodium.

4. The quinoline-sulfur poison of Rosenmund and Zetzsche [2] is prepared by refluxing 1 g. of sulfur with 6 g. of quinoline for 5 hours and diluting the resultant dark brown liquid to 70 ml. with the purified xylene. The literature on the Rosenmund reduction contains many conflicting reports concerning the necessity for a catalyst poison; however, the work of Zetzsche and collaborators [3,4] indicates that the purity of the solvent is the determining factor. These workers found that by using technical xylene without added poison a good yield of aldehyde could usually be obtained but after the xylene had been purified by distilling over anhydrous aluminum chloride practically no aldehyde was obtained under the same conditions. Instead, products arising from further reduction of the aldehyde were obtained. In view of these results the use of a poison is recommended in order to ensure controlled conditions. The submitters claim that the use of twice the ratio of poison specified has no effect except slowing up the reaction; the yield and quality of the product remain the same.

5. The rapid rate of stirring desirable for maximum reaction rate often causes spraying of fine droplets of mercury from the seal. This can be prevented by a layer of paraffin oil over the mercury. It is important for the gas-inlet tube to extend below the surface of the stirred liquid, for absorption of hydrogen occurs chiefly at the rapidly agitated surface.

6. The palladium may be recovered from used catalyst by ignition and solution in aqua regia.[5]

7. According to the submitters, this reaction is quite satisfactory on a small scale and can be used with other acid chlorides. In a 0.05-mole run carried out in the same manner, an 83% yield of β-naphthaldehyde was obtained. 1-Acetoxy-3-naphthaldehyde, m.p. 112–114°, was obtained in 70% yield from 0.85 g. of the corresponding acid chloride. Methyl β-formylpropionate, b.p. 69–70°/14 mm., was also obtained in 65% yield from the acid chloride; reduction proceeds rapidly at 110° in this case.

3. Methods of Preparation

β-Naphthaldehyde has been prepared from β-chloromethylnaphthalene by the use of hexamethylenetetramine in ethanol,[6] or by oxidation with lead nitrate;[7] from β-bromomethylnaphthalene by the use of hexamethylenetetramine in ethanol[8] or in acetic acid,[9] or by oxidation with lead nitrate;[7] by distillation of a mixture of calcium formate and calcium β-naphthoate;[10,11] by reduction of β-naphthoic acid with sodium amalgam;[12] from β-naphthylcarbinol by oxidation with chromic acid;[13] from β-naphthylglyoxylic acid anil;[14] from β-naphthylmagnesium iodide and methyl orthoformate;[15] from β-naphthylmagnesium bromide and ethoxymethylaniline[16] or orthoformic ester;[14,17] by treatment of β-naphthylmagnesium bromide with carbon disulfide, followed by conversion of the dithioacid to a semicarbazone and hydrolysis;[18] from β-naphthonitrile by Stephen reduction;[19,20] from β-naphthoyl chloride by Rosenmund reduction;[3,21,22] and from 2-methylnaphthalene by oxidation with selenium dioxide.[23]

[1] Stephen, *J. Chem. Soc.*, **1930**, 2786.
[2] Rosenmund and Zetzsche, *Ber.*, **54**, 436 (1921).
[3] Zetzsche and Arnd, *Helv. Chim. Acta,* **9**, 173 (1926).
[4] Zetzsche, Enderlin, Flütsch, and Menzi, *Helv. Chim. Acta,* **9**, 177 (1926).
[5] Fröschl, Maier, and Heuberger, *Monatsh.*, **59**, 256 (1932).
[6] Wahl, Goedkoop, and Heberlein, *Bull. soc. chim. France,* (5) **6**, 533 (1939).
[7] Schulze, *Ber.*, **17**, 1530 (1884); Kikkoji, *Biochem. Z.*, **35**, 71 (1911).
[8] Mayer and Sieglitz, *Ber.*, **55**, 1857 (1922).
[9] Badger, *J. Chem. Soc.*, **1941**, 536.
[10] Battershall, *Ann.*, **168**, 116 (1873).
[11] Sah, *Rec. trav. chim.*, **59**, 461 (1940).
[12] Weil, *Ber.*, **44**, 3058 (1911); Weil and Ostermeier, *Ber.*, **54**, 3217 (1921).
[13] Bamberger and Boekmann, *Ber.*, **20**, 1118 (1887).
[14] Rousset, *Bull. soc. chim. France,* (3) **17**, 305 (1897).
[15] Sah, *Rec. trav. chim.*, **59**, 1021 (1940).
[16] Monier-Williams, *J. Chem. Soc.*, **89**, 275 (1906); Gattermann, *Ann.*, **393**, 228 (1912).

[17] Tschitschibabin, *Ber.*, **44**, 447 (1911).
[18] Wuyts, Berman, and Lacourt, *Bull. soc. chim. Belg.*, **40**, 665 (1931).
[19] Fulton and Robinson, *J. Chem. Soc.*, **1939**, 200; Williams, *J. Am. Chem. Soc.*, **61**, 2248 (1939).
[20] Stephen, *J. Chem. Soc.*, **127**, 1874 (1925).
[21] Rosenmund et al., *Ber.*, **51**, 585, 594 (1918); **54**, 2888 (1921); **55**, 609 (1922).
[22] Fieser and Hershberg, *J. Am. Chem. Soc.*, **62**, 52 (1940).
[23] Sultanov, Rodionov, and Shemyakin, *J. Gen. Chem. U.S.S.R.*, **16**, 2072 (1946) [*C. A.*, **42**, 880 (1948)].

α-NAPHTHONITRILE *

(1-Naphthonitrile)

$$\alpha\text{-}C_{10}H_7Br + CuCN \rightarrow \alpha\text{-}C_{10}H_7CN + CuBr$$

Submitted by M. S. Newman.
Checked by R. L. Shriner and F. J. Wolf.

1. Procedure

In a dry 200-ml. flask fitted with a ground-joint reflux condenser and protected from moisture by a calcium chloride tube are put 66 g. (0.32 mole) of α-bromonaphthalene (Note 1), 35 g. (0.39 mole) of dry powdered cuprous cyanide (Note 2), and 30 ml. of pyridine (Note 3) in the order mentioned. This mixture is heated in a Wood's metal bath (Note 4) at 215–225° for 15 hours. The resulting dark brown solution is poured while still hot (about 100°) into a flask containing 150 ml. of aqueous ammonia (sp. gr. 0.90) and 150 ml. of water. About 140 ml. of benzene is added, and the flask is stoppered and shaken until all the lumps have disintegrated. After the mixture has cooled to room temperature, 100 ml. of ether is added and the mixture is filtered (Note 5). The filtrate is transferred to a 1-l. separatory funnel, and the aqueous layer is separated (Note 6). The ether-benzene layer is washed successively with (*a*) four 100-ml. portions of dilute aqueous ammonia (Note 7), (*b*) two 100-ml. portions of 6 *N* hydrochloric acid (Note 8), (*c*) two 100-ml. portions of water, and (*d*) two 100-ml. portions of saturated sodium chloride solution. The ether and benzene are removed by distillation from a water bath, and the residue is distilled under reduced pressure from a 125-ml. modified Claisen flask. The temperature rises rapidly, and the yield of colorless α-naphthonitrile, b.p. 173–174°/27 mm. (166–169°/18 mm.), is 40–44 g. (82–90%) (Notes 9 and 10).

2. Notes

1. The α-bromonaphthalene was redistilled just before use and the fraction boiling at 153–154°/22 mm. taken.

2. "Baker's Analyzed" cuprous cyanide (powdered) was used. The cuprous cyanide must be dry.

3. Pure pyridine dried over barium oxide was used. A considerable amount of heat is liberated on addition of the pyridine to the mixture.

4. A fused salt bath consisting of 8.5 parts (by weight) of sodium nitrite and 10 parts of potassium nitrate has a melting point of about 140° and may replace the metal bath.

5. The cuprammonium solution attacks filter paper and hence this solution is best filtered through a fritted-glass filter. Two layers of ordinary toweling cloth on a Büchner funnel may also be used.

6. It is occasionally necessary to add an additional 50–100 ml. of ether to facilitate separation.

7. If the final ammoniacal wash solution is not colorless, the ether-benzene solution should be washed with additional 100-ml. portions of dilute aqueous ammonia.

8. If a precipitate separates during this operation it should be removed by filtration.

9. The yields vary with the quality of the cuprous cyanide. One lot of this reagent gave yields of only 60–75%. Larger amounts may be run, but the amount of pyridine used should be reduced to decrease the reflux. For a 2-mole run, 160 ml. of pyridine was satisfactory. The yield of the large runs was 81%.

10. α-Chloronaphthalene (b.p. 144–146.5°/29 mm.) may be used in place of α-bromonaphthalene, but 24 hours' heating at 245–250° is required. The yields are about the same.

3. Methods of Preparation

α-Naphthonitrile has been prepared from α-naphthylamine by the Sandmeyer reaction,[1] from α-chloronaphthalene,[2] from α-bromonapththalene [3] by heating with cuprous cyanide, by heating the sodium salt of α-naphthalenesulfonic acid with sodium cyanide [4] or with potassium ferrocyanide, iron, and ferric oxide,[5] by heating α-naphthamide with phosphorus pentachloride [6] or with sodium chloroaluminate,[7] and by distilling α-naphthoic acid with dicyandiamide.[8]

[1] Rupe and Brentano, *Helv. Chim. Acta*, **19**, 581 (1936); McRae, *J. Am. Chem. Soc.*, **52**, 4550 (1930); Clarke and Read, *J. Am. Chem. Soc.*, **46**, 1001 (1924).

[2] Ger. pat. 293,094 (1916) [*Frdl.*, **13**, 269 (1923)].
[3] Newman, *J. Am. Chem. Soc.*, **59**, 2472 (1937).
[4] Whitmore and Fox, *J. Am. Chem. Soc.*, **51**, 3363 (1929).
[5] Wahl, Goedkoop, and Heberlein, *Bull. soc. chim. France*, (5) **6**, 533 (1939).
[6] Blicke, *J. Am. Chem. Soc.*, **49**, 2848 (1927).
[7] Norris and Klemka, *J. Am. Chem. Soc.*, **62**, 1432 (1940).
[8] Dangyan, Danielyan, and Akopyan, *Bull. Armenian Branch Acad. Sci. U.S.S.R.*, Ser. II, **1943**, 57 [*C. A.*, **40**, 3421 (1946)].

1,2-NAPHTHOQUINONE-4-SULFONATE, AMMONIUM AND POTASSIUM

(1-Naphthalenesulfonic acid, 3,4-dihydro-3,4-dioxo, ammonium and potassium salts)

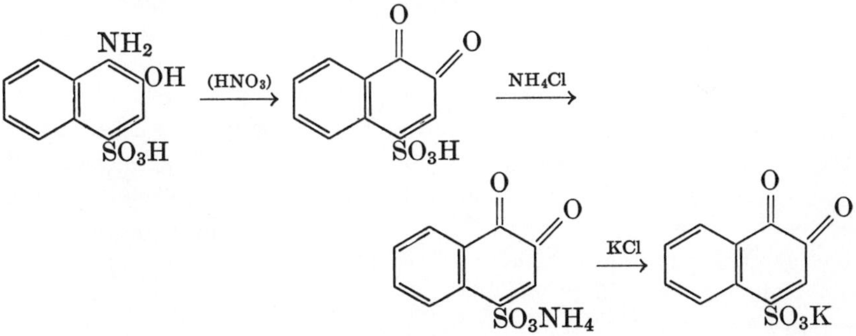

Submitted by E. L. Martin and L. F. Fieser.
Checked by R. L. Shriner and Eldred Welch.

1. Procedure

A mixture of 145 ml. of nitric acid (sp. gr. 1.42) and 400 ml. of water in a 2-l. beaker is cooled to 30° in a slush of ice and water, and 350 g. (1.46 moles) of pure, anhydrous 1-amino-2-naphthol-4-sulfonic acid (Note 1) is weighed into a separate 2-l. beaker. The beaker is removed from the bath; a 10-g. portion of the 1-amino-2-naphthol-4-sulfonic acid is stirred into the solution, and the liquid is then allowed to become entirely motionless. Generally, oxidation starts during this addition; if not, 2 ml. of concentrated nitric acid is poured carefully down the side of the beaker without stirring. Oxidation commences in 1–2 minutes and the mixture turns yellow (Note 2).

The beaker is replaced in the ice bath, and 20–25 g. of 1-amino-2-naphthol-4-sulfonic acid is stirred into the mixture by hand. A second portion of 20–25 g. is added immediately and stirred. The mixture begins to froth and is covered with a layer of 100 ml. of ether which serves as an efficient subsident (Note 3). The remainder of the 1-amino-2-naphthol-4-sulfonic acid is added in 20–25 g. portions during the course of 3–4 minutes, the mixture being stirred well after each addition. Oxides of nitrogen are freely evolved, and a stiff yellow-orange paste is formed. The temperature is maintained between 25° and 30° by vigorous stirring and by controlling the rate of addition of the compound (Note 4). The oxidation is complete within 3–4 minutes after the last addition and gas is then no longer evolved. The thick mass is stirred until the temperature has dropped to 5–10° and then 175 ml. of saturated ammonium chloride solution (30°) is added.

After the mixture has been cooled to 0°, the ammonium 1,2-naphthoquinone-4-sulfonate is collected on a 20-cm. Büchner funnel and as much of the mother liquor is removed as possible by pressing the cake with a porcelain spatula or glass stopper. The product is washed with three equal portions of a cold mixture of 150 ml. of saturated ammonium chloride solution and 100 ml. of water. The wash solution is removed as completely as possible, and the product is washed twice with 50-ml. portions of ethanol, followed by 300 ml. of ether in small portions (Note 5). The ammonium 1,2-naphthoquinone-4-sulfonate is spread out in a thin layer and dried to constant weight at 35–40°. An orange, microcrystalline product of bright appearance is thus obtained. The yield is 350–365 g. (94–98%). The ammonium salt is of good quality and is sufficiently pure for many purposes. No satisfactory method has been devised for its further purification, but it can be converted into a pure potassium salt as follows.

Seventeen hundred milliliters of water containing 0.3 ml. of liquid bromine is heated in a 2-l. Erlenmeyer flask to 50° on a steam bath (Note 6). The flask is removed, 50 g. of ammonium 1,2-naphthoquinone-4-sulfonate is added, and the mixture is shaken for a few minutes until solution is complete. Three grams of Norit is added, and the solution is stirred for 2–3 minutes. It is then filtered by suction and the clear orange filtrate is transferred to a 4-l. Erlenmeyer flask. Four hundred milliliters of saturated potassium chloride solution (30°) is added rapidly in one portion, and the flask is allowed to stand undisturbed. Orange crystals of potassium salt begin to separate immediately, and after standing for 30 minutes the contents of the flask are cooled to 0° in an ice-salt bath and the potassium 1,2-naphthoquinone-4-sulfonate is collected on a 15-cm. Büchner funnel. The solid is washed with 150 ml. of cold, dilute potassium chloride solution (30 ml. of saturated potassium chloride solution added to 120 ml. of water), and then with 150 ml. of ethanol in small portions followed by 300 ml. of ether. The potassium salt is dried to constant weight at 40–50°. The yield is 48–50 g. (90–92% based on the ammonium salt used). The product consists of orange needles free of colored decomposition products but still containing traces of ammonium salts. An ammonium-free compound is obtained by a second crystallization. Seventeen hundred milliliters of water containing 0.2 ml. of liquid bromine is heated to 60° in a 2-l. Erlenmeyer flask on a steam bath; the flask is removed, and 50 g. of the potassium salt is added. The salt dissolves rapidly; the solution is filtered with suction; and the clear orange filtrate is transferred to a 4-l. Erlenmeyer flask. Three hundred milliliters of saturated potassium chloride solution is added rapidly in one portion, and the flask is allowed to stand undisturbed. Orange needles of potassium salt appear in a moment or two, and after standing for 30 minutes the mixture is cooled to 0° in an ice-salt bath. The crystals are collected, washed, and dried as outlined above. The salt thus obtained weighs 45–49 g. (90–98% based on the potassium 1,2-naphthoquinone-4-sulfonate used). This product compared favorably with the sodium salt prepared by the more elaborate borax process [1] with respect to color, colored decomposition products and ammonia content (Notes 7 and 8).

2. Notes

1. The 1-amino-2-naphthol-4-sulfonic acid earlier described [*Org. Syntheses* Coll. Vol. **2,** 42 (1943)] is not quite pure and gives an oxidation product of somewhat inferior quality. The procedure in question yields a gray product containing water of hydration (the percentage yield reported is thus in error). On reinvestigating the matter it has been found that the darkening of the material can be minimized if not prevented entirely, and that the colored impurity can be removed completely by extraction with ethanol. The colorless, anhydrous sulfonic acid required for the present preparation is thus prepared by modifying the earlier procedure in the following respects. By stirring the mixture of nitroso-β-naphthol and sodium bisulfite solution vigorously by hand (wooden paddle), all the soluble product can be dissolved in 3–4 minutes. The solution is then filtered as rapidly as possible, using two 15-cm. Büchner funnels and changing filter papers frequently. The clear, golden-yellow filtrate is acidified immediately on completion of the filtration. The product is then light gray, whereas, if much time elapses before the bisulfite solution is acidified, the solution turns red and the aminonaphtholsulfonic acid may be deep purple-gray. After the product has been collected and washed with water, it is

washed with warm ethanol until the filtrate is colorless; 1.5–2 l. of ethanol is required. The product is washed with two 100-ml. portions of ether and dried to constant weight at 60–80° in the absence of light. A pure white, dust-dry product which weighs 370–380 g. (75–78% based on the β-naphthol) is thus obtained. This product is used in the present procedure.

If technical 1-amino-2-naphthol-4-sulfonic acid is used, the yield of ammonium 1,2-naphthoquinone-4-sulfonate is 313 g. (84%) of a rust-colored product. If the technical material is washed with 2 l. of warm 95% ethanol the yield is 337 g. (90%) of an orange-colored ammonium salt.

2. It is important that the oxidation start at this point. Should oxidation not begin, a second 1–2 ml. portion of concentrated nitric acid is added; the acid is introduced along the side of the beaker with as little disturbance as possible.

3. It is necessary to add more ether from time to time to replace that which evaporates.

4. The temperature must be kept within this range. At lower temperatures the oxidation is slow and not satisfactory; at higher temperatures the quinone is slightly decomposed.

5. The ethanol and ether remove the small amount of decomposition products and assist in the rapid drying of the product. It is advisable to dry the preparation as rapidly as possible.

6. Aqueous solutions of 1,2-naphthoquinone-4-sulfonate begin to decompose when heated to this temperature in the absence of bromine.

7. Tests for the purity of the product were devised by Folin and later improved by Danielson.[1] A direct comparison of the two preparations revealed no difference in the degree of purity of the product.

8. This purified salt is suitable for use in the procedures of Folin and of Sullivan for the determination of amino acids.

3. Methods of Preparation

Salts of 1,2-naphthoquinone-4-sulfonate have been prepared by the oxidation of 2-amino-1-naphthol-4-sulfonic acid with nitric acid,[2] or by the oxidation of the more readily available 1-amino-2-naphthol-4-sulfonic acid with the same reagent.[3–5]

[1] Danielson, *J. Biol. Chem.*, **101**, 507 (1933).
[2] Witt and Kaufman, *Ber.*, **24**, 3162 (1891).
[3] Folin, *J. Biol. Chem.*, **51**, 386 (1922).
[4] Böniger, *Ber.*, **27**, 24 (1894).
[5] Fieser, *J. Am. Chem. Soc.*, **48**, 2929 (1926).

NAPHTHORESORCINOL *

(1,3-Naphthalenediol)

$$C_2H_5OH + Mg + CH_2(COOC_2H_5)_2 \rightarrow C_2H_5OMgCH(COOC_2H_5)_2 + H_2$$

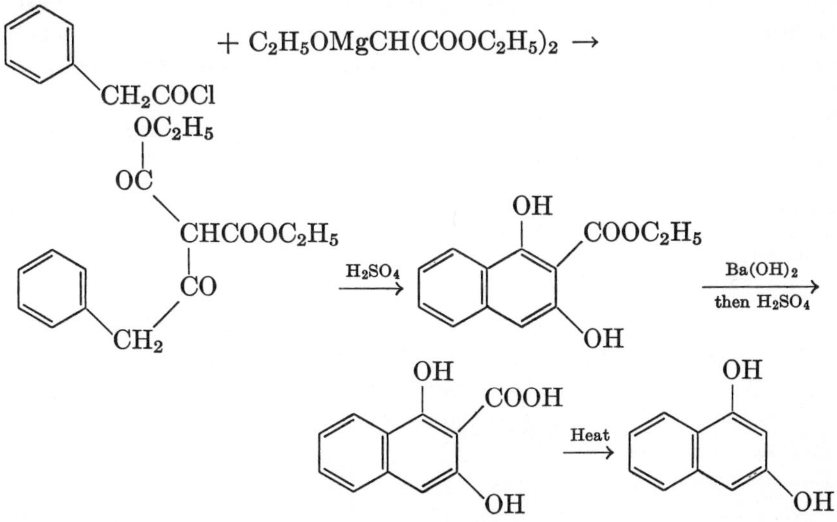

Submitted by Karl Meyer and Henry S. Bloch.
Checked by H. R. Snyder and Curtis W. Smith.

1. Procedure

A. *Ethyl phenylacetylmalonate.* A 1-l. three-necked round-bottomed flask is equipped with a dropping funnel (Note 1) and a reflux condenser provided with a calcium chloride tube. Before the apparatus is assembled, it is washed in the following manner with absolute ethanol which has been prepared in a distilling flask (Note 2). Ninety milliliters of the anhydrous ethanol is distilled through the dropping funnel into the apparatus, used for rinsing, and discarded. In the reaction flask are placed 12.5 g. (0.52 atom) of magnesium turnings, 1 ml. of carbon tetrachloride (Note 3), and 40 g. (0.25 mole) of malonic ester (Note 3); then 90 ml. of absolute ethanol is distilled into the dropping funnel and transferred to the flask. The reaction is started by heating and is controlled by applying an ice bath when the condenser begins to flood. When the reaction has subsided, 40 g. (0.25

mole) more of malonic ester is added at one time. After this reaction has subsided, the flask is cooled slightly, 180 ml. of dry ether is added, and the reaction mixture is heated on a steam bath for 1 hour. Then 88 g. (74 ml., 0.57 mole) of phenylacetyl chloride, diluted with 90 ml. of dry ether, is added slowly (in about 30 minutes) from the dropping funnel in portions at such a rate that the vigorous reaction subsides between additions. When the addition has been completed, the flask is warmed for 10 minutes on a steam bath. After the mixture has been cooled, 100 ml. of water is added dropwise in about 30 minutes. The oily layer is washed with two 100-ml. portions of water and dried over sodium sulfate, and the ether is removed by distillation under reduced pressure (water pump). The crude phenylacetylmalonic ester (87 ml.) which remains is a reddish, somewhat viscous oil.

B. *Ethyl* 1,3-*dihydroxy*-2-*naphthoate.* In a 1-l. flask 1 volume (87 ml.) of the crude ester is added in one lot to 3 volumes (261 ml.) of concentrated sulfuric acid without cooling, and the solution is allowed to stand for about 1 week (Note 4). The whole mixture, including any precipitate which may have formed, is poured slowly with stirring into a mixture of 1 kg. of ice and 500 ml. of water (Note 5); the solid yellow ester is filtered with suction, washed with a small amount of cold water, pressed in the filter for 30 minutes with a rubber dam, and finally dried in a vacuum desiccator. The yield of ester melting at 80° is 58–68 g. (50–59% based on the malonic ester used).

Recrystallization of 50 g. of the ester from 300 ml. of 70% ethanol (Note 6) yields 45 g. of yellow, needle-like crystals melting at 82°.

C. 1,3-*Dihydroxy*-2-*naphthoic acid.* In a 1-l. three-necked round-bottomed flask, fitted with a mechanical stirrer, a nitrogen inlet tube, and a condenser carrying a separatory funnel attached by means of a notched cork, is placed 20.9 g. (0.09 mole) of the recrystallized ester dissolved in 300 ml. of dioxane. The apparatus is flushed with nitrogen, the solution is stirred and heated on a steam bath, and a solution of 40 g. (0.13 mole) of barium hydroxide octahydrate in 500 ml. of water is added in the course of 1 hour. Heating and stirring are continued for 3 hours. The precipitated barium salt, which may vary in color from yellow to greenish gray, is filtered from the hot solution and transferred immediately to a cool (25°) solution of 11 ml. of concentrated sulfuric acid in 185 ml. of water in a 500-ml. three-necked round-bottomed flask fitted with a nitrogen inlet tube, a mechanical stirrer, and a condenser. The mixture is stirred in an atmosphere of nitrogen for 5 minutes without heating (Note 7). The precipitate is removed by filtration, washed with a small amount of distilled water, transferred to a beaker, and extracted with two 100-ml. portions of hot absolute ethanol. To the warm filtrate is added 200 ml. of water, and the solution is allowed to cool in an ice bath for 2 hours. The yield of the crude acid which is obtained by filtration as a fine, yellow powder melting at 135° (Note 8) is 16.5 g. (90%).

D. *Naphthoresorcinol.* In a 200-ml. three-necked round-bottomed flask, equipped with a nitrogen inlet tube, a stirrer, and a condenser, is placed 65 ml. of water; the water is then boiled to expel dissolved oxygen. To the boiling water is added 16.5 g. of the crude acid (Note 9), and the mixture is boiled with stirring for 2 hours in an atmosphere of nitrogen. The water is decanted from any pasty residue; the residue is boiled with 35 ml. of water for 2 hours, and the supernatant liquid is decanted through a filter and combined with the first solution. Just enough sodium hydrosulfite (about 0.3–0.5 g.) is added to decolorize the red solution, 15 g. of sodium chloride is dissolved in the solution, and the mixture is allowed to stand at 5° for 24 hours. The naphthoresorcinol precipitates first as an oil, which solidifies to a reddish solid on further cooling, and then as colorless or slightly yellow plates. The yield is 7.0–7.3 g. (54–56%). After standing over calcium chloride in a vacuum desiccator for 1 day, the white or slightly yellow plates melt at 119–122° and the reddish solid melts at 110–118°.

To 100 ml. of hot water are added 3.5 g. of the reddish solid, just enough sodium hydrosulfite to remove the red coloration, and 0.3 g. of Norit. The mixture is filtered, 7 g. of sodium chloride is dissolved in the filtrate, and the solution is allowed to stand in a closed container at 5° for 24 hours. The yield of naphthoresorcinol, which separates as large, nearly transparent plates melting at 122–124° (Note 10), is 2.6 g. (75% recovery).

2. Notes

1. In order to measure the volume of ethanol to be used, an equal volume of water (90 ml.) is placed in the dropping funnel and a label is pasted at the level of the water. The funnel is then emptied and carefully dried.

2. A liter of commercial absolute ethanol is dried by means of sodium and ethyl phthalate [1] in a 2-l. distilling flask equipped with a reflux condenser. A short section of rubber tubing closed by a screw clamp is attached to the side arm of the flask. In the next step the condenser is set for downward distillation.

3. The carbon tetrachloride, ether, and malonic ester are dried over anhydrous magnesium sulfate. The malonic ester is distilled before use.

4. The yield of ethyl 1,3-dihydroxy-2-naphthoate is decreased to 45% if the sulfuric acid solution is allowed to stand for only 2 days.

5. After about 25 ml. of the sulfuric acid solution has been added, the mixture is stirred until a yellow solid separates. Precipitation of the main portion of the product then occurs very smoothly as the remainder of the sulfuric acid solution is added.

6. The ester may be recrystallized also by solution in dioxane and precipitation by water.

7. An alternative procedure consists in decomposing the barium salt with hydrochloric acid. However, in the checkers' hands this procedure did not yield a barium-free product.

8. The submitters obtained the pure acid as follows: To 0.5 g. of the crude acid was addded 200 ml. of boiling water, and the mixture was heated for an instant to effect solution and filtered immediately. The filtrate was cooled by vigorous shaking in an ice bath. Addition of 1 ml. of concentrated hydrochloric acid to the solution caused the separation of crystals, which were collected after 30 minutes' additional cooling in ice, washed with ice water, and dried in vacuum over phosphorus pentoxide. The recovery of material melting in the range of 138.5–140.5° to 146.5° was 62%.

9. The solid has a tendency to assume a semi-plastic consistency, but the size of the mass decreases as decarboxylation occurs.

10. If part of the material separates as an oil and solidifies, the appearance of the product may be poor, although the melting point is 122–124°. Separation as an oil may be prevented by slow cooling with shaking and seeding. An alternative procedure for final purification of naphthoresorcinol consists in sublimation at 120–130°/5×10^{-4} mm.

3. Methods of Preparation

Naphthoresorcinol has been prepared by heating 1-amino-3-hydroxy-4-naphthalenesulfonic acid or its salt in water or slightly acidic solution,[2] by cyclization of ethyl phenylacetylmalonate[3] or of ethyl phenylacetoacetate,[4] and by alkaline fusion of 1,3-naphtholsulfonic acid or 1,3-naphthalenedisulfonic acid.[5] Phenylacetyl malonic ester has been prepared by condensing phenylacetyl chloride and malonic ester with sodium in ether,[3] or with magnesium.[6]

[1] *Org. Syntheses* Coll. Vol. **2**, 155 (1943).

[2] Friedlander and Rüdt, *Ber.*, **29**, 1609 (1896); Ger. pat. 84,990 [*Frdl.*, **4**, 229 (1894–1897)]; Ger. pat. 87,429 [*Frdl.*, **4**, 584 (1894–1897)]; Ger. pat. 90,096 [*Frdl.*, **4**, 585 (1894–1897)].

[3] Metzner, *Ann.*, **298**, 374 (1897); Wagreich, Roberts, and Harrow, *Chemist-Analyst,* **31**, 59 (1942).

[4] Soliman and West, *J. Chem. Soc.*, **1944**, 53.

[5] Kozlov and Odintosov, *J. Appl. Chem. U.S.S.R.*, **17**, 719 (1944) [*C. A.*, **39**, 2744 (1945)].

[6] Ogata, Nosaki, and Takagi, *J. Pharm. Soc. Japan,* **59**, 105 (1939) [*C. A.*, **33**, 4230 (1939)].

o-NITROBENZALDEHYDE *

(Benzaldehyde, *o*-nitro-)

$$o\text{-}O_2NC_6H_4CH_3 + 2(CH_3CO)_2O \xrightarrow{CrO_3} o\text{-}O_2NC_6H_4CH(OCOCH_3)_2 + 2CH_3COOH$$

$$o\text{-}O_2NC_6H_4CH(OCOCH_3)_2 + H_2O \xrightarrow{HCl} o\text{-}O_2NC_6H_4CHO + 2CH_3COOH$$

Submitted by S. M. Tsang, Ernest H. Wood, and John R. Johnson.
Checked by Lee Irvin Smith and J. W. Opie.

1. Procedure

A. *o-Nitrobenzaldiacetate.* In a 2-l. three-necked round-bottomed flask equipped with an effective mechanical stirrer and a thermometer, and surrounded by an ice bath, are placed 600 g. (570 ml.) of glacial acetic acid, 565 ml. (612 g., 6 moles) of acetic anhydride, and 50 g. (0.36 mole) of *o*-nitrotoluene. To this solution is added slowly, with stirring, 156 g. (85 ml., 1.5 moles) of concentrated sulfuric acid. When the mixture has cooled to 5°, 100 g. (1 mole) of chromium trioxide is added in small portions at such a rate that the temperature does not rise above 10°; about 2 hours is required for the addition (Note 1). Stirring is continued for 5 hours after the chromium trioxide has been added (Note 2). The reaction mixture is poured into a large jar two-thirds filled with chipped ice, and cold water is added until the total volume is about 6 l. The mixture is then stirred vigorously for at least 15 minutes to promote solidification of the oily layer (Note 3). The oily solid is filtered with suction on a Büchner funnel, washed with cold water, and then stirred mechanically with 500 ml. of cold 2% sodium carbonate solution. The solid is then collected on a filter, washed with cold water, and dried in the air. The crude substance is digested with 150 ml. of petroleum ether (b.p. 60–70°) for 30 minutes, cooled, and filtered (Note 4). After having been dried in a vacuum desiccator, the product weighs 21–22 g. (23–24%) and melts at 87–88° (Note 5).

B. *o-Nitrobenzaldehyde.* A suspension of 51.6 g. of the diacetate in a mixture of 500 g. (272 ml.) of concentrated hydrochloric acid, 450 ml. of water, and 80 ml. of ethanol is stirred and refluxed for 45 minutes. The mixture is then cooled to 0°, and the solid is filtered with suction and washed with water (Notes 6 and 7). The crude product is purified by rapid steam distillation through a 100-cm. Allihn condenser into a cooled receiver. About 3.5 l. of distillate is collected during 30 minutes; the cooled distillate is filtered, and the *o*-nitrobenzaldehyde is dried in a desiccator over calcium chloride. It weighs 23.7 g. (74%) and melts at 44–45°.

2. Notes

1. Proper control of the temperature is very important during addition of the chromium trioxide. If the temperature rises very much, the reaction may become violent.

2. A hard, tarry mass forms during the later stages of this oxidation. For this reason stirring becomes increasingly difficult and a rather powerful motor must be used. After decantation of the liquid portion of the reaction mixture, ice and a little water (total volume 1 l. for the amounts of materials specified) are added to the reaction flask, and the mass is broken up with a spatula. The suspension is then combined with the mixture in the large jar.

3. It is essential that the oily layer should be stirred vigorously until it has solidified; otherwise large losses will be incurred during the subsequent filtration.

4. This treatment removes unchanged *o*-nitrotoluene.

5. The checkers have carried out this preparation using three times the amounts of material specified here. The percentage yields were the same.

6. In one run, the submitters substituted an equivalent amount of sulfuric acid for the hydrochloric acid, omitted the filtration, and steam-distilled the mixture directly. The yield of *o*-nitroaldehyde was 74.7%; this product also melted at 44–45°.

7. The crude aldehyde may be dried and then purified by distillation under reduced pressure. For this purpose, an ordinary Claisen flask is used. No fractionation is necessary; the distillation serves only to remove a small amount of tarry material. The boiling range is rather wide (120–144°/3–6 mm.), but the entire distillate (30 g. from 66 g. of the diacetate) solidifies and melts at 44–45°. The aldehyde may also

be recrystallized, although with some loss. For this purpose 11 g. of the solid is dissolved in toluene (2–2.5 ml. per gram) at room temperature, the solution is diluted with petroleum ether (b.p. 30–60°; 7 ml. per ml. of solution) and cooled to −10° or below. The pale yellow needles weigh 9 g. and melt at 44–45°.

3. Methods of Preparation

o-Nitrobenzaldehyde has been prepared by action of oxides of nitrogen upon *o*-nitrobenzyl alcohol;[1] by oxidation of *o*-nitrocinnamic acid or ester with permanganate or nitrous acid;[2] by oxidation of *o*-nitrophenylpyruvic acid with permanganate;[3] and from *o*-nitrotoluene by a number of methods. These include: action of isoamyl nitrite and sodium ethoxide, followed by hydrolysis of the resulting *o*-nitrobenz-*anti*-aldoxime with concentrated hydrochloric acid;[4] direct action of a number of oxidizing agents including potassium dichromate and sulfuric acid,[5] manganese dioxide,[6] nickel oxide,[7] and cerium oxide;[8] action of mercury oxide in the presence of alkali, followed by hydrolysis of the resulting *o*-nitrobenzalmercurioxide with nitrous or nitric acid;[9] and action of chromic acid in acetic acid and acetic anhydride, followed by hydrolysis of the resulting *o*-nitrobenzaldiacetate.[10] *o*-Nitrobenzaldehyde is also formed, along with much *m*-nitrobenzaldehyde, by direct nitration of benzaldehyde.[11]

1 Cohen and Harrison, *J. Chem. Soc.*, **71**, 1057 (1897).

2 Friedländer and Henriques, *Ber.*, **14**, 2803 (1881).

3 Reissert, *Ber.*, **30**, 1042 (1897).

4 Lapworth, *J. Chem. Soc.*, **79**, 1274 (1901); Farbwerke vorm. Meister, Lucius, and Brüning, Ger. pat. 107,095 (*Chem. Zentr.*, **1900**, I, 886).

5 Lauth, *Bull. soc. chim. France,* (3) **31**, 133 (1904).

6 Gilliard, Monnet, and Cartier, Ger. pat. 101,221 (*Chem. Zentr.*, **1899**, I, 959); Badische Anilin- u. Sodafabr., Ger. pat. 175,295 (*Chem. Zentr.*, **1906**, II, 1589).

7 Badische Anilin- u. Sodafabr., Ger. pat. 127,388 (*Chem. Zentr.*, **1902**, I, 150).

8 Farbwerke vorm. Meister, Lucius, and Brüning, Ger. pat. 174,238 (*Chem. Zentr.*, **1906**, II, 1297).

9 Reissert, *Ber.*, **40**, 4216, 4220 (1907); Ger. pat. 186,881 (*Chem. Zentr.*, **1907**, I, 1295).

10 Thiele and Winter, *Ann.*, **311**, 356 (1900).

11 Friedländer and Henriques, *Ber.*, **14**, 2801 (1881).

m-NITROBENZALDEHYDE DIMETHYLACETAL

(Benzaldehyde, *m*-nitro-, dimethylacetal)

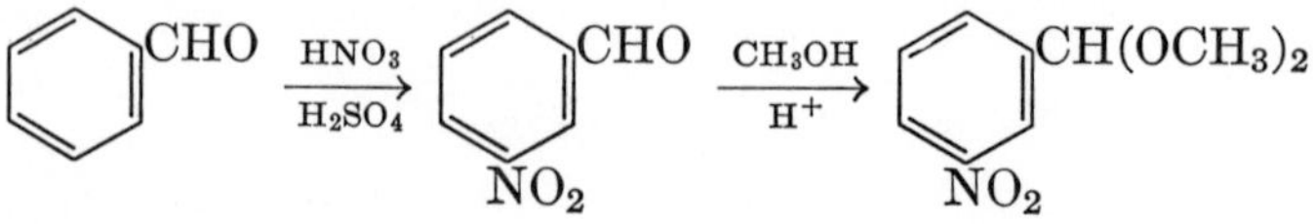

Submitted by ROLAND N. ICKE, C. E. REDEMANN, BURNETT B. WISEGARVER, and GORDON A. ALLES.
Checked by H. R. SNYDER and FRANK X. WERBER.

1. Procedure

A. *m-Nitrobenzaldehyde.* In a 3-l. three-necked flask provided with a dropping funnel, a thermometer, and an efficient mechanical stirrer is placed 1.25 l. of technical concentrated sulfuric acid (sp. gr. 1.84). The acid is stirred during the fairly rapid addition of 167 ml. (250 g., 4 moles) of technical fuming nitric acid (sp. gr. 1.49–1.50), with sufficient cooling by an ice bath so that the temperature of the mixed acids does not exceed 10°. The mixture is maintained at 5–10° while 213 g. (2 moles) of U.S.P. benzaldehyde (Note 1) is added with good stirring over a period of 2–3 hours. At the end of the addition the ice bath is removed and the mixture is allowed to stand overnight at room temperature.

The mixture is poured carefully onto about 3.25 kg. of cracked ice in a 2-gal. crock (Note 2) with manual stirring. The yellow precipitate is collected on a large Büchner funnel, washed with cold water, and pressed as dry as possible (Note 3). Further removal of water is accomplished by dissolving the moist product in about 100 ml. of warm benzene and separating the water layer in a separatory funnel. The benzene solution is filtered (Note 4) into a 1-l. distilling flask and concentrated on a steam bath. Distillation is discontinued when benzene no longer comes over or when nitrous fumes (from incompletely removed nitric acid) appear. The residual *m*-nitrobenzaldehyde is sufficiently pure for use in the next step (Note 5).

B. *m-Nitrobenzaldehyde dimethylacetal.* The above crude product is dissolved in 750 ml. of technical anhydrous methanol, 1 ml. of concentrated hydrochloric acid is added if necessary (Note 6), and the solution is allowed to stand at room temperature for 5 days (Note 7). A solution of sodium methoxide in methanol is added until the solution is just alkaline to moistened litmus paper. The methanol is removed by distillation on a steam bath; the residue is cooled to room temperature and treated with cool water to dissolve the inorganic salts. The aqueous solution is extracted with two 50-ml. portions of ether, and the extracts are added to the crude acetal. The solution is subjected to a preliminary drying over anhydrous magnesium sulfate (or sodium sulfate), filtered, and then dried for at least 12 hours over anhydrous sodium carbonate. After the ether has been removed by distillation on a steam bath the residue is distilled under reduced pressure from a 500-ml. Claisen flask. The yield of the light-yellow liquid acetal, boiling at 141–143°/8 mm. (Note 8), is 300–335 g. (76–85%) (Note 9).

2. Notes

1. Commercial benzaldehyde may contain some benzoic acid. When taken from a previously unopened bottle, U.S.P. benzaldehyde usually is quite satisfactory. If there is any doubt concerning its purity, the aldehyde may be washed with dilute sodium carbonate solution, dried over anhydrous sodium carbonate, and distilled under diminished pressure just before use.

2. Two 4-l. beakers may be used.

3. A small amount of an oily liquid passes through the filter; this material is a mixture of the *ortho* and *meta* isomers. If a very pure product is desired, this oil may be discarded, although some of the *meta* isomer will be lost. It is usually satisfactory to combine it with the solid material, the *m*-nitro acetal being obtained in a state of purity sufficient for most purposes in the final distillation.

4. A small amount of a solid by-product is removed.

5. If pure *m*-nitrobenzaldehyde is desired, the benzene solution is diluted with additional benzene and is washed well with aqueous sodium bicarbonate solution until the washings are alkaline. (*Warning! It has been reported* [*K. Ueno*] *that incomplete washing of the product in solution may result in an explosion during the distillation; all remaining acids must be removed before distillation is attempted.*) The solution is dried (with anhydrous sodium or magnesium sulfate), and the solvent is removed. The aldehyde is distilled under reduced pressure from a Claisen flask connected to an air-cooled condenser. The yield of product boiling at 119–123°/4 mm. is 226–254 g. (75–84%). The product readily crystallizes in the receiver.

6. When nitrous fumes are observed, there usually is enough acid present to catalyze the acetal formation.

7. The yield is not increased by extending the reaction period beyond 5 days.

8. The acetal boils at 116–119°/1 mm., 122–126°/2 mm., 130–132°/3 mm. The boiling range depends somewhat upon the rate of distillation.

9. The submitters obtained yields in the upper part of this range from runs twice the size described. They have stored samples of the product for as long as 4 years without deterioration.

3. Methods of Preparation

The only satisfactory method for the preparation of *m*-nitrobenzaldehyde is the direct nitration of benzaldehyde with a mixture of nitric and sulfuric acids. The procedure described is essentially that of Baker and Moffitt.[1]

m-Nitrobenzaldehyde dimethylacetal has been prepared from the aldehyde and methanol in the presence of formiminomethyl ether hydrochloride [2] or Twitchell's reagent [3] as the acid catalyst.

[1] Baker and Moffitt, *J. Chem. Soc.*, **1931,** 314.
[2] Claisen, *Ber.*, **31,** 1016 (1898).
[3] Zaganiaris, *Ber.*, **71,** 2002 (1938).

p-NITROBENZONITRILE *

(Benzonitrile, *p*-nitro-)

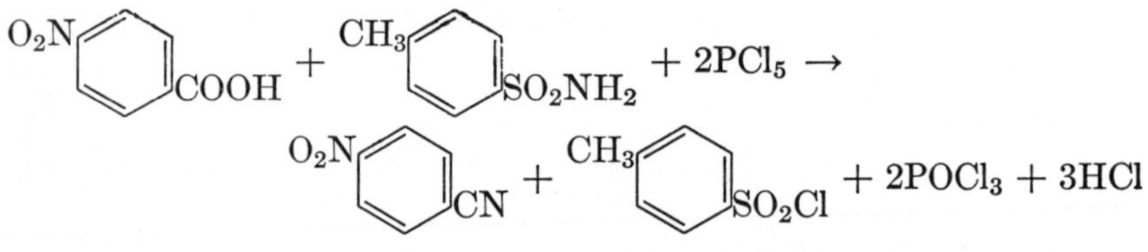

Submitted by Charles S. Miller.
Checked by Arthur C. Cope and James J. Ryan.

1. Procedure

In a 1-l. round-bottomed flask with a ground-glass joint are mixed 100.3 g. (0.6 mole) of *p*-nitrobenzoic acid and 109.9 g. (0.64 mole) of *p*-toluenesulfonamide (Note 1). To this mixture is added 262.2 g. (1.26 moles, 5% excess) of phosphorus pentachloride, which is manually stirred into the mixture while the flask is rotated to facilitate mixing (Note 2). The flask is then attached to a short column fitted with a thermometer and side arm for distillation. The thermometer is adjusted so that the bulb is immersed in the reaction mixture, and the mixture is then warmed gently in a hot-air bath to initiate the reaction (Note 3). In some cases this initial reaction must be controlled by a cold-water bath which temporarily replaces the hot-air bath. After the first reaction has subsided and the contents of the flask have become almost liquid, the temperature is gradually raised to 205° and maintained at 200–205° until no more material distils (Note 4). The cooled reaction mixture is treated with 240 ml. of pyridine and warmed gently with manual stirring until solution is complete. To this solution is added cautiously 1.1 l. of water with manual stirring (Note 5), and the cooled suspension is filtered. The solid is washed with water and suspended in 400 ml. of 5% sodium hydroxide solution. After this mixture has been stirred for 30 minutes, the solid is removed, washed thoroughly with water, and dried. The yield of the light-tan product is 74.5–80 g. (85–90%); m.p. 146–147° (cor.) (Notes 6 and 7). This product may be further purified with 91% recovery by recrystallization from 50% acetic acid (6.5 ml. per gram). The recrystallized material melts at 147–148° (cor.).

2. Notes

1. Benzenesulfonamide and *p*-nitrobenzenesulfonamide have also been used in place of *p*-toluenesulfonamide.

2. This preparation is carried out in a hood.

3. The air bath consists of a hemispherical metal pan about 5 in. in diameter fitted to the bottom of the flask so as to leave an air space of ⅛ in. to ¼ in. between the metal and the flask. The bath is heated with a small flame.

4. This operation requires about 30 minutes, and the phosphorus halides that distil over weigh 145–155 g.

5. The water should be added about 1 ml. at a time at the beginning. Since this decomposition is very vigorous, the solution is stirred until the reaction subsides before the next addition is made. Later in the addition the water may be poured in slowly.

6. Alternatively, according to the submitter, the cooled reaction mixture may be treated with crushed ice to decompose the remaining phosphorus halides, after which the oily solid is separated from the water by filtration and washed thoroughly with water. The oily solid is added to a mixture of 350 ml. of 14% ammonium hydroxide and 350 ml. of acetone and stirred for 15–30 minutes to ensure complete reaction between the *p*-toluenesulfonyl chloride and ammonia. The ace-

tone is evaporated on a steam bath, and, after cooling, the solid is filtered and ground in a mortar with a pestle. The finely divided solid is then stirred with 300 ml. of 10% sodium hydroxide until all the *p*-toluenesulfonamide is dissolved (0.5 to 1 hour). The solid *p*-nitrobenzonitrile is filtered, washed with water, and dried. The alkaline filtrate is treated with charcoal, filtered and acidified with dilute hydrochloric acid to yield about 75% of the *p*-toluenesulfonamide that was used as a reactant.

7. The submitter states that a number of aromatic mono- and dinitriles have been prepared by this procedure with slight or no modification in the temperature. The reaction mixtures were usually worked up by the acetone-ammonia method described in Note 6. Among the compounds prepared by this method are *o*-nitrobenzonitrile, *o*-bromobenzonitrile, *m*-methoxybenzonitrile, 4,4′-dicyanodiphenylsulfone, 4,4′-dicyanostilbene, α,γ-di-(4-cyanophenoxy)propane. With the last, a temperature of 185–190° for 20 minutes gave the best results. The yields before purification ranged between 75% and 95% and after purification between 63% and 79%. Aliphatic acids give low yields of the corresponding nitriles and, in some cases, chlorinated by-products.

3. Methods of Preparation

p-Nitrobenzonitrile has been prepared by the action of phosphorus pentoxide on *p*-nitrobenzamide,[1,2] by the Sandmeyer reaction on *p*-nitroaniline,[3,4] by thermal decomposition of *p*-nitrophenylglyoxylic acid oxime,[5] by the thermal decomposition of *o*-benzoyl-*p*-nitrobenzaldoxime,[6] by the action of acetic anhydride and phosphorus pentoxide on *p*-nitrobenzaldoxime,[7] by the action of fused sodium chloride-aluminum chloride on *p*-nitrobenzamide;[8] by the action of tertiary amines on *syn*-*p*-nitrobenzaldoxime benzoate,[9,10] by heating *p*-nitrobenzoic acid or *p*-nitrobenzoic anhydride with benzenesulfonamide or *p*-toluenesulfonamide,[11] and by heating *p*-nitrobenzamide with sulfamic acid.[12]

[1] Engley, *Ann.*, **149**, 298 (1869).
[2] Fricke, *Ber.*, **7**, 1321 (1874).
[3] Sandmeyer, *Ber.*, **18**, 1492 (1885).
[4] Bogert and Kohnstamm, *J. Am. Chem. Soc.*, **25**, 479 (1903).
[5] Borsche, *Ber.*, **42**, 3597 (1909).
[6] Neber, Hartung, and Ruopp, *Ber.*, **58B**, 1234 (1925).
[7] Bogoslovskii, *J. Gen. Chem. U.S.S.R.*, **8**, 1784 (1938) [*C. A.*, **33**, 4973 (1939)].
[8] Norris and Klemka, *J. Am. Chem. Soc.*, **62**, 1432 (1940).
[9] Hauser and Vermillion, *J. Am. Chem. Soc.*, **63**, 1224 (1941).

[10] Vermillion and Hauser, *J. Am. Chem. Soc.*, **63**, 1227 (1941).
[11] Oxley, Partridge, Robson, and Short, *J. Chem. Soc.*, **1946**, 763; Brit. pat. 583,586 [*C. A.*, **41**, 2750 (1947)].
[12] Kirsanov and Zolotov, *J. Gen. Chem. U.S.S.R.*, **20**, 284 (1950) [*C. A.*, **44**, 6384 (1950)].

p-NITROBENZOYL PEROXIDE

[Peroxide, bis(*p*-nitrobenzoyl)]

$$2p\text{-}O_2NC_6H_4COCl + Na_2O_2 \rightarrow [p\text{-}O_2NC_6H_4COO]_2 + 2NaCl$$

Submitted by Charles C. Price and Edwin Krebs.
Checked by R. L. Shriner and C. H. Tilford.

1. Procedure

A 600-ml. beaker containing 100 ml. of water and equipped with an efficient stirrer, a thermometer, and a 200-ml. separatory funnel is immersed in an ice-water bath. When the temperature of the water has fallen to 0–5°, 10 g. (0.13 mole) of sodium peroxide (Note 1) is added. Then, with vigorous stirring, a solution of 37 g. (0.2 mole) of *p*-nitrobenzoyl chloride in 100 ml. of dry toluene is added dropwise over a period of about 30 minutes. After the mixture has been stirred for an additional 1.5 hours, the precipitate is filtered and washed with 200 ml. of cold water (Note 2). The yield of *p*-nitrobenzoyl peroxide is 28.5–29 g. (86–88%). It melts at 155–156°.

The product may be recrystallized most conveniently by dissolving it as rapidly as possible (Note 3) in 500 ml. of dry toluene which has been preheated to 80–85°. As soon as the solid is completely dissolved (2–3 minutes of stirring), the solution is filtered through a warm Büchner funnel and the filtrate is immediately cooled in an ice-water bath. The yield of very pale yellow glistening needles is 25 to 26 g. (86–89% recovery); they melt at 156° with vigorous decomposition (Note 4).

2. Notes

1. In a parallel experiment in which 25 g. of sodium peroxide was used, the precipitate was not the peroxide but evidently consisted of sodium *p*-nitroperbenzoate.

2. *p*-Nitrobenzoic acid may be recovered by acidification of the filtrate.

3. Excessive heating during recrystallization leads to extensive decomposition of the peroxide.

4. By a procedure similar to that described for *p*-nitrobenzoyl peroxide, the following peroxides have been prepared: *m*-nitrobenzoyl peroxide (m.p. 136–137°) in 90% yield; anisoyl peroxide (m.p. 126–127°) in 86–89% yields; *p*-bromobenzoyl peroxide (m.p. 144°) in 73% yield; and 3,4,5-tribromobenzoyl peroxide (m.p. 186°) in 40% yield.[1] The procedure is not satisfactory for preparation of acetylsalicylyl peroxide, which is more conveniently prepared by action of hydrogen peroxide upon an acetone solution of the acid chloride.

3. Methods of Preparation

p-Nitrobenzoyl peroxide,[2] as well as *m*-nitrobenzoyl peroxide [2] and anisoyl peroxide,[3] have been prepared in about 50% yields from the acid chlorides and an acetone solution of hydrogen peroxide, in the presence of a basic substance such as pyridine, sodium acetate, or sodium hydroxide. *m*-Nitrobenzoyl peroxide has also been prepared by nitration of benzoyl peroxide with cold concentrated nitric acid,[4] or with a cold [5] or hot [2] mixture of nitric and sulfuric acids.

[1] Price, Kell, and Krebs, *J. Am. Chem. Soc.*, **64**, 1104 (1942).
[2] Vanino and Uhlfelder, *Ber.*, **33**, 1046 (1900).
[3] Vanino and Uhlfelder, *Ber.*, **37**, 3624 (1904).
[4] Brodie, *J. Chem. Soc.*, **17**, 271 (1864).
[5] Vanino, *Ber.*, **30**, 2004 (1897); Gelissen and Hermans, *Ber.*, **58**, 285 (1925).

p-NITROBENZYL ACETATE

(Acetic acid, *p*-nitrobenzyl ester)

$$p\text{-}O_2NC_6H_4CH_2Cl + NaOCOCH_3 \rightarrow p\text{-}O_2NC_6H_4CH_2OCOCH_3 + NaCl$$

Submitted by W. W. Hartman and E. J. Rahrs.
Checked by R. L. Shriner and Charles Russell.

1. Procedure

A mixture of 250 g. (1.46 moles) of *p*-nitrobenzyl chloride, 225 g. (2.74 moles) of fused sodium acetate, and 375 g. (6.25 moles) of glacial acetic acid is refluxed for 8–10 hours in a 2-l. flask heated by an oil bath, the temperature of which is maintained at 160–170° (Note 1). After this time the bath is allowed to cool to about 125°, and the acetic acid is removed by distillation under reduced pressure. Care must be taken not to reduce the pressure too rapidly in the early stages of the distillation. As the distillation slows down, the pressure is further reduced until it reaches 50 mm. or lower, and the temperature is slowly raised to 160°. From 2.5 to 3 hours is required for the complete removal of the acetic acid. About 500 ml. of water is added, and the hard cake is broken up with a stirring rod (Note 2).

The entire contents of the flask are then transferred to a 1.5-l. beaker and stirred with a mechanical stirrer for about 30 minutes or long enough to break up all the lumps. The finely divided material is filtered on a Büchner funnel and washed with 200 ml. of cold water. The above process of washing, stirring, and filtering is repeated twice. The product is then transferred to a 1.5-l. beaker, and 500 ml. of methanol is added and heated to boiling in order to effect solution. The hot solution is filtered through a heated Büchner funnel and allowed to cool slowly. When the solution has cooled to 20°, the product is collected on a filter and air-dried. This first crop consists of yellow needles which melt at 74–77° and weigh 215–225 g. The filtrate is evaporated to 100 ml. and cooled. An additional 25–30 g. of solid separates. Both crops are again washed with cold water and recrystallized from 500 ml. of hot methanol. The first crop from this second crystallization weighs 210–218 g. and melts at 77–78°. Evaporation of the filtrate to 100 ml. and cooling yield an additional 15–18 g. of acetate which is purified by recrystallization from methanol; about 12 to 15 g. of product is obtained. The combined yield of pale yellow crystals which melt at 77–78° amounts to 222–233 g. (78–82%).

2. Notes

1. *p*-Nitrobenzyl chloride obtained from the Eastman Kodak Company was used.

2. The hard cake may be removed by the addition of 500 ml. of boiling water; the ester melts and forms an oily layer. The molten ester solidifies in small lumps when the flask is cooled rapidly in an ice bath while the molten mixture is stirred vigorously. Purification in this manner by repeating the process two or three times does not seem to produce better results than purification with cold water.

3. Methods of Preparation

p-Nitrobenzyl acetate has been prepared by the action of ethanolic potassium acetate on *p*-nitrobenzyl chloride [1] or bromide,[2] and by the nitration of benzyl acetate.[3,4]

[1] Grimaux, *Zeit. für Chem.*, **1867**, 562.

[2] Reid, *J. Am. Chem. Soc.*, **39**, 130 (1917).

[3] Beilstein and Kuhlberg, *Ann.*, **147**, 340 (1868).
[4] Hurd, Fancher, and Bonner, *J. Org. Chem.*, **12**, 369 (1947).

p-NITROBENZYL ALCOHOL *

(Benzyl alcohol, *p*-nitro)

$$p\text{-}O_2NC_6H_4CH_2OCOCH_3 + NaOH \rightarrow p\text{-}O_2NC_6H_4CH_2OH + CH_3CO_2Na$$

Submitted by W. W. Hartman and E. J. Rahrs.
Checked by R. L Shriner and Charles Russell.

1. Procedure

A solution of 218 g. (1.12 moles) of *p*-nitrobenzyl acetate (p. 650) in 500 ml. of hot methanol is prepared in a 2-l. flask. To the hot solution is added 380 g. (1.43 moles) of a 15% solution of sodium hydroxide. The alkali should be added slowly at first with shaking to prevent too vigorous boiling. After standing for 5 minutes, the mixture is poured with vigorous hand stirring into 4.5 kg. of a mixture of cracked ice and water. The precipitate is collected on a Büchner funnel and recrystallized from 3 to 3.7 l. of hot water with the aid of 15 g. of Norit. The alcohol is dried at 60° to 65° in an oven for several hours and bottled (Note 1). The yield of slender, nearly colorless needles amounts to 110–121 g. (64–71%); the product melts at 92–93° (Note 2).

2. Notes

1. The product turns yellow if it is spread out to dry in the air.
2. The submitters report that *p*-iodobenzyl alcohol can be prepared in a similar manner, and that it is unnecessary to isolate the acetate. A mixture of 148 g. of *p*-iodobenzyl bromide, 52 g. of potassium acetate, and 750 ml. of 95% ethanol is refluxed for 8 hours, cooled, and filtered from the salt. To the filtrate is added 33.6 g. of potassium hydroxide, and the solution is refluxed for 6 hours. It is then diluted with 2 l. of water, and the oil taken up in 500 ml. of chloroform. After concentration to 150 ml. and chilling, *p*-iodobenzyl alcohol crystallizes. It is filtered and the solid rinsed with petroleum ether; the rinse added to the mother liquor causes the separation of an additional amount of *p*-iodobenzyl alcohol. The combined yield is 95–100 g. (81–86%); the product melts at 70–72°.

3. Methods of Preparation

p-Nitrobenzyl alcohol has been secured by the oxidation of *p*-nitrotoluene electrolytically [1] or chemically,[2] by the reduction of *p*-nitrobenzaldehyde with formaldehyde in alkaline solution,[3] and by the hydrolysis of the acetate.[4,5] *p*-Iodobenzyl alcohol has been prepared by hydrolysis of *p*-iodobenzyl bromide and p-iodobenzyl acetate.[6,7]

[1] Elbs, *Ber.* (Ref.), **29,** 1122 (1896).

[2] Dieffenbach, Ger. pat. 214,949 [*Frdl.,* **9,** 156 (1908–10)].

[3] B. M. Bogoslovskii and Z. S. Kazakova, *Zhur. Priklad. Khim.,* **21,** 1183 (1948) [*C. A.,* **43,** 6182 (1949)].

[4] Beilstein and Kuhlberg, *Ann.,* **147,** 343 (1868).

[5] Basler, *Ber.,* **16,** 2715 (1883).

[6] Mabery and Jackson, *Ber.,* **11,** 56 (1878).

[7] Jackson and Mabery, *Am. Chem. J.,* **2,** 251 (1880).

2-NITRO-*p*-CYMENE *

(*p*-Cymene, 2-nitro-)

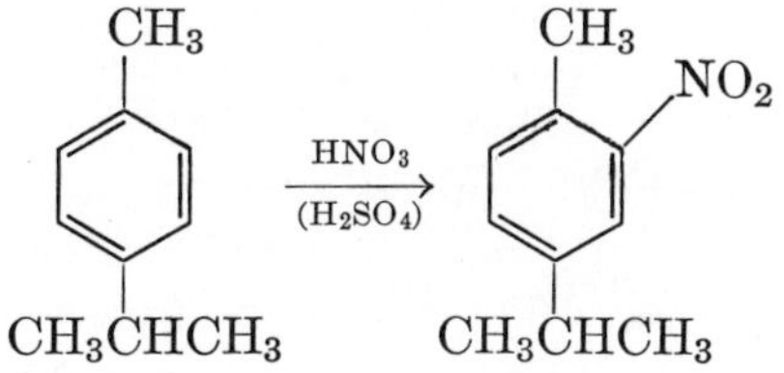

Submitted by KENNETH A. KOBE and THOMAS F. DOUMANI.
Checked by C. F. H. ALLEN and JAMES VANALLAN.

1. Procedure

A 5-qt. enamel pail or bain-marie jar is fitted with two mechanical stirrers (Note 1) placed off center, a thermometer for reading low temperatures, and a dropping funnel, the lower end of which is placed just over the vortex created by one of the stirrer blades, so that each drop of added liquid is immediately mixed with and diluted by the chilled reaction mixture.

In the pail are placed 1 kg. (544 ml.) of concentrated sulfuric acid and 300 ml. of glacial acetic acid (Note 2), and this mixture is chilled by Dry Ice until the temperature is 0° to −5°. Next, 500 g. (585 ml.) of *p*-cymene (Note 3) is added from the dropping funnel with vigorous stirring, any rise of temperature being prevented by the addition of

Dry Ice (Note 4). Concurrently the nitrating mixture is prepared from 369 g. (262 ml.) of nitric acid (sp. gr. 1.42) and 1 kg. (544 ml.) of concentrated sulfuric acid, and cooled to 0–5° by the direct addition of small pieces of Dry Ice (Note 4).

The hydrocarbon emulsion is then cooled to −15° to −10° and the nitrating mixture is admitted dropwise from the dropping funnel over a period of about 2 hours, avoiding any temperature rise (Notes 4 and 5). Stirring is continued for 10 minutes, after which the entire contents of the pail are poured, with adequate stirring, into a mixture of 1 kg. of cracked ice and 1 l. of water. Separation into layers occurs after about 2 hours, whereupon the lower acid layer is drawn off by siphoning or by means of a large separatory funnel (Note 6) and extracted twice, using 50-ml. portions of petroleum ether or 500-ml. portions of ether (Note 7). The extracts and crude nitrocymene are combined and washed three times with 300-ml. portions of water (Note 8), and the ether layer is dried by 50 g. of calcium chloride. After the ether is filtered and distilled, the residual liquid is fractionated under reduced pressure, using a 1-l. modified Claisen flask with indented neck. Fractions are collected at 15–17 mm. as follows: up to 125°, 25–30 g. (Note 9); 125–139°, 520–550 g. (78–82%), n_D^{26} 1.5280; tarry residue, 25–30 g. The product is mainly 2-nitro-*p*-cymene but contains about 8% of *p*-nitrotoluene; it is suitable for reduction [1] to 2-amino-*p*-cymene, from which the *p*-toluidine is readily removed. Upon refractionation of the mixed nitro compounds, the last half is essentially pure 2-nitro-*p*-cymene, b.p. 137–139°/17 mm.; 128–131°/12 mm.; 108°/3 mm., n_D^{25} 1.5290. The properties of pure 2-nitro-*p*-cymene are as follows: b.p. 126°/10 mm.; 142°/20 mm.; n_D^{20} 1.5287.

2. Notes

1. It is advisable to use more powerful stirring motors than those usually found in the organic laboratory, since the success of the preparation is largely dependent on the production of a good emulsion. The checkers employed two "Lightnin" stirrers (Model L mixers) having shafts and blades of Monel metal.

2. On mixing the acids, the temperature rises to about 60°.

3. The authors recommended that a technical grade of cymene be used, since the terpenic impurities present facilitated emulsification. The checkers used both a technical product and one free from terpenes; They encountered no difficulties with emulsification and secured a slightly higher yield (3–4%) with the purified grade. Cymene from different sources varies considerably in its composition.

4. The amounts of Dry Ice given are approximate. The actual quantities depend on the initial temperature and the rapidity of cooling required. Ordinarily 200 g. is required to cool the nitrating mixture, 500 g. to chill the cymene-sulfuric-acetic acid mixture, and 1.5 kg. during the actual nitration. The nitrating acid will solidify if chilled too far. It is extremely important to keep the temperatures indicated to secure the best yields, and Dry Ice should be used freely. The carbon dioxide evolved aids in mixing.

5. If the mixture is cooled too much, the emulsion is broken.

6. The separation into layers is not sharp; 10–15 g. of product may be lost with the acid layer at this point (Note 7).

7. The extractions can be omitted with a smaller yield of product. Thus from the first extraction there is obtained approximately 15 g. of nitro compound, and 7 g. from the second.

8. Sometimes an emulsion is formed during washing; this can be broken by the addition of a few milliliters of sulfuric acid and shaking.

9. The fore-run, on refractionation at atmospheric pressure, yields 4–6 g. of impure *p*-cymene, b.p. 160–165°, and 10–13 g. of nitro compound. This latter amount is included in the total yields given in the procedure.

3. Methods of Preparation

2-Nitro-*p*-cymene has always been secured by the nitration of the hydrocarbon, either by the use of mixed sulfuric and nitric acids with [2] or without [3] the addition of acetic acid, or by fuming nitric acid in acetic acid.[4]

[1] Doumani and Kobe, *Ind. Eng. Chem.*, **31**, 264 (1939).

[2] Söderbaum and Widman, *Ber.*, **21**, 2126 (1888); Andrews, *J. Ind. Eng. Chem.*, **10**, 453 (1918); Wheeler and Smithey, *J. Am. Chem. Soc.*, **43**, 2613 (1921); Phillips, *J. Am. Chem. Soc.*, **44**, 1777 (1922); Demonbreun and Kremers, *J. Am. Pharm. Assoc.*, **12**, 296 (1926) [*C. A.*, **17**, 3906 (1923)]; Kobe and Doumani, *Ind. Eng. Chem.*, **31**, 257 (1939).

[3] Andrews, U. S. pat. 1,314,920 [*C. A.*, **13**, 2765 (1919)]; Selden, Brit. pat. 142,226 [*C. A.*, **14**, 2645 (1920)].

[4] Schumow, *J. Russ. Phys. Chem. Soc.*, (1) **19**, 119 (1887); *Ber. Ref.*, **20**, 218 (1887).

5-NITRO-2,3-DIHYDRO-1,4-PHTHALAZINEDIONE

(1,4-Phthalazinedione, 5-nitro-2,3-dihydro-)

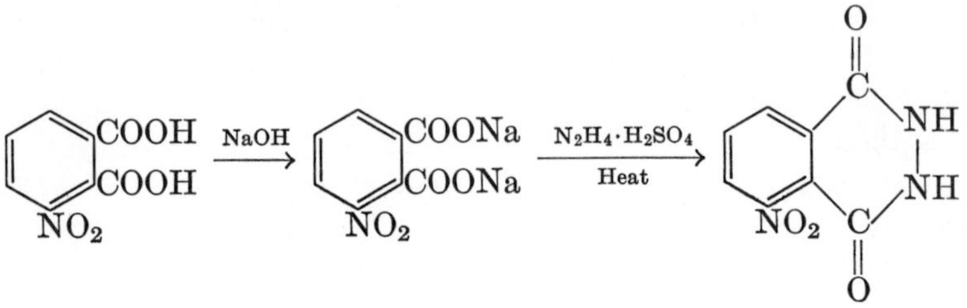

Submitted by CARL T. REDEMANN and C. ERNST REDEMANN.
Checked by CLIFF S. HAMILTON and CARL L. CARLSON.

1. Procedure

In a 16-cm. evaporating dish are placed 42.2 g. (0.2 mole) of 3-nitrophthalic acid,[1] 50 ml. of water, and a few drops of phenolphthalein indicator solution. The mixture is made faintly alkaline to phenolphthalein with about 66 ml. (0.4 mole) of 6 *N* sodium hydroxide solution (Note 1), the last portions of which are added slowly with good stirring so that the end point may be observed. The color of the phenolphthalein is discharged by the addition of 0.2–0.3 g. of the 3-nitrophthalic acid, and 26.0 g. (0.2 mole) of hydrazine sulfate[2] is added. The solution is evaporated to dryness over a sand bath, with stirring at the latter part of the evaporation (Note 2), and the residual solid is cooled, ground in a mortar to a fine powder, and placed in a 200-ml. conical flask with 25 ml. of tetralin (Note 3). The mixture is heated at 160–170° for 3 hours and allowed to cool. After the addition of 50 ml. of benzene the solid is collected on a Büchner funnel and pressed down well, and most of the benzene is removed by suction. The solid is then washed with two 25-ml. portions of ether and allowed to stand in the air until the odor of ether is no longer detectable. The resulting material, weight 62–68 g., is an equimolar mixture of sodium sulfate and 5-nitro-2,3-dihydro-1,4-phthalazinedione, and it may be used directly for preparing 5-amino-2,3-dihydro-1,4-phthalazinedione (luminol, p. 69).

For purification, the crude material is suspended in 600–700 ml. of boiling water, and solid sodium carbonate is added in portions until the nitro compound has dissolved. Decolorizing carbon is added cautiously, and the suspension is boiled for a few minutes and filtered. The clear red-brown filtrate is acidified with concentrated hydrochloric acid, and the 5-nitro-2,3-dihydro-1,4-phthalazinedione is precipitated as a cream-colored flocculent solid. The solution is allowed to cool to room temperature, and the solid is separated by filtration and dried. The product thus obtained weighs 31–32 g. (75–78%) and has a melting point of 315–316° (dec.) when determined with the Kullman copper block (Note 4).

2. Notes

1. The strength of the sodium hydroxide solution is not critical; the equivalent amount of a solution of a different concentration may be used.

2. It is wise to cover the hand with a cloth or a glove while stirring, for the hot mixture may spatter if the heating is too rapid. The evaporation may be finished by transferring the evaporating dish to an oven shortly after the solid begins to separate. It is necessary to continue the evaporation until the solid is completely dry.

3. In an optional procedure the dry powder is heated in an oven at 160–170° for 3 hours.

4. No satisfactory solvent has been found for the recrystallization of 5-nitro-2,3-dihydro-1,4-phthalazinedione, but this precipitation gives a satisfactory product, free of sodium sulfate. The melting points given in the literature range from 297° to 320°.

3. Methods of Preparation

5-Nitro-2,3-dihydro-1,4-phthalazinedione has been prepared by heating 3-nitrophthalic acid with a large excess of hydrazine hydrate,[3] by heating 3-nitrophthalic acid with hydrazine sulfate[4] and sodium acetate, and by heating the nitro acid with hydrazine hydrate in ethanol at 150°.[5]

[1] *Org. Syntheses* Coll. Vol. **1**, 408 (1941).
[2] *Org. Syntheses* Coll. Vol. **1**, 309 (1941).
[3] Bogert and Boroschek, *J. Am. Chem. Soc.,* **23,** 740 (1901).
[4] Huntress, Stanley, and Parker, *J. Am. Chem. Soc.,* **56,** 241 (1934).
[5] Radulescu and Alexa, *Z. physik. Chem.,* **B8,** 393 (1930).

m-NITRODIMETHYLANILINE

(Aniline, N,N-dimethyl-*m*-nitro-)

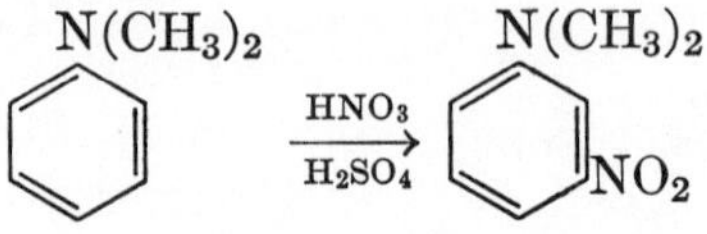

Submitted by HOWARD M. FITCH.
Checked by C. F. H. ALLEN and J. A. VANALLAN.

1. Procedure

In a 3-l. three-necked round-bottomed flask fitted with an effective mechanical stirrer, a dropping funnel, and a thermometer, and surrounded by an ice bath, is placed 1270 ml. (23.0 moles) of concentrated sulfuric acid (sp. gr. 1.84). Then 363 g. (3.0 moles) of dimethylaniline (Note 1) is slowly added with stirring and cooling so that the temperature remains below 25°. Stirring and cooling are continued until the temperature falls to 5°. A nitrating mixture is prepared by adding 366 g. (200 ml., 3.6 moles) of concentrated sulfuric acid (sp. gr. 1.84) to 286 g. (200 ml., 3.15 moles) of concentrated nitric acid (sp. gr. 1.42), with cooling and stirring. This is placed in the dropping funnel and added, drop by drop, the end of the dropping funnel being kept beneath the surface of the solution of dimethylaniline sulfate. The temperature is regulated between 5° and 10°, best by addition of small pieces of Dry Ice (Note 2); about 1.5 hours is required (Note 3). When the addition has been completed, the solution is stirred at 5–10° for 1 hour, and then poured, with stirring (Note 4), into an enameled pail containing 6 l. of ice and water (Note 5). Concentrated ammonium hydroxide is added slowly, with good stirring and using a dropping funnel that extends beneath the surface of the liquid, until the color of the precipitate changes to a light orange (Note 6). During this step the temperature is kept below 25°, using Dry Ice. From 1900 to 2050 ml. of ammonium hydroxide is required. The crude *p*-nitrodimethylaniline is collected on a filter and washed with 200 ml. of water (Note 7). Concentrated ammonium hydroxide (sp. gr. 0.90) is again added to the combined filtrate and washings, with good agitation and cooling to keep the temperature below 25°, until the liquid gives a purple color on Congo red paper (Note 8). From 1500 to 1650 ml. of ammonium hydroxide is required. The product is collected on a Büchner funnel and washed with 500 ml. of water. The crude *m*-nitrodimethylaniline (316–342 g. dry weight) is recrystallized from 400 ml. of hot 95% ethanol and carefully washed on the filter with 100 ml. of cold 95% ethanol. The yield is 280–316 g. (56–63%) of bright orange crystals that melt at 59–60°.

2. Notes

1. The dimethylaniline should be of good quality. Dimethylaniline (free from mono), m.p. 1°, of the Eastman Kodak Company, and dimethylaniline, Merck, are satisfactory.

2. A cooling bath of Dry Ice and acetone can be used, but addition of the solid directly to the mixture is more satisfactory and less is required.

3. The mixture may be left overnight at room temperature. It is advisable to complete the subsequent neutralization in 1 day.

4. A powerful stirrer is required to mix the large volumes.

5. If large equipment is not available, it is convenient to divide the nitration mixture into four parts (about 490 ml. each) and to pour each portion into 1.5 l. of ice and water in a 4-l. beaker. After neutralization, the precipitates from the four lots are combined on the filter.

6. It is essential that all the *para* isomer be precipitated, since it is not easily removed by recrystallization from solvents. The point at which the yellow *p*-nitrodimethylaniline ceases to precipitate and the orange *m*-nitrodimethylaniline begins to precipitate is difficult to judge without experience, since the solution itself has an orange color. If doubt exists, a small sample may be filtered so that the color of the precipitate can be clearly seen. If very much of the *para* isomer remains in solution, a yellow precipitate or turbidity is obtained on diluting the filtrate with several volumes of water.

7. If the *para* isomer is to be recovered, the precipitate should be washed well with water to remove all acid and dried. The solid (143–155 g.) is recrystallized from 400 ml. of benzene and carefully washed on the filter with 125 ml. of cold benzene. The yield is 74–92 g. (14–18%) of bright yellow crystals that melt at 163–164°.

8. Substantially all the *m*-nitrodimethylaniline is precipitated at a *p*H of about 3, while any dimethylaniline present remains in solution.

3. Methods of Preparation

The above procedure is a modification of the method of Groll;[1] a more laborious method of purification has been described.[2] *m*-Nitrodimethylaniline can also be prepared by methylation of *m*-nitroaniline using methyl sulfate [3] or methyl *p*-toluenesulfonate.[4]

[1] Groll, *Ber.*, **19**, 198 (1886).

[2] Bobranskiĭ and Eker, *J. Applied Chem. U.S.S.R.*, **14**, 524 (1941) [*C. A.*, **36**, 3159 (1942)].

[3] Ullmann, *Ann.*, **327**, 112 (1903).

[4] Simonov, *J. Gen. Chem. U.S.S.R.*, **13**, 51 (1943) [*C. A.*, **38**, 338 (1944)].

5-NITROINDAZOLE *

(Indazole, 5-nitro-)

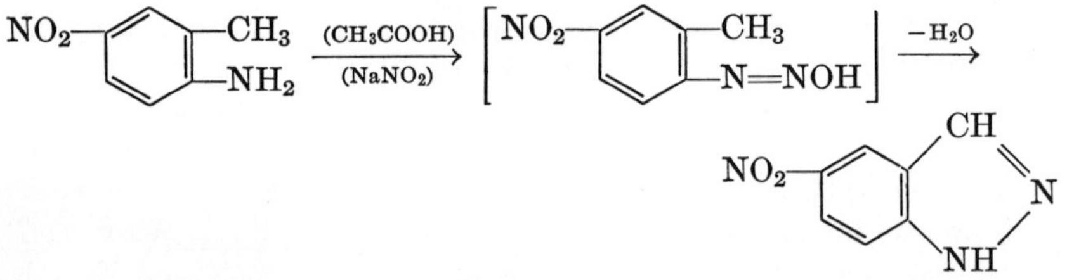

Submitted by H. D. Porter and W. D. Peterson.
Checked by N. L. Drake and A. F. Freeman.

1. Procedure

To a solution of 55 g. (0.36 mole) of 2-amino-5-nitrotoluene (m.p. 129–132°) in 2.5 l. of glacial acetic acid in a 5-l. round-bottomed flask, provided with an efficient mechanical stirrer, is added all at once (Note 1) a solution of 25 g. (0.36 mole) of sodium nitrite in 60 ml. of water. During this addition the temperature is not allowed to rise above 25° (Note 2). After the nitrite solution has been added, stirring is continued for 15 minutes to complete the diazotization. Any yellow precipitate formed during the next few hours is filtered and discarded (Note 3).

The solution is allowed to stand for 3 days at room temperature, and is then concentrated on the steam bath under reduced pressure (water pump) until spattering makes further evaporation impossible. Two hundred milliliters of water is added to the residue, and the contents of the flask are washed into a small beaker where they are stirred to a smooth slurry. The product is filtered, washed thoroughly on the funnel with cold water, and dried in an oven at 80–90°. The crude material melts at 204–206° and weighs 47–57 g. (80–96%). It is purified by recrystallization from 650 ml. of boiling methanol using 5 g. of decolorizing charcoal. The recrystallized, pale yellow needles of 5-nitroindazole melt at 208–209°. The yield is 42–47 g. (72–80%) (Note 4). Further recrystallization does not raise the melting point.

2. Notes

1. If the sodium nitrite solution is added slowly, a considerable quantity of a yellow precipitate, presumably the diazoamino compound, is formed.

2. If the solution is cooled in an ice bath to 15–20° before addition of the nitrite solution, the temperature of the mixture will not rise above 25° during the diazotization.

3. This is presumably the diazoamino compound, insoluble in most organic solvents. It melts at about 200°.[1]

4. The unsubstituted *o*-toluidine gives indazole itself, but the yield is very low (3–5%).

3. Method of Preparation

The procedure is essentially that of Noelting.[2]

[1] Meunier, *Bull. soc. chim. France,* (3) **31,** 641 (1904).
[2] Noelting, *Ber.,* **37,** 2584 (1904).

2-NITRO-4-METHOXYANILINE

(*p*-Anisidine, 2-nitro-)

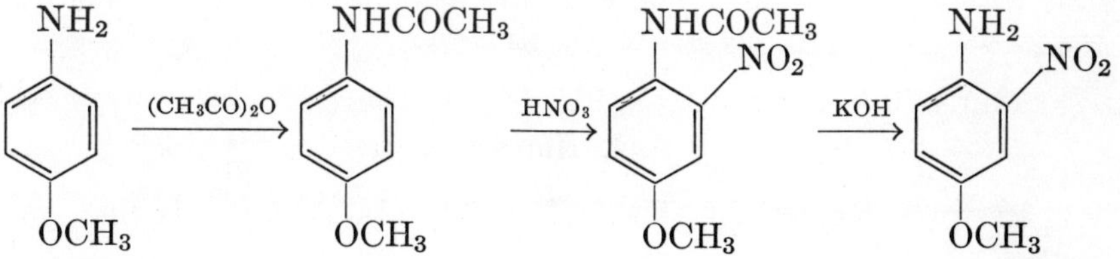

Submitted by Paul E. Fanta and D. S. Tarbell.
Checked by W. E. Bachmann and N. C. Deno.

1. Procedure

A. 2-*Nitro*-4-*methoxyacetanilide.* In a 2-l. three-necked, round-bottomed flask equipped with a mechanical stirrer and a thermometer are placed 123 g. (1 mole) of *p*-anisidine (Note 1), 300 ml. of glacial acetic acid, and 217 ml. of water. Stirring is started, and, when the *p*-anisidine has dissolved, 350 g. of ice is added. When the temperature reaches 0–5°, 103 ml. (1.1 moles) of acetic anhydride is added all at once with rapid stirring. Within several seconds the contents of the flask set to a crystalline mass and the temperature rises to 20–25°. The flask is heated on a steam bath until the crystalline material dissolves and is then cooled with stirring to 45°, at which temperature crystals begin to separate. An ice bath is applied, and 100 ml. (55% excess) of concentrated nitric acid (sp. gr. 1.42) is added all at once. The temperature rises rapidly to 70° and soon begins to fall. By suitable adjustment of the cooling bath the temperature is maintained at 60–65° for 10 minutes and then brought down to 25° in the course of 10 minutes (Note 2).

The solution is chilled overnight in an ice chest, and the precipitated yellow crystals are collected on a 19-cm. Büchner funnel. The crystals are washed with 270 ml. of ice-cold water and pressed as dry as possible with a rubber dam. The filter cake can be dried (Note 3) in air or in a vacuum desiccator over calcium chloride and soda lime. The yield of 2-nitro-4-methoxyacetanilide melting at 116–116.5° is 158–168 g. (75–79%).

B. 2-*Nitro*-4-*methoxyaniline*. A mixture of 160 g. of 2-nitro-4-methoxyacetanilide and 250 ml. of cold Claisen's alkali (Note 4) in a 2-l. beaker is stirred and warmed on a steam bath for 15 minutes; it first becomes liquid and then sets to a thick, red paste. After the addition of 250 ml. of hot water the mixture is stirred and digested on a steam bath for an additional 15 minutes and then cooled to 0–5°. The product is collected on a 19-cm. Büchner funnel, washed with three 160-ml. portions of ice-cold water, and pressed as dry as possible with a rubber dam. The yield of vacuum-dried product melting at 122.5–123° is 122–124 g. (95–97%).

2. Notes

1. Eastman Kodak Company's Practical grade of *p*-anisidine was used.

2. In a run in which the mixture was allowed to cool spontaneously from 65° the product became dark.

3. Drying is unnecessary if the 2-nitro-4-methoxyacetanilide is used in step B. The material may be recrystallized with 97% recovery from dilute aqueous ethanol (2 ml. of 95% ethanol and 4 ml. of water per gram). The product so obtained melts at 116.5–117°.

4. Claisen's alkali is prepared by dissolving 88 g. of potassium hydroxide in 63 ml. of water, cooling, and diluting to 250 ml. with methanol.

3. Methods of Preparation

2-Nitro-4-methoxyacetanilide has been prepared by the nitration of *p*-acetaniside.[1] The procedure described for the nitration is essentially that used by Lothrop.[1d]

2-Nitro-4-methoxyaniline has been prepared by heating nitrohydroquinone dimethyl ether with aqueous ammonia;[2] by heating the tetramethylammonium salt of 3-nitro-4-aminophenol;[3] by the hydrolysis of 2-nitro-4-methoxyacetanilide[1a] with alcoholic potassium hydroxide[1b,1c] or hydrochloric acid;[1d] and by the hydrolysis of the *p*-toluenesulfonamide,[4] the 3-nitrobenzenesulfonamide,[5] the 3-nitro-*p*-toluenesulfonamide,[5] the benzenesulfonamide,[6] and the acetyl derivative of the *p*-toluenesulfonamide[7] of 2-nitro-4-methoxyaniline with concentrated sulfuric acid.

[1] (*a*) Hinsberg, *Ann.*, **292**, 249 (1896); (*b*) Reverdin, *Ber.*, **29**, 2595 (1896); *Jahresb.*, **1896**, 1156; *Bull. soc. chim. France,* (3) **17**, 115 (1897); (*c*) Hata, Tatematsu, and Kubota, *Bull. Chem. Soc. Japan,* **10**, 425 (1935) [*C. A.*, **30**, 1056 (1936)]; (*d*) Lothrop, *J. Am. Chem. Soc.*, **64**, 1698 (1942).

[2] Scheidel, Ger. pat. 36,014 [*Frdl.*, **1**, 221 (1888)].

[3] Hahle, *J. prakt. Chem.*, (2) **43**, 66 (1891).

[4] Reverdin, *Ber.*, **42**, 1525 (1909); Simonov, *J. Gen. Chem. U.S.S.R.*, **10**, 1580 (1940) [*C. A.*, **35**, 2870 (1941)].

[5] Reverdin and de Luc, *Ber.*, **45**, 352 (1912).

[6] Elderfield, Gensler, and Birstein, *J. Org. Chem.*, **11**, 812 (1946).

[7] Reverdin and de Luc, *Ber.*, **43**, 3461 (1910).

4-NITRO-1-NAPHTHYLAMINE

(1-Naphthylamine, 4-nitro-)

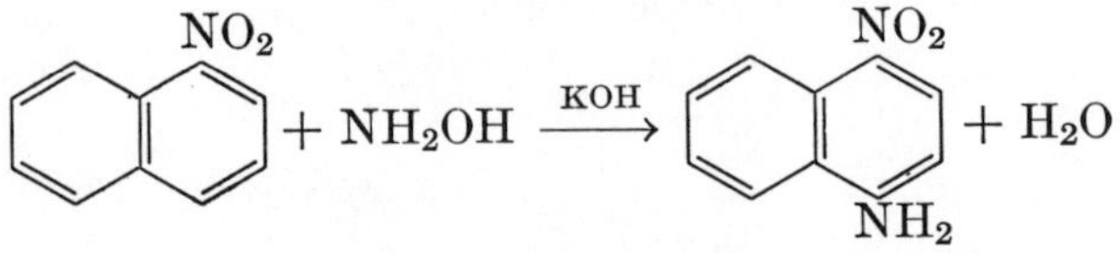

Submitted by CHARLES C. PRICE and SING-TUH VOONG.
Checked by RICHARD T. ARNOLD and JAY S. BUCKLEY, JR.

1. Procedure

Twenty grams (0.115 mole) of α-nitronaphthalene (Note 1) and 50 g. (0.72 mole) of powdered hydroxylamine hydrochloride are dissolved in 1.2 l. of 95% ethanol contained in a 3-l. flask which is heated in a bath maintained at 50–60°. A filtered solution of 100 g. of potassium hydroxide in 500 g. (630 ml.) of methanol is added gradually with vigorous mechanical stirring (Note 2) over a period of 1 hour. Stirring is continued for an additional hour, and the warm solution is poured slowly into 7 l. of ice water. After the solid has coagulated, it is collected on a filter and washed thoroughly with water. The crude 4-nitro-1-naphthylamine is purified by recrystallization from 500 ml. of 95% ethanol (Note 3). About 12–13 g. (55–60%) of long golden-orange needles, m.p. 190.5–191.5°, is obtained.

2. Notes

1. The α-nitronaphthalene, m.p. 56–57°, was obtained from Eastman Kodak Company.

2. The color changes from yellow to orange, and potassium chloride separates.

3. In some experiments a few milliliters of dilute hydrochloric acid (1:1) or sulfuric acid (1:1) was added to facilitate the crystallization of the 4-nitro-1-naphthylamine.

3. Methods of Preparation

This method is essentially that described by Goldhahn.[1] 4-Nitro-1-naphthylamine has also been prepared by the nitration of α-naphthylamine,[2] acetyl-α-naphthylamine,[3,4] and ethyl-1-naphthyloxamate,[5,6] by the oxidation of 4-nitroso-1-naphthylamine,[7] and by reaction of 4-chloro-1-nitronaphthalene with ammonia.[8]

[1] Goldhahn, *J. prakt. Chem.*, **156**, 315 (1940); **157**, 96 (1940).
[2] Meldola and Streatfeild, *J. Chem. Soc.*, **63**, 1055 (1893).
[3] Lellmann and Remy, *Ber.*, **19**, 796 (1886).
[4] Hodgson and Walker, *J. Chem. Soc.*, **1933**, 1205.
[5] Lange, Ger. pat. 58,227 [*Frdl.*, **3**, 509 (1890–1894)].
[6] Sergievskaya, Russ. pat. 50,696 (1937); *J. Gen. Chem. U.S.S.R.*, **10**, 55 (1940)
[7] Vorozhtsov and Kozlov, *J. Gen. Chem. U.S.S.R.*, **9**, 587 (1939).
[8] Ger. pat. 117,006 [*Frdl.*, **6**, 176 (1900–1902)].

p-NITROPHENYLARSONIC ACID

(Benzenearsonic acid, p-nitro-)

$$p\text{-}NO_2C_6H_4N_2BF_4 + NaAsO_2 + 2NaOH \xrightarrow{Cu_2Cl_2} p\text{-}NO_2C_6H_4AsO_3Na_2 + NaBF_4 + N_2 + H_2O$$

$$p\text{-}NO_2C_6H_4AsO_3Na_2 + 2HCl \rightarrow p\text{-}NO_2C_6H_4AsO_3H_2 + 2NaCl$$

Submitted by A. Wayne Ruddy and Edgar B. Starkey.
Checked by Cliff S. Hamilton and Richard E. Benson.

1. Procedure

In a 2-l. beaker provided with an efficient mechanical stirrer, 52 g. (0.4 mole) of sodium metaarsenite and 16 g. (0.4 mole) of sodium hydroxide are dissolved in 600 ml. of water (Note 1) and 6 g. of cuprous chloride is suspended in the solution.

A mixture of 300 ml. of water and the p-nitrobenzenediazonium borofluoride [1] obtained from 0.25 mole of p-nitroaniline (Note 2) is added during a period of 1 hour to the sodium arsenite solution. The foaming that accompanies the evolution of nitrogen is readily controlled by the occasional addition of small amounts of ether or benzene (Note 3). As the reaction proceeds, 100 ml. of 10% sodium hydroxide solution (0.25 mole) is added in 20-ml. portions. Stirring is continued for another hour, and then the mixture is warmed to 60° for 30 minutes and filtered with suction through a sintered-glass funnel. The residue on the funnel is washed with two 50-ml. portions of water. To the combined filtrate and washings is added concentrated hydrochloric acid (sp. gr. 1.19) until the solution is acid to litmus paper. The mixture is filtered, activated charcoal is added to the filtrate, and the solution is concentrated to about 350 ml. After the hot solution has been filtered with suction, concentrated hydrochloric acid is added until the solution is acid to Congo red paper. The solution is placed in a refrigerator overnight, and the crystals are collected on a Büchner funnel and washed twice with 20-ml. portions of ice water. The combined filtrates and washings are concentrated to about 150 ml., chilled, and filtered. The total quantity of crystals is dissolved in 10% ammonium hydroxide solution; the solution is filtered and again made acid to Congo red paper with concentrated hydrochloric acid. After the solution has been thoroughly chilled (preferably overnight), the *p*-nitrophenylarsonic acid is filtered on a Büchner funnel and washed with small portions of ice water until free of ammonium chloride. After drying, the yellow crystals melt with decomposition at 298–300.° The yield is 44–48.5 g. (71–79%) (Note 4).

2. Notes

1. The sodium arsenite solution may also be prepared by dissolving 39.6 g. (0.2 mole) of arsenious oxide and 32 g. (0.8 mole) of sodium hydroxide in 600 ml. of water.

2. According to the submitters *p*-nitrobenzenediazonium borofluoride may also be prepared as follows:

p-Nitroaniline (34.5 g.) is dissolved in 63 ml. of concentrated hydrochloric acid (sp. gr. 1.19) and 50 ml. of water. The solution is cooled to 0°, and 17.3 g. of sodium nitrite in 40 ml. of water is added slowly with vigorous stirring. After the diazotization is complete, as indicated by a positive test with starch iodide paper, a solution of 55 g. of sodium fluoborate in 110 ml. of water is added. The thick slurry is stirred for 15 minutes and then filtered with suction and washed with ice water, twice with methanol, and twice with ether. The solid should be sucked as free as possible from liquid after each washing. The compound may be kept in an evacuated desiccator until needed.

3. The checkers found that the foaming is more readily controlled by amyl alcohol than by the addition of ether or benzene.

4. This is a general method for preparing arylarsonic acids. The melting points and yields of other arsonic acids prepared by the submitters are as follows: phenyl, 156°, 58%; *o*-nitrophenyl, 232–234° dec., 67%; *m*-nitrophenyl, 182°, 47%; *o*-tolyl, 159–160°, 63%; *m*-tolyl, 150°, 54%; *p*-tolyl, 300° dec., 73%; *o*-chlorophenyl, 182°, 52%; *p*-chlorophenyl, above 300°, 63%; *o*-carboxyphenyl, above 300°, 65%; *p*-carboxyphenyl, 232° dec., 67%; *p*-carbethoxyphenyl, 260°, 60%; *p*-acetophenyl, 175°, 70%.

3. Methods of Preparation

p-Nitrophenylarsonic acid has been prepared by heating *p*-nitrobenzenediazonium chloride with arsenious acid in hydrochloric acid,[2] by the action of *p*-nitrobenzenediazonium chloride on sodium arsenite,[3] by the action of sodium arsenite on sodium *p*-nitrobenzeneisodiazo oxide,[4] by the diazotization of *p*-nitroaniline in acetic acid in the presence of arsenic chloride and cuprous chloride,[5] and by the reaction of *p*-nitrobenzenediazonium borofluoride with sodium arsenite in the presence of cuprous chloride.[6]

[1] *Org. Syntheses* Coll. Vol. **2,** 225 (1943).

[2] Bart, Ger. pat. 250,264 [*Frdl.,* **10,** 1254 (1910–12)].

[3] Bart, *Ann.,* **429,** 95 (1922).

[4] Bart, *Ann.,* **429,** 103 (1922); Jacobs, Heidelberger, and Rolf, *J. Am. Chem. Soc.,* **40,** 1580 (1918).

[5] Scheller, U. S. pat. 1,704,106; Fr. pat. 624,028 [*Chem. Zentr.,* **98,** II, 2229 (1927)]; Brit. pat. 261,026; Ger. pat. 624,028 [*Frdl.,* **17,** 2372 (1930)].

[6] Ruddy, Starkey, and Hartung, *J. Am. Chem. Soc.,* **64,** 828 (1942).

p-NITROPHENYL SULFIDE

[Sulfide, bis-(*p*-nitrophenyl)]

$$2p\text{-}NO_2C_6H_4Cl \xrightarrow{KSCSOC_2H_5} p\text{-}NO_2C_6H_4SC_6H_4NO_2\text{-}p$$

Submitted by Charles C. Price and Gardner W. Stacy.
Checked by Cliff S. Hamilton and Paul D. Berry.

1. Procedure

In a 1-l. round-bottomed flask equipped with a reflux condenser are placed 157.5 g. (1 mole) of *p*-chloronitrobenzene, 160 g. (1 mole) of potassium xanthate (Note 1), and 450 ml. of 95% ethanol. This reaction mixture is heated under reflux on a steam bath for 48 hours. The crystalline product, which deposits from solution during the course of the reaction, is collected by filtration, crushed into small particles in a mortar, and washed twice with hot ethanol and once with hot water. The yield of *p*-nitrophenyl sulfide melting at 158–160° is 105–113 g. (76–82%). This product is pure enough for most purposes. Recrystallization from glacial acetic acid (15 ml. per gram) raises the melting point to 160–161°.

2. Note

1. Potassium xanthate may be prepared in the following manner: With heating, 300 g. (5.36 moles) of potassium hydroxide is dissolved in 3 l. of absolute ethanol. The solution is then cooled in an ice bath, and the temperature is kept below 10° while carbon disulfide is added in portions with stirring until the solution is no longer alkaline; about 456 g. (360 ml., 5.95 moles) of carbon disulfide is required. The potassium xanthate is collected by suction filtration and air-dried on large sheets of filter paper; yield, 429–472 g. (50–55%).

3. Methods of Preparation

p-Nitrophenyl sulfide has been prepared by the reaction between *p*-chloronitrobenzene and sodium sulfide.[1] This is not a practical means of preparation, however, because of the variety of substances formed.[2] The method described has been published.[3] A preparation of pure *p*-nitrophenyl sulfide has been reported from *p*-chloronitrobenzene and sodium thiosulfate.[4]

[1] Nietski and Bothof, *Ber.*, **27**, 3261 (1894).
[2] Kehrmann and Bauer, *Ber.*, **29**, 2362 (1896).
[3] Price and Stacy, *J. Am. Chem. Soc.*, **68**, 498 (1946).
[4] Dall'Olio, *Chimica e industria,* **28**, 73 (1946) [*C. A.*, **41**, 380 (1947)].

NITROSOBENZENE

(Benzene, nitroso-)

$$C_6H_5NO_2 \xrightarrow[NH_4Cl]{Zn} C_6H_5NHOH \xrightarrow[H_2SO_4]{Na_2Cr_2O_7} C_6H_5NO$$

Submitted by George H. Coleman, Chester M. McCloskey, and Frank A. Stuart.
Checked by W. E. Bachmann, N. C. Deno, and R. F. Edgerton.

1. Procedure

A mixture of 250 ml. (2.44 moles) of nitrobenzene (Notes 1 and 2) and a solution of 150 g. of ammonium chloride in 5 l. of water in a 5-gal. crock (Note 3) is stirred vigorously (Note 4), and 372 g. (5.15 moles) of zinc dust (90% zinc) is added in small portions over a period of 5 minutes. About 5 minutes after the addition of the zinc, the main reaction occurs and the temperature rises. When the temperature reaches about 65°, enough ice is added to the stirred mixture to bring the temperature down to 50–55° (Note 5). Twenty minutes after the addition of zinc was started, the solution is filtered through a 24-cm. Büchner funnel and the zinc oxide residues are washed with 3 l. of boiling water. The filtrate and washings are combined in a 6-gal. crock (Note 6) and cooled immediately (Note 7) by the addition of enough ice to bring the temperature to 0° to −2° and leave at least 1 kg. of ice unmelted.

To this cold solution or suspension of β-phenylhydroxylamine, a cold solution of sulfuric acid (750 ml. of concentrated acid and sufficient ice to bring the temperature down to −5°) is added with stirring. An ice-cold solution of 170 g. of sodium dichromate dihydrate in 500–750 ml. of water is added as rapidly as it can be poured into the mixture, which is stirred or swirled (Note 8). After 2 to 3 minutes, the straw-colored precipitate of nitrosobenzene is collected on a Büchner funnel and washed with 1 l. of water (Note 9).

The crude nitrosobenzene is steam-distilled (Note 10), and the distillate is collected in a receiver cooled by ice (Note 11). The nitrosobenzene is finely ground in a mortar, transferred to a Büchner funnel, and washed with water until the washings are no longer brown. After it has been sucked as dry as possible on the filter the nitrosobenzene is pressed between layers of filter paper (or other porous paper). One or two changes of paper may be necessary. The yield of nitrosobenzene melting at 64–67° is 128–138 g. (49–53%) (Notes 12 and 13). If a purer product is desired, the crude nitrosobenzene can be recrystallized from a small amount of ethanol with good cooling, and the product dried over calcium chloride at atmospheric pressure.

2. Notes

1. Commercial nitrobenzene of good quality is satisfactory.
2. Contact with nitrobenzene, phenylhydroxylamine, and nitrosobenzene or prolonged breathing of the vapors should be avoided.
3. The checkers used a 12-l. round-bottomed flask.
4. Vigorous stirring is necessary in order to prevent the zinc dust from caking on the bottom of the crock. The submitters employed two mechanically driven stirrers in order to keep the zinc dust in suspension. The checkers used a single paddle stirrer successfully.
5. At higher temperatures secondary reactions take place.
6. The checkers used a 22-l. round-bottomed flask.

7. Since β-phenylhydroxylamine decomposes on standing, the filtrate should be cooled as soon as the filtration is completed, and the oxidation to nitrosobenzene should be carried out immediately. An excess of ice should be present since the oxidation reaction is exothermic and the temperature of the solution should still be near 0° after the oxidation. At higher temperatures oxidation products other than nitrosobenzene are produced.

8. It is important to add the dichromate solution rapidly in order to obtain a good yield of easily filterable product. In a run in which the dichromate solution was added over a period of 25 minutes, only a 10% yield of nitrosobenzene was obtained.

9. This product is stable for about a week if kept at 0°.

10. Connections should be of glass since cork and rubber are attacked readily. Since nitrosobenzene decomposes at the elevated temperature, it should be steam-distilled as rapidly as possible. If several runs are made, not more than four or five should be combined for distillation. A Hopkins still head is effective for preventing contamination.

The nitrosobenzene condenses to a green liquid which solidifies to a white solid. Care should be taken that the solid does not clog the condenser. Distillation is stopped when yellow oily material appears in the condenser.

11. Cooling of the receiver is necessary because the nitrosobenzene has a very high vapor pressure at room temperature.

12. Yields as high as 65–70% have been obtained in smaller runs.

13. Nitrosobenzene can be kept at room temperature in a closed container for 1–2 days. Over longer periods it should be stored at 0°.

3. Methods of Preparation

Nitrosobenzene can be prepared by the oxidation of aniline with permonosulfuric acid [1] or peracetic acid [2] and by the oxidation of β-phenylhydroxylamine,[3] which is prepared by the reduction of nitrobenzene. Purification by sublimation has been recommended.[4]

[1] Caro, *Z. angew. Chem.*, **11,** 845 (1898).

[2] D'Ans and Kneip, *Ber.*, **48,** 1144 (1915).

[3] Bamberger, *Ber.*, **27,** 1555 (1894); Parsons and Bailar, *J. Am. Chem. Soc.*, **58,** 268 (1936).

[4] Robertson and Vaughan, *J. Chem. Education,* **27,** 605 (1950).

OLEYL ALCOHOL

(9-Octadecen-1-ol)

$$CH_3(CH_2)_7CH{=}CH(CH_2)_7COOC_2H_5 + 4Na + 3C_2H_5OH \rightarrow CH_3(CH_2)_7CH{=}CH(CH_2)_7CH_2OH + 4C_2H_5ONa$$

Submitted by Homer Adkins and R. H. Gillespie.
Checked by Cliff S. Hamilton, C. W. Winter, and Alfred Stepan.

1. Procedure

In a 5-l. round-bottomed flask fitted with a large-bore reflux condenser are placed 200 g. (230 ml., 0.65 mole) of ethyl oleate (Note 1) and 1.5 l. of absolute ethanol (Note 2). Through the reflux condenser is added 80 g. (3.5 gram atoms) of sodium rapidly enough to keep up a vigorous reaction. The flask is shaken occasionally. After the initial reaction has subsided, about 200 ml. more of absolute ethanol is added, and the mixture is heated on a steam bath until all the sodium has reacted. Then 500 ml. of water is added, and the mixture is refluxed for 1 hour to saponify the unreacted ester. The mixture is cooled, and 1.2 l. of water is added. The unsaponifiable fraction is extracted with ether, and the extracts are washed with 1% potassium hydroxide solution and then with water till free of alkali when tested externally with phenolphthalein. The ether extract is dried over sodium sulfate, the ether removed by distillation, and the residue distilled through an efficient column (Note 3). A yield of 84–89 g. (49–51%), b.p. 150–152°/1 mm., is obtained.

Fifty grams of the crude oleyl alcohol is crystallized from 400 ml. of acetone at −20° to −25° (Note 4) in a jacketed sintered-glass funnel (Note 5). The residue, amounting to about 25 g., is then recrystallized from 250 ml. of acetone at −5° (Note 6) to remove saturated alcohols. The oleyl alcohol in the filtrate is recrystallized at −60° to −65° (Note 7) and then distilled from a 25-ml. flask with a thermometer well and a low side arm. The yield is 13–16 g., b.p. 148–150°/less than 1 mm.; n_D^{25} 1.4590.

2. Notes

1. The ethyl oleate used by the submitters was prepared by esterification of commercial U.S.P. grade oleic acid and fractionated through

a Widmer column. The fraction boiling at 160–170°/1 mm., n_D^{25} 1.4482–1.4570, iodine value 83.6 (calculated 81.8), was used. The checkers distilled ethyl oleate (technical, Eastman Kodak Company) and used the fraction boiling at 164–166°/1 mm., n_D^{25} 1.4522.

2. The purity of the absolute ethanol is of prime importance. Ethanol dried only by a lime process gives a low yield (20–25%). The ethanol used in this procedure was dried over lime and then over aluminum *tert*-butoxide, after which it was distilled directly into the flask used for the reaction.

3. The submitters used a modified Widmer column; the checkers, a Claisen flask modified with a short Vigreux column.

4. The temperature is controlled by selecting a liquid whose freezing point is approximately the temperature at which one wishes to carry out the crystallization, and crushed Dry Ice is added with stirring to keep just a slush. There must be an intimate mixture of solid and liquid at all times. Carbon tetrachloride was used for temperatures of −20° to −25°.

5. The apparatus was as follows: A 50-ml. Pyrex sintered-glass funnel, with a sealed-on extension tube, was jacketed with an inverted 500-ml. wide-mouth bottle with the bottom cut off and the edges ground smooth. The jacketed funnel was then placed in a 250-ml. suction flask, the side arm of which was connected to a three-way stopcock. One arm of the stopcock was connected to a water aspirator; the other, to a large test tube (1⅛ in. by 8 in.) fitted with a two-hole stopper. From the second hole in this stopper was led a tube extending to the bottom of a smaller test tube partly filled with mercury. During crystallization, this large test tube was filled three-quarters full of Dry Ice, the carbon dioxide gas evolved being allowed to bubble up through the bottom of the sintered-glass funnel. This prevents the solution from filtering through and also stirs the solution by the rise of small carbon dioxide bubbles, and hence induces crystallization. The porosity of the sintered-glass funnel is a factor in the size and number of bubbles formed. The mercury trap may be raised or lowered to adjust the pressure, and to act as a vent to prevent the accumulation of carbon dioxide to give excessive pressures. After crystallization had taken place (about 1 hour) the stopcock was turned to the aspirator and the solution immediately filtered. With this apparatus, about 80 ml. of solution may be crystallized at a time, and no transferring from one container to another is necessary. The checkers found it more convenient to use a 500-ml. sintered-glass funnel along with a proportionally larger bottle, flask, etc. The 400 ml. of

acetone solution was then poured into the funnel in 50-ml. portions with adequate cooling intervals.

6. A crushed ice-salt bath was used.

7. A Dry Ice-chloroform bath was used.

3. Methods of Preparation

Oleyl alcohol has been prepared by the catalytic reduction of ethyl oleate [1] and heavy-metal oleates; [2] by the action of sodium and absolute ethanol on ethyl [3] or butyl [4] oleate; by the action of sodium and absolute butyl alcohol on butyl oleate; [5] and by the action of sodium and *tert*-butyl alcohol on ethyl oleate.[6] The above procedure is essentially that of Kass and Burr,[7] who prepared linoleyl alcohol. The purification step was derived from the low-temperature crystallization technique of Hartsuch.[8]

[1] Sauer and Adkins, *J. Am. Chem. Soc.*, **59,** 1 (1937).

[2] Brit. pat. 584,939 [*C. A.*, **41,** 3812 (1947)].

[3] Bouveault and Blanc, *Bull. soc. chim. France,* (3) **31,** 1210 (1904).

[4] Palfray and Anglaret, *Compt. rend.*, **224,** 404 (1947); *Bull. mens. ITERG,* **1947,** No. 9, 3 [*C. A.*, **42,** 3310 (1948)].

[5] *Org. Syntheses* Coll. Vol. **2,** 468 (1943).

[6] Hausley, *Ind. Eng. Chem.*, **39,** 54 (1947).

[7] Kass and Burr, *J. Am. Chem. Soc.*, **62,** 1796 (1940).

[8] Hartsuch, *J. Am. Chem. Soc.*, **61,** 1142 (1939).

OZONE

(A laboratory ozonizer)

$$3O_2 \xrightarrow[\text{discharge}]{\text{Silent electrical}} 2O_3$$

Submitted by L. I. Smith, F. L. Greenwood, and O. Hudrlik.
Checked by R. L. Shriner, C. E. Kaslow, and R. D. Stayner.

1. Description of Apparatus and Procedures

A diagrammatic sketch of the complete apparatus for the laboratory production and use of ozone in organic reactions is shown in Fig. 21.

A. *Purification train.* Oxygen from a cylinder (*A*) fitted with a reducing valve (*B*) is led to a pressure release tube (*C*). This is a T-tube, the long arm of which dips into a test tube of mercury. The height of the mercury column should be about 2 to 3 cm. The release tube is connected through a stopcock (1) to a 40-cm. condenser jacket (*D*). This is filled with 4-mesh anhydrous calcium chloride held in

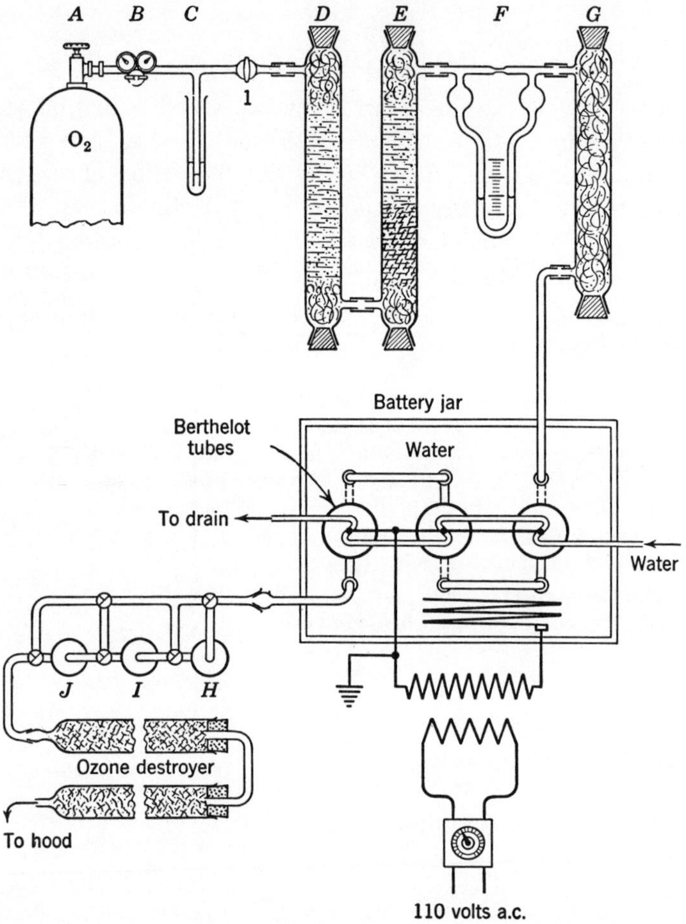

FIG. 21.

place by plugs of glass wool at the ends. A second condenser jacket (*E*) is filled about halfway with 4-mesh soda lime, then a 4- to 6-in. layer of anhydrous calcium chloride, and the remainder of the tube is packed with glass wool. The ends of the condensers are closed with rubber stoppers. Part *F* is a flowmeter, the U-tube of which should be about 20 cm. long. The bore of the capillary should be about 0.5 mm. in diameter. The flowmeter tube is filled about half full with Hyvac pump oil. Part G is a condenser jacket (40 cm.) loosely packed with glass wool.

All parts up to the ozonizer may be connected with heavy rubber tubing or, preferably, Tygon tubing. If rubber tubing is used, it is desirable to place this purification train several feet from the ozonizer proper in order to prolong the life of the rubber connectors (Note 1). The sizes of parts C through G are not critical. After this part of the apparatus has been assembled, a very rapid stream of oxygen should be passed through it for 30 minutes in order to blow out all dust particles.

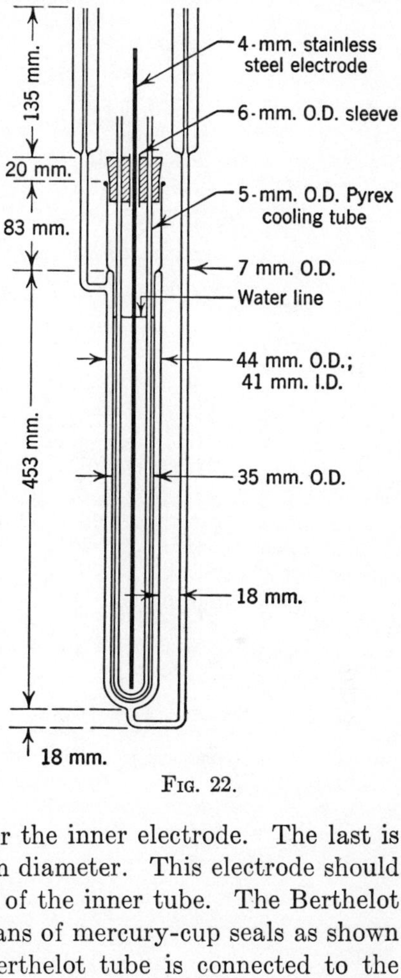

Fig. 22.

B. *Ozonizer.* The conversion of oxygen to ozone is accomplished by means of three Berthelot tubes constructed of soft lime glass (Note 2) with the dimensions shown in Fig. 22. It is important that the glass be thoroughly cleaned and that the annular space through which the gases pass be as uniform as possible. Each tube has a cooling coil (Note 3) consisting of a long U-shaped piece of 6-mm. Pyrex glass tubing which fits into the innermost space of the Berthelot tube and reaches nearly to the bottom. The top of the Berthelot tube is closed by a cork with three holes, two for the cooling tubes and one for the inner electrode. The last is a stainless steel rod about 2–4 mm. in diameter. This electrode should extend to about 1 cm. of the bottom of the inner tube. The Berthelot tubes (Fig. 22) are connected by means of mercury-cup seals as shown in Fig. 23 (Note 4). The third Berthelot tube is connected to the reaction vessels by means of either mercury-cup seals or ground-glass joints. No rubber connectors can be used for gases containing ozone.

The three Berthelot tubes are mounted vertically in a large battery jar about 13 in. by 9 in. by 18 in. (Note 5). They are held in place by a wooden top (see Fig. 23) which has holes and slots cut to fit the tubes. A second wooden block with holes and slots to fit the bottom of the Berthelot tubes (Fig. 22) is placed in the bottom of the battery jar. It should be weighted with a few pieces of lead. Both the top and the bottom blocks should be soaked in hot molten paraffin wax. The holes in the top should be directly above those in the bottom so that the Berthelot tubes will be as nearly vertical as possible to facili-

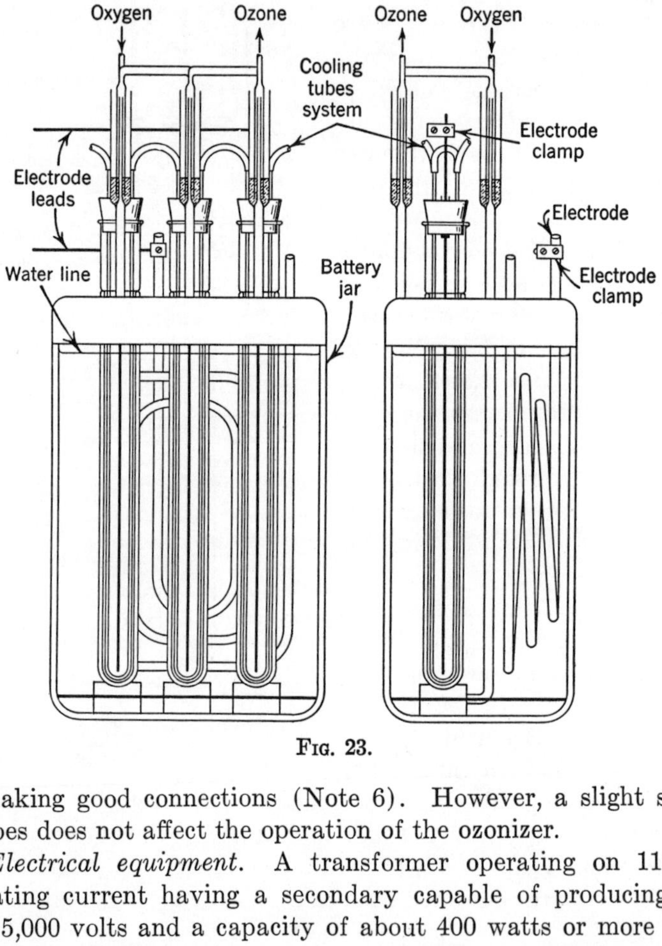

FIG. 23.

tate making good connections (Note 6). However, a slight slant of the tubes does not affect the operation of the ozonizer.

C. *Electrical equipment.* A transformer operating on 110 volts alternating current having a secondary capable of producing about 8000–15,000 volts and a capacity of about 400 watts or more is suitable. It is important that provision be made for varying the secondary voltage. This may be done by purchasing a transformer provided with taps on the primary windings or by inserting a variable transformer in series with the primary (Note 7). During the initial test runs for calibration it is desirable to connect an ammeter in the primary circuit in order to make certain the transformer is not overloaded.

The inner electrodes of the Berthelot tubes are connected to one of the secondary terminals of the transformer. *This wire must be grounded* (Note 8); otherwise the cooling tube inside the Berthelot tube will act as a conductor to the water line and the laboratory water line would be charged. The other wire from the secondary is connected to the electrode in the battery jar. This outer electrode con-

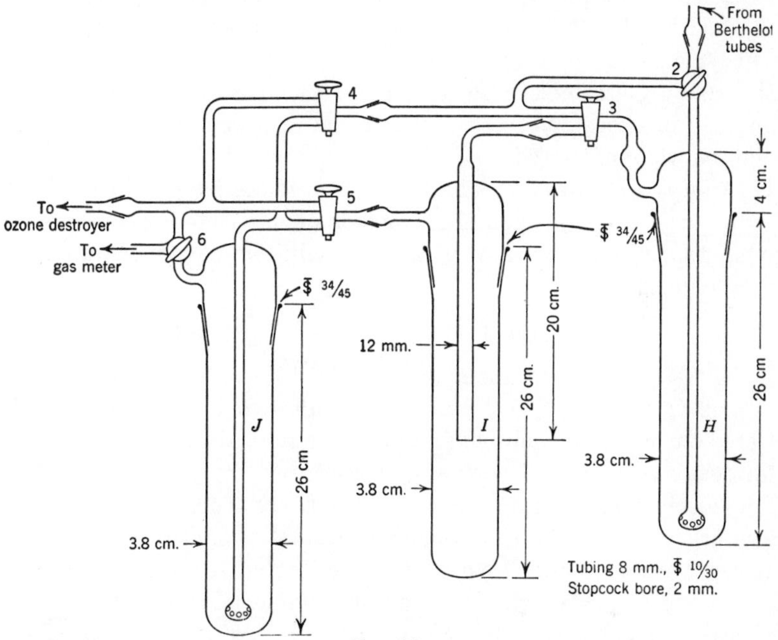

FIG. 24.

sists of a grid of stainless-steel wire (No. 8 or 10 B. and S. gauge). Connecting the inner electrode to the ground results in a charge on the battery jar. *Precautions* must be taken so that the jar is not touched while the current is passing. Also the battery jar must be kept away from any of the plumbing so that it will not be grounded. It is best to arrange a wooden shield so that the operator cannot come in contact with any of the high-voltage connections of the secondary.

D. *Ozone absorption assembly.* One useful combination of reaction flasks is shown in Fig. 24. It is best that the parts be connected by ground-glass joints. Tube *H* is the main reaction flask (Note 9), *I* is a condensing tube surrounded by acetone and Dry Ice to trap any volatile compounds, and *J* is an auxiliary analyzing tube for determining the amount of ozone not absorbed by the compound in tube *H*. The tubes should be assembled and clamped high enough above the desk for cooling baths to be placed around tubes *H* and *I* in order that ozonizations may be carried out at temperatures below that of the surroundings.

E. *Ozone destroyer.* Ozone is a powerful irritant. The maximum possible working concentration has been reported to be 0.15 to 1.0 part per million of air.[1] It is necessary to destroy any excess ozone and to see that the exit tube from the above absorption assembly is connected to a good hood. It is safer to incorporate an ozone destroyer in the set-up. One such destroyer consists of two tall towers (30 in.) filled with broken glass moistened with 5% aqueous sodium hydroxide and connected in series (Note 10). The room in which an ozonizer is used should be well ventilated.

F. *Analysis for ozone.* The analysis is made by passing a definite amount of oxygen through the ozonizer at a selected secondary voltage and then through a neutral 2% solution of potassium iodide. Iodine is liberated,

$$O_3 + 2KI + H_2O \rightarrow I_2 + 2KOH + O_2$$

and the resulting solution is acidified with 10% sulfuric acid (about 15 ml.) and titrated with a previously standardized 0.1 *N* sodium thiosulfate solution using soluble starch as the indicator.

G. *Procedure and calibration of apparatus.* For satisfactory use of an ozonizer it is desirable to obtain data relating rate of flow and secondary voltage with the amount of ozone produced.

A wet-type gas meter (Note 11) is connected to the tube *J* as indicated in Fig. 24. Enough 2% potassium iodide solution (about 70 ml.) is added to absorption tubes *H* and *J* to fill them to a depth of about one-third to one-half their height, and the tubes are marked at this level (Note 12). The reducing valve of the oxygen tank is opened far enough to show a pressure of 3–4 lb. per sq. in.; stopcocks 2, 3, 4, 5, and 6 are turned so as to by-pass *H, I,* and *J* and direct the gases through the ozone destroyer. Stopcock 1 is opened far enough to give a flowmeter reading of about 2–3 mm.; the transformer is turned on and adjusted to a definite voltage (Note 13). After the apparatus has been swept out for 5 minutes, the stopcocks are turned so as to pass the ozonized oxygen successively through tubes *H, I, J,* and the gas meter. The ozonized oxygen is passed through the potassium iodide solution for 5 minutes (stopwatch), and notes are made of the exact flowmeter reading and the volume of gas passing through the wet meter. At the end of 5 minutes the stopcocks are turned so as to by-pass tubes *H, I, J,* and the gas meter; and the ozonized oxygen stream is directed through the ozone destroyer. (*Caution: do not pass the ozone through the wet gas meter.*) The solutions in *H* and *J* are then combined, acidified, and titrated as described in paragraph F above. Practically all the ozone is absorbed in tube *H*. The second tube *J* is to prevent any damage to the wet gas meter.

The procedure is then repeated, the secondary voltage being kept constant but the rate of flow of oxygen being increased so that the flowmeter readings are about 5 mm. and 10 mm. The voltage is then changed so that the general range of 7000–12,000 volts in steps of about 1000 volts is obtained (Note 13). Determinations of ozone produced are made for each voltage and each rate of flow. These data may be plotted on coordinate paper in order to determine the performance of the ozonizer. A portion of such data is summarized in Table I, which represents the results from two different ozonizers constructed by different workers in different laboratories.

TABLE I

VARIATION IN PERCENTAGE (BY VOLUME) OF OZONE (FROM OXYGEN) WITH RATE OF FLOW AND SECONDARY VOLTAGE

Flowmeter Reading, mm.	*Rate of Flow,* l./hr.	*Secondary Voltage* 7000	8000	9000	10,000	11,000	12,000
		Ozonizer 1					
2.0	5.6	1.3%	3.1%	4.5%	6.0%	...	...
6.0	10.6	0.9	2.1	4.2	5.6	...	...
12.0	15.4	0.5	1.7	3.6	4.9	...	...
		Ozonizer 2					
2.0	5.7	6.7%	7.0%	7.1%	7.3%	7.2%	7.1%
5.0	9.8	6.0	6.5	6.7	7.2	6.9	6.8
10.0	14.4	5.5	6.2	6.5	6.7	6.8	6.9

For practical use in ozonolysis of compounds it is convenient to recalculate these data to show the time required to produce 0.1 mole of ozone at a specified rate of flow and voltage. This is illustrated by Table II.

TABLE II

HOURS REQUIRED TO PRODUCE 0.1 MOLE OZONE FROM OXYGEN

Flowmeter, mm.	*Rate of Flow,* l./hr.	*Secondary Voltage* 7000	8000	9000	10,000	11,000	12,000
		hr.	hr.	hr.	hr.	hr.	hr.
2.0	5.7	5.97	5.55	5.55	5.48	5.48	5.48
5.0	9.8	3.60	3.55	3.42	3.32	3.33	3.33
10.0	14.4	2.90	2.57	2.43	2.32	2.29	2.26

All ozonizers have individual characteristics since it is difficult if not impossible to build two absolutely identical ozonizers. Variations in the composition of the glass and the annular space in the Berthelot tubes cause considerable changes in the amount of ozone produced, as shown by Table I. Hence the performances indicated by the data in Table I would be duplicated only under very fortuitous circumstances. The production of ozone from the same ozonizer may vary from time to time; hence it is desirable to check the percentage of ozone occasionally. If the production of ozone drops markedly, the Berthelot tubes should be carefully cleaned with hot nitric acid, thoroughly rinsed, and dried. Higher percentages of ozone may be obtained by having more Berthelot tubes. With six tubes, concentrations of 9–11% may be obtained.

H. *Ozonization of organic compounds.* The ozonization of each unsaturated organic compound is more or less an individual problem, but some general comments may be made. *Organic ozonides are highly explosive,* and hence it is safest to carry out the ozonization in a solvent which dissolves both the original compound and the ozonide. *In all cases, a shatter proof screen* of laminated safety glass should be placed between the operator and the tubes *H, I,* and *J*. A second screen should be placed back of the tubes to protect other pieces of the apparatus.

The general procedure consists in dissolving a weighed amount of the compound in a suitable solvent such as glacial acetic acid, methyl chloride, ethyl chloride, or carbon tetrachloride. The solution is placed in tube *H*, and such an amount is used that the same hydrostatic head is obtained as when the 2% potassium iodide solution was used in both tubes *H* and *J*. Usually both tubes *H* and *I* are surrounded by cooling baths. The ozonizer is started and the gases by-passed through the destroyer for about 5 minutes while the apparatus is attaining equilibrium. The ozonized oxygen is then passed through the solution of the compound for the calculated time. Since all organic compounds do not absorb ozone rapidly enough for a quantitative absorption it is frequently necessary to run the ozonization longer. The presence of unsaturation may often be detected by testing a small portion of the reaction mixture with a dilute solution of bromine in carbon tetrachloride. The ozonization is continued until the test with the bromine solution is negative.

It is also possible to place a solution of the compound in tube *H* to the proper height and use a 2% potassium iodide solution in tube *J*. At the end of the calculated time the contents of tube *J* are titrated in order to determine the amount of ozone that failed to react with the compound. From these data the additional time necessary to complete the ozonization is calculated and the apparatus operated accordingly (Notes 14 and 15).

The procedures for the decomposition of the ozonide and separation of the reaction products will vary according to the nature of the compounds and must be designed and selected accordingly.

Selective ozonization of compounds containing different types of unsaturated linkages is possible by choosing a proper concentration of ozone and by stopping the ozonization at the right time. It is for this reason that the calibration described above is carried out, since for certain ozonizations it is important to have the correct concentration for best results.

1. *Use of air in place of oxygen.* When air is the source of oxygen the exit gases from the Berthelot tubes contain lower percentages of ozone than when pure oxygen is used. The gases also contain small amounts of nitric anhydride,[2] the presence of which may cause some side reactions to occur. However, a considerable number of compounds may be satisfactorily ozonized with air as the source of oxygen. The above apparatus will operate satisfactorily with compressed air. It is essential to bubble the compressed air through three 5-l. flasks half filled with concentrated sulfuric acid before passing the air into the purification train *B*, *C*, *D*, *E*, *F*, *G*. Some data on the production of ozone from air at a secondary potential of 10,000 volts are given in Table III.

TABLE III

PRODUCTION OF OZONE FROM AIR AT SECONDARY VOLTAGE OF 10,000

	Rate of Flow	
	15.3 l./hr.	21.8 l./hr.
Ozone, % by volume	3.0	2.9
Hours for production of 0.1 mole ozone	4.8	3.4
N_2O_5, % by volume	0.13	0.08
Moles HNO_3 formed per 0.1 mole ozone	0.009	0.006

At higher secondary voltages the percentage of nitrogen pentoxide rises. A decrease in the rate of flow also increases the amount of nitrogen pentoxide. It is obvious from the data in Table III that the amounts of nitrogen pentoxide are very small and need to be considered only when examining a reaction mixture for small amounts of by-products or when the presence of this oxide of nitrogen would act as a catalyst for the oxidation of the organic compounds by oxygen or ozone.

2. Notes

1. If necessary, the rubber connecting tubes in the purification train may be protected by painting them with molten paraffin wax.

2. Berthelot tubes constructed of Pyrex may fail to produce ozone or may give a low yield.[3] However, Henne and Perilstein[4] have described a satisfactory ozonizer constructed of Pyrex glass.

With Pyrex tubes, voltages of 14,000–18,000 volts are needed. The power consumption is 15–25 watts per tube. The gap should be 1.5–2.00 mm.; this may be obtained through the use of 22-mm. o.d. and

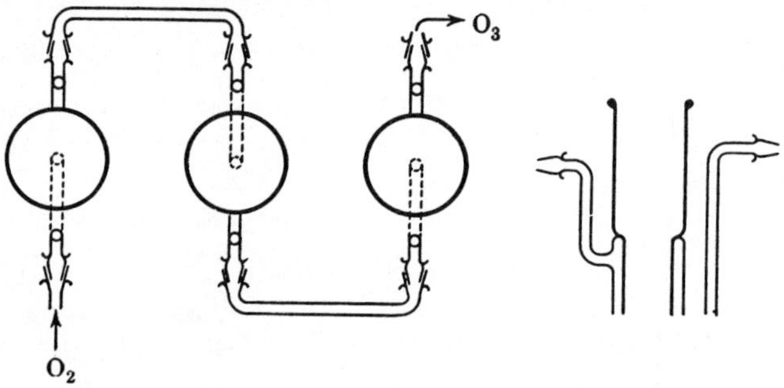

Fig. 25.

28-mm. o.d. Pyrex tubing. The transformer which is recommended (E. O. Brimm and G. A. Cook, private communication) for use with Pyrex equipment is catalog No. 969-001-395, 15,000 volts, 30 M.A., of the Jefferson Electric Company.

3. Some form of cooling is necessary. The equipment described here involves the use of a cooling coil in the inner tubes. It is also possible to cool the entire unit by means of a copper cooling coil inserted in the outer bath. When the latter method is used, the inner tubes are not cooled and the outer bath is grounded. This is reported to be a less hazardous method of operation (E. O. Brimm and G. A. Cook).

4. Mercury cup seals have been recommended in the past, but satisfactory results have been obtained with ground joints lubricated with a very small amount of 85% phosphoric acid. It is also possible to use Tygon connections with the tube ends butted together. If ground-glass connections are used the inlet and outlet tubes should be bent to facilitate union. Figure 25 shows top and side views of such Berthelot tubes. The holes in the wooden supports should be large enough to permit the tubes to fit loosely.

5. Any large commercial laboratory battery jar is suitable. The Exide types F-9 or F-11 are satisfactory.

6. The dimensions of these wooden supports will be determined by the battery jar. The position of the holes and slots should be arranged so that the tubes may be spaced at convenient distances as shown in Fig. 23.

7. The submitters used a transformer purchased from the Franklin Transformer Company, Minneapolis, Minnesota. It was provided with taps so that secondary voltages of 5500, 6600, 7700, 8800, 9900, and 11,000 could be obtained. These transformers do not always deliver the rated voltage and hence should be calibrated by actual measurement. The checkers used a luminous tube transformer obtained from the Jefferson Electric Company, Bellwood, Illinois, catalog No. 721-411. Cap. 825 V.A. Primary 115 V.A.C. 60 cycles. Secondary 15,000 volts, 60 M.A. The variable transformer used to regulate the voltage should be rated at 7.5 amperes.

8. It may make a difference which side of the transformer is grounded, depending on the construction of the transformer; the maker of the transformer should be queried on this point.

9. The reaction flask *H* can be made of different sizes depending on the amount of solution being ozonized. The dimensions of the parts in Fig. 24 are not critical; any tubes and stopcocks of convenient size are suitable.

10. Many reagents act as contact catalysts for the destruction of ozone; a study of some of them has been made.[5]

11. A wet gas meter such as No. S-39465, E. H. Sargent and Company, is satisfactory. This meter need not be a permanent part of the apparatus. It is used only to calibrate the flowmeter.

12. In order that the flowmeter may be used during an ozonization when tube *J* is empty it is necessary to mark the heights of liquid on both tubes. The same liquid head (height $\times$ density) must be present in the absorption tubes during calibration and ozonization. For very accurate work a pressure regulator should be incorporated in the oxygen train and a gas meter made part of the set-up. For ordinary preparative work, however, these are not necessary.

13. If a transformer with taps is used the voltage chosen is determined by each tap. With a luminous tube transformer and Variac, a voltmeter is connected across the primary of the transformer. Since these transformers have a fixed ratio of primary to secondary windings, the secondary voltage will be nearly proportional to the impressed primary voltage. Thus a transformer designed to deliver

15,000 volts with a primary voltage of 115 volts will deliver approximately 12,000 volts when the primary voltage is $\frac{4}{5}$ of 115 or 92 volts.

14. It is evident that for ordinary preparative work the careful calibration given in section G is not essential. It is only necessary to adjust the voltage of the transformer to about 10,000 to 11,000 volts and turn on the flow of oxygen to as rapid a rate as the absorption tubes will handle when surrounded by cooling baths. The amount of ozone produced in 5 minutes at the observed flowmeter reading is determined as in section F. By operating the ozonizer at this rate of flow and voltage the ozonization of organic compounds can be carried out.

15. The ozonizer may be operated at higher rates of flow of oxygen than shown in Tables I and II, provided that the organic compound reacts with ozone at a reasonably rapid rate. Some data on high rates of flow are given in Table IV.

TABLE IV

Production of Ozone from Oxygen at High Rates of Flow and Secondary Voltage of 10,000

Rate, l./hr.	*% Ozone* (by volume)	*Hours to Produce* 0.1 *Mole Ozone*
27	4.7	1.76
32	4.5	1.55
45	4.1	1.23

16. A laboratory ozonizer is manufactured by the Welsbach Corporation, Philadelphia, Pennsylvania.

3. Methods of Preparation

Ozone for laboratory use has always been prepared by the action of a silent electric discharge upon a stream of air or oxygen. Although dielectrics other than glass are used in commercial ozonizers, they do not give a percentage of ozone high enough for laboratory use, and practically all laboratory ozonizers employ the Berthelot tube and are modeled after the one originally constructed by Harries.[6] Good ozonizers of this type have been described by Briner, Patry, and de Luserna,[7] and by Church, Whitmore, and McGrew.[8] The ozonizer described above is a modification of the one described by Smith,[9] as improved by Henne,[10] and by Smith and Ullyot[3] and Greenwood.[11] Henne and Perilstein[4] described a modification of their ozonizer in which the inner electrode is a tube filled with mercury; the outer electrode is water-cooled. More recently a 4% concentration of ozone has been prepared using a compressed-air-cooled inner electrode.[12] Ozone has also been produced in good concentrations (over 12%) by electrolysis methods.[13]

[1] Jacobs, *Analytical Chemistry of Industrial Poisons, Hazards, and Solvents,* p. 292, Interscience Publishers, New York, 1941.
[2] Kaslow and Stayner, *Proc. Ind. Acad. Sci.,* **54,** 107 (1945).
[3] Smith and Ullyot, *J. Am. Chem. Soc.,* **55,** 4327 (1933).
[4] Henne and Perilstein, *J. Am. Chem. Soc.,* **65,** 2183 (1943).
[5] Smith, *J. Am. Chem. Soc.,* **47,** 1850 (1925).
[6] Harries, *Ann.,* **343,** 339, 343, 376 (1905).
[7] Briner, Patry, and de Luserna, *Helv. Chim. Acta,* **7,** 62 (1924).
[8] Church, Whitmore, and McGrew, *J. Am. Chem. Soc.,* **56,** 176 (1934).
[9] Smith, *J. Am. Chem. Soc.,* **47,** 1844 (1925).
[10] Henne, *J. Am. Chem. Soc.,* **51,** 2676 (1929).
[11] Greenwood, *Ind. Eng. Chem., Anal. Ed.,* **17,** 446 (1945).
[12] Vigroux, *Bull. soc. chim. France,* **1949,** 402.
[13] Boer, *Rec. trav. chim.,* **67,** 217 (1948).

PALLADIUM CATALYSTS *

$$Na_2PdCl_4 + CH_2O + 3NaOH \rightarrow Pd + HCOONa + 4NaCl + 2H_2O$$

$$PdCl_2 + H_2 \rightarrow Pd + 2HCl$$

Submitted by RALPH MOZINGO.
Checked by HOMER ADKINS and JAMES E. CARNAHAN.

1. Procedures (Note 1)

A. *Palladium on barium sulfate catalyst* (5% *Pd*). A solution (Note 2) of 8.2 g. of palladium chloride (0.046 mole) in 20 ml. (0.24 mole) of concentrated hydrochloric acid and 50 ml. of water is prepared. To a rapidly stirred, hot (80°) solution of 126.2 g. (0.4 mole) of reagent barium hydroxide octahydrate in 1.2 l. of distilled water contained in a 4-l. beaker (Notes 3 and 4) is added all at once 120 ml. (0.36 mole) of 6 *N* sulfuric acid. More 6 *N* sulfuric acid is added to make the suspension just acid to litmus (Note 5). To this hot barium sulfate suspension (Note 6) are added the palladium solution and 8 ml. (0.1 mole) of 37% formaldehyde solution. The suspension is then made slightly alkaline to litmus with 30% sodium hydroxide solution, constant stirring being maintained. The suspension is stirred 5 minutes longer, and then the catalyst is allowed to settle (Note 7). The clear supernatant liquid is decanted and replaced by water, and the catalyst is resuspended. The catalyst is washed by decantation eight to ten times. After the final decantation, the catalyst is collected on a 90-mm. medium-porosity sintered-glass funnel (Note 8). Most of the water is removed from the cake, but not enough to cause the cake to break or channel. The filter cake is washed with 250 ml. of water in five portions, the last being removed as completely as possible by filtration. The funnel and its contents are then placed in an oven at 80° until the catalyst is dry. The catalyst (93–98 g.) is powdered and stored in a tightly closed bottle (Note 9).

B. *Palladium on carbon catalyst* (5% *Pd*). A suspension of 93 g. of nitric acid-washed Darco G-60 (Note 10) in 1.2 l. of water contained in a 4-l. beaker (Notes 3 and 4) is heated to 80°. To this is added a solution of 8.2 g. (0.046 mole) of palladium chloride in 20 ml. (0.24 mole) of concentrated hydrochloric acid and 50 ml. of water (Note 2). Eight milliliters (0.1 mole) of 37% formaldehyde solution is added. The suspension is made slightly alkaline to litmus with 30% sodium hydroxide solution, constant stirring being maintained. The suspension is stirred 5 minutes longer. The catalyst is collected on a filter and washed ten times with 250-ml. portions of water. After removal of as much water as possible by filtration, the filter cake is dried (Note 11), first in air at room temperature, and then over potassium hydroxide in a desiccator. The dry catalyst (93–98 g.) is stored in a tightly closed bottle.

C. *Palladium chloride on carbon* (5% *Pd*). A solution of 8.2 g. (0.046 mole) of palladium chloride in 20 ml. (0.24 mole) of concentrated hydrochloric acid and 50 ml. of water is prepared (Note 2). The solution is diluted with 140 ml. of water and poured over 92 g. of nitric acid-washed Darco G-60 (Note 10) in an 8-in. evaporating dish (Note 3). After the palladium chloride solution has been thoroughly mixed with the carbon, the whole mixture is dried, first on a steam bath and then in an oven at 100°, with occasional mixing until completely dry. The mass (98–100 g.) is powdered and stored in a closed bottle.

The required quantity of palladium chloride on carbon is transferred to a hydrogenation bottle and reduced with hydrogen in the solvent to be used for the hydrogenation (Notes 12 and 13). When no more hydrogen is absorbed by the catalyst, it is collected (Note 14) on a sintered-glass funnel and washed with more of the solvent to remove the hydrogen chloride, and then returned to the reduction bottle, the last being washed in with the solvent. The material to be hydrogenated is then added and the hydrogenation is completed in the usual way.

D. *Palladium on carbon catalyst* (10% *Pd*). A solution of 8.33 g. of palladium chloride in 5.5 ml. of concentrated hydrochloric acid and 40 ml. of water is prepared by heating the mixture on a steam bath (Notes 2 and 15). The resulting solution is poured into a solution of 135 g. of sodium acetate trihydrate in 500 ml. of water contained in a 1-l. reduction bottle (Note 16). Forty-five grams of Norit (Note 10) is added, and the mixture is hydrogenated until absorption ceases after 1–2 hours. The catalyst is collected on a Büchner funnel and washed with 2 l. of water in five portions. The filter cake, after removal of most of the water, is dried in air and then in a desiccator over calcium chloride (Note 11). The catalyst (48–50 g.) is stored, after being powdered, in a tightly closed bottle.

2. Notes

1. The four procedures given for the preparation of palladium catalysts differ in that in A the support is barium sulfate or barium carbonate whereas in the others the support is carbon. In procedures A and B, alkaline formaldehyde is the reducing agent; in C and D, hydrogen is used. The catalysts A, B, and D are prepared and stored until required with the palladium in the reduced form ready for use. In C,[1] the palladium salt is reduced to the metal as needed, so that there is no loss of activity during storage. Catalyst A is similar to that usually recommended for Rosenmund reductions; D is essentially that developed by Hartung[2] and extensively used by Cope[3] and others. Catalyst D carries twice as much palladium per unit weight as the others.

Catalysts reduced with formaldehyde carry no adsorbed hydrogen and are less pyrophoric. Barium carbonate as a support may sometimes be advantageous in that the neutrality of the hydrogenation mixture may be maintained. Barium sulfate or barium carbonate may be a better support than carbon, which may, in some instances, so strongly adsorb the derived product that recovery is difficult or incomplete. Palladium may be more completely and easily recovered from a spent catalyst where carbon rather than barium sulfate is the support. In general, the submitter prefers a catalyst prepared according to procedure C.

2. Since palladium chloride dissolves rather slowly in aqueous acid, the mixture is heated on a steam bath for about 2 hours, or until solution is complete. If the dihydrate of palladium chloride is used the quantity should be increased to 9.9–10.0 g.

3. The entire preparation is carried out with all-glass or porcelain equipment in order to prevent contamination with iron or other metals.

4. The catalyst may be prepared in ten times the amount given here, with a 20-l. battery jar in place of the beaker.

5. The rapid addition of sulfuric acid is made to give finely divided barium sulfate.

6. An equal weight of precipitated barium carbonate (93 g.) may be substituted for the barium hydroxide and sulfuric acid to give a palladium on barium carbonate catalyst. The amount of hydrochloric acid should then be reduced to 8.2 ml.

7. After 5 minutes, the solution is colorless and free of palladium chloride.

8. A Büchner funnel may be used, but filtration through paper is very slow. The washing process may be carried out by centrifugation instead of filtration.

9. The palladium may be conveniently separated from the barium sulfate by solution in aqua regia. The used catalyst is collected from the reaction mixture on a sintered-glass funnel. The organic material is removed with suitable solvents, and the solvents are replaced by water. The palladium is dissolved in aqua regia and is washed out with dilute hydrochloric acid, the solutions being collected for recovery of the metal. For recovery of the palladium from carbon, the mass is ignited and the ash is extracted with aqua regia for several hours. The palladium solution is filtered, and any residue is reignited and then treated with alkaline formaldehyde solution to reduce any oxides of palladium which may have been formed, and which are only slowly soluble in aqua regia. The solids are collected on a filter, and the palladium is extracted with aqua regia.

10. Norit, Darco, or other carbons may be used. The carbon is heated on a steam bath with 10% nitric acid for 2–3 hours, washed free of acid with water, and dried at 100–110° before use.

11. The palladium on carbon catalysts should be dried at room temperature or the carbon may ignite. These catalysts are first dried in air and then over potassium hydroxide (or calcium chloride) in a desiccator.

12. The solvent is conveniently that in which the hydrogenation is to be done. During the reduction of the palladium chloride, a neutral solvent is to be preferred; any acid or alkali needed for the hydrogenation is added after reduction of the catalyst.

13. The presence of hydrogen chloride during the hydrogenation of many organic compounds is desirable or without effect, so that the washing operations may be omitted in such cases. Thus, the palladium

chloride on carbon may be used in the same manner as the prereduced catalysts, i.e., simply added before reduction to the solvent and the hydrogen acceptor.

14. The catalyst should be kept wet with the solvent during the washing process, as it is pyrophoric.

15. The resulting solution is approximately equivalent to 50 ml. of the commercial palladium chloride solution (p. 385) suggested by Hartung and Cope.[3]

16. The checkers reduced the palladium chloride, in three batches, in a 500-ml. bottle. The bottle was not shaken, but the contents were rapidly stirred under a pressure of 1.1 atmospheres of hydrogen. The reduction of each batch required about 5 hours.

3. Methods of Preparation

Palladium catalysts have been prepared by fusion of palladium chloride in sodium nitrate to give palladium oxide;[4,5] by reduction of palladium salts by alkaline formaldehyde[6-8] or sodium formate,[9] by hydrazine,[10] and by the reduction of palladium salts with hydrogen.[11] The metal has been prepared in the form of palladium black,[6,9] and in colloidal form in water containing a protective material,[10] as well as upon supports. The supports commonly used are asbestos[12] barium carbonate,[13] barium sulfate,[1,7,8,14] calcium carbonate,[15] carbon,[1,11,14,16] kieselguhr,[14,16] silica-gel,[17] and strontium carbonate.[18] The catalysts described here are prepared by modifications of the methods of Schmidt,[8] Rosenmund and Langer,[14] Mannich and Thiele,[11] and Hartung.[2,3]

Polyvinyl alcohol[19] and aluminum oxide[20] have been used as supports for palladium catalysts.

[1] Mozingo, Harris, Wolf, Easton, Hoffhine, and Folkers, *J. Am. Chem. Soc.*, **67**, 2092 (1945).

[2] Hartung, *J. Am. Chem. Soc.*, **50**, 3373 (1928).

[3] Alexander and Cope, *J. Am. Chem. Soc.*, **66**, 886 (1944).

[4] Shriner and Adams, *J. Am. Chem. Soc.*, **46**, 1683 (1924).

[5] *Org. Syntheses* Coll. Vol. **1**, 470 (1941).

[6] Willstätter and Waldschmidt-Leitz, *Ber.*, **54**, 123 (1921).

[7] Houben, *Die Methoden der organischen Chemie,* 3rd ed., Vol. II, p. 500, Verlag Georg Thieme, Leipzig, 1930.

[8] Schmidt, *Ber.*, **52**, 409 (1919).

[9] Zelinsky and Glinka, *Ber.*, **44**, 2309 (1911).

[10] Paal and Amberger, *Ber.*, **37**, 124 (1904).

[11] Mannich and Thiele, *Ber. deut. pharm. Ges.*, **26**, 36 (1916).

[12] Zelinsky and Borisoff, *Ber.*, **57**, 150 (1924); Zelinsky and Turowa-Pollak, *Ber.*, **58**, 1295 (1925).

[13] Harris, Stiller, and Folkers, *J. Am. Chem. Soc.*, **61**, 1242 (1939).
[14] Rosenmund and Langer, *Ber.*, **56**, 2262 (1923).
[15] Busch and Stöve, *Ber.*, **49**, 1063 (1916).
[16] Sabalitschka and Moses, *Ber.*, **60**, 786 (1927).
[17] Fester and Brude, *Ber.*, **56**, 2247 (1923).
[18] Martin and Robinson, *J. Chem. Soc.*, **1943**, 491.
[19] Kavanagh and Nord, *J. Am. Chem. Soc.*, **66**, 2126 (1944).
[20] U. S. pat. 2,366,409 [*C. A.*, **39**, 2001 (1945)].

PENTAACETYL *d*-GLUCONONITRILE

(Glucononitrile, D-, pentaacetate)

$$H(CHOH)_5CHO \rightarrow H(CHOH)_5CH{=}NOH \rightarrow H(CHOCOCH_3)_5CN$$

Submitted by H. T. Clarke and S. M. Nagy.
Checked by W. E. Bachmann and Wayne Cole.

1. Procedure

To 350 ml. of anhydrous methanol contained in a 1-l. three-necked round-bottomed flask, to which is attached a reflux condenser protected by a drying tube, is added 20 g. (0.87 gram atom) of sodium (Note 1) in large pieces. The reaction is kept under control by cooling the flask in a pan of ice water. To the resulting solution of sodium methoxide is added a solution (Note 2) of 61 g. of hydroxylamine hydrochloride (0.88 mole) in 20 ml. of water; during the addition the mixture is swirled in order to avoid spattering. After 20 minutes the mixture is cooled to 0° and filtered with suction. The sodium chloride is washed with 350 ml. of anhydrous methanol. The combined filtrate and washings are warmed to 65° in a 3-l. round-bottomed flask, and a solution of 100 g. of finely powdered commercial crystalline glucose monohydrate (0.50 mole) in 200 ml. of warm 25% aqueous methanol is added, with stirring. The resulting solution is held at 65° for 2 hours and then concentrated under reduced pressure until no further distillate is obtained; the residue weighs 155–160 g. The resulting syrup (Note 3) is diluted with 300 ml. of methanol and again distilled, and this process is repeated once (Note 4).

A mixture of 100 g. of powdered, anhydrous sodium acetate and 677 ml. of 90% acetic anhydride (Note 5) is heated on a steam bath in a 3-l. round-bottomed flask under an efficient reflux condenser. Without interrupting the heating a solution of the syrupy glucose oxime in 50 ml. of glacial acetic acid and 100 ml. of cold acetic anhydride is added through a dropping funnel to the hot mixture (Note 6); this requires about 1 hour (Note 7). Heating is continued for another hour, and the bulk (380–420 ml.) of the acetic acid and any unchanged acetic anhydride is distilled under reduced pressure from a water bath. The residue is immediately stirred into 2 l. of cold water, stirred occasionally during the first 3 hours, and allowed to stand overnight.

After the mixture has been chilled to 0°, the brown, crystalline mass is filtered with suction and washed with 500 ml. of water. The solid is dissolved in 300 ml. of hot 95% ethanol, and the solution is heated with 10–15 g. of Norit for 5 minutes and filtered with suction. The filtrate is gradually cooled to 0°; the crystals are filtered with suction and washed with 20 ml. of cold ethanol. The weight of the first crop is 90–93 g. A second crop is obtained by concentrating the mother liquor under reduced pressure to 25 ml., boiling the solution with Norit, filtering, and chilling the filtrate to 0°. The total yield of colorless pentaacetyl *d*-gluconónitrile melting at 82.5–83.5° is 95–96 g. (50%).

2. Notes

1. In preparing free hydroxylamine, a little less than the theoretical amount of sodium is employed to avoid the presence of free alkali in the reaction mixture.

2. The hydroxylamine hydrochloride dissolves in the small amount of water when the mixture is warmed to about 125°.

3. The isolation of glucose oxime is unnecessary in this preparation.

4. The distillation with methanol serves to remove water almost completely.

5. If acetic anhydride of a higher concentration is available, correspondingly smaller quantities may be employed.

6. The viscous, syrupy glucose oxime dissolves with difficulty, and it may be necessary to warm the mixture slightly. If this is done, a pan of ice water should be at hand in order to cool the mixture should the temperature begin to rise rapidly.

7. In the process described in the literature, the oxime, sodium acetate, and acetic anhydride are allowed to react without dilution, a condition which frequently leads to an uncontrollably violent reaction.

3. Methods of Preparation

The above method for preparing glucose oxime is the modification of that of Jacobi [1] developed by Wohl,[2] who first converted the oxime into the pentaacetyl gluconónitrile by means of acetic anhydride. The latter reaction was later employed for the same purpose by Zemplén and Kiss.[3]

Pentaacetyl gluconitrile has also been prepared by dehydration of pentaacetyl gluconamide with phosphorus oxychloride.[4]

[1] Jacobi, *Ber.*, **24**, 696 (1891).

[2] Wohl, *Ber.*, **26**, 730 (1893).

[3] Zemplén and Kiss, *Ber.*, **60**, 165 (1927).

[4] Ladenburg, Tischler, Wellman, and Babson, *J. Am. Chem. Soc.*, **66**, 1217 (1944).

PENTAMETHYLENE BROMIDE *

(Pentane, 1,5-dibromo-)

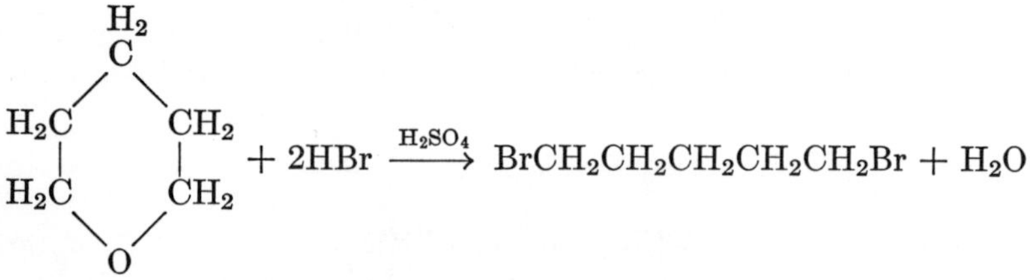

Submitted by D. W. Andrus.
Checked by Nathan L. Drake and Charles M. Eaker.

1. Procedure

A hydrobromic acid solution [*Org. Syntheses* Coll. Vol. **1,** 26 (1941)] is prepared in a 500-ml. round-bottomed flask by passing sulfur dioxide into a mixture of 120 g. (37.7 ml., 0.75 mole) of bromine, 50 ml. of water, and 150 g. of crushed ice. This is equivalent to a mixture of 253 g. (1.5 moles) of 48% hydrobromic acid and 74 g. of concentrated sulfuric acid. To the mixture 21.5 g. (0.25 mole) of tetrahydropyran (p. 794) is added, a reflux condenser is attached to the flask, and the light-brown homogeneous mixture is refluxed for 3 hours (Note 1).

The heavy lower layer is separated (Note 2), washed once with a saturated solution of sodium bicarbonate and once with water, and then dried over 4–5 g. of anhydrous calcium chloride. The crude product is decanted from the calcium chloride, and the drying agent is rinsed once or twice with a small quantity of ethyl bromide which is added to the main product. The mixture is distilled under reduced pressure, and the pentamethylene bromide, which weighs 46–47 g. (80–82%), is collected at 104–106°/19 mm.

2. Notes

1. The submitter refluxed the mixture for 10 hours, but the checkers obtained equally good yields in 3 hours.

2. The upper aqueous layer contains considerable unchanged hydrobromic acid. If this layer is distilled, about 150 g. (0.9 mole) of constant-boiling hydrobromic acid (b.p. 123–124°/748 mm.) may be recovered.

3. Methods of Preparation

The methods of preparing pentamethylene bromide are given in *Org. Syntheses* Coll. Vol. **1,** 428 (1941), where the preparation of the dihalide from benzoylpiperidine and phosphorus pentabromide is described in detail. The procedure given above is based upon the work of Paul.[1]

[1] Paul, *Bull. soc. chim. France,* (4) **53,** 1489 (1933).

1,5-PENTANEDIOL *

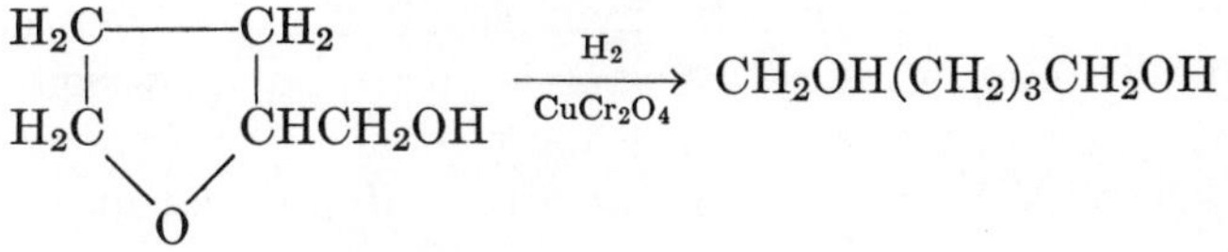

Submitted by Daniel Kaufman and Wilkins Reeve.
Checked by Homer Adkins and Harry Billica.

1. Procedure

Five hundred and ten grams (5 moles) of pure tetrahydrofurfuryl alcohol (Note 1) and 50 g. of copper chromite (Note 2) are placed in a hydrogenation bomb having a void of 1.4 l. (Note 3). The bomb, in a suitable rocker assembly, is filled with hydrogen to a pressure of 3300 to 3500 lb. per sq. in. (Note 4). The bomb is rocked and heated. *The pressure must not at any time be allowed to go much above* 6000 *lb. per sq. in.* If the pressure rises higher than 6200 lb., the heating of the bomb should be stopped (Note 5). In a typical run, the pressure reaches a maximum of about 6100 lb. at 255° after 2 hours. The pressure then slowly falls to about 4400 lb. during another hour, as the temperature rises to a little over 300°. During the ensuing 6 hours, the pressure falls to about 3000 lb. while the temperature is held at 300–310°.

The bomb is allowed to cool to room temperature, where the pressure should be 1000–1100 lb. The contents of the bomb are poured into a beaker, and the bomb is rinsed twice with 100-ml. portions of acetone. The catalyst is removed by centrifuging, and the reaction mixture is distilled through a fractionating column (Note 6). A fraction boiling in the range 60–140°, containing α-methyltetrahydrofuran, water, and *n*-amyl alcohol, is distilled at atmospheric pressure (Note 7). Tetrahydrofurfuryl alcohol (80–110 g.) is recovered, boiling at 65–70°/10 mm.; the 1,5-pentanediol boiling at 118–120°/6 mm. is obtained in amounts varying from 200 g. to 244 g. The yield is 40–47% of the theoretical, without allowance being made for the recovered tetrahydrofurfuryl alcohol. The residue of products boiling above the glycol amounts to 25–35 g.

2. Notes

1. The tetrahydrofurfuryl alcohol available from the Quaker Oats Company, or the practical grade from the Eastman Kodak Company, has been used. If the material available does not hydrogenate satisfactorily, it may be purified by hydrogenation over Raney nickel at 150°/100–200 atm. pressure. A sample of good quality boils at 177–178°/740 mm. and does not become dark-colored when a few milliliters are shaken with 1 drop of concentrated sulfuric acid at room temperature.

2. The catalyst is prepared as described in Note 11 for the preparation of copper chromite.[1]

3. The hydrogenation bomb supplied under catalog No. 406–21, by the American Instrument Company, Silver Spring, Maryland, is satisfactory. The internal volume of the empty bomb should be at least 2.7 times that of the volume of tetrahydrofurfuryl alcohol added, as otherwise not enough hydrogen can be added for the completion of the reaction, without the use of excessively high pressures. The submitters maintained the pressure above 5500 lb. per sq. in. by intermittently adding hydrogen.

4. Temperatures higher than about 300° and pressures higher than 6000 lb. per sq. in. should not be used in the vessels and with the gauges ordinarily supplied for "high-pressure" hydrogenations. *Only clean equipment, in first-class condition, and under careful control, can be used safely and successfully in carrying out reactions under the conditions described.*

5. If hydrogenolysis does not occur, the pressure would be about 7000 lb. when a temperature of 300° is reached. If the pressure rises above 6200, starting with 3500 lb. at room temperature, it is evident that the quality of the catalyst or the alcohol is not satisfactory. Further attempts to prepare the glycol should be made with alcohol and catalyst of better quality, or starting with a pressure sufficiently low (3000 lb.) so that a safe operating pressure will not be exceeded.

6. A Vigreux-type column 60 cm. in length and 2 cm. in outside diameter was used for the fractionation. The submitters state that they have also used a column of similar dimensions, packed with glass helices, for a fractionation at atmospheric pressure. They recommended the Vigreux column under reduced pressure, as used by the checkers.

7. When acetone was not used for washing out the bomb, a fraction weighing 70–75 g. was obtained by the submitters. After drying over anhydrous potassium carbonate, they obtained, by fractional distillation, 3–6 g. of α-methyltetrahydrofuran, b.p. 80–81°, and 23–28 g. of *n*-amyl alcohol, b.p. 137–138°. The alcohol obtained was pure, and neither secondary amyl nor butyl alcohols could be detected.

3. Methods of Preparation

1,5-Pentanediol has been prepared by the reduction of methyl [2,3] or ethyl [4] glutarate, of 2,3-dihydropyran,[5] of δ-hydroxyvaleraldehyde,[6–10] and of furfural.[11] It has also been prepared from pentamethylene bromide by conversion to the diacetate and subsequent hydrolysis.[12] The procedure described is a modification of that of Connor and Adkins.[13]

[1] *Org. Syntheses* Coll. Vol. **2**, 144 (1943).
[2] Wojcik and Adkins, *J. Am. Chem. Soc.*, **55**, 4939 (1933).
[3] Bennett and Reynolds, *J. Chem. Soc.*, **1935**, 138.
[4] Adkins and Billica, *J. Am. Chem. Soc.*, **70**, 3121 (1948).
[5] U. S. pat. 2,440,929 [*C. A.*, **42**, 5466 (1948)].
[6] Paul, *Bull. soc. chim. France*, (5) **1**, 978 (1934).
[7] Schniepp and Geller, *J. Am. Chem. Soc.*, **68**, 1646 (1946).
[8] Woods and Saunders, *J. Am. Chem. Soc.*, **68**, 2111 (1946).
[9] Ital. pat. 439,947 [*C. A.*, **44**, 5915 (1950)].
[10] U. S. pat. 2,497,812 [*C. A.*, **44**, 6428 (1950)].
[11] Brit. pat. 627,293 [*C. A.*, **44**, 2564 (1950)].
[12] Bennett and Heathcoat, *J. Chem. Soc.*, **1929**, 273.
[13] Connor and Adkins, *J. Am. Chem. Soc.*, **54**, 4678 (1932).

3-PENTEN-2-OL

$$CH_3Cl + Mg \xrightarrow{\text{Ether}} CH_3MgCl$$

$$CH_3CH{=}CHCHO + CH_3MgCl \rightarrow$$
$$CH_3CH{=}CHCH(OMgCl)CH_3 \xrightarrow[NH_4Cl]{H_2O}$$
$$CH_3CH{=}CHCHOHCH_3 + MgCl_2 + NH_4OH$$

Submitted by E. R. Coburn.
Checked by Homer Adkins, B. W. Winner, John Woods, R. F. McElwee and Robert M. Ross.

1. Procedure

A 5-l. round-bottomed three-necked flask is equipped with a mechanical stirrer in a suitable seal (Note 1), a reflux condenser of the cold-finger type (Note 2) protected from moisture in the air by a drying tube, and a gas delivery tube extending nearly to the bottom of the flask. The flask is surrounded by an ice bath, and the cold finger is filled with solid carbon dioxide in acetone. Approximately 1.7 l. of dry ether and 61 g. (2.5 gram atoms) of magnesium are placed in the flask and cooled to about 0°. Methyl chloride (130 ml., 130 g., 2.6 moles) is condensed (Note 3) in a 250- to 300-ml. test tube held in a bath of solid carbon dioxide in acetone (Note 4). About 50 ml. of methyl chloride is allowed to distil from the test tube containing the methyl chloride through the gas delivery tube into the rapidly stirred mixture of ether and magnesium. The reaction mixture is then warmed until the reaction of methyl chloride and magnesium is under way. A crystal of iodine may be added if the reaction does not start readily. The methyl chloride is then allowed to distil into the reaction mixture during a period of about 3 hours. The reaction mixture and the tube containing the methyl chloride are cooled if the refluxing of the reaction mixture becomes so vigorous that the reflux condenser does not condense the methyl chloride. After all the methyl chloride has been added, the reaction mixture is warmed for 1 hour so that there is a gentle reflux.

At the end of this time, when almost all the magnesium has reacted, the Dry Ice cold finger is replaced with a water condenser and the gas delivery tube with a dropping funnel. A solution of 142 g. (2.02

moles) of freshly distilled crotonaldehyde in 300 ml. of dry ether is added dropwise while the reaction mixture is stirred vigorously and cooled. The reaction mixture is allowed to stand at room temperature for 1 hour.

The Grignard addition compound is decomposed by adding 435 ml. of a saturated ammonium chloride solution (Note 5) dropwise, with vigorous stirring, to the thoroughly cooled reaction mixture. A dense white precipitate, too heavy to be stirred mechanically, forms and settles to the bottom of the flask. After the reaction mixture has been allowed to stand for 1 hour, the ether solution is poured off and the precipitate washed by decantation with two 300-ml. portions of ether.

The ether is removed by distillation, and the residual 3-penten-2-ol is distilled through a short column (Note 6) at atmospheric pressure. The yield of material boiling at 119–121° is 140–150 g. (81–86%). Pure 3-penten-2-ol boils at approximately 120°/740 mm.

2. Notes

1. A Hershberg type of stirrer is preferred. The submitter used a stirrer of tantalum with a mercury seal; the checkers used a Nichrome stirrer in a simple rubber seal.

2. The reflux condenser must be of very high capacity, as otherwise methyl chloride may be lost. The checkers used a cold-finger type of condenser in which the dimensions of the finger or container for the refrigerant were 30 cm. in length and 3 cm. in outside diameter. The glass jacket surrounding the finger was 4.5 cm. in outside diameter, but was drawn down to 1.3 cm. below the finger for convenience of insertion into one of the necks of the reaction flask.

3. Methyl chloride is led from a commercial cylinder to the bottom of the test tube used for measuring and storing the reagent. The tube is previously marked so that the desired volume (130 ml.) of methyl chloride may be readily measured.

4. The checkers also obtained equally good results with less effort by allowing a slow stream of dry methyl chloride to pass directly from the commercial cylinder into the reaction mixture until practically all the magnesium had reacted.

5. Hydrolysis of the Grignard complex with saturated ammonium chloride solution possesses the advantage that the resulting ethereal solution of the alcohol is neutral and sufficiently dry so that it need not be dried before distillation. The alcohol is dehydrated if it is distilled from a mixture containing even a trace of a mineral acid.

Approximately 125 g. of ammonium chloride and 345 ml. of water are required for the saturated solution referred to above.

6. The submitter used a Hempel column. The checkers used a modified Widmer or Vigreux column, 1.3 cm. in diameter and 15 cm. in length.

3. Methods of Preparation

3-Penten-2-ol has been prepared by the addition of methylmagnesium iodide [1,2] or bromide [3–6] to crotonaldehyde and by the partial dehydration of pentanediol,[7] and by the hydrolysis of 2-chloropentene-3.[8]

1 Courtot, *Bull. soc. chim. France,* (3) **35,** 983 (1906).
2 Kyriakides, *J. Am. Chem. Soc.,* **36,** 663 (1914).
3 Reif, *Ber.,* **39,** 1603 (1906); **41,** 2739 (1908).
4 Mulliken, Wakeman, and Gerry, *J. Am. Chem. Soc.,* **57,** 1605 (1935).
5 Auwers and Westermann, *Ber.,* **54,** 2996 (1921).
6 Hurd and Cohen, *J. Am. Chem. Soc.,* **53,** 1917 (1931).
7 Kyriakides, *J. Am. Chem. Soc.,* **36,** 996 (1914).
8 U. S. pat. 2,464,244 [*C. A.,* **43,** 4288 (1949)].

4-PENTEN-1-OL

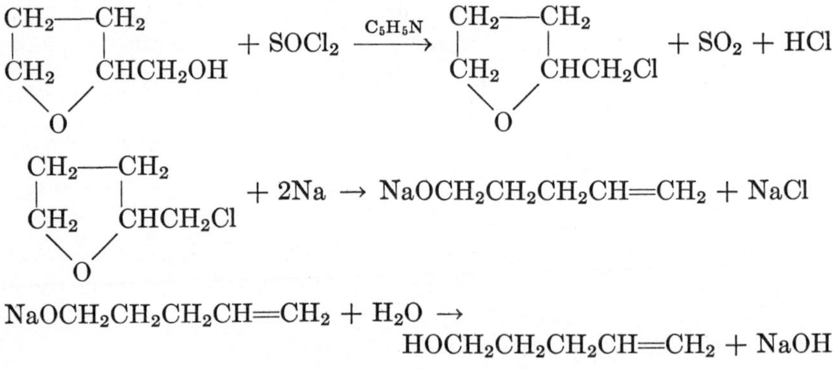

Submitted by L. A. Brooks and H. R. Snyder.
Checked by Nathan L. Drake and W. Mayo Smith.

1. Procedure

A. *Tetrahydrofurfuryl chloride.* In a 2-l. three-necked flask, fitted with a mechanical stirrer, a dropping funnel, and a thermometer, are placed 408 g. (4 moles) of freshly distilled tetrahydrofurfuryl alcohol

(Note 1) and 348 g. (4.4 moles) of pyridine. To the rapidly stirred mixture, which is cooled in an ice bath, 500 g. (4.2 moles) of freshly distilled thionyl chloride (Note 1) is added from the dropping funnel at the rate of 3–5 drops per second. When one-third to one-half of the thionyl chloride has been added, a pasty crystalline mass begins to separate and the temperature begins to rise rapidly. The temperature should not be allowed to go above 60°. As more thionyl chloride is added the mass redissolves and a dark brown liquid forms. When the addition is complete, the bath is removed and the mixture is stirred for 3–4 hours. The liquid (or the slurry, if some crystallization has occurred) is poured into a beaker (Note 2) and extracted seven times with 500-ml. portions of ether (Note 3); the ether extracts are decanted and combined. The ether is removed by distillation, and the residue is washed three times with 100-ml. portions of water, dried over anhydrous magnesium sulfate, and distilled under reduced pressure. The yield of tetrahydrofurfuryl chloride boiling at 41–42°/11 mm. (47–48°/15 mm.) is 354–360 g. (73–75%).

B. 4-*Penten*-1-*ol*. A 2-l. three-necked flask containing 112 g. (4.87 moles) of powdered sodium (Note 4) under 700 ml. of anhydrous ether is fitted with a mechanical stirrer, a separatory funnel, and a reflux condenser with a drying tube. A few milliliters (2–3) of a mixture of 300 g. (2.5 moles) of tetrahydrofurfuryl chloride and 300 ml. of anhydrous ether is added to the rapidly stirred suspension. A vigorous reaction occurs, and the solution turns blue. The remainder of the solution of the chloride is then added dropwise over a period of 5 hours, during which time the flask is cooled in an ice bath (Note 5). When the addition is complete, stirring is continued for 2 hours. The suspension is decanted from any sodium that remains (Note 6) into a dry beaker and decomposed with sufficient ice water to give two liquid layers. The ether layer is separated and dried over magnesium sulfate. After the removal of the ether by distillation on a steam cone, the residue is distilled. The yield of 4-penten-1-ol boiling at 134–137° is 161–178 g. (76–83%).

2. Notes

1. Undistilled commercial thionyl chloride and Eastman practical tetrahydrofurfuryl alcohol may be used, but the yields are slightly lower (65–70%).

2. The checkers found it easier to separate the ethereal extract from the residue when the mixture was in a large separatory funnel.

3. The yield will be low if extraction is incomplete. It is advisable to stir with a heavy glass rod and break up any lumps that have formed.

4. The powdered sodium is prepared under hot xylene with the aid of a Hershberg stirrer; the xylene is decanted and replaced with ether.

5. The ice bath is not used until the reaction has definitely started.

6. Occasionally a little sodium is left on the bottom of the flask. This is destroyed with ethanol and the flask is washed with ice water.

3. Methods of Preparation

Tetrahydrofurfuryl chloride has been prepared from the alcohol and thionyl chloride [1] or phosphorus trichloride. 4-Penten-1-ol has been prepared from tetrahydrofurfuryl bromide or chloride and magnesium,[2,3] sodium,[4–6] sodium-potassium,[6] or lithium; [6] and by the reaction of allylmagnesium chloride with ethylene oxide, followed by hydrolysis.[7]

[1] Kirner, *J. Am. Chem. Soc.,* **52,** 3251 (1930).
[2] Paul, *Bull. soc. chim. France,* (4) **53,** 424 (1933).
[3] Robinson and Smith, *J. Chem. Soc.,* **1936,** 195.
[4] Paul, *Bull. soc. chim. France,* (5) **2,** 745 (1935).
[5] Gaubert, Linstead, and Rydon, *J. Chem. Soc.,* **1937,** 1971.
[6] Paul and Normant, *Bull. soc. chim. France,* (5) **10,** 484 (1943).
[7] Kharasch and Fuchs, *J. Org. Chem.,* **9,** 359 (1944).

PHENANTHRENE-9-ALDEHYDE

(9-Phenanthrenecarboxaldehyde)

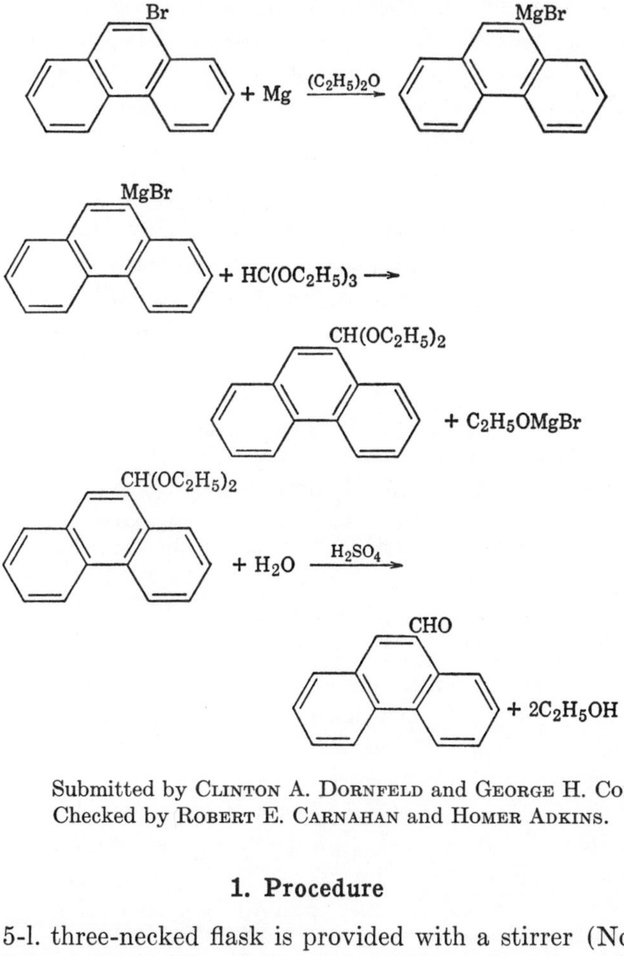

Submitted by CLINTON A. DORNFELD and GEORGE H. COLEMAN.[1]
Checked by ROBERT E. CARNAHAN and HOMER ADKINS.

1. Procedure

A dry 5-l. three-necked flask is provided with a stirrer (Note 1), a nitrogen inlet tube, a 500-ml. Pyrex separatory funnel, and a large Allihn reflux condenser. To the upper end of the condenser are attached an outlet tube and a 1-l. separatory funnel. Both separatory funnels and the outlet tube are provided with calcium chloride drying tubes. To the flask is added 50.3 g. (2.07 gram atoms) of magnesium turnings (Note 2). Nitrogen gas, dried by bubbling through concentrated sulfuric acid, is passed in to displace the air in the flask. The nitrogen atmosphere is maintained until the hydrolysis of the Grignard addition product is completed. Five hundred and fourteen grams (2 moles) of crude 9-bromophenanthrene (p. 134) (Note 3) is melted and poured into the Pyrex separatory funnel (Note 4). One liter of anhydrous ether (dried over sodium wire) is placed in the upper separatory funnel. About 200 ml. of the ether and 10 ml. of the melted bromophenanthrene are allowed to run into the reaction flask. The reaction of the bromophenanthrene with magnesium is initiated by the addition of a few crystals of iodine and 1 ml. of ethyl bromide; the reaction begins after the mixture is stirred for a few minutes without external heating. As the reaction proceeds, the ether and the bromo compound are added at rates sufficient to maintain gentle refluxing. The relative rates of addition should be such that the two separatory funnels will be emptied at about the same time. After the additions are complete, but while the reaction is still in progress, the Grignard reagent begins to precipitate on the sides of the flask. One liter of dry, thiophene-free benzene is added from the Pyrex separatory funnel at such a rate as to keep the Grignard reagent in solution. When refluxing due to the exothermic reaction stops, the mixture is heated at gentle reflux with stirring for 4 hours.

The mixture is allowed to cool until refluxing ceases, and 296.4 g. (2 moles) of ethyl orthoformate (Note 5) is added from the lower separatory funnel over a period of about 30 minutes. The mixture is then refluxed gently for 6 hours.

The reaction mixture is cooled with stirring in an ice bath, and 1 l. of cold 10% hydrochloric acid (Note 6) is added from the separatory funnel; the acid is added dropwise at first and more rapidly after the reaction subsides. The benzene-ether layer is separated from the aqueous layer and concentrated under reduced pressure in a 5-l. round-bottomed flask on a steam bath. One liter of 25% sulfuric acid is added to the residue, and the mixture is refluxed gently for 12 hours.

The mixture is then cooled in an ice bath, the acid is decanted, and the residue is washed twice by decantation with water. The residue is dissolved in 1 l. of benzene in the same flask, and 1.5 l. of water and 1.2 kg. of sodium bisulfite are added. The flask is fitted with a stirrer, and the mixture is stirred vigorously overnight. The mixture is filtered through an 8-in. Büchner funnel, and the bisulfite addition product is washed on the funnel with 500 ml. of benzene.

The filter cake is broken up and returned to the same 5-l. flask. A saturated solution of sodium bicarbonate is added slowly (Note 7) with stirring until there is no further evidence of decomposition. The mixture is stirred for 2 hours longer. The solution is kept alkaline to litmus throughout by the addition of more sodium bicarbonate if necessary. The crude aldehyde is collected on an 8-in. Büchner funnel, washed with water, and allowed to dry as completely as possible. The product is dissolved in 1 l. of chloroform, the small aqueous layer is separated (Note 8), and the solution is dried with Drierite or another suitable drying agent.

A 250-ml. modified Claisen flask, equipped with a dropping funnel, a thermometer, a water-cooled condenser, and a receiver, is arranged for distillation. The chloroform solution is filtered into the dropping funnel, from which it is admitted to the flask slowly as the solvent is distilled (Note 9). When the solvent has been removed, the dropping funnel is replaced by a stopper and the condenser by a 250-ml. distilling flask as a receiver. The residue is distilled at 160–170°/1 mm. The distillate weighs 206–216 g. (50–52%). This material is recrystallized once from glacial acetic acid (approximately 1 g. to 0.9 ml.) and then from ethanol (about 1 g. to 3 ml.) to give 166–174 g. (40–42% over-all yield) of phenanthrene-9-aldehyde melting at 100–101°.

2. Notes

1. A mercury seal may be used, but a glycerol-rubber tube seal is adequate.

2. The checkers operated on one-fifth the scale specified.

3. Crude bromophenanthrene prepared by the bromination of technical (90%) phenanthrene and purified by distillation only was used by the submitters in this preparation. The anthracene-9-aldehyde, which may be formed from the anthracene present as an impurity in "90% phenanthrene," does not form a sodium bisulfite addition product and so will not contaminate the phenanthrene-9-aldehyde. The checkers used 9-bromophenanthrene, m.p. 54–56° (p. 134), exclusively, but without any advantage in yield. The submitters report yields of 55–60% from pure 9-bromophenanthrene.

4. It is not feasible to add the 9-bromophenanthrene as an ether solution because of the limited solubility of the substance in this solvent. Since the melting point of the crude 9-bromophenanthrene is about 50° it is desirable to heat the melted material to 70° in order to prevent crystallization in the funnel. If the bromo compound begins to solidify in the funnel it may be melted again by careful heating with a microburner.

5. The ethyl orthoformate should be freshly distilled with rejection of the fraction boiling below 140°.

6. If this procedure is used for the preparation of the acetal instead of the aldehyde, it may be preferable to use ammonium chloride solution for hydrolysis instead of 10% hydrochloric acid.

7. The alkaline solution must be added carefully to avoid excessive foaming.

8. The water in the filter cake is removed with difficulty by drying in air or even in an oven under reduced pressure. If the water is not removed as indicated in the procedure, difficulty may be encountered in the early part of the distillation.

9. The distillation of the solvent may be carried out at reduced pressure if desired.

3. Methods of Preparation

Phenanthrene-9-aldehyde has been obtained by the Sonn and Müller synthesis from 9-phenanthroyl chloride,[2] by the Rosenmund reduction of 9-phenanthroyl chloride,[3] by the Gattermann hydrogen cyanide synthesis from phenanthrene,[4] and by the reaction of 9-phenanthrylmagnesium bromide with ethyl formate.[5] The procedure described above is an adaptation of the method of Miller and Bachman.[6] This Grignard method has also been used by others.[7]

[1] Work done under contract with the Office of Scientific Research and Development.

[2] Shoppee, *J. Chem. Soc.*, **1933**, 40.

[3] Mosettig and van de Kamp, *J. Am. Chem. Soc.*, **55**, 2996 (1933).

[4] Hinkel, Ayling, and Beynon, *J. Chem. Soc.*, **1936**, 344.

[5] Bergmann and Israelashwili, *J. Am. Chem. Soc.*, **67**, 1955 (1945).

[6] Miller and Bachman, *J. Am. Chem. Soc.*, **57**, 768 (1935).

[7] Hewett, *J. Chem. Soc.*, **1938**, 195; Weizmann, Bergmann, and Berlin, *J. Am. Chem. Soc.*, **60**, 1332 (1938).

dl-PHENYLALANINE *

(Alanine, phenyl-)

$$CH_2(CO_2C_2H_5)_2 + C_6H_5CH_2Cl \xrightarrow{(NaOC_2H_5)} C_6H_5CH_2CH(CO_2C_2H_5)_2$$

$$C_6H_5CH_2CH(CO_2C_2H_5)_2 \xrightarrow[\text{(then HCl)}]{(KOH)} C_6H_5CH_2CH(CO_2H)_2$$

$$C_6H_5CH_2CH(CO_2H)_2 + Br_2 \rightarrow C_6H_5CH_2CBr(CO_2H)_2 + HBr$$

$$C_6H_5CH_2CBr(CO_2H)_2 \xrightarrow{\text{Heat}} C_6H_5CH_2CHBrCO_2H + CO_2$$

$$C_6H_5CH_2CHBrCO_2H + 2NH_3 \rightarrow C_6H_5CH_2CHNH_2CO_2H + NH_4Br$$

Submitted by C. S. Marvel.[1]
Checked by Lee Irvin Smith, R. T. Arnold, and K. L. Howard.

1. Procedure

A. *Diethyl benzylmalonate.* To 2.5 l. of absolute ethanol in a 5-l. three-necked flask set on a steam cone and equipped with a mercury-sealed stirrer, reflux condenser, and a 500-ml. dropping funnel is added 115 g. (5 gram atoms) of sodium cut in small slices. When all the sodium has reacted, a calcium chloride tube is placed on the condenser and 830 g. (5.18 moles) of diethyl malonate is added through the separatory funnel in a steady stream. This is followed by the drop-wise addition of 632 g. (5 moles) of benzyl chloride over a period of 2–3 hours. The mixture is refluxed, with stirring, until neutral to moist litmus paper (about 8–11 hours). The reflux condenser is then exchanged for a downward condenser, and the ethanol is distilled into another 5-l. three-necked flask equipped with a reflux condenser (Note 1). About 3 hours is required to remove the ethanol, and slightly more than 2 l. is recovered.

The residue is then treated with no more than 2 l. of water (Note 2) and shaken; if necessary, salt is added to make the ester layer separate sharply from the aqueous layer. The combined ester layers from two such runs are distilled from a 5-l. two-necked flask fitted with a well-wrapped 18-mm. Vigreux column. The fraction distilling at 145–155°/5 mm. is collected; it amounts to 1265–1420 g. (51–57%). The residue is chiefly diethyl dibenzylmalonate.

B. *α-Bromo-β-phenylpropionic acid.* Eight hundred and sixty grams of technical potassium hydroxide is dissolved in 850 ml. of water in a 12-l. round-bottomed flask equipped with a stirrer and set on a large steam cone. While the solution is still hot, 1 kg. (4 moles) of diethyl benzylmalonate is added from a dropping funnel over a period of 1 hour. The removal of alcohol vapors is facilitated by placing a tube connected to the water pump in the mouth of the flask. Heating and stirring are continued for 3 hours, and more water is added, if necessary, to keep the mass from solidifying. The flask is then cooled and the contents poured into a crock surrounded by an ice bath and equipped with a stirrer. Five hundred grams of ice is added to lower the temperature; when it reaches 20°, technical hydrochloric acid is added at such a rate that the temperature does not rise. The addition is slow at first, but more rapid when the excess alkali has been neutralized. The solid monopotassium salt that separates is returned to solution by adding the acid more rapidly and stirring by hand. When the reaction mixture is acid to Congo red paper, an excess of 150 ml. of acid is added and the contents of the crock are transferred to a 12-l. round-bottomed flask fitted with a stopper containing a large stopcock which barely pierces the stopper, and a glass tube which reaches to the bottom of the flask. In this way the flask may be used as a large separatory funnel if the stopper is wired in tightly.

The benzylmalonic acid is extracted with four 1-l. portions of ether; the ether extracts are combined in a 5-l. flask and allowed to stand over 150 g. of calcium chloride overnight. The ether layer is then decanted into a 5-l. flask equipped with an efficient reflux condenser, mercury-sealed stirrer, and dropping funnel. Two hundred and twenty-five milliliters of dry bromine is dropped in at such a rate that the ether refluxes (Note 3). The time required is about 4 hours. After the complete addition of bromine, 1 l. of water is added through the dropping funnel at such a rate that the ether merely refluxes (Note 4).

The ether layer of bromobenzylmalonic acid is separated by decantation and the ether removed by distillation. The residue is then decarboxylated by heating to a temperature of 130–135° in a 3-l. flask in an oil bath for 5 hours.

C. *dl-Phenylalanine.* The crude bromo acid is divided into four portions and each portion is added to 2 l. of technical ammonium hydroxide (sp. gr. 0.90) in a 3-l. round-bottomed flask. The flask is well shaken, a rubber stopper is wired in, and the mixture is allowed to stand for a week. The contents of the four amination flasks (Note 5) are then combined in a 12-l. flask, 20 g. of Norit is added, and the flasks are heated on a steam cone overnight. The ammonia which is evolved is conducted into a gas-absorption trap or merely led into water by a tube from the flask. The solution is filtered while still hot; on cooling most of the phenylalanine precipitates. This is filtered, washed with 250 ml. of methanol, and the filtrate evaporated under the pressure of a water pump until more crystals form. The solution is then cooled and an additional crop of phenylalanine obtained, which is also washed with methanol. The yield of crude product is 500 g., but it is slightly wet; if it is dried overnight in an oven at about 80°, it will weigh 460 g. This need not be done, however, as the yield of pure product is the same whether or not the crude product is dried.

The phenylalanine is recrystallized as follows: the crude product is dissolved in 9 l. of water heated to 95° on a steam cone, treated with 15 g. of Norit, and filtered. Three liters of alcohol is added and the solution cooled in the ice chest overnight. The yield of pure product amounts to 367 g. An additional 45 g. may be obtained by evaporating the mother liquor under reduced pressure until crystals separate, adding an amount of alcohol equivalent to one-third the volume of the concentrated mother liquor, and cooling. Additional material may be obtained by continuing to work down all mother liquors. The yield, 412 g., is 62.4% based on the diethyl benzylmalonate. The white crystals decompose at 271–273° (uncor.) in a closed capillary (Note 6).

2. Notes

1. The sodium required for the next run can be added to the second flask as the alcohol distils into it.

2. Usually 1.5 l. of water suffices, and it is not necessary to add salt.

3. To start the bromination 10 ml. of bromine is added and the solution stirred until decolorized. After the reaction has been started in this manner, it runs very smoothly. When large amounts of the amino acid are to be prepared, it is more convenient to double the portion used in the bromination step.

4. Care must be taken not to add the water too fast, as the reaction mixture will foam out of the flask.

5. The aminations are usually colored, and frequently the flasks contain small deposits of oil on the bottom. This oil disappears in subsequent treatment.

6. The product thus obtained has the calculated amino nitrogen content.

3. Methods of Preparation

Methods of making *dl*-phenylalanine have been summarized [*Org. Syntheses* Coll. Vol. **2,** 489 (1943)]. The method described here is essentially the one originally described by Fischer.[2] For the preparation of large amounts of amino acids it is undoubtedly the cheapest and best procedure.

Additional preparative methods include the hydrolysis and decarboxylation of benzyl formamidomalonic ester,[3] benzylacetamidomalonic ester,[4] or benzylacetamidocyanoacetic ester.[5]

[1] These directions are the result of the efforts of many men who have worked on the preparation of phenylalanine at the University of Illinois.

[2] Fischer, *Ber.*, **37**, 3064 (1904).

[3] Galat, *J. Am. Chem. Soc.*, **69**, 965 (1947).

[4] Albertson and Archer, *J. Am. Chem. Soc.*, **67**, 308 (1945).

[5] Albertson and Tullar, *J. Am. Chem. Soc.*, **67**, 502 (1945).

1-PHENYL-3-AMINO-5-PYRAZOLONE

(5-Pyrazolone, 3-amino-1-phenyl-)

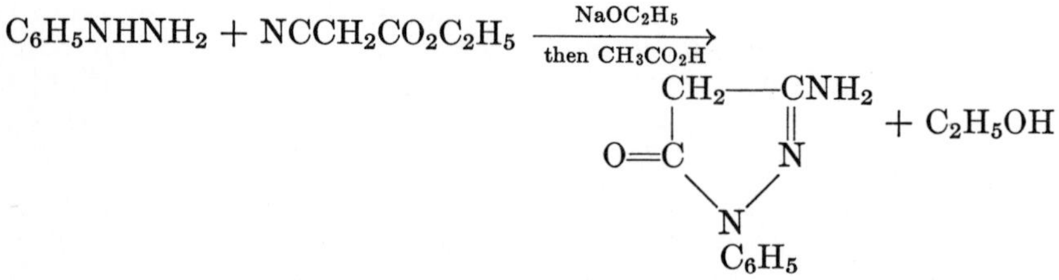

Submitted by H. D. Porter and A. Weissberger.
Checked by R. T. Arnold and K. Murai.

1. Procedure

Sodium ethoxide is prepared from 46 g. (2 gram atoms) (Note 1) of sodium and 800 ml. of absolute ethanol in a 2-l. three-necked flask equipped with stirrer and a reflux condenser. To the hot solution is added 113 g. (106 ml., 1 mole) of ethyl cyanoacetate followed by 108 g. (98 ml., 1 mole) of phenylhydrazine (Note 2), and the mixture is stirred and heated in an oil bath at 120° for 16 hours. Then most of the ethanol is removed under reduced pressure and the residue is dissolved in 1 l. of water; the mixture is warmed to about 50° and stirred to facilitate solution. After being cooled to room temperature, the solution is extracted with three 100-ml. portions of ether (Note 3). The aqueous phase is acidified by the addition of 100 ml. of glacial acetic acid, cooled in ice, and filtered. The crude product is washed on the filter with 100 ml. of 95% ethanol; it is then transferred to a flask and boiled with 500 ml. of the same solvent, and this mixture is cooled and filtered. The solid is washed with ethanol and dried. The tan crystalline 1-phenyl-3-amino-5-pyrazolone, melting with decomposition at 216–218°, weighs 76–82 g. (43–47%) (Note 4).

2. Notes

1. At least two equivalents of sodium ethoxide are necessary for the reaction, but larger amounts do not improve the yield. Sodium hydroxide in ethanol, or sodamide in benzene, cannot be substituted for the sodium ethoxide solution.

2. All the chemicals used were obtained from the Eastman Kodak Company.

3. Instead of the isolation of the product by concentration, solution in water, and extraction with ether, water may be added directly to the ethanolic solution and the resulting solution acidified. This procedure, however, often leads to a more highly colored product.

4. This material is sufficiently pure for most purposes. If desired, an almost white product may be obtained by two recrystallizations from dioxane (Norit). This treatment entails a 40% loss of the product and raises the melting point by only 2° (to 218–220°).

3. Methods of Preparation

The procedure is essentially that given by Conrad and Zart for the original preparation of the substance,[1] to which they assigned the incorrect structure of 1-phenyl-3-hydroxy-5-pyrazolone imide.[2] The compound may also be prepared in about the same yield by the reaction of phenylhydrazine with ethyl malonate monoimidoester.[3]

The procedure given has been applied with varying success to a number of aromatic and heterocyclic hydrazines.[4]

[1] Conrad and Zart, *Ber.*, **39**, 2282 (1906).

[2] Weissberger and Porter, *J. Am. Chem. Soc.*, **64**, 2133 (1942).

[3] Weissberger, Porter, and Gregory, *J. Am. Chem. Soc.*, **66**, 1851 (1944).

[4] Weissberger and Porter, *J. Am. Chem. Soc.*, **66**, 1849 (1944).

PHENYL AZIDE

(Benzene, azido-)

$$C_6H_5NHNH_2 + HCl + NaNO_2 \rightarrow C_6H_5N_3 + NaCl + 2H_2O$$

Submitted by R. O. Lindsay and C. F. H. Allen.
Checked by R. L. Shriner and J. C. Lawler.

1. Procedure

In a 1-l. three-necked flask fitted with a stirrer, a thermometer, and a dropping funnel are placed 300 ml. of water and 55.5 ml. of concentrated hydrochloric acid. The flask is surrounded by an ice-salt bath, the stirrer is started, and 33.5 g. (0.31 mole) of phenylhydrazine (Note 1) is added dropwise (5–10 minutes is required). Phenylhydrazine hydrochloride separates as fine white plates. Stirring is continued, and, after the temperature has fallen to 0°, 100 ml. of ether is added, after which a previously prepared solution of 25 g. of technical sodium nitrite in 30 ml. of water is added from the dropping funnel at such a rate that the temperature *never* rises above 5°. This requires 25–30 minutes.

The reaction mixture is subjected to steam distillation until about 400 ml. of distillate is obtained. The ether layer is removed from the distillate, and the aqueous layer is extracted once with 25 ml. of ether. The combined ethereal solutions are dried over 10 g. of anhydrous calcium chloride. The dried solution is placed in a 200-ml. ordinary Claisen flask arranged for vacuum distillation. *The flask must be surrounded by a cylindrical wire screen, and a laminated glass screen must be interposed between the operator and the apparatus* (Note 2). The flask is immersed in a water bath at 25–30°, and the ether is removed under reduced pressure. Then the temperature of the water bath is raised to 60–65°, and the product is distilled under reduced pressure. Phenyl azide boils at 49–50° at 5 mm. (Note 3). A yield of 24–25 g. (65–68%) of the pungent, pale yellow, oily azide is obtained (Note 4).

2. Notes

1. The phenylhydrazine used was the best grade supplied by the Eastman Kodak Company. With technical material, or a preparation

that was appreciably discolored, the yield was much less (45–50%), and a considerable amount of tar was formed.

2. Care must be exercised during the distillation. Phenyl azide explodes when heated at ordinary pressure, and occasionally at lower pressures. The water-bath temperature should never be permitted to rise above 80° at any time.

3. Phenyl azide boils at 66–68°/21 mm. with a bath temperature of 70–75°. It is advisable to use as low a bath temperature as possible and a pressure of 5 mm. or less. The checkers have used these directions repeatedly without any explosions.

4. The product should be stored in a brown glass bottle. It will keep for a month in a cool, dark place.

3. Methods of Preparation

Phenyl azide has been prepared by the action of nitrous acid upon phenylhydrazine hydrochloride;[1] of ammonia upon diazobenzene perbromide;[2] and by the reaction between a diazo salt and sodium azide,[3] hydroxylamine,[4] or *p*-toluenesulfonamide.[5]

[1] Dimroth, *Ber.*, **35**, 1032 (1902).

[2] Griess, *Ann.*, **137**, 68 (1866).

[3] Nölting, *Ber.*, **26**, 86 (1893).

[4] Fischer, *Ann.*, **190**, 96 (1877); Mai, *Ber.*, **25**, 372 (1892); **26**, 1271 (1893); Forster and Fierz, *J. Chem. Soc.*, **91**, 855, 1350 (1907).

[5] Bretschneider and Rager, *Monatsh.*, **81**, 970 (1950).

p-PHENYLAZOBENZOIC ACID *

(Benzoic acid, *p*-phenylazo-)

$$C_6H_5NO + H_2NC_6H_4COOH \rightarrow C_6H_5N{=}NC_6H_4COOH + H_2O$$

Submitted by Harry D. Anspon.
Checked by W. E. Bachmann and N. C. Deno.

1. Procedure

Fifty-four grams (0.39 mole) of *p*-aminobenzoic acid is dissolved in 390 ml. of warm glacial acetic acid in a 1-l. Erlenmeyer flask. The solution is cooled to room temperature, 42 g. (0.39 mole) of nitrosobenzene (p. 668) is added, and the mixture is shaken until the nitrosobenzene dissolves. The flask is stoppered, and the solution is allowed to stand for 12 hours at room temperature. The product begins to crystallize after about 15 minutes.

The *p*-phenylazobenzoic acid is collected on a Büchner funnel (Note 1) and washed with acetic acid and with water. The yield of air-dried acid melting at 245–247° cor. is 62 g. (70%). By recrystallization from 95% ethanol (60 ml. per g.) the acid is obtained as orange-gold plates which melt at 248.5–249.5° cor.; the yield is 54 g. (61%).

2. Note

1. The solution is not cooled below room temperature before filtering; cooling below 20° brings down impurities.

3. Methods of Preparation

The method employed here is essentially the one described by Angeli and Valori.[1]

[1] Angeli and Valori, *Atti accad. Lincei,* **22,** I, 132 (1913) [*C. A.,* **7,** 2223 (1913)].

p-PHENYLAZOBENZOYL CHLORIDE *

(Benzoyl chloride, *p*-phenylazo-)

$$C_6H_5N{=}NC_6H_4CO_2H + SOCl_2 \rightarrow C_6H_5N{=}NC_6H_4COCl + HCl + SO_2$$

Submitted by George H. Coleman, Gust Nichols, Chester M. McCloskey, and Harry D. Anspon.
Checked by W. E. Bachmann and N. C. Deno.

1. Procedure

Fifty grams (0.22 mole) of recrystallized *p*-phenylazobenzoic acid (p. 711) (Note 1) and 50 g. (0.47 mole) of anhydrous sodium carbonate (Note 2) are placed in a 1-l. flask and thoroughly mixed by shaking. To this mixture is added 250 ml. (3.5 moles) of thionyl chloride (Note 3); a reflux condenser with a drying tube is fitted to the flask, and the mixture is refluxed for 1.5 hours (Note 4). A condenser is set for distillation, and as much as possible of the thionyl chloride is distilled on a steam bath (Note 5).

The acid chloride is dissolved by refluxing with 500 ml. of 90–100° ligroin, and the hot solution is decanted from the sodium carbonate onto a fluted filter. This process is repeated with three 150-ml. portions of ligroin. The combined filtrates are concentrated to 500 ml., filtered if necessary, and cooled to 0°. The acid chloride is collected on a Büchner funnel and pressed as dry as possible. It is washed twice with 30–60° petroleum ether and stored in a vacuum desiccator over phosphorus pentoxide and paraffin shavings (Note 6). The yield of orange-red crystals melting at 94.5–95.5° is 48 g. (89%).

2. Notes

1. The checkers found that 50 g. of unrecrystallized acid gave 45 g. of acid chloride melting at 92–94°. When this was recrystallized from 90–100° petroleum ether (10 ml. per g.), 41.5 g. of orange-red crystals melting at 94.5–95.5° was obtained. The over-all yield of pure acid chloride on the basis of unrecrystallized acid is the same. The use of unrecrystallized acid has an advantage in that the volume of solvent required for recrystallization of the acid chloride is much less than for the acid.

2. The use of sodium carbonate is unusual. It is claimed that the sodium carbonate prevents decomposition and tar formation during the reaction.

3. The checkers used Eastman Kodak Company's best grade of thionyl chloride.

4. Refluxing should be carried out under a hood or in an apparatus provided with a gas trap.

5. About 170 ml. of good thionyl chloride can be recovered.

6. The acid chloride holds the petroleum ether tenaciously. About a week is required for complete removal of the petroleum ether.

3. Methods of Preparation

p-Phenylazobenzoyl chloride (azoyl chloride) has been prepared by the action of thionyl chloride on the acid.[1,2] The method used is a modification of that of Ladenburg, Fernholz, and Wallis.[2]

[1] Reich, *Biochem. J.*, **33**, 1001 (1939).

[2] Ladenburg, Fernholz, and Wallis, *J. Org. Chem.*, **3**, 294 (1938).

PHENYL CINNAMATE

(Cinnamic acid, phenyl ester)

$$C_6H_5CH{=}CHCO_2H + SOCl_2 \rightarrow C_6H_5CH{=}CHCOCl + HCl + SO_2$$

$$C_6H_5CH{=}CHCOCl + C_6H_5OH \rightarrow C_6H_5CH{=}CHCO_2C_6H_5 + HCl$$

Submitted by Ennis B. Womack and J. McWhirter.
Checked by R. L. Shriner, N. S. Moon, and S. C. Kelton, Jr.

1. Procedure

A mixture of 148 g. (1 mole) of cinnamic acid and 119 g. (1 mole) of thionyl chloride (Note 1) is placed in a 500-ml. Claisen flask. The side arms are stoppered, and the flask is fitted with a reflux condenser. The apparatus is mounted at an angle so that the condensate will not run into the side arm. To the top of the condenser is attached an exit tube, for evolved hydrogen chloride and sulfur dioxide, leading to a gas-absorption trap. The mixture is heated on a steam bath, cautiously at first, until no further evolution of hydrogen chloride is noted (45–60 minutes), and then allowed to cool, and 94 g. (1 mole) of phenol (Note 2) is added. The mixture is again heated on the steam bath until the evolution of hydrogen chloride has ceased (about 1 hour). It is then placed on a sand bath and brought just to the reflux temperature in order to complete the reaction and remove the hydrogen chloride more completely (Note 3).

The reaction mixture is cooled and distilled under reduced pressure (Note 4). The fraction boiling at 190–210° at 15 mm. is collected. The distillate solidifies to a pale yellow solid melting at 64° to 69° and weighing 186–200 g. (83–89%). It is purified by grinding in a mortar to a powder and washing with 500 ml. of cold 2% sodium bicarbonate solution. The residue is recrystallized from 300 ml. of 95% ethanol. The recovery is 141–168 g. (63–75%) of pure white crystals melting at 75–76°.

2. Notes

1. The thionyl chloride should be redistilled before use. The material used in this preparation boiled at 75.0–75.5°.
2. The phenol used was Mallinckrodt's analytical reagent grade.

3. A sand-bath temperature of about 350° will effect refluxing. Prolonged heating on the sand bath causes considerable loss of product due to decomposition and polymerization and to the conversion of the acid to stilbene by the loss of carbon dioxide.

4. In carrying out the vacuum distillation, it is well not to include the manometer in the system until the unchanged phenol and most of the hydrogen chloride have been removed. Bumping during the distillation may be minimized by holding the burner in the hand and directing the free flame at the surface of the boiling liquid. If the vapors are superheated too much, the boiling-point range may be 190–220° at 15 mm.

3. Methods of Preparation

Phenyl cinnamate and other phenolic esters have been prepared by heating the acid and phenol in the presence of phosphorus oxychloride,[1] and by heating the acid anhydride and phenol together in the presence of a dehydrating agent such as fused zinc chloride or anhydrous sodium acetate.[2] Phenyl cinnamate has also been prepared by the careful distillation of phenyl fumarate.[3]

1 Nencki, *Compt. rend.*, **108**, 254 (1889).
2 Franchimont, *Ber.*, **12**, 2059 (1879); Liebermann, *Ber.*, **21**, 1172 (1888).
3 Anschütz, *Ber.*, **18**, 1945 (1885).

α-PHENYLCINNAMONITRILE

(Acrylonitrile, α-β-diphenyl-)

$$C_6H_5CHO + C_6H_5CH_2CN \xrightarrow{NaOC_2H_5} C_6H_5CH{=}C(C_6H_5)\text{—}CN + H_2O$$

Submitted by Stanley Wawzonek and Edwin M. Smolin.
Checked by Cliff S. Hamilton and Karl W. R. Johnson.

1. Procedure

In a 2-l. beaker fitted with a strong, efficient, mechanical stirrer is placed a mixture of 106 g. (101 ml., 1 mole) of freshly distilled benzaldehyde and 117 g. (1 mole) of purified dry benzyl cyanide (Note 1), in 650 ml. of 95% ethanol (Note 2). To this mixture is added dropwise, with stirring, a solution of 7 g. of sodium ethoxide in 50 ml. of absolute ethanol (Note 3). When 40–50 ml. has been added, the mixture becomes warm, turns cloudy, and solidifies. Mechanical stirring is continued as long as possible, and then the mixture is stirred by hand with a thick stirring rod in order to break up the lumps that form. The mixture is cooled in an ice bath (Note 4), and the product is separated by filtration. The filtrate is removed and may be saved (Note 5). The white mass is washed first with 200 ml. of distilled water, then with 50 ml. of 95% ethanol to remove unchanged reagents. The nitrile is dried at 25° and melts at 86–88°. The yield is 178–199 g. (87–97%) of product sufficiently pure for most purposes. Recrystallization from 700 ml. of 95% ethanol gives 170–187 g. (83–91%) of a pure, white product melting at 88° (Note 6).

2. Notes

1. The benzyl cyanide may be purified by a procedure described earlier.[1] If commercial benzyl cyanide is used, the yield is between 80% and 90% of a slightly yellow product. Two recrystallizations are necessary for purification.

2. Denatured alcohol (Formula 3A) containing 10% absolute methanol is a satisfactory solvent for recrystallization.

3. A 40% solution of sodium hydroxide may also be used as the condensing agent; 35–60 ml. is required. With this reagent, the product is less pure and needs an additional recrystallization. The yields range from 70% to 82%.

4. This preliminary cooling helps to prevent clogging of the 20-cm. Büchner funnel used during filtration.

5. An additional 10–21 g. of crude nitrile melting at 84–86° can be obtained by evaporating the combined alcoholic filtrates to a volume of 300 ml. Two or three recrystallizations from 95% ethanol are necessary to raise the melting point to 88°.

6. Similar yields of substituted α-phenylcinnamonitriles can be obtained using *p*-methoxybenzyl cyanide and anisaldehyde, or benzyl cyanide and anisaldehyde.[2,3]

3. Methods of Preparation

α-Phenylcinnamonitrile can be prepared from benzaldehyde and benzyl cyanide with no solvent and with sodium ethoxide as a catalyst.[4] Sodium hydroxide [5] (40%) or piperidine [6] may also be used as catalysts. The nitrile has been made by the condensation of benzyl cyanide and excess benzyl chloride with strong sodium hydroxide at 170° [7,8] and by heating α,β-diphenylsuccinonitrile with alcohol at 180° in a sealed tube [9] or at 230–250° under 100–110 mm. pressure with palladium.[10]

[1] *Org. Syntheses* Coll. Vol. **1,** 108 (1941).
[2] Wawzonek, *J. Am. Chem. Soc.,* **68,** 1157 (1946).
[3] Niederl and Ziering, *J. Am. Chem. Soc.,* **64,** 885 (1942).
[4] De Schuttenbach, *Ann. chim.,* (11) **6,** 90 (1936).
[5] Walther, *J. prakt. Chem.,* (2) **53,** 454 (1896).
[6] Knoevenagel, Ger. pat. 94,132 [*Chem. Zentr.,* **69,** 228 (1898)].
[7] Neure, *Ann.,* **250,** 155 (1888).
[8] Janssen, *Ann.,* **250,** 129 (1888).
[9] Chalaney and Knoevenagel, *Ber.,* **25,** 297 (1892).
[10] Knoevenagel and Bergdott, *Ber.,* **36,** 2861 (1903).

α-PHENYLETHYLAMINE *

(Benzylamine, α-methyl-)

$$C_6H_5COCH_3 + NH_3 + H_2 \xrightarrow[150°]{\text{Raney nickel}} C_6H_5\underset{\displaystyle NH_2}{\underset{|}{C}}HCH_3 + H_2O$$

Submitted by John C. Robinson, Jr., and H. R. Snyder.
Checked by Nathan L. Drake and Daniel Draper.

1. Procedure

In a 2-l. bomb are placed 720 g. (6 moles) of pure acetophenone and 1 tablespoon of Raney nickel catalyst (p. 181). After the cap and gauge block are securely fastened, 700 ml. (30 moles) of liquid ammonia is introduced (Note 1). The mixture is hydrogenated at 150° under 5000–3500 lb. (Note 2). The reaction is allowed to continue as long as hydrogen is absorbed, generally 4–6 hours. The bomb is cooled, the excess ammonia is allowed to escape, and the contents are filtered from the catalyst. The mixture is cooled in an ice bath, acidified to Congo red with concentrated hydrochloric acid (200–300 ml.), and steam-distilled for 10–12 hours to remove excess acetophenone (Note 3). The residue is then cooled and added slowly to 200 g. of solid sodium hydroxide in a flask surrounded by an ice bath. The amine is separated, and the aqueous layer is extracted with three 150-ml. portions of benzene. The extracts and amine are combined and dried over solid sodium hydroxide. After removal of the benzene, the residue is fractionated under diminished pressure. The yield of α-phenylethylamine (Note 4), b.p. 80–81°/18 mm., is 320–380 g. (Note 5) (44–52%).

2. Notes

1. Liquid ammonia is introduced into the large bomb as follows: The cap and gauge block of the large bomb are tightened in place. The inner gas inlet tube is removed from the cap assembly of a smaller bomb (capacity about 250 ml.). This bomb is equipped with a test-tube-type liner which is kept chilled in a bath of Dry Ice while it is filled with liquid ammonia. This test tube is then placed in the small bomb, and the cap and gauge block are quickly (15–30 seconds) tightened. The bomb is then filled with hydrogen under high pressure and connected with the larger bomb by means of a short length of the conventional steel pressure tubing. The smaller bomb is inverted, and the valves are opened. This operation will introduce about 150 ml. of liquid ammonia at one time and may be repeated as often as necessary.

2. A booster pump is required, for it is quite important to keep the pressure above the minimum value of about 3500 lb. The temperature of the reduction is above the critical temperature of ammonia, and the pressure will not fall much below 3500 lb. At this point hydrogen must be pumped into the bomb until the pressure is about 5000 lb.; this process is repeated until the reaction is complete. If a safety disk is to be incorporated into the line, it *must not* be made of *copper,* as ammonia, even under 2–3 atm., rapidly attacks copper. A *special* disk of steel, nickel, or other suitable material is required.

3. It is necessary to heat the flask externally with a flame or the volume of the solution will greatly increase during the lengthy steam distillation.

4. According to the submitters, methyl amyl ketone (800 g.) and ammonia (600 ml.) have been converted to 2-aminoheptane, b.p. 139–141°, in exactly the same manner, in 50–55% yields. A slightly modified procedure was used in the preparation of *n*-heptylamine and furfurylamine. Heptaldehyde (320 g.) was dissolved in 500 ml. of methanol, and 150 ml. of liquid ammonia was added; the reduction was conducted as above. *n*-Heptylamine, b.p. 57–58°/23 mm., was obtained in yields of 53–63%. Freshly distilled furfural (290 g.) was dissolved in 500 ml. of methanol, 150 ml. of liquid ammonia was introduced, and the reduction carried out as usual. The product was removed, filtered, and fractionated directly. Furfurylamine, b.p. 144–146°, was obtained in 50% yield.

5. The yields are based upon the amount of acetophenone initially used and do not make allowances for the material recovered from the steam distillation. A small amount of di-(α-phenylethyl)amine, b.p. 61–62°/2 mm., may be recovered from the residues.

3. Methods of Preparation

α-Phenylethylamine has been prepared by reducing acetophenone with hydrogen at high pressures over nickel catalysts in the presence of ammonia;[1,2] with hydrogen at low pressures over a nickel catalyst in the presence of ammonia-saturated ethanol;[3] and with hydrogen at low pressures over a platinum catalyst in the presence of ammonia-saturated methanol containing ammonium chloride (69% yield).[4]

l-α-Phenylethylamine has been prepared through the oxime of *d*-α-phenylethyl methyl ketone by the Beckmann rearrangement;[5] from *d*-phenylmethylacethydroxamic acid by the Lossen rearrangement;[5] from *d*-hydratropic azide;[6,7] from *d*-hydratropic acid by the Schmidt reaction;[5] from *d*-hydratropamide by treatment with alkaline hypobromite;[8] and by the reduction of acetophenone oxime with lithium aluminum hydride.[9]

Other methods of preparing α-phenylethylamine are reviewed in *Org. Syntheses* Coll. Vol. **2,** 503 (1943), where detailed directions are given for the preparation of this amine from acetophenone and ammonium formate. The procedure given above was based upon that of Schwoegler and Adkins.[2]

Methods of preparing *d*- and *l*-α-phenylethylamine, based on the resolution of *dl*-α-phenylethylamine, are reviewed in *Org. Syntheses* Coll. Vol. **2,** 506 (1943), where detailed directions are given for the resolution of this amine by *l*-malic and *d*-tartaric acids.

[1] Couturier, *Ann. chim.,* (11) **10,** 610 (1938).

[2] Schwoegler and Adkins, *J. Am. Chem. Soc.,* **61,** 3499 (1939).

[3] Mignonac, *Compt. rend.,* **172,** 223 (1921).

[4] Alexander and Misegades, *J. Am. Chem. Soc.,* **70,** 1315 (1948).

[5] Campbell and Kenyon, *J. Chem. Soc.,* **1946,** 25.

[6] Bernstein and Whitmore, *J. Am. Chem. Soc.,* **61,** 1324 (1939).

[7] Kenyon and Young, *J. Chem. Soc.,* **1941,** 263.

[8] Arcus and Kenyon, *J. Chem. Soc.,* **1939,** 916.

[9] Larsson, *Trans. Chalmers Univ. Technol., Gothenburg,* **94,** 15 (1950) [*C. A.,* **45,** 1494 (1951)].

β-PHENYLETHYLAMINE *

(Phenethylamine)

$$C_6H_5CH_2CN + 2H_2 \xrightarrow{\text{Raney nickel}} C_6H_5CH_2CH_2NH_2$$

Submitted by John C. Robinson, Jr., and H. R. Snyder.
Checked by Nathan L. Drake and Daniel Draper.

1. Procedure

In a 2-l. bomb are placed 1 kg. (8.55 moles) of pure (Note 1) benzyl cyanide and 1 tablespoon of Raney nickel catalyst (p. 181). After the cap is securely fastened down, 150 ml. of liquid ammonia is introduced (Note 2). Hydrogen is introduced until the pressure is about 2000 lb. The bomb is then heated to 120–130° and shaking is begun. The reduction is complete well within an hour (Note 1). The bomb is cooled and opened, and the contents are removed. The bomb is rinsed with a little ether, and the combined liquids are filtered from the catalyst. The ether is removed, and the residue is fractionated under reduced pressure. The yield is 860–890 g. (83–87%) of β-phenylethylamine, b.p. 90–93°/15 mm. (Notes 3, 4, and 5).

2. Notes

1. Benzyl cyanide, prepared according to *Org. Syntheses* Coll. Vol. **1,** 107 (1941), should be distilled from Raney nickel. Minute traces of halide have a strong poisoning effect on the catalyst. If the reduction does not occur within an hour, the contents of the bomb should be removed and filtered. New catalyst is then added and the process is repeated.

2. The presence of ammonia in the reduction mixture reduces the amount of secondary amine formed. For directions for introducing the liquid ammonia, see Note 1 to the preparation of α-phenylethylamine (p. 718).

3. If several runs are made, a small amount of the secondary amine may be recovered from the combined residues. Di-(β-phenylethyl)-amine boils at 155–157°/4 mm.

4. Similarly *n*-amyl cyanide has been converted to *n*-hexylamine, b.p. 128–130°, in 67–70% yields.

5. It has been reported (R. N. Icke and C. E. Redemann, private communication) that β-phenylethylamine, as well as several substituted β-phenylethylamines, may be prepared in excellent yields by catalytic reduction of the corresponding cyanides in 10 *N* methanolic ammonia. An example of this procedure follows. Commercial anhydrous methanol is saturated with ammonia gas at 0°; this solution is approximately 10 *N*. A solution of 58.5 g. (0.5 mole) of benzyl cyanide in 300 ml. of 10 *N* methanolic ammonia (the ratio of ammonia to benzyl cyanide should be at least 5:1 in order to minimize the formation of the secondary amine) is placed in a high-pressure hydrogenation bomb, 5–10 ml. of settled Raney nickel catalyst (p. 181) is added, the bomb is closed, and hydrogen is introduced until the pressure is 500–1000 lb. The bomb is shaken and heated to 100–125° until absorption of hydrogen ceases (about 2 hours). The bomb is cooled and opened, and the contents are removed. The bomb is rinsed with two or three 100-ml. portions of methanol, and the combined liquids are poured through a fluted filter to remove the catalyst. (*Caution! If the catalyst becomes dry, it is likely to ignite.*) The solvent and the ammonia are removed by distillation, and the residue is fractionated through a short column. The yield of β-phenylethylamine boiling at 92–93°/19 mm. (62–63°/4 mm.) is 51–54.5 g. (84–90%). The hydrochloride, after crystallization from dry ethanol, melts at 218–219°. This procedure has also been used for preparation of the following β-phenylethylamines from the cyanides; the yields of amines were uniformly high: 3,4-dimethoxyphenylethylamine, b.p. 119–119.5°/1 mm.; *o*-methylphenylethylamine, b.p. 67°/0.5 mm.; *m*-methylphenylethylamine, b.p. 68°/2 mm.; *p*-methylphenylethylamine, b.p. 71°/2 mm.; and 3,4-methylenedioxyphenylethylamine, b.p. 109°/2 mm.

3. Methods of Preparation

β-Phenylethylamine has been made by a number of reactions, many of which are unsuitable for preparative purposes. Only the most important methods, from a preparative point of view, are given here. The present method is adapted from that of Adkins,[1] which in turn was based upon those of Mignonac,[2] von Braun and coworkers,[3] and Mailhe.[4] Benzyl cyanide has been converted to the amine by catalytic reduction with palladium on charcoal,[5] with palladium on barium sulfate,[6] and with Adams' catalyst;[7] by chemical reduction with sodium and ethanol,[8] and with zinc dust and mineral acids.[9] Hydrocinnamic acid has been converted to the azide and thence by the Curtius rearrangement to β-phenylethylamine;[10] also the Hofmann degradation of hydrocinnamide has been used successfully.[11] β-Nitrostyrene,[12] phenylthioacetamide,[13] and the benzoyl derivative of mandelonitrile[14] all yield β-phenylethylamine upon reduction. The amine has also been prepared by cleavage of N-(β-phenylethyl)phthalimide[15] with hydrazine; by the Delépine synthesis from β-phenylethyl iodide and hexamethylenetetramine;[16] by the hydrolysis of the corresponding urethan and urea;[17] by reduction of phenylacetaldoxamine;[18] and by catalytic reduction of O-carbethoxymandelonitrile in the presence of acids.[19]

More recent methods for preparation of the amine include the lithium aluminum hydride reduction of β-nitrostyrene[20] and of phenylacetamide.[21]

The Raney nickel reduction of the nitrile in the presence of formamide is reported to give an 87% yield of the formylated primary amine.[22]

1 Adkins, *The Reaction of Hydrogen with Organic Compounds over Copper-Chromium Oxide and Nickel Catalysts,* University of Wisconsin Press, Madison, Wisconsin, **1937**, pp. 53–54.

2 Mignonac, French pat. 638,550 [*C. A.*, **23**, 154 (1929)]; Brit. pat. 282,038 [*C. A.*, **22**, 3668 (1928)].

3 von Braun, Blessing, and Zobel, *Ber.*, **56**, 1988 (1923).

4 Mailhe, *Bull. soc. chim. France,* (4) **23**, 237 (1918).

5 Strack and Schwaneberg, *Ber.*, **65**, 710 (1932).

6 Rosenmund and Pfannkuch, *Ber.*, **56**, 2258 (1923).

7 Carothers and Jones, *J. Am. Chem. Soc.*, **47**, 3051 (1925).

8 Wohl and Berthold, *Ber.*, **43**, 2184 (1910).

9 Bernthsen, *Ann.*, **184**, 304 (1877).

10 Sah and Kao, *Science Repts. Natl. Tsing Hua Univ., Ser. A,* **3**, 525 (1936) [*C. A.*, **31**, 3889 (1937)].

11 McRae and Vining, *Can. J. Research,* **6**, 409 (1932).

12 Kindler, Brandt, and Gehlhaar, *Ann.*, **511**, 209 (1934).

13 Kindler, *Ber.*, **57**, 775 (1924).

14 Hartung, *J. Am. Chem. Soc.*, **50**, 3373 (1928).

15 Ing and Manske, *J. Chem. Soc.*, **1926**, 2348.

16 Galat and Elion, *J. Am. Chem. Soc.*, **61**, 3585 (1939).

17 Curtius and Jordan, *J. prakt. Chem.*, (2) **64**, 308 (1901).

18 Bischler and Napieralski, *Ber.*, **26**, 1905 (1893).

19 Indler and Schrader, *Ann.*, **564**, 49 (1949).

20 Nystrom and Brown, *J. Am. Chem. Soc.*, **70**, 3738 (1948).

21 Uffer and Schlittler, *Helv. Chim. Acta,* **31**, 1397 (1948).

22 Sekiya, *J. Pharm. Soc. Japan,* **70**, 520 (1950).

β-PHENYLETHYLDIMETHYLAMINE

(Phenethylamine, N,N-dimethyl-)

$$C_6H_5CH_2CH_2NH_2 + 2HCHO + 2HCOOH \rightarrow C_6H_5CH_2CH_2N(CH_3)_2 + 2CO_2 + 2H_2O$$

Submitted by Roland N. Icke and Burnett B. Wisegarver with Gordon A. Alles.
Checked by H. R. Snyder and James H. Saunders.

1. Procedure

To 51.2 g. (1 mole) of 90% formic acid in a 500-ml. round-bottomed flask (Note 1), cooled in running tap water, is added slowly 24.2 g. (0.2 mole) of β-phenylethylamine. To the resulting clear solution are added 45 ml. (0.6 mole) of formaldehyde solution (concentration, 37%) (Note 2) and a small boiling stone. The flask is connected to a reflux condenser and is placed in an oil bath which has been heated to 90–100°. A vigorous evolution of carbon dioxide begins after 2–3 minutes, at which time the flask is removed from the bath until the gas evolution notably subsides (15–20 minutes); then it is returned to the bath and heated at 95–100° for 8 hours.

After the solution has been cooled, 100 ml. of 4 *N* hydrochloric acid is added and the solution is evaporated to dryness under reduced pressure (water pump) from a water bath; the receiver is cooled in an ice bath. The pale yellow syrupy residue (or crystalline solid) is dissolved in 60–75 ml. of water, and the organic base is liberated by the addition of 50 ml. of 18 *N* sodium hydroxide solution. The upper (organic) phase is separated, and the lower (aqueous) phase is extracted with two 30-ml. portions of benzene. The combined organic base and benzene extracts are dried over 10 g. of anhydrous granular potassium carbonate (Note 3). After the benzene has been distilled slowly under slightly reduced pressure from a 125-ml. Claisen flask, the pressure is lowered further, and the product is distilled. The yield of colorless β-phenylethyldimethylamine boiling at 97–98°/22 mm. (Note 4) is 22–24.7 g. (74–83%) (Note 5).

2. Notes

1. A flask of this size is used because of the tendency of the solution to froth during the gas evolution. Frothing usually is not bad with this amine but is quite bothersome when the higher aliphatic amines (decylamine to octadecylamine) are methylated.

2. U.S.P. formaldehyde was used. The commercial aqueous-methanolic solution contains 37% formaldehyde by weight. It is sometimes called "40% formalin" because 100 ml. of the solution contains 40 g. of formaldehyde.

3. If complete separation of the benzene extracts from the aqueous solution is difficult, it is advantageous to dry the benzene solution roughly over 10 g. of the anhydrous potassium carbonate and to decant the resulting clear solution into another flask where it may be dried over 5 g. of fresh drying agent. The spent drying agent is rinsed with 15–20 ml. of benzene, and the rinsings are added to the main solution.

4. Another boiling point is 66–68°/6 mm. If the product is distilled through a short column (12–15 cm.) packed with glass helices, it boils constantly at 98°/22 mm. The recovery is somewhat lower when a column is used.

The product gives a negative carbylamine test and hence contains no significant amount of unchanged primary amine.

5. This methylation procedure is quite generally satisfactory for simple primary and secondary amines. For methylation of a secondary amine only half as much formaldehyde is required, although a larger amount does no harm. The submitters also have prepared, in uniformly good yields, benzyldodecylmethylamine (b.p. 180–182°/4 mm.) from benzyldodecylamine, and α-amylhexyldimethylamine (b.p. 115°/16 mm.) from α-amylhexylamine. It is reported [1] that the reaction can be successfully applied to the methylation of butylamine, benzylamine, tetramethylenediamine, piperidine, and α-phenyl-α-aminobutyric acid.

3. Methods of Preparation

The procedure given above is an adaptation of the methylation method first used by Sommelet and Ferrand [2] and developed more fully by Clarke, Gillespie, and Weisshaus.[1] β-Phenylethyldimethylamine has been prepared from β-phenylethylamine by alkylation with dimethyl sulfate; [3] by the reaction of β-phenylethylamine and of N-methyl-β-phenylethylamine with formaldehyde;[4] by catalytic reduction of phenylacetonitrile in the presence of dimethylamine;[5] by the reaction of dimethylamine with β-phenylethyl chloride[6–8] and with β-phenylethyl bromide;[8] and by the reaction of phenylacetaldehyde with dimethylamine.[9]

[1] Clarke, Gillespie, and Weisshaus, *J. Am. Chem. Soc.*, **55**, 4571 (1933).
[2] Sommelet and Ferrand, *Bull. soc. chim. France*, (4) **35**, 446 (1924).
[3] Johnson and Guest, *J. Am. Chem. Soc.*, **32**, 761 (1910).
[4] Decker and Becker, *Ber.*, **45**, 2404 (1912); *Ann.*, **395**, 344 (1913).
[5] Buck, Baltzly, and Ide, *J. Am. Chem. Soc.*, **60**, 1789 (1938).
[6] Barger, *J. Chem. Soc.*, **95**, 2193 (1909).
[7] Tiffeneau and Fuhrer, *Bull. soc. chim. France*, (4) **15**, 173 (1914).
[8] v. Braun, *Ber.*, **43**, 3209 (1910).
[9] Ger. pat. 291,222 [*Frdl.*, **12**, 802 (1914–1916)].

2-PHENYLINDOLE

(Indole, 2-phenyl-)

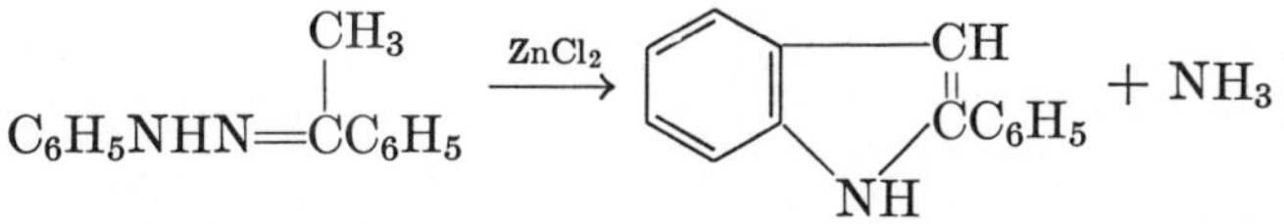

Submitted by R. L. Shriner, W. C. Ashley, and E. Welch.
Checked by Homer Adkins, Alan K. Roebuck, and Harry Coonradt.

1. Procedure

In a tall 1-l. beaker is placed an intimate mixture of 53 g. (0.25 mole) of freshly prepared acetophenone phenylhydrazone (Note 1) and 250 g. of powdered anhydrous zinc chloride (Note 2). The beaker is immersed in an oil bath at 170°, and the mixture is stirred vigorously by hand. The mass becomes liquid after 3–4 minutes, and evolution of white fumes begins. The beaker is removed from the bath, and the mixture is stirred for 5 minutes. In order to prevent solidification to a hard mass, 200 g. of clean sand is thoroughly stirred into the reaction mixture. The zinc chloride is dissolved by digesting the mixture overnight on the steam cone with 800 ml. of water and 25 ml. of concentrated hydrochloric acid (sp. gr. about 1.2). The sand and crude 2-phenylindole are removed by filtration, and the solids are boiled with 600 ml. of 95% ethanol. The hot mixture is decolorized with Norit and filtered through a hot 10-cm. Büchner funnel, and the sand and

Norit are washed with 75 ml. of hot ethanol. After the combined filtrates are cooled to room temperature, the 2-phenylindole is collected on a 10-cm. Büchner funnel and washed three times with small amounts (15–20 ml.) of cold ethanol. The first crop is quite pure; after drying in a vacuum desiccator over calcium chloride it weighs 30–33 g. and melts at 188–189° (cor.) (Notes 3 and 4).

A little Norit is added to the combined filtrate and washings, which are then concentrated to a volume of 200 ml. and filtered. The filtrate, on cooling, deposits a second crop of 5–6 g. (Note 5) of impure product, which melts at 186–188°. The total yield of 2-phenylindole is 35–39 g. (72–80%).

2. Notes

1. Acetophenone phenylhydrazone is prepared [1] by warming a mixture of 40 g. (0.33 mole) of acetophenone and 36 g. (0.33 mole) of phenylhydrazine on the steam cone for 1 hour. The hot mixture is dissolved in 80 ml. of 95% ethanol, and crystallization is induced by agitation. The mixture is then cooled in an ice bath, and the product is removed and washed with 25 ml. of ethanol. There is obtained 54–57 g. of white product. A second crop of 4–10 g. is obtained by concentrating the combined filtrate and washings. The combined solids are dried under reduced pressure, over calcium chloride, for 30 minutes. The total yield of acetophenone phenylhydrazone, m.p. 105–106°, is 61–64 g. (87–91%).

2. It has been reported [2] that such a large amount of zinc chloride is not necessary, but the submitters found that equal parts of acetophenone phenylhydrazone and zinc chloride gave lower yields.

3. Using 3.2 times the quantities specified above, except that no sand was added in separating the product, the checkers have obtained yields of 75–80% of 2-phenylindole.

4. While drying, the surface of the product becomes very light green.

5. The filtrate still contains some 2-phenylindole, but great difficulty is encountered in trying to purify the crude material.

3. Methods of Preparation

2-Phenylindole has been prepared by heating benzoyl-*o*-toluidine in an atmosphere of hydrogen,[3] the reaction being improved by the use of sodium amyloxide as the condensing agent; by heating phenylacetaldehydephenylhydrazone with 5 parts of anhydrous zinc chloride; [4] by warming phenacyl bromide [5,6] or N-phenacylaniline with aniline,[7] aniline hydrochloride, or zinc chloride;[8] by elimination of water from *o*-aminodesoxybenzoin,[9] prepared in turn from *o*-nitrodesoxybenzoin; by dehydration of benzal-*o*-toluidine;[10] by the action of anhydrous zinc chloride[11] or boron trifluoride[12] upon the phenylhydrazone of acetophenone; by the action of potassium amide and potassium nitrate upon 2-phenyl-4-quinolinecarboxylic acid.[13] N^{15}-2-Phenylindole has been prepared from the α-N^{15}-phenylhydrazone of acetophenone.[14]

[1] Reisenegger, *Ber.*, **16**, 662 (1883).
[2] Arbuzov and Tichvinshi, *J. Russ. Phys. Chem. Soc.*, **45**, 70 (1913); [*C. A.*, **7**, 2225, 3599 (1913)].
[3] Madelung, *Ber.*, **45**, 1131 (1912).
[4] Fischer and Schmitt, *Ber.*, **21**, 1072 (1888).
[5] Mohlau, *Ber.*, **15**, 2480 (1882).
[6] Bischler, *Ber.*, **25**, 2860 (1892).
[7] Mentzer, Molho, and Berguer, *Bull. soc. chim. France,* **1950**, 555.
[8] Verkade and Janetsky, *Rec. trav. chim.*, **62**, 763 (1943).
[9] Pictet, *Ber.*, **19**, 1064 (1886).
[10] Etard, *Compt. rend.*, **95**, 730 (1882).
[11] Fischer, *Ann.*, **236**, 133 (1886).
[12] Snyder and Smith, *J. Am. Chem. Soc.*, **65**, 2452 (1943).
[13] White, Doctorate thesis, Stanford University, 1940; White and Bergstrom, *J. Org. Chem.*, **7**, 497 (1942).
[14] Allen and Wilson, *J. Am. Chem. Soc.*, **65**, 611 (1943).

PHENYLMETHYLGLYCIDIC ESTER

(Hydrocinnamic acid, α,β-epoxy-β-methyl-, ethyl ester)

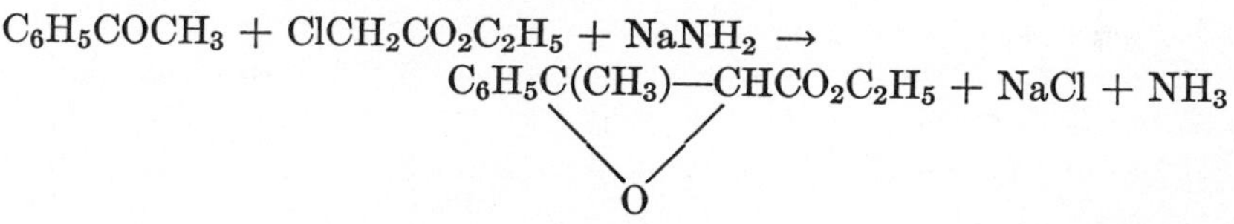

Submitted by C. F. H. Allen and J. VanAllan.
Checked by Nathan L. Drake and Carl Blumenstein.

1. Procedure

To a mixture of 120 g. (118.5 ml., 1 mole) of acetophenone (Note 1), 123 g. (109 ml., 1 mole) of ethyl chloroacetate, and 200 ml. of dry benzene in a 1-l. three-necked round-bottomed flask, fitted with a stirrer and low-temperature thermometer, is added, over a period of 2 hours, 47.2 g. (1.2 moles) of finely pulverized sodium amide. The temperature is kept at 15–20° by external cooling (Note 2). After the addition has been completed, the mixture is stirred for 2 hours at room temperature, and the reddish mixture is poured upon 700 g. of cracked ice, with hand stirring. The organic layer is separated and the aqueous layer extracted once with 200 ml. of benzene. The combined benzene solutions are washed with three 300-ml. portions of water, the last one containing 10 ml. of acetic acid. The benzene solution is dried over 25 g. of anhydrous sodium sulfate, filtered, the drying agent rinsed with a little dry benzene, and, after removal of the solvent, the residue is fractionated under reduced pressure, using a modified Claisen flask. The fraction boiling at 107–113°/3 mm. is collected separately and used for the preparation of hydratropaldehyde (p. 733). Redistillation yields a product which boils at 111–114°/3 mm. (Note 3). The yield is 128–132 g. (62–64%) (Note 4).

2. Notes

1. The practical grades of ketone and ester were used.

2. The reaction is strongly exothermic, and much ammonia is evolved.

3. Other boiling points are 272–275°/760 mm. with decomposition; 153–159°/20 mm.; and 147–149°/12 mm.

4. There are several side reactions which reduce the yield. There are always unchanged ketone and ester in the low-boiling fraction, and also some chlorocinnamic ester.

3. Methods of Preparation

This is an example of a general reaction by which haloesters are condensed with ketones by means of sodium,[1] sodium ethoxide,[2,3] or sodium amide.[4–6]

[1] Erlenmeyer, *Ann.*, **271,** 161 (1892).

[2] Darzens, *Compt. rend.*, **139,** 1215 (1904).

[3] Dutta, *J. Indian Chem. Soc.*, **18,** 235 (1941) [*C. A.*, **36,** 761 (1942)].

[4] Claisen and Feyerabend, *Ber.*, **38,** 702 (1905).

[5] I. G. Farbenind. A.-G., Ger. pat. 591,452 [*Frdl.*, **19,** 288 (1934)] [*C. A.*, **28,** 2367 (1934)].

[6] I. G. Farbenind. A.-G., Ger. pat. 602,816 [*Frdl.*, **20,** 217 (1935)] [*C. A.*, **29,** 1438 (1935)].

1-PHENYLNAPHTHALENE

(Naphthalene, 1-phenyl-)

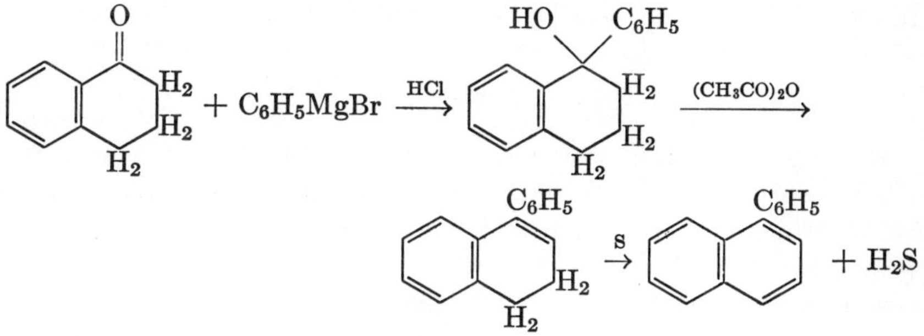

Submitted by RICHARD WEISS.
Checked by C. F. H. ALLEN and F. P. PINGERT.

1. Procedure

A. 1-*Phenyldialin.* A solution of phenylmagnesium bromide is prepared in the usual manner [1] from 11 g. (0.45 gram atom) of magnesium, 75 g. (0.48 mole) of bromobenzene, and 175 ml. of ether. As soon as the metal has reacted, a solution of 58.4 g. (0.4 mole) of α-tetralone in 60 ml. of ether is added from a dropping funnel as rapidly as possible so that vigorous refluxing is maintained; about 30 minutes is required. The reaction mixture is then heated under reflux for an additional 30 minutes and allowed to stand for 1 hour. The magnesium complex is decomposed by about 250 g. of ice and 40 ml. of concentrated hydrochloric acid. The ether layer is separated and distilled with steam to remove impurities (Note 1); about 6 hours is necessary, and approximately 4.5 l. of distillate is collected. The heavy residual oil is separated from the water, and 80 ml. of ether and 10 g. of calcium chloride are added. After 4–5 minutes, the calcium chloride is removed by filtration, and the ether is driven off by distillation on a steam bath. Twenty milliliters of acetic anhydride is then added, and the flask and contents are heated in a steam bath (Note 2) for 20–25 minutes. The mixture is finally distilled under reduced pressure through a 15-in. Widmer column (Note 3). The fraction which boils at 135–140°/2 mm. is collected; the yield is 35–40 g. (42–48%).

B. 1-*Phenylnaphthalene.* A mixture of 6 g. (0.18 mole) of powdered sulfur and 35 g. (0.17 mole) of 1-phenyldialin in a 200-ml. Claisen flask having a modified side arm is heated for 30 minutes in a metal bath the temperature of which is 250–270° (Note 4). At the end of this time the evolution of hydrogen sulfide will have ceased. The heavy oil (Note 5) is then distilled from the same flask; the yield is 32–33 g. (91–94%); the boiling range of the product is 134–135°/2 mm. or 189–190°/12 mm. (Notes 6 and 7).

2. Notes

1. Steam distillation removes unused bromobenzene, biphenyl, etc.
2. The reaction is best carried out in a 500-ml. Erlenmeyer flask or beaker which can be lowered into the steam pot and held by a cloth wrap.
3. If a fractionating column is not used, several distillations are necessary to secure a product with the boiling range given. When this distillation is done as described, subsequent operations are facilitated, and a column is not required for the final distillation of the hydrocarbon. Ebullition tubes are desirable in both distillations.
4. The temperature inside the flask should be 250°.
5. Significant amounts of hydrocarbon are lost with each change of flask. For this reason it is advisable to perform the dehydrogenation and distillation in the same flask.
6. Sometimes the hydrocarbon is blue, probably owing to traces of an azulene. The contaminant is easily removed by dissolving the hydrocarbon in an equal volume of hexane or petroleum ether and shaking this solution with an equal volume of 85% syrupy phosphoric acid until the color has been removed. The hydrocarbon is then obtained on evaporation of the solvent; it does not need redistillation.
7. The over-all yield is 40%, based on the α-tetralone.

3. Methods of Preparation

1-Phenylnaphthalene has been prepared by the reaction of α-halonaphthalenes with mercury diphenyl [2] or with benzene in the presence of aluminum chloride,[2] and by means of a Grignard synthesis, starting with either bromobenzene, or cyclohexyl chloride and α-tetralone [3,4] or with α-bromonaphthalene and cyclohexanone.[3,5–7] A 60% over-all yield has been reported for the Grignard preparation from cyclohexanone.[7] The formation of the hydrocarbon through the diazo reaction [8–11] appears to be less attractive than the method described. Dehydrogenation of the reduced naphthalene has been accomplished by the use of sulfur,[3] bromine,[5] platinum black, or selenium.[4]

[1] *Org. Syntheses* Coll. Vol. **1**, 226 (1941).
[2] Chattaway, *J. Chem. Soc.*, **63**, 1187 (1893).
[3] Weiss and Woidich, *Monatsh.*, **46**, 455 (1925).
[4] Cook and Lawrence, *J. Chem. Soc.*, **1936**, 1432.
[5] Sherwood, Short, and Clausfield, *J. Chem. Soc.*, **1932**, 1834.
[6] Veselý and Štursa, *Collection Czechoslov. Chem. Commun.*, **5**, 344 (1933) [*C. A.*, **28**, 144 (1934)].
[7] Orchin and Reggel, *J. Am. Chem. Soc.*, **69**, 505 (1947).
[8] Grieve and Hay, *J. Chem. Soc.*, **1938**, 108.
[9] Waters, *J. Chem. Soc.*, **1939**, 864.
[10] Hodgson and Marsden, *J. Chem. Soc.*, **1940**, 208.
[11] Bachmann and Hofmann, *Org. Reactions*, **2**, 248 (1944).

PHENYLPROPARGYL ALDEHYDE

(Propiolaldehyde, phenyl-)

$$C_6H_5CH{=}CHCHO \xrightarrow{Br_2} C_6H_5CHBrCHBrCHO \xrightarrow{K_2CO_3}$$
$$C_6H_5CH{=}CBrCHO \xrightarrow{HC(OC_2H_5)_3} C_6H_5CH{=}CBrCH(OC_2H_5)_2 \xrightarrow{KOH}$$
$$C_6H_5C{\equiv}CCH(OC_2H_5)_2 \xrightarrow[H_2SO_4]{H_2O} C_6H_5C{\equiv}CCHO$$

Submitted by C. F. H. Allen and C. O. Edens, Jr.
Checked by C. D. Heaton and C. R. Noller.

1. Procedure

A. *α-Bromocinnamic aldehyde.* A mixture of 44 g. (0.33 mole) of cinnamic aldehyde and 167 ml. of acetic acid in a 500-ml. three-necked round-bottomed flask, surrounded by a cold-water bath and fitted with a stirrer, reflux condenser, and dropping funnel, is stirred vigorously while 17.1 ml. (53.5 g., 0.33 mole) of bromine is added. This is followed by the addition of 23 g. (0.17 mole) of anhydrous potassium carbonate. When the evolution of gas has ceased, the mixture is refluxed for 30 minutes, then cooled and poured into 435 ml. of water in a 1-l. flask; a lower, reddish layer of crude α-bromoaldehyde separates. The flask is stoppered, cooled under running water, and shaken vigorously. The resulting granular solid is filtered with suction and dissolved without drying by warming with 220 ml. of 95% ethanol. After the addition of 50 ml. of water, the solution is warmed until it becomes clear and is then set aside to crystallize at room temperature and finally in a refrigerator. α-Bromocinnamic aldehyde separates as nearly colorless needles, which are filtered with suction, rinsed with 17 ml. of 80% ethanol, and air-dried. The yield of product melting at 72–73° is 52–60 g. (75–85%) (Note 1).

B. *α-Bromocinnamic aldehyde acetal.* In a 250-ml. flask are placed 45 g. (0.21 mole) of α-bromocinnamic aldehyde, 50 ml. (0.3 mole) of ethyl orthoformate, 40 ml. of absolute ethanol, and 0.5 g. of ammonium chloride. After the mixture has been refluxed for 30 minutes, it is transferred to a modified Claisen flask, and the low-boiling constituents are removed at atmospheric pressure and a bath temperature up to 150° (Note 2). The yield of acetal boiling at 146–149°/5 mm. and with n_D^{22} 1.5500 is 50–52 g. (82–86%).

C. *Phenylpropargyl aldehyde acetal.* A solution of 20.7 g. (0.25 mole) of potassium hydroxide in 200 ml. of absolute ethanol is added to 50 g. (0.18 mole) of α-bromocinnamic aldehyde acetal in a 500-ml. flask. The mixture is refluxed for 1.5 hours and poured into a 3-l. separatory funnel containing 1.5 l. of water. The oil that separates is extracted with three 170-ml. portions of chloroform, and the combined chloroform extracts are washed with three 75-ml. portions of water and then dried over 15 g. of anhydrous sodium sulfate. The chloroform is removed by distillation, and the residual oil is distilled from a modified Claisen flask. The yield of phenylpropargyl aldehyde acetal boiling at 153–156°/19 mm. is 28–31 g. (80–86%).

D. *Phenylpropargyl aldehyde.* Twenty-nine grams (0.14 mole) of the acetal is added to 140 ml. of water containing 10 ml. of concentrated sulfuric acid, and the mixture is heated on a steam bath with occasional shaking for 30 minutes. The aldehyde is then steam-distilled and extracted from the distillate with two 250-ml. portions of ether. The ethereal solution is dried over 20 g. of anhydrous sodium sulfate, the solvent is removed, and the residue is distilled from a 100-ml. modified Claisen flask. The yield of phenylpropargyl aldehyde boiling at 114–117°/17 mm. and with n_D^{25} 1.6032 is 13–15 g. (70–81%).

2. Notes

1. The same percentage yield is obtained in runs three times as large.

2. Both the reaction and the distillation can be carried out in the same flask.

3. Methods of Preparation

The procedure given for the preparation of phenylpropargyl aldehyde is a modification of Claisen's directions [1] in part due to Kalff.[2] The monobromocinnamic aldehyde was described by Zincke.[3] Other methods of possible preparative value for the aldehyde are the interaction of sodium phenylacetylene,[4–7] or the Grignard reagent from phenylacetylene with ethyl orthoformate. Sodium[5] and lithium[8] phenylacetylene react with ethyl formate to yield the aldehyde.

[1] Claisen, *Ber.*, **31,** 1020 (1898).
[2] Kalff, *Rec. trav. chim.*, **46,** 594 (1927).
[3] Zincke and Hagen, *Ber.*, **17,** 1815 (1884).
[4] Moureu and Delange, *Compt. rend.*, **133,** 106 (1901).
[5] Moureu and Delange, *Bull. soc. chim. France,* (3) **31,** 1329 (1904).
[6] Brachim, *Bull. soc. chim. France,* (3) **35,** 1165 (1906).
[7] Charon and Dugoujon, *Compt. rend.*, **137,** 126 (1903).
[8] Nightingale and Wadsworth, *J. Am. Chem. Soc.*, **69,** 1181 (1947).

α-PHENYLPROPIONALDEHYDE *

(Hydratropaldehyde)

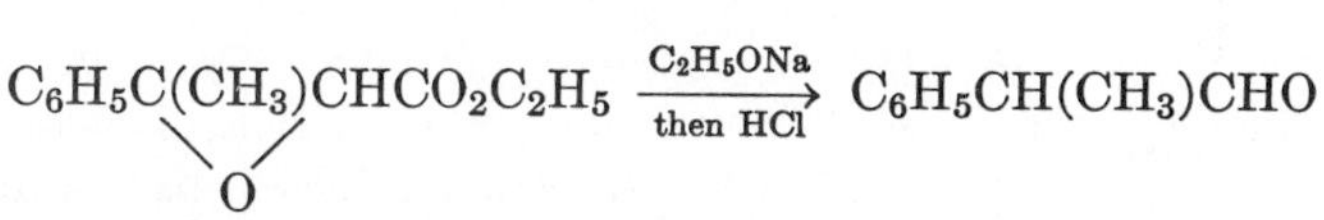

Submitted by C. F. H. Allen and J. VanAllan.
Checked by Nathan L. Drake and Carl Blumenstein.

1. Procedure

An ethanolic solution of sodium ethoxide is prepared in a 1-l. round-bottomed flask from 15.5 g. (0.67 gram atom) of sodium and 300 ml. of absolute ethanol (Note 1). One hundred thirty-three grams (0.64 mole) of phenylmethylglycidic ester (p. 727) is added to this solution slowly and with shaking. The flask is then cooled externally to 15°, and 15 ml. of water is slowly added; considerable heat is evolved, and the sodium salt soon begins to separate. After the mixture has stood overnight, the salt is filtered by suction and rinsed with one 50-ml. portion of ethanol and a similar amount of ether. The dried salt weighs 102–108 g. (80–85%) and melts at 255–256° with decomposition.

The salt is added to a dilute solution of hydrochloric acid prepared by mixing 300 ml. of water and 56 ml. of concentrated acid (sp. gr. 1.18); the acid should be contained in a 1-l. flask under a reflux condenser. The mixture is warmed gently, whereupon carbon dioxide is evolved and an oil separates. The flask is heated on a steam bath for

1.5 hours, and the oil is then extracted from the cooled mixture with 150 ml. of benzene. The extract is washed once with 200 ml. of water and distilled under reduced pressure, using an ordinary 500-ml. Claisen flask. The hydratropaldehyde distils at 90–93°/10 mm. or 73–76°/4 mm. (oil bath at 120–130°), leaving only a slight residue (3–5 g.). The yield is 56–60 g. (65–70%) (Notes 2 and 3).

2. Notes

1. The metal is placed in the flask, and the ethanol is added through the condenser as fast as refluxing will allow (p. 215, Note 1).

2. The generality of this reaction is limited only by the availability of the glycidic esters.

3. The 2,4-dinitrophenylhydrazone forms yellow prisms that melt at 135°.

3. Methods of Preparation

Hydratropaldehyde has been prepared by hydrolysis of phenylmethylglycidic ester,[1–3] by chromyl chloride oxidation of cumene,[4] by the elimination of halogen acid or water from halohydrins or glycols,[5–7] and by the distillation at ordinary pressure of methylphenylethylene oxide.[8,9]

Hydratropaldehyde has also been prepared by the catalytic addition of hydrogen and carbon monoxide to styrene.[10]

1 Claisen, *Ber.*, **38**, 703 (1905).

2 I. G. Farbenind. A.-G., Ger. pat. 602,816 [*Frdl.*, **20**, 217 (1935)].

3 Dutta, *J. Indian Chem. Soc.*, **18**, 235 (1941) [*C. A.*, **36**, 161 (1942)].

4 v. Miller and Rohde, *Ber.*, **24**, 1359 (1889).

5 Bougault, *Ann. chim. phys.*, (7) **25**, 548 (1902).

6 Tiffeneau, *Bull. soc. chim. France*, (3) **27**, 643 (1902); *Compt. rend.*, **134**, 846 (1902); **137**, 1261 (1903); **142**, 1538 (1906); *Ann. chim. phys.*, (8) **10**, 353 (1907).

7 Stoermer, *Ber.*, **39**, 2298 (1906).

8 Klages, *Ber.*, **38**, 1971 (1905).

9 Tiffeneau, *Compt. rend.*, **140**, 1459 (1905); *Ann. chim. phys.*, (8) **10**, 192 (1907).

10 Adkins and Krsek, *J. Am. Chem. Soc.*, **70**, 383 (1948).

α-PHENYLTHIOUREA *

(Urea, 1-phenyl-2-thio-)

$$NH_4SCN + C_6H_5COCl \longrightarrow C_6H_5CONCS + NH_4Cl$$

$$C_6H_5CONCS + C_6H_5NH_2 \longrightarrow C_6H_5CONHCSNHC_6H_5$$

$$C_6H_5CONHCSNHC_6H_5 \xrightarrow{NaOH} C_6H_5NHC(=S)NH_2$$

Submitted by Robert L. Frank and Paul V. Smith.[1]
Checked by Richard T. Arnold and Sherman Sundet.

1. Procedure

In a 500-ml. three-necked flask fitted with a reflux condenser, a mechanical stirrer, and a 100-ml. dropping funnel are placed 17 g. (0.22 mole) of ammonium thiocyanate and 100 ml. of dry acetone (Note 1). Through the dropping funnel is added, with stirring, 28.2 g. (0.2 mole) of benzoyl chloride. After the addition is complete, the mixture is refluxed for 5 minutes. Then a solution of 18.6 g. (0.2 mole) of aniline in 50 ml. of dry acetone is added at such a rate that the solution refluxes gently. The mixture is poured carefully with stirring into 1.5 l. of water, and the resulting yellow precipitate (α-benzoyl-β-phenylthiourea) is separated by filtration. The crystals are heated for 5 minutes with a boiling solution of 30 g. of sodium hydroxide in 270 ml. of water. After the removal of a small amount of insoluble material by filtration, the solution is acidified with concentration hydrochloric acid and then made slightly basic with ammonium hydroxide. Upon standing, the solution deposits the crystalline product. The yield of oven-dried material (m.p. 151–153°) is 25.8 g. (85%). Recrystallization from ethanol yields 23.2 g. (76%) of white plates melting at 152.5–153°.

2. Note

1. The acetone is dried for at least 48 hours over anhydrous calcium sulfate (Drierite) and distilled just before it is used.

3. Methods of Preparation

α-Phenylthiourea has been prepared by the action of ammonium thiocyanate,[2] thiocyanic acid,[3] thiuramdisulfide,[4] or silicon thiocyanate [5] on aniline; by the action of ammonium thiocyanate on aniline hydrochloride; [6] by the action of ammonia on phenyl isothiocyanate,[7] 1-phenyl-2-thiobiuret,[8] thiocarbanilide,[9] phenyldithiocarbamazide,[10] or phenyl isothiocyanate hexasulfide; [11] by the addition of hydrogen sulfide to monophenylcyanamide; [12] by the decomposition of salts of phenyldithiocarbamic acid in the presence of lead carbonate,[13] ammonium polysulfide,[14] or ammonium carbonate; [15] by the reaction of thiophosgene, aniline, and ammonia; [16] and by the action of hydrazine hydrate on phenyldithiobiuret.[17]

The preparation of α-phenylthiourea by the procedure described herein has not been reported, although Douglass and Dains [18] have applied the method to the preparation of various substituted phenyl derivatives.

1 Work done under contract with the Office of Rubber Reserve.

2 Schiff, *Ann.*, **148**, 338 (1868); Rathke, *Ber.*, **18**, 3102 (1885).

3 Salkowski, *Ber.*, **24**, 2724 (1891); Challenger and Collins, *J. Chem. Soc.*, **125**, 1377 (1924); DeBeer, Buck, Ide, and Hjort, *J. Pharmacol.*, **57**, 19 (1936).

4 Klason, *J. prakt. Chem.*, [2] **36**, 57 (1887); Fromm, *Ber.*, **42**, 1955 (1909).

5 Reynolds, *J. Chem. Soc.*, **89**, 397 (1906).

6 deClermont, *Ber.*, **9**, 446 (1876); Liebermann, *Ann.*, **207**, 122 (1881); Bertram, *Ber.*, **25**, 48 (1892); Ger. pat. 604,639, French pat. 762,310 [*C. A.*, **29**, 819 (1935)]; Krall and Gupta, *J. Indian Chem. Soc.*, **12**, 629 (1935).

7 Hofmann, *Compt. rend.*, **47**, 424 (1858); Otterbacher and Whitmore, *J. Am. Chem. Soc.*, **51**, 1909 (1929).

8 Birckenbach and Kraus, *Ber.*, **71**, 1492 (1938).

9 Gebhardt, *Ber.*, **17**, 3043 (1884); v. Walther and Stenz, *J. prakt. Chem.*, [2] **74**, 223 (1906).

10 Oliveri-Mandala, *Gazz. chim. ital.*, **51**, II, 195 (1921).

11 Levi, *Gazz. chim. ital.*, **61**, 619 (1931).

12 Weith, *Ber.*, **9**, 810 (1876).

13 Heller and Bauer, *J. prakt. Chem.*, [2] **65**, 365 (1902).

14 Azzalin, *Gazz. chim. ital.*, **55**, 895 (1925).

15 Drozdov, *J. Gen. Chem. U.S.S.R.*, **6**, 1368 (1936).

16 Hutin, *Rev. gén. mat. plastiques*, **7**, 95 (1931).

17 Fromm, *Ann.*, **426**, 326 (1922).

18 Douglass and Dains, *J. Am. Chem. Soc.*, **56**, 1408 (1934).

PHTHALALDEHYDIC ACID

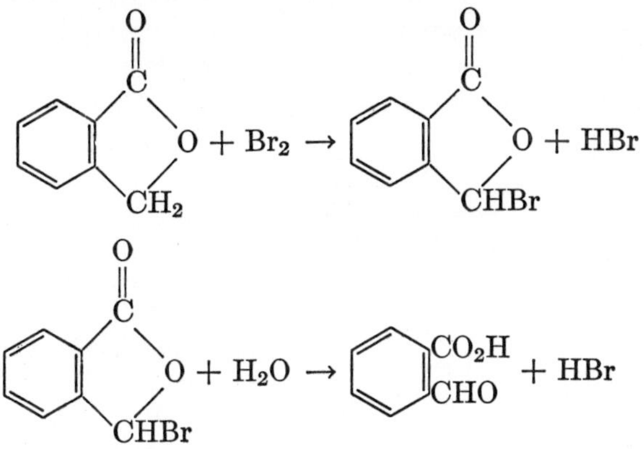

Submitted by R. L. SHRINER and F. J. WOLF.
Checked by LEE IRVIN SMITH, R. T. ARNOLD, and WALTER FRAJOLA.

1. Procedure

A. 2-*Bromophthalide.* The apparatus shown in Fig. 26 is used for the bromination. Flasks *A* and *B* are of 200-ml. capacity. Bromine is introduced by means of a current of carbon dioxide, which passes through mineral oil or sulfuric acid in the bubble counter, then through the bromine in flask *B*, and finally through the drying tower. The tower is conveniently made from a condenser jacket and is filled with anhydrous calcium chloride. Flask *A* is surrounded by an oil bath and is equipped with a thermometer, an outlet tube of wide bore connected to a gas-absorption trap, and a gas-inlet tube having an inside diameter of 2 mm. The inlet tube reaches to the bottom of the flask.

In the reaction flask *A* is placed 134 g. (1 mole) of phthalide (Note 1). In flask *B* is placed 160 g. (53.5 ml., 1 mole) of bromine. The oil bath is maintained at 140–155°, and the stream of carbon dioxide is started when the temperature of the phthalide has reached 140°. The temperature inside flask *A* is maintained at 135–150° (oil bath 140–155°) during the course of the reaction (Note 2). Carbon dioxide is introduced at such a rate that no bromine vapor is observed in the outlet tube (5–8 bubbles per second). The stream of carbon dioxide is continued for 30 minutes after all the bromine color has disappeared from the train. The reaction is complete in 10–13 hours, depending upon the rate at which the bromine has been introduced into the reaction mixture.

While still warm, the reaction mixture is transferred to a 250-ml. modified Claisen flask fitted for distillation under reduced pressure. Any hydrogen bromide remaining in the reaction mixture is removed by heating at 120° under the vacuum of a water pump. The product is then distilled under reduced pressure. The fore-run of less than

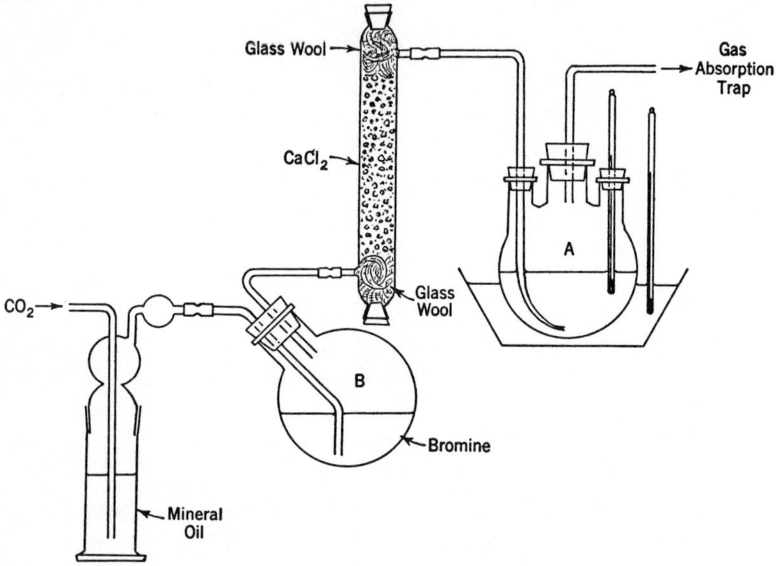

Fig. 26.

15 g. is largely phthalide (Note 3). The 2-bromophthalide, which distils at 138–142°/4 mm. (128–132°/2 mm.), weighs 175–178 g. (82–83% based on the phthalide) (Note 4). It is water-clear and solidifies to a solid which melts at 69–73°.

The distilled product is pure enough for use in the subsequent hydrolysis, but it may be purified by recrystallization from 100 ml. of carbon tetrachloride. Upon cooling, 100 g. of pure 2-bromophthalide melting at 75° is obtained. An additional 30–40 g. of slightly yellow material is obtained by concentrating the mother liquor.

B. *Phthalaldehydic acid.* The entire distillate is placed in a 500-ml. flask and covered with 230 ml. of water. The flask is equipped with a mechanical stirrer and is heated on a steam cone. The hydrolysis is complete when the layer of 2-bromophthalide has disappeared (about 30 minutes). The reaction mixture is then placed in a refrigerator overnight, during which time the entire mass solidifies. The product is filtered, washed with two 50-ml. portions of ice water, and dried in the air. The yield of crude product melting at 60–65° (Note 5) is 140–160 g.

The crude product is recrystallized from 400 ml. of hot water and dried in the air. The recrystallized product is white, melts at 95–96°, and weighs 97–102 g. (78–83% based on the bromo compound or 65–68% based on the phthalide).

2. Notes

1. The phthalide used by the submitters and by the checkers was a commercial product, obtained from E. I. du Pont de Nemours and Company, Wilmington, Delaware. This product is no longer available. Phthalide may be prepared in 82.5% yield by hydrogenation of phthalic anhydride in benzene at 270° under 3000 lb. pressure in the presence of copper chromite [1] or, in yields of 61–71%, from phthalimide according to the procedure given in *Org. Syntheses* Coll. Vol. **2,** 526 (1943).

2. At a temperature below 135°, bromination does not take place readily. Above 155°, the reaction mixture becomes considerably darker, and the yield is lower.

3. A change in crystalline structure of the distillate is observed when all the phthalide has been removed.

4. The checkers consistently obtained yields of at least 87%, and in one run the yield of product melting at 78° was 95%.

5. In one run, the checkers obtained 180 g. of product which melted at 67°. This, after recrystallization, melted at 94.5–95° and weighed 121 g. This yield is 84.9% of the theoretical amount based on the bromo compound, or 80% based on the phthalide. The crude product holds water tenaciously, but this is removed by allowing the product to stand in a vacuum desiccator over Drierite.

3. Methods of Preparation

To the methods listed in *Org. Syntheses* Coll. Vol. **2,** 523 (1943) may be added the hydrolysis of 2-chlorophthalide; [2] the action of carbon dioxide and sodium, under pressure, upon *o*-chlorobenzaldehyde; [3] and from the hydrolysis of methyl phthalaldehydate, obtained by the Rosenmund reduction of the acid chloride of methyl hydrogen phthalate.[4] The procedure described above is essentially that of Racine.[5]

[1] Austin, Bousquet, and Lazier, *J. Am. Chem. Soc.,* **59,** 864 (1937).

[2] Austin and Bousquet, U. S. pat. 2,047,946 [*C. A.,* **30,** 6011 (1936)].

[3] Morton, LeFevre, and Hechenbleikner, *J. Am. Chem. Soc.*, **58**, 754 (1936).
[4] Eliel and Burgstahler, *J. Am. Chem. Soc.*, **71**, 2251 (1949).
[5] Racine, *Ann.*, **239**, 79 (1887).

PICOLINIC ACID HYDROCHLORIDE *

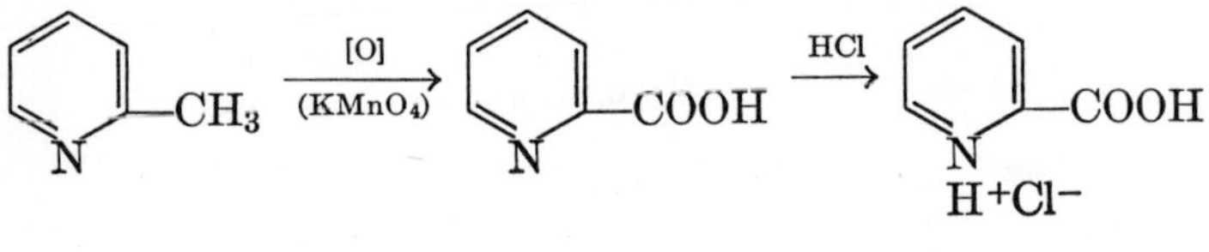

Submitted by ALVIN W. SINGER and S. M. MCELVAIN.
Checked by C. F. H. ALLEN and ALAN BELL.

1. Procedure

In a 5-l. three-necked flask, fitted with a reflux condenser and stirrer, are placed 2500 ml. of water and 50 g. of α-picoline (0.54 mole) (Note 1). Ninety grams (0.57 mole) of potassium permanganate is added, and the solution is heated on a steam bath until the purple color has practically disappeared (about 1 hour). A second 90-g. portion of permanganate is then introduced, followed by 500 ml. of water, and the heating is continued until the purple color is destroyed (2–2.5 hours). The reaction mixture is allowed to cool slightly, and the precipitated oxides of manganese are filtered and washed well with 1 l. of hot water (Note 2). The filtrate is concentrated under reduced pressure to 150–200 ml., filtered, if necessary, and acidified to Congo red with concentrated hydrochloric acid (65–70 ml., sp. gr. 1.19). This acid solution is then evaporated to dryness under reduced pressure. The solid residue is refluxed for one hour with 250 ml. of 95% ethanol and filtered, and the extraction is repeated with 150 ml. of 95% ethanol. Dry hydrogen chloride is passed into the combined ethanolic filtrates until crystals begin to separate. The solution is then chilled to about 10° in a freezing mixture, the addition of hydrogen chloride being continued until the solution is saturated. The crystals of picolinic acid hydrochloride which separate are filtered and air-dried. The yield is 43–44 g. (50–51%), m.p. 228–230° (Note 3).

This hydrochloride may contain traces of potassium chloride, which can be removed by dissolving the hydrochloride in hot absolute ethanol (50 g. requires 1 l.) and filtering from insoluble material. An equal volume of dry ether is then added to the warm ethanolic solution, and, after cooling, the crystallized product is filtered. The recovery is 40–43 g., m.p. 210–212° (230°) (Notes 3 and 4).

2. Notes

1. The submitters used a picoline fraction, b.p. 128–132°, whereas the checkers used the practical product, Eastman Kodak Company, b.p. 128–134°. The same yield was obtained with a carefully fractionated picoline, boiling over the 1° range 128–129°.

2. The washed manganese dioxide does not contain an appreciable amount of acid.

3. When the melting point is determined in the ordinary manner, it is found to be 228–230° with decomposition. If, on approaching the melting point, the rate of heating is reduced to 1° in 5 minutes, the value 210–212° can be observed.

4. About 2 g. of potassium chloride is removed by this procedure. The unpurified material can be used for most purposes.

3. Methods of Preparation

Picolinic acid has been generally prepared by the permanganate oxidation of α-picoline and isolated through the copper salt.[1–5] In one instance,[6] it was isolated directly as in the present procedure. It has also been secured by the hydrolysis of ω-trichloropicoline,[7,8] and by the nitric acid oxidation of α-picoline.[9]

[1] Weidel, *Ber.,* **12,** 1992 (1879).
[2] Pinner, *Ber.,* **33,** 1226 (1900).
[3] Camps, *Arch. Pharm.,* **240,** 345 (1902).
[4] Ley and Ficken, *Ber.,* **50,** 1132 (1917).
[5] Clemo and Ramage, *J. Chem. Soc.,* **1931,** 440.
[6] Mende, *Ber.,* **29,** 2887 (1896).
[7] Dyson and Hammick, *J. Chem. Soc.,* **1939,** 781.
[8] Ochiai and Okuda, *J. Pharm. Soc. Japan,* **70,** 156 (1950) [*C. A.,* **44,** 5878 (1950)].
[9] U. S. pat. 2,505,568 [*C. A.,* **44,** 6443 (1950)].

γ-*n*-PROPYLBUTYROLACTONE AND β-(TETRAHYDROFURYL)PROPIONIC ACID

(Enanthic acid, γ-hydroxy-, lactone, and 2-furanpropionic acid, tetrahydro-)

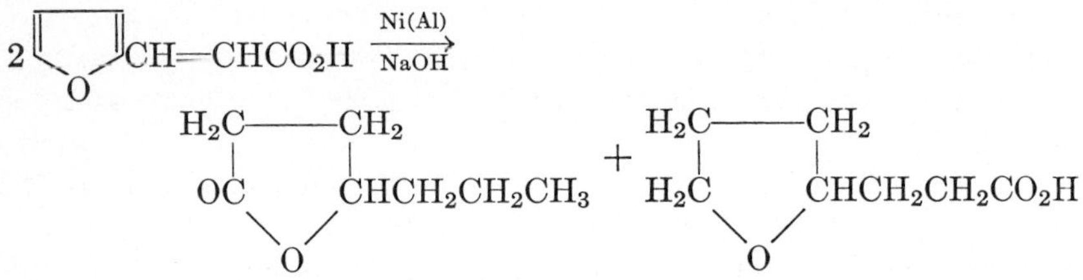

Submitted by ERWIN SCHWENK, DOMENICK PAPA, HILDA HANKIN and HELEN GINSBERG.
Checked by CLIFF S. HAMILTON and FLAVEN E. JOHNSON.

1. Procedure

In a 2-l. beaker resting on a hot plate and equipped with a mechanical stirrer are placed 150 g. of sodium hydroxide (Note 1) dissolved in 800 ml. of water and 41.4 g. (0.3 mole) of furylacrylic acid (p. 425). The stirrer is started, and to the warm solution (Note 2) 100 g. of Raney nickel-aluminum alloy is added, in small portions, over a period of 4–4.5 hours (Notes 3 and 4). During the addition of the alloy, the temperature of the mixture is kept at 60–70° and then is raised to approximately 95° where it is held for an additional 2–3 hours, with stirring. From time to time sufficient water is added to the reaction mixture to maintain approximately the original volume. The hot solution is filtered by decantation, and the nickel residue (Note 5) is washed with two 50-ml. portions of hot 2% sodium hydroxide solution. The combined filtrates and washings are cooled and then immediately added slowly (Note 6), with good stirring, to 800 ml. of concentrated hydrochloric acid (Note 7). The solution at this point is strongly acid to Congo red paper (Note 8). The acidified solution is cooled and thoroughly extracted with three 200-ml. portions of ether (Note 9). The ether extracts are combined and washed once with 100 ml. of 10% sodium chloride solution, the ether is evaporated, and the residue is fractionally distilled in a modified Claisen flask. The yield of γ-*n*-propylbutyrolactone boiling at 84–85°/5 mm., n_D^{25} 1.4385, is 13.0–14.9 g. (33–37%), while the yield of β-(tetrahydrofuryl)propionic acid boiling at 123–124°/5 mm., n_D^{25} 1.4578, is 15.0–17.0 g. (34–39%) (Notes 10, 11).

2. Notes

1. Solid sodium hydroxide is used in order to take advantage of the heat of solution. If the mixture at this stage is allowed to cool, it will be necessary to heat the solution to 50–60° before the addition of the alloy is begun.

2. Some of the furylacrylic acid remains undissolved but goes into solution readily after the addition of the nickel-aluminum alloy is begun.

3. If any excessive frothing occurs during the addition of the alloy, a few drops of octyl alcohol can be added from time to time.

4. During the addition of the alloy, it is advantageous to stir the solution efficiently, the alloy being added in the vortex of the solution. If the addition of the alloy is made on the surface of the liquid, most of the hydrogen developed is lost without entering into the reaction.

5. The Raney nickel residue is quite active and will ignite if allowed to become dry. It may readily be disposed of by pouring into dilute mineral acid. This nickel residue is sufficiently active for various types of catalytic hydrogenations requiring the use of Raney nickel catalyst.

6. If the acidified mixture is allowed to become too hot, some of the material may be lost by steam distillation.

7. The alkaline solution is added to the hydrochloric acid in the manner described, since the reverse order of addition usually results in the precipitation of aluminum salts which dissolve only after considerable heating and stirring. The alkaline solution is added at such a rate that at no time is there any appreciable amount of solid present.

8. If insufficient hydrochloric acid is used at this stage, aluminum salts will precipitate and will make the ether extractions difficult because of the formation of emulsions.

9. The acidified solution has also been extracted with ether in a continuous liquid-liquid extractor, this extraction requiring approximately 20 hours. Only slightly higher yields of the two products are obtained by this modification.

10. According to the submitters an alternative procedure for the separation of γ-n-propylbutyrolactone and β-(tetrahydrofuryl)propionic acid is the following:

The ether extracts, after washing with salt solution, are extracted with one 100-ml. and two 50-ml. portions of 5% sodium carbonate solution. It is important that the ether solution be free of any mineral acid before the carbonate extractions are made. The residual ether solution, after being washed with 10% sodium chloride solution, is dried over sodium sulfate. The ether is removed by distillation, and the residue is distilled under reduced pressure. The yield of γ-*n*-propylbutyrolactone is 14.5 g. (36%), b.p. 78–80°/2 mm. The combined sodium carbonate extracts are acidified to Congo red paper with concentrated hydrochloric acid and then thoroughly extracted with one 100-ml. and two 50-ml. portions of ether. The combined ether extracts are washed with 10% sodium chloride solution and dried; and, after the ether is removed by distillation, the residue is distilled under reduced pressure. The yield of β-(tetrahydrofuryl)propionic acid is 15.9 g. (36%), b.p. 118–120°/2 mm.

11. The submitters have used this procedure for the reduction of β-(α-thenoyl)propionic acid[1] from which ω-hydroxycaprylic acid[2] has been obtained in a yield of 38%, melting at 54–55°, and δ-*n*-propylvalerolactone in a yield of 31%, boiling at 116–117°/10 mm.

3. Methods of Preparation

γ-*n*-Propylbutyrolactone has been obtained from γ-bromoenanthic acid on boiling with water,[3] from β,γ-hepteneoic acid and sulfuric acid,[4] and by distilling γ-propylparaconic acid.[5]

β-(Tetrahydrofuryl)propionic acid has been obtained from furylacrylic acid by catalytic reduction with platinum oxide,[6] with nickel on kieselguhr,[7] with platinum black,[8] or with Raney nickel.[9] It has also been obtained from β-furyl-β-propiolactone by reduction,[10] and from 2-tetrahydrofurylpropanol by oxidation.[11]

The reduction procedure described has been published.[12]

[1] Fieser and Kennelly, *J. Am. Chem. Soc.*, **57**, 1615 (1935).
[2] Chuit and Hausser, *Helv. Chim. Acta*, **12**, 466 (1929).
[3] Fittig and Schmidt, *Ann.*, **255**, 80 (1889).
[4] Rupe, *Ber.*, **35**, 4272 (1902).
[5] Fittig, *Ann.*, **255**, 76 (1889).
[6] Kaufmann and Adams, *J. Am. Chem. Soc.*, **45**, 3042 (1923).
[7] Burdick and Adkins, *J. Am. Chem. Soc.*, **56**, 440 (1934).
[8] Windaus and Dalmer, *Ber.*, **53**, 2304 (1920).
[9] Paul and Hilly, *Compt. rend.*, **208**, 359 (1939).
[10] U. S. pat. 2,484,497 [*C. A.*, **44**, 6426 (1950)].
[11] Szarvasi, Dupont, and Dulou, *Chimie & industrie*, **62**, 143 (1949).
[12] Papa, Schwenk, and Ginsberg, *J. Org. Chem.*, **16**, 253 (1951).

PROTOCATECHUIC ACID

(Benzoic acid, 3,4-dihydroxy-)

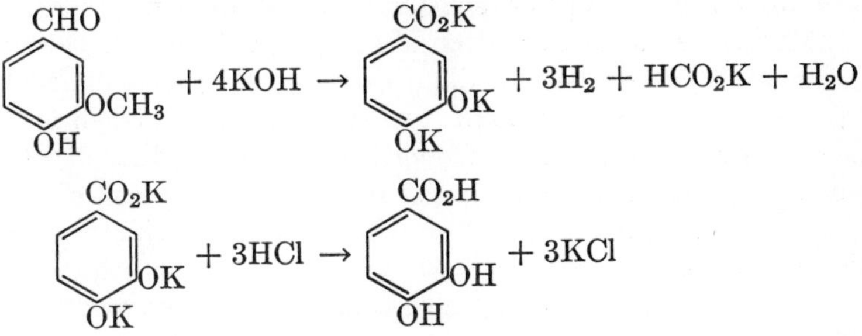

Submitted by Irwin A. Pearl.
Checked by C. F. H. Allen and Calvin Wolf.

1. Procedure

In a stainless-steel beaker of approximately 3-l. volume (180 mm. by 150 mm.) (Note 1), equipped with an efficient mechanical stainless-steel stirrer and heated by an electric hot plate, are placed 84 g. (2 moles) of 97% sodium hydroxide pellets, 332 g. (5 moles) of 85% potassium hydroxide pellets (Note 2), and 50 ml. of water. The mixture is stirred and heated. When the temperature of the fluid mixture reaches 160°, 152 g. (1 mole) of vanillin is added in portions over a period of 2.5–3 minutes at a rate sufficient to maintain the reaction (Note 3). The temperature at this point is 190–195°. Stirring is continued, and heat is applied until the temperature reaches 240–245° (Note 4). The temperature is maintained at 240–245° for 5 minutes. The hot plate is removed, and the mixture is allowed to cool with stirring. When the mixture has cooled to about 150–160°, 1 l. of water is added, and the mixture is stirred until all the fusion mixture is dissolved. The solution is transferred to a 4-l. beaker, another 500 ml. of water is added, and sulfur dioxide gas is introduced for 2 minutes (Note 5); the mixture is then completely acidified with 1.5 l. of 6 N hydrochloric acid. The acidified mixture is cooled in an ice bath (5°) for 2 hours, and the crystalline precipitate is filtered, washed on the filter with two 100-ml. portions of ice water, and air dried. The tan crystals of protocatechuic acid melting at 196–198° weigh 90–100 g. Extraction of the filtrate and washings with two 1-l. portions of ether yields an additional 45–55 g. of protocatechuic acid melting at 190°. The total yield of crude protocatechuic acid amounts to 135–153 g. (89–99%) (Note 6).

2. Notes

1. In the checkers' opinion a 2-l. beaker is sufficiently large, and the contents are easier to stir. Iron or nickel pots have also been used.

2. The exact proportion of sodium hydroxide to potassium hydroxide is not too critical as long as the total amount of alkali is more than 7 moles. Mixtures containing from 10% to 60% sodium hydroxide become fluid between 120° and 130°. Increased percentages of sodium hydroxide in the mixture result in darker protocatechuic acid, but yields are not affected until 70% sodium hydroxide is reached.

3. This reaction is the oxidation of vanillin to vanillic acid with the liberation of hydrogen.

4. The demethylation of vanillic acid to protocatechuic acid takes place to a slight degree between 210° and 235° but goes to completion only at temperatures above 240–245°.

5. The sulfur dioxide treatment prevents the formation of a very dark-colored product when the reaction mixture is acidified with a strong acid.

6. The first crop of acid is light tan and is suitable for most purposes. It can be improved slightly by recrystallizing from hot water, with 3 ml. per g. and 1 g. of Norit for every 10 g. of acid; the recovery is 75%, the remainder being retained by the charcoal. This recrystallized acid is a cream color and melts at 199–200°. If the Norit is omitted, the recovery is 90%, m.p. 198–199°, and color unchanged.

The second crop is of decidedly inferior quality. It is easily purified as follows: Fifteen grams of crude product is dissolved in 100 ml. of 10% sodium hydroxide solution at room temperature, 2 g. of Norit is added, and the mixture is stirred for 10 minutes and filtered. Sulfur dioxide is then passed in for 2 minutes, after which 60 ml. of 6 N hydrochloric acid is added. After chilling and standing, 10 g. (67%) of purified protocatechuic acid, m.p. 196–198°, is recovered.

3. Methods of Preparation

The only practical method for the preparation of protocatechuic acid is by the alkaline fusion of vanillin.[1–3]

[1] Pearl, *J. Am. Chem. Soc.*, **68**, 2180 (1946).
[2] Tiemann and Haarman, *Ber.*, **7**, 617 (1874).
[3] U. S. pat. 2,547,920 [*C. A.*, **45**, 8042 (1951)].

PSEUDOIONONE

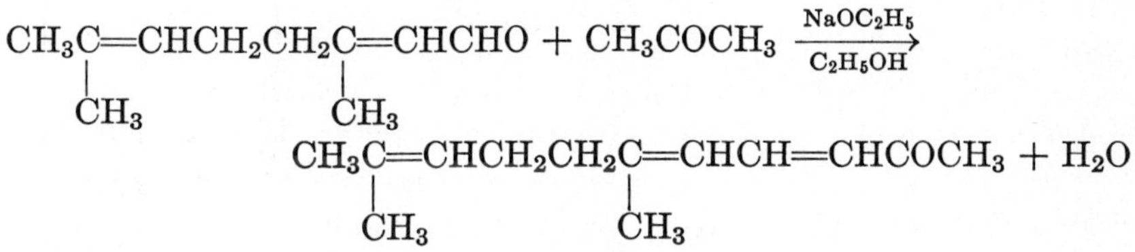

Submitted by ALFRED RUSSELL and R. L. KENYON.
Checked by LEE IRVIN SMITH, W. B. RENFROW, JR., and LOUIS E. DEMYTT.

1. Procedure

A. *Purification of citral.* In a 4-l. bottle are placed 1 l. of water, 1 kg. of crushed ice, 450 g. of anhydrous sodium sulfite (or an equivalent amount of hydrated sodium sulfite), 320 g. of sodium bicarbonate, and 270 g. (304 ml., 1.78 moles) of commercial citral (Note 1). A tightly fitting stopper is securely wired into place, and the bottle is shaken thoroughly for 5–6 hours. The solution, which contains very little unchanged citral, is extracted twice with 300-ml. portions of ether (Note 2).

A 5-l. round-bottomed flask is fitted with a 1-l. dropping funnel, the mouth of which is connected to a tube passing through the stopper of the flask, and the whole apparatus is rigidly attached to a shaker (Note 3). About one-half of the sulfite solution is placed in the flask and covered with 800 ml. of ether. In the dropping funnel is placed 800 ml. of 10% aqueous sodium hydroxide. Shaking is begun, and when the contents of the flask are thoroughly mixed, the sodium hydroxide solution is admitted in a thin continuous stream over a period of about 1 hour. After all the sodium hydroxide solution is added, shaking is continued for not longer than 5 minutes, and the mixture is then poured into a large separatory funnel. The aqueous layer is drawn off and the ether layer is set aside. The aqueous layer is returned to the separatory funnel, covered with 300 ml. of ether, and shaken with an additional 200 ml. of 10% aqueous sodium hydroxide. After separation, the aqueous layer is extracted once with 150 ml. of ether, and the combined ether extracts are then dried over anhydrous sodium sulfate.

The remaining half of the sulfite solution is subjected to the same treatment, and the total ether extract, after drying, is evaporated on a water bath. The residue is then distilled under reduced pressure, yielding 200–215 g. of almost colorless citral which boils at 84–85°/2 mm. (93–95°/5 mm.).

B. *Pseudoionone.* In a 2-l. round-bottomed flask, fitted with a mechanical stirrer, a dropping funnel, and a thermometer, are placed 203 g. (230 ml., 1.33 moles) of pure citral and 800 g. (110 ml., 13.8 moles) of acetone (C.P., dried over anhydrous potassium carbonate). The mixture is cooled to −5° (or below) in an ice-salt bath, vigorous stirring is begun, and there is added through the dropping funnel a cold solution of 9.2 g. (0.4 mole) of sodium in 200 ml. of absolute ethanol. The solution of sodium ethoxide is added at the maximum rate which will permit maintenance of the temperature at −5° or below (rapid dropping). After the addition is complete, stirring is continued for 3–4 minutes. A solution of 30 g. (0.2 mole) of tartaric acid in 200 ml. of water is added, and the mixture is *immediately* steam-distilled to remove the excess acetone. A white precipitate may form when the tartaric acid is added, but this disappears during the steam distillation (Note 4). The mixture in the distilling flask is cooled in an ice bath, and the upper layer (about 380 ml.) is then removed and refluxed vigorously for 5–6 hours with three times its volume of 25% aqueous sodium bisulfite.

After cooling, the mixture is extracted twice with 200-ml. portions of ether to remove any material that has not reacted with the bisulfite. One-half of the aqueous solution is placed in the shaking extractor together with 650 ml. of ether. The calculated amount of 10% aqueous sodium hydroxide (1 mole of sodium hydroxide per mole of sodium bisulfite used) is now added as described for purification of the citral. Shaking is continued for 15 minutes after the addition of the alkali is complete (Note 5). The layers are separated (Note 6); the aqueous layer is returned to the separatory funnel and covered with 200 ml. of ether. After 100 ml. of 10% sodium hydroxide has been added, the mixture is vigorously shaken, and then the aqueous layer is removed and extracted with another 200-ml. portion of ether. The two ether extractions, one with addition of sodium hydroxide and one without, are repeated. The second half of the bisulfite solution is subjected to the same treatment as the first half, and all the ether extracts are combined (Note 7) and dried over anhydrous sodium sulfate. The ether is removed on the water bath, and the residual yellow-green oil is distilled under reduced pressure, yielding 120–130 g. (45–49% based upon 210 g. of pure citral) of pale yellow pseudoionone, boiling at 114–116°/2 mm. (124–126°/4 mm.).

2. Notes

1. Unless a pure pseudoionone free from isomers is wanted, it is not necessary to perform the elaborate purification of citral, or the elaborate purification of pseudoionone. The checkers, using commercial citral obtained from the Florasynth Laboratories, Inc., New York, found 90% of it to boil over a 3° range between 100° and 103°/7 mm. Using a solution of 203 g. of this distilled citral in 1 l. of commercial acetone, the checkers proceed as follows: The citral solution is cooled in an ice-salt bath to −10°, and one-fourth of the solution is forced, by dry compressed air, into a 500-ml. round-bottomed three-necked flask, fitted with a stirrer, dropping funnel, and an adjustable outlet tube long enough to reach to the bottom of the flask. The temperature is maintained at 0° to −5° while one-fourth of a solution of 9.2 g. of sodium in 200 ml. of absolute ethanol is added dropwise. After all the base is added, stirring is continued for 3 minutes, and then the reaction mixture is forced over into one-fourth of a solution of 33 g. of tartaric acid in 200 ml. of water. The elapsed time, from the addition of the first drop of base until the mixture is forced into the acid, is 14 minutes. The condensation is repeated three times more, and the combined acidified mixture is steam-distilled until 1 l. of distillate is obtained. It is important that the solution should remain slightly acid during steam distillation. The material remaining in the flask is cooled, the layers are separated, and the aqueous layer is extracted with ether. The combined organic layers are dried over sodium sulfate and then distilled. The yield of pseudoionone boiling at 123–124°/2.5 mm. is 177.8 g. (70% based upon the citral). By carrying out the condensation in small batches, the temperature is much more easily controlled and the yields are greatly improved. This pseudoionone, when catalytically reduced in a bomb, gave hexahydropseudoionol, boiling at 124–128°/10 mm., in a yield of more than 90%.

2. In handling these large quantities, much better results are obtained if extractions, etc., are carried out in portions.

3. A pressure outlet may be used, but the arrangement outlined here is quite sufficient. Citral and pseudoionone are both rapidly polymerized by contact with aqueous sodium hydroxide. This apparatus continuously provides an intimate mixture of the sulfite solution with the ether. On decomposition, the free carbonyl compound is immediately extracted and prolonged contact with sodium hydroxide is thus avoided.

4. From the completion of addition of the sodium ethoxide solution to the steam distillation, fast work is advantageous, and all equipment should be in readiness before the addition is started. Allowing the solution to stand before addition of the tartaric acid may cause darkening and formation of gummy material. It is important that the solution should remain slightly acid during the steam distillation.

5. The pseudoionone bisulfite addition product, unlike that of citral, apparently decomposes rather slowly; if the separation is made too soon, some undecomposed bisulfite compound is left in solution. This is later decomposed and, in contact with the alkali, polymerizes to a dark red gum.

6. The checkers obtained three layers at this point—a lower aqueous layer, a dark red oily layer, and the upper ether layer. The ether layer was removed and the other two layers were returned for further treatment. The dark oily layer gradually disappeared during the subsequent extractions.

7. The checkers washed these ether extracts with a little water to ensure removal of any basic material.

3. Methods of Preparation

Pseudoinone has been prepared by the condensation of citral and acetone, using as condensing agent a saturated solution of barium hydroxide,[1] a solution of sodium ethoxide in ethanol,[2,3] or metallic sodium.[4] Impure products have been obtained from citral and acetone, using alcoholic sodium hydroxide as the condensing agent,[4] and by treatment of oil of lemon grass and acetone with bleaching powder, cobalt nitrate, and alcohol.[5–7]

Pseudoionone, together with methylpseudoionone, has been obtained by oxidation of geraniol by acetone or methyl ethyl ketone in the presence of an aluminum alcoholate.[8] It has also been prepared from linaloöl.[9]

[1] Tiemann and Krüger, *Ber.*, **26,** 2692 (1893).
[2] Stiehl, *J. prakt. Chem.*, (2) **58,** 84 (1898).
[3] Tiemann, *Ber.*, **32,** 115 (1899).
[4] Hibbert and Cannon, *J. Am. Chem. Soc.*, **46,** 119 (1924).
[5] Ziegler, *J. prakt. Chem.*, (2) **57,** 493 (1898).
[6] Tiemann, *Ber.*, **31,** 2313 (1898).
[7] Haarmann and Reimer Company, Ger. pat. 73,098 [*Frdl.*, **3,** 889 (1890–1894)].
[8] Yamashita and Honjo, *J. Chem. Soc. Japan*, **63,** 1335 (1942) [*C. A.*, **41,** 3041 (1947)].
[9] Tavel, *Helv. Chim. Acta*, **33,** 1266 (1950).

PSEUDOTHIOHYDANTOIN

(4-Thiazolidone, 2-imino-)

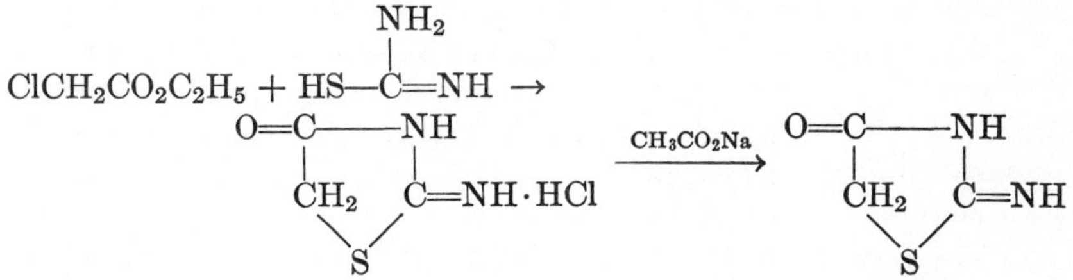

Submitted by C. F. H. Allen and J. A. VanAllan.
Checked by H. R. Snyder and Frank X. Werber.

1. Procedure

In a 1-l. flask surmounted by a reflux condenser, 76 g. (1 mole) of thiourea, m.p. 174–176°, is dissolved in 500 ml. of 95% ethanol by refluxing for 10–15 minutes. Then 125 g. (108 ml., 1.02 moles) of ethyl chloroacetate is added slowly (15–20 minutes) through the condenser while gentle refluxing is continued. After the mixture has been refluxed for 3 hours longer, it is allowed to cool to room temperature and the solid is filtered by suction on a 14-cm. Büchner funnel. The filtrate is used to rinse any solid adhering to the walls of the reaction flask onto the filter. The crude product is pressed down firmly and washed with 50 ml. of ethanol (Note 1).

The crude hydrochloride is dissolved in 1.2 l. of hot, freshly boiled water (Note 2) in a 2-l. beaker, a boiling solution of 121 g. of sodium acetate trihydrate in 150 ml. of water is added, and the mixture is heated to boiling (Note 3). The resulting clear solution is stored in the ice chest overnight. The crystalline pseudothiohydantoin is filtered and dried to constant weight at 60°. The product weighs 92–95 g. (79–82%). A reproducible decomposition point of 255–258° can be determined with the aid of a melting-point bar (Note 4).

2. Notes

1. The yield of the crude hydrochloride, dried to constant weight, is about 126 g.; the decomposition range is 210–255°.

2. The water is boiled before use to expel dissolved oxygen, the presence of which in the solution may cause the final product to have a yellow color.

3. Prolonged boiling should be avoided, as hydrolysis of the imino group occurs very easily. When sodium carbonate is substituted for sodium acetate the yield drops 10%.

4. The decomposition point as measured in the ordinary manner varies with the temperature of the bath at the time of the introduction of the sample. When the melting-point bath is heated so that the temperature rises by 2° per minute, a sample introduced just as the temperature reaches 200° develops an observable brown color at 206° and then darkens so rapidly that no liquid phase can be identified; if, under the same conditions, the sample is introduced when the bath temperature is 238° the decomposition begins immediately but a liquid phase can be recognized when the bath temperature is in the range 241–243°.

3. Methods of Preparation

Pseudothiohydantoin has been obtained from thiourea and ethyl chloroacetate [1] and from thiourea and chloroacetic [2,3] or dichloroacetic acids.[4]

[1] Klason, *Ber.*, **10,** 1352 (1877).

[2] Andreasch, *Monatsh.*, **8,** 424 (1887).

[3] Schmidt, *Arch. Pharm.*, **258,** 229 (1920).

[4] Dixon, *J. Chem. Soc.*, **63,** 816 (1893).

PYOCYANINE

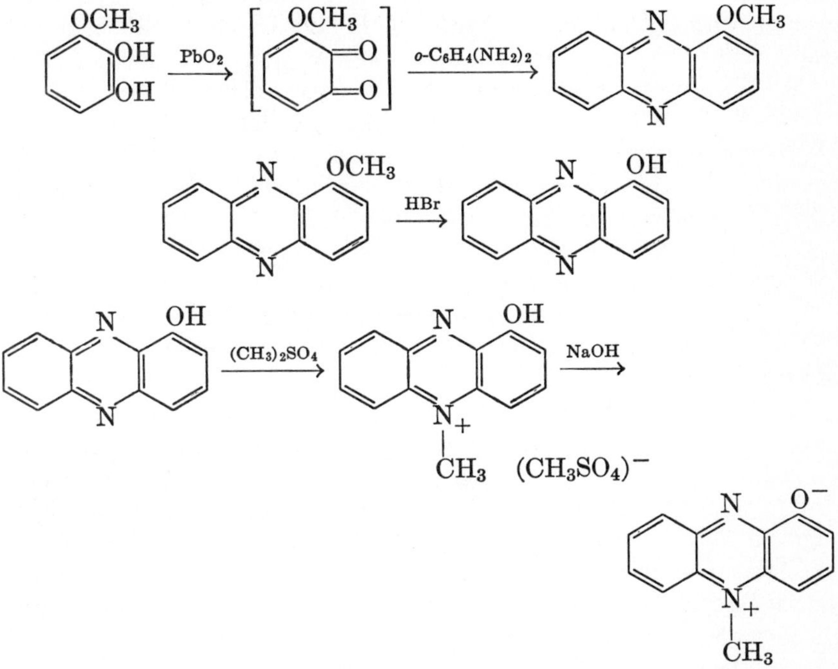

Submitted by Alexander R. Surrey.
Checked by Lee Irvin Smith and Chien-Pen Lo.

1. Procedure

A. *α-Methoxyphenazine* (*condensation*). Two hundred grams (0.42 mole) of powdered lead dioxide (Note 1) is added to a solution of 10 g. (0.07 mole) of pyrogallol monomethyl ether (p. 759) in 3 l. of dry benzene in a 1-gal. narrow-necked acid bottle. The bottle and contents are placed in a shaking machine and shaken for 10–20 minutes (Note 2). The reddish brown solid is filtered through an 11-cm. Büchner funnel, and the filter cake is washed once with 400 ml. of benzene. To this filtrate there is added, immediately and with mechanical stirring, a solution of 6 g. (0.06 mole) of *o*-phenylenediamine (Note 3) in a mixture of 80 ml. of glacial acetic acid and 200 ml. of benzene. The solution, which becomes dark brown, is allowed to stand at room temperature for 1.5 hours; it is then divided into two portions and each portion is washed, in a 3-l. separatory funnel, three times with water, twice with 5% sodium hydroxide solution, and finally twice with water, 100-ml. portions being taken each time. Each of the benzene solutions is shaken with 50 g. of anhydrous sodium carbonate and 5 g. of Norit and filtered through an 11-cm. Büchner funnel. Each filtrate is stirred with 50–60 g. of activated alumina (Note 4) until a filtered sample shows a light-yellow color. The alumina is filtered on a folded filter and washed with benzene until the washings are almost colorless. The benzene is removed from the combined filtrates by distillation under reduced pressure on a water bath at 40–50°. The residual light-yellow solid (Note 5) is recrystallized by dissolving it in the least possible amount of hot pyridine, adding water to the point of incipient precipitation, and cooling. The light-yellow crystals are filtered on a 7-cm. Büchner funnel, washed with water, and air-dried. The yield is about 5 g. (33%) of a product that melts at 167–169° (Note 6).

B. *α-Hydroxyphenazine* (*demethylation*). A solution of 4.2 g. (0.02 mole) of α-methoxyphenazine, from A above, in 125 ml. of 55% hydrobromic acid (Note 7) is placed in a 250-ml. round-bottomed flask fitted with a reflux condenser. The flask is immersed in an oil bath, and the solution is heated to 110–120° for 5 hours; the evolved gases are absorbed with water in a trap. The reaction mixture is cooled to room temperature, diluted with about 125 ml. of water, almost neutralized with sodium hydroxide (Note 8), and extracted six times with 30- to 40-ml. portions of ether. The combined ether extracts are extracted with 25-ml. portions of 10% sodium hydroxide solution (Note 9) until no more purple sodium salt is removed from the ether. The aqueous extracts are combined, made acid to litmus with dilute acetic acid, and re-extracted four times with 50-ml. portions of ether. The combined ether extracts are dried over anhydrous sodium sulfate, and the ether is removed by distillation on a steam bath. The residue is recrystallized as follows: It is dissolved in the least possible amount of hot ethanol, water is added to the point of incipient precipitation, then 0.5 g. of Norit is added, and the hot solution is filtered. The filtrate is cooled in ice water, and the orange solid is filtered on a 7-cm. Büchner funnel, washed with water, and dried in an oven at 100°. The yield is 2.7 g. (70%) of a product that melts at 153–155°.

C. *Pyocyanine* (*alkylation*). A solution of 2 g. (0.011 mole) of α-hydroxyphenazine in 13.4 g. (10 ml., 0.1 mole) of methyl sulfate (Note 10) is placed in a 250-ml. Erlenmeyer flask fitted with a calcium chloride drying tube and heated at 100° (oil bath) for 10 minutes. The solution is allowed to cool to room temperature, and about 75 ml. of dry ether is added. The dark brown solid is filtered on a 7-cm. Büchner funnel and washed with about 150 ml. of dry ether in several portions (Note 11).

The dry methosulfate, dissolved in about 30 ml. of water, is made alkaline with 2–3 ml. of 10% sodium hydroxide, and the solution is then extracted *exhaustively* with successive 15-ml. portions of chloroform until no more blue substance is removed from the aqueous solution (Note 12). The combined chloroform solutions are extracted three times with 20-ml. portions of 5% hydrochloric acid. The combined acid extracts are made alkaline to phenolphthalein with 10% sodium hydroxide and re-extracted *exhaustively* with 25-ml. portions of chloroform until no more blue substance is removed from the aqueous solution (Note 12). The combined chloroform solutions are dried over anhydrous sodium sulfate and decanted, and the chloroform is removed by distillation under reduced pressure. The blue crystalline residue is recrystallized by dissolving it in the least possible amount of water at 60° and then cooling the solution in an ice bath. The product is filtered on a 5-cm. Büchner funnel and dried in the dark in a vacuum desiccator over calcium chloride. The yield is 1.35 g. (58%) of dark blue needles that melt at 133° (Note 13).

2. Notes

1. Lead peroxide "analytical reagent" of low manganese content was used.

2. On a scale twelve times as large, a "Lightnin" stirrer was used for 20 minutes.

3. *o*-Phenylenediamine, m.p. 99–101°, from Eastman Kodak Company was used.

4. The amount of alumina needed depends upon the color of the benzene solution. Sufficient alumina (activated alumina, 80 mesh, from Aluminum Corporation of America) is taken so that the filtrate is only light yellow in color. On the larger scale only 100–200 g. of alumina was needed.

5. The residue, without recrystallization, is pure enough for the preparation of α-hydroxyphenazine.

6. The submitter reports yields somewhat higher (33–40%) and states that, on a larger scale, the yields are slightly better than 40%.

7. The concentration of hydrobromic acid used ranged from 50% to 55%.

8. Concentrated sodium hydroxide solution (about 100 ml. of 35%) is used at the beginning of the neutralization, and dilute sodium hy-

droxide solution is added towards the end. The reaction mixture should be just faintly acid to litmus.

9. If any solid sodium salt separates during the extraction, it can be redissolved by adding water.

10. The methyl sulfate should be freshly distilled under reduced pressure.

11. The yield of methosulfate is practically quantitative.

12. The solution must be extracted *exhaustively;* the checkers found that thirty such extractions were required to remove all the product. A continuous extractor might be used at this point and certainly would be necessary for large-scale runs.

13. The same percentage yield was obtained with double the amounts specified above. The pyocyanine thus obtained can be stored in a vacuum desiccator in the dark for several weeks without appreciable decomposition. It slowly decomposes on longer standing to give the yellow α-hydroxyphenazine.

3. Methods of Preparation

This series of reactions is essentially the one described by Wrede and Strack.[1] Pyocyanine can also be prepared by the photochemical oxidation of phenazine methosulfate.[2]

A different synthesis of 1-hydroxyphenazine, starting with 2,6-dinitroaniline and *o*-iodonitrobenzene, has also been described.[3]

[1] Wrede and Strack, *Ber.*, **62**, 2053, 2054 (1929); *Z. physiol. Chem.*, **74**, 181, 184, 185 (1929).

[2] McIlwain, *J. Chem. Soc.*, **1937**, 1708.

[3] Hegedüs, *Helv. Chim. Acta,* **33**, 766 (1950).

1-(α-PYRIDYL)-2-PROPANOL

[2-(β-Hydroxypropyl)pyridine]

$$C_6H_5Br + 2Li \rightarrow C_6H_5Li + LiBr$$

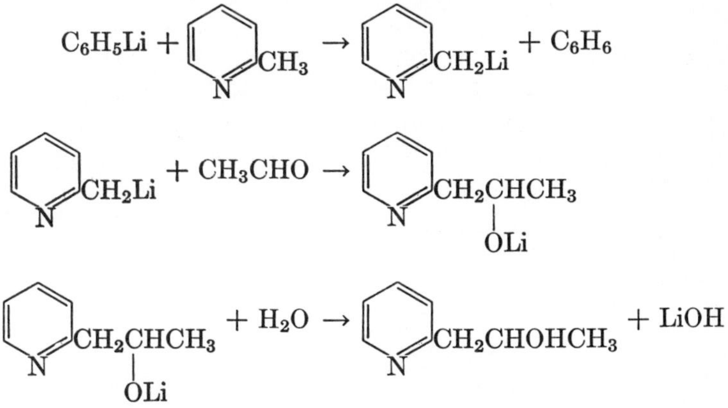

Submitted by L. A. Walter.
Checked by C. F. H. Allen and James VanAllan.

1. Procedure

While a current of dry nitrogen is passed through the apparatus, 400 ml. of dry ether and 6.9 g. (1 gram atom) of lithium (in small pieces) (Note 1) are placed in a 1-l. three-necked flask fitted with a dropping funnel, mechanical stirrer, and reflux condenser protected from moisture. The stirrer is started, and 10–15 ml. of a solution of 79 g. (0.5 mole) of dry bromobenzene in 100 ml. of dry ether is added from the dropping funnel. The reaction usually starts immediately; if not, the flask may be warmed, and the remainder of the mixture is then added at such a rate that the ether refluxes gently. The mixture is stirred until the lithium disappears (Note 2).

Forty-six grams (0.5 mole) of α-picoline (Note 3) is then added, and the mixture is stirred at room temperature for 1 hour, during which time the dark red solution of picolyllithium is formed. The flask is then immersed in an ice-salt bath, and when the mixture is thoroughly chilled the nitrogen train is disconnected. Then 20 g. of dry acetaldehyde in 50 ml. of dry ether (Note 4) is slowly dropped into the mixture over a period of 20 minutes. The red color entirely disappears. After

15 minutes, 100 ml. of water is slowly added and then 100 ml. of concentrated hydrochloric acid (sp. gr. 1.2). The aqueous layer is removed and poured, with stirring, into a warm solution of 300 g. of sodium carbonate decahydrate in 100 ml. of water (Note 5). The crude reaction product separates as an oil and is taken up in 300 ml. of chloroform. The precipitated lithium carbonate is filtered, transferred to a beaker, and stirred with four 200-ml. portions of chloroform. The chloroform extracts are decanted or filtered, and all the chloroform solutions are combined (Note 6). The chloroform is removed by distillation, and the residue is fractionated under reduced pressure through a good column. The 1-(α-pyridyl)-2-propanol boils sharply at 116–117°/17 mm. (124–125°/20 mm.). A small fore-run and a considerable amount of high-boiling residue are discarded. The yield is 30–34 g. (44–50% based on the α-picoline) (Notes 7 and 8). This product darkens on exposure to light, and it should be preserved in a brown glass bottle.

2. Notes

1. The most unsatisfactory operation of this preparation is cutting the lithium. It may be finely divided by rubbing the metal against a coarse wood rasp and allowing the filings to drop through a large paper funnel directly into the ether while a rapid stream of dry nitrogen is passed through the flask. This procedure is most convenient when a large piece of lithium is available and the amount of filings can be determined by the loss in weight. Larger pieces, cut with a knife [*Org. Syntheses* Coll. Vol. **2,** 518 (1943)], can be used equally well (Note 2). Bartlett, Swain, and Woodward [1] have published a convenient method for the preparation of lithium sand.

2. The time depends upon the size of the pieces; the solution may be stirred for 24 hours without affecting the yield.

3. The submitters used α-picoline, b.p. 128–130°, obtained from the Barrett Company. The checkers used practical α-picoline, b.p. 128–134°, freshly distilled under reduced pressure. Samples of picoline containing water should be carefully fractionated to remove the water as the α-picoline-water azeotrope, b.p. 93°.

4. Alternatively, the acetaldehyde may be distilled into the flask through the nitrogen inlet. During this operation, the outlet must be kept well above the surface, to prevent clogging. The introduction of aldehyde is stopped when the red color has disappeared.

5. A solution of 111 g. of anhydrous sodium carbonate in 150 ml. of water may be substituted.

6. If sodium hydroxide is used to liberate the amino alcohol from its salt, extraction with chloroform produces unworkable emulsions.
7. Equally good yields of 1-(α-pyridyl)-3-propanol, b.p. 116–118°/4 mm., may be obtained by using ethylene oxide in place of acetaldehyde.
8. The submitters obtained yields as high as 43 g. (60%).

3. Methods of Preparation

This procedure is based upon that of Ziegler.[2] 1-(α-Pyridyl)-2-propanol has also been prepared in poor yields (4–6 per cent) by heating α-picoline with aqueous acetaldehyde in sealed tubes.[3,4]

[1] Bartlett, Swain, and Woodward, *J. Am. Chem. Soc.*, **63**, 3230 (1941).
[2] Ziegler and Zeiser, *Ann.*, **485**, 174 (1931).
[3] Ladenburg, *Ann.*, **301**, 140 (1898).
[4] Meisenheimer and Mahler, *Ann.*, **462**, 308 (1928).

PYROGALLOL 1-MONOMETHYL ETHER

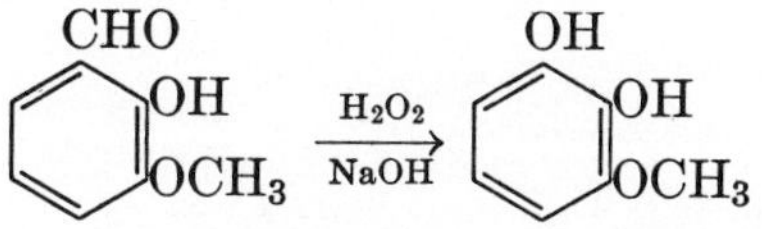

Submitted by Alexander R. Surrey.
Checked by Lee Irvin Smith and Chien-Pen Lo.

1. Procedure

The apparatus consists of a 1-l. three-necked flask fitted with a gas inlet tube extending about 3 cm. into the flask and connected to the flask through a bubbler, a thermometer extending to the bottom, a mechanical stirrer, and a reflux condenser connected at the upper end with an exit tube leading to the hood. The reaction is carried out in an atmosphere of illuminating gas (Note 1).

In the flask are placed 60.8 g. (0.4 mole) of 2-hydroxy-3-methoxybenzaldehyde (Note 2) and 200 ml. of 2 *N* sodium hydroxide (0.4 mole). The mixture is stirred until almost all the solid has dissolved. The stirrer is replaced by a dropping funnel which contains 284 ml. (0.5 mole) of 6% hydrogen peroxide (Note 3). With occasional shaking, the hydrogen peroxide is added in portions of 20–25 ml. About 1 hour is required for the addition; the temperature is kept between 40° and 50°. After the addition of the first portion of hydrogen peroxide, the temperature rises to about 45° and a dark solution results. The temperature is allowed to fall to 40° before the next portion of the peroxide is added.

After all the hydrogen peroxide is added, the reaction mixture is allowed to cool to room temperature and is then saturated with sodium chloride, after which it is extracted four times with 100-ml. portions of ether. The combined extracts are dried over sodium sulfate. The ether is removed by distillation on a steam bath, and the residue is then distilled under reduced pressure. Pyrogallol monomethyl ether is collected at 136–138°/22 mm. The yield is 38–44.5 g. (68–80%) of a colorless to light yellow oil which solidifies on standing (Note 4).

2. Notes

1. Nitrogen can be used in place of illuminating gas. The gas is introduced at the rate of about 3 bubbles per second.

2. Practical 2-hydroxy-3-methoxybenzaldehyde (Eastman Kodak Company) was used in this preparation. Care should be taken when working with this material as its vapors are irritating and will cause sneezing.

3. The hydrogen peroxide solution was prepared by diluting 63 g. of a solution containing 27% hydrogen peroxide with water to 284 ml.

4. The procedure described is similar to that of Dakin[1] for the preparation of catechol.[2] The reaction has been carried out using four times the quantities specified here; the yield was 81% (C. F. H. Allen, private communication).

3. Methods of Preparation

Pyrogallol monomethyl ether has been prepared by the methylation of pyrogallol with dimethyl sulfate[3] or methyl iodide;[4] by the decarboxylation of 2,3-dihydroxy-4-methoxybenzoic acid;[5] and by the methylation of pyrogallol carbonate with diazomethane and subsequent hydrolysis.[6] The method described is taken from the improved procedure of Baker and Savage[7] for the preparation of pyrogallol monomethyl ether from *o*-vanillin by oxidation with hydrogen peroxide.

[1] Dakin, *Am. Chem. J.*, **42**, 477 (1909).
[2] Dakin, *Org. Syntheses* Coll. Vol. **1**, 149 (1941).
[3] Graebe and Hess, *Ann.*, **340**, 232 (1905).
[4] Herzig and Pollak, *Monatsh.*, **25**, 811 (1904).
[5] Herzig and Pollak, *Monatsh.*, **25**, 507 (1904).
[6] Hillemann, *Ber.*, **71**, 41 (1938).
[7] Baker and Savage, *J. Chem. Soc.*, **1938**, 1607.

RESACETOPHENONE *

(Acetophenone, 2,4-dihydroxy-)

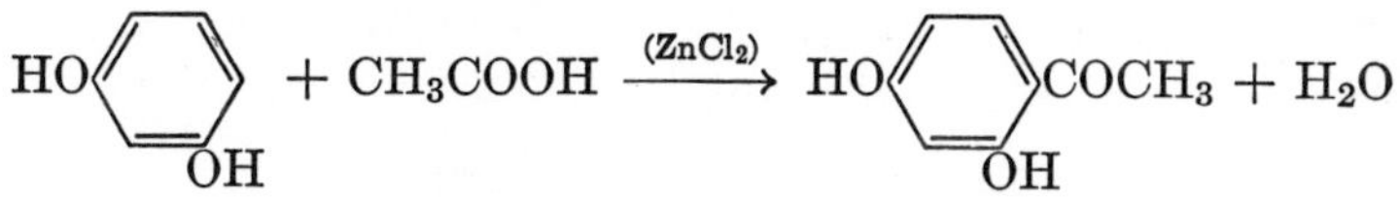

Submitted by S. R. Cooper.
Checked by Nathan L. Drake, Harry D. Anspon, and Ralph Mozingo.

1. Procedure

One hundred and sixty-five grams (1.2 moles) of anhydrous zinc chloride (Note 1) is dissolved with the aid of heat in 165 g. (158 ml., 2.7 moles) of glacial acetic acid, which has been placed in a 1-l. beaker. To this hot mixture (about 140°), 110 g. (1 mole) of resorcinol is added with constant stirring. The solution is heated on a sand bath until it *just* begins to boil (about 152°). The flame is then removed and the reaction allowed to complete itself at a temperature not in excess of 159° (Note 2). After standing on the sand bath without further heating for 20 minutes, the solution is diluted with a mixture of 250 ml. of concentrated hydrochloric acid and 250 ml. of water. The dark red solution is placed in an ice bath and cooled at 5°. The resulting precipitate is collected on a filter and washed free from zinc salts with 1 l. of dilute (1:3) hydrochloric acid in 200-ml. portions. This orange-red product, after drying, weighs 104–110 g. and melts at 141–143°. It is distilled under reduced pressure (Note 3), and boils at 180–181° at 10 mm. (147–152° at 3–4 mm.). After most of the product has distilled, the temperature rises sharply, and the operation is discontinued when the temperature reaches 190°. The light-yellow distillate is removed from the receiver with hot ethanol and the ethanol is evaporated (Note 4). This product weighs 100–106 g. It is further purified as follows: the substance is dissolved in 1.8 l. of hot dilute (1:11) hydrochloric acid, filtered hot, and cooled to 5°. The crystals are removed by filtration, washed with two 200-ml. portions of ice water, and dried. The yield of tan-colored resacetophenone, melting at 142–144°, is 93–99 g. (61–65%).

2. Notes

1. Although finely ground zinc chloride dissolves more rapidly, lumps or sticks are satisfactory.

2. If the temperature rises much above the initial boiling point, the yield of red product increases at the expense of resacetophenone formation. The boiling point also may vary slightly.

3. Any convenient apparatus for distillation of a solid may be used.

4. It is convenient to remove most of the material by melting and pouring it out of the receiver. The remainder is removed with hot ethanol.

3. Methods of Preparation

Resacctophenone has been prepared by heating resorcinol with zinc chloride and acetic acid,[1,2] with zinc chloride and acetic anhydride,[2] with zinc chloride and acetyl chloride,[3] with boron trifluoride and acetic anhydride,[4] and with aluminum chloride and acetyl chloride.[5] It has been prepared by the action of zinc chloride on resorcinol diacetate,[2] by heating 4-methylumbelliferone with potassium hydroxide,[6] by heating resacetophenone carbonic acid,[7] and by the action of acetyl chloride on resorcinol.[8]

[1] Robinson and Shah, *J. Chem. Soc.,* **1934,** 1491.
[2] Nencki and Sieber, *J. prakt. Chem.,* (2) **23,** 147 (1881).
[3] Eijkman, *Chem. Weekblad,* **1,** 453 (1904) [*Chem. Zentr.,* **75,** II, 1597 (1904)].
[4] Killelea and Lindwall, *J. Am. Chem. Soc.,* **70,** 428 (1948).
[5] Desai and Ekhlas, *Proc. Indian Acad. Sci.,* **8A,** 194 (1938).
[6] Pechmann and Duisberg, *Ber.,* **16,** 2119 (1883).
[7] Liebermann and Lindenbaum, *Ber.,* **40,** 3570 (1907).
[8] Cox, *Rec. trav. chim.,* **50,** 848 (1931).

RHODANINE *

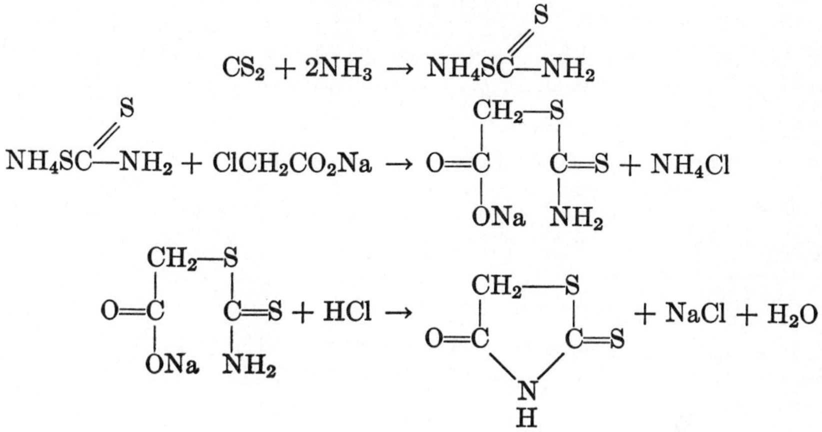

Submitted by C. Ernest Redemann, Roland N. Icke, and Gordon A. Alles.
Checked by H. R. Snyder and John H. Johnson.

1. Procedure

A. *Ammonium dithiocarbamate.* Gaseous ammonia is passed into 250 ml. of 95% ethanol (Note 1) contained in a 1-l. Erlenmeyer flask immersed in an ice bath until the gain in weight is 39 g. (2.3 moles). To this solution, still cooled by the ice bath, is added a well-cooled mixture of 76 g. (60 ml., 1 mole) of carbon disulfide and 200 ml. of ether. The flask is stoppered loosely (Note 2) and allowed to remain in the ice bath for 2–3 hours and then at room temperature overnight (Note 3). The mixture is again cooled in an ice bath or refrigerator, and the crystals are collected by filtration (*hood*), sucked dry, and washed on the filter with two 50-ml. portions of ether. Air is drawn through the crystals for 5 minutes (Note 4) to effect removal of most of the ether. The product is used promptly without further treatment; the weight of the lemon-yellow solid varies between 80 and 90 g., depending principally on the completeness of the removal of the solvent.

B. *Rhodanine.* Just before the filtration of the ammonium dithiocarbamate, a solution of sodium chloroacetate is prepared by dissolving 71 g. (0.75 mole) of chloroacetic acid in 150 ml. of water contained in a 1-l. wide-mouthed round-bottomed flask and neutralizing the acid with 40 g. (0.38 mole) of anhydrous sodium carbonate (or an equivalent amount of the hydrate) while stirring the solution mechanically. This solution is cooled in an ice bath, and the ammonium dithiocarbamate from the preceding preparation is added during 5 minutes with continual stirring. As soon as the first portion of ammonium dithiocarbamate is added the solution becomes very dark in color. After all the dithiocarbamate has been added the ice bath is removed and stirring is discontinued. The solution is allowed to stand for 20–30 minutes longer, during which time the color changes to a clear yellow. In a 1-l. beaker 400 ml. of 6 *N* hydrochloric acid is heated to boiling, and the above solution (Note 5) is poured slowly with stirring into the hot acid. Heating is continued until the solution has attained a temperature of 90–95°, after which the solution is allowed to cool slowly to room temperature. The rhodanine separates as nearly colorless long blades which are collected by filtration, washed well with water, and dried. The product weighs 83–89 g. (83–89% based on the chloroacetic acid), and melts at 167–168°. Recrystallization from boiling glacial acetic acid (2 ml. per gram) raises the melting point to 168–168.5° (Note 6); the recovery, without reworking of the mother liquor, is 87%.

2. Notes

1. Absolute ethanol is satisfactory. Methanol may be used, but ammonium dithiocarbamate is much more soluble in methanol than in ethanol and the recovery will be lower.

2. The flask should be stoppered loosely to retard the escape of ammonia. It is not wise to stopper it tightly, as some gas is evolved and pressure may develop. Hydrogen sulfide is present in the gases evolved.

3. The success of the preparation depends upon securing ammonium dithiocarbamate of good quality. Although much solid sometimes separates within 1–2 hours, it contains a large amount of the very unstable ammonium trithiocarbamate.[1] After prolonged standing, the solid is nearly pure ammonium dithiocarbamate.

4. Ammonium dithiocarbamate is relatively unstable, and no attempt should be made to dry the compound thoroughly before use.

5. If this solution is not clear and free from solid impurities it should be filtered before addition to the acid.

6. The melting points observed with the aid of a hot-stage microscope are slightly higher—170–170.5° for the unrecrystallized material, and 170.5–171° for the purified product. Although rhodanine usually is described in the literature as melting with decomposition, the checkers observed no evidence of decomposition during melting under the

microscope and found the melting point unchanged when molten samples in ordinary melting-point tubes were cooled and remelted.

3. Methods of Preparation

Rhodanine has been prepared by the reaction of chloroacetic acid with ammonium thiocyanate;[2] by the action of ethyl chloroacetate upon ammonium dithiocarbamate in the presence of alcohol and hydrogen chloride;[3] by saturating a solution of thioglycolic acid and potassium thiocyanate in absolute ethanol with hydrogen chloride;[4] by ring closure of thiocarbamylthioglycolic acid in various ways.[5–7] The present method of preparation is adapted from that of Julian and Sturgis.[7]

[1] Miller, *Contrib. Boyce Thompson Inst.*, **5**, 31 (1933).
[2] Nencki, *J. prakt. Chem.*, (2) **16**, 2 (1877).
[3] Miolati, *Ann.*, **262**, 85 (1891).
[4] Fredyl, *Monatsh.*, **10**, 82 (1889).
[5] Holmberg, *Ber.*, **39**, 3069 (1906); *J. prakt. Chem.*, (2) **79**, 261, 265 (1909).
[6] Granacher, *Helv. Chim. Acta*, **5**, 610 (1920).
[7] Julian and Sturgis, *J. Am. Chem. Soc.*, **57**, 1126 (1935).

SALICYL-*o*-TOLUIDE

(*o*-Salicylotoluide)

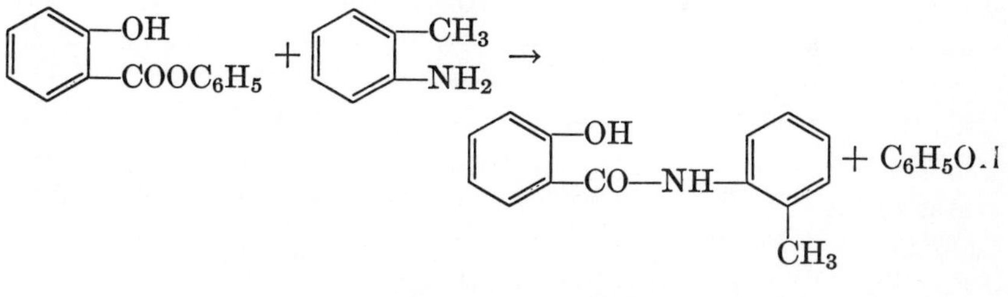

Submitted by C. F. H. Allen and J. VanAllan.
Checked by H. R. Snyder and R. L. Rowland.

1. Procedure

In a 250-ml. flask attached to a Vigreux column 30 cm. in over-all length (Note 1), a mixture of 42.8 g. (0.2 mole) of phenyl salicylate ("Salol," m.p. 42–43°), 26.7 g. (0.25 mole) of *o*-toluidine, and 60 g. of 1,2,4-trichlorobenzene (m.p. 15–16°), is heated at the boiling point, so that the phenol formed slowly distils. The temperature rises from 183° to 187° during the first hour, and 22–23 g. of distillate is collected. Heating is continued until the temperature rises to 202° and a total of 45–46 g. of distillate has been collected (Note 2). The flask is then removed, and to it are added 3 g. of Norit and 10 ml. of trichlorobenzene. The mixture is heated to boiling and filtered hot by suction. The filtrate is allowed to stand in the ice chest overnight. The crystalline amide is filtered by suction, slurried with 75 ml. of ligroin (b.p. 90–120°) at 35–40°, and filtered. After being dried to constant weight, the salicyl-*o*-toluide, m.p. 143–144°, amounts to 33–35 g. (73–77%) (Notes 3, 4, and 5).

2. Notes

1. These are measurements of standard ground-glass equipment.

2. The checkers preferred α-methylnaphthalene (Eastman, practical) as the diluent. When it is used in the apparatus described the phenol does not distil, so that a reaction flask fitted with an air-cooled condenser is more convenient. The reactants in 60 g. of α-methylnaphthalene are heated in an oil bath at 230° for 1.5–2 hours. Three grams of Norit and 20 g. more of α-methylnaphthalene are then added, and the mixture is treated as described under Procedure. The yield and melting point of the product are identical with those described.

3. Since the pure material[1] melts at 144°, the product needs no further purification for ordinary purposes. It has a faint odor, which can be removed by recrystallization from aqueous ethanol (suitably by solution in 8–10 ml. of ethanol per gram of material and the addition of water just short of precipitation from the hot solution). The loss in recrystallization is only about 5%; the melting point is unchanged.

4. By working up the filtrate, only about 2 g. of material of poor quality can be secured.

5. Other anilides can be prepared by this procedure.

Amine Used	*M.p. of Anilide*	*Yield, %*
Aniline	131–132°	70
2,5-Dichloroaniline	228°	89
β-Naphthylamine	188°	84
2-Aminopyridine	206°	86
1-Aminoanthraquinone	278–284°	79

When the boiling point of the amine is widely different from that of phenol, a diluent is unnecessary though it facilitates purification. In the absence of a diluent, the hot melt is poured into ethanol and recrystallized, using a decolorizing carbon; there is considerable tendency towards crystallization during the filtration from the charcoal.

Salicylanilide (m.p. 131–132°) is obtained by heating the mixture without a diluent for 3 hours at 180–200° with a short air condenser, pouring the melt into 100 ml. of ethanol, and working up as above. The persistent pink color is not easily removed.

3. Methods of Preparation

Salicyl-*o*-toluide has been prepared only by the action of phosphorus oxychloride upon a mixture of salicylic acid and *o*-toluidine. The useful methods of preparation of salicylanilide are by the interaction of salicylic acid and aniline in the presence of phosphorus trichloride,[2–4] by heating phenyl salicylate and aniline,[5] and from *o*-hydroxybenzamide and bromobenzene in the presence of small amounts of sodium acetate and metallic copper.[6] A number of these and other anilides have been described.[7]

1 Pictet and Hubert, *Ber.,* **29,** 1191 (1896).

2 Wanstrat, *Ber.,* **6,** 336 (1873).

3 Kupferberg, *J. prakt. Chem.,* [2] **16,** 442 (1877).

4 Hubner and Mensching, *Ann.,* **210,** 342 (1881).

5 Cohn, *J. prakt. Chem.,* [2], **61,** 547 (1900).

6 Goldberg, *Ber.,* **39,** 1692 (1906).

7 Fargher, Galloway, and Probert, *J. Textile Inst.,* **21,** 245T (1930) [*C. A.,* **24,** 6026 (1930)].

SEBACONITRILE AND ω-CYANOPELARGONIC ACID

(Sebaconitrile and pelargonic acid, θ-cyano-)

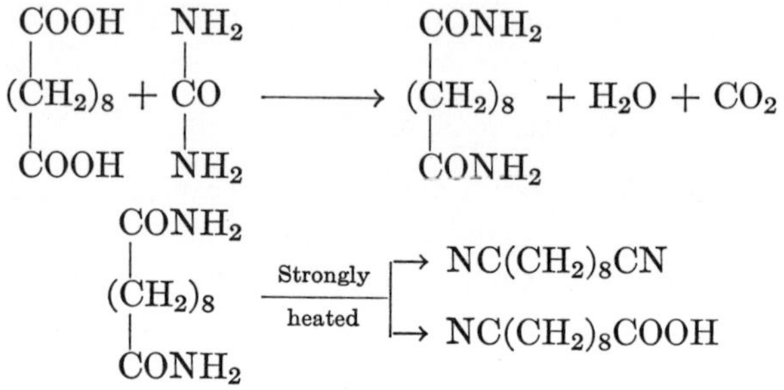

Submitted by B. S. Biggs and W. S. Bishop.
Checked by C. F. H. Allen and J. VanAllan.

1. Procedure

A. *Sebaconitrile.* A 3-l. three-necked flask (Note 1), equipped with a mechanical stirrer (Notes 2 and 3) and a thermometer which dips into the liquid, is heated in an oil bath to 160°. In the flask are placed 505 g. (2.5 moles) of commercial sebacic acid (Note 4) and 180 g. (3 moles) of urea (Note 5), and the melt is heated with stirring for 4 hours at about 160° (Note 6). The oil bath is removed, the surplus oil is wiped off, the flask is insulated (Note 7), and the temperature is then raised, as rapidly as foaming permits, to 220° by means of a triple burner and wire gauze. It is important to continue the stirring for at least 5 minutes after 220° is attained; otherwise the mass will foam over during the subsequent distillation.

The stirrer is then replaced by a short still head connected to a long (90-cm.) air condenser and receiver (Note 8), and the product is distilled at atmospheric pressure as long as any distillate is obtained. The temperature of the vapor rises gradually to 340°. The distillate, which consists chiefly of water, dinitrile, acid nitrile, and sebacic acid, is poured into a large (2-l.) separatory funnel and, after the addition of 500 ml. of ether (Note 9), is extracted three times with 650-ml. portions of 5% ammonium carbonate (Note 10). The crude dinitrile which remains after the removal of the ether is distilled under reduced pressure; after a small fore-run (20–25 ml.) the main product is collected at 185–188°/12 mm. (Note 11). The yield of sebaconitrile is 190–200 g. (46–49%) (Note 12).

B. *ω-Cyanopelargonic acid.* The combined ammoniacal extracts are heated nearly to boiling in a large enameled pot or 4-l. beaker and neutralized to phenolphthalein with concentrated hydrochloric acid (about 120 ml.) (Note 13). A hot solution of 400 g. of barium chloride is then added slowly with stirring, and the hot solution is filtered from the precipitated barium sebacate (Note 14) through a fluted filter paper or "Shark Skin" filter paper (Note 15) on a 20-cm. Büchner funnel. The barium salt of the cyano acid that separates on cooling is filtered on the same sized funnel and dissolved in 1 l. of hot water, and the solution is acidified to litmus with concentrated hydrochloric acid (about 30 ml.). The cyano acid separates as a clear oil. The filtrate from which the barium salt was filtered is acidified likewise to litmus with concentrated hydrochloric acid (about 50 ml.). The two oils are combined. The cyano acid is washed by decantation with three 200-ml. portions of hot water, separated completely from water, and dried in a vacuum desiccator over calcium chloride. The yield of ω-cyanopelargonic acid melting at 48–49° is 146–155 g. (32–34%) (Note 16).

2. Notes

1. Ground-glass equipment is preferable, since the final temperature is high enough to decompose rubber or cork stoppers.

2. The stirrer is a bent glass rod that will break up the foam produced.

3. Attention is called to the powerful but compact compressed-air stirrer, catalog No. 9224, of the A. H. Thomas Company. When a compressed-air line is available, such a stirrer is preferable to an electric motor.

4. Commercial grades of sebacic acid (m.p. 127–130°) and urea (m.p. 132–134°) were used; pure sebacic acid (m.p. 132–133°) gives only a slightly (13%) higher yield. Since the acid is light and bulky, it is convenient to melt it in an enameled pot (p. 614, Note 2) and pour the liquid into the preheated flask.

5. It is important to use the amount of urea which is specified. When 2 moles of urea per mole of sebacic acid was employed, the yield of sebaconitrile was only 27%.

6. At the end of this time the product consists almost entirely of sebacamide, which may be isolated if desired.

7. This insulation is easily accomplished by wrapping with asbestos rope, starting at the neck and winding as far down the bulge of the flask as possible. After the flask is clamped in place on the wire gauze,

asbestos paper is bent around to enclose the bottom. Magnesia can also be used.

8. If the condenser is short, it is advisable to have a second receiver with a reflux condenser attached, because some dinitrile may be carried through by uncondensed steam. A water-cooled condenser should not be placed ahead of the first receiver, however, since the distillate contains some solid products (sebacic acid and products from urea).

9. Although not absolutely essential, the ether facilitates the operation. Benzene is not satisfactory.

10. Ammonium hydroxide (about 3%) is equally satisfactory. The aqueous extracts are saved.

11. Another boiling point is 201–203°/16 mm.

12. The submitters obtained 410–450 g. (50–55%) in runs twice this size.

13. The checkers used a *p*H meter and neutralized to *p*H 7.8.

14. This amounts to 43–45 g.

15. S & S "Shark Skin" filter paper has "high wet-strength" and a high resistance to acids and alkalies.

16. If the cyano acid obtained from the precipitated barium salt is worked up separately, it melts at 51–52°.

3. Methods of Preparation

Sebaconitrile has been obtained by heating sebacic acid in a stream of ammonia,[1] and from sebacamide by pyrolysis [2,3] or by dehydration with phosphorus pentachloride [4] or phosphorus oxychloride.[5] Sebacamide has been prepared from ethyl sebacate,[6,7] sebacoyl chloride,[7] or polysebacic anhydride [8] and ammonia; by heating sebacic acid with urea [2,3,9] or ammonium thiocyanate; [10] and by the electrolysis of adipic acid monoamide.[11]

[1] Greenewalt and Rigby, U. S. pat. 2,132,849 [*C. A.*, **33**, 178 (1939)].
[2] Biggs and Bishop, *J. Am. Chem. Soc.*, **63**, 944 (1941).
[3] Biggs, U. S. pat. 2,322,914 [*C. A.*, **38**, 118 (1944)].
[4] Phookan and Krafft, *Ber.*, **25**, 2252 (1892).
[5] Wilcke, U. S. pat. 1,828,267 [*C. A.*, **26**, 735 (1932)].
[6] Rowney, *Ann.*, **82**, 123 (1852).
[7] Aschan, *Ber.*, **31**, 2350 (1898).
[8] Coffman, Cox, Martin, Mochel, and Van Natta, *J. Polymer Sci.*, **3**, 85 (1948).
[9] U. S. pat. 2,109,941 [*C. A.*, **32**, 3419 (1938)].
[10] Ssolonina, *J. Russ. Phys. Chem. Soc.*, **28**, 558 (1896).
[11] Offe, *Z. Naturforsch.*, **2B**, 185 (1947) [*C. A.*, **42**, 4548 (1948)].

SELENOPHENOL

(Benzeneselenol)

$$C_6H_5MgBr \xrightarrow[\text{then HCl}]{\text{Se}} C_6H_5SeH + MgClBr$$

Submitted by Duncan G. Foster.
Checked by C. F. H. Allen and H. W. J. Cressman.

1. Procedure

Most selenium compounds are toxic, and many have a vile odor. It is frequently advisable to work with them on alternate days. All manipulations should be done in a good hood. Rubber gloves should be worn, and it is well to keep the window of the hood down so that the glass is between the apparatus and the face of the operator.

Hydrogen selenide, a possible by-product, is very toxic, being comparable with hydrogen cyanide. Its accidental inhalation in small amounts may produce a sore throat.

A 500-ml. three-necked round-bottomed flask is fitted with an efficient reflux condenser, a glycerol-sealed mechanical stirrer, a dropping funnel, and a gas inlet tube extending nearly to the blades of the stirrer (Note 1). An absorption train,[1] with the addition in *J* of a safety tube which extends nearly to the bottom, is connected to the upper end of the reflux condenser (Note 2). A 2-cm. layer of water in *J* allows it to serve as a bubble counter; *K* is one-third filled with a 50% potassium hydroxide solution. The entire apparatus is set up in subdued light in a hood and swept with dry hydrogen (Notes 3 and 4). Phenylmagnesium bromide is prepared in the flask by the usual procedure[2] from 78.5 g. (0.5 mole) of bromobenzene, 12 g. (0.5 gram atom) of magnesium, and 500 ml. of dry ether. The dropping funnel is then replaced by an addition flask (p. 550) which contains 38 g. (0.48 gram atom) of dry powdered black selenium (Note 5). The solution is warmed sufficiently to bring about gentle refluxing, and the selenium is then added gradually over a period of 30 minutes at such a rate as to maintain gentle refluxing without heating. Stirring is continued for an additional 30 minutes (Note 6).

The contents of the flask are then poured upon 600 g. of cracked ice, and, with hand stirring, 75 ml. of hydrochloric acid (sp. gr. 1.18) is added. The cold mixture is now filtered through glass wool in an ordinary funnel into a 2-l. separatory funnel. The aqueous layer is separated and extracted once with 250 ml. of ether (Note 7). The combined extract and main product are dried over 30 g. of calcium chloride, the ether is removed on a steam bath, and the residue is distilled using a 500-ml. modified Claisen flask. The selenophenol is collected at 57–59°/8 mm. or 84–86°/25 mm.; the yield is 43–54 g. (57–71%) (Notes 6 and 9 through 14). The product should be sealed at once in a glass vial (Note 8).

2. Notes

1. Dry hydrogen or dry nitrogen may be used. The gas must be oxygen-free. Hydrogen tends to decrease the amount of oxidation to diselenide.

2. The submitter used a simpler train consisting of two small wide-mouthed bottles in series closed by stoppers bearing the necessary inlet and outlet tubes constructed of 10-mm. or larger glass tubing. In each flask was a shallow layer of 50% potassium hydroxide solution; the outlet tube of the first bottle extended nearly to the surface of the solution and dipped *below* the surface of the solution in the second bottle.

3. Most selenium compounds are affected by sunlight, and many by any bright light. It is often essential to use amber glassware or to wrap the flasks in light-proof paper.

4. Selenium compounds which contain an —SeH group are easily oxidizable in the air to diselenides. It is advantageous to replace the air by an inert gas, and to work as rapidly as possible.

5. The selenium is dried overnight in a vacuum desiccator over concentrated sulfuric acid.

6. The higher yields of selenophenol are favored by exclusion of air, rapid stirring, and not too rapid addition of selenium.

7. An alternative procedure is to extract the selenophenol by sodium hydroxide solution, and subsequently to acidify and extract the liberated substance. The yield of selenophenol is not improved by employing such a procedure, but it may be of value with some compounds.

8. The selenophenol is water-white but rapidly turns yellow in contact with the air.

9. The residue in the still (or alkali-insoluble material in the ether layer if Note 6 has been employed) contains diphenyl selenide (b.p. 167°/16 mm.) and diphenyl diselenide (m.p. 63°). It can be separated by a combination of distillation and crystallization from ethanol, but

the amounts are small, and, unless the residues from several runs are combined, the procedure is not economical.

10. This is a general reaction. It can be used for preparing other selenophenols whenever the desired Grignard reagent can be obtained. The submitter has made the three selenocresols, *p*-bromophenylselenophenol, and *n*-butylselenol by this procedure. He has also obtained thiophenols by the substitution of sulfur for selenium.

11. The submitter has synthesized nine alkyl phenyl selenides in yields of 85–95% by treating ethanol solutions of the sodium salt (the selenophenol is dissolved in the calculated amount of 50% aqueous sodium hydroxide diluted with ethanol) with the appropriate alkyl halide or sulfate.

12. Many selenophenols are more advantageously prepared by hydrolysis of the aryl selenocyanate.

13. It is well to have a hot sulfuric-nitric acid cleaning bath in the same hood, so that apparatus need not be handled in the open laboratory. The large separatory funnel is conveniently cleaned by pouring into it 50 ml. of concentrated nitric acid. After a few minutes, the acid reacts violently with selenides remaining on the sides. The oxides of nitrogen produced effectively clean the funnel, which can then be rinsed with water.

14. Water-soluble selenium compounds are poured down the sink in the hood and flushed with much water. Rubber stoppers and gloves can be freed from toxic compounds by soaking them for a few minutes in bromine or chlorine water and then in dilute sodium hydroxide solution.

3. Methods of Preparation

Selenophenol has been prepared from selenium tetrachloride and benzene in the presence of anhydrous aluminum chloride,[3] and by the procedure described,[4] which is a development of Taboury's.[5]

[1] *Org. Syntheses* Coll. Vol. **1**, 266, Fig. 15 (1941).

[2] *Org. Syntheses* Coll. Vol. **1**, 226 (1941).

[3] Chabrie, *Bull. soc. chim. France,* (2) **50**, 133 (1888); *Ann. chim. phys.,* (6) **20**, 229 (1890).

[4] Foster and Brown, *J. Am. Chem. Soc.,* **50**, 1184 (1928).

[5] Taboury, *Bull. soc. chim. France,* (3) **29**, 762 (1903); *Ann. chim. phys.,* (8) **15**, 36, 38 (1908).

dl-SERINE *

$$CH_2{=}CHCO_2CH_3 \xrightarrow[CH_3OH]{Hg(OAc)_2} CH_3OCH_2CH(HgOAc)CO_2CH_3 \xrightarrow{KBr}$$
$$CH_3OCH_2CH(HgBr)CO_2CH_3 \xrightarrow{Br_2}$$
$$CH_3OCH_2CHBrCO_2CH_3 \xrightarrow[(H_2SO_4)]{NaOH} CH_3OCH_2CHBrCO_2H \xrightarrow{NH_3}$$
$$CH_3OCH_2CH(NH_2)CO_2H \xrightarrow{HBr} HOCH_2CH(NH_3Br)CO_2H \xrightarrow{NH_3}$$
$$HOCH_2CH(NH_2)CO_2H$$

Submitted by HERBERT E. CARTER and HAROLD D. WEST.
Checked by NATHAN L. DRAKE and WILLIAM A. STANTON.

1. Procedure

A. *Methyl α-bromo-β-methoxypropionate.* In a 5-l. flask is placed 450 g. of a 60% solution of methyl acrylate in methanol (Note 1) containing 3.1 moles of methyl acrylate (Note 2). To this solution are added 180 g. of methanol and 960 g. (3 moles) of mercuric acetate (Note 3). The mixture is allowed to stand at room temperature for 3 days with occasional shaking (Note 4). The flask is cooled in an ice bath, and a solution of 360 g. (3 moles) of potassium bromide in 1.2 l. of water is added with stirring during 15 minutes. The heavy oil that separates is extracted with 2.4 l. of chloroform (Note 5), and the aqueous layer is again extracted with 600 ml. of chloroform. The chloroform solutions are combined, washed three times with water, and carefully dried over anhydrous magnesium sulfate (Note 6).

The solution, filtered to remove the drying agent, is placed in a 4-l. beaker and warmed to 50°. The beaker is then exposed to direct sunlight (Note 7), and 450 g. (2.81 moles) of bromine is added with stirring (Note 8) as fast as it is used. The reaction starts slowly but accelerates rapidly, and considerable heat is evolved. The temperature of the solution should be kept below 55° (Note 9). The bromination usually requires 20–30 minutes. At the conclusion of the reaction, the flask is cooled for 15 minutes in an ice-salt bath, and the mercuric bromide is separated by filtration. The chloroform is removed by distillation under reduced pressure through a 20-in. column, and the residue is fractionally distilled under reduced pressure from a modified Claisen flask. The yield is 480–510 g. (81–86% based on the mercuric acetate) of a product boiling at 70–80° at 6 mm. This material contains 5–10% of methyl α,β-dibromopropionate, which is not readily removed. Since the impurity causes no trouble in the synthesis of serine, the crude bromo ester is used in succeeding steps (Note 10).

B. *α-Bromo-β-methoxypropionic acid.* Eight hundred grams of the bromo ester and 1 l. of 0.5 *N* sodium hydroxide are placed in a 5 l. three-necked flask equipped with an efficient stirrer and a separatory funnel, and cooled with running tap water. The stirrer is started, and 800 ml. of 5 *N* sodium hydroxide is added during the course of 2 hours. After the addition is complete, the solution is stirred for 1 hour and then neutralized with an equivalent quantity of sulfuric acid (Note 11). The neutralized solution is extracted once with a 1-l. portion, and three times with 500-ml. portions, of ether. The ether extracts are combined, washed once with a cold saturated solution of sodium sulfate, and dried over anhydrous sodium sulfate; the ether is removed by distillation (Note 12). There remains 700–750 g. of crude bromo acid which is used without purification in the preparation of serine (Note 13).

C. *dl-Serine.* One-half (350–375 g.) of the crude bromomethoxypropionic acid prepared from the saponification of 800 g. of methyl α-bromo-β-methoxypropionate is heated with 3.5 l. of concentrated ammonium hydroxide in a glass-lined autoclave (Note 14) for 10–15 hours at 90–100°, and the solution from the bomb is concentrated to a thick syrup under reduced pressure (Note 15). Two liters of water is added, and the solution is concentrated to dryness. The cake is dissolved in 1.5 l. of 48% hydrobromic acid and refluxed for 2.5 hours. The resulting dark solution is concentrated to a volume of about 500 ml. and then cooled under the tap. The precipitated ammonium bromide is removed by filtration, and the filtrate is concentrated to a thick syrup (Note 16). One liter of water is added, and the solution is again concentrated to a thick syrup. The syrupy residue is dissolved in 375 ml. of warm water, and ammonium hydroxide is added carefully until a faint odor of ammonia persists after vigorous shaking. One and one-half liters of absolute ethanol is then added slowly (Note 17), and the mixture is allowed to stand overnight. The crude serine is filtered and the filtrate discarded. The precipitate is dissolved in 500 ml. of boiling water (Note 18), heated on the steam cone for 10 minutes with 10–15 g. of Darco (Note 19), and then filtered. Five hundred milliliters of absolute ethanol is added to the filtrate slowly with stirring. The mixture is cooled to 0° and is kept at this temperature for 1 hour with occasional stirring to ensure formation of a finely divided precipitate (Note 20). The serine is filtered, washed with ethanol and ether, and air-dried. The precipitates from several runs may be combined and recrystallized as above from 50% ethanol until a white product is obtained (one or two recrystallizations).

All the filtrates should be combined and evaporated to dryness under reduced pressure, and the residue should be recrystallized as above. The last concentrates (including filter paper and Darco washings), if quite dark, may be precipitated from 70% ethanol before recrystallizing. By this process a nearly quantitative recovery may be effected.

The over-all yield is 30 to 40% of the theoretical amount based on the mercuric acetate (Note 21).

2. Notes

1. The solution of methyl acrylate in methanol was obtained from Rohm and Haas Company.

2. If the methyl acrylate solution has stood for some time, it is advisable to use a 10% excess, since the solution slowly deteriorates. A low yield of bromo ester is usually caused by insufficient methyl acrylate.

3. Six hundred and forty-eight grams (3 moles) of mercuric oxide and 345 ml. (360 g., 6 moles) of acetic acid may be used just as satisfactorily as mercuric acetate. The solution of these substances should be cooled in an ice bath and vigorously shaken for several minutes after the reagents have been mixed, since the reaction of mercuric oxide with acetic acid produces considerable heat.

4. At this time there is still a small amount of undissolved mercuric acetate (and mercuric oxide, if used). Longer standing does not improve the yield.

5. All solutions of the mercury compounds should be handled with care, as they are extremely vesicant to some individuals.

6. The brominations proceed more smoothly if the chloroform solution is washed free of methanol and dried.

7. The bromination may be carried out just as satisfactorily with the light from two No. 2 Photoflood lamps. These lamps should be mounted in suitable reflectors as near the surface of the solution as possible. The time required may be slightly longer.

8. An efficient mechanical stirrer is required. Near the end of the bromination the precipitate of mercuric bromide becomes quite heavy.

9. The temperature may be controlled by passing tap water through an 8-mm. glass cooling coil mounted in the beaker. The bromination proceeds rapidly at 45–55°, rather slowly at 35–40°, and practically ceases below 30°.

10. The crude ester may be used directly in the preparation of serine. However, greater ease of purification and slightly better yields make it advantageous to prepare the free acid. Pure methyl α-bromo-β-methoxypropionate is obtained by fractionally distilling the crude material under reduced pressure through a Widmer column. From 100 g. of crude material there is obtained 80–90 g. of pure ester, b.p. 73–75°/6 mm., n_D^{20} 1.4586.

11. It is advisable to keep the temperature below 30° during both the hydrolysis and neutralization.

12. The acid should not be heated for long periods. The last traces of ether should be removed under reduced pressure.

13. The crude bromo acid on standing 2–3 days may deposit an amorphous solid which is probably a polymerization product of α-bromoacrylic acid. If this is removed before the bromo acid is aminated, the subsequent purification of serine is much easier.

14. The checkers employed a simple autoclave, constructed as follows: A piece of 10-in. steam pipe 11 in. long and threaded at both ends was closed at one end by a standard pipe cap. A standard flange was screwed to the other end, and a blank companion flange seating on an "ammonia gasket" provided the closure for the top of the autoclave. A Pyrex battery jar 10 in. high and 8.5 in. in diameter fitted loosely in the pipe and was protected from breakage due to contact with the sides and bottom of the pressure vessel by rings made of rubber tubing. The battery jar was covered with a loosely fitting germinating dish to prevent condensate from the top from dripping into the reaction mixture. After the bromo acid had been introduced into the battery jar, concentrated ammonia was poured into the annular space between the jar and the pipe to provide better heat transfer. The autoclave was heated in a wash tub containing water which was kept boiling by a steam coil.

15. All the concentrations under reduced pressure required in this preparation were carried out with the aid of an efficient water pump.

16. The distillate may be redistilled at atmospheric pressure to recover the hydrobromic acid.

17. Absolute ethanol (200 ml.) is added to the warm (60–70°) solution, and the sides of the flask are scratched to induce crystallization; the remainder of the ethanol is then added in large portions with stirring over a period of 1 hour.

18. Pure *dl*-serine is soluble in water to the extent of 50 g. per liter at 25°, 200 g. at 80°, and approximately 30 g. at 5–10°. The presence of impurities, however, will increase its solubility greatly.

19. Norit is not so satisfactory as Darco.

20. Slow precipitation tends to give large crystals with resulting inclusion of colored impurities.

21. The yield is directly dependent upon the quality of the methyl acrylate used.

3. Methods of Preparation

Serine has been prepared by the Strecker method from glycol aldehyde [1] and from ethoxyacetaldehyde [2–4] by the condensation of ethyl formate with ethyl hippurate followed by reduction and hydrolysis,[5,6] from the reaction product of chloromethyl ether with ethyl sodium phthalimidomalonate,[7] by amination of α-bromo-β-methoxypropionic acid with subsequent demethylation,[8] by hydrolysis of 5-(methoxymethyl)hydantoin,[9] by hydrolysis of ethyl α-acetamido-β-hydroxypropionate,[10] by hydrogenolysis of N-benzylserine,[11] and by reduction of the 2,4-dinitrophenylhydrazone of hydroxypyruvic acid.[12]

[1] E. Fischer and Leuchs, *Ber.*, **35**, 3787 (1902).
[2] Leuchs and Geiger, *Ber.*, **39**, 2644 (1906).
[3] Dunn, Redemann, and Smith, *J. Biol. Chem.*, **104**, 511 (1934).
[4] Redemann and Icke, *J. Org. Chem.*, **8**, 159 (1943).
[5] Erlenmeyer, *Ber.*, **35**, 3769 (1902).
[6] Erlenmeyer and Stoop, *Ann.*, **337**, 236 (1904).
[7] Mitra, *J. Indian Chem. Soc.*, **7**, 799 (1930).
[8] Schiltz and Carter, *J. Biol. Chem.*, **116**, 793 (1936).
[9] Nadeau and Gaudry, *Can. J. Research*, **27B**, 421 (1949).
[10] King, *J. Am. Chem. Soc.*, **69**, 2738 (1947); U. S. pat. 2,530,065 [*C. A.*, **45**, 2971 (1951)].
[11] U. S. pat. 2,446,651 [*C. A.*, **42**, 8211 (1948)].
[12] Sprinson and Chargaff, *J. Biol. Chem.*, **164**, 417 (1946).

SODIUM AMIDE *

$$2Na + 2NH_3 \rightarrow 2NaNH_2 + H_2$$

Submitted by F. W. BERGSTROM.
Checked by C. F. H. ALLEN and C. V. WILSON.

1. Procedure

The apparatus is assembled as shown in Fig. 27. Ammonia gas from a commercial cylinder (Note 1) enters the system at *K*. *R* is a mercury trap which would serve as a safety valve if the system should become blocked by solidification of the amide owing to an accidental drop in temperature. *J* is a U-tube containing just enough mercury to seal the bend, and it serves to estimate the rate of ammonia flow. *I* is a Kjeldahl trap which prevents any mercury from being thrown into the fusion pot *A*, which (Note 2) is conveniently supported on a tripod set on bricks to raise it to a convenient height above the burner *M*. Through the cover of the fusion pot passes an outlet tube *B*, a thermometer well *T*, and the combined inlet tube *CDE*. The thermometer well is welded shut at the bottom and projects about 6 mm.

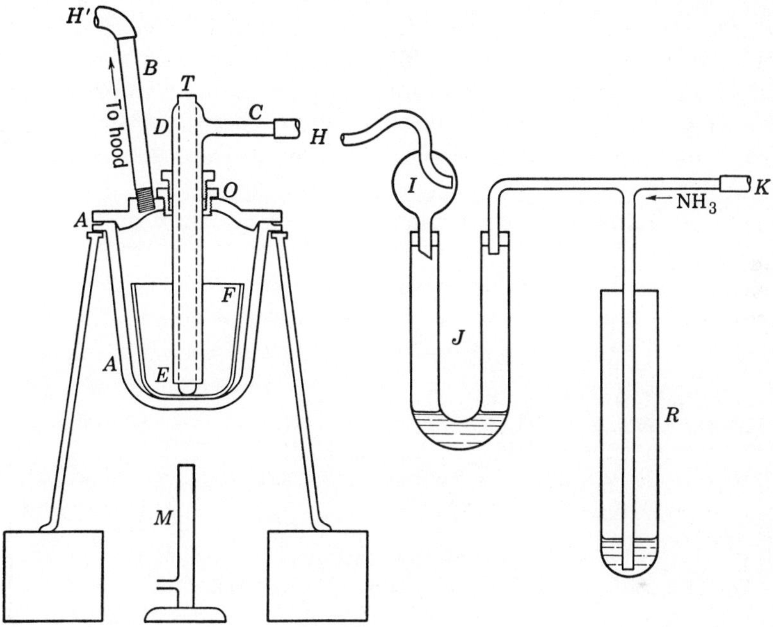

FIG. 27.

below the wider inlet tube, to which it is welded at the top. A gland or packed joint *O*, through which the inlet passes, is packed with a few turns of asbestos cord, the upper hexagonal nut being turned down with a wrench so that ammonia will not escape past the packing and so that there will be sufficient resistance to hold *CDE* in any position to which it may be raised (Note 3). The rubber tubes *H*, *H'* should be of sufficient length (5–7 cm.) to be very flexible and facilitate manipulation of the hot cover. The outlet tube *B* is at least 10 mm. in diameter.

At the outset of the run the pot *A*, with the thermometer well in the position shown in the diagram, is heated to about 120° for 10 minutes in a slow stream of ammonia (Note 4). This serves to sweep the air and any traces of moisture from the system. The apparatus is then allowed to cool to 70–80°, the cover is removed, and a 250-ml. nickel crucible *F* is placed in the pot. The ammonia inlet *CDE* is raised to just above the top of the nickel crucible, in which is then placed approximately 175 g. (7.6 gram atoms) of clean sodium (Note 5). The pot is now heated with the full flame of the burner; the sodium melts in 5–10 minutes, whereupon the inlet tube *CDE* is pushed through the gland until it rests on the bottom of the crucible. When the temperature has reached 320°, the burner is turned down and adjusted to maintain the temperature at 350–360°. The ammonia flow is regulated so that the bubbles in *J* are just too rapid to count. After 3.5–4 hours (Notes 6 and 7), the temperature is lowered to about 320°, and the cover is lifted enough to see whether any unreacted metal remains; if none remains, the flame is removed and the crucible allowed to cool to 230–240°; at this temperature the burner is replaced and heating continued for 30 minutes to ensure removal of the bulk of the sodium hydride.

The burner is now extinguished, the ammonia shut off, and the pot cover removed by disconnecting at *H*, *H'*. The crucible is removed from the pot with tongs, and the molten amide is poured into a shallow iron tray, which has previously been heated to remove traces of moisture (Note 8). At this point it is essential to work rapidly to avoid solidification of the amide in the crucible (Note 9). As soon as the product has solidified sufficiently, the tray is transferred to a large desiccator to cool. When cold enough to handle, the tray is inverted on a clean heavy paper; the amide is removed by rapping the bottom of the pan and is at once transferred to convenient wide-mouthed bottles and covered with a petroleum fraction (Notes 10 and 11). The yields vary from 267 to 282 g. (90–95%) (Notes 12, 13, and 14).

The sodium amide thus prepared is easily pulverized; it may be ground in a mortar under any hydrocarbon solvent. It is safer, though not necessary, where ether is to be used as a reaction medium, to grind the amide first under a hydrocarbon, the mixture being transferred to the reaction flask and then replaced by ether in the usual way (Notes 15, 16, and 17).

2. Notes

1. Ordinary commercial cylinders of ammonia are used; it is unnecessary to dry the gas.

2. The fusion pot is obtainable on the market from the Denver Fire Clay Company (cast-iron crucible and cover, 0.25 gal., catalog No. 2136).

3. The gland *O* may be replaced by a sleeve or bushing and held in place by a set screw or by a clamp at any position desired.

4. Considerable time is saved by using a Meker or triple burner for raising the apparatus to reaction temperature, but an ordinary Tirrill burner is sufficient for the reaction.

5. The oxide coating or oil on commercial sodium should be removed before using. It is more convenient to use approximately 175 g. of sodium than to cut this exact amount.

6. The reaction time depends largely upon the rate at which the ammonia is admitted. If the current is too rapid there will be considerable splashing, and much of the molten amide will collect in the iron pot. Ordinarily the quantity of sodium specified will react completely in the time indicated. The total time for a run is slightly under 6 hours, of which not more than 2–2.5 hours of actual attention are required.

7. Any unreacted metal is easily visible as a globule, floating on the surface of the darker liquid. A flashlight aids in rapid inspection.

8. A pan 2 cm. high and 13 cm. in diameter is suitable for a run of this size. Any oxide coat should be removed by heating to redness, cooling, and polishing with emery paper; otherwise the product is deeply colored where it comes into contact with the pan. The same pan may be used repeatedly without further treatment other than cleaning and drying.

9. Some of the product invariably splashes out of the crucible onto the walls of the pot. If the quantity should be large (too rapid current of ammonia) it can be poured out, but if small it is chipped out after cooling.

10. Commercial "heptane" from petroleum, b.p. 90–100°, is preferable, but other fractions may be used. A 750-ml. bottle will hold the product from one run.

11. Alternatively, the amide is allowed to cool completely in the nickel crucible in a slow current of ammonia and removed when cold. *CDE* is raised above the melt before cooling.

12. The chief variation in the yield is due to loss by splashing; it is difficult to remove the amide that has soldified on the walls of the iron pot. Some loss is accounted for by the sodium hydride carried away with the effluent gas.

13. Runs of other sizes may be made in the same apparatus. With half the quantity of sodium specified, temperature control demands much more attention. With larger quantities, the nickel crucible is dispensed with and the carefully cleaned pot is used. The checkers used 260–270 g. of sodium and averaged a yield of 94%; the reaction

time was increased by only 30 minutes. By arranging two series of apparatus in parallel, but connected to the same cylinder of ammonia, one operator can prepare twice as much amide in almost the same time.

14. The product is nearly white if the iron vessels are carefully cleaned but may be considerably on the gray side.

15. *Caution.* Sodium amide is a very reactive substance; it combines with oxygen and reacts explosively with water. The submitters recommended keeping the amide in sealed glass containers in an atmosphere of ammonia. The checkers preferred the use of petroleum fractions for greater convenience in handling; they have kept specimens under this solvent for 3 years without appreciable loss in activity.

When exposed to the atmosphere, sodium amide rapidly takes up moisture and carbon dioxide. When exposed to only limited amounts, as in imperfectly sealed containers, products are formed which render the resulting mixture highly explosive.[1] The formation of oxidation products is accompanied by the development of a yellow or brownish color. If such a change is noticed, the substance should be destroyed at once. This is conveniently accomplished by covering with much benzene, toluene, or kerosene and slowly adding dilute ethanol with stirring.

16. After the preparation is completed, the cooled reactor should be dismantled and the parts immediately washed with ethanol, care being taken that all traces of sodium amide have been destroyed before water is brought into contact with any part of the equipment.

17. The submitters have prepared potassium amide in yields of 95% in the same manner, but maintaining the temperature at 350–360° for the entire run. The apparatus should be rinsed with an ethanol-benzene mixture.

3. Methods of Preparation

Sodium amide has been prepared by the action of gaseous[2] or liquid[3] ammonia on sodium, by the action of ammonia on alloys of sodium,[4] and by the electrolysis of a solution of sodium cyanide[5] in liquid ammonia with a sodium amalgam electrode. A summary of the chemistry of alkali amides is given by Bergstrom and Fernelius.[1]

[1] Bergstrom and Fernelius, *Chem. Revs.*, **12**, 43 (1933); **20**, 413 (1937).

[2] Wislicenus, *Ber.*, **25**, 2084 (1892); Titherly, *J. Chem. Soc.*, **65**, 504 (1894); *Inorg. Syntheses*, **1**, p. 74, McGraw-Hill Book Company, New York, 1939; De Forcrand, *Compt. rend.*, **121**, 66 (1895); Dennis and Browne, *J. Am. Chem. Soc.*, **26**, 587 (1904); Winter, *J. Am. Chem. Soc.*, **26**, 1484 (1904); Ruff and Geisel, *Ber.*,

39, 828 (1906); Kraus and Cuy, *J. Am. Chem. Soc.*, **45**, 712 (1923); Guntz and Benoit, *Bull. soc. chim. France,* (4) **41**, 434 (1927); Fernelius and Bergstrom, *J. Phys. Chem.*, **35**, 740 (1931); Gilbert, Scott, Timmerli, and Hausley, *Ind. Eng. Chem.*, **25**, 740 (1933); Shreve, Riechers, Rubenkoenig, and Goodman, *Ind. Eng. Chem.*, **32**, 173 (1940).

[3] Joannis, *Compt. rend.*, **112**, 392 (1891); McGee, *J. Am. Chem. Soc.*, **43**, 586 (1921); Brit. pat. 222,718 (1923) [*C. A.*, **19**, 1143 (1925)]; Vaughn, Vogt, and Nieuwland, *J. Am. Chem. Soc.*, **56**, 2120 (1934); *Inorg. Syntheses,* **2**, 128 (1946).

[4] U. S. pat. 1,359,080 [*C. A.*, **15**, 415 (1921)].

[5] Brit. pat. 222,718 (1923) [*C. A.*, **19**, 1143 (1925)]; U. S. pat. 1,570,467 [*C. A.*, **20**, 714 (1926)].

SORBIC ACID *

$$CH_3CH{=}CHCHO + H_2C(CO_2H)_2 \xrightarrow{\text{Pyridine}} CH_3CH{=}CHCH{=}CHCO_2H + CO_2 + H_2O$$

Submitted by C. F. H. Allen and J. VanAllan.
Checked by C. S. Hamilton and R. A. Alberty.

1. Procedure

A mixture of 80 g. (93.2 ml., 1.14 moles) of crotonaldehyde (b.p. 102–103°), 120 g. (1.15 moles) of malonic acid (m.p. 134–135°), and 120 g. (122 ml., 1.52 moles) of pyridine (b.p. 113–115°) is heated for 3 hours in a 1-l. flask on a steam bath under a reflux condenser. At the end of this period the evolution of carbon dioxide will have practically ceased. The flask and contents are then cooled in ice, and a solution of 42.5 ml. (0.76 mole) of concentrated sulfuric acid in 100 ml. of water is added with shaking. Most of the sorbic acid separates at once; the remainder is obtained by chilling the solution in an ice bath for 3 hours. The crude acid is filtered by suction and washed once with a small amount of ice water; it is recrystallized at once from 250 ml. of boiling water. The purified acid, which separates on standing overnight in the ice chest, is filtered; it melts at 134°. The yield is 36–41 g. (28–32%) (Notes 1, 2, and 3).

2. Notes

1. The submitters have found that the percentage yield is the same when double the above quantities are used.

2. This is an example of a general reaction. If acetic acid is used as a solvent, the substituted malonic acids can be secured, whereas

organic bases facilitate the loss of carbon dioxide. The product is generally a mixture from which but a single substance can be isolated.

3. The use of simple aldehydes gives better yields of unsaturated acids; this is especially noticeable when aromatic aldehydes are employed.[1] Mixtures of α,β- and β,γ-unsaturated acids have been reported when aliphatic aldehydes and certain basic catalysts are used.[2,3]

3. Methods of Preparation

Sorbic acid has been prepared from crotonaldehyde[4] or aldol[5] and malonic acid in pyridine solution; by hydrogen peroxide oxidation of the condensation product of crotonaldehyde and pyruvic acid;[6] and by the action of alkali on 3-hydroxy-4-hexenoic acid,[7,8] β,δ-disulfo-*n*-caproic acid,[9] parasorbic acid,[10,11] and 3,5-hexadienoic acid.[12]

[1] K. A. Pandya and R. B. Pandya, *Proc. Indian Acad. Sci.*, **14A**, 112 (1941) [*C. A.*, **36**, 1599 (1942)].

[2] Auwers, *Ann.*, **432**, 58 (1923).

[3] Boxer and Linstead, *J. Chem. Soc.*, **1931**, 740.

[4] Doebner, *Ber.*, **33**, 2140 (1900).

[5] Riedel, *Ann.*, **361**, 90 (1908).

[6] Smedley and Lubrzynska, *Biochem. J.*, **7**, 370 (1913).

[7] Jaworski, *J. Russ. Phys. Chem. Soc.*, **35**, 274 (1903) [*Chem. Zentr.*, **1903**, II, 555].

[8] Jaworski and Reformatski, *Ber.*, **35**, 3636 (1902).

[9] Notebohn, *Ann.*, **412**, 77 (1916).

[10] Hofmann, *Ann.*, **110**, 132 (1859).

[11] Doebner, *Ber.*, **27**, 351 (1894).

[12] Paul and Tchelitcheff, *Bull. soc. chim. France*, **1948**, 108.

STEAROLIC ACID

$$CH_3(CH_2)_7CH{=}CH(CH_2)_7CO_2CH_3 + Br_2 \rightarrow CH_3(CH_2)_7CHBrCHBr(CH_2)_7CO_2CH_3$$

$$CH_3(CH_2)_7CHBrCHBr(CH_2)_7CO_2CH_3 + 3KOH \rightarrow CH_3(CH_2)_7C{\equiv}C(CH_2)_7CO_2K + CH_3OH + 2KBr + 2H_2O$$

$$CH_3(CH_2)_7C{\equiv}C(CH_2)_7CO_2K + HCl \rightarrow CH_3(CH_2)_7C{\equiv}C(CH_2)_7CO_2H + KCl$$

Submitted by HOMER ADKINS and R. E. BURKS, JR.
Checked by ARTHUR C. COPE and HAROLD R. NACE.

1. Procedure

Bromine is added dropwise with stirring to 35 g. (0.118 mole) of methyl oleate (Note 1) in a 500-ml. round-bottomed flask. The mixture is kept below 50° throughout the addition, which is continued until a slight excess of bromine is present; approximately the theoretical amount (18.9 g.) is decolorized. Methyl oleate (2–3 drops) is then added until the bromine color just disappears. *n*-Amyl alcohol (50 ml.) (Note 2) and potassium hydroxide pellets (40 g., 0.61 mole assuming 85% purity) are added to the flask, and the mixture is heated under reflux for 4 hours in an oil bath at 150°. Then approximately 50 ml. of the *n*-amyl alcohol is distilled at atmospheric pressure (Note 3). The residue on cooling solidifies into a tan-colored mass. Phenolphthalein is added as an indicator, the mixture is cooled in an ice bath, and concentrated hydrochloric acid is added in portions until the red color disappears but reappears on stirring of the viscous mass. This process is continued until the mixture remains colorless. Water (approximately 200 ml.) is added, and the mixture is allowed to come to room temperature. Concentrated hydrochloric acid is added until the mixture is acid to methyl orange. The mixture is again cooled in an ice bath; the oily layer solidifies into a wax, and the acidic water solution is decanted. The wax is dissolved in 100 ml. of 95% ethanol at room temperature, and water is added until the solution becomes turbid. The mixture is heated on a steam bath until a clear solution is formed, and then it is cooled in an ice bath and stirred while the product crystallizes. The semisolid mass is filtered, and the product is recrystallized three times from an ethanol-water mixture. After drying in a vacuum desiccator the yield is 11–14 g. (33–42%), m.p. 46–46.5°; neutral equivalent 279.7–281.4 (calcd. 280.4); hydrogen absorbed by catalytic reduction 94–100%.

2. Notes

1. The methyl oleate was prepared by esterification[1] of commercial U.S.P. grade oleic acid and fractionated through a Widmer column. The fractions used boiled at 140–144°/0.5 mm., 175–179°/2 mm., and had n_D^{25} 1.4500–1.4527 and an iodine number of 93–97 (calcd. 85.6) by iodine bromide titration.[2]

2. *n*-Amyl alcohol was selected as a solvent with a convenient boiling point for the dehydrohalogenation. Acetylenic triple bonds are attacked by water in the presence of strong alkali or strong acid.

3. Part of dehydrohalogenation occurs during the distillation of *n*-amyl alcohol at atmospheric pressure. If the solution was concentrated by distillation under reduced pressure, the product contained bromine and failed to crystallize.

3. Methods of Preparation

Stearolic acid has been prepared by the dehydrohalogenation of brominated olive or almond oil,[3] dibromostearic acid,[4] or dichlorostearic acid.[5] The procedure described is a modification of one used by Kino.[4]

[1] Skraup and Schwamberger, *Ann.*, **462**, 155 (1928).

[2] Mahin, *Quantitative Analysis*, 4th ed., pp. 416–420, McGraw-Hill Book Co., New York, 1932.

[3] Hoffmann-La Roche, Ger. pat. 243,582 [*Chem. Zentr.*, **1912**, I, 695].

[4] Kino, *J. Soc. Chem. Ind. Japan*, **32**, 187 (1929) [*C. A.*, **24**, 1998 (1930)].

[5] Inoue and Suzuki, *Proc. Imp. Acad. Tokyo*, **7**, 261 (1931) [*C. A.*, **26**, 87 (1932)].

trans-STILBENE *

$$C_6H_5CHOHCOC_6H_5 + 4H \xrightarrow[HCl]{Zn(Hg)_x} C_6H_5CH{=}CHC_6H_5 + 2H_2O$$

Submitted by R. L. Shriner and Alfred Berger.
Checked by W. E. Bachmann and Charles E. Maxwell.

1. Procedure

In a 4-l. beaker, equipped with a powerful mechanical stirrer which reaches nearly to the bottom, are placed 500 ml. of water and 50 g. of mercuric chloride. The stirrer is started, and 200 g. (3.06 gram atoms) of zinc dust (Note 1) is rapidly sifted into the suspension. Stirring is continued until the mercuric chloride dissolves (about 20–30 minutes). The zinc is then allowed to settle, the supernatant liquid is removed by decantation, and the amalgam is filtered and washed with 200 ml. of water. The zinc amalgam is returned to the beaker, which is now surrounded by an ice bath, and 500 ml. of 95% ethanol and 100 g. (0.48 mole) of benzoin are added. The stirrer is started, and 500 ml. of concentrated hydrochloric acid is added through a dropping funnel at such a rate that addition is complete in about 2 hours; throughout the reaction the temperature is maintained below 15°. Stirring is continued for about 2 hours more.

About 2 l. of cold water is added to the reaction mixture, and the insoluble material is collected on a Büchner funnel. The precipitate is transferred to a 2-l. beaker and extracted with two 600-ml. portions of hot ethanol. The combined extracts, on cooling, deposit long needles of stilbene which weigh 55–59 g. and melt at 116–121°. These are filtered with suction and are recrystallized from 600 ml. of 95% ethanol. The yield is 45–48 g. (53–57%) of colorless needles melting at 123–124°.

2. Note

1. The zinc dust used was obtained from J. T. Baker and Company.

3. Methods of Preparation

The procedure described is essentially that of Ballard and Dehn.[1] Stilbene has also been prepared by reduction of desoxybenzoin,[2a] benzaldehyde,[2b] and benzil;[2a,c] by dehydrogenation of ethylbenzene,[3a] toluene,[3a,b,c] and bibenzyl;[3b,d] by alkaline reduction of phenylnitromethane,[4a] phenylnitroacetonitrile,[4a] and desoxybenzoin;[4b] by distillation of benzyl sulfone,[5a] benzyl sulfide,[5a,b] calcium cinnamate,[5c] cinnamic acid,[5d] phenyl cinnamate,[5e,f] and diphenyl fumarate,[5e] by dehydrohalogenation of α,α′-dichlorobibenzyl[6a] and benzyl chloride;[6b] by dehalogenation of α,α,α′,α′-tetrachlorobibenzyl[7a] and benzal chloride;[7b] by the coupling of cinnamic acid and phenyldiazonium chloride;[8] by dehydration of benzylphenylcarbinol,[6a,9a] benzyl ether,[9b] and benzyl alcohol;[9b,c] by treatment of benzaldehyde ethyl mercaptal with Raney nickel;[10] by pyrolysis of methyl benzyl dithiocarbonate;[11] and by catalytic desulfurization of trithiobenzaldehyde.[12] The diazonium coupling reaction[8] appears to have the widest applicability for the synthesis of substituted stilbenes.

[1] Ballard and Dehn, *J. Am. Chem. Soc.*, **54**, 3969 (1932).

[2] (*a*) Irvine and Weir, *J. Chem. Soc.*, **91**, 1384 (1907); (*b*) Williams, *Jahresb.*, **1867**, 672; Barbaglia and Marquardt, *Gazz. chim. ital.*, **21**, 195 (1891); Baumann

and Klett, *Ber.*, **24**, 3307 (1891); Law, *J. Chem. Soc.*, **91**, 748 (1907); Schepps, *Ber.*, **46**, 2564 (1913); (*c*) Jena and Limpricht, *Ann.*, **155**, 89 (1870); Blank, *Ann.*, **248**, 1 (1888).

[3] (*a*) Meyer and Hofmann, *Monatsh.*, **37**, 681 (1916); (*b*) Behr and van Dorp, *Ber.*, **6**, 753 (1873); Aronstein and van Nierop, *Rec. trav. chim.*, **21**, 448 (1902); (*c*) Lorenz, *Ber.*, **7**, 1096 (1874); *Ber.*, **8**, 1455 (1875); Lange, *Ber.*, **8**, 502 (1875); Michaelis and Lange, *Ber.*, **8**, 1313 (1875); Michaelis and Paneck, *Ann.*, **212**, 203 (1882); (*d*) Dreher and Otto, *Ann.*, **154**, 171 (1870); Barbier, *Compt. rend.*, **78**, 1769 (1874); Radziszewski, *Ber.*, **8**, 756 (1875); Kade, *J. prakt. Chem.*, (2) **19**, 461 (1879).

[4] (*a*) Wislicenus and Endes, *Ber.*, **36**, 1194 (1903); (*b*) Sudborough, *J. Chem. Soc.*, **67**, 601 (1895).

[5] (*a*) Fromm and Achert, *Ber.*, **36**, 534 (1903); (*b*) Märcker, *Ann.*, **136**, 75 (1865); Limpricht and Schwanert, *Ann.*, **145**, 330 (1868); Forst, *Ann.*, **178**, 370 (1875); (*c*) Engler and Leist, *Ber.*, **6**, 254 (1873); (*d*) von Miller, *Ann.*, **189**, 338 (1877); (*e*) Anschütz, *Ber.*, **18**, 1945 (1885); (*f*) Skraup and Beng, *Ber.*, **60**, 942 (1927).

[6] (*a*) Meisenheimer and Heim, *Ann.*, **355**, 249 (1907); (*b*) Tschitschibabin, *J. Russ. Phys. Chem. Soc.*, **34**, 130 (1902); Loeb, *Ber.*, **36**, 3059 (1903); Acree, *Am. Chem. J.*, **29**, 593 (1907).

[7] (*a*) Liebermann and Homeyer, *Ber.*, **12**, 1971 (1879); (*b*) Limpricht, *Ann.*, **139**, 303 (1866); Lippmann and Hawliczek, *Jahresb.*, **1877**, 405.

[8] Meerwein, Büchner, and van Emster, *J. prakt. Chem.*, (2) **152**, 237 (1939).

[9] (*a*) Limpricht and Schwanert, *Ann.*, **155**, 59 (1870); Hell, *Ber.*, **37**, 453 (1904); (*b*) Szperl, *Chemik Polski*, **15**, 23 (1917) [*C. A.*, **13**, 2865 (1919)]; (*c*) Guerbet, *Compt. rend.*, **146**, 298 (1908); *Bull. soc. chim. France*, (4) **3**, 500 (1908).

[10] Hauptmann and Camargo, *Experientia*, **4**, 385 (1948).

[11] Laakso, *Suomen Kemistilehti*, **16B**, 19 (1943); [*C. A.*, **40**, 4687 (1946)].

[12] Wood, Bacon, Meibohm, Throckmorton, and Turner, *J. Am. Chem. Soc.*, **63**, 1334 (1941).

TEREPHTHALALDEHYDE

$$\text{A.}\quad C_6H_4\begin{cases}CH_3\\ CH_3\ (p)\end{cases} + 4Br_2 \rightarrow C_6H_4\begin{cases}CHBr_2\\ CHBr_2\ (p)\end{cases} + 4HBr$$

$$\text{B.}\quad C_6H_4\begin{cases}CHBr_2\\ CHBr_2\ (p)\end{cases} + 2H_2O \xrightarrow{(H_2SO_4)} C_6H_4\begin{cases}CHO\\ CHO\ (p)\end{cases} + 4HBr$$

Submitted by J. M. Snell and A. Weissberger.
Checked by R. L. Shriner and N. S. Moon.

1. Procedure

A. *α,α,α′,α′-Tetrabromo-p-xylene.* A 1-l. three-necked flask is fitted with an oil-sealed mechanical stirrer, a 500-ml. dropping funnel, and a reflux condenser, the top of which is connected to a gas-absorption trap.

A 300-watt tungsten lamp is clamped in such a position that its bulb is within 1 in. of the flask (Note 1). Into the flask is introduced 100 g. (0.94 mole) of dry *p*-xylene (m.p. 11–12°), and the flask is heated in an oil bath maintained between 140° and 160°. The stirrer is started, and, when the xylene starts boiling, 700 g. (224 ml., 4.38 moles) of dry bromine (Note 2) is gradually added through the dropping funnel at such a rate that there is never any large amount of unreacted bromine in the flask. Stirring and heating are continued throughout the reaction, which requires 6–10 hours. After all the bromine has reacted, the mixture is cooled and dissolved in 1 l. of warm chloroform. The chloroform solution is cooled in an ice bath and the product removed by filtration. It is light gray, melts at 165–169°, and weighs 250–258 g. After a second recrystallization from 1 l. of chloroform, 190–200 g. of light-gray crystals melting at 168–170° is obtained. An additional 15–20 g. may be obtained by concentrating the chloroform filtrate to 250 ml. and recrystallizing the precipitate from fresh chloroform. The total yield is 205–220 g. (51–55%).

B. *Terephthalaldehyde.* A 2-l. round-bottomed flask is fitted with a still head, capillary ebullition tube, and receiver in an assembly for a vacuum distillation. Into the flask are introduced 84.3 g. (0.2 mole) of finely powdered tetrabromo-*p*-xylene and 200 ml. of concentrated sulfuric acid (95%). The reactants are thoroughly mixed by shaking. A vacuum is applied by means of a water pump, and a stream of air is allowed to pass through the capillary tube in order to facilitate the rapid removal of hydrogen bromide. The flask is heated in an oil bath to 70°, and, as the evolution of gas becomes less vigorous, the temperature of the bath is gradually raised to 110° (Note 3). The reaction is complete when a perfectly clear solution is obtained and no more gas is evolved (about 2.5 hours). The flask is then cooled and the contents are poured on 600 g. of crushed ice. The crystalline solid is collected on a filter, washed with a little water, and recrystallized from 600 ml. of 10% methanol with the aid of 1 g. of decolorizing carbon to remove the yellow color. The small amount of tetrabromoxylene which remains undissolved is separated with the carbon and washed on the filter with two 100-ml. portions of hot 10% methanol. The yield of pure product, m.p. 115–116°, is 21.7–22.5 g. (81–84%).

2. Notes

1. It is important to have light shining on the reaction mixture throughout the bromination. The bromination may also be accomplished by placing the flask in direct sunlight.

2. The bromine is dried by shaking it with two 200-ml. portions of concentrated sulfuric acid.

3. The mixture foams considerably and must be watched. The foaming can be controlled by raising the temperature slowly and regulating the vacuum. For this reason it is difficult to hydrolyze larger amounts of the tetrabromide.

3. Methods of Preparation

Terephthalaldehyde has been made by the action of lead nitrate on α,α′-dichloro-*p*-xylene [1] or α,α′-dibromo-*p*-xylene; [2] by the action of fuming nitric acid on dibromo-*p*-xylene; [3] by the hydrolysis of terephthaldehyde tetraacetate; [4] by the action of phosphorus pentachloride on *p*-xylyleneglycol monoethyl ether; [5] by the hydrolysis of α,α,α′,α′-tetrachloro-*p*-xylene; [6] by the hydrolysis of α,α,α′,α′-tetrabromo-*p*-xylene; [7] and by the Sommelet procedure.[8] The method described here is essentially the modification described by Weissberger and Bach.[9]

[1] Grimaux, *Bull. soc. chim. France,* (2) **25,** 337 (1876); *Compt. rend.,* **83,** 825 (1877).

[2] Löw, *Ann.,* **231,** 361 (1885).

[3] Löw, *Ber.,* **18,** 2072 (1885).

[4] Thiele and Winter, *Ann.,* **311,** 341 (1900).

[5] Colson, *Bull. soc. chim. France,* (2) **42,** 152 (1884).

[6] Colson and Gautier, *Bull. soc. chim. France,* (2) **45,** 506 (1886); *Ann. chim.,* (6) **11,** 25 (1887).

[7] Hönig, *Monatsh.,* **9,** 1150 (1888); Thiele and Gunther, *Ann.,* **347,** 106 (1906).

[8] Angyal, Morris, Tetaz, and Wilson, *J. Chem. Soc.,* **1950,** 2141.

[9] Weissberger and Bach, *Ber.,* **65,** 24 (1932).

TEREPHTHALIC ACID *

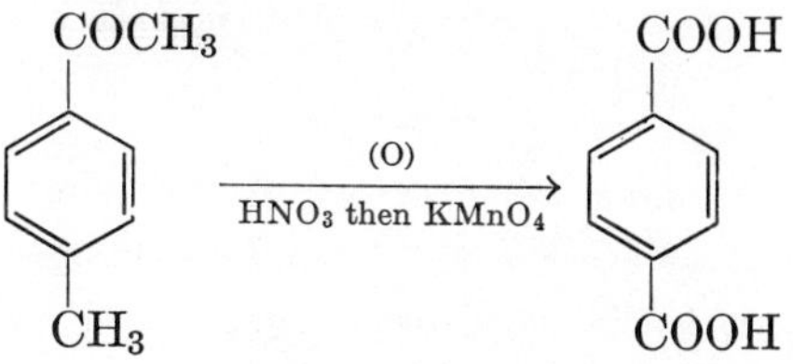

Submitted by C. F. KOELSCH.
Checked by C. F. H. ALLEN and J. VANALLAN.

1. Procedure

One hundred grams (0.75 mole) of *p*-methylacetophenone [1] is added to a mixture of 250 ml. (4 moles) of concentrated nitric acid (sp. gr. 1.42) and 1 l. of water in a 3-l. flask, and the mixture is refluxed in a hood for 4 hours. After the mixture has been cooled, the sticky, yellow solid is collected on a 14-cm. Büchner funnel, pressed down well, and washed with 300 ml. of cold water.

The moist solid is mixed with 1 l. of water and 35 g. of sodium hydroxide in a 3-l. three-necked flask, fitted with a mechanical stirrer and a reflux condenser, and the stirred mixture is heated almost to its boiling point. Through the momentarily opened third neck of the flask is added, in portions of about 20 g., 300 g. (1.9 moles) of potassium permanganate, at such a rate that the boiling of the stirred mixture is maintained without external heating. After the addition has been completed, the mixture is refluxed for 2 hours; if any permanganate remains it is destroyed by the addition of 25 ml. of ethanol. The mixture is then filtered through a 14-cm. Büchner funnel, and the manganese dioxide is washed by removing it from the funnel, slurrying with 500 ml. of hot water (Note 1), and filtering.

The combined filtrates are heated nearly to boiling and acidified with a solution of 108 ml. of concentrated sulfuric acid (sp. gr. 1.84) in 400 ml. of water. After the mixture has been cooled to room temperature, the terephthalic acid is filtered and washed by stirring on the filter with three successive 100-ml. portions of cold water. The product is dried in an evaporating dish on a steam bath. The yield of terephthalic acid which sublimes at 300° or higher without melting is 105–109 g. (84–88%) (Notes 2 and 3).

2. Notes

1. This extraction yields 1–2 g. of the product.

2. The acid is analytically pure. There is no satisfactory solvent for the recrystallization of large amounts of terephthalic acid. Small quantities may be recrystallized from acetic acid, but the purity of a properly precipitated and washed sample is not thereby improved.

3. The high yields of ethyl ester obtainable from the product attest its purity. A mixture of 50 g. of terephthalic acid, 500 ml. of absolute ethanol, and 25 ml. of sulfuric acid was boiled for 16 hours and then distilled to half its volume and poured into dilute aqueous sodium carbonate. There was obtained 56.7 g. of diethyl terephthalate (m.p. 42–44°), and from the wash water there was recovered 4.6 g. of terephthalic acid; these materials account for 93.3% of the original substance.

3. Methods of Preparation

Terephthalic acid has been obtained from a great many *p*-disubstituted derivatives of benzene or cyclohexane by oxidation with permanganate, chromic acid, or nitric acid. The following routes appear to have preparative value: from *p*-toluic acid,[2] *p*-methylacetophenone,[3] or dihydro-*p*-tolualdehyde[4] by oxidation with permanganate; from *p*-cymene by oxidation with sodium dichromate and sulfuric acid;[5] from *p*-dibromobenzene or from *p*-chloro- or *p*-bromobenzoic acid by heating at 250° with potassium and cuprous cyanides;[6] and from *p*-dibromobenzene[7] or *p*-iodobenzoic acid,[8] butyllithium, and carbon dioxide. It has been obtained also by the air oxidation of *p*-xylene in the presence of cobalt naphthenates.[9] Oxidations of *p*-toluic acid,[10] *p*-tolualdehyde,[11] and α-chloro-*p*-xylene[12] have been patented.

[1] *Org. Syntheses* Coll. Vol. **1,** 111 (1941).

[2] Frey and Horowitz, *J. prakt. Chem.,* (2) **43,** 116 (1891).

[3] Randall, Benger, and Groocock, *Proc. Roy. Soc.,* **A165,** 432 (1938) [*C. A.,* **32,** 5373 (1938)].

[4] Allen, Ball, and Young, *Can. J. Research,* **9,** 169 (1933).

[5] Bogert and Harris, *J. Am. Chem. Soc.,* **41,** 1680 (1919).

[6] Rosenmund and Struck, *Ber.,* **52,** 1752 (1919).

[7] Gilman, Langham, and Moore, *J. Am. Chem. Soc.,* **62,** 2331 (1940).

[8] Gilman and Arntzen, *J. Am. Chem. Soc.,* **69,** 1537 (1947).

[9] Brit. pat. 623,836 [*C. A.,* **44,** 4035 (1950)].

[10] U. S. pat. 2,529,448 [*C. A.,* **45,** 2981 (1951)].

[11] Brit. pat. 644,667 [*C. A.,* **45,** 4743 (1951)].

[12] Brit. pat. 644,707 [*C. A.,* **45,** 4744 (1951)].

TETRAHYDROFURFURYL BROMIDE

(Furan, 2-(bromomethyl)tetrahydro-)

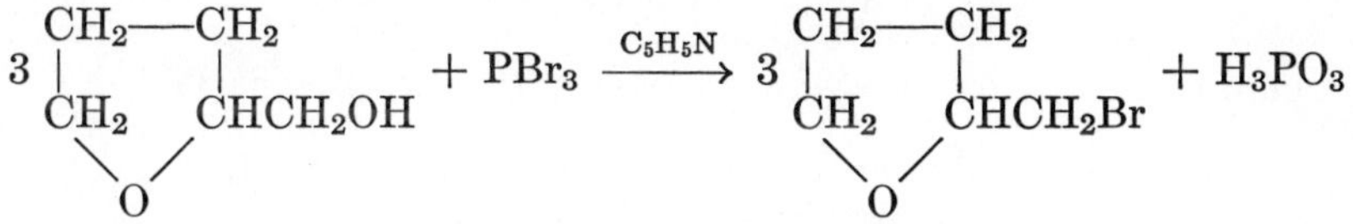

Submitted by L. H. Smith.
Checked by John R. Johnson and R. L. Sawyer.

1. Procedure

In a 500-ml. three-necked flask, fitted with a mechanical stirrer, thermometer, separatory funnel, and calcium chloride tube, are placed 96 g. (56.5 ml., 0.36 mole) of redistilled phosphorus tribromide (b.p. 174–175°/740 mm.) and 50 ml. of dry benzene. From the separatory funnel, 15 g. of dry pyridine is added with stirring over a period of 15 minutes. The flask is then surrounded by an ice-salt mixture, and the contents are cooled to −5°. A mixture of 102 g. (1 mole) of redistilled tetrahydrofurfuryl alcohol (b.p. 79–80°/20 mm.) and 5 g. of dry pyridine (total pyridine, 20 g., 0.25 mole) is added slowly from the dropping funnel with stirring over a period of 4 hours. During this time the internal temperature is kept at −5° to −3°. Stirring is continued for 1 hour longer, and the cooling bath is then allowed to warm up to room temperature.

The mixture is allowed to stand for 24–48 hours (Note 1) and is then transferred to a 500-ml. Claisen flask. Two small portions of benzene are used for rinsing the flask. The benzene is distilled by reducing the pressure gradually to about 60 mm. and heating the flask gently in an oil bath (not above 90°). After the benzene has been removed, the pressure is reduced to 5–10 mm. and the bath is heated slowly to 150–155° until no more material distils (Note 2). The crude distillate (110–126 g.) is redistilled through an efficient column, and the purified tetrahydrofurfuryl bromide is collected at 69–70°/22 mm. (61–62°/13 mm., 49–50°/4 mm.) (Note 3). The yield is 90–102 g. (53–61%).

2. Notes

1. The yields were slightly higher when the mixture was allowed to stand for 48 hours.

2. Most of the material distils while the bath is at 100–120°. When the bath reaches 155–160° the mixture begins to decompose and white fumes are copiously evolved.

3. Unless a good fractionation is obtained, the material will contain some pyridine and will discolor in a few days. Carefully fractionated material will remain colorless for 2 months or more. A considerable amount of dark, viscous residue remains in the distilling flask.

3. Methods of Preparation

Tetrahydrofurfuryl bromide has been obtained in low yields by the action of hydrobromic acid, or of phosphorus tribromide, on the corresponding alcohol.[1] The yield is improved markedly by use of phosphorus tribromide and pyridine.[2,3] The bromide has also been prepared by the action of potassium hydroxide on 4,5-dibromopentanol-1.[4]

[1] Dox and Jones, *J. Am. Chem. Soc.*, **50**, 2033 (1928).
[2] Paul, *Bull. soc. chim. France,* (4) **53**, 417 (1933).
[3] Robinson and Smith, *J. Chem. Soc.*, **1936**, 195.
[4] Paul, *Ann. chim.*, (10) **18**, 303 (1932).

TETRAHYDROPYRAN *

(Pyran, tetrahydro-)

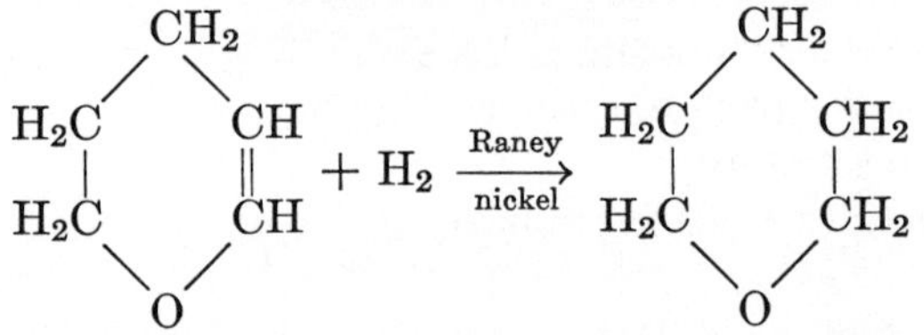

Submitted by D. W. Andrus and John R. Johnson.
Checked by Nathan L. Drake and Charles M. Eaker.

1. Procedure

The hydrogenation is carried out in a low-pressure catalytic hydrogenation apparatus. Raney nickel catalyst (p. 181) is washed with ether three times on a Büchner funnel, then 8 g. of it is transferred under ether to the hydrogenation bottle. The bottle is fitted with a rubber stopper bearing a small dropping funnel and a glass tube that leads to one arm of a three-way stopcock. The other arms of the stopcock are connected respectively to a water pump and a source of inert gas (Note 1) in such a way that the ether can be pumped off and the bottle can then be filled with the inert gas. When this has been done, 50.5 g. (0.6 mole) of dihydropyran (p. 276) is introduced through the dropping funnel.

The bottle is connected to the hydrogenation apparatus and alternately evacuated and filled with hydrogen twice. Hydrogen is then admitted to the system until the pressure gauge reads 40 lb. The shaker is started, and the pressure drops to the theoretical value for absorption of 0.6 mole in 15–20 minutes; beyond this point shaking causes no further absorption of hydrogen (Note 2). The bottle is removed and the nickel catalyst is allowed to settle. The tetrahydropyran is decanted, but enough is left in the bottle to cover the catalyst (Note 3). The product boils at 85–86°, but it need not be distilled for many purposes. The yield is practically quantitative.

2. Notes

1. Purified nitrogen is a convenient inert gas, but natural gas containing no oxygen, or sulfur compounds, is equally suitable; the checkers used hydrogen.

2. As the catalyst becomes older, it loses its activity somewhat and a longer time is required for the pressure to drop to the theoretical value.

3. The same catalyst may be used many times. For the next run the dihydropyran is merely poured into the bottle containing the catalyst which is wet with the product of the previous run.

3. Methods of Preparation

Tetrahydropyran has been prepared by hydrogenation of dihydropyran using a platinum black catalyst;[1] by heating pentamethylene bromide with water;[2,3] or with water and zinc oxide in a sealed tube;[4] or by heating pentamethylene glycol with 3 volumes of 60% sulfuric acid in a pressure tube,[5] or dehydration of this glycol in the vapor phase over kaolin or aluminum oxide.[6]

[1] Paul, *Bull. soc. chim. France,* (4) **53,** 1489 (1933).

[2] Hochstetter, *Monatsh.,* **23,** 1073 (1902).

[3] Demjanow, *J. Russ. Phys. Chem. Soc.,* **45,** 169 (1913) [*C. A.,* **7,** 2226 (1913)].

[4] Clarke, *J. Chem. Soc.,* **101,** 1802 (1912); Allen and Hibbert, *J. Am. Chem. Soc.,* **56,** 1398 (1934).

[5] Demjanow, *J. Russ. Phys. Chem. Soc.,* **22,** 389 (1890) [*J. Chem. Soc.,* **62,** 1292 (1892)].

[6] Beati and Mattei, *Ann. chim. appl.,* **30,** 21 (1940).

TETRAIODOPHTHALIC ANHYDRIDE *

(Phthalic anhydride, tetraiodo-)

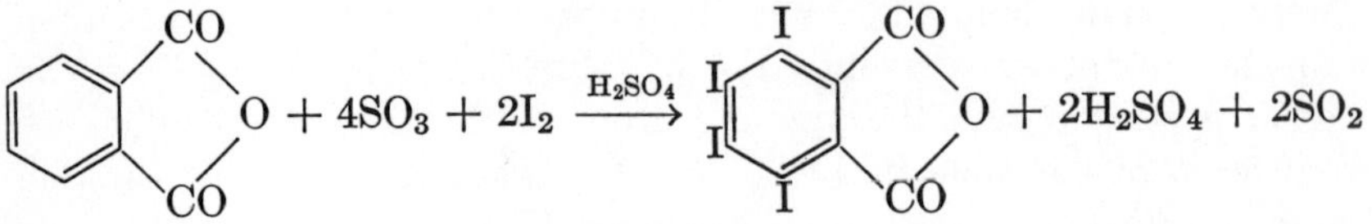

Submitted by C. F. H. Allen, Homer W. J. Cressman, and H. B. Johnson.
Checked by H. R. Snyder and E. Van Heyningen.

1. Procedure

In a 3-l. three-necked flask with ground-glass joints are placed 148 g. (1.0 mole) of phthalic anhydride, 324 g. (60% of the total of 2.12 moles to be added) of iodine, and 600 ml. of 60% fuming sulfuric acid. The flask, fitted with an air condenser 90–100 cm. in length (Note 1), is arranged for heating by a water bath. A tube leads from the condenser to a gas trap.

The temperature of the water bath is raised cautiously to 45–50°, at which point the reaction begins (Note 2). Heating is continued until visible reaction has almost completely stopped. During this time (Note 3), the temperature has been raised gradually to 65°. The flask is then cooled by flooding the bath with ice water, and a second portion, 162 g. (30% of the total), of iodine is added. The flask is heated again at 65° until the reaction ceases. After cooling as before, the last portion, 54 g. (10% of the total), of iodine is added.

When reaction at 65° has ceased and the flask has been cooled, the condenser is removed, and the flask is arranged for heating on an oil bath in a well-ventilated area (Note 4). The temperature of the bath is raised to 175° and maintained at 170–180° until the evolution of sulfur trioxide and iodine fumes has slowed considerably (about 2 hours at 170°). The flask is allowed to cool to about 60° before the contents are poured into a beaker, which is allowed to stand overnight at room temperature.

The solid is filtered by suction on a glass filter cloth and is washed with two 100-ml. portions of concentrated sulfuric acid and then with three 200-ml. portions of water (Note 5). The light-yellow crystalline material is then transferred to a 3-l. beaker where it is stirred for 30 minutes with a solution of 20 g. of sodium bisulfite in 1.5 l. of water to remove the last traces of iodine. The heavy solid is allowed to settle to the bottom of the beaker, and the bisulfite solution is poured off. The crystals are washed with three 1-l. portions of water, each portion being decanted as before (Note 6), and then are transferred to a funnel. The product is washed with an additional 1-l. portion of water and two 200-ml. portions of acetone and dried in an oven at 60°. The product amounts to 520–535 g. (80–82% based on the phthalic anhydride) (Notes 7 and 8) and melts at 327–328° (Note 9).

2. Notes

1. A water-cooled condenser maintained at 20–25° may be used.

2. Temporary cooling by cold water may be necessary to keep the reaction in check. It must be closely watched at this stage.

3. The lengths of the periods of heating are approximately 5, 3, and 1 hour, respectively. In a run of one-half the size described, the checkers found the periods required to be about 4, 1.5, and 1 hour, respectively.

4. This operation is carried on most conveniently out-of-doors. For the 0.5-mole run, the checkers fitted the flask to the gas trap with a U-tube of 2-cm. bore, having a ground-glass joint for the connection to the flask. The large-diameter tube permitted the operation to be carried out in a hood with no danger of clogging of the apparatus by sublimed iodine.

5. This method of working up the product is superior to the usual one of pouring on ice in that a purer material is obtained and a difficult recrystallization is thus avoided.

6. Some impurities tend to float and are removed with the wash water.

7. It was observed that sunlight or illumination by a Photoflood lamp tended to make the reaction more vigorous but did not produce any significant increase in yield.

8. Yields of the same order were obtained starting with 0.5, 1.0, and 10.0 moles of phthalic anhydride. In the runs with 10.0 moles of phthalic anhydride, a greater excess of iodine was found necessary; i.e., the weights of iodine added were 3240 g., 1620 g., 540 g., 500 g., 300 g. The lengths of the heating periods were approximately 72, 48, 24, 24, and 12 hours, respectively.

9. The melting point may vary from 325° to 332° but usually falls within a 2° range. No suitable solvent for recrystallization has been

found. The recrystallized product always has a lower melting point than the original material.

3. Methods of Preparation

The method given is the only satisfactory one so far reported.[1-5]

[1] Juvalta, Ger. pat. 50,177 [*Frdl.*, **2,** 94 (1887–1890)].
[2] Rupp, *Ber.*, **29,** 1634 (1896).
[3] Pratt and Shupp, *J. Am. Chem. Soc.*, **40,** 254 (1918).
[4] Perkins and Quimba, *Am. J. Pharm.*, **106,** 467 (1934).
[5] Twiss and Heinzelmann, *J. Org. Chem.*, **15,** 496 (1950).

α-TETRALONE

(1(2H)-Naphthalenone, 3,4-dihydro-)

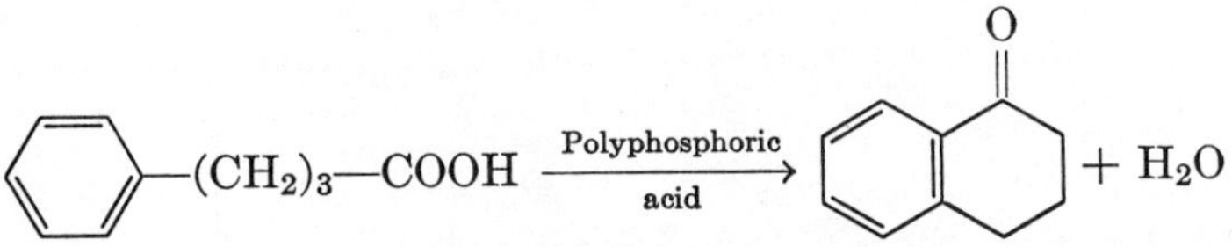

Submitted by H. R. Snyder and Frank X. Werber.
Checked by E. C. Horning and G. N. Walker.

1. Procedure

In a 600-ml. beaker, 80 g. of polyphosphoric acid (Note 1) is warmed on a steam bath to 90°. There is then added in one lot, with manual stirring, 22.0 g. (0.134 mole) of molten γ-phenylbutyric acid (Note 2), previously warmed to 65–70°. The mixture is removed from the steam bath and is stirred steadily by hand for 3 minutes; during this time a temperature of 90° or slightly higher is maintained by the heat of the reaction. An additional 70 g. of polyphosphoric acid is then added, and the mixture is again warmed on a steam bath for 4 minutes while being stirred vigorously by hand. The resulting solution is allowed to cool to 60°, and is hydrolyzed by addition to 200 g. of ice and water.

When the hydrolysis is complete, as indicated by the disappearance of orange, viscous material and the appearance in its stead of yellow oil, the aqueous mixture is transferred to a 500-ml. separatory funnel and extracted with two portions of ether, 200 ml. and 100 ml., respectively. The extracts are combined and washed successively with 200 ml. of water, two 100-ml. portions of 5% sodium hydroxide solution

(Note 3), 200 ml. of water, 100 ml. of 3% aqueous acetic acid, 100 ml. of 5% sodium bicarbonate solution, and 100 ml. of water. The ether solution is dried over anhydrous magnesium sulfate. The ether is removed either by evaporation on a steam bath or by distillation from a 50-ml. flask into which the solution is introduced continuously, as fast as the solvent is distilled, by means of a dropping funnel reaching below the side arm of the distilling head. The flask is fitted with a 6-in. Vigreux column, and the residual oil is distilled under reduced pressure. There is obtained 14.8–16.9 g. (75–86%) of colorless oil, b.p. 90–91°/0.5–0.7 mm. or 94–95°/2.1–2.3 mm. (Note 4).

2. Notes

1. Polyphosphoric acid obtained from the Victor Chemical Company, Chicago, was used.
2. γ-Phenylbutyric acid was obtained by catalytic reduction (palladium-charcoal) of β-benzoylpropionic acid and was purified by distillation; the yield of material, b.p. 133–135°/1.0–1.3 mm., was 96%.
3. The amount of unchanged γ-phenylbutyric acid obtained by acidification of the alkaline solution is negligible. When only 72 g. of polyphosphoric acid in one lot was employed in the reaction, there was recovered 1.5 g. of γ-phenylbutyric acid, and the yield of α-tetralone was 15.9 g. (81%).
4. The submitters have used polyphosphoric acid in the cyclization of hydrocinnamic acid, 2-bromo-5-methoxyhydrocinnamic acid, and *o*-benzoylbenzoic acid.

3. Methods of Preparation

Other methods of preparation have been reviewed in earlier volumes.[1,2] The method consisting of air oxidation of tetralin[2–6] has in some cases been aided by the use of copper oxide[4] or sulfate,[5] and has been said to yield a product which is contaminated with α-tetralol.[7] The ketone has also been prepared from γ-phenylbutyric acid by cyclization with phosphoric acid-phosphoric anhydride[8] and with hydrofluoric acid,[9] and from γ-phenylbutyryl chloride by the action of aluminum chloride.[10] The method described is essentially that of Synder and Werber.[11]

[1] *Org. Syntheses,* **15,** 77 (1935).
[2] *Org. Syntheses,* **20,** 97 (1940).
[3] Hock and Susemihl, *Ber.,* **66,** 61 (1933).
[4] Brown, Widiger, and Letang, *J. Am. Chem. Soc.,* **61,** 2601 (1939).

[5] U. S. pat. 2,462,103 [*C. A.*, **43**, 3848 (1949)].

[6] Medvedev and Pod'yapol'skaya, *J. Phys. Chem. U.S.S.R.*, **13**, 719 (1939).

[7] Mentzer and Billet, *Bull. soc. chim. France*, **1948**, 835; Beyerman, *Rec. trav. chim.*, **72**, 550 (1953).

[8] Birch, Jaeger, and Robinson, *J. Chem. Soc.*, **1945**, 582.

[9] Fieser and Herschberg, *J. Am. Chem. Soc.*, **61**, 1272 (1939).

[10] Johnson and Glenn, *J. Am. Chem. Soc.*, **71**, 1092 (1949).

[11] Snyder and Werber, *J. Am. Chem. Soc.*, **72**, 2962 (1950).

2,3,4,6-TETRAMETHYL-*d*-GLUCOSE

(D-Glucose, 2,3,4,6-tetramethyl-)

$$HOCH_2\underbrace{CH(CHOH)_3CH}_{O}OH + 5(CH_3)_2SO_4 + 5NaOH \rightarrow$$

$$CH_3OCH_2\underbrace{CH(CHOCH_3)_3\text{—}CH}_{O}OCH_3 + 5CH_3NaSO_4 + 5H_2O$$

$$CH_3OCH_2\underbrace{CH(CHOCH_3)_3\text{—}CH}_{O}OCH_3 + H_2O \rightarrow$$

$$CH_3OCH_2\underbrace{CH(CHOCH_3)_3\text{—}CH}_{O}OH + CH_3OH$$

Submitted by EDWARD S. WEST and RAYMOND F. HOLDEN.
Checked by W. W. HARTMAN and A. J. SCHWADERER.

1. Procedure

This preparation must be carried out in a hood having good ventilation. Methyl sulfate has a high vapor pressure in spite of its high boiling point and is very poisonous. Ammonia is a specific antidote and should be kept on hand to destroy any of the ester accidentally spilled. It is advisable to wash the hands in dilute ammonium hydroxide frequently.

In a 2-l. distilling flask immersed in a 4-l. water bath are placed 25 g. (0.14 mole) of anhydrous glucose (Note 1) and 15 ml. of water. The flask is fitted with a cork through which passes a strong mechanical stirrer (Note 2) and a dropping tube connected by rubber tubing and a screw clamp to a 500-ml. reservoir flask. The side tube of the flask is connected to a water condenser, fitted with a suction flask, from which fumes are led to a flue by a piece of tubing. The temperature of the water bath is raised to 55°, and the glucose is brought into solution with rapid stirring, which is maintained throughout the process. A mixture of 90 ml. (120 g., 9.5 moles) of methyl sulfate and 125 ml. of carbon tetrachloride is added from the reservoir as quickly as possible to the flask. The clamp is closed, and 400 ml. (580 g.) of 40% (by weight) sodium hydroxide is placed in the reservoir. The alkali is admitted to the flask at the rate of 1 drop in 2 seconds for 5 minutes, then 1 drop per second for 5 minutes, and then 3 drops per second until the distillation of carbon tetrachloride slackens or ceases. This is usually accomplished in 15–20 minutes and after the addition of 70–90 ml. of alkali. The heat of reaction generally maintains the proper temperature of 50–55° throughout this stage without the necessity of heating the water bath externally. The remainder of the alkali is added as quickly as possible, and the bath temperature raised to and maintained at 70–75°. Then 160 ml. (208 g., 1.65 moles) of methyl sulfate is placed in the reservoir immediately and added at the rate of 3–4 drops per second (slower if the mixture foams seriously).

After all the methyl sulfate has been added, the bath is boiled for 30 minutes with continued stirring. The contents of the flask are cooled, diluted with sufficient water to dissolve most of the separated sodium sulfate, and extracted four times with 150-ml. portions of chloroform; the chloroform and water layers are separated carefully. The combined chloroform extracts (Note 3) are placed in a 2- or 3-l. distilling flask (with condenser attached) with 400 ml. of 2 *N* hydrochloric acid, and the chloroform is removed by distillation. A rapid current of steam is then passed through the solution for 1 hour, care being taken to maintain the volume approximately constant by heating the flask. Five grams of Norite is added to the hot solution, which is then cooled and filtered. The filtrate is saturated with sodium sulfate and extracted four times with 150-ml. portions of chloroform. The combined chloroform extracts are dried with sodium sulfate, 1 g. of Norit is added, and the mixture is filtered. The chloroform is removed as completely as possible in a boiling water bath without vacuum and finally at the water pump. The heavy syrup is treated with 40–50 ml. of petroleum ether (30–60°) and shaken for a short time, whereupon it sets to a mass of crystals. After cooling in an ice bath for 30 minutes, the crystals are filtered, washed with a little cold petroleum ether, and dried over calcium chloride under reduced pressure. The yield is 15–18 g. (46–55%) (Note 3). The specific rotation, $[\alpha]_D^{26}$, in water was found to be about +79–79.5°, when $c = 4$; a drop of 15% ammonia was added to speed equilibration, and the readings were made 1 hour after the solution was prepared in a 2-dm. tube.

The slightly impure product is recrystallized from petroleum ether (b.p. 30–60°) containing 0.5% of anhydrous ether in a continuous extractor, using 100 ml. of solvent per 6–7 g. of compound. One crystallization is generally sufficient to give a pure product. The specific rotation of pure tetramethyl-*d*-glucose prepared by this method is approximately $[\alpha]_D^{20}$ +81.3° (Note 4).

2. Notes

1. α-Methylglucoside may be methylated in the same way as glucose, using four-fifths of the reagents and hydrolyzing the tetramethyl-α-methylglucoside (contained in the chloroform solution) for 3 hours instead of the 1 hour required for the mixture of α- and β-tetramethylmethylglucosides obtained directly from glucose. Tetramethylglucose prepared in this way may show the correct rotation without recrystallization.

2. The stirrer consists of a 3/16-in. Monel rod formed into a 1-in. flattened loop at one end for stirring. The stirrer is passed through a closely fitting 3/8 by 5 in. brass bearing with a packing nut at the lower end, the top of the bearing being countersunk to facilitate oiling with heavy engine oil. This metal stirrer permits much more vigorous stirring (which is essential) than the ordinary glass apparatus.

3. The tetramethylmethylglucosides may be isolated from the chloroform solution by drying with sodium sulfate, distilling off the chloroform, and then vacuum-distilling the syrup. The distillation can be easily carried out in a distilling flask (100-ml.) filled with glass wool (which effectively prevents bumping) and having a low side arm. The glass wool materially lowers the distilling temperature. The mixture of α- and β-glucosides (approximately 85% β form) and the pure α form distil at about 88–90°/0.15 mm. under the above conditions, and they constitute the entire distillate after the solvent has been removed. The specific rotation, $[\alpha]_D$, lies between +9° and +12°, depending upon the proportion of isomers in the distillate; for the pure α form it is 151°/25°. The refractive index is 1.4445_D^{20} for the α,β mixture and 1.4460_D^{20} for the pure α isomer.

4. The correct value for any temperature between 5° and 37° may be calculated from the equation $[\alpha]_D^T = 85 - 0.1846t$. The melting point of the compound varies according to the proportion of α and β isomers present. After prolonged digestion with petroleum ether, the α isomer predominates and the melting point rises. A sample of the product recrystallized once melted at 90–93°, and after five recrystallizations it melted at 98°, without change in the optical rotation.

3. Methods of Preparation

Tetramethyl-*d*-glucose has been prepared by the action of methyl iodide and silver oxide on methylglucoside,[1] by the action of methyl sulfate and alkali on both methylglucoside and glucose,[2] and also by hydrolysis of various methylated polysaccharides, for example methyl decamethyl maltotrioside.[3]

[1] Purdie and Irvine, *J. Chem. Soc.*, **83**, 1021 (1903).
[2] Haworth, *J. Chem. Soc.*, **107**, 8 (1915); **113**, 188 (1918).
[3] Sugihara and Wolfrom, *J. Am. Chem. Soc.*, **71**, 3357 (1949).

TETRANITROMETHANE *

(Methane, tetranitro-)

$$4(CH_3CO)_2O + 4HNO_3 \rightarrow C(NO_2)_4 + 7CH_3CO_2H + CO_2$$

Submitted by Poe Liang.
Checked by Nathan L. Drake and Ralph Mozingo.

1. Procedure

Caution! The product is toxic. The reaction should be carried out in a hood.

In a 250-ml. Erlenmeyer flask provided with a two-holed stopper which has a slit cut in one edge to serve as an air vent and which holds a thermometer, the bulb of which reaches almost to the bottom of the flask, is placed 31.5 g. (0.5 mole) of anhydrous nitric acid (Note 1). The flask is cooled below 10° in ice water, and 51 g. (0.5 mole) of acetic anhydride (Note 2) is slowly added from a buret through the second hole in the stopper in portions of about 0.5 ml. at a time. The temperature of the reaction mixture is never allowed to rise above 10° (Note 3). After about 5 ml. of the acetic anhydride has been added the reaction becomes less violent, and larger portions, increasing gradually from 1 to 5 ml., may be introduced at a time with constant shaking. After all the acetic anhydride has been added, the stopper and the thermometer are removed. The neck of the flask is wiped clean with a towel, and the flask is then covered with an inverted beaker and allowed to come up to room temperature in the original ice bath (Note 4).

The mixture is allowed to stand at room temperature for 7 days (Note 5), and the tetranitromethane is separated by pouring the mixture into 300 ml. of water in a 500-ml. round-bottomed flask and steam distilling (Note 6). The tetranitromethane passes over with the first 20 ml. of the distillate. The heavy product is separated from the upper layer of water, washed first with dilute alkali, finally with water, and dried over anhydrous sodium sulfate. The yield of tetranitromethane is 14–16 g. (57–65%). *The product should not be distilled, as it may decompose with explosive violence.* Tetranitromethane must be kept out of contact with aromatic compounds except in *very small* test portions, since violently explosive reactions can occur (Note 7).

2. Notes

1. The anhydrous nitric acid (sp. gr. $>$ 1.53) is most easily obtained by slowly distilling ordinary fuming nitric acid from its own bulk of concentrated sulfuric acid. If ordinary concentrated nitric acid (sp. gr. 1.42) is used, it is advisable to distil twice from equal volumes of sulfuric acid. A technical grade of fuming nitric acid having a specific gravity of 1.60 was found to give satisfactory yields when used without further treatment; but an equivalent amount of a weaker commercial acid, corresponding to 98% nitric acid by gravity, gave considerably lower yields. The use of more than the calculated amount of nitric acid decreases the yield of tetranitromethane.

2. According to the submitter, the purity of acetic anhydride is not so important as that of the nitric acid. Equivalent amounts of 99–100% and 94–95% acetic anhydride gave practically the same yield of tetranitromethane.

3. If the flask is not cooled, the reaction proceeds more and more vigorously as the temperature rises and may, if unchecked, become violent.

4. If the flask is removed from the ice bath after the addition of the anhydride, and allowed to stand at room temperature, the reaction may become violent with great loss of product.

5. The yield is only 35% after 2 days and no greater than 65% after 10–15 days. If it should be necessary to obtain the product in shorter time, the reaction mixture may be allowed to stand at room temperature for 48 hours and then slowly heated to 70° during an interval of 3 hours and maintained at 70° for 1 hour longer before pouring into water. The yield of a run carried out in this way was, according to the submitter, 40%.

6. It is most convenient to use a 50-ml. graduated separatory funnel as the receiver during the steam distillation, if it is desired to estimate roughly the yield of tetranitromethane, the density of which is 1.65 at 15°.

7. Tetranitromethane is a valuable reagent for detecting the presence of double bonds, especially those which do not give the ordinary reactions of such linkages.[1]

3. Methods of Preparation

The procedure described is essentially that of Chattaway.[2] Tetranitromethane has also been prepared by nitrating nitroform,[3] from acetic anhydride by the action of diacetylorthonitric acid,[4] from iodopicrin and silver nitrite,[5] from acetyl nitrate by heating with acetic anhydride or glacial acetic acid,[6] from nitrobenzene by distilling with a mixture of nitric acid and fuming sulfuric acid,[7] by adding acetic anhydride to nitrogen pentoxide or a mixture of nitrogen pentoxide and nitrogen peroxide,[8] by the action of acetic anhydride on highly concentrated nitric acid,[9] from toluene by nitration,[10] from acetylene by the action of nitric acid,[11] from nitrobenzene and a mixture of nitric and fuming nitric acids,[12] and from acetylene and ethylene by the action of nitric acid in the presence of a catalyst.[13]

[1] Meyer, H., *Analyse und Konstitutionsermittlung organischer Verbindungen,* 6th ed., p. 773, Julius Springer, Vienna, 1938; Ruzicka, Huyser, Pfeiffer, and Seidel, *Ann.,* **471,** 21 (1929).

[2] Chattaway, *J. Chem. Soc.,* **1910,** 2099; *Chem. News,* **102,** 307 (1910).

[3] Schischkoff, *Ann.,* **119,** 247 (1861).

[4] Pictet and Genequand, *Ber.,* **36,** 2225 (1903).

[5] Hantzsch, *Ber.,* **39,** 2478 (1906).

[6] Pictet and Khotinsky, *Compt. rend.,* **144,** 210 (1907); *Ber.,* **40,** 1163 (1907).

[7] Claessen, Ger. pat. 184,229 [*C. A.,* **1,** 2524 (1907)].

[8] Schenck, Ger. pats, 211,198, 211,199 [*C. A.,* **3,** 2205 (1909)].

[9] Bayer and Co., Brit. pat. 24,299 [*C. A.,* **5,** 2305 (1911)] and Ger. pat. 224,057; Berger, *Compt. rend.,* **151,** 813 (1910) and *Bull. soc. chim. France,* (4) **9,** 26 (1911).

[10] Will, *Ber.,* **47,** 704 (1914).

[11] Orton, Brit. pat. 125,000; Orton and McKie, *J. Chem. Soc.,* **1920,** 283.

[12] McKie, *J. Soc. Chem. Ind.,* **44,** 430T (1925).

[13] McKie, *J. Chem. Soc.,* **1927,** 962.

TETRAPHENYLCYCLOPENTADIENONE

(Cyclopentadienone, tetraphenyl-)

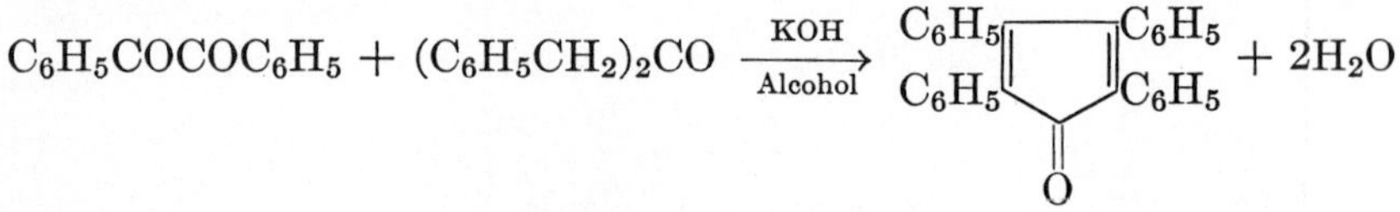

Submitted by John R. Johnson and Oliver Grummitt.
Checked by Nathan L. Drake and Stuart Haywood.

1. Procedure

In a 500-ml. round-bottomed flask, 21 g. (0.1 mole) of benzil and 21 g. (0.1 mole) of dibenzyl ketone (Note 1) are dissolved in 150 ml. of hot ethanol. The flask is fitted with a reflux condenser, the temperature of the solution is raised nearly to the boiling point, and a solution of 3 g. of potassium hydroxide in 15 ml. of ethanol is added slowly in two portions through the condenser. When the frothing has subsided the mixture is refluxed for 15 minutes and then cooled to 0°. The dark crystalline product is filtered with suction and washed with three 10-ml. portions of 95% ethanol. The product melts at 218–220° and weighs 35–37 g. (91–96%) (Note 2).

2. Notes

1. The dibenzyl ketone should melt at 34–35°.
2. This product is sufficiently pure for most purposes. It may be crystallized from a mixture of ethanol and benzene, using 155–160 ml. solvent for 5 g. of tetraphenylcyclopentadienone; the melting point of the recrystallized material is 219–220°.

3. Methods of Preparation

Tetraphenylcyclopentadienone has been prepared by the action of phenylmagnesium bromide on benzaldiphenylmaleide,[1] and by reduction, dehydration, and oxidation of the methylenedesoxybenzoin obtained by condensing formaldehyde with desoxybenzoin.[2] The present procedure is essentially that of Dilthey.[3]

[1] Lowenbein and Uhlich, *Ber.*, **58**, 2662 (1925).
[2] Ziegler and Schnell, *Ann.*, **445**, 266 (1925).

[3] Dilthey and Quint, *J. prakt. Chem.*, (2) **128**, 146 (1930); Ger. pat. 575,857 [*Frdl.*, **20**, 503 (1933); *C. A.*, **28**, 1356 (1934)].

TETRAPHENYLPHTHALIC ANHYDRIDE

(Phthalic anhydride, tetraphenyl-)

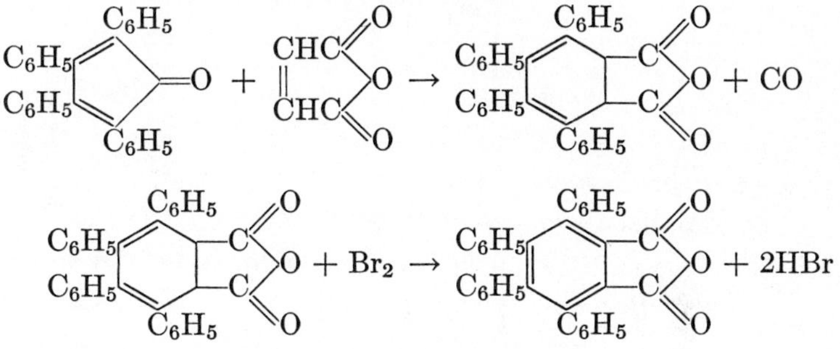

Submitted by OLIVER GRUMMITT.
Checked by NATHAN L. DRAKE and CHARLES M. EAKER.

1. Procedure

An intimate mixture of 35 g. (0.094 mole) of tetraphenylcyclopentadienone (p. 806) and 9.3 g. (0.095 mole) of maleic anhydride is placed in a 200-ml. round-bottomed flask (Note 1), and to it is added 25 ml. of bromobenzene. After the mixture has been refluxed gently for 3.5 hours (Note 2), it is cooled (Note 3), a solution of 7 ml. of bromine in 10 ml. of bromobenzene is added through the condenser, and the flask is shaken until the reagents are thoroughly mixed. After the first exothermic reaction has subsided, the mixture is refluxed gently for 3 hours (Note 4). The flask is then immersed in a cooling bath and the temperature of the mixture is held at 0–10° for 2–3 hours. The mixture is filtered with suction, and the crystalline product is washed three times with 10-ml. portions of petroleum ether (b.p. 60–68°). After the product has been dried in the air, it weighs 37–38 g. (87–89%) and melts at 289–290°. It is light brown, but when pulverized it is almost colorless. The filtrate, when diluted with an equal volume of petroleum ether and cooled to 0–10°, yields an additional 2–3 g. of a less pure product which melts at 285–288°. The impure material may be purified by recrystallization from benzene, using 8–9 ml. of benzene per gram of solid (Note 5).

2. Notes

1. Ground-glass equipment is preferred; corks are attacked by the bromine used later.

2. The operation should be carried out in a hood because of the carbon monoxide evolved.

3. The tetraphenyldihydrophthalic anhydride may be isolated at this point in practically quantitative yields by cooling the mixture, filtering with suction, and washing the solid with three 10-ml. portions of petroleum ether (b.p. 60–68°). The yield is 41–42 g.; the product melts at 235–240°.

4. This operation should also be performed in a hood, or the top of the condenser should be connected to a suitable trap.

5. The crystallized product must be dried at 110° for 1–1.5 hours; otherwise benzene, possibly benzene of crystallization, will be retained by the solid indefinitely.

3. Methods of Preparation

Tetraphenylphthalic anhydride has been prepared by condensation of tetraphenylcyclopentadienone and maleic anhydride in nitrobenzene,[1] followed by dehydrogenation of the tetraphenyldihydrophthalic anhydride with sulfur.[2,3]

Tetraphenylphthalic anhydride has also been prepared by condensation of tetraphenylcyclopentadienone with chloromaleic anhydride;[4] and by oxidation of tetraphenylhydrindene with chromic acid[5] or oxidation of 7-carboxy-4,5,6,7-tetraphenyl-3*a*,4,7,7*a*-tetrahydroindene with potassium permanganate.[6]

1 Dilthey, Schommer, and Trosken, *Ber.*, **66**, 1627 (1933).
2 Dilthey, Thewalt, and Trosken, *Ber.*, **67**, 1959 (1934).
3 Allen and Sheps, *Can. J. Research,* **11**, 171 (1934).
4 Synerholm, *J. Am. Chem. Soc.*, **67**, 1229 (1945).
5 Grummitt, Klopper, and Blenkhorn, *J. Am. Chem. Soc.*, **64**, 604 (1942).
6 Allen, Jones, and VanAllan, *J. Am. Chem. Soc.*, **68**, 708 (1946).

m-THIOCRESOL

(m-Toluenethiol; m-tolyl mercaptan)

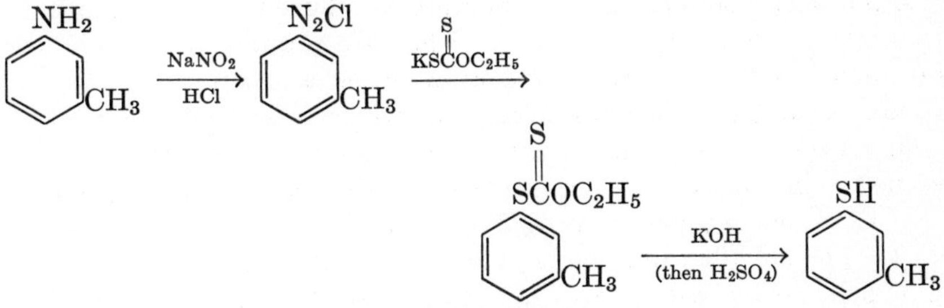

Submitted by D. S. Tarbell and D. K. Fukushima.
Checked by C. F. H. Allen and John R. Byers, Jr.

1. Procedure

All the steps in this preparation, including sealing the product in bottles or ampoules, should be carried out under a good hood. Care should be exercised to avoid contact with m-thiocresol or its solutions since it is a skin irritant.

In a 1-l. flask, equipped with a mechanical stirrer and thermometer for reading low temperatures, and immersed in an ice bath, are placed 150 ml. of concentrated hydrochloric acid (sp. gr. 1.18) and 150 g. of crushed ice. The stirrer is started, and 80 g. (0.75 mole) of *m*-toluidine (b.p. 92–93°/15 mm.) is slowly added. The mixture is cooled to 0°, and a cold solution of 55 g. (0.8 mole) of sodium nitrite in 125 ml. of water is slowly added, the temperature being kept below 4°.

In a 2-l. flask equipped with a thermometer, dropping funnel, and stirrer is placed a solution of 140 g. of potassium ethyl xanthate (Note 1) in 180 ml. of water. This mixture is warmed to 40–45° and kept in that range during the slow addition of the cold diazonium solution (Note 2); about 2 hours is required (Note 3). After an additional 30 minutes at this temperature to ensure complete decomposition of the intermediate compound, the red, oily *m*-tolyl ethyl xanthate is separated and the aqueous layer is extracted twice, using 100-ml. portions of ether. The combined oil and extracts are washed once with 100 ml. of 10% sodium hydroxide solution (Note 4) and then with several portions of water until the washings are neutral to litmus.

The ether solution is dried over 25 g. of anhydrous calcium chloride, and the ether is removed by distillation. The crude residual *m*-tolyl ethyl xanthate is dissolved in 500 ml. of 95% ethanol, the solution brought to boiling, and the source of heat removed. To this hot solution is added slowly 175 g. of potassium hydroxide pellets so that the solution keeps boiling, and the mixture is refluxed until a sample is completely soluble in water (about 8 hours). Approximately 400 ml. of ethanol is then removed by distillation on a steam bath, and the residue is taken up in the minimum of water (about 500 ml.). The aqueous solution is extracted with three 100-ml. portions of ether, the extract being discarded. The aqueous solution is now made strongly acid to Congo red paper, using 6 *N* sulfuric acid (Note 5) (625–650 ml.). The acidified solution is placed in a 3-l. flask, 2 g. of zinc dust is added, and the *m*-thiocresol is distilled with steam. The lower layer of the *m*-thiocresol is separated; the aqueous layer is extracted with three 100-ml. portions of ether, the extracts being added to the oil. After drying with 50 g. of Drierite, the ether is removed by distillation, and the oily residue is distilled under reduced pressure. The yield of colorless *m*-thiocresol, b.p. 90–93°/25 mm., is 59–69 g. (63–75%) (Notes 6 and 7). It is best preserved in sealed glass bottles because of its disagreeable odor.

2. Notes

1. Eastman Kodak Company technical potassium ethyl xanthate was used.

2. The diazonium solution is left in the ice bath, and only 10- to 15-ml. portions are placed in the dropping funnel at one time.

3. Many diazonium solutions have been reported to react explosively with solutions of metallic polysulfides even at low temperatures.[1,2] A violent reaction with xanthates is mentioned only in one report.[3] Neither the authors nor the checkers observed any unusual reactivity during this preparation or with the procedure given for dithiosalicylic acid.[4] On a large scale (100 moles of *m*-toluidine) flashes of light have been occasionally observed (private communication, L. J. Roll).

4. This wash serves to remove any *m*-cresol present.

5. This acidification liberates carbon oxysulfide, which has a very disagreeable odor.

6. The refractive index is n_D^{25} 1.568–1.571.

7. Other boiling points are 195°/760 mm.; 120°/100 mm.; 107°/50 mm.

3. Methods of Preparation

The only practical laboratory procedure for preparing *m*-thiocresol is by the alkaline hydrolysis of *m*-tolyl ethyl xanthate, obtained from *m*-toluenediazonium chloride and potassium ethyl xanthate.[3,5] The procedure described is essentially that of Bourgeois.[5]

[1] Nawiasky, Ebersole, and Werner, *Chem. Eng. News,* **23,** 1247 (1945).
[2] Hodgson, *Chemistry & Industry,* **1945,** 362.
[3] Leuckart, *J. prakt. Chem.,* [2] **41,** 189 (1890).
[4] *Org. Syntheses* Coll. Vol. **2,** 580 (1943).
[5] Bourgeois, *Rec. trav. chim.,* **18,** 447 (1899).

2-THIOPHENEALDEHYDE *

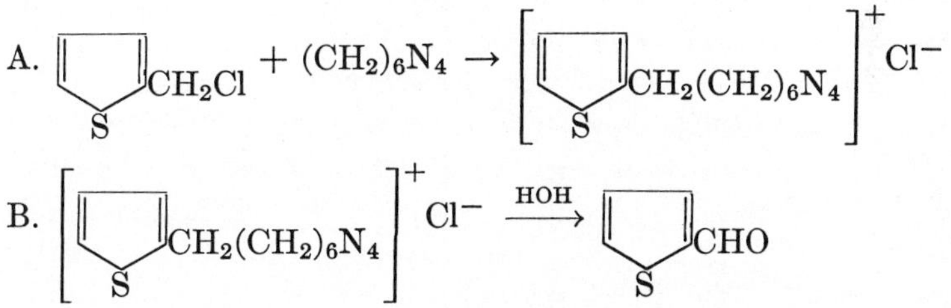

Submitted by Kenneth B. Wiberg.
Checked by Cliff S. Hamilton and J. L. Pauley.

1. Procedure

A. 2-*Thienylmethylhexamethylenetetrammonium chloride.* In a 1-l. round-bottomed flask are placed 67 g. (0.5 mole) of 2-chloromethylthiophene (p. 197), 400 ml. of chloroform, and 70 g. (0.5 mole) of hexamethylenetetramine. The flask is fitted with a reflux condenser, and the mixture is boiled gently for 30 minutes. The mixture is cooled, and filtered on a Büchner funnel. The precipitate is washed with 100 ml. of cold chloroform, drained thoroughly, and air-dried. The yield is 128–136 g. (94–99%) of a white powder.

B. 2-*Thiophenealdehyde.* The hexamethylenetetrammonium salt is placed in a 2-l. round-bottomed flask containing 400 ml. of warm water. The flask is fitted for steam distillation, and steam is passed in until all the aldehyde has distilled (Note 1). The distillate is cooled, 10 ml. of 6 *N* acetic acid is added (Note 2), and the aldehyde is extracted with two 100-ml. portions of ether. The ether solution is dried over anhydrous calcium chloride, and the ether is evaporated on a steam bath until the volume of the solution has decreased to about 50 ml. The solution is placed in a 100-ml. Claisen flask, the ether is removed by distillation, and the aldehyde distilling at 89–91°/21 mm., n_D^{25} 1.5880, is collected. The yield is 27–30 g. (48–53%) of a colorless oily liquid which darkens slowly on standing (Notes 3 and 4).

2. Notes

1. About 1.5 l. of distillate is collected before all the thiophenealdehyde is distilled over.

2. The acetic acid is added to remove traces of amines that come over in the steam distillation. This method of purification must be used because of the high solubility of the bisulfite addition compound of the aldehyde.

3. If the aldehyde is to be stored for any period of time, the addition of a small amount of hydroquinone is advisable.

4. It has been reported (S. J. Angyal and D. F. Penman) that the procedure for 1-naphthaldehyde [*Org. Syntheses,* **30,** 67 (1950)] may also be followed with slight modification for the preparation of 2-thiophenealdehyde. A mixture of 41.5 g. (0.31 mole) of 2-chloromethylthiophene, 88 g. (0.60 mole) of hexamethylenetetramine, 130 ml. of glacial acetic acid, and 130 ml. of water is swirled until, with considerable evolution of heat, the mixture has become homogeneous. The mixture is heated under reflux for 4 hours; at the end of this period, 125 ml. of concentrated hydrochloric acid is added, and heating under reflux is continued for 5 minutes. After cooling, the mixture is extracted with three 100-ml. portions of ether. The combined ether extracts are dried (with anhydrous sodium or magnesium sulfate), and the ether is removed. The crude product is distilled through a short column under reduced pressure; after a fore-run of acetic acid the product is collected at 63–66°/6 mm. or 115–118°/65 mm. The yield is 25–26 g. (71–74%).

3. Methods of Preparation

2-Thiophenealdehyde has been prepared by the decarboxylation of 2-thienylglyoxalic acid,[1] by the action of 2-thienylmagnesium iodide on ethyl orthoformate followed by hydrolysis of the acetal,[2] in small yields by the Rosenmund reduction of 2-thiophenecarboxylic acid chloride,[3] in small yields by the action of hydrogen cyanide, hydrogen chloride, and aluminum chloride on thiophene, using benzene as a solvent,[4] by a series of reactions from 1-chloro-2,3-diketocyclopentane,[5] by the hydrolysis of 2-thenylmethylhexamethylenetetrammonium chloride in neutral solution,[6] by the action of N-methylformanilide on thiophene in the presence of phosphorus oxychloride,[7] by the acid hydrolysis of N-(2-thenyl)-2′-thenaldimine,[8] and by the oxidation of N,N-di-(2-thenyl)hydroxylamine with alkaline potassium permanganate.[9]

[1] Biedermann, *Ber.*, **19**, 636 (1886).
[2] Grischkewitsch-Trochimowski, *J. Russ. Phys. Chem. Soc.*, **44**, 570 (1912).
[3] Barger and Easson, *J. Chem. Soc.*, **1938**, 2100.
[4] Reichstein, *Helv. Chim. Acta*, **13**, 349 (1930).
[5] Hantzsch, *Ber.*, **22**, 2838 (1889).
[6] Dunn, Waugh, and Dittmer, *J. Am. Chem. Soc.*, **68**, 2118 (1946).
[7] King and Nord, *J. Org. Chem.*, **13**, 635 (1948); *Org. Syntheses*, **31**, 108 (1951).
[8] Hartough, *J. Am. Chem. Soc.*, **69**, 1355 (1947).
[9] U. S. pat. 2,463,500 [*C. A.*, **43**, 4302 (1949)].

dl-THREONINE

$$CH_3CH{=}CHCO_2H \xrightarrow[CH_3OH]{Hg(OAc)_2} \text{Addition product} \xrightarrow{KBr}$$

$$CH_3CH(OCH_3)CH(HgBr)CO_2K \xrightarrow[Br_2]{KBr}$$

$$CH_3CH(OCH_3)CHBrCO_2K \xrightarrow{HBr} CH_3CH(OCH_3)CHBrCO_2H \xrightarrow{NH_3}$$

$$CH_3CH(OCH_3)CH(NH_2)COOH \xrightarrow[(CH_3CO)_2O]{HCO_2H}$$

$$CH_3CH(OCH_3)CH(NHCHO)CO_2H \xrightarrow{HBr}$$

$$CH_3CHOHCH(NH_3Br)CO_2H \xrightarrow{NH_3} CH_3CHOHCH(NH_2)CO_2H$$

Submitted by HERBERT E. CARTER and HAROLD D. WEST.
Checked by NATHAN L. DRAKE and WILLIAM A. STANTON.

1. Procedure

A. *α-Bromo-β-methoxy-n-butyric acid.* In a 5-l. flask are placed 3 l. of methanol, 640 g. (2 moles) of mercuric acetate (Note 1), and 172 g. (2 moles) of crotonic acid (Note 2). The flask is warmed on a steam cone and shaken vigorously until the mercuric acetate dissolves (about 10 minutes). The solution is allowed to stand at room temperature for 48 hours (Note 3), then the precipitate is filtered, washed twice with 300-ml. portions of methanol, and air-dried. The yield is 625–650 g. (Note 4). This material is powdered and dissolved in a solution of 360 g. (3 moles) of potassium bromide in 2 l. of water, and the solution is placed in a 4-l. beaker, which is cooled in an ice bath and exposed to direct sunlight (Note 5). A solution of 320 g. (2 moles) of bromine and 360 g. (3 moles) of potassium bromide in 600 ml. of water is added with stirring during 20–40 minutes, a large excess of bromine being avoided. After 10–15 minutes' standing, any excess bromine is destroyed with sodium bisulfite. The bromo acids are isolated as follows: The solution is extracted once with 300 ml. of ether to remove a small amount of lachrymatory material and, after acidification with 400 ml. of 40% hydrobromic acid, is extracted again with six 800-ml. portions of ether. The ether extracts are combined, washed once with a small volume of cold water, and dried over anhydrous sodium sulfate. The ether is removed by distillation, leaving the crude bromo acid mixture.

The yield of crude bromo acids is 350–370 g. (88–93% based on the crotonic acid used in the first step). This material is used without purification in the preparation of aminomethoxybutyric acid.

The crude bromo acids may be freed from impurities by fractionation under reduced pressure. The fraction distilling below 125°/10 mm. (105°/3 mm.) is discarded; the remainder distils at 125–128°/10 mm. (105–107°/3 mm.) and consists of a mixture of stereoisomeric acids. The yield is 75–85% based on the crotonic acid used in the first step.

B. *dl-Threonine.* One hundred and seventy-five grams of crude bromomethoxybutyric acid is heated with 2 l. of concentrated ammonium hydroxide for 6 hours at 90–100° in a glass-lined autoclave (p. 777) (Note 6). The solution is concentrated to a thick gum under reduced pressure (Note 7), water is added, and the solution is reconcentrated under reduced pressure. The residue is allowed to stand under acetone with frequent shaking (Note 8) until the material has crystallized completely (1–2 days). The acetone is decanted, and the residue dissolved in 1 l. of 85–90% formic acid (Note 9). The solution is warmed to 45°, and 350 g. (330 ml.) of acetic anhydride is added with stirring during 10 minutes. The heat of reaction causes the temperature of the solution to rise to 70–80°, and the temperature of the mixture is maintained within this range for about 15 minutes. The solution is next evaporated to dryness under reduced pressure. The residue is dissolved, while being warmed on the steam bath, in the minimum amount of water (Note 10), and the solution is cooled overnight in the icebox. The crystals are filtered and air-dried. This material is a mixture of formyl derivatives (Note 11). One recrystallization from 150 ml. of hot water yields about 25 g. of practically pure formyl-*dl*-O-methylthreonine melting at 174–176°. An additional 3–5 g. is obtained by working up the filtrates. The yield is 25% (Note 12).

Twenty-five grams (0.16 mole) of formyl-*dl*-O-methylthreonine is refluxed for 2 hours with 360 ml. of constant-boiling hydrobromic acid. The solution is concentrated under reduced pressure (Note 13). Sufficient water is added to dissolve all the residue, and the solution is reconcentrated under reduced pressure. The gummy residue is next dissolved in 450 ml. of absolute ethanol, and concentrated ammonium hydroxide is added until the odor of ammonia persists after vigorous shaking. The solution is cooled in the icebox overnight, and the crystals are filtered and dissolved in 3 volumes of hot water (about 5 ml. of water per gram of crude material). Seven volumes of absolute ethanol are added, and the solution is cooled to room temperature with scratching of the flask to induce crystallization. It is then cooled in an icebox overnight. The crystals are filtered and washed twice with 90-ml. portions of absolute ethanol and once with ether. The yield is 18–20 g. (85–90% based on the formyl-*dl*-O-methylthreonine) of pure *dl*-threonine, melting with decomposition at 234–235° (Note 14).

2. Notes

1. The mercuric acetate may be replaced by equivalent amounts of mercuric oxide and glacial acetic acid.

2. The crotonic acid was obtained from the Niacet Chemical Company and used without purification.

3. It is advisable to scratch the flask with a glass rod after 3–4 hours. This usually initiates precipitation of the addition product in a finely divided form. If this is not done, the addition product may crystallize slowly on the sides of the flask in a cake which is removed only with the greatest difficulty. It is also advantageous to stir the mixture mechanically for several hours after crystallization begins in order to prevent caking.

4. The exact yield cannot be calculated, since the structure of the addition product is unknown. The yield is almost quantitative, however, since only a small amount of mercury remains in the filtrate.

5. The bromination can be carried out equally successfully under the illumination of two No. 2 Photoflood lamps in suitable reflectors placed directly above the surface of the liquid. Under these conditions, however, the addition of the bromine requires 10–15 minutes longer.

6. According to the submitters the amination can be carried out in ordinary 500-ml. glass bottles if the temperature does not exceed 85°. The time of heating should then be extended to 8–10 hours.

7. All the concentrations under reduced pressure required in this preparation may be carried out at the pressure provided by an efficient water pump.

8. The material, if allowed to stand without shaking, solidifies to a hard cake. The shaking furthers extraction of certain gummy impurities which interfere with the separation to be carried out later.

9. If a mixture of *dl*-threonine and *dl*-allothreonine is desired instead of *dl*-threonine alone, the residue may be dissolved directly in 1.2 l. of 48% hydrobromic acid and refluxed for 2 hours. After removal of the hydrobromic acid under reduced pressure, the gummy residue is dissolved in warm water, and concentrated ammonium hydroxide is added slowly until a faint odor of ammonia persists after vigorous shaking. The solution is concentrated until crystals appear, and 3–4 volumes of ethanol is added. The acids are recrystallized by dissolving in the minimum amount of hot water (4–5 ml. per gram) and adding 4–5 volumes of ethanol. The solution is allowed to cool and permitted to stand overnight at room temperature.

10. If, after solution is complete on the steam bath, 10% more water is added, the quality of the product is better, but the yield is slightly less.

11. Formyl-*dl*-O-methylthreonine melts at 173–174°. Formyl-*dl*-O-methylallothreonine melts at 152–153°.

12. Crude *dl*-allothreonine may be obtained from the mother liquors by concentrating them to dryness, refluxing the residue with 10 volumes of 48% hydrobromic acid, and working up the solution as described for *dl*-threonine. The product contains a small amount of *dl*-threonine which can be largely removed by three or four recrystallizations from 50% ethanol. *dl*-Allothreonine of this purity melts at 242–243°.

13. Constant-boiling hydrobromic acid can be recovered by fractionating the distillate at atmospheric pressure.

14. The melting points obtained by the checkers were consistently 3–4° above those given. The melting points of these compounds vary with the method of determination.

3. Methods of Preparation

α-Amino-β-hydroxybutyric acid has been prepared by a procedure similar to the one described, using ethyl crotonate as the starting material.[1] A mixture of the α-amino-β-hydroxy- and α-hydroxy-βaminobutyric acids has been secured by treating crotonic acid with hypochlorous acid and heating the resulting product with dry ammonia under pressure.[2] A mixture containing threonine has been obtained by treatment of acetoacetic ester with sodium nitrite and acetic acid; the resultant ethyl oximinoacetoacetate was then converted by means of diethyl sulfate into ethyl O-ethyloximinoacetoacetate. This product was reduced by hydrogen and Raney nickel to an impure ethyl α-amino-β-hydroxybutyrate, which was then hydrolyzed to a mixture of *dl*-threonine and *dl*-allothreonine.[3] The method described has been published.[4]

Improvements in this method have been reported,[5] and an excellent synthesis from ethyl acetamidoacetoacetate has been described.[6]

Interconversion methods for obtaining *dl*-threonine from *dl*-allothreonine via the azlactone [7] or the oxazoline [6] have been reported.

1 Abderhalden and Heyns, *Ber.*, **67**, 530 (1934).

2 Burch, *J. Chem. Soc.*, **137**, 310 (1930).

3 Adkins and Reeve, *J. Am. Chem. Soc.*, **60**, 1330 (1938).

4 West and Carter, *J. Biol. Chem.*, **119**, 109 (1937).

5 Pfister, Howe, Robinson, Shabica, Pietrusza, and Tishler, *J. Am. Chem. Soc.*, **71**, 1096 (1949).

6 Pfister, Robinson, Shabica, and Tishler, *J. Am. Chem. Soc.*, **71**, 1101 (1949).

7 Carter, Handler, and Melville, *J. Biol. Chem.*, **129**, 362 (1939).

o-TOLUALDEHYDE *

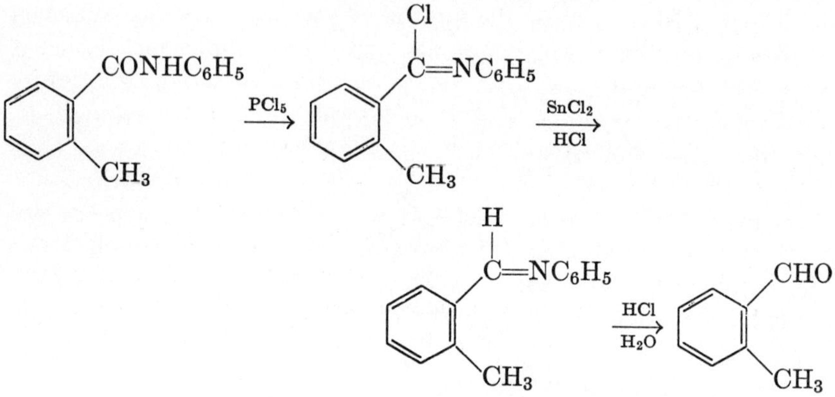

Submitted by JONATHAN W. WILLIAMS, CHARLES H. WITTEN, and JOHN A. KRYNITSKY.
Checked by W. E. BACHMANN and N. C. DENO.

1. Procedure

Thirty grams (0.14 mole) of o-toluanilide (Note 1) and 20 ml. of dry benzene are placed in a 125-ml. Claisen flask which is fitted with a condenser for distillation under reduced pressure and set in a water bath. The bath is heated to 50°, and 30 g. (0.14 mole) of phosphorus pentachloride is added to the mixture over a period of 10 minutes. The bath is then heated to 75°, and this temperature is maintained for 15 minutes. The benzene and most of the phosphorus oxytrichloride are then removed by distillation at 20 mm. from a bath at 95°. The crude N-phenyl-o-toluimidyl chloride which remains as a viscous liquid is sufficiently pure for the next step.

A mixture of 50 g. (0.26 mole) of anhydrous stannous chloride (p. 627) and 225 ml. of dry ether is placed in a 1-l. three-necked round-bottomed flask fitted with a rubber-tube sealed stirrer, an inlet tube reaching nearly to the bottom of the flask, and a reflux condenser (Note 2) protected by a calcium chloride drying tube. The mixture is saturated with dry hydrogen chloride (Note 3) with continuous stirring. Within 3 hours all the stannous chloride dissolves, forming a clear viscous lower layer. The source of hydrogen chloride is then disconnected, and the freshly prepared imidyl chloride is transferred into the mixture with the aid of 25 ml. of dry ether (Note 4). Stirring is continued for

1 hour, and then the reactants are allowed to stand at room temperature for 12 hours.

Ice (about 100 g.) and then 100 ml. of cold water are added, the mixture is stirred for 10 minutes, and then the ether is removed by distillation (Note 5). To the residue in the flask is added sufficient water to make the total volume about 300 ml. Steam distillation is then carried out until the distillate comes over clear. This process requires about 1 hour. The aldehyde is extracted from the steam distillate with three equal portions of the ether recovered above, and the ether solution is dried for several hours over anhydrous sodium sulfate.

The ether is removed by dropping the solution slowly from a separatory funnel into a 50-ml. Claisen flask fitted with a condenser for downward distillation and heated on a steam bath (Note 6), and the residual aldehyde is distilled under reduced pressure. The yield of *o*-tolualdehyde boiling at 90–93°/19 mm. is 10.5–12 g. (62–70%).

2. Notes

1. The *o*-toluanilide was prepared by the reaction of *o*-tolylmagnesium bromide with phenyl isocyanate.[1] If the *o*-toluanilide has been recrystallized from petroleum ether, it may be necessary to melt the light fluffy crystals (to a more compact form) in order to get all the compound into the flask.

2. A reflux condenser is not necessary if the mixture is kept ice-cold during the addition of hydrogen chloride.

3. The checkers used hydrogen chloride from one of the cylinders of hydrogen chloride now available commercially.

4. At this point considerable evolution of heat occurs. A cold-water bath should be kept in readiness for application to the reaction flask.

5. The distillate, which may contain a small amount of *o*-tolualdehyde, is collected and used for the subsequent extraction.

6. In this step the recovered ether is collected in a dry atmosphere, and about 100 ml. of this ether is used in two portions to extract the sodium sulfate residue in order to transfer into the flask any small quantities of the aldehyde that may have been trapped by the drying agent.

3. Methods of Preparation

o-Tolualdehyde has been prepared by the oxidation of *o*-xylene,[2] by the oxidation of *o*-xylyl chloride,[3] by the oxidation of *o*-tolylcarbinol,[4] by the reaction of *o*-tolylmagnesium bromide with ethyl formate,[5] ethyl orthoformate,[6,7] ethoxymethyleneaniline,[7,8] N-methylformanilide,[6] or with carbon disulfide followed by treatment with semicarbazide,[7] and by the aluminum amalgam reduction of N,N′-diphenyl-N-carbethoxy-*o*-toluamidine.[9] The procedure described is based on the method of Sonn and Müller,[10] as studied for *o*-tolualdehyde by King, L'Ecuyer, and Openshaw.[11]

An excellent preparative method from α-bromo-*o*-xylene has been published.[12]

[1] Schwartz and Johnson, *J. Am. Chem. Soc.*, **53**, 1066 (1931).
[2] (*a*) Fournier, *Compt. rend.*, **133**, 635 (1901); (*b*) Law and Perkin, *J. Chem. Soc.*, **91**, 263 (1907); (*c*) Bornemann, *Ber.*, **17**, 1467 (1884).
[3] Rayman, *Bull. soc. chim. France*, (2) **27**, 498 (1877).
[4] Fournier, *Compt. rend.*, **137**, 717 (1903).
[5] Gattermann and Maffezzoli, *Ber.*, **36**, 4152 (1903).
[6] Smith and Bayliss, *J. Org. Chem.*, **6**, 437 (1941).
[7] Smith and Nichols, *J. Org. Chem.*, **6**, 489 (1941).
[8] Monier-Williams, *J. Chem. Soc.*, **89**, 273 (1906).
[9] Shirsat and Shah, *J. Indian Chem. Soc.*, **27**, 13 (1950).
[10] Sonn and Müller, *Ber.*, **52**, 1929 (1919).
[11] King, L'Ecuyer, and Openshaw, *J. Chem. Soc.*, **1936**, 352.
[12] *Org. Syntheses*, **30**, 99 (1950).

o-TOLUIC ACID *

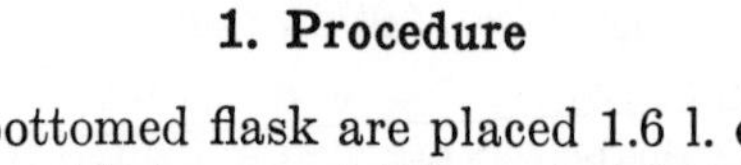

Submitted by Harold E. Zaugg and Richard T. Rapala.
Checked by R. L. Shriner and Curtis M. Snow.

1. Procedure

In a 5-l. round-bottomed flask are placed 1.6 l. of water, 800 ml. of concentrated nitric acid (sp. gr. 1.42), and 400 ml. (364 g., 3 moles) of commercial 90% *o*-xylene (Note 1). A reflux condenser (40 cm. or longer) is fitted to the flask with a cork, and a gas-absorption trap is attached to the top of the condenser. The mixture is refluxed by heating in an oil bath kept between 145° and 155° for 55 hours (Note 2). At the end of this time the organic layer has settled to the bottom of the flask. The hot reaction mixture is poured with stirring into 1 kg. of ice in a 4-l. beaker. After cooling, the solid product is filtered, suspended in 2 l. of cold water, and filtered again. The wet product is dissolved by warming in 1 l. of 10% sodium hydroxide solution. After cooling, any unreacted xylene is separated by extraction with

250 ml. of ether. The aqueous layer is then heated on the steam bath with 5–10 g. of Norit and filtered hot with suction through a layer of Norit. The warm, clear red alkaline solution is added with vigorous mechanical stirring to 225 ml. of concentrated hydrochloric acid. The product is filtered from the warm solution, washed with cold water, and sucked as dry as possible on a Büchner funnel (Note 3).

The crude product is dissolved in 350 ml. of 95% ethanol and heated on the steam bath with 5 g. of Darco for 1 hour. The hot solution is filtered by gravity through a heated funnel. The filter paper is heated with an additional 70-ml. portion of 95% ethanol, and the mixture is filtered hot through a fresh paper, into the main solution. To the ethanolic solution, adjusted to a temperature of 55–60°, is added 480 ml. of warm (55–60°) water. After being cooled first to room temperature and then to ice-bath temperature, the crystallized product is filtered, washed with 250 ml. of cold 50% ethanol, and dried (Note 4). The yield of light-tan product melting at 99–101° amounts to 218–225 g. (53–55%) (Notes 5 and 6).

2. Notes

1. The *o*-xylene (90–92% purity) may be obtained from the Barrett Division of the Allied Chemical and Dye Corporation, 40 Rector Street, New York 6, New York.

2. It is not necessary that this heating be continuous. Longer times of refluxing or mechanical stirring do not improve the yield.

3. The crude product after air drying for 24–30 hours weighs between 250 g. and 300 g. and melts at 94–98°.

4. If the humidity is low, the product may be air-dried; otherwise, it is best to dry in a vacuum desiccator overnight at room temperature. *o*-Toluic acid sublimes at higher temperatures.

5. The yield depends on the *o*-xylene content of the starting material. The yields stated are based on 90% *o*-xylene.

6. This product is pure enough for most purposes. Pure white needles of *o*-toluic acid may be obtained by recrystallization from water (2 l. of water for 20 g. of acid), Darco being used for decolorization. The recovery is about 85% of white needles melting at 101–103°.

3. Methods of Preparation

References to methods of preparation are given in earlier volumes.[1] *o*-Toluic acid has been prepared by the catalytic oxidation of *o*-xylene.[2]

[1] *Org. Syntheses,* **11,** 97 (1931); Coll. Vol. **2,** 589 (1943).

[2] Emerson, Lucas, and Heimsch, *J. Am. Chem. Soc.,* **71,** 1742 (1949).

p-TOLUIC ACID *

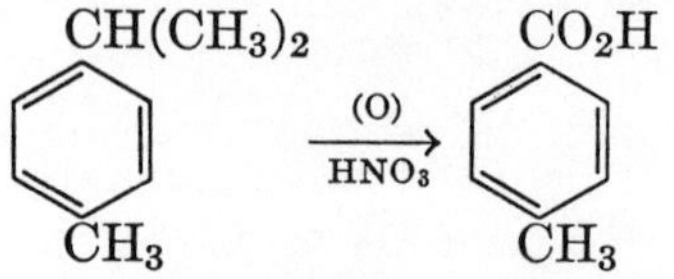

Submitted by W. F. Tuley and C. S. Marvel.
Checked by H. T. Clarke and D. Blumenthal.

1. Procedure

In a 5-l. round-bottomed flask are mixed 2.7 l. of water and 750 ml. of concentrated nitric acid (sp. gr. 1.42). The flask is fitted with an efficient stirrer (Note 1) and a reflux condenser whose outlet is connected with a trap to remove oxides of nitrogen. One hundred and five grams (125 ml., 0.78 mole) of *p*-cymene (Note 2) is added, the stirrer is started, and the reaction mixture is boiled gently for 8 hours. It is then allowed to cool, and the solid which crystallizes is collected on a hardened filter paper in a Büchner funnel (Note 3). The crude product (Note 4) is washed with 200 ml. of water in small portions and then dissolved in 850 ml. of 1 *N* sodium hydroxide. The alkaline solution is placed in a 2-l. flask with 20 g. of zinc dust (Note 5) and distilled until the distillate runs clear (Note 6). The undissolved zinc is removed by filtration, and the yellowish filtrate is poured in a thin stream with vigorous stirring into 500 ml. of boiling 5 *N* hydrochloric acid. After cooling, the precipitated acid is filtered, washed with cold water until substantially free of chloride, and dried. About 80 g. of a light-brown powder is thus obtained.

The product is extracted for 6 hours with 300 ml. of toluene in the apparatus described in a previous volume [1] (Note 7). The toluene extract is chilled to 0°, and the light-brown crystals of *p*-toluic acid are filtered. This material weighs 56–58 g.; an additional 5 g. is obtained by concentrating the filtrate to 100 ml. The total yield of product melting at 174–177° is 60–63 g. (56–59%). The acid may be purified further with very little loss by dissolving it in 0.5 *N* sodium hydroxide, treating the solution with Norit, precipitating the acid by pouring the alkaline solution into excess hot hydrochloric acid, and recrystallizing the product from toluene (Note 8). The purified *p*-toluic acid melts at 176–177° and weighs about 55 g. (51%).

2. Notes

1. A stirrer of the tubular type, running in a bearing consisting of a glass tube which extends well below the surface of the liquid, is recommended.

2. The fraction of spruce turpentine which boils at 175–178° is satisfactory.

3. The filtrate contains too little dissolved product (about 4 g.) to repay extraction. It can be employed for a subsequent run by adding sufficient concentrated nitric acid (about 300 ml.) to restore the specific gravity to its initial value, 1.115.

4. The crude product consists of *p*-toluic acid contaminated with small amounts of terephthalic acid, methyl *p*-tolyl ketone, and nitration products.

5. The zinc serves to reduce nitration products that are otherwise difficult to remove. The resulting amines remain in the filtrate after acidification.

6. About 300 ml. of distillate is collected, of which 5 ml. consists of methyl *p*-tolyl ketone.

7. About 4 g. of light-tan terephthalic acid remains on the filter paper.

8. The last traces of color are removed only with considerable difficulty by Norit. An alternative procedure consists in distilling the toluic acid under reduced pressure from a two-bulbed flask with a wide connecting tube and crystallizing the distillate from toluene.

3. Methods of Preparation

p-Toluic acid has been prepared by the oxidation of cymene,[2] *p*-xylene,[3,4] or dihydro-*p*-tolualdehyde;[5] by reaction of *p*-chlorotoluene and metallic sodium[6] or *p*-bromotoluene and butyllithium followed by carbonation;[7] by hydrolysis of *p*-tolunitrile;[8] by fusing *p*-tolyl phenyl ketone or di-*p*-tolyl ketone with potassium hydroxide;[9] and by reaction of oxalyl chloride with toluene in the presence of aluminum chloride.[10]

[1] *Org. Syntheses* Coll. Vol. **1,** 375 (1941).

[2] Noad, *Ann.,* **63,** 287 (1847).

[3] Yssel de Schepper and Beilstein, *Ann.,* **137,** 302 (1866).

[4] Emerson, Lucas, and Heimsch, *J. Am. Chem. Soc.,* **71,** 1742 (1949).

[5] Allen, Ball, and Young, *Can. J. Research,* **9,** 169 (1933).

[6] Morton, LeFevre, and Hechenbleikner, *J. Am. Chem. Soc.,* **58,** 754 (1936).

[7] Gilman, Wright, and Moore, *J. Am. Chem. Soc.,* **62,** 2330 (1940).

[8] *Org. Syntheses* Coll. Vol. **2**, 589 (1943).
[9] Kozlov, Fedoseev, and Lazarev, *J. Gen. Chem. U.S.S.R.*, **6**, 485 (1936); [*C. A.*, **30**, 5574 (1936)].
[10] Fahim, *Nature*, **162**, 526 (1948).

o-TOLUIDINESULFONIC ACID

(*m*-Toluenesulfonic acid, 6-amino-)

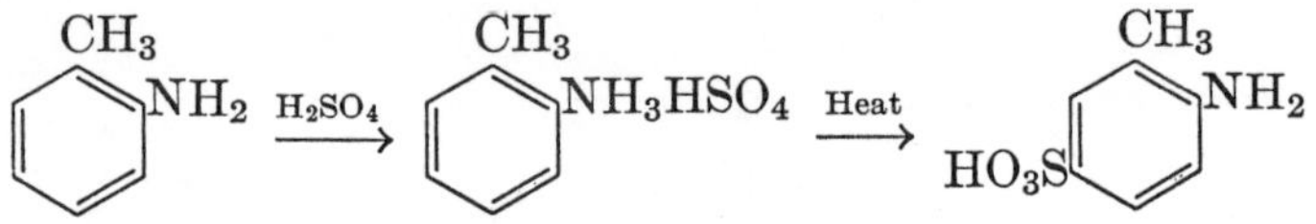

Submitted by C. F. H. Allen and J. A. VanAllan.
Checked by R. L. Shriner and Charles R. Russell.

1. Procedure

In a 500-ml. three-necked round-bottomed flask, fitted with a stirrer, thermometer, and 250-ml. dropping funnel, are placed 19 ml. of water and 27.8 ml. (0.5 mole) of concentrated sulfuric acid (sp. gr. 1.84), and the whole is warmed to 70°. From the dropping funnel is added, dropwise, over a 30-minute period, 53.5 g. (0.5 mole) of *o*-toluidine (Note 1). The temperature is allowed to reach about 85° in order to prevent the separation of solid before all the amine has been added. When addition is complete, the temperature is raised to 100–105° for 5 minutes. The flask is then placed on a steam bath, the stirrer and dropping funnel are removed, and the flask is evacuated by means of a water pump to about 15 mm. (Note 2); after 21 hours all the water has been removed. The flask is then immersed in an oil bath, and the internal temperature is raised to 180° over a period of 2 hours and kept at 180–195° (bath temperature about 190–205°) for 7 hours while the same diminished pressure is maintained (Note 3). The crude product is dissolved in 1.3 l. of boiling water, and 99 g. of barium hydroxide octahydrate is added with stirring; the solution is now alkaline to litmus. After the addition of 3 g. of Norit, the solution is heated for 1.5 hours on a steam bath, and 17.5 ml. of concentrated sulfuric acid (sp. gr. 1.84) diluted with 20 ml. of water is added with good stirring. The solution is heated to 96° and filtered through a large, preheated Büchner funnel; the filter cake is rinsed with 100 ml. of boiling water (Note 4).

The filtrate is transferred to a 3-l. round-bottomed flask and concentrated under a pressure of about 50 mm. to a volume of 200 ml. (Note 5). After the mixture has been chilled overnight in a refrigerator, the crystalline acid (68–74 g.) is filtered by suction and dried (Note 6). An additional small amount (5–8 g.) can be obtained by evaporating the filtrate to dryness. This residue is allowed to remain in contact with 75 ml. of methanol for about an hour with occasional shaking. The solid is filtered, dried, and weighed. The total yield of *o*-toluidinesulfonic acid is 74–79 g. (79–83%) (Notes 7 and 8).

2. Notes

1. The amine used should boil over a range of not more than 2°.

2. If time is an important consideration, and a smaller yield (60%) is acceptable, the procedure may be modified: the acid sulfate is heated for 1 hour under reduced pressure; the temperature is then raised to 180° for 1 hour and finally to 180–190° for 3 hours. The solid residue is taken up in 1 l. of hot water and 55 ml. of 40% sodium hydroxide, a small amount of unchanged amine is distilled with steam, and the clear solution is treated with Norit and filtered as above. The acid is then precipitated by adding 65 ml. of concentrated hydrochloric acid.

3. During this time the amine acid sulfate melts and loses water, and the product solidifies to a white crystalline mass.

4. Low yields may result from crystallization of some of the acid in the funnel at this point if the funnel is preheated insufficiently. Some acid may also be lost in the large bulk of inorganic solid. Filtration through the finely divided filter cake is often so slow that the under side and stem of the funnel collect large amounts of crystalline acid which is not rinsed out by the 100 ml. of boiling water. The use of a steam-heated Büchner funnel or prolonged preheating of an ordinary Büchner funnel does not seem to minimize this effect. It is advisable to remove the funnel after filtration is complete, and to take out the filter cake and rinse the funnel with two 100-ml. portions of boiling water. The combined washings are heated to boiling and filtered, and the filtrate is added to that obtained in the original filtration.

5. The flask should be equipped with a still head to prevent carry-over of the solution during distillation; a capillary jet is also an aid in this evaporation. Much of the acid separates during this concentration; the crystals are easily filterable.

6. The acid was spread out on a large filter paper and dried above a steam radiator for 48 hours. The literature reports the acid to be a monohydrate which loses its water of crystallization by drying at 120° for 3 hours. However, titration of the air-dried acid showed that it was not hydrated. A sample of the acid heated to 120° for 3 hours showed no change in neutral equivalent.

7. This aminosulfonic acid has no definite melting point. On a Maquenne block a decomposition range of 335–380° was observed.

8. The product secured by this procedure is free from sodium sulfate and isomeric sulfonic acids. Titrations of the first crop of crystals gave neutral equivalents ranging from 187.0 to 187.6; the calculated neutral equivalent is 187.2. The second crop is less pure since it had a neutral equivalent of about 177.

3. Methods of Preparation

o-Toluidinesulfonic acid has been obtained by a variety of reactions. Of these the ones of preparative value are the sulfonation of *o*-toluidine by fuming sulfuric acid;[1–3] the rearrangement of the amine sulfate or acid sulfate;[4,5] and the reduction of the corresponding nitrotoluenesulfonic acid.[6] The above procedure is based on the directions of Huber,[5] who has described the preparation of ten aminoarylsulfonic acids by this method.

[1] Gerver, *Ann.*, **169**, 374 (1873).
[2] Claus and Immel, *Ann.*, **265**, 72 (1891).
[3] Schultz and Lucas, *J. Am. Chem. Soc.*, **49**, 299 (1927).
[4] Neville and Winther, *Ber.*, **13**, 1941 (1880).
[5] Huber, *Helv. Chim. Acta*, **15**, 1372 (1932).
[6] Foth, *Ann.*, **230**, 306 (1885).

m-TOLYLBENZYLAMINE

(*m*-Toluidine, N-benzyl)

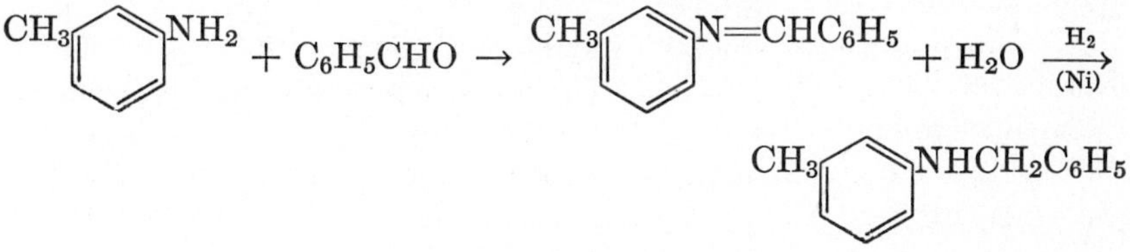

Submitted by C. F. H. Allen and James VanAllan.
Checked by Nathan L. Drake and Ralph Mozingo.

1. Procedure

One hundred and six grams (1 mole) of benzaldehyde (Note 1) and 107 g. (1 mole) of *m*-toluidine are mixed in a suitable flask; the temperature rises to about 60° (Notes 2 and 3). The mixture is cooled below 35° in cold water, 200 ml. of ether is added, and the solution is placed in the steel reaction vessel of a high-pressure hydrogenation apparatus. Eight to ten grams of Raney nickel catalyst (p. 181) is added, the bomb is closed, and hydrogen is admitted up to 1000 lb. pressure (Note 4). The bomb is shaken continuously at room temperature for 15 minutes (Note 5). The contents are removed, and the bomb is washed out with two 200-ml. portions of ether. After the catalyst has been separated by filtration (Note 6), the ether is removed by distillation and the product is distilled from a modified Claisen flask (Note 7). After a small fore-run the N-benzyl-*m*-toluidine boils at 153–157°/4 mm.; 315–317°/760 mm. The yield is 175–185 g. (89–94%) (Notes 8 and 9).

2. Notes

1. Technical grades of benzaldehyde (b.p. 57–59°/8 mm.) and *m*-toluidine (b.p. 76–77°/7 mm.) are satisfactory.

2. It is unnecessary to isolate the Schiff base.

3. The ether used as solvent may be replaced by an equal amount of 95% ethanol. If ethanol is used, cooling is unnecessary.

4. The initial pressure of hydrogen used is determined by the apparatus available. The drop in pressure depends upon the size of the bomb.

5. With a less active Raney nickel catalyst it may be necessary to carry out the reduction at a somewhat higher temperature. If necessary, the reaction mixture may be heated slowly to 60° and the reduction carried out at this temperature. The yield is then somewhat lower (82–84%).

6. The catalyst is filtered through paper on a Büchner funnel. Owing to its activity, suction should be discontinued as soon as all the liquid has passed through. If this precaution is not observed, a fire may result.

7. In some instances (*o*-tolylbenzylamine), a part of the product may crystallize and can be filtered.

8. The hydrochloride, m.p. 198–199°, may serve for characterization.

9. According to the submitters, other aromatic aldehydes and amines may be used in a similar manner with essentially the same yield of product. The exact procedure for isolation of the amine will depend upon its physical properties. Benzaldehyde and *o*-toluidine yield *o*-tolylbenzylamine, m.p. 56–57° (hydrochloride, m.p. 165–166°), while *p*-tolylbenzylamine, obtained from *p*-toluidine, has a b.p. 162–163°/5 mm. (hydrochloride, m.p. 181–182°).

3. Methods of Preparation

N-Benzyl-*m*-toluidine has been prepared by the electrolytic reduction of benzal-*m*-toluidine,[1] and by hydrolysis of N-benzyl-N-formyl-*m*-toluidine, which was prepared in turn by the reaction of sodium N-formyl-*m*-toluidine with benzyl chloride.[2]

[1] Law, *J. Chem. Soc.*, **1912,** 154.

[2] Grunfeld, *Bull. soc. chim. France,* (5) **4,** 654 (1937).

1,3,5-TRIACETYLBENZENE

(Benzene, 1,3,5-triacetyl)

$$HCO_2C_2H_5 + CH_3COCH_3 + NaOC_2H_5 \rightarrow CH_3COCH{=}CH(ONa) + 2C_2H_5OH$$

$$3CH_3COCH{=}CH(ONa) + 3CH_3CO_2H \rightarrow$$

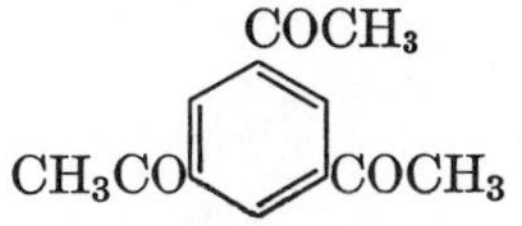

$$+ 3CH_3CO_2Na + 3H_2O$$

Submitted by Robert L. Frank and Robert H. Varland.
Checked by Arthur C. Cope and W. H. Jones.

1. Procedure

Sixty-nine grams (3.0 gram atoms) of freshly cut sodium is placed in a 1-l. round-bottomed flask with 400 ml. of dry xylene (Note 1) and heated until all the sodium is melted. The flask is closed with a rubber stopper and shaken vigorously to form finely powdered sodium (Note 2). When cool, the contents are transferred to a 5-l. three-necked round-bottomed flask, and the xylene is decanted. The powdered sodium is then washed with two 100-ml. portions of anhydrous ether by decantation, after which 1 l. of anhydrous ether is added. The flask is placed on a steam bath and fitted with a Hershberg stirrer, an upright condenser, and a 500-ml. dropping funnel. The condenser and funnel are protected from moisture by calcium chloride tubes. Through the dropping funnel is then added with stirring 138 g. (175 ml., 3.0 moles) of absolute ethanol at such a rate that gentle refluxing occurs. The mixture is refluxed and stirred for 6 hours after the addition is complete.

The reaction mixture is diluted with 1.5 l. of anhydrous ether, and a mixture of 174 g. (220 ml., 3.0 moles) of acetone (Note 3) and 222 g. (241 ml., 3.0 moles) of ethyl formate (Note 4) is placed in the dropping funnel and added to the flask over a period of 2 hours. Stirring is continued for 2 hours after the acetone and ethyl formate have been added (Note 5).

The reaction mixture is then rapidly extracted with five 1-l. portions of water (Note 6). Acetic acid is added to the water solution until it is acid to litmus. The acidified solution is warmed to 50° on a steam bath and maintained at approximately that temperature for 2 hours. It is subsequently allowed to stand at room temperature for 48 hours, during which time the triacetylbenzene crystallizes.

The crude yellow crystalline solid (m.p. in the range 150–162°) is collected on a filter; the yield is 84–94 g. (41–46%). It is recrystallized by dissolving in hot ethanol (18 ml. of ethanol per gram), adding 2 g. of Norit, filtering through a steam-heated funnel (Note 7), and cooling the filtrate in an ice bath. The yield of shiny white crystals, m.p. 162–163°, is 62–79 g. (30–38%) (Note 8).

2. Notes

1. The first portion of this synthesis is very similar to the preparation of acetylacetone by the method of Adkins and Rainey (p. 17). The procedure and Notes 1–6 and 8 for that synthesis may therefore be helpful in the present preparation.

2. A more finely powdered sodium can be obtained if paraffin oil ("Stanolind") is used in place of xylene (Gilbert Ashburn, private communication).

3. Acetone dried over calcium sulfate and distilled from phosphorus pentoxide is satisfactory.

4. The ethyl formate is dried over calcium sulfate and distilled.

5. If the solution becomes so thick as to make stirring difficult, it is advisable to add more anhydrous ether.

6. It is important that the extraction and acidification be carried out without delay. Oxidation of the intermediate product takes place readily on exposure to air and causes the solution to become dark in color.

7. If a steam-heated funnel is not used, a larger volume of ethanol should be employed to prevent crystallization during filtration.

8. The higher yield (38%) was obtained if the commercial absolute ethanol was dried by treatment with sodium and ethyl phthalate.[1]

3. Method of Preparation

Triacetylbenzene has been prepared only by the condensation of acetone with ethyl formate followed by the trimerization of the intermediate acetylacetaldehyde.[2,3]

[1] *Org. Syntheses* Coll. Vol. **2**, 155 (1943).

[2] Claisen and Stylos, *Ber.*, **21**, 1145 (1888).

[3] Kaushal, Sovani, and Deshapande, *J. Indian Chem. Soc.*, **19**, 107 (1942).

TRIBIPHENYLCARBINOL

[Carbinol, tris-(4-phenylphenyl-)]

$$3C_6H_5C_6H_4Cl(p) + (C_2H_5O)_2CO + 6Na \rightarrow (p\text{-}C_6H_5C_6H_4)_3CONa + 3NaCl + 2NaOC_2H_5$$

$$(p\text{-}C_6H_5C_6H_4)_3CONa + H_2O \rightarrow (p\text{-}C_6H_5C_6H_4)_3COH + NaOH$$

Submitted by A. A. MORTON, J. R. MYLES, and W. S. EMERSON.
Checked by R. L. SHRINER, JOSEPH DEC, and J. HARKEMA.

1. Procedure

In a 3-l. three-necked round-bottomed flask, equipped with a mercury-sealed stirrer and a 12-bulb reflux condenser, are placed 188.5 g. (1 mole) of *p*-chlorobiphenyl (Note 1), 47.2 g. (0.4 mole) of ethyl carbonate (Note 2), and 1.5 l. of anhydrous, thiophene-free benzene. The side tube is closed with a cork, stirring is begun, and the mixture is heated by means of a hot plate (Note 3). As soon as the mixture begins to boil, 2 g. of powdered sodium (Note 4) is added. The reaction starts in 1–2 minutes, as indicated by more vigorous refluxing and a change in color from yellow to brown (Note 5). Powdered sodium is then introduced in small portions, over a period of 1 hour, until an additional 28 g. has been added (total amount of sodium, 30 g.). The reaction mixture is refluxed and vigorously stirred (Note 6) for 2 hours. When the mixture has cooled somewhat, 75 to 100 ml. of absolute ethanol is added. After all the particles of sodium have reacted, 500 ml. of water is added, the condenser is arranged for downward distillation, and the benzene and unchanged *p*-chlorobiphenyl are removed by distillation from a steam bath (Note 7). The crude product remaining in the flask is separated by filtration, washed with 100–200 ml. of water, and pressed as dry as possible. The solid is dissolved in 600 ml. of xylene in a 1-l. distilling flask, and the solution is subjected to distillation until 25 to 50 ml. of distillate (xylene and water) has been collected. The solution is cooled somewhat, 1 to 2 g. Norit is added, and then the mixture is boiled for 5 minutes. The hot solution is filtered rapidly and the filtrate is cooled. The product is collected by filtration, washed with 25 to 50 ml. of cold xylene then with 200 ml. of petroleum ether, and dried. The tribiphenylcarbinol forms small white crystals which melt at 207–208°. The product weighs 57–65 g. (35–40%) (Note 8).

2. Notes

1. Technical *p*-chlorobiphenyl was purified by recrystallizing 300 g. of it from 750 ml. of petroleum ether. The recovery was 240–250 g.

2. Technical ethyl carbonate was washed successively with 10% aqueous sodium carbonate and saturated calcium chloride solution. It was then dried with magnesium sulfate and distilled. The fraction boiling at 124–125° was used.

3. A hot plate should be used to reduce the fire hazard. All flames in the vicinity should be extinguished.

4. The powdered sodium is prepared by melting 30 g. of the clean metal under 1 l. of dry xylene. The mixture is then stirred vigorously with a metal stirrer until the sodium is powdered.

5. If anhydrous reagents are used and the benzene solution is boiling, the reaction starts immediately. The reaction must start before more sodium is added.

6. The stirring must be sufficiently vigorous to prevent caking on the sides of the flask.

7. Unchanged *p*-chlorobiphenyl steam-distils along with the benzene. It is essential that it be removed, since a pure product is difficult to obtain by crystallization if unchanged chloro compound is present.

8. Better yields may be obtained in small-scale preparations if sodium chips are used. A solution of 19 g. of *p*-chlorobiphenyl, 5 g. of ethyl carbonate, and 50 ml. of anhydrous thiophene-free benzene is heated on a steam bath until refluxing begins. During 30 minutes 5 g. of sodium, cut into particles about 0.5 mm. square and 2 mm. thick, is dropped through the condenser. After the mixture has refluxed for 12 hours, it is decomposed and the product is isolated as described above, but with one-tenth the amounts of solvents. The yields range from 7 to 9 g. (47 to 55%), but the long period of refluxing required and the tedious cutting of the sodium chips make this method of value for small-scale preparations only. A large-scale run by this method gave only a 31% yield.

3. Methods of Preparation

Tribiphenylcarbinol has been prepared by the action of 4-xenylmagnesium iodide upon 4,4′-diphenylbenzophenone, or upon the methyl ester of *p*-phenylbenzoic acid;[1] and by the action of *p*-chlorobiphenyl upon ethyl carbonate in the presence of powdered sodium [2] or sodium wire.[3]

[1] Schlenk, *Ann.*, **368**, 295 (1909).

[2] Morton and Stevens, *J. Am. Chem. Soc.*, **53**, 4028 (1931); Morton and Emerson, *J. Am. Chem. Soc.*, **59**, 1947 (1937); Morton and Wood, private communication.

[3] Bachmann and Wiselogle, *J. Org. Chem.*, **1**, 372 (1936).

1,2,5-TRIHYDROXYPENTANE

(1,2,5-Pentanetriol)

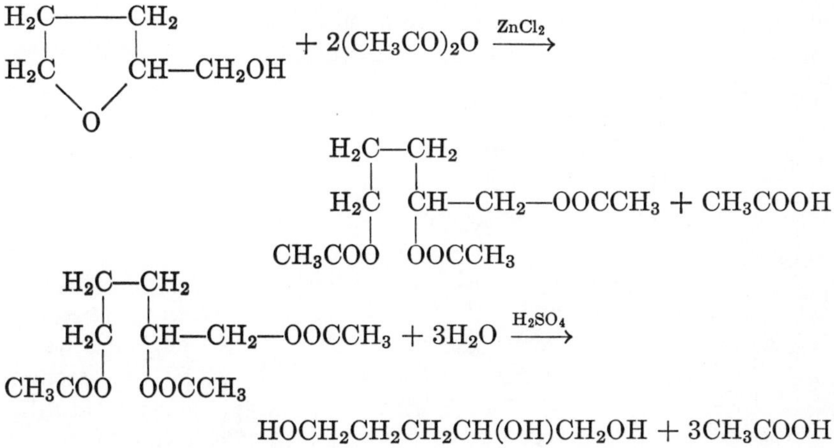

Submitted by Oliver Grummitt, James A. Stearns, and A. A. Arters.
Checked by Eugene Dineen, Christopher L. Wilson, and Charles C. Price.

1. Procedure

In a 5-l. three-necked flask fitted with a mercury-sealed mechanical stirrer, a 500-ml. dropping funnel, and an efficient reflux condenser are placed 2630 ml. (2856 g., 28 moles) of acetic anhydride and 57 g. of crushed anhydrous zinc chloride. This mixture is stirred and brought to gentle boiling. After 5–10 minutes at reflux, 680 ml. (714 g., 7 moles) of redistilled tetrahydrofurfuryl alcohol (Note 1) is added from the dropping funnel at such a rate that the mixture refluxes vigorously. This addition requires about 30 minutes. The mixture is then stirred and refluxed gently for 24 hours. After being cooled, about one-half of the liquid is decanted from the zinc chloride residue through a glass wool plug in a funnel into a 2-l. Claisen flask and then distilled under reduced pressure. The second portion is likewise distilled. The total weight of fore-run is 1740–1780 g. (Note 2). The 1,2,5-triacetoxypentane is collected at 155–165°/14 mm. The yield is 1500–1550 g. (87–90%).

Fifteen hundred grams (6.1 moles) of the ester and 1.2 l. of 1% sulfuric acid (6.9 ml. of concentrated acid in 1.2 l. of solution) are added to a 5-l. round-bottomed flask fitted with a reflux condenser. The mixture is heated strongly and agitated by tilting the ring stand back and forth until the two layers become homogeneous. This initial hydrolysis step requires 15–30 minutes. When homegenity is attained, the flask and condenser are set for steam distillation. The flask is heated by a Meker burner to maintain approximately constant volume, and the mixture is vigorously steam-distilled until 20 l. of distillate has been collected (Note 3). The residue is allowed to cool and then made basic to litmus paper by the addition of 10–20 g. of lime in small portions with vigorous agitation. The calcium sulfate is removed by suction filtration, the filtrate tested again with litmus paper to make certain that it is basic (Note 4), and the filtrate distilled under reduced pressure. This is most conveniently done by placing about one-half of the solution in a 2-l. Claisen flask, heating the flask in an oil bath at 100–110°, and distilling the water under the pressure of the water pump. The remainder of the filtrate is added, the water is removed, and the distillation is continued with an oil or mercury-vapor pump. The yield of 1,2,5-trihydroxypentane collected at 167–170°/0.5–1.0 mm. is 460–520 g. (63–71%); n_D^{25} 1.4730 (Note 5).

2. Notes

1. Tetrahydrofurfuryl alcohol from the Quaker Oats Company was redistilled, and the portion boiling at 175–177° was used. If the 1001 industrial grade of alcohol is used without purification, the yield of triacetate is about 70% of the theoretical.

2. This consists mostly of acetic acid, acetic anhydride, and some tetrahydrofurfuryl acetate. By fractional distillation through a 15-in. glass-helix-packed column, ½ in. in diameter, about 700 g. of acetic acid boiling at 117–125° and about 1075 g. of acetic anhydride boiling at 137–142° can be recovered.

3. With 10-mm. glass tubing for the steam inlet and outlet and an efficient condenser the maximum rate of steam distillation is 3 l. per hour. The 20 l. of distillate collected contains 95% of the calculated yield of acetic acid. The removal of the acetic acid is slow because

the rate of hydrolysis in the final stages appears to be slow and the rate of acetic acid distillation is slow. The remainder of the acid is collected so slowly that it is not considered worth while to distil further, although the yield of product can be slightly increased.

4. If an acidic solution is distilled, the yield of product is decreased, probably because of dehydration and etherification reactions.

5. Analysis of this product for hydroxyl group content by the acetic anhydride-pyridine method gave 40.0%, compared to a calculated value of 42.5%, indicating a purity of 94%.

3. Methods of Preparation

1,2,5-Trihydroxypentane has been made by saponification of 1,2,5-triacetoxypentane with barium hydroxide[1,2] and by hydrolysis with 0.1 *N* hydrochloric acid.[3] The triacetoxypentane has been made by the action of potassium acetate and acetic anhydride on 4,5-dibromopentanol-1[1] or 1,2,5-tribromopentane[2,4] or 1-chloro-4,5-diacetoxypentane,[5] but it is most conveniently made by the reaction of acetic anhydride with tetrahydrofurfuryl alcohol.[3,4] The present method is a modification of that described by Wilson.[3]

[1] Paul, *Compt. rend.*, **192**, 1574 (1931).
[2] Paul, *Ann. chim.*, (10) **18**, 303 (1932).
[3] Wilson, *J. Chem. Soc.*, **1945**, 48.
[4] Paul, *Bull. soc. chim. France*, **53**, 417 (1933).
[5] Paul, *Compt. rend.*, **211**, 645 (1940).

TRIMETHYLENE OXIDE

(Oxetane)

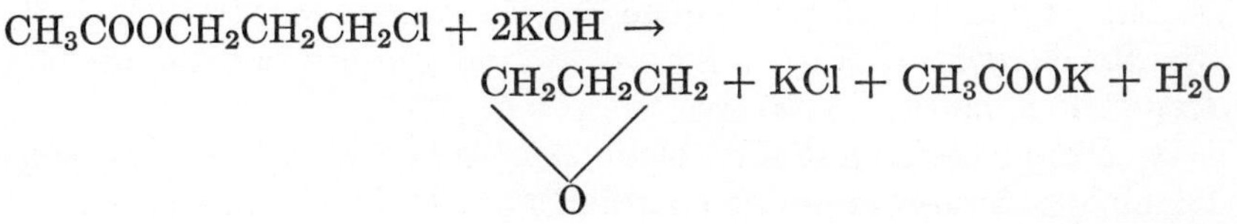

Submitted by C. R. Noller.
Checked by C. F. H. Allen and F. W. Spangler.

1. Procedure

A 3-l. Pyrex flask (Note 1) is fitted with an efficient mechanical stirrer, a dropping funnel, a thermometer, and an upright tube (Note

2) to act as an air condenser and to carry the trimethylene oxide vapors through a spiral condenser. In the flask are placed 1344 g. (24 moles) of potassium hydroxide and 120 ml. of water. The contents are heated to about 140° with stirring (Note 3). The upper part of the flask should be kept sufficiently hot so that the potassium hydroxide mush which is thrown up on the sides does not solidify. The spiral condenser is then packed with an ice-salt mixture, and 1092 g. (8 moles) of γ-chloropropyl acetate (p. 203) (Note 4) is slowly run into the flask by means of the dropping funnel so that the distillate comes over at the rate of about 1 drop a second. The time required for the addition is 1.5–3 hours. After all the ester has been added, the mixture is stirred, with continued heating at 140–150°, until the distillate amounts to 250–400 g., depending on how efficiently the air condenser has separated the trimethylene oxide from higher-boiling substances. About 100 g. of potassium hydroxide pellets is added to the crude distillate, which is then distilled through a 25-cm. fractionating column packed with ¼-in. Berl Saddles. By careful distillation the trimethylene oxide can be obtained in a single fractionation, the portion boiling at 45–50° being collected. The higher-boiling residue still contains some trimethylene oxide, but consists mostly of water, allyl alcohol, and unchanged γ-chloropropyl acetate. The yield is 195–205 g. (42–44%). Redistillation over freshly fused potassium hydroxide gives a product boiling at 47–48°; n_D^{23} 1.3905.

2. Notes

1. A half-gallon jacketed iron autoclave may be employed.
2. A plain 90-cm. 10-mm. tube or a 30-cm. Vigreux column is equally satisfactory.
3. An electric heating mantle is advisable. The reaction is exothermic, and if it appears to become too vigorous the current is temporarily discontinued.
4. The submitter states that a yield of 20–25% may be obtained when trimethylene chlorohydrin is substituted for the acetate, with 2 moles of potassium hydroxide and 10 ml. of water per mole of chlorohydrin.

3. Methods of Preparation

This procedure is an improvement upon that of Reboul,[1] and of Derick and Bissell.[2] Reboul, as also did Ipatow,[3] used trimethylene chlorohydrin, but the yield is higher from the acetate.

[1] Reboul, *Ann. chim.*, (5) **14**, 495 (1878).
[2] Derick and Bissell, *J. Am. Chem. Soc.*, **38**, 2478 (1916).

[8] Ipatow, *J. Russ. Phys. Chem. Soc.*, **46**, 67 (1914) [*C. A.*, **8**, 1965 (1914); *Chem. Zentr.*, **85**, I, 2161 (1914)].

2,4,7-TRINITROFLUORENONE *

(9-Fluorenone, 2,4,7-trinitro-)

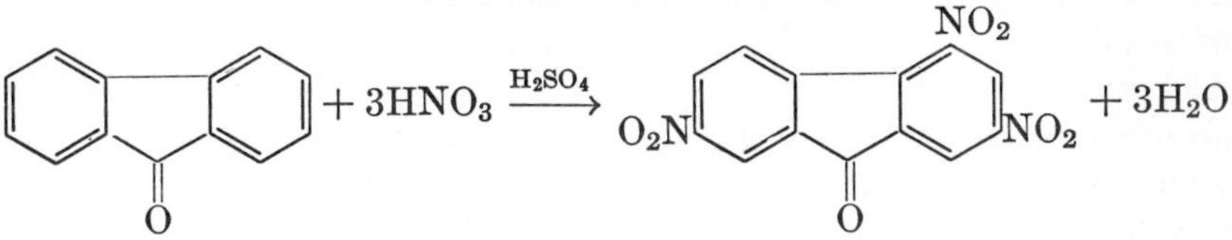

Submitted by E. O. Woolfolk and Milton Orchin.
Checked by Charles C. Price and Benjamin D. Halpern.

1. Procedure

In a well-ventilated hood, a 250-ml. dropping funnel, a thermometer, and a mechanical stirrer are connected by ground-glass joints to a 3-l. three-necked round-bottomed flask. Nine hundred milliliters of red fuming nitric acid (sp. gr. 1.59–1.60) is placed in the flask, and 675 ml. of concentrated sulfuric acid (sp. gr. 1.84) is added with stirring. The acid mixture is cooled to 20°, and the cooling bath is removed. A solution of 45 g. (0.25 mole) of fluorenone (Note 1) in 135 ml. of glacial acetic acid is added dropwise to the mixed acids over a period of about 40 minutes (Note 2). At the end of the addition, the temperature of the reaction mixture is about 45°. The stirrer, thermometer, and funnel are removed, the flask is attached to a reflux condenser with a ground-glass joint, and ground-glass stoppers are placed in the other two openings. The reaction mixture is refluxed for 1 hour and then poured (in the hood) slowly with shaking onto 7 kg. of cracked ice in a 12-l. flask. The product separates as a yellow solid and is filtered by suction and washed with 5 l. of water. The product and 2 l. of water are placed in a 5-l. round-bottomed flask, and steam is passed into the mixture for 1 hour to dissolve and remove acidic impurities (Note 3). The product is filtered by suction, washed with water until the washings are no longer acid to Congo red paper, and air-dried overnight. The material is further dried in a 1-l. round-bottomed flask connected to a water pump and heated in a water bath at 80–90° for several hours. The product is a yellow powder weighing 72–74 g. (91–94%) and melting at 166–171° (Note 4).

The crude product is dissolved in 350 ml. of boiling glacial acetic acid, and the hot solution is filtered by suction (Note 5). Any crystals that separate during the filtration are redissolved by heating the suction flask, and the solution is allowed to cool slowly. The small yellow needles that separate are filtered with suction and washed successively with small quantities of ethanol (30 ml.), water (50 ml.), and ethanol (30 ml.). The yield is 59–61 g. (75–78%) of 2,4,7-trinitrofluorenone, melting at 175.2–176.0°. Additional material can be recovered from the mother liquor by dilution with water, drying the precipitate so formed, and recrystallization from acetic acid. The second crop consists of about 5 g. of pure material, melting at 175.2–176.0°, which usually has a slightly darker color than the first crop.

2. Notes

1. Commercial fluorenone was recrystallized from benzene-ethanol to give material melting at 83–84°. Fluorenone can also be prepared conveniently from commercial fluorene according to the procedure of Huntress, Hershberg, and Cliff.[1] When the checkers nitrated unrecrystallized commercial fluorenone, m.p. 79–82°, the product obtained (yield 55.5 g.) was of a duller yellow color but melted at 175.5–176.5°.

2. Nitrogen oxides are evolved vigorously if the acetic acid solution of fluorenone is added to the acid mixture too rapidly.

3. If this treatment is omitted the recrystallized product has a brown color which is not removed by repeated recrystallizations.

4. The checkers observed melting points of 165–169° for the crude product.

5. Filter paper is not satisfactory for the filtration of hot acetic acid; a sintered-glass Büchner funnel is recommended.

3. Methods of Preparation

2,4,7-Trinitrofluorenone has been prepared by the nitration of fluorenone,[2,3] 2,7-dinitrofluorenone,[4] and 2,5-dinitrofluorenone.[4] The present method has been published [5] in connection with reports of the use of the reagent for the conversion of aromatic compounds to solid derivatives.

[1] Huntress, Hershberg, and Cliff, *J. Am. Chem. Soc.*, **53**, 2720 (1931).

[2] Schmidt and Bauer, *Ber.*, **38**, 3760 (1905). The material reported by these authors as 2,6,7-trinitrofluorenone was subsequently shown to be the 2,4,7-isomer (Bell, *J. Chem. Soc.*, **1928**, 1990).

[3] Schmidt, Retzlaff, and Haid, *Ann.*, **390**, 231 (1912).

[4] Ray and Francis, *J. Org. Chem.*, **8**, 58 (1943).

[5] Orchin, Reggel, and Woolfolk, *J. Am. Chem. Soc.*, **69**, 1225 (1947); Orchin and Woolfolk, *J. Am. Chem. Soc.*, **68**, 1727 (1946).

TRIPHENYLCARBINOL *

(Carbinol, triphenyl-)

$$2C_6H_5MgBr + C_6H_5CO_2C_2H_5 \rightarrow (C_6H_5)_3COMgBr + MgBrOC_2H_5$$

$$(C_6H_5)_3COMgBr + H_2O \rightarrow (C_6H_5)_3COH + MgBrOH$$

Submitted by W. E. Bachmann and H. P. Hetzner.
Checked by R. L. Shriner and P. L. Southwick.

1. Procedure

In a 2-l. three-necked flask, fitted with a separatory funnel, a reflux condenser, and a mercury-sealed stirrer, 27 g. (1.1 gram atoms) of magnesium, 181 g. (122 ml., 1.15 moles) of bromobenzene, and a total of 450 ml. of anhydrous ether (Note 1) are converted into phenylmagnesium bromide by the procedure described in *Org. Syntheses* Coll. Vol. **1,** 226 (1941). Calcium chloride tubes are used to prevent the entrance of moisture through the condenser and the separatory funnel during the addition. In order to dissolve all the magnesium it may be necessary, in some instances, to continue stirring for longer than the prescribed 10 minutes after the addition is complete.

To the Grignard solution, 75 g. (71.5 ml., 0.5 mole) of ethyl benzoate in 200 ml. of dry benzene (Note 2) is added at such a rate that the mixture refluxes gently. The flask is cooled in a pan of cold water during the addition, which requires about an hour. After the addition is complete, the mixture is refluxed for an hour on a steam bath. The reaction mixture is cooled in an ice-salt bath and then poured slowly, with constant stirring, into a mixture of 1.5 kg. of cracked ice and 50 ml. of concentrated sulfuric acid. The mixture is stirred at intervals until all the solid that separates at the benzene-water interface has dissolved. If necessary, 50 g. of ammonium chloride is added to facilitate the decomposition of the magnesium salt, and additional benzene may be added if the amount present is insufficient to dissolve all the product. When the solids have disappeared, the benzene layer is separated and washed, successively, with 200 ml. of water, 200 ml. of a 5% solution of sodium bicarbonate, and finally with 200 ml. of water. The solvents are removed as completely as possible by distillation on a steam bath, and the remaining solution or solid mass is steam-distilled to remove biphenyl and unchanged bromobenzene. The product is filtered, washed with water, and dried. The crude triphenylcarbinol, which weighs 120–125 g., is recrystallized from carbon tetrachloride (4 ml. of solvent per gram of solid). The weight of the first crop of colorless triphenylcarbinol is 110–115 g. (Note 3). A second crop may be obtained by heating the filtrate with a gram of Norit, then concentrating to about 125 ml. and cooling the solution. The total yield of triphenylcarbinol melting at 161–162° is 116–121 g. (89–93%) (Note 4).

2. Notes

1. Commercial anhydrous ether, before it is used, should be dried over slices of sodium, and the bromobenzene should be dried by distillation.
2. The benzene may be dried over anhydrous magnesium sulfate.
3. Triphenylcarbinol separates from carbon tetrachloride with solvent of crystallization. The solvent is readily lost on exposure of the crystals to air. The weight mentioned is that of the solvent-free product.
4. Practically the same yield of triphenylcarbinol was obtained from benzophenone and phenylmagnesium bromide; in this reaction only one-half as much Grignard reagent is required as that needed for the reaction with ethyl benzoate. To the cooled Grignard reagent prepared from 13.5 g. (0.55 gram atom) of magnesium, was added a solution of 91 g. (0.5 mole) of benzophenone in 200 ml. of dry benzene at such a rate that the mixture refluxes gently. After the mixture had been refluxed for 1 hour, the isolation of the triphenylcarbinol was carried out in the manner described above.

3. Methods of Preparation

Triphenylcarbinol has been obtained by the reaction between phenylmagnesium bromide and benzophenone,[1] methyl benzoate,[2] phosgene,[3] or ethyl or butyl pyrocarbonate;[4] by action of phenylsodium on benzophenone, benzoyl chloride, ethyl chlorocarbonate, or ethyl benzoate;[5] by hydrolysis of triphenylchloromethane;[6] and by oxidation of triphenylmethane.[7]

[1] Acree, *Ber.*, **37**, 2755 (1904).
[2] Ullmann and Münzhuber, *Ber.*, **36**, 406 (1903).
[3] Sachs and Loevy, *Ber.*, **36**, 1588 (1903).
[4] Shamshurin, *J. Gen. Chem.*, *U.S.S.R.*, **13**, 569 (1943); [*C. A.*, **39**, 700 (1945)].
[5] Acree, *Am. Chem. J.*, **29**, 594 (1903).
[6] Meissel, *Ber.*, **32**, 2422 (1899).
[7] Law and Perkin, *J. Chem. Soc.*, **93**, 1637 (1908); Schwarz, *J. Am. Chem. Soc.*, **31**, 848 (1909).

TRIPHENYLCHLOROMETHANE *

(Methane, chlorotriphenyl-)

A. FROM TRIPHENYLCARBINOL AND ACETYL CHLORIDE

$$(C_6H_5)_3COH + CH_3COCl \rightarrow (C_6H_5)_3CCl + CH_3COOH$$

Submitted by W. E. BACHMANN.
Checked by R. L. SHRINER and ELDRED WELCH.

1. Procedure

A mixture of 250 g. of pure triphenylcarbinol (p. 839) and 80 ml. of dry benzene is placed in a 1-l. round-bottomed flask provided with a reflux condenser. The condenser is provided with a calcium chloride tube at the top (Note 1). The mixture is heated on a steam bath; when it is hot, 50 ml. of acetyl chloride (Note 2) is added through the top of the condenser. Heating is continued while the mixture is shaken vigorously. In about 5 minutes all the solid triphenylcarbinol disappears and a clear solution results. In the course of 10 minutes, an additional 100 ml. of acetyl chloride is added in 10-ml. portions. The solution is then refluxed for 30 minutes longer.

The solution is cooled by shaking the flask under running water, and during this operation 200 ml. of petroleum ether (Note 3) is added through the top of the condenser; the triphenylchloromethane separates in sugarlike crystals. The mixture is cooled in an ice bath for 1–2 hours, and the product is filtered and washed with 100–150 ml. of petroleum ether (Note 4). The colorless solid, after drying in a desiccator over soda-lime and paraffin (Note 5), weighs 212–224 g. (79–83%) and melts at 111–112° with slight previous softening. An additional 30–37 g. of colorless material may be obtained by concentrating the filtrate to a volume of about 75 ml. For this purpose, the petroleum ether and acetyl chloride are distilled at ordinary pressure, and the acetic acid under reduced pressure. The warm solution is treated with 2 g. of Norit, filtered, and again warmed with 2 g. of Norit. The mixture is filtered, 50 ml. of petroleum ether is added to the filtrate, and the solution is then cooled in an ice bath. The crystals of triphenylchloromethane are collected on a filter and washed with 70 ml. of cold petroleum ether. This second crop of material is pure white, like the first crop; the solid melts at 110.5–112° with previous softening (Note 6). The total yield is 249–254 g. (93–95%).

2. Notes

1. It is best to carry out the reaction under a hood.

2. The acetyl chloride should be of good quality. Good results were obtained with practical acetyl chloride which was distilled just before use.

3. Petroleum ether having a boiling point of 30–60° should be used because it is easily removed from the product.

4. The filtration should be rapid. Triphenylchloromethanc is hydrolyzed by moisture in the air.

5. The final product is perfectly colorless and should have no sharp odor. This product is sufficiently pure for most purposes. It may be recrystallized by dissolving it in 100 ml. of hot benzene and cooling the solution after diluting it with 200 ml. of petroleum ether. If the product is kept in a bottle, the stopper should be coated with paraffin in order to keep out the moisture of the air.

6. Triphenylbromomethane may be made in a similar manner by substituting acetyl bromide for the acetyl chloride.

B. FROM BENZENE AND CARBON TETRACHLORIDE, BY THE FRIEDEL-CRAFTS REACTION

$$3C_6H_6 + CCl_4 + AlCl_3 \longrightarrow (C_6H_5)_3CCl \cdot AlCl_3 + 3HCl$$

$$(C_6H_5)_3CCl \cdot AlCl_3 + xH_2O \xrightarrow{HCl} (C_6H_5)_3CCl + AlCl_3 \cdot xH_2O$$

Submitted by C. R. HAUSER and BOYD E. HUDSON, JR.
Checked by R. L. SHRINER and FLAVIUS W. WYMAN.

1. Procedure

A mixture of 2 kg. (2.28 l., 25.6 moles) of dry thiophene-free benzene (Note 1) and 800 g. (470 ml., 5.2 moles) of dry sulfur-free carbon tetrachloride (Note 2) is placed in a 5-l. three-necked flask immersed in an ice bath and equipped with a mercury-sealed mechanical stirrer and a reflux condenser which is connected to a trap for the absorption of hydrogen chloride. The third neck of the flask is connected to a 1-l. Erlenmeyer flask by means of short pieces of large-bore (15-mm.) glass and rubber tubing (see Fig. 19, p. 550). In the Erlenmeyer flask is placed 600 g. (4.51 moles) of fresh resublimed aluminum chloride (Note 3). The Erlenmeyer flask is tilted and gently tapped, so that the aluminum chloride is added in small portions to the reaction mixture, and at such a rate that addition is completed in 1.5–2 hours. The reaction mixture is not allowed to reflux during the addition of the aluminum chloride. Fifteen minutes after all the aluminum chloride has been added, the ice bath is removed, and the reaction is allowed to proceed without further cooling. When no further heat is evolved, the mixture is refluxed until the evolution of hydrogen chloride subsides (about 2 hours). The mixture is then allowed to cool to room temperature.

A 10-l. copper can (22.5 cm. in diameter and 30 cm. deep) (Note 4) is equipped with a powerful mechanical stirrer and a thermometer and is immersed to a depth of 20 cm. in an ice bath. A mixture of 1 l. of thiophene-free benzene and 2 l. of 6 *N* hydrochloric acid is placed in the copper vessel. The above reaction mixture is added to the vigorously stirred contents of the copper can at such a rate that the temperature does not rise above 25° (about 2 hours is required). When the addition is complete, the reaction flask is rinsed with a little ice water, and the rinsings are added to the hydrolysis mixture. Stirring is continued for 10 minutes longer. The benzene layer is decanted, and the aqueous layer is diluted with 1 l. of ice water. The aqueous layer is extracted with 500 ml. of benzene and is discarded. The combined benzene solutions are washed with 250 ml. of ice-cold hydrochloric acid and are dried for 2 hours in a stoppered flask over 250 g. of anhydrous calcium chloride.

The dried benzene solution is filtered through glass wool into a 5-l. flask fitted with a cork carrying a delivery tube and a thermometer which reaches within 5 cm. of the bottom of the flask. Several boiling chips are added, and the benzene is removed by distillation, slowly towards the last, until the temperature of the boiling residue reaches 120°. The residue is transferred to a 2-l. Erlenmeyer flask with the aid of 40–50 ml. of dry benzene and is cooled to about 40°. Twenty-five milliliters of acetyl chloride (Note 5) is added, and the mixture is heated nearly to the boiling point. The solution is vigorously shaken while it is cooled rapidly to room temperature; it is then chilled in ice water for 2 hours. The solid triphenylchloromethane is crushed thoroughly with a porcelain spatula and is filtered with suction. The filtrate is set aside, and the crystals are washed with three 300-ml. portions of ligroin (b.p. 70–90°) (Note 6). The solvent is removed from the product by allowing it to stand in a vacuum desiccator over mineral oil or paraffin shavings for 24 hours. The crystals are stirred occasionally, and the desiccator is evacuated to 5–10 mm. (Note 7). The light-greenish yellow crystals melt at 111–112° and weigh 870–940 g. (69–75% based on the aluminum chloride).

Second crops of crystals may be obtained by distilling, separately, the filtrate and washings until the temperature reaches 110° and 100°, respectively, and cooling the residues rapidly while they are shaken. The crystals are washed with minimum quantities of ligroin (b.p. 70–90°) and are dried as described for the first crop. The second crop from the filtrate weighs 70–80 g. and melts at 111–112° with sintering at 108°; that from the washings weighs 50–135 g. and melts at 110–111° with sintering at 107°. These second crops are darker in color than the first crop, and of the two second crops, that from the washings is the darker greenish yellow. The total yield of product varies from 1060 g. to 1085 g. (84–86% based on the aluminum chloride) (Note 8).

2. Notes

1. The presence of sulfur compounds, especially thiophene, in the reagents leads to low yields and a dark-colored product. A good grade of thiophene-free benzene should be used. The benzene may be dried by distillation or by allowing it to stand over calcium chloride. The benzene which is recovered in this preparation may be used in future runs after it has been extracted with alkali, washed, and dried. Small amounts of carbon tetrachloride in the benzene do not interfere, since the quantities of benzene and carbon tetrachloride used are in excess of the relative quantity of the aluminum chloride.

2. A good grade of commercial carbon tetrachloride (b.p. 75–76°) may be used without a noticeable decrease in yield. However, if a colorless product is desired, sulfur-free C.P. carbon tetrachloride is recommended. The carbon tetrachloride may be dried in the same manner as the benzene.

3. The yield of triphenylchloromethane depends very much upon the quality of the aluminum chloride used. Fresh commercial resublimed aluminum chloride, in powder form, has been used in the preparation described.

4. The high thermal conductivity of copper makes possible rapid removal of heat during the hydrolysis; the presence of copper in the mixture does not appear to produce any undesirable effects. If a metal can is not available, a 3-gal. enameled bucket may be used.

5. A technical grade of acetyl chloride may be used, since the presence of small amounts of acetic acid does not interfere. The acetyl chloride serves to convert into triphenylchloromethane any triphenylcarbinol which may be present.

6. Technical ligroin (b.p. 90–120°) should be extracted several times with concentrated sulfuric acid, washed, and dried before use.

7. The vacuum desiccator shown in Fig. 28 has been in use in the research laboratories of the Eastman Kodak Company for some time, and has been found to be of great convenience when large amounts of materials must be handled. One of two similar round-bottomed flasks (*B*) is provided with a tube for connection to the vacuum pump. Although not necessary, a glass stopcock in this tube is a convenience, and it enables the flask to be used as a large separatory funnel. A glass sleeve *C*, which must not fit too loosely, serves as a guide for the union between the two flasks *A* and *B*. The sleeve carries a rubber gasket (*D*) which is most conveniently cut from an old inner tube.

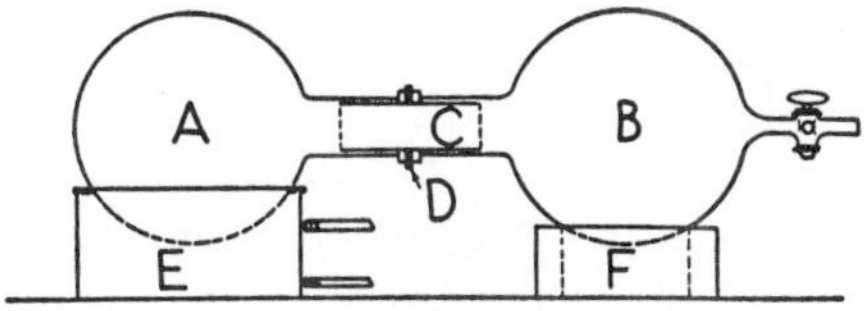

FIG. 28.

The gasket provides a tight seal for the vacuum, and the sleeve prevents the contact surfaces from slipping sidewise. *Without the guiding sleeve, the assembly is not safe.* Special guiding clamps may be used, but they offer no particular advantage. When in use, flask *A* contains the product and *B* contains the drying agent. *E* is a steam bath, and *F* is any convenient support. The apparatus, assembled from 5-l. stock flasks chosen at random, holds a vacuum of 2 mm. for 24 hours with the aid of a high-grade vacuum grease. This "dumb-bell desiccator" is particularly useful for synthetic work; frequently one operation may be eliminated, for flask *A* may be the vessel in which a reaction, or process of concentration, was performed and the product may be dried without transferring it, or flask *A* may be the vessel in which a subsequent reaction is to be carried out; while the product is drying, it may be stirred by rotating the assembly; and, within limits, the desiccator may be of any size desired. It is desirable to coat the flasks on the outside with a thin film of some soluble plastic, for the hazards are thereby greatly reduced; very little shattering occurred when evacuated flasks so coated were experimentally "imploded." A comparative test, in which a highly hygroscopic syrup was used, indicated that, with this form of vacuum desiccator and with equal amounts of drying agent, operations may be performed about twice as rapidly as in the conventional model. When the flask *A* is warmed gently, the efficiency of the assembly is at least four times that of a conventional desiccator. (F. P. Pingert, private communication.)

8. The product may be stored satisfactorily in ordinary screw-top bottles, provided that they are well sealed with paraffin or some similar material. Triphenylchloromethane is slowly hydrolyzed to triphenylcarbinol by the moisture of the air. Partially hydrolyzed triphenylchloromethane may be purified by recrystallization from one-third its weight of benzene containing 5–25% of acetyl chloride. The product is washed with ligroin containing a little acetyl chloride.

3. Methods of Preparation

Triphenylchloromethane has been made by the action of phosphorus pentachloride,[1] hydrogen chloride,[2] or acetyl chloride [3] upon triphenylcarbinol. It has also been made by the reaction between carbon tetrachloride and benzene in the presence of aluminum chloride [4] or ferric chloride; [5] and by the action of phosphorus pentachloride,[6] or of sulfuryl chloride in the presence of peroxides,[7] upon triphenylmethane. Procedure A is based upon that of Gomberg and Davis; [3] procedure B is a modification of that originally reported by Gomberg [4] and later described by Gattermann and Wieland.[8]

[1] Hemilian, *Ber.*, **7**, 1207 (1874); Fischer and Fischer, *Ann.*, **194**, 257 (1878).
[2] Gomberg, *Ber.*, **35**, 2401 (1902).
[3] Gomberg and Davis, *Ber.*, **36**, 3925 (1903).
[4] Gomberg, *Ber.*, **33**, 3147 (1900).
[5] Wertyporoch, Kowalski, and Roeske, *Ber.*, **66**, 1237 (1933).
[6] Cone and Robinson, *Ber.*, **40**, 2163 (1907).
[7] Kharasch and Brown, *J. Am. Chem. Soc.*, **61**, 2148 (1939).
[8] Gattermann and Wieland, *Laboratory Methods of Organic Chemistry*, p. 346, The Macmillan Company, New York, 1937.

UNDECYL ISOCYANATE

$$n\text{-}C_{11}H_{23}COCl + NaN_3 \rightarrow n\text{-}C_{11}H_{23}CON_3 \xrightarrow{\text{Heat}} n\text{-}C_{11}H_{23}NCO + N_2$$

Submitted by C. F. H. Allen and Alan Bell.
Checked by Nathan L. Drake and John Sterling.

1. Procedure

In a 1-l. three-necked flask, equipped with a stirrer and a thermometer and immersed in an ice bath, is placed 46 g. (0.7 mole) of sodium azide (Note 1) in 150 ml. of water. A mixture of 109 g. (0.5 mole) of lauroyl chloride (b.p. 134–137°/11 mm.) and 150 ml. of acetone is then added from a separatory funnel to the well-stirred solution of the azide at such a rate that the temperature remains at 10–15°. After the mixture has been stirred at this temperature for an hour, the stirrer is stopped and, when the layers have separated, the lower water layer is removed carefully by suction through a glass capillary tube (Note 2). The upper layer is then added slowly to 500 ml. of benzene which has been warmed to 60° (Note 3). A rather rapid evolution of gas results, and the mixture is kept at 60–70° (Note 4) until no more nitrogen is evolved; the conversion of azide to isocyanate requires about an hour. The solution is filtered to remove any insoluble matter, and the benzene is removed by distillation from a modified Claisen flask. Distillation of the residue yields 80–85 g. of ester (81–86%) (Notes 5 and 6).

2. Notes

1. A practical grade of sodium azide such as that obtained from the Eastman Kodak Company is satisfactory.

2. It is important that the water be removed as completely as possible before the azide is added to the warm benzene. Failure to remove the water causes formation of the *sym*-disubstituted urea during decomposition of the azide. If the water is separated carefully, there will be no need to filter the benzene solution before the final distillation.

3. If the azide is added too rapidly, the solution may froth over; it is best to carry out this reaction in a 1-l. beaker.

4. The heat of reaction is usually sufficient to maintain the temperature at 60–70°.

5. On redistillation all the product boils at 103°/3 mm. A second distillation is unnecessary; the original ester is pure enough for all practical purposes.

6. This method is a general one for the preparation of isocyanates.[1]

3. Method of Preparation

This procedure is one used by Schröter for preparing alkyl isocyanates.[1]

[1] Schröter, *Ber.*, **42**, 3356 (1909).

dl-VALINE *

(Isovaleric acid, α-amino)

$$(CH_3)_2CHCH_2CO_2H + Br_2 \xrightarrow{(PCl_3)} (CH_3)_2CHCHBrCO_2H + HBr$$

$$(CH_3)_2CHCHBrCO_2H + 2NH_3 \rightarrow (CH_3)_2CHCH(NH_2)CO_2H + NH_4Br$$

Submitted by C. S. Marvel.[1]
Checked by C. F. H. Allen and J. VanAllan.

1. Procedure

A. *α-Bromoisovaleric acid.* One kilogram of commercial isovaleric acid monohydrate is placed in a 3-l. round-bottomed flask together with 500 ml. of benzene. The water and benzene are distilled, using a short column, until the temperature of the vapor reaches 100°. The temperature rises rapidly when the benzene is removed. The residue is cooled, and 878 g. (934 ml., 8.6 moles) of it is placed in a 3-l. round-bottomed flask fitted with a long reflux condenser. The top of the condenser is connected by glass tubing to an empty 500-ml. Erlenmeyer flask which acts as a safety trap. A second outlet on the Erlenmeyer flask connects to a gas-absorption trap (Note 1). One and five-tenths kilograms (480 ml.) of dry bromine (Note 2) is added to the acid, and then 15 ml. of phosphorus trichloride is added through the top of the condenser.

The mixture is heated on an oil bath at 70–80° for 10–20 hours or until the condenser no longer shows the deep red color of bromine. Another 25-ml. portion of bromine is added and the flask heated as before. When the color has again disappeared, the temperature of the bath is slowly raised to 100–105° and maintained there for 1.5–2 hours.

The crude bromo acid is placed in a 2-l. modified Claisen flask and distilled under reduced pressure. The low-boiling fraction is mainly unbrominated acid (Note 3). The fraction boiling at 110–125°/15 mm. is collected. The yield is 1364–1380 g. (87.5–88.6%).

B. *dl-Valine.* To 2 l. of technical ammonium hydroxide (sp. gr. 0.90) in a 3-l. round-bottomed flask is added 330 g. (1.82 moles) of α-bromoisovaleric acid. A stopper is wired in and the flask allowed to stand at room temperature for a week. The contents from three such amination flasks are combined in a 12-l. flask, and the ammonia is removed by heating on the steam cone overnight. The solution is then concentrated to a thin paste (about 800 ml.) by means of a water pump (Note 4). The solid material is collected on a filter and, when dry, amounts to 470 g. This is recrystallized by dissolving in 2.4 l. of water heated to 95° on a steam cone, treating with 10 g. of Norit for 30 minutes, filtering hot, adding an equal volume of 95% ethanol, and cooling overnight in the icebox. The valine is collected on a filter and washed with 150 ml. of cold absolute ethanol. The yield is 200–235 g. A second crop is obtained by evaporating the filtrate from the recrystallization on the water pump until crystals form, adding an equal volume of 95% ethanol, and cooling as before. The amino acid obtained in this way amounts to 34 g.

The second filtrate from the recrystallization together with the filtrate from the original concentration is evaporated to dryness and extracted with 500 ml. of glacial acetic acid on a steam cone. The inorganic salts are filtered and the acetic acid is removed by distillation under reduced pressure. One liter of water is added, and it, too, is removed by distillation under reduced pressure. This operation is repeated. These three distillations require 1 day. The residue is dissolved in the minimum amount of hot water (about 300 ml.). The solution is then treated with Norit as before and filtered hot, and an equal volume of 95% ethanol is added. The yield on cooling overnight in the icebox is 34 g. An additional 8 g. can be obtained from the mother liquor by concentration and addition of ethanol as was done with the original mother liquor. The total yield of valine is 300–311 g. (47–48%), which decomposes at 280–282° in a sealed capillary (Notes 5 and 6).

2. Notes

1. The hydrogen bromide may be absorbed in water and constant-boiling hydrobromic acid formed (*Org. Syntheses* Coll. Vol. **1,** 26 (1941).

2. The bromine is dried by shaking with 1 l. of concentrated sulfuric acid.

3. The low-boiling fraction (56–80 g.) may be combined with the next portion of acid to be brominated, or several such fractions are collected and brominated together. In this latter case, only three-fifths as much bromine is used as in the original run.

4. This condition is attained when 1850–1950 ml. of distillate has been collected (5 hours).

5. The quantities stated for each fraction are approximate. If separations are incomplete, the melting point will be 275–280°.

6. Valine prepared in this manner has the calculated amino nitrogen content.

3. Methods of Preparation

Valine has been prepared by the action of ethanolic ammonia on α-chloroisovaleric acid;[2] by the action of aqueous ammonia on α-bromoisovaleric acid;[3] through the reaction of hexamethylenetetramine with α-bromoisovaleric acid in dioxane or dioxane-xylene (91% yield);[4] by the action of ammonia and ammonium carbonate,[5] or ammonium carbonate alone on α-bromoisovaleric acid;[6] by heating isopropylmalonazidic acid[7] or isopropylcyanoacetazide[8] in ethanol and subsequent hydrolysis; by the action of ammonia and hydrogen cyanide[9] or ammonium chloride and potassium cyanide[10] on isobutyraldehyde, followed by hydrolysis; by reaction of α-hydroxyisovaleronitrile with ammonia under pressure, followed by hydrolysis;[11] by reaction of isopropyl bromide with the sodium salt of acetamidomalonic ester or acetamidocyanoacetic ester,[12] and subsequent hydrolysis; by reduction of dimethylpyruvic acid phenylhydrazone;[13] by hydrolysis of α-carbomethoxy-α-phenylacetamidoisovaleronitrile;[14] and by the reaction of isobutyraldehyde with ammonium carbonate and sodium cyanide in aqueous methanol followed by hydrolysis.[15]

α-Bromoisovaleric acid has been prepared in several ways as described in an earlier volume.[16]

[1] These directions are the result of the efforts of a large number of men who have worked on the preparation of valine at the University of Illinois.

[2] Schlebusch, *Ann.*, **141**, 326 (1867).

[3] Clark and Fittig, *Ann.*, **139**, 202 (1866); Schmidt and Sachtleben, *Ann.*, **193**, 105 (1878).

[4] Hillman, *Z. physiol. Chem.*, **283**, 71 (1948).

[5] Slimmer, *Ber.*, **35**, 401 (1902).

[6] Cheronis and Spitzmueller, *J. Org. Chem.*, **6**, 349 (1945).

[7] Curtius, *J. prakt. Chem.*, **125**, 228 (1930).

[8] Gaudry, Gagnon, King, and Savard, *Ann. ACFAS*, **7**, 85 (1941) [*C. A.*, **40**, 1452 (1946)]; Gagnon, Gaudry, and King, *J. Chem. Soc.*, **1944**, 13.

[9] Lipp, *Ann.*, **205**, 18 (1880).

[10] Gaudry, *Can. J. Research*, **24B**, 301 (1946).

[11] Gresham and Schweitzer, U. S. pat. 2,520,312 [*C. A.*, **44**, 10732 (1950)].

[12] Albertson and Tullar, *J. Am. Chem. Soc.*, **67**, 502 (1945); Brit. pat. 621,477; [*C. A.*, **43**, 6653 (1949)]; U. S. pat. 2,479,662 [*C. A.*, **44**, 1133 (1950)].

[13] Feofilaktov and Zaïtseva, *J. Gen. Chem. U.S.S.R.*, **10**, 1391 (1940) [*C. A.*, **35**, 3606 (1941)].

[14] Ehrhart, *Chem. Ber.*, **82**, 60 (1949).

[15] U. S. pat. 2,480,644 [*C. A.*, **44**, 2017 (1950)]; U. S. pat. 2,527,366 [*C. A.*, **45**, 3415 (1951)].

[16] *Org. Syntheses* Coll. Vol. **2**, **95** (1943).

VINYLACETIC ACID

(3-Butenoic acid)

$$CH_2{=}CHCH_2CN + 2H_2O + HCl \rightarrow CH_2{=}CHCH_2CO_2H + NH_4Cl$$

Submitted by Edward Rietz.
Checked by C. F. H. Allen and James VanAllan.

1. Procedure

In a 500-ml. flask attached to a reflux condenser is placed a mixture of 67 g. (80 ml., 1 mole) of allyl cyanide[1] (Note 1) and 100 ml. (1.2 moles) of concentrated hydrochloric acid (sp. gr. 1.19). The mixture is heated by a small flame and is shaken frequently. After 7–8 minutes, the reaction begins, a voluminous precipitate of ammonium chloride appears, the temperature rises rapidly, and the mixture refluxes. After 15 minutes the flame is removed, 100 ml. of water is added, and the upper layer of the acid is separated (Note 2). The aqueous layer is extracted with two 100-ml. portions of ether. The extracts and the acid are combined and distilled. Most of the ether is removed at atmospheric pressure (Notes 3, 4, and 5), and the remainder is removed as the pressure is diminished. The acid is collected at 70–72°/9 mm. after a fore-run of approximately 40 g. The yield of crude acid is 50–53 g. (52–62%) (Note 5).

Although this product is pure enough for most purposes, it contains small amounts of by-products which cannot be removed by distillation. Further purification is accomplished by the following procedure: In a 250-ml. three-necked round-bottomed flask fitted with a stirrer, a small dropping funnel, and a thermometer for reading low temperatures, 24 g. of sodium hydroxide is dissolved in 80 ml. of water. While the temperature *of the solution* is maintained at 8–15° by external cooling, 45 g. (40.5 ml.) of the impure vinylacetic acid is added; this operation requires 25 minutes. This solution is transferred to a 600-ml. conical separatory funnel and extracted with 50 ml. of chloroform (Notes 6 and 7). The alkaline solution is *immediately* transferred to a 1-l. beaker, and 300 ml. of dilute sulfuric acid (Notes 8 and 9) is added with stirring. This acid solution is *at once* extracted with three 100-ml. portions of chloroform (Notes 7 and 10). The solvent is then removed by distillation, first at atmospheric pressure and then at reduced pressure, from a 200-ml. modified Claisen flask. The residue is distilled under reduced pressure. Almost all the material boils at 69–70°/12 mm. (163°/760 mm.). The recovery is 30–33 g. (75–82%) (Note 11).

2. Notes

1. An improved procedure for preparation of allyl cyanide (3-butenenitrile) is as follows: In a dry (washed with absolute ethanol and absolute ether) 500-ml. three-necked flask, equipped with a sealed mechanical stirrer and a 90-cm. bulb condenser set vertically and protected by a calcium chloride tube, are placed 85 g. of dry cuprous cyanide (commercial, or prepared as previously described;[1] dried for 72 hours in an oven at 110° just before use), 0.25 g. of potassium iodide, and 72.5 g. of allyl chloride (dried over calcium chloride and freshly distilled; b.p. 45–47°). The stirrer is started, and the mixture is heated on a water bath. After about 6 hours the reaction is substantially complete, as indicated by cessation of the refluxing; heating is continued for 1 hour beyond this point. With larger runs it may be necessary to moderate the reaction by removing the water bath when vigorous refluxing sets in. This usually occurs about 3–5 hours after heating is started. Runs of the size described do not require any special attention. The water bath is replaced by an oil bath, the condenser is set downward for distillation, and stirring is continued while the allyl cyanide is distilled into a 100-ml. distilling flask. Near the end of the distillation it may be necessary to discontinue stirring, and it is advisable to reduce the pressure somewhat to aid in the removal of the last portion of the product. The distillate is redistilled, and 50–53 g. (79–84%) of allyl cyanide boiling at 116–122° is collected. (Private communication from Curtis W. Smith and H. R. Snyder; checked by W. E. Bachmann and G. Dana Johnson.)

2. This amounts to 90–95 g. A separatory funnel of about 600-ml. capacity is the most convenient size.

3. The first ether extract contains 10 g. of nitrile; the second extract contains 4.5 g.

4. This procedure is shorter and less tedious than the more common method of drying and fractionating the ethereal solution.

5. A Dry Ice trap, inserted between the oil pump and the apparatus, condenses 5–10 g. of unchanged allyl cyanide. The yield of crude acid, after allowing for the recovered nitrile, is 62–72%.

6. This extraction removes about 1 g. of non-acidic impurity.

7. Ether can be substituted for chloroform without materially decreasing the yield.

8. This is prepared by diluting 16.5 ml. of concentrated sulfuric acid to 300 ml.

9. Contact with alkali results in the isomerization of vinylacetic acid to crotonic acid.

10. The first two extractions remove 27–28 g.; the last one, 3–4 g.

11. When cooled in a Dry Ice-acetone bath the acid should remain clear until it crystallizes at −36° to −35°. This indicates freedom from crotonic acid.[2]

3. Methods of Preparation

The only practical methods for the preparation of vinylacetic acid involve hydrolysis of allyl cyanide;[3,4] carbonation of allylmagnesium bromide[5,6] or an alkali salt of propylene.[7] The malonic acid synthesis is less satisfactory than the hydrolysis of allyl cyanide.[4]

[1] *Org. Syntheses* Coll. Vol. **1**, 46 (1941).

[2] Fichter, *Ber.*, **35**, 938 (1902).

[3] Falaise and Frognier, *Bull. soc. chim. Belg.*, **42**, 433 (1933) [*C. A.*, **28**, 2329 (1934)].

[4] Linstead, Noble, and Boorman, *J. Chem. Soc.*, **1933**, 560.

[5] Houben, *Ber.*, **36**, 2897 (1903).

[6] Gilman and McGlumphy, *Bull. soc. chim. France*, (4) **43**, 1327 (1928).

[7] Morton, Brown, Holden, Letsinger, and Magat, *J. Am. Chem. Soc.*, **67**, 2224 (1945).

VINYL CHLOROACETATE

(Acetic acid, chloro-, vinyl ester)

$$ClCH_2CO_2H + HC{\equiv}CH \xrightarrow{HgO} ClCH_2CO_2CH{=}CH_2$$

Submitted by Richard H. Wiley.
Checked by Maynette Vernsten and Homer Adkins.

1. Procedure

Vinyl chloroacetate is lachrymatory.

A 1-l. three-necked flask is equipped with an efficient stirrer, a thermometer, a gas inlet tube 10 mm. in diameter, and a reflux condenser. The bulb of the thermometer and the lower end of the gas inlet tube are sufficiently close to the bottom of the flask to be covered by the reactants (Note 1). The upper end of the condenser is attached to a small gas-washing bottle containing enough water so that the rate of passage of the exit gases may be noted. The flask is charged with 200 g. (2.12 moles) of monochloroacetic acid, 0.2 g. of hydroquinone, and 20 g. of yellow mercuric oxide (Note 2). A slow stream of acetylene is passed through a spiral trap cooled in Dry Ice-acetone mixture, a mercury safety valve, an empty wash bottle, a sulfuric acid wash bottle, a soda-lime tower, and then into the reaction flask through the gas inlet tube. The stirrer is started, and the contents of the flask are heated gently with steam until the chloroacetic acid just melts (Note 3). The temperature of the reaction mixture is lowered to 40–50° after 30 minutes or as soon as the melting point of the mixture permits the lower temperature to be attained without solidification. An ice bath is used for cooling the mixture when necessary (Note 4). The stirrer is operated fast enough to throw the contents of the flask vigorously against the walls, in order to obtain the most rapid absorption. The absorption of acetylene, very rapid at first, becomes very slow after about 3 hours, and the introduction of the gas is discontinued (Note 5).

The contents of the flask are decanted and filtered or centrifuged to remove as much as possible of the finely divided mercury salt (Note 6). The filtrate is distilled from a Claisen flask, and 117–135 g. of material boiling at 45–55°/20 mm. is collected (Note 7). The distillate is fractionated, and 107–125 g. (42–49%) of vinyl chloroacetate, b.p. 37–38°/16 mm., n_D^{25} 1.4422, is collected (Notes 8 and 9).

2. Notes

1. The apparatus is similar to that previously shown [1] except that one of the inlet tubes, T_1 or T_2, is replaced by a thermometer and the stirrer is equipped with a short sleeve which is connected to the shaft by a rubber tube moistened with glycerol.

2. Merck or Mallinckrodt Reagent grade or Baker C.P. monochloroacetic acid was used without further purification. Distillation of the acid did not improve the yield. Baker or Mallinckrodt yellow mercuric oxide, C.P., was used. The hydroquinone may be added at this point or before the distillation.

3. The submitter suggested that the mixture be heated to 50–55°. The melting point of the chloroacetic acid used by the checkers required that the temperature at the beginning of the reaction be somewhat above 60°. The chief difficulty in carrying out the reaction is due to solidification of chloroacetic acid in the inlet tube if the temperature of the reaction mixture is allowed to fall to the point where

the mixture begins to crystallize. If the inlet tube becomes plugged, the pressure in the system will be relieved by the mercury safety valve. The safety valve may consist of a small bottle containing a layer of mercury and carrying a stopper fitted with two glass tubes; one of the tubes, extending just below the stopper, is connected to the acetylene line, and the other, which extends *just* beneath the surface of the mercury, is open to the air.

4. Higher temperatures are said to facilitate the formation of the ethylidene compound. Similar yields have been obtained, however, when the temperature was maintained at 50–55° during the entire reaction period.

5. Addition of more mercuric oxide at this time or after 1½ hours' operation does not affect the rate of absorption or increase the yield. If less than 20 g. of mercuric oxide is used the yield is poorer; for example, an experiment with 10 g. of mercuric oxide resulted in a 37% yield of crude ester.

6. It is difficult to remove the finely divided mercury salt completely by decantation or filtration. All the suspended salt can be removed by centrifuging, but dissolved salt remains and separates from the solution on distillation. The suspended salt does not interfere if the product is distilled rapidly. Long-continued heating of the crude product, such as would be required in a careful fractionation, has resulted in a vigorous decomposition of the mixture.

7. Hydroquinone should be added to the crude ester immediately to stabilize it against polymerization. If it is to be kept for any length of time before refractionation it should be stored in a cold chest.

8. The submitter observed a somewhat higher boiling point (41–42°/15 mm.) and did not report the refractive index of the product. Both the submitter and checkers used a column 20 cm. in length, 12 mm. in diameter, packed with glass helices, and equipped with a partial take-off head. If the fractionation is carefully conducted, the product is sufficiently pure for most polymerization work. The compound should be stabilized with hydroquinone and stored in a cold chest if it is not to be used immediately.

9. According to the submitter the crude ester may also be purified as follows: The centrifuged reaction mixture is placed in a 1-l. separatory funnel with 500 ml. of ether and washed with 200-ml. portions of a 5% sodium carbonate solution until the unchanged acid is removed. Difficulty with emulsions is sometimes encountered at this point. The ether layer is dried with anhydrous sodium sulfate, and the ether is evaporated on a water bath. The residue is fractionated as above.

3. Methods of Preparation

Vinyl chloroacetate has been prepared from acetylene and chloroacetic acid in the vapor phase at 250° with a zinc-cadmium catalyst,[2] and in the liquid phase with a mercury salt catalyst.[3] The procedure described is an adaptation of that employed by Klatte,[4] by Skirrow and Morrison,[5] and by others.[6]

[1] *Org. Syntheses* Coll. Vol. **2,** 363 (1943).

[2] Hermann, Deutsch, and Baum, U. S. pat. 1,822,525 [*C. A.,* **25,** 5900 (1931)].

[3] Ger. pat. 271,381 [*Frdl.,* **11,** 54 (1912–1914)].

[4] U. S. pat. 1,084,581 [*C. A.,* **8,** 991 (1914)].

[5] U. S. pat. 1,710,197 [*C. A.,* **23,** 2724 (1929)].

[6] The preparation of vinyl esters has been reviewed by Ellis, *The Chemistry of Synthetic Resins,* Vol. II, p. 1017, Reinhold Publishing Corporation, New York, 1935.

TYPE OF REACTION INDEX

This index lists the preparations contained in this volume in accordance with general types of reactions. Only those preparations are included which can be classified under the selected headings with definiteness. The arrangement of types and of preparations is alphabetical.

TYPE OF COMPOUND INDEX

Preparations are listed by functional groups or by ring systems. Phenyl, ethylenic, and acetylenic groups are not considered as substituents unless otherwise stated. Salts are included with the corresponding acids and bases.

FORMULA INDEX

All preparations are listed in this index. The system of indexing is that used by *Chemical Abstracts*. The essential principles involved are as follows: (1) The arrangement of symbols in formulas is alphabetical except that in carbon compounds C always comes first, followed immediately by H if hydrogen is also present. (2) The arrangement of formulas is also alphabetical except that the number of atoms of any specific kind influences the order of compounds: e.g., all formulas with one carbon atom precede those with two carbon atoms, thus: CH_2I_2, CH_3NO_2, CH_5N, C_2H_2O. (3) The arrangement of entries under any heading is strictly alphabetical according to the names of the isomers. (4) Inorganic salts of organic acids and inorganic addition compounds of organic compounds are listed under the formulas of the compounds from which they are derived.

INDEX TO PREPARATION OR PURIFICATION OF SOLVENTS AND REAGENTS

ORGANIC SYNTHESES procedures frequently include notes describing the purification of solvents and reagents. These have been placed in a single index for convenience. The preparation of useful reagents and catalysts is also included.

ILLUSTRATION INDEX

GENERAL INDEX

The name of a compound in Small Capital Letters together with a number in bold-faced type indicates complete preparative directions for the substance named. The name of a compound in ordinary light-faced type together with a number in bold-faced type indicates directions, usually adequate but not in full detail, for preparing the substance named. A name in light-faced type together with a number in light-faced type indicates a compound or an item mentioned in connection with a preparation.